Lecture Notes in Electrical Engineering

Volume 397

About this Series

"Lecture Notes in Electrical Engineering (LNEE)" is a book series which reports the latest research and developments in Electrical Engineering, namely:

- Communication, Networks, and Information Theory
- Computer Engineering
- Signal, Image, Speech and Information Processing
- Circuits and Systems
- Bioengineering

LNEE publishes authored monographs and contributed volumes which present cutting edge research information as well as new perspectives on classical fields, while maintaining Springer's high standards of academic excellence. Also considered for publication are lecture materials, proceedings, and other related materials of exceptionally high quality and interest. The subject matter should be original and timely, reporting the latest research and developments in all areas of electrical engineering.

The audience for the books in LNEE consists of advanced level students, researchers, and industry professionals working at the forefront of their fields. Much like Springer's other Lecture Notes series, LNEE will be distributed through Springer's print and electronic publishing channels.

More information about this series at http://www.springer.com/series/7818

Rachid El-Azouzi · Daniel Sadoc Menasché
Essaïd Sabir · Francesco De Pellegrini
Mustapha Benjillali
Editors

Advances in Ubiquitous Networking 2

Proceedings of the UNet'16

Springer

Editors
Rachid El-Azouzi
Computer Science Laboratory (LIA)
University of Avignon
Avignon
France

Daniel Sadoc Menasché
Federal University of Rio de Janeiro
Rio de Janeiro
Brazil

Essaïd Sabir
Hassan II University of Casablanca
ENSEM
Casablanca
Morocco

Francesco De Pellegrini
CREATE-NET
Trento
Italy

Mustapha Benjillali
INPT
Rabat
Morocco

ISSN 1876-1100 ISSN 1876-1119 (electronic)
Lecture Notes in Electrical Engineering
ISBN 978-981-10-1626-4 ISBN 978-981-10-1627-1 (eBook)
DOI 10.1007/978-981-10-1627-1

Library of Congress Control Number: 2016943864

Printed on acid-free paper

This Springer imprint is published by Springer Nature
The registered company is Springer Science+Business Media Singapore Pte Ltd.

Committee

Program Committee

Mohamed-Slim Alouini, King Abdullah University of Science and Technology (KAUST), Saudi Arabia
Eitan Altman, INRIA, Sophia-Antipolis, France
Yacine Atif, College of Information Technology, UAE University, UAE
Amar Prakash Azad, Samsung Advanced Institute of Technology, Bangalore, India
Leila Azouz Saidane, ENSI, Tunisia
Elarbi Badidi, Faculty of Information Technology, UAEU
Abdelmajid Badri, FST, Hassan II University of Casablanca, Morocco
Mohamed Bakhouya, University of Technology of Belfort Montbeliard
Haythem Bany Salameh, Yarmouk University, Irbid, Jordan
Mohammed Baslam, FST, Beni Mellal, Morocco
Hicham Behja, ENSEM, Hassan II University of Casablanca, Morocco
Abdelhamid Belmekki, INPT, Rabat, Morocco
Yann Ben Maissa, INPT, Rabat, Morocco
Jalel Ben Othmane, University of Paris 13, France
Nabil Benamar, Moulay Ismail University, Morocco
Abderrahim Beni Hssane, Chouaib Doukkali University, Morocco
Hassan Bennani, ENSIAS, Mohammed V University of Rabat, Morocco
Mehdi Bennis, Centre of Wireless Communications, University of Oulu, Finland
Karim Benzidane, FSAC, Hassan II University of Casablanca, Morocco
Abdelmajid Berdai, ENSEM, Hassan II University of Casablanca, Morocco
Ismail Berrada, L3I and LIMS, Sidi Mohammed Ben Abdellah University, Fez, Morocco
Noureddine Boudriga, Sup'COM, Cartage University, Tunisia
Mohammed Boulmalf, International University of Rabat, Morocco
Jaouad Boumhidi, Sidi Mohammed Ben Abdellah University, Fez, Morocco
Olivier Brun, LAAS-CNRS, Toulouse, France
Tijani Chahed, TELECOM & Management SudParis, France

Abdelaali Chaoub, INPT, Rabat, Morocco
Reda Chbihi, FSAC, Hassan II University of Casablanca, Morocco
Hatim Chergui, Télécom-Bretagne, France
Adel Omar Dahmane, University of Québec à Trois-Rivières, Canada
Hamza Dahmouni, INPT, Rabat, Morocco
Imane Daoudi, ENSEM, Hassan II University of Casablanca, Morocco
Francesco De Pellegrini, Creat-Net, Italy
Mérouane Debbah, Centrale-Supelec/Huawei, Paris, France
Loubna Echaabi, INPT, Rabat, Morocco
Rachid El-Azouzi, LIA-CERI, University of Avignon, France
Mohamed El-Kamili, LiM, FSDM, Sidi Mohammed Ben Abdellah University, Fez
Mourad El Yadari, Moulay Ismail University, Errachidia, Morocco
Halima Elbiaze, University of Quebec, Montreal, Canada
Hajar Elhammouti, INPT, Rabat, Morocco
Oussama Elissati, INPT, Rabat, Morocco
Mohammed Elkoutbi, ENSIAS, Mohammed V University of Rabat, Morocco
Mohamed Emad Eldin, Canadaian University, Dubai, United Arab Emirates
Mohammed Erradi, ENSIAS, Mohammed V University of Rabat, Morocco
Larbi Esmahi, Athabasca University, Canada
Mohammed Essaaidi, ENSIAS, Mohammed V University of Rabat, Morocco
Mohamed Et-Tolba, INPT, Rabat, Morocco
Mohamed Fizari, University of BAdji Mokhtar Annaba, Algeria
Hicham Ghennioui, Sidi Mohammed Ben Abdellah University, Fez
Mounir Ghogho, UIR, Rabat, Morocco; University of Leeds, UK
Oussama Habachi, University of Limoges, France
Majed Haddad, LIA-CERI, University of Avignon
Hatim Hafiddi, INPT, Rabat, Morocco
Harroud Hamid, Al Akhawayn University, Ifrane, Morocco
Tembiné Hamidou, New York University, UAE/USA
Kenza Hamidouche, Centrale-Supelec, France
Abdelkarim Haqiq, FST, Hassan First University, Settat, Morocco
Moulay Lahcen, Hasnaoui, Faculty of Sciences Dhar Mahraz, Sidi Mohammed Ben Abdellah University, Fez
Yezekael Hayel, LIA-CERI, University of Avignon, France
Mostafa Hefnawi, Royal Military College, University of Canada, Canada
Khalil Ibrahimi, University IBN Tofail, Faculty of Sciences
Said Jai-Andaloussi, Faculty of sciences, Hassan II University of Casablanca, Morocco
Tania Jiminèz, LIA-CERI, University of Avignon, France
Min Ju, LIA-CERI, University of Avignon, France
Hamoudi Kalla, University of Batna, Batna, Algeria
Vijay Kamble, University of California, Berkeley, USA
Abdelilah Karouit, University of Avignon, France
Mohammed Khaldoun, Université Hassan Hassan Ain-Cock
Ismail Khriss, University of Rimouski, Canada

Abdellatif Kobbane, ENSIAS, Mohammed V University of Rabat, Morocco
Mohammed-Amine Koulali, ENSA, University Mohammed First of Oujda, Morocco
Rim Koulali, Mohammed First University of Oujda, Morocco
Moez Krichen, National School of Engineers of Sfax, Tunisia
Marwan Krunz, University of Arizona, Arizona, USA
Adlen Ksentini, University of Rennes/IRISA, Rennes, France
Sujit Kumar Samanta, National Institute of Technology Raipur, India
Latif Ladid, University of Luxembourg, Luxembourg
Jaime Lloret, Polytechnic University of Valencia, Spain
Zaixin Lu, Marywood University, Scranton, Pennsylvania, USA
Issame Mabrouki, University of Sousse, Tunisia
Sujith Mathew, UAE University, UAE
Daniel Sadoc Menasché, Department of Computer Science at Federal University of Rio de Janeiro, Brazil
Pascale Minet, INRIA, Paris, France
Orazio Mirabella, University of Catania, Italy
Lynda Mokdad, University of Paris-Est, France
El Habib Nfaoui, Sidi Mohamed Ben Abdellah University, Fez
Luis Orosco Barbosa, Albacete University, Spain
Ouail Ouchetto, University Paris-Sud, France
Mohamed Ouzzif, ESTC, Hassan II University of Casablanca, Morocco
Sofie Pollin, Faculty of Engineering Science, Ku Leuven University, Belgium
Guy Pujolle, LIP6-CNRS, University of Pierre and Marie Curie, Paris, France
Mohammed Raiss El Fenni, NPT, Rabat, Morocco
Sreenath Ramanath, Systems R&D, Lekha Wireless, Bangalore, India
Fernando Ramirez-Mireles, Instituto Tecnologico Autonomo, Mexico
Alexandre Reiffers, University of Avignon, France
Slim Rekhis, Sup'COM, Cartage University, Tunisia
Zouheir Rezki, King Abdullah University of Science and Technology, Saoudi Arabia
Mounir Rifi, ESTC, Hassan II University of Casablanca, Morocco
Julio Rojas-Mora, School of Informatics Engineering, Universidad Católica de Temuco, Chile
Rachid Saadane, LETI, EHTP, Casablanca, Morocco
Mohamed Nabil Saidi, INSEA, Rabat, Morocco
Aziz Salah, Quebec University of Montreal, Canada
Habib Sidi, Orange Labs., France Telecom, France
Alonso Silva, Alcatel-Lucent Bell Labs, Paris, France
Nabil Tabbane, Sup'COM, University of Carthage, Tunisia
Tarik Taleb, Aalto University, Aalto, Finland
Alpcan Tansu, the University of Melbourne, Australia
Corinne Touati, LIG-INRIA, Grenoble, France
Abdelwahed Tribak, INPT, Rabat, Morocco
Kavitha Veeraruna, Indian Institute of Technology of Bombay, India

Mohamed Wahbi, Insight centre, University College Cork, Cork, Ireland
Xiaoyan Wang, National Institute of Informatics, China
Carlos Becker Westphall, Universidade Federal de Santa Catarina, Brazil
Sulan Wong, Faculty of Law Universidad Católica de Temuco, Temuco, Chile
Li Xu, College of Mathematics and Computer Science, Fujian University, China
Yuedong Xu, SIST, Fudan University, China
Iraqi Youssef, Khalifa University, Emirates Arab United
Quanyan Zhu, New York University, USA

Steering Committee

Mohamed-Slim Alouini, KAUST University, Saudi Arabia
Eitan Altman, INRIA, France
Rachid El-Azouzi, University of Avignon, France
Mounir Ghogho, UIR/Morocco, University of Leeds/UK
Marwan Krunz, University of Arizona, USA
Francesco De Pellegrini, Create-net/INSPIRE, Italy
Essaïd Sabir, ENSEM, Hassan II University of Casablanca, Morocco

Organizing Committee

Honorary Chairs
Latif Ladid, Founder and President of IPv6 FORUM; Chair of 5G World Alliance; Luxembourg
Idriss Mansouri, President Hassan II University of Casablanca, Morocco
Driss Aboutajdine, Hassan II Academy, CNRST, Morocco
Haris Hassabis, President International University of Casablanca, Morocco

General Chairs
Rachid El-Azouzi, University of Avignon, France
Daniel Sadoc Menasché, Federal University of Rio de Janeiro, Brazil

Local Chairs
Hicham Medromi, ENSEM, Hassan II University of Casablanca
Essaïd Sabir, ENSEM, Hassan II University of Casablanca

Technical Program Chairs
Mehdi Bennis, Centre for Wireless Communication, University of Oulu, Finland
Francesco De Pellegrini, Create-net/INSPIRE, Italy
Mustapha Benjillali, INPT, Morocco

Publication Chairs
Quanyan Zhu, New York University, USA
Mohamed Sadik, ENSEM, Hasssan II University of Casablanca

Industry Panel Chair
Mounir Ghogho, UIR, Rabat; University of Leeds, UK

Special Sessions Chairs
Elarbi Badidi, Faculty of Information Technology, Emirates Arab United
Mohammed-Amine Koulali, Mohammed 1st University of Oujda, Morocco
Fouad Moutaouakkil, ENSEM, Hassan II University of Casablanca
Said Jai-Andaloussi, Hassan II University of Casablanca, Morocco

Local Arrangement Chairs
Mounire Trifess, International University of Casablanca, Morocco
Zineb Chraibi, International University of Casablanca, Morocco
Mama Alaoui, International University of Casablanca, Morocco

Publicity and Patron Chairs
Abdelilah Esmili, International University of Casablanca, Morocco
Yuedong Xu, SIST, Fudan University, China
Alonso Silva, Alcatel-Lucent Bell Labs, Paris, France
Hamidou Tembine, University of New York, USA
Ahmed Errami, ENSEM, Hassan II University of Casablanca

Local Organizing Committee
Mounir Tantaoui El Araki, International University of Casablanca
Abdellah Moujahid, International University of Casablanca
Othmane Benhmamouch, International University of Casablanca
Mohammed Boutabia, International University of Casablanca
Hajar Iguer, International University of Casablanca
Saida Tallal, ENSEM, Hassan II University of Casablanca
Abdelmajid Berdai, ENSEM, Hassan II University of Casablanca
Sihame Benhaddou, ENSEM, Hassan II University of Casablanca
Imane Daoudi, ENSEM, Hassan II University of Casablanca
Adil Sayouti, Royal Naval School, Casablanca
Mohamed El-Kamili, FSDM, Sidi Mohammed Ben Abdellah University of Fez
Abdellatif Kobbane, ENSIAS, Mohammed V University of Rabat
Khalil Ibrahimi, Faculty of Sciences, Ibn Totail University, Kénitra
Mohamed Baslam, FST, Sultan My Ismail University, Béni Mellal
Mohammed Raïs-EL-Fenni, INPT, Rabat

Webmaster and Social Media
Sidi Ahmed Ezzahidi, Mohammed V University of Rabat
Sara Handouf, ENSEM, Hassan II University of Casablanca

A Welcome Message From the General Chairs

It is our pleasure to welcome you to the 2016 edition of the International Symposium on Ubiquitous Networking, UNet'16. The conference will be held in the city of Casablanca, Morocco, from May 30 to June 1, following the success of last year's first edition. Morocco counts a growing and active community of networking researchers and the choice of Casablanca for UNet'16 allows its attendants, coming from all parts of the globe, to interact in a fascinating environment.

The growth of pervasive and ubiquitous networking in the past few years is unprecedented. Nowadays, a significant portion of the world's population is connected to the Internet most of the time through smart phones, and the Internet of Things promises to broaden the impact of the Internet to encompass devices ranging from electric appliances and medical devices to unmanned vehicles. The goal of UNet is to be a premier forum to discuss technical challenges and solutions related to such a widespread adoption of networking technologies, including broadband multimedia, machine-to-machine applications, Internet of Things, sensor networks, and RFID technologies. To this aim, we count with a main technical track of papers, together with three special sessions on smart cities, big data, and unmanned aerial vehicles.

The UNet'16 program features five special talks addressed by distinguished keynote speakers: Prof. Mohamed-Slim Alouini from KAUST (Saudi Arabia), Prof. Eitan Altman from INRIA (France), Prof. Mehdi Bennis from Oulu University (Finland), Prof. Mohammed Essaaidi from Mohammed V University (Morocco), and Prof. Marwan Krunz from University of Arizona (USA). It also counts with an industrial panel lead by Prof. Latif Ladid, Founder and Chair of the IPv6 Forum and the 5G World Alliance, about new trends and industrial efforts in IPv6, 5G, Internet of Things, and SDN.

With a rich program that reflects on the most recent advances in ubiquitous computing, involving a broad range of theoretical tools (e.g., game theory, mechanism design theory, learning theory, etc.) and practical methodologies (e.g., SDR/SDN platforms, embedded systems, etc.) to study modern technologies (e.g., LTE-A, LTE-B, 5G), we are very pleased to welcome you to this second edition of UNet.

We are very grateful to our technical sponsors, without whom UNet would have not been viable. We thank the IPv6 Forum, 5G World Alliance, IEEE COMSOC 5G Mobile Wireless Internet Emerging Technologies Subcommittee, IEEE COMSOC Internet of Things Emerging Technologies Subcommittee and the IEEE COMSOC Software Defined Networking and Network Functions Virtualization Emerging Technologies Subcommittee. Among our Morocco collaborators, we are especially thankful to the IEEE Morocco Section, the COMSOC Morocco Chapter and the International University of Casablanca.

Enjoy the conference!

Sincerely
UNet'16 General Chairs
Rachid El-Azouzi, Daniel Sadoc Menasché
Essaïd Sabir and Hicham Medromi

A Welcome Message From the TPC Chairs

It is with great pleasure that we welcome you to the 2016 International Symposium on Ubiquitous Networking (UNet 2016) in Casablanca, Morocco. You will find an interesting technical program of three technical sessions reporting on recent advances in context-awareness, autonomy paradigms, mobile edge networking, virtualization, and discussing the enablers, the challenges, and the applications of ubiquitous communications and networking in today's and future contexts. UNet'16 also features five keynote speeches by world-class experts, an industry panel covering the new trends and the industrial efforts in IPv6, 5G, Internet of Things, and software-defined networking, and three special sessions on smart cities and urban informatics for sustainable development, unmanned aerial vehicles theory and applications, and big data applications and solutions.

We have received 144 paper submissions from 15 countries. From those, 44 were accepted as main track papers and 12 were accepted as special session papers, after a careful review process to be included in UNet'16 proceedings. The overall acceptance rate of full papers in the UNet'16 main tracks is 36 %, 38 % including special sessions.

The preparation of this excellent program would not have been possible without the dedication and the hard work of the different chairs, the keynote speakers, and all technical program committee (TPC) members and reviewers. We grasp the opportunity to acknowledge their valuable work, and sincerely thank them for their help in ensuring that UNet 2016 will be remembered as a high-quality event.

We hope that you will enjoy this edition's technical program, and we look forward to meeting you in Casablanca.

Sincerely
UNet'16 TPC Chairs
Mustapha Benjillali, Francesco De Pellegrini and Mehdi Bennis

Contents

Part I
Main Track 1: Context-Awareness and Autonomy Paradigms

The Allocation in Cognitive Radio Network: Combined Genetic Algorithm and ON/OFF Primary User Activity Models

Yasmina El Morabit, Fatiha Mrabti and El Houssein Abarkan

Abstract Cognitive radio (CR) has appeared as a promising solution to the problem of spectrum underutilization. Cognitive radio user (CU) is an intelligent equipment who scent the spectrum which is licensed to primary radio users (PUs) when it is idle and use it with other CUs for their communication. Thus by modeling PUs activity, CUs can predict the future state ON or OFF (busy or idle) of PUs by learning from the history of their spectrum utilization. In this manner, CUs can select the best available spectrum bands. On this point, many PU ON/OFF activity models have been proposed in the literature. Among this models, Continuous Time Markov chain, Discrete Time Markov chain, Bernoulli and Exponential models. In this paper, we firstly compare these four models in term of better numbers of OFF slots to deduce which model give best performance of available resources. Then, the activity history patterns generated from each model are combined with the genetic algorithm as sensing vectors to select the best available channel in terms of quality and least PU arrivals.

Keywords Cognitive radio · Allocation · Primary user ON/OFF models · Continuous time markov chain · Discrete time markov chain · Bernoulli model · Exponential model · Genetic algorithm · Optimization

---

Y. El Morabit (✉) · F. Mrabti · E.H. Abarkan
Laboratory SSC, Faculty of Sciences and Technology, Sidi Mohamed Ben Abdellah University, Fez, Morocco
e-mail: elmorabityasmina@gmail.com

F. Mrabti
e-mail: f_mrabti@yahoo.fr

E.H. Abarkan
e-mail: habarkan@yahoo.fr

R. El-Azouzi et al. (eds.), *Advances in Ubiquitous Networking 2*, Lecture Notes in Electrical Engineering 397, DOI 10.1007/978-981-10-1627-1_1

1 Introduction

The rapid development of wireless communication technologies is seriously challenged by the spectrum scarcity and spectrum underutilization problem. According to some studies sponsored by Federal Communications Commission, in certain geographical areas, many frequency bands that are regulated by the traditional fixed spectrum allocation policy are not occupied in most of the time. However, the Cognitive Radio (CR) has appeared as the promising technology to improve the spectrum usage efficiency.

In cognitive radio network (CRN), two types of users exist, one is primary radio user (PU) and the other is cognitive radio user (CU) which is also called secondary user (SU). The CU is considered to be intelligent and self-managing radio user that operate in a decentralized pattern without the need of a central base station. PU has licensed spectrum on which it operates while CU has no licensed spectrum and it operates either on unlicensed spectrum or on PU licensed spectrum when it is idle. If PU arrives on its spectrum band (channel) while CU is utilizing it, then CU has to vacate this spectrum immediately without causing interference to PU and to switch to another available idle channel. Thus, by modeling PU activity with ON/OFF models, In such ON/OFF models a given channel is either occupied (ON state) by the PUs and is unavailable for CUs, or vacant (OFF state) indicates that a channel is free, so it can be utilized by CU. CU can predict the future state (ON or OFF) of the PUs by learning from the history of their spectrum utilization. In this manner, CUs can assign best available channel for their communication. Due to this fact, PU activity modeling is very important for the performance of CRN. In this point of view, many PU ON-OFF activity models have been proposed in the literature [1–3]. Among these models, there are four (4) we're going to study: Continuous Time Markov chain (**CTMC**) [2, 4–6], Discrete Time Markov chain (**DTMC**) [7, 8], Bernoulli (**BN**) process [9–11] and Exponential (**Exp**) model [3]. The lengths of the ON and OFF periods being random variables following some specified distributions.

In **CTMC** model, the time spent is modeled as continuous random variables between transition states. It makes two hypotheses, the first, if the current state is i, the time will be exponentially distributed until the next state transition, the second, if the current state is i, the next state will be j with probability P_{ij} which will be independent of past history of previous state and process until the next transition. In **DTMC**, the state changes at discrete time intervals and they are determined by transition matrix and probability matrix [1]. **Exp** process is a random process which depends on a particular factor called ON-OFF mean time and tells how long ON/OFF time will be. It serves as input to the function in generating random ON-OFF period time in depicting PU behavior [3]. Lastly, **BN** process is a sequence (finite or infinite) of binary random variables which takes only two values, 0 or 1 corresponds to OFF and ON states respectively. These states change according to independent and identically distributed Bernoulli process.

In **CTMC** based modeling it is assumed that CUs observe the system state continuously and can detect randomly arriving PUs. On the contrary, in **DTMC** model, CUs perform the sensing periodically relying in discrete time instants to observe the system state and therefore they cannot instantly detect PUs arriving between sensing instants. Similarly, in the **Exp** and **BN** models, the sensing is done in discrete and probabilistic manner. However, **CTMC** model will not lose the listening of the channel during the sensing process contrary to other models. So, it must yield better results in OFF slots.

The purpose of these models is to provide a realistic model of PU activity pattern which is considered in the network by CUs in taking decisions about spectrum by selecting the best available channel in terms of quality (such as transmission power (POW), bit error rate (BER), etc.), and least PU arrivals. These diverse factors with contradictory objectives bring the channel selection problem inside the domain of multi-objective optimization problem. The most popular approach to solve such problems is the genetic algorithm (GA). The GA optimizes the multiple parameters of a problem in parallel and provides the optimal solution for a given problem.

The GA is a search algorithm based on the principles of natural selection and genetics. It relies upon evolving a set of solutions, represented by the so-called chromosomes, over a period of time. Eventually, through the GA operators (selection, crossover and mutation) a good solution will be found by combining different possible solutions [12].

In this paper, we have firstly performed a comparative study of these four ON/OFF PU activity models in term of the average of ON and OFF state durations of each model, to deduce which model gives better resources of idle slots to give the CU the leverage to utilize them. The history channel patterns maintained from each model is used as a sensing vector to calculate the Cognitive User Opportunity Index (CUOI), and we used it as the new information gene to the same structure of chromosome used in our previous work [13]. Thus, our proposal takes into account the behavior of PU in the channel to find the optimal channel with least PU activities and with the quality required by the CU. The comparison of the performance of each model combined with the GA is showed in Matlab. The simulation results show that a combined allocation model based CTMC and GA outperforms other models in term of the average fitness value.

2 Problem Formulation

We consider a system with M channels; each channel is divided into N slots. Let $Z^k(t)$ denote the state ON or OFF of a channel k at time t and a sample ON/OFF period correspond to the value 0/1. The sensing produces a binary random sequence for each channel (Fig. 1).

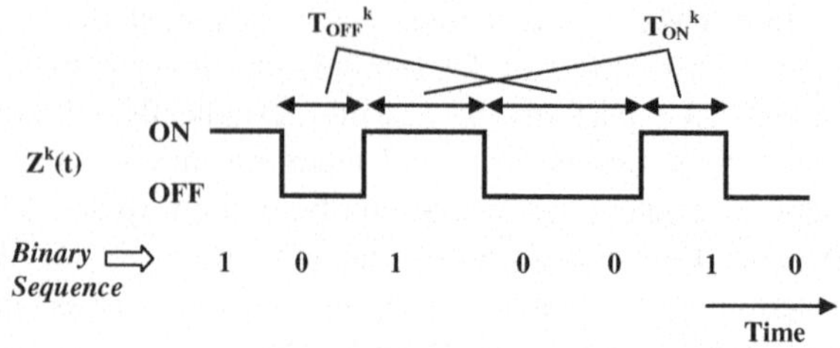

Fig. 1 ON/OFF PU activity

2.1 *PU ON-OFF Models*

The ON/OFF channel usage model specifies a time slot in which the PU signal is occupying or not occupying a channel. Source alternating between states ON (busy) and OFF (idle).

ON-OFF DTMC Model. The DTMC is a stochastic process $\{X_t, t = 0, 1, 2\ldots\}$ takes a finite number of possible values, represented by a transition matrix P. If $X_t = i$, then the process is said to be in state i at time t, and the probability P_{ij} that it will next be in state j is $P_{ij} = P(X_1 = j \mid X_0 = i)$. So we have that $P_{ij} = P(X_1 = j \mid X_0 = i) = P(X_{t+1} = j \mid X_t = i)$ for all $t \geq 0$ [14, 15].

The transition probability matrix P of this channel is assumed to be known and does not change, which can be represented by a 2 × 2 matrix as shown in Eq. (1) for the case of a single channel [7].

$$P = \begin{bmatrix} P_{00} & P_{01} \\ P_{10} & P_{11} \end{bmatrix} \quad (1)$$

Where P_{00} is the probability of the channel to remain in the busy state, P_{01} is the probability of the channel to change from busy state to idle state, P_{11} is the probability that the channel remain in the idle state and P_{10} is the probability that the channel change from idle to busy state (Fig. 2).

The cumulative distribution function (CDF) of the ON and OFF durations adopted from the Ref. [3] and expressed by Eqs. (2) and (3) respectively:

$$T_{ON}(n) = \lim_{n \to N} \sum_{i=n}^{N} P_{01} P_{00}^{i} \quad (2)$$

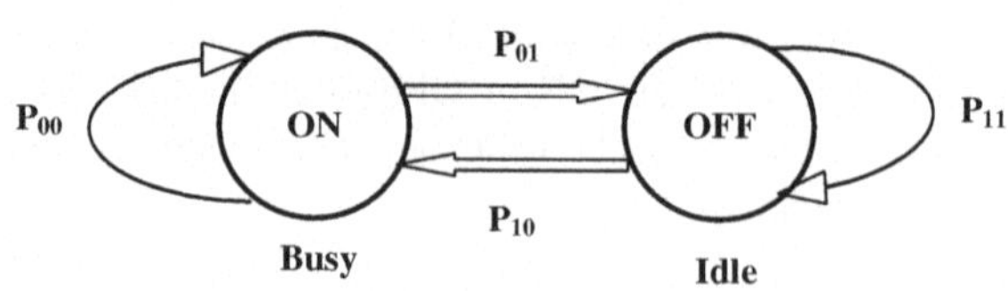

Fig. 2 ON-OFF two states discrete Markov model for a single channel

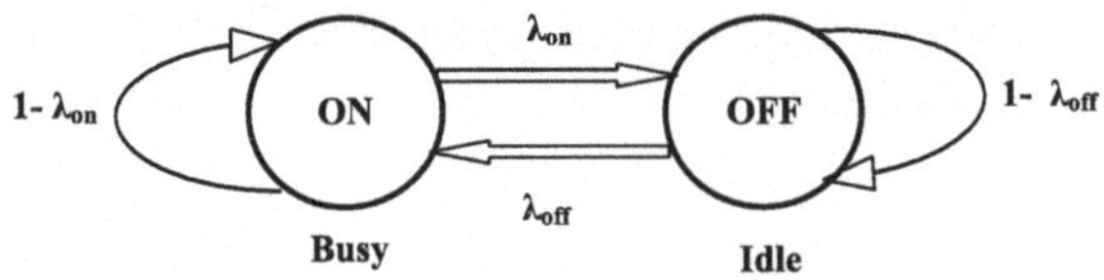

Fig. 3 ON-OFF two states continuous Markov model

$$T_{OFF}(n) = \lim_{n \to N} \sum_{j=n}^{N} P_{10} P_{11}^{j} \tag{3}$$

With $n = \{1, 2 \ldots N\}$.

ON-OFF CTMC Model. CTMC behave much like DTMC, with the key difference that the jumps between states take place at random times rather than at fixed steps. Characterized by:

- The amount of time spent in state i is an exponential distribution with mean λ.
- When the process leaves state i, it next enters state j with transition rate q_{ij} (Fig. 3).

Let λ_{on} be the flow rate from zero to one and λ_{off} be the flow rate from one to zero. The transition rate matrix is given by:

$$Q = \begin{bmatrix} -\lambda_{on} & \lambda_{on} \\ \lambda_{off} & -\lambda_{off} \end{bmatrix} \tag{4}$$

Mathematically, the durations for PU to stay in ON state and in the OFF state follow two independent random variables which are given to be exponentially distributed, with probability distribution functions [16]:

$$f_{on}\left(T_{on}^{k}\right) = \lambda_{on} e^{-\lambda_{on} T_{on}^{k}} \tag{5}$$

$$f_{off}\left(T_{off}^{k}\right) = \lambda_{off} e^{-\lambda_{off} T_{off}^{k}} \tag{6}$$

where $(\lambda_{on}, \lambda_{off})$ are the mean parameters for the exponential distributions.

ON-OFF Bernoulli Model. The probability density function for Bernoulli process given by:

$$f(x) = p^{x}(1-p)^{(1-x)} \tag{7}$$

where $x \in \{0,1\}$, p is the probability of channel in ON state and $(1 - p)$ is the probability of channel in OFF state.

ON-OFF Exponential Model. Mathematically the ON and OFF durations are represented as [3]:

$$f_{on}\left(T_{on}^{k}\right) = \lambda_{on} e^{-\lambda_{on} T_{on}^{k}} \tag{8}$$

$$f_{off}\left(T_{off}^{k}\right) = \lambda_{off} e^{-\lambda_{off} T_{off}^{k}} \tag{9}$$

2.2 Genetic Algorithm

The structure chromosome of our scheme consists of 5 genes with different size and characteristics. We utilize the same genes from our previous work [13]: FB (frequency band), PWR (power), BER (bite error rate) and MOD (modulation). In this study we added a new gene CUOI (Table 1). The CUOI is represented between 0 to 1 considering 4 bits and 16 possible levels ranging from 0000 to 1111.

Table 1 Structure of the chromosome

FB	PW	BER	MOD	CUOI
5bits	4bits	4bits	2bits	4bits

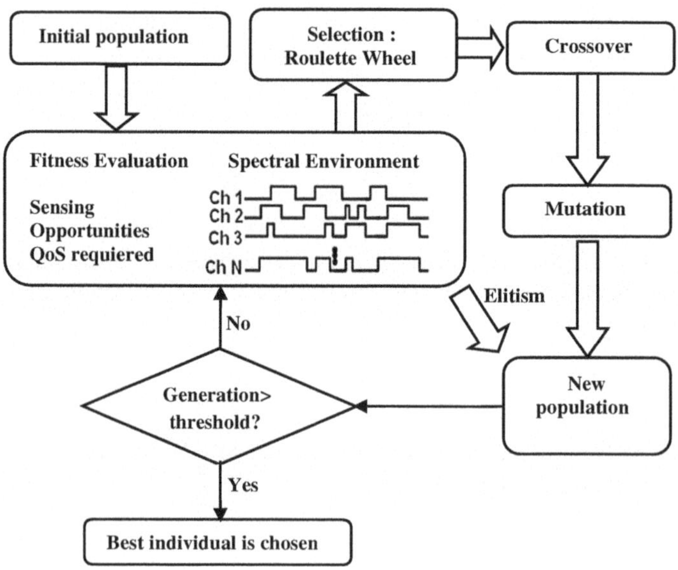

Fig. 4 Flowchart of GA scheme

The Fig. 4, show the flowchart of the proposed genetic algorithm (GA) formulation.

The fitness function of the new added gene can be represented using Eq. (10) as given in [17]:

$$f_\gamma = \frac{\gamma}{\gamma_{max}} \tag{10}$$

Where:

- γ is future channel usage opportunity for the CU, is presented by the following equation:

$$\gamma(t) = e^{-Z_k^t} \tag{11}$$

- Z_k^t The history channel vector of K_{th} channel at the instant t, which maintain the idle/busy state of a channel k. It can be represented as follows:

$$\mathrm{Z_k^t} = \{\mathrm{Z_k^1}, \mathrm{Z_k^2}, \ldots, \mathrm{Z_k^t}\} \tag{12}$$

- γ_{max} Indicates the maximum value of the CUOI.

The overall fitness function value of chromosome F can be calculated as cumulative sum of individual fitness value f_i of all the genes. This value is obtained by:

$$F = \sum_{i=1}^{5} w_i f_i \tag{13}$$

where, $w = [w_{FB}\ w_{PWR}\ w_{BER}\ w_{MOD}\ w_{CUOI}]$ is a weight vector. In current article, we utilize the following weights, $w_{CUOI} = 0.5$ and remaining 0.5 is equally divided among other four objective functions.

3 Simulation Results

In order to analyze the performance of the ON/OFF behavioral pattern of primary users (PUs) in the investigated models: CTMC, DTMC, Exponential (Exp) and Bernoulli (BN) models, five simulation scenarios was carried out using Matlab software. These simulations show the availability of resources in terms of unused slots.

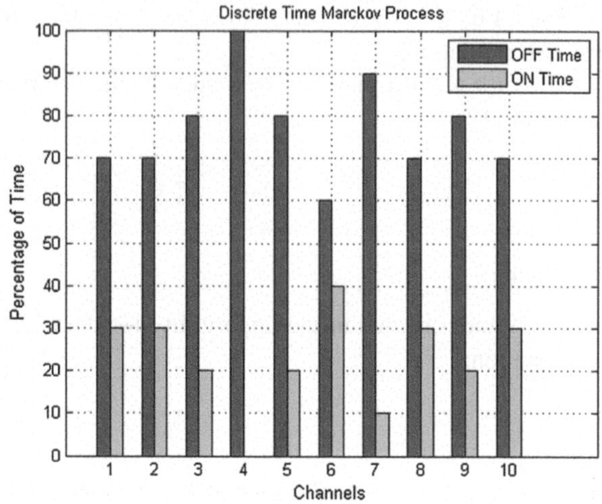

Fig. 5 PUs ON-OFF DTMC behavior

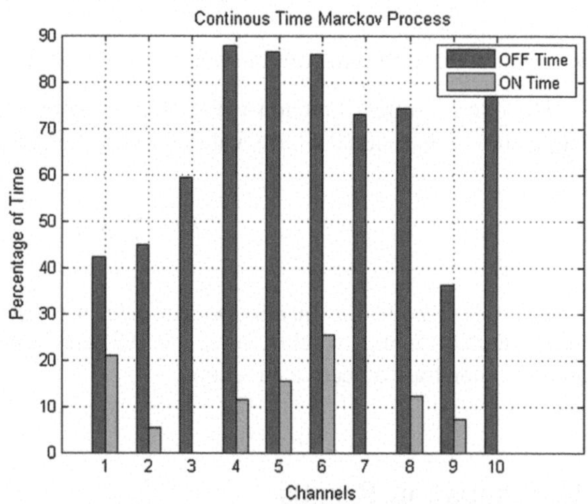

Fig. 6 PUs ON-OFF CTMC behavior

In the simulation we consider the parameters: $M = 10$ channels, $N = 10$ slots, the transition matrix P for the Markov models was chosen from [3], $P = [0.8866, 0.1134; 0.5309, 0.4691]$, exponential parameters rate $\lambda_{ON} = 0.2$ and $\lambda_{OFF} = 0.5$ and the probability $p = 0.5$ for the Bernoulli model. The Figs. 5, 6, 7 and 8 gave a percentage usage of ON/OFF time for each of the PUs slots in CTMC, DTMC, Exponential and Bernoulli models respectively.

The simulation results show that the percentage usage of the channels revolves between 10 and 40 % in DTMC. While in CTMC model it revolves around 0 and

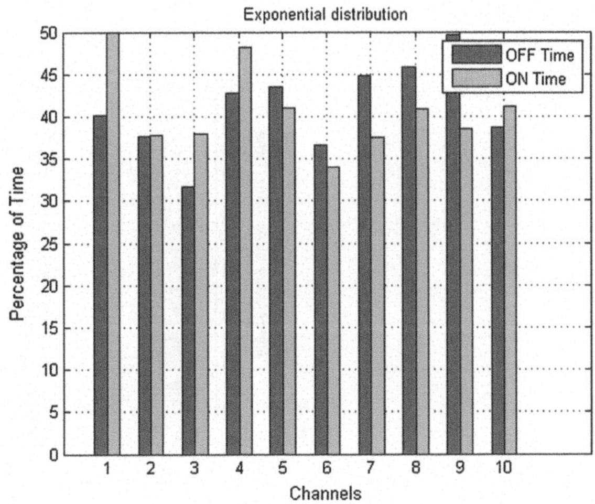

Fig. 7 PUs ON-OFF exponetial behavior

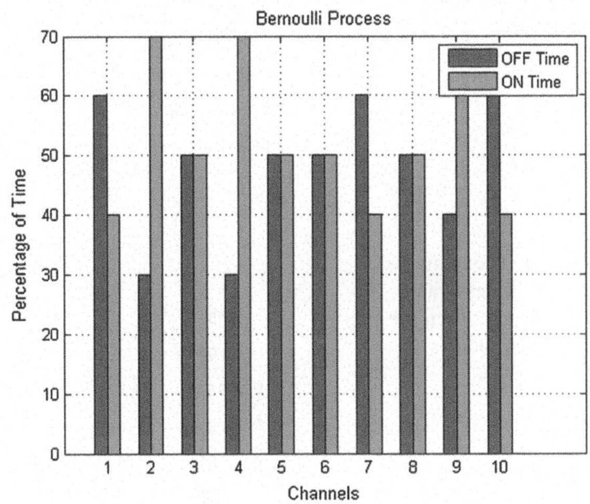

Fig. 8 PUs ON-OFF Bernoulli behavior

25 %. Contrariwise in Exp and BN models, the usage exceeds 40 %. However CTMC and DTMC are relatively reliable in terms of OFF slots than Exp and BN models with some difference between the Markov models, which is explained by the continuous sensing in the case of CTMC model (Fig. 6).

Figure 9 shows an average time for each model: an average of 85 % of available resources found in CTMC model, and 80 % in DTMC model, which agrees with [3] views about DTMC behavior. However, for both models, Exponential and Bernoulli there is no stable average of available resources.

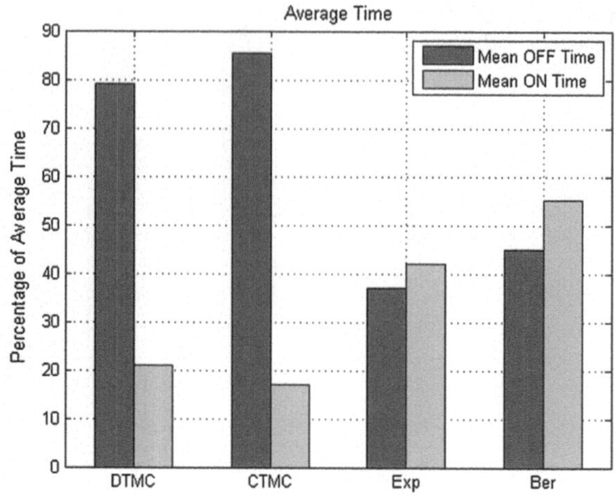

Fig. 9 Average PUs ON-OFF time for four behaviors

Table 2 Simulation parameters

Parameter	Value
Population size	20
Crossover rate	0.8
Mutation rate	0.003
Iterations	300

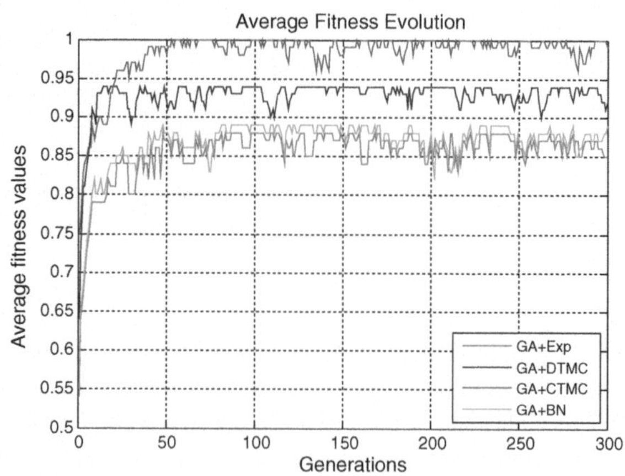

Fig. 10 The compare of average fitness value evolution between the four combined models

These results show that Markov models based on CTMC yielding potentially better results in OFF slots. Because in CTMC based modeling it is assumed that CUs observe the system state continuously and can detect randomly arriving PUs without lose of the listening contrary to other models. So, CTMC can be used for the prediction of the activities of PUs.

Each ON/OFF model is combined with the genetic algorithm (GA) to improve the channel allocation problem. Each model generates the sensing vector which describes idle/busy state of the channels. Table 2 summarizes the GA parameters employed in this work.

We compute and plot the average objective function for such combined model. The Fig. 10 shows that the average values of the given models are quickly converge to optimal value for very less number of iterations. From the results, GA converges to the best value with CTMC which is close to 1. These results confirm previous ones that CTMC based modeling can be used for prediction of the activities of PUs.

4 Conclusion

This paper gives in the first hand a comparative overview of the performance of four ON/OFF models, Continuous Time Markov chain (CTMC), Discrete Time Markov chain (DTMC), Exponential and Bernoulli models, in term of best available resources (OFF slots). The simulations show that CTMC based modeling gives better and stable results in OFF slots which can be used by cognitive radio users (CUs). So the CTMC based modeling can be used for the prediction of the activities and the behaviors of primary radio users (PUs). In the second hand, we used the genetic algorithm (GA) in the context of channel allocation problem combined with these models. Each model performs the sensing task, and enhanced the chromosome with the factor of PU activity. We added the gene, cognitive user opportunity index, to quantify the PU activity on a given channel. The performances of these models are compared in term of average fitness value. The simulation results show that the average fitness value is better by considering PU behaviors based CTMC which confirming previous results.

The multi-tier heterogeneous networks (HetNets), which consist of different types of base stations (BSs) (such as macro BSs, micro BSs, pico BSs and femto BSs), can effectively improve network capacity and thus are expected to be the dominant scenarios in the 5G era. However, the huge energy consumption of HetNets brings heavy burdens to the network operators [18]. In future work, we will implement the ON/OFF CTMC model in small cell heterogeneous networks, in order to reduce the energy consumption by switching OFF some underutilized cells during off peak hours.

References

1. Saleem, Y., Rehmani, M.H.: Primary radio user activity models for cognitive radio networks: A survey. J. Netw. Comput. Appl. **43**, 1–16 (2014)
2. Horvath, L.C., Bito, J.: Primary and secondary user activity models for cognitive wireless network. In: Proceedings of the 11th International Conference on Telecommunications, pp. 301–306 (2011)

3. Ebenezer, E., Tom, W.: Primary users ON/OFF behaviour models in cognitive radio networks. In: Proceedings of International Conference for Wireless and Mobile Communication System, Lisbon-Portugal (2014)
4. Bayhan, S., Alagoz, F.: A Markovian approach for best-fit channel selection in cognitive radio networks. J. Ad Hoc Netw. **12**, 165–177 (2014)
5. Li, Y., Dong, Y., Zhang, H., Zhao, H., Shi, H., Zhao, X.: Spectrum usage prediction based on high-order Markov model for cognitive radio networks. In: Proceedings of 10th International Conference on Computer and Information Technology, pp. 2784–2788 (2010)
6. Min, A.W. Shin, K.G.: Exploiting multi-channel diversity in spectrum-agile networks. In: Proceedings of 27th International Conference on Computer Communications, pp. 1921–1929 (2008)
7. John, A.M., Hongjun, X.: A POMDP Framework for Throughput Optimization MAC Scheme in Presence of Sensing Errors for Cognitive Radio Networks. J. Comput. Sci. Appl. **1**(4), pp. 205–216. (October 2014)
8. Kanan, E., Husari, G. Al-Ayyoub, M., Jararweh, Y.: Towards improving channel switching in cognitive radio networks. In: Proceedings of 6th International Conference on Information and Communication Systems, pp. 280–285 (April 2015)
9. Ganti, A., Modiano, E., Tsitsiklis, J.N.: Optimal transmission scheduling in symmetric communication models with intermittent connectivity. J. Trans. Inf. Theory, **53**, pp. 998–1008 (2007)
10. Banaei, A., Georghiades, C.N.: Throughput analysis of arandomized sensing scheme in cell-based ad-hoc cognitive networks. In: Proceedings of International Conference on Communications, pp. 1–6 (2009)
11. Jonathan, G., Simeone, O., Bar-Ness, Y., Spagnolini, U., Yu, T.: Packet-wise vertical handover for unlicensed multi-standard spectrum access with cognitive radios. J. IEEE Trans. Wirel. **7**, pp. 5172–5176 (2008a)
12. Balieiro, A., Yoshioka, P., Dias, K., Cavalcanti, D., Cordeiro, C.: A multi-objective genetic optimization for spectrum sensing in cognitive radio. J. Expert Syst. Appl. **41**, pp. 3640–3650 (2014)
13. El Morabit, Y., Mrabti, F., Abarkan, H.: Spectrum allocation using genetic algorithm in cognitive radio networks. In: Proceedings of 3rd International Workshop on RFID and Adaptive Wireless Sensor Network, pp. 90–93. IEEE Xplorer (2015)
14. Tim, B.: Stochastic Simulation of Processes. Fields and Structures. Ulm University Institute of Stochastics (2014)
15. Yigit, S.: Introduction to Probability Theory for Graduate Economics (2008)
16. Wang, Z., Chew, Y.H., Yuen, C.: On discretizing the exponential ON-OFF primary radio activities in simulations. In: Proceedings of 22nd International Symposium on Personal, Indoor and Mobile Radio Communications, pp. 556–560 (2011)
17. Aslam, S., Shahid, A., Lee, K.: GA-CSS: genetic algorithm based control channel selection scheme for cognitive radio networks. In: Proceedings of 7th International Conference on Next Generation Mobile Apps, Services and Technologies, pp. 232–236 (2013)
18. Zhang, S., Gong, J., Zhou, S., Niu, Z.: How Many Small Cells Can be Turned Off via Vertical Offloading Under a Separation Architecture? J. IEEE Trans. Wirel. Commun. **14**, 5440–5453 (2015)

An Optimized Vertical Handover Approach Based on M-ANP and TOPSIS in Heterogeneous Wireless Networks

Mohamed Lahby, Abdelbaki Attioui and Abderrahim Sekkaki

Abstract Due to a deployment of different networks technologies such as 3G (UMTS, IEEE 802.11), 4G (LTE, IEEE 802.16) and 5G, the users have the opportunity to be connected to Internet at any time and any where. This ability to be quickly and easily connected is ensured by using the intelligent mobile terminal multi-modes such as mobile phones, smart-phones, IPAD, etc. These equipments mobiles have enabled users also to handle simultaneously various applications by using different access networks. The most issue in this heterogeneous wireless network is enabling for users to continuously choose the most appropriate access network during their communication. To deal with this task, we propose a new approach for network selection based on two multi attribute decision making (MADM) methods namely multiple analytic network process (M-ANP) and technique for order preference by similarity to ideal solution (TOPSIS) method. The M-ANP is used to weigh each criterion and TOPSIS is applied to rank the alternatives. The simulation results illustrate the effectiveness of our optimized approach in terms of reducing of the reversal phenomenon and the ping-pong phenomenon.

Keywords Heterogeneous wireless networks · IEEE 802.21 · Vertical handover · Multi Attribute Decision Making · M-ANP · TOPSIS

M. Lahby (✉) · A. Attioui
Laboratory of Mathematics and Applications, University Hassan II,
Ecole Normale Suprieure (ENS), Casablanca, Morocco
e-mail: mlahby@gmail.com

A. Attioui
e-mail: abdelbaki.attioui@gmail.com

A. Sekkaki
Laboratory of Computer Science and Decision Support, Faculty of Sciences Ain Chock,
University Hassan II, Casablanca, Morocco
e-mail: sekkabd@gmail.com

R. El-Azouzi et al. (eds.), *Advances in Ubiquitous Networking 2*, Lecture Notes in Electrical Engineering 397, DOI 10.1007/978-981-10-1627-1_2

1 Introduction

Nowadays, several wireless technologies such as 3G (UTMS, IEEE 802.11a, IEEE 802.11b, etc.), and 4G (IEEE 802.16, LTE, LTE-A) have already deployed by different telecommunication operator's. Moreover, this heterogeneous environment, can ensure diversity for multimedia applications and provide to mobile user the ability to be connected by using the mobile Internet. In addition theses application, taking advantage of the advanced features of the mobile devices which are equipped with several wireless interfaces. These diversity of interfaces allow the users not only to be connected at any access network, but also he can benefit simultaneously from variety of services delivered by these technologies.

The most important issue concerning this heterogeneous networks, is to ensure ubiquitous access for the end users, under the principle "Always Best Connected" (ABC) [1]. For that, the IEEE 802.21 standard [2] is intended to determine whether a vertical handoff should be initiated, and to choose the most suitable network in terms of quality of service (QoS) for mobile users.

The standard IEEE 802.21 defines three parts in order to manage the vertical handover process [3]. These parts are:

- Handover initiation: in this step, the terminal discovers available networks.
- Handover decision: it's namely also network selection decision. In this step the mobile terminal evaluates the reachable wireless networks to make a decision according some criteria such as battery, velocity, QoS level, security level, users preferences, perceived QoS, etc.
- Handover execution: it consists on establishing the target access network by using mobile IP protocol.

However, the network selection algorithm is not specified in IEEE 802.21 which is important role in the vertical handover process. To cope with this issue, our objective in this paper is to optimize this step by proposing a new approach for network selection decision which allows to the user to choose the most suitable network in terms of QoS.

During recent years, different algorithms were proposed in order to solve and to optimize the network selection problem. According to [3], we can categorize the network selection algorithms into four kinds such handover based RSS, handover based bandwidth, cost function and combination algorithms. The last category includes handover algorithms that use fuzzy logic, neural networks, genetic algorithms and MADM methods. Based on the literature review, the MADM methods represent a promising solution to choose dynamically the optimal access network, which can satisfying the QoS from the available networks.

This paper is organized as follows. Section 2 describes Multi Attribute Decision Making methods (MADM). Section 3 presents our access network selection algorithm based on M-ANP and TOPSIS two MADM methods. Section 4 includes the simulations and results. Section 5 concludes this paper.

2 MADM-based Network Selection

2.1 Related Work and Problem Statement

Several network selection algorithms based on MADM methods have been proposed and developed exhaustively in the literature in the last decade. In [4] the authors have evaluated the performance of eight MADM methods namely SAW, MEW, TOPSIS, GRA, VIKOR, DIA, E-TOPSIS and FADM. This comparison study allows to identify a suitable MADM algorithm which can be used in the context of vertical handover decision. In [5, 6] the network selection algorithm is based on Analytic Hierarchy Process (AHP) and Gray Relation Analysis (GRA) two MADM methods. The AHP method is used to determine weights for each criterion and GRA method is applied to rank the alternatives. In [7, 8] the network selection algorithm combines two MADM methods AHP and TOPSIS. The AHP method is used to get weights of the criteria and TOPSIS method is applied to determine the ranking of access network.

In addition, there are several methods used to assign weights for the criteria such as analytic hierarchy process (AHP), fuzzy analytic hierarchy process (FAHP), analytic network process (ANP), fuzzy analytic network process (FANP) and random weighting. Determining the most suitable weights for different criteria for each traffic classes is one of the main problems in the network selection decision. The work in [9] studied and compared five weighting algorithms namely AHP, FAHP, ANP, FANP and RW for all four traffic classes namely, conversational, streaming, interactive and background. According to reference [9], the ANP method is the appropriate algorithm which should be used to weigh the criteria. In this context, the work in [10] proposed intelligent network selection strategy which combines two MADM algorithms the ANP method to the TOPSIS technique. The ANP method is used to find the differentiate weights of available networks by considering each criterion and the TOPSIS method is applied to rank the alternatives.

However, one of the major limitations of the ANP method is that in the majority of situations necessitate to re-establish the pairwise comparison matrix in cases, where the judgment matrix is inconsistent. This weakness is due to the decision markers, ANP method is based only on the experience of one expert to build the matrix decision which can not reflect the real user's preferences. To deal with these weakness we propose Multiple Analytic Network Process (M-ANP) method, this one takes into account the experiences of multiple experts to build the matrix decision and to determine the weights of criteria. On the other hand TOPSIS method suffers from ranking abnormality [11].

The goal of this paper, is providing an optimal network selection algorithm, which can deal with the ranking abnormality of TOPSIS method. For that, we propose a new approach which combines two MADM methods, the multiple analytic network process (M-ANP) and TOPSIS method. The M-ANP is applied to determine the suitable weights for different criteria and TOPSIS method is used to rank the alternatives.

2.2 The ANP Method

The ANP method is proposed by Saaty [12], in order to extend the AHP approach to problems with dependence and feed beck within clusters (inner dependence) and between clusters (outer dependence). The ANP method is based on six steps:

1. Model construction: A problem is decomposed into a network in which nodes corresponds to components. The elements in a component can interact with some or all of the elements of another component. Also, relationships among elements in the same component can exist. These relationships are represented by arcs with directions.
2. Construct of the pairwise comparisons: To establish a decision, ANP builds the pairwise matrix comparison such as

$$A = (x_{ij})\ where\ x_{ji} = \begin{cases} 1 & \text{si } i = j; \\ \frac{1}{x_{ij}} & \text{si } i\#j. \end{cases} \tag{1}$$

   Elements x_{ij} are obtained from the Table 1, it contains 1–9 preference scales.
3. Construct the normalized decision matrix: A_{norm} is the normalized matrix of A(1), where $A(x_{ij})$ is given by, $A_{norm}(a_{ij})$ such:

$$a_{ij} = \frac{x_{ij}}{\sum_{i=1}^{n} x_{ij}} \tag{2}$$

4. Calculating the weights of criterion: The weights of the decision factor i can be calculated by

$$W_i = \frac{\sum_{j=1}^{n} a_{ij}}{n}\ and\ \sum_{j=1}^{n} W_i = 1 \tag{3}$$

   With n is the number of the compared elements.
5. Calculating the coherence ratio (CR): To test consistency of a pairwise comparison, a consistency ratio (CR) can be introduced with consistency index (CI) and random index (RI).

Table 1 Saaty's scale for pairwise comparison

Saaty's scale	The relative importance of the two sub-elements
1	Equally important
3	Moderately important with one over another
5	Strongly important
7	Very strongly important
9	Extremely important
2, 4, 6, 8	Intermediate values

Table 2 Value of random consistency index RI

Criteria	3	4	5	6	7	8	9	10
RI	0.58	0.90	1.12	1.24	1.32	1.41	1.45	1.49

Let us define consistency index CI

$$CI = \frac{\lambda_{max} - n}{n - 1} \tag{4}$$

Also, we need to calculate the λ_{max} by the following formula:

$$\lambda_{max} = \frac{\sum_{i=1}^{n} b_i}{n} \quad such \; b_i = \frac{\sum_{j=1}^{n} W_i * a_{ij}}{W_i} \tag{5}$$

We calculate the coherence ratio CR by the following formula:

$$CR = \frac{CI}{RI} \tag{6}$$

The various values of RI are shown in Table 2. If the CR is less than 0.1, the pairwise comparison is considered acceptable.

6. Construct the super-matrix formation: the local priority vectors are entered into the appropriate columns of a super-matrix, which is a partitioned matrix where each segment represents a relationship between two components.

2.3 *The TOPSIS Technique*

The TOPSIS technique is known as a classical MADM method, has been developed in 1981 [13]. The basic principle of the TOPSIS is that the chosen alternative should have the shortest distance from the positive ideal solution and the farthest distance from the negative ideal solution.

The procedure can be categorized in six steps:

1. Construct of the decision matrix: the decision matrix is expressed as

$$D = (d_{ij}) \tag{7}$$

where d_{ij} is the rating of the alternative A_i with respect to the criterion C_j

2. Construct the normalized decision matrix: each element r_{ij} is obtained by the euclidean normalization.

$$r_{ij} = \frac{d_{ij}}{\sqrt{\sum_{i=1}^{m} d_{ij}^{2}}}, i = 1, ..., m, j = 1, ..., n. \quad (8)$$

3. Construct the weighted normalized decision matrix: The weighted normalized decision matrix v_{ij} is computed as:

$$v_{ij} = W_i * r_{ij} \ \ where \sum_{i=1}^{m} W_i = 1 \quad (9)$$

4. Determination of the ideal solution A^* and the anti-ideal solution A^-:

$$A^* = [V_1^*, ..., V_m^*] \ \ and \ \ A^- = [V_1^-, ..., V_m^-], \quad (10)$$

- For desirable criteria:

$$V_i^* = max\{v_{ij}, j = 1, ..., n\} \ and \ V_i^- = min\{v_{ij}, j = 1, ..., n\} \quad (11)$$

- For undesirable criteria:

$$V_i^* = min\{v_{ij}, j = 1, ..., n\} \ and \ V_i^- = max\{v_{ij}, j = 1, ..., n\} \quad (12)$$

5. Calculation of the similarity distance:

$$S_j^* = \sqrt{\sum_{i=1}^{m} (V_i^* - v_{ji})^2}, j = 1, ..., n \quad (13)$$

and

$$S_j^- = \sqrt{\sum_{i=1}^{m} (v_{ji} - V_i^-)^2}, j = 1, ..., n \quad (14)$$

6. Ranking:

$$C_j^* = \frac{S_j^-}{S_j^* + S_j^-}, j = 1, ..., n. \quad (15)$$

A set of alternatives can be ranked according to the decreasing order of C_j^*.

3 Our Optimized Vertical Handover Algorithm

In order to provide an optimal network selection algorithm, we propose a new approach which combines two MADM methods such as M-ANP and TOPSIS. The M-ANP method, takes into consideration the experiences of multiple experts to build the matrix decision and to weigh each criterion. In this work, M-ANP method is based on the experience of three experts. The basic principle of M-ANP as follows:

Let us define the weight vector W_{ANP_i}, obtained by ANP based only on the experience of one expert i:

$$W_{ANPi} = [a_{i1}, a_{i2}, ...a_{im}] \ where \ \sum_{j=1}^{m} a_{ij} = 1 \ and \ i = 1, ..., 3 \tag{16}$$

The weight vector $W_{M\text{-}ANP}$, can be calculated by using geometric mean:

$$W_{M\text{-}ANP} = [c_1, c_2, ...c_m], \ c_j = \sqrt[3]{\prod_{i=1}^{3} a_{ij}} \ where \ j = 1, ..., m \tag{17}$$

In addition, the algorithm assumes wireless overlay networks which entails three heterogeneous networks such as UMTS, IEEE 802.11 and IEEE 802.16. The six attributes associated in this heterogeneous environment are: Cost per Byte (CB), Available Bandwidth (AB), Security (S), Packet Delay (D), Packet Jitter (J) and Packet Loss (L).

The M-ANP algorithm based network selection contain three level in order to weigh the criteria. The first level includes three criteria QoS, security and cost, the second level includes four QoS parameters such as AB, D, J and L and the level 3 includes three available networks UMTS, IEEE 802.11 and IEEE 802.16.

Our new approach for network selection based on M-ANP and TOPSIS consists of the four following steps:

1. Assign weights to level-1: the M-ANP method is used to get a weight of the decision criteria of level 1.
2. Assign weights to level-2: the M-ANP method is used to get a weight of the decision criteria of level 2.
3. Assign weights to level-3: the weight vector of each available network is calculated by multiplication of the weight vector obtained in level 1 with the weight vector obtained in level 2.
4. Select the best access network: the method TOPSIS is applied to rank the available networks and select the access network that has the highest value of C_j^* (see the steps of TOPSIS method).

4 Simulation and Results

In order to validate our optimized vertical handover approach which based on M-ANP to weigh different criteria and TOPSIS to rank available networks, we present the performance comparison between four algorithms:

Table 3 Attribute values for the candidate networks

Nework/criteria	CB (%)	S (%)	AB (mbps)	D (ms)	J (ms)	L ($per10^6$)
UMTS	60	70	0.1–2	25–50	5–10	20–80
IEEE 802.11	10	50	1–11	100–150	10–20	20–80
IEEE 802.16	50	60	1–60	60–100	3–10	20–80

- TOPSIS-ANP 1: this algorithm is applied by the first expert, it's based on the ANP method which used to get the weights of criteria and TOPSIS algorithm which applied to rank each access network.
- TOPSIS-ANP 2: this algorithm is applied by the second expert, it's based on the ANP method to weigh criteria and TOPSIS algorithm.
- TOPSIS-ANP 3: this algorithm is applied by the third expert, it's based on the ANP method and TOPSIS algorithm.
- TOPSIS-M-ANP: this algorithm represents our optimized strategy for network selection. Firstly the M-ANP is used to weigh each criterion. While the TOPSIS is applied to get the ranking of different networks.

We perform four simulations according to four traffic classes [14] namely background, conversational, interactive, and streaming. For each simulation, we provided the values for average of ranking abnormality and the number of handoffs.

We execute these algorithms in 1000 decision points by using MATLAB simulator. During the simulation, the measures of each criterion for candidate networks are randomly varied according to the ranges shown in Table 3.

4.1 Simulation 1

In this simulation, the traffic analyzed is background traffic. The set of importance weights of the criteria based on each algorithm are displayed in Fig. 1.

4.1.1 Ranking Abnormality

Figure 2 shows that TOPSIS-ANP 1, TOPSIS-ANP 2, TOPSIS-ANP 3 and TOPSIS-M-ANP reduce the risk to have an abnormality problem with the values of 33, 35, 32.5 and 25.42 % respectively. For background traffic, our strategy TOPSIS- M-ANP can reduce the ranking abnormality problem better than TOPSIS based on one decision maker.

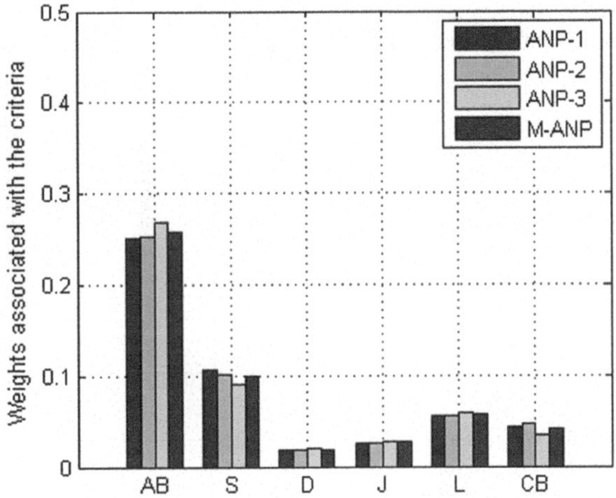

Fig. 1 Weights associated with the criteria for background traffic

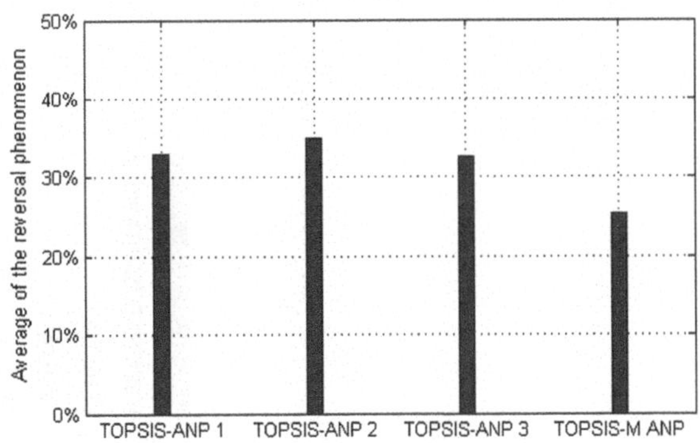

Fig. 2 Average of Ranking abnormality for background traffic

4.1.2 Number of Handoffs

Figure 3 shows that TOPSIS-ANP 1, TOPSIS-ANP 2, TOPSIS-ANP 3 algorithms diminish the number of handoffs with the values of 42, 46 and 42.50 % respectively. While the TOPSIS-M-ANP method provides a value of 33.35 %. We deduce that for background traffic, TOPSIS-M-ANP method provides better performances concerning the number of handoffs than all TOPSIS based on one expert to weigh the criterion.

4.2 Simulation 2

In this simulation, the traffic analyzed is conversational traffic. The weights of the criteria based on each algorithm are displayed in Fig. 4.

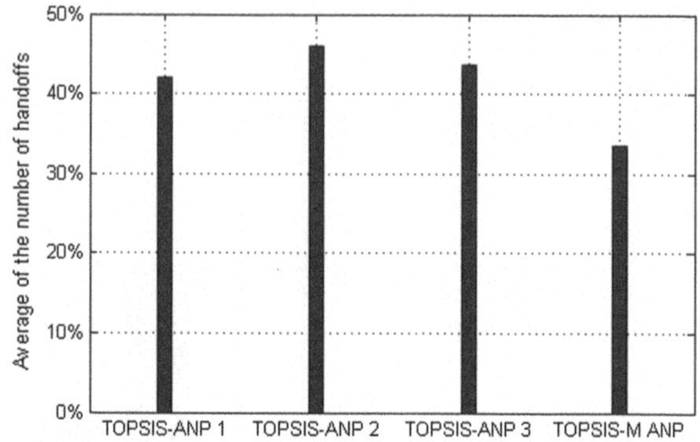

Fig. 3 Average of number of handoffs for background traffic

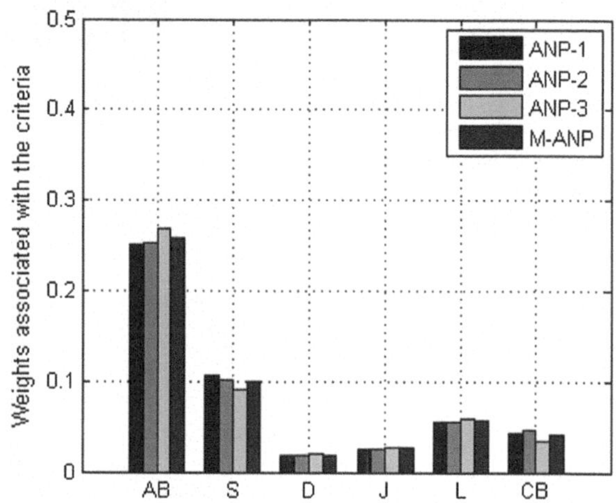

Fig. 4 Weights associated with the criteria for conversational traffic

4.2.1 Ranking Abnormality

Figure 5 shows that the three methods TOPSIS-ANP 1, TOPSIS-ANP 2, TOPSIS-ANP 3 reduces the risk of the abnormality phenomenon with the values of 25.5, 23.33 and 26.66 % respectively. While our TOPSIS based on M-ANP reduces the risk with a value of 20.5 %. For conversational traffic, our approach TOPSIS-M-ANP can reduce the ranking abnormality problem better than all algorithms which based on TOPSIS and one expert using ANP method.

4.2.2 Number of Handoffs

Figure 6 shows that the TOPSIS-ANP 1 method diminishes the number of handoffs with a value of 37.5 %, the TOPSIS-ANP 2 provides a value of 36 % and the TOPSIS-ANP 3 provides a value of 38.66 %. While the TOPSIS-M-ANP method

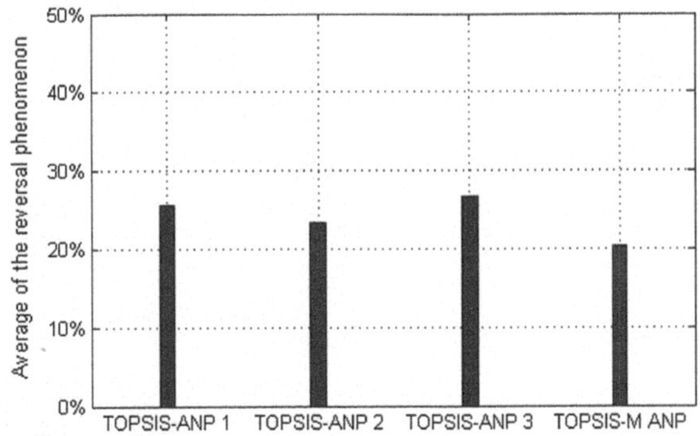

Fig. 5 Average of Ranking abnormality for conversational traffic

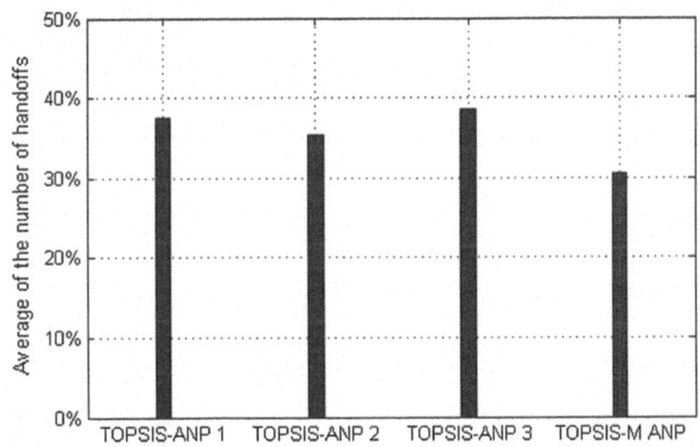

Fig. 6 Average of number of handoffs for conversational traffic

provides a value of 30.44 %. We deduce that for conversational traffic, TOPSIS based on M-ANP provides better performances concerning the number of handoffs than all algorithms.

4.3 Simulation 3

This simulation consists in analyzing interactive traffic, the weights of the criteria based on each algorithm are displayed in Fig. 7.

4.3.1 Ranking Abnormality

Figure 8 shows that the four algorithms TOPSIS-ANP 1, TOPSIS-ANP 2, TOPSIS-ANP 3 and TOPSIS-M-ANP reduce the risk of ranking abnormality with the values of 18.33, 19.67, 17.57 and 14.33 % respectively. For interactive traffic, our strategy TOPSIS-M-ANP can reduce the ranking abnormality problem better than TOPSIS based on one decision maker.

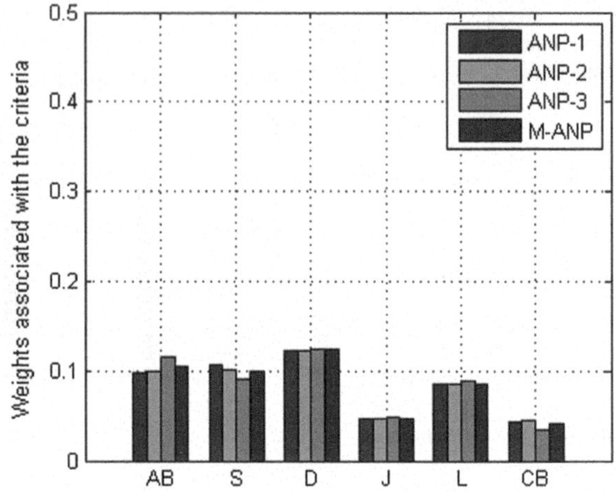

Fig. 7 Weights associated with the criteria for interactive traffic

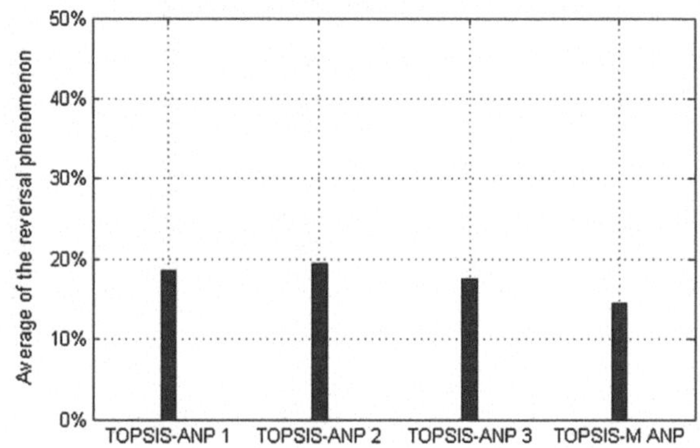

Fig. 8 Average of ranking abnormality for interactive traffic

4.3.2 Number of Handoffs

Figure 9 shows that TOPSIS-ANP 1, TOPSIS-ANP 2, TOPSIS-ANP 3 algorithms diminish the number of handoffs with the values of 25.5 %, 26.5 % and 24.66 % respectively. While the TOPSIS-M-ANP method provides a value of 18.33 %. We deduce that for interactive traffic, TOPSIS-M-ANP method provides better performances concerning the number of handoffs than all TOPSIS based on one expert to weigh the criterion.

4.4 Simulation 4

This simulation consists in analyzing streaming traffic, the weights of the criteria based on each algorithm are displayed in Fig. 10.

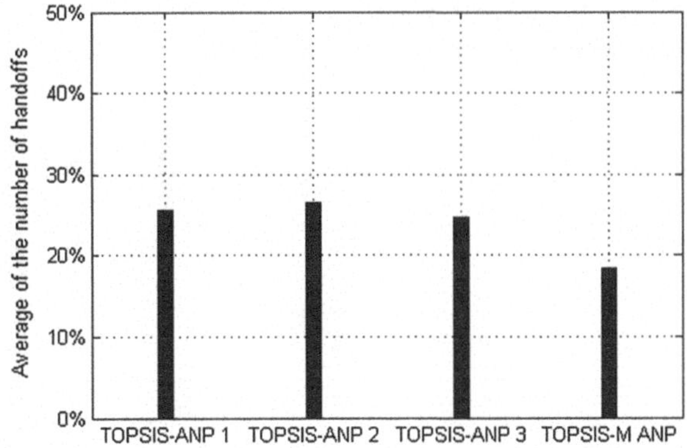

Fig. 9 Average of number of handoffs for interactive traffic

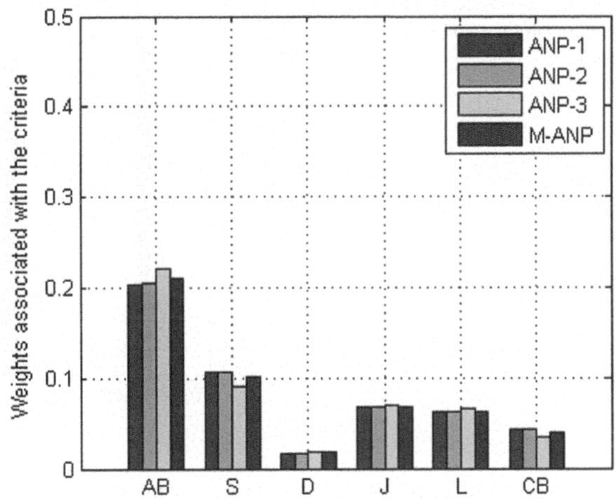

Fig. 10 Weights associated with the criteria for streaming traffic

4.4.1 Ranking Abnormality

Figure 11 shows that the three methods TOPSIS-ANP 1, TOPSIS-ANP 2, TOPSIS-ANP 3 reduces the risk of the abnormality phenomenon with the values of 35 %, 36.5 % and 35.66 % respectively. While our TOPSIS based on M-ANP reduces the risk with a value of 28.5 %. For streaming traffic, our approach TOPSIS-M-ANP can reduce the ranking abnormality problem better than all algorithms which based on TOPSIS and one expert using ANP method.

4.4.2 Number of Handoffs

Figure 12 shows that the TOPSIS-ANP 1 method diminishes the number of handoffs with a value of 45.5 %, the TOPSIS-ANP 2 provides a value of 46.33 % and the TOPSIS-ANP 3 provides a value of 45.44 %. While the TOPSIS-M-ANP method

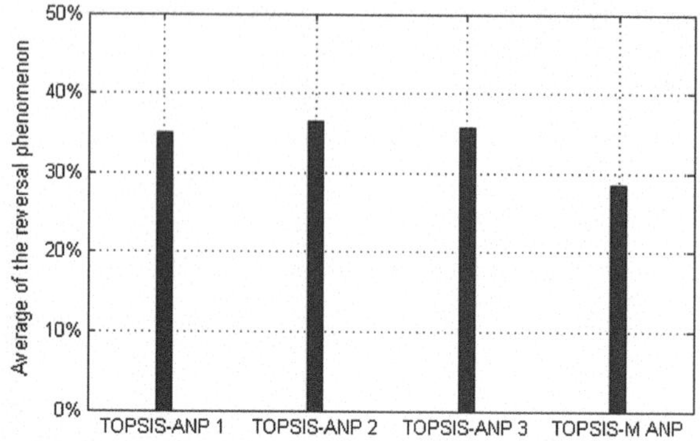

Fig. 11 Average of Ranking abnormality for streaming traffic

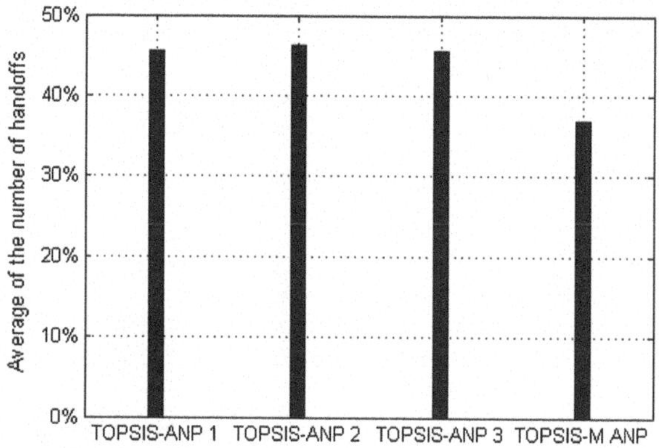

Fig. 12 Average of number of handoffs for streaming traffic

provides a value of 36 %. We deduce that for streaming traffic, TOPSIS based on M-ANP provides better performances concerning the number of handoffs than all algorithms.

5 Conclusion

In this work, we have proposed a new approach based on multiple analytic network process (M-ANP) method and TOPSIS method. The M-ANP method, allows to assign a suitable weights of different criteria better than ANP method.

The simulation shows that, for each traffic classes, our method based on M-ANP and TOPSIS can reduce the ranking abnormality problem better than ANP and TOPSIS method for all traffic classes.

In the other hand our optimized algorithm which combine M-ANP and TOPSIS two MADM methods provides best performance concerning the number of handoffs than the classical algorithm based on ANP and TOPSIS for each traffic.

References

1. Gustafsson, E., Jonsson, A.: Always best connected. IEEE Wirel. Commun. Mag. **10**(1), 49–55, Feb 2003
2. IEEE 802.21. Ieee standard for local and metropolitan area networks, part 21: media independent handover services, 21 Jan 2009
3. Lahby, M., et al.: A novel ranking algorithm based network selection for heterogeneous wireless access. J. Netw. **8**(2), 263–272 (2013)
4. Lahby, M., Silki, B., Sekkaki, A.: Survey and comparison of MADM methods for network selection access in heterogeneous networks. In: 7th IFIP International Conference on New Technologies, Mobility and Security (NTMS), pp. 1–6 (2015)
5. Lahby, M., et al.: Network selection mechanism by using M-AHP/GRA for heterogeneous networks. In: the Sixth Joint IFIP Wireless and Mobile Networking Conference (WMNC), pp. 1–6 (2013)
6. Fu, J., et al.: Novel AHP and GRA based handover decision mechanism in heterogeneous wireless networks. Information Computing and Applications, pp. 213–220. Springer, Berlin (2010)
7. Lahby, M., Leghris, C., Adib, A.: A hybrid approach for network selection in heterogeneous multi-access environments. In: 4th IFIP International Conference on New Technologies, Mobility and Security (NTMS), pp. 1–5 (2011)
8. Sgora, A., et al.: An access network selection algorithm for heterogeneous wireless environments. In: The IEEE symposium on Computers and Communications (2010)
9. Lahby, M., et al.: A Survey and comparison study on weighting algorithms for access network selection. In: the Proceedings of the 9th Annual Conference on Wireless On-Demand Network Systems and Services, pp. 35–38 (2012)
10. Lahby, M., et al.: An intelligent network selection strategy based on MADM methods in heterogeneous networks. Int. J. Wirel. Mob. Netw. (IJWMN) **4**(1), 83–96 (2012)
11. Bari, F., Leung, V.: Multi-attribute network selection by iterative TOPSIS for heterogeneous wireless access. In: 4th IEEE Consumer Communications and Networking Conference, pp. 808–812, Jan 2007
12. Lee, J., Kim, S.: Using analytic network process and goal programming for interdependent information system project selection. Comput. Oper. Res. **27**(4), 367–382, Apr 2000
13. Triantaphyllou, E.: Multi-Criteria Decision Making Methods: A Comparative Study. Applied optimization series. Kluwer Academic Publishers (2002)
14. 3GPP, QoS Concepts and Architecture tS 22.107 (v 6.3.0) (2005)

Performance Analysis of Routing Protocols in Vehicular Ad Hoc Network

Bouchra Marzak, Hicham Toumi, Elhabib Benlahmar and Mohamed Talea

Abstract Vehicular Ad Hoc Network (VANET) networks are very likely to be deployed in the coming years and thus become the most relevant form of mobile ad hoc networks. They provide wireless communication among vehicles and vehicle-to-road side equipment. The communication between vehicles is used for safety, comfort and for entertainment as well. The performance of communication depends on how better the routing takes place in the network. Routing of data depends on the routing protocols being used in network. In this article, we investigated different routing protocols for VANET. The main aim of our study was to identify which routing method has better performance in highly mobile environment of VANET.

Keywords VANET · Mobile ad hoc network · Wireless · Routing protocols · Performance

1 Introduction

Vehicular Ad hoc Networks (VANET) is the subclass of Mobile Ad Hoc Networks. VANET is one of the influencing areas for the improvement of Intelligent Transportation System (ITS) in order to provide safety and comfort to the road users. VANET assists vehicle drivers to communicate and to coordinate among them-

B. Marzak (✉) · H. Toumi · M. Talea
Laboratory of Information Processing, University Hassan II, Cdt Driss El Harti, BP 7955 Sidi Othman, 20702 Casablanca, Morocco
e-mail: marzak8bouchra@gmail.com

H. Toumi
e-mail: toumi.doc@gmail.com

E. Benlahmar
Information Technology and Modeling Laboratory, University Hassan II, Cdt Driss El Harti, BP 7955 Sidi Othman, 20702 Casablanca, Morocco
e-mail: h.benlahmer@gmail.com

R. El-Azouzi et al. (eds.), *Advances in Ubiquitous Networking 2*, Lecture Notes in Electrical Engineering 397, DOI 10.1007/978-981-10-1627-1_3

selves in order to avoid any critical situation through Vehicle to Vehicle communication e.g. road side accidents, traffic jams, speed control, free passage of emergency vehicles and unseen obstacles etc. Besides safety applications VANET also provide comfort applications to the road users. VANET are self-organizing network. It does not rely on any fixed network infrastructure. Although some fixed nodes act as the roadside units to facilitate the vehicular networks for serving geographical data or a gateway to internet etc. [1]. Higher node mobility, speed and rapid pattern movement are the main characteristics of VANET. This also causes rapid changes in network topology [2].

VANET is a particular type of MANET, in which vehicles act as nodes. Contrary to MANET, vehicles move on predefined roads, vehicles velocity depends on the speed signs and in addition these vehicles also have to follow traffic signs and traffic signals [3]. There are many challenges in VANET that are needed to be solved in order to provide reliable services. Stable and reliable routing in VANET is one of the major issues [4]. Therefore, more research is needed to be conducted in order to make VANET more applicable. As vehicles have dynamic behavior, high speed and mobility that make routing even more challenging.

VANET applications are based on Vehicle-to-Vehicle (V2V) and Vehicle-to-Infrastructure (V2I) communications. Vehicles become smarter with the installation of embedded systems and sensors. Sensors collect crucial data about the situation on the road and this information is exchanged in order to help the driver make appropriate decisions. The driver receives information about a local anomaly, a too short inter-distance with the leading vehicle, lane departure etc. Exchange of this information among neighboring vehicles is crucial for VANET applications to be efficient. Communication between vehicles can be used to inform drivers about congested roads ahead, a car accident, parking facilities etc. As a result, Inter-vehicle communications (IVC) may help drivers avoid dangerous situations, decrease driver time, fuel consumption and have an overall better driving satisfaction. Most of these applications demand data dissemination among vehicles. VANET has the communication type: Vehicle-to-Vehicle (V2V) and Vehicle-to-Infrastructure (V2I).

Data dissemination generally refers to the process of spreading data or information over distributed wireless networks. This dissemination uses one of the two available communication modes. The message will be disseminated in a multi-hop fashion when V2V communication is enabled and will be broadcasted by all the roadside units (RSU) when V2I communications are used instead. A hybrid version is also possible, RSUs broadcast the messages and, as they do not cover the whole network, some vehicles are selected to forward the message to complete the dissemination. These messages can be flooded at a certain number of hops or in a given area depending on the application purposes [5]. In V2V mode, the tasks of a dissemination protocol consist in selecting a pertinent set of vehicles to disseminate the message, and defining retransmission procedures to ensure the entire applications requirements on reliability, delay, etc. [6].

Several routing protocols have been proposed to make routing more efficient and reliable in VANET [7]. These protocols looking to maximize throughput while

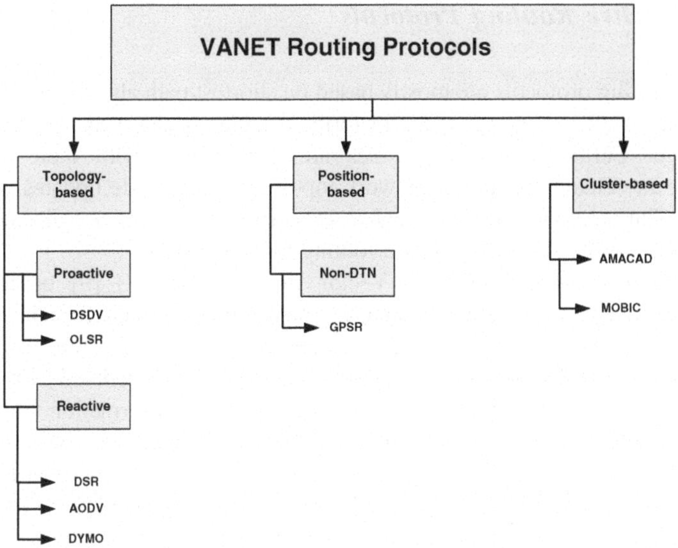

Fig. 1 Taxonomy of VANET routing protocols

minimizing packet loss, security, interference and controlling overhead. However, routing protocols in VANETs are very complicated and challenging task due to the high mobility of nodes making topology of the network dynamic and causing frequent links disconnections. For this, the selection of routing protocol heavily depends on the nature of the network. Consequently, single routing protocol is not sufficient enough in meeting all the different types of networks.

Many routing protocols have been developed to answer for VANET routing requirements. These routing protocols are classified into category based on following: topology-based routing, position-based routing, cluster-based routing, geocast-routing and broadcast-routing as shown in Fig. 1.

This paper focuses on the routing protocol for VANETs and briefly describes some of those protocols. Furthermore, we present a thorough evaluation of routing protocols. All the protocols have been examined with varying mobility and offered load using the NS-2. The comparison focuses on the following performance metrics: average end-to-end delay and bandwidth. The contribution of the paper is to demonstrate the advantages and limitations of different routing protocols in VANET environment.

2 Topology-Based Routing Protocols

Topology based routing protocols is based on links information within the network to send the data packets from source to destination. It can be classified into proactive, reactive and hybrid protocols.

2.1 Proactive Routing Protocols

Proactive routing protocols are mostly based on shortest path algorithms. They keep information of all connected nodes in form of tables because these protocols are table based. Furthermore, these tables are also shared with their neighbors. Whenever any change occurs in network topology, every node updates its routing table [8]. The strategies implemented in proactive algorithms are distance-vector routing such as DSDV and link-state routing such as OLSR.

Destination-Sequenced Distance-Vector Routing (DSDV) [9] based on the Bellman–Ford algorithm, which is a table-driven routing scheme for ad hoc mobile networks. The main contribution of the algorithm was to solve the routing loop problem, increases the convergence speed, and minimizes overhead of the control message. In DSDV [10], all the nodes sustain a next-hop information table and are exchanged table's information with their neighbors. Each entry in the routing table contains a sequence number, the sequence numbers are generally even if a link is present. Further, an odd number is used. The number is generated by the destination, and the emitter needs to send out the next update with this number. The DSDV provides a loop-free single path to the destination and sends two types of packets: full dump and incremental. In full dump packets, all the routing information is sent, whereas in the incremental type, only updates are sent. This function decreases bandwidth utilization by sending only updates instead of complete routing information. The incremental packet still increases the overhead in the network because the packets are so frequent and are therefore unsuitable for large-scale networks.

Optimized link state routing (OLSR) [11] includes of well-known unicast routing protocols adapted to VANETs which does periodic flooding of control information using special nodes that act as Multi Point Relays (MPRs). OLSR maintains routing information by sending link state information. After each change in the topology every node sends updates to selective nodes. Thus, every node in the network receives updates only once. Unselected packets cannot retransmit updates; they can only read updated information.

2.2 Reactive Routing Protocols

Reactive routing protocols are also known as on-demand driven, they regularly renew the routing table. It was designed in such a manner to overcome the overhead that was created by proactive routing protocols. This is overcome by maintaining only those routes that are currently active [8]. However, reactive protocols use a flooding method for route discovery that initiates more routing overhead and also suffer from the initial route discovery process. Many reactive protocols have been proposed so far but in this section we briefly described about AODV, DYMO and DSR. Then we check the suitability of these protocols for VANET.

Ad Hoc On-Demand Distance Vector (AODV) [12] algorithm enables dynamic, self-starting, multi hop routing between participating mobile nodes wishing to establish and maintain an ad hoc network. AODV functions on demand basis when it is required by network, which is fulfilled by nodes within the network. Route discovery and route maintenance is also carried out on demand basis even if only two nodes need to communicate with each other. AODV reduces the need of nodes in order to always remain active and to continuously update routing information at each node. In other words, AODV maintains and discovers routes only when there is a need of communication among different nodes.

Dynamic source routing protocol (DSR) [13] is an on-demand, whereby all the routing information is maintained at mobile nodes. DSR enables the network to be completely self-organizing and self-configuring, without any existing network infrastructure or administration. This protocol is composed of two operations Route Discovery and Route Maintenance. In route discovery DSR discovers for the routes from source node to destination. In DSR, data packets stored the routing information of all intermediate nodes in its header to reach at a particular destination. Routing information for every source node can be change at any time in the network and DSR updates it after each change occurs. Intermediate routers don´t need to have routing information to route the passing traffic, but they save routing information for their future use.

DYMO [14] is a new reactive (on demand) routing protocol, which is currently developed in the scope of the IETF's MANET working group. DYMO builds upon experience with previous approaches to reactive routing, especially with the routing protocol AODV. It aims at a somewhat simpler design, helping to reduce the system requirements of participating nodes, and simplifying the protocol implementation. DYMO retains proven mechanisms of previously explored routing protocols like the use of sequence numbers to enforce loop freedom. At the same time, DYMO provides enhanced features, such as covering possible MANET–Internet gateway scenarios and implementing path accumulation. Besides route information about a requested target, a node will also receive information about all intermediate nodes of a newly discovered path. Therein lies a major difference between DYMO and AODV, the latter of which only generates route table entries for the destination node and the next hop, while DYMO stores routes for each intermediate hop.

3 Position-Based Routing Protocols

In position based routing, each node has knowledge about its geographical position by GPS or by some other position determining services. It used geographical information of nodes in the communication selection process. In it all nodes also has the knowledge of source, destination and other neighboring nodes. Position based routing provides hop-by-hop communication to vehicular networks. A position based routing protocol consists of many major components such as "beaconing", "location service and servers" and "recovery and forwarding strategies" [15].

The position-based routing can be classified as non-delay tolerant network (non-DTN) routing protocols and delay tolerant network (DTN) routing protocols. In this category, we find several protocols have been proposed so far but in this section we briefly described about GPSR.

Greedy Perimeter Stateless Routing (GPSR) [16] is one of the most known position-based protocols in literature aimed at handling mobile environments. GPSR uses closest neighbor's information of destination in order to forward packet. This method is also known as greedy forwarding. In GPSR each node has knowledge of its current position and also the neighboring nodes. The knowledge about node positions provides better routing and also provides knowledge about the destination. On the other hand neighboring nodes also assists to make forwarding decisions more correctly without the interference of topology information. GPSR protocol depends on two modes: Greedy mode and Perimeter mode.

4 Cluster-Based Routing Protocols

In cluster-based routing protocols, the node possess the similar characteristics can form a cluster and chose a cluster-head which is responsible for intra-and inter-cluster management purposes. The intra-cluster nodes interact with one another through direct links, whereas inter-cluster interaction is performed through the cluster-headers [10]. In this category we find: AMACAD and MOBIC.

Adaptable Mobility-Aware Clustering Algorithm based on Destination in vehicular networks (AMACAD) [17] takes into account the destination of the vehicles to arrange the clusters and implements an efficient message mechanism to respond in real time and avoid global re-clustering. It algorithm tries to accurately follow the mobility pattern of the network and prolong the cluster lifetime and reduce the global overhead. There might be a problem with knowing the final destination a priori as drivers usually do not use navigation system for known routes. Cluster size is variable according to vehicle density, speed and required minimum bandwidth or QOS where parameters can be predefined or provided on the fly from vehicle sensors and application profiles [18].

MOBIC [19] is a relative mobility metric for nodes in a MANET based on the ratio of the power levels due to successive packet receptions at a node that works also in VANETs. It is based on the Lowest-ID algorithm but uses signal power levels mobility metric derived from successive receptions instead. MOBIC calculate the variance of relative mobility of a mobile node with each of its neighbors, where a small value of variance indicates the mobile node is moving relatively less than its neighborhood. However, in the case in which few neighbor nodes move differently, the method still results in dramatic increase in the variance [17].

5 Performance Evaluation

5.1 Simulation

The performance of routing protocols is evaluated through simulation using NS2 [20] and MOVE [21] tool to generate the motion of the nodes. The positions of vehicles are then made by the simulator SUMO [22]. The VANET network has been simulated with 20–100 vehicles. Note that the number of vehicles is variable in our simulations. Moreover, the position of vehicles is selected randomly. Constant bit rate (CBR) traffic was used in simulation. Table 1 illustrates the characteristics of the environment in which the simulation is experimented. For network simulation, there are several performances metrics which is used to evaluate the performance such as Packet Delivery Ratio, Average End-to-End Delay and Bandwidth are investigated based on a Number of Vehicle and Vehicle Speed. The following is the definition of the performance metrics we shall evaluate in the following subsections:

- Packet delivery ratio is the ratio of total number of data packets received at the destination to the total data packets sent from the source. The performance of the protocol depends on several parameters chosen for the simulation. The major parameters are packet size, no of nodes, transmission range and the structure of the network. The performance is better when packet delivery ratio is high.

Table 1 Simulation scenario

Parameters	Values
Highway length	10000 m
Number of lanes	2
Speed of vehicles	90–130 km/h
Transmission rate	6 Mbps
Packet size	512 bytes
MAC protocol	IEEE802.11
The maximum size of packet in the queue	70
Traffic type	CBR
Number of vehicles	20-100
Channel capacity	2 Mbps
Simulation time	900 s
Channel	Channel/WirelessChannel
Antenna type	Antenna/OmniAntenna
Propagation model	Propagation/TwoRayGround
Network interface	Phy/WirelessPhy
Interface queue type	Queue/DropTailQueue
Routing protocol	DSDV, OLSR, AODV, DSR, GPSR, DYMO, AMACAD, MOBIC

- Average End-to-End Delay is the average time that a packet takes to traverse from a source node to a destination node. This is the time from the generation of the packet in the sender up to its reception at the destination's application layer and it is measured in seconds. It therefore includes all the delays in the network such as buffer queues, transmission time and delays induced by routing activities and MAC control exchanges.
- Bandwidth is the total link capacity of a link to carry information (bits/second).

5.2 Simulation Results

The simulation environment includes 20–100 vehicles. We executed two sets of experiments. The first set evaluates how the variation of the number of vehicles affects the 8 routing protocols performance. The second set evaluates the vehicle speed effects on the performance. The performance is measured by the average end-to-end delay, packet delivery ratio and bandwidth.

5.2.1 Number of Vehicles

Figure 2 shows the Packet Delivery Ratio of all 8 routing protocols. The behavior of Packet Delivery Ratio of every routing protocol with respect to Number of Vehicles is shown. We observe that AMACAD and MOBIC with short interval values is able to give better Packet Delivery Ratio consistently as compared to other protocols. AODV, DSR, GPSR, OLSR and DYMO in this scenario perform well. DSDV has lower Packet Delivery Ratio throughout this scenario. Overall it can be concluded that cluster-based routing protocols with short interval values provide

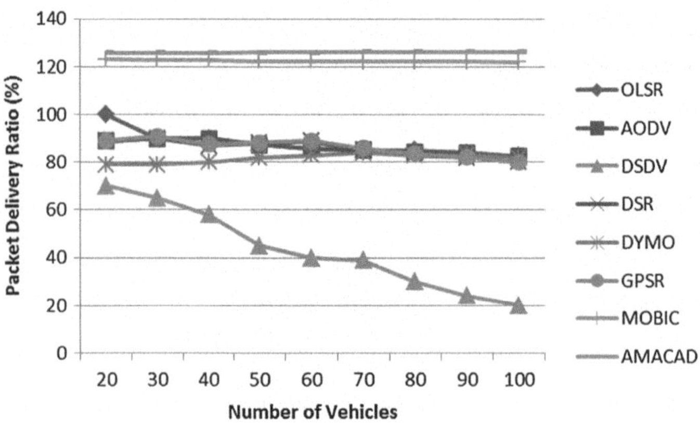

Fig. 2 Packet delivery ratio versus number of vehicles

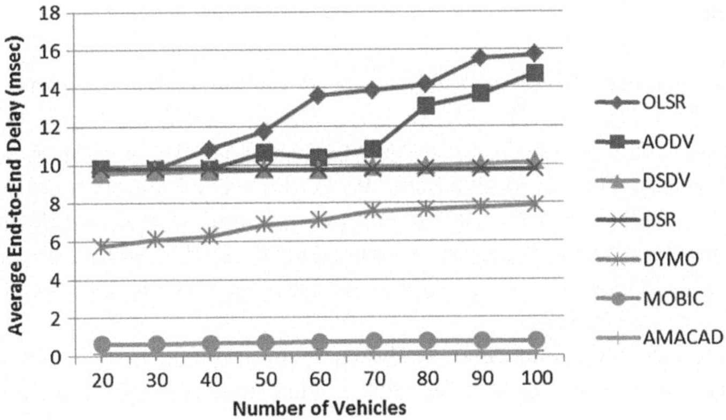

Fig. 3 Average end-to-end delay versus number of vehicles

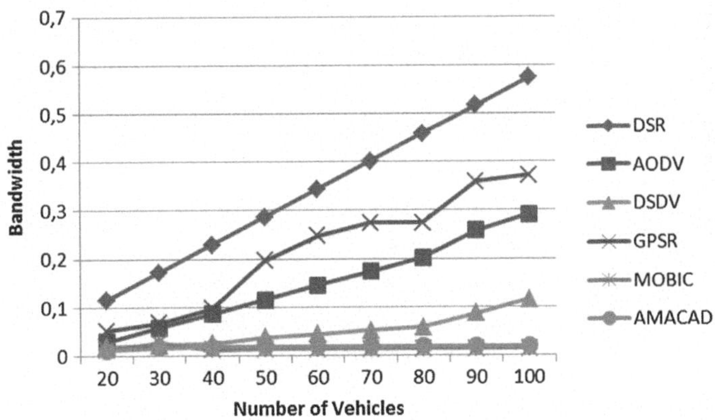

Fig. 4 Bandwidth versus number of vehicles

better Packet Delivery Ratio, as their routing table is updated quickly and one node is responsible to deliver messages to all nodes of the clusters.

Figure 3 shows the end-to-end delay of AODV, DSR, DYMO, GPSR, AMACAD, MOBIC, DSDV and OLSR in Highway Scenario against increasing number of vehicles. We observe that OLSR and AODV routing protocols have higher end-to-end delay than DYMO. MOBIC and AMACAD again outperforms the other routing protocols when the node density increases in this scenario.

Figure 4 depicts the bandwidth values associated with mentioned routing protocols. It turns out that the AMACAD and MOBIC routing protocols has the less consumption of bandwidth in comparison with other routing protocol, followed by DSDV routing protocol in number of vehicles diagram. DSR owns the highest bandwidth value in this scenario.

5.2.2 Vehicle Speed

Figure 5 shows the effects of Vehicle Speed on the Packet Delivery Ratio of AODV, DYMO, DSR, GPSR, AMACAD, MOBIC, DSDV and OLSR in highway scenario. Here again we observe that AMACAD and MOBIC with shorter intervals of control messages is able to give better Packet Delivery Ratio with the increase in node density. AODV and OLSR have shown the similar result while the GPSR protocol decreased at the speed of 90. The DSDV and DYMO routing protocols represent a significant downward trend of nearly 9 % while the vehicles speed varies from 90 to 130 km/h.

In Fig. 6, the end-to-end delay is shown. It is observed that variant of OLSR consistently have more end-to-end delay than reactive and position routing

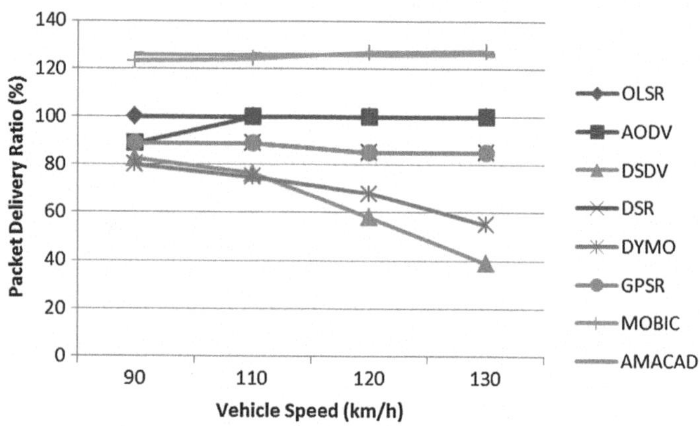

Fig. 5 Packet delivery ratio versus vehicle speed

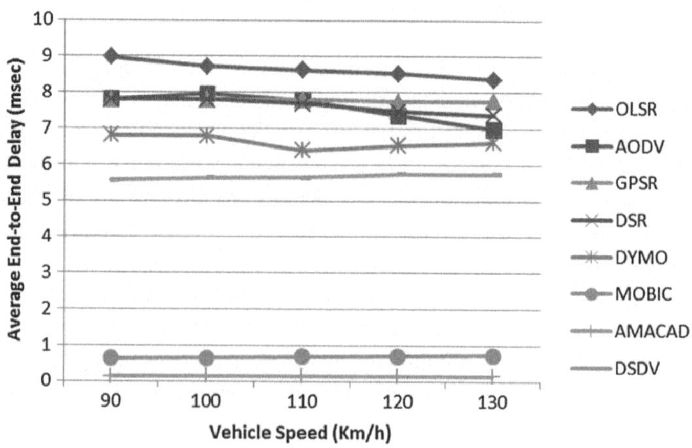

Fig. 6 Average end-to-end delay versus vehicle speed

protocols. This is because the proactive routing protocols generate sustained control traffic, in order to have updated information about the topology and routing paths. This process is done irrespective of user communication. AMACAD and MOBIC have shown the lowest Average End-to-End delays in speed and density diagrams respectively while the DYMO routing protocol plays a role in between position and proactive protocols according to both diagrams.

6 Conclusion

The VANET has to overcome the challenges of communication delay, low delivery rate, reliability, scalability and congestion. This paper reveals the performance analysis of reactive, proactive and position routing protocols in comparison with cluster-based routing protocols AMACAD and MOBIC. Cluster-based routing protocols represent some similarities in terms of average end-to-end delay, packet delivery ratio and bandwidth. Reactive routing protocols represent also some similarities in terms of packet delivery ratio. However difference among reactive routing protocols due to the different approach of routing storage and maintenance. Simulation of fundamental yet major parameters such as packet delivery ratio, Average End-to-End delay and bandwidth based on vehicle speed and number of vehicles for cluster-based routing protocols in VANET results in useful information. Simulation results show that the cluster-based routing protocols could be applied on VANET although in increasing the speed and density of vehicles, in most cases, the performance of cluster-based routing protocols will increase, which makes use of clustering routing protocol is suitable for VANET.

References

1. Bernsen, J., Manivannan, D.: Routing protocols for vehicular ad hoc networks that ensure quality of service. The Fourth International Conference on Wireless and Mobile Communications, 2008. ICWMC'08. IEEE (2008)
2. Wex, P., et al.: Trust issues for vehicular ad hoc networks. Vehicular Technology Conference, 2008. VTC Spring 2008. IEEE. IEEE (2008)
3. Taleb, T., et al.: A stable routing protocol to support ITS services in VANET networks. IEEE Trans. Veh. Technol. **56**(6), 3337–3347 (2007)
4. Marzak, B., et al.: Cluster head selection algorithm in vehicular Ad Hoc networks. 2015 International Conference on Cloud Technologies and Applications (CloudTech). IEEE (2015)
5. Ferreira, M.C.P., et al.: Methods and systems for coordinating vehicular traffic using in-vehicle virtual traffic control signals enabled by vehicle-to-vehicle communications. U.S. Patent No. 8,972,159. 3 March 2015
6. Johnson, D.B., Maltz, D.A. Broch, J.: The dynamic source routing protocol for multihop wireless Ad Hoc Networks. Ad Hoc Netw. 139–172

7. Ding, Y., Wang, C., Xiao, L.: A static-node assisted adaptive routing protocol in vehicular networks. In: Proceedings of the fourth ACM International Workshop on Vehicular Ad Hoc Networks. ACM (2007)
8. Abolhasan, Mehran, Wysocki, T., Dutkiewicz, E.: A review of routing protocols for mobile ad hoc networks. Ad Hoc Netw. **2**(1), 1–22 (2004)
9. Nithya, S., Kumar, G.A., Adhavan, P.: Destination-sequenced distance vector routing (DSDV) using clustering approach in mobile adhoc network. 2012 International Conference on Radar, Communication and Computing (ICRCC). IEEE (2012)
10. Sharef, B.T., Alsaqour, R.A., Ismail, M.: Vehicular communication ad hoc routing protocols: A survey. J. Netw. Comput. Appl. **40**, 363–396 (2014)
11. Toutouh, J., García-Nieto, J., Alba, E.: Intelligent OLSR routing protocol optimization for VANETs. IEEE Trans. Veh. Technol. **61**(4), 1884–1894 (2012)
12. Perkins, C., Belding-Royer, E., Das, S.: Ad hoc on-demand distance vector (AODV) routing. No. RFC 3561. (2003)
13. Maltz, D.B., Johnson, D.A., Broch, J.: DSR: The dynamic source routing protocol for multi-hop wireless ad hoc networks. Computer Science Department Carnegie Mellon University Pittsburgh, PA, pp. 15213–3891 (2001)
14. Sommer, C., Dressler, F.: The DYMO routing protocol in VANET scenarios. Vehicular Technology Conference, 2007. VTC-2007 Fall. 2007 IEEE 66th. IEEE (2007)
15. Liu, J., et al.: A survey on position-based routing for vehicular ad hoc networks. Telecommun. Syst. 1–16 (2015)
16. Karp, B., Kung, H.-T.: GPSR: Greedy perimeter stateless routing for wireless networks. In: Proceedings of the 6th annual international conference on Mobile computing and networking. ACM (2000)
17. Morales, M.M.C., Hong, C.S., Bang, Y.-C.: An adaptable mobility-aware clustering algorithm in vehicular networks. In: Network Operations and Management Symposium (APNOMS), 2011 13th Asia-Pacific. IEEE (2011)
18. Vodopivec, S., Bešter, J., Kos, A.: A survey on clustering algorithms for vehicular ad-hoc networks. 2012 35th International Conference on Telecommunications and Signal Processing (TSP). IEEE (2012)
19. Basu, P. Khan, N. Little, T.D.C.: A mobility based metric for clustering in mobile ad hoc networks. 2001 International Conference on. Distributed Computing Systems Workshop. IEEE (2001)
20. Issariyakul, T., Hossain, E.: Introduction to network simulator NS2. Springer Science & Business Media (2011)
21. Karnadi, F.K., Mo, Z.H., Lan, K.-C.: Rapid generation of realistic mobility models for VANET. Wireless Communications and Networking Conference, 2007. WCNC 2007. IEEE. IEEE (2007)
22. Behrisch, M., et al.: SUMO–simulation of urban mobility. The Third International Conference on Advances in System Simulation (SIMUL 2011), Barcelona, Spain (2011)

A Message Removal Mechanism for Delay Tolerant Networks

Elenilson da Nóbrega Gomes, Carlos Alberto V. Campos, Sidney C. de Lucena and Aline Carneiro Viana

Abstract The dissemination of redundant copies of a message is one of the techniques used in Delay/Disruption Tolerant Networks (DTN) to improve the delivery rate and to decrease the incurred delay. Nevertheless, many of these copies remain in the buffer of intermediate nodes even after the message is delivered to the destination. In this paper, we propose mechanism to remove obsolete messages for DTN routing protocols. Furthermore, we compare its performance with other state-of-art techniques under two different realistic scenarios and using two datasets of human mobility traces. Through the obtained results, we observed a better performance of the tested DTN routing protocols in most scenarios in terms of delivery ratio and overhead messages, when compared to others related works.

1 Introduction

In mobile wireless networks, the end-to-end communication between users can not be available all time, or may never becomes available. For this scenario, we use a network approach known as Delay/Disruption Tolerant Networks—DTN [1]. In these networks, the message storage is persistent due to the use of the *store-carry-and-forward* mechanism, in which a node stores a message for later in a new contact, forward it [2]. To increase the chances of delivery and reduce the time that a message

E. da Nóbrega Gomes · C.A.V. Campos (✉) · S.C. de Lucena
Federal University of State of Rio de Janeiro (UNIRIO), Rio de Janeiro, Brazil
e-mail: beto@uniriotec.br

E. da Nóbrega Gomes
e-mail: elenilson@gmail.com

S.C. de Lucena
e-mail: sidney@uniriotec.br

A. Carneiro Viana
Institut National de Recherche en Informatique et en Automatique (INRIA),
Palaiseau, France
e-mail: aline.viana@inria.fr

R. El-Azouzi et al. (eds.), *Advances in Ubiquitous Networking 2*, Lecture Notes in Electrical Engineering 397, DOI 10.1007/978-981-10-1627-1_4

takes to reach the destination, many forwarding protocols have been proposed where multiple copies of a message are disseminated in the network.

Unfortunately, many of these copies remain stored in the intermediate nodes after their delivery to the destination. The use of message removal mechanisms aims thus to reduce the occupancy rate of such copies in the buffer of nodes and enable more efficient data exchange on the network.

In this context, this paper proposes a mechanism that uses the acknowledgment information of a message by the destination to remove from buffers of intermediate nodes the possible obsolete copies of this message. This information (copies of messages) is stored in a list that is exchanged at every contact between nodes, allowing an update of both.

The main contributions of our paper are threefold: (i) the proposal of an obsolete message removal mechanism; (ii) the implementation of the proposed mechanism and some state-of-art mechanisms as IMMUNE, IMMUNE-TX, and TTL in the ONE (Opportunistic Networking Environment) simulator; (iii) the evaluation of such mechanisms when combined with the Epidemic, Spray and Wait, and Bubble Rap forwarding protocols under two different realistic scenarios and two human mobility datasets.

This paper is organized as follows. In Sect. 2, related works are presented and discussed. The proposed mechanism will be described in Sect. 3. An evaluation of the mechanisms is presented in Sect. 4. In Sect. 5, the results are presented and discussed. Finally, Sect. 6 concludes this work.

2 Related Works

The obsolete message removal problem has attracted significant attention of the networking community in the literature. We discuss some of these works in the following. In the work [3], the TTL (time-to-live) is used to limit the number of message copies disseminated in the network. Thus, the TTL is set based on time and network hops. A node has initially a defined TTL_{max} value. This value is decremented every second until zero. The messages are replicated while the TTL value is greater than zero. When the time expires (i.e., TTL equals zero), all messages are simultaneously deleted.

The work at [4] presents a mechanism for removing copies of messages. The presented approach is composed into two parts: (i) distribution of acknowledgment messages (ACK) and (ii) the use of auxiliary nodes to retransmit ACK messages. The ACKs are also divided into active ACK and passive ACK. The passive ACK will only be forwarded to a node that receives a copy of a message that has been delivered to the destination. The passive ACK messages are distributed slowly and copies are deleted accordingly. In active ACK, a node that has an ACK of any message transfers this ACK to any neighbor node (broadcast-like). The concepts of passive and active ACK are used by three message discarding strategies proposed at [5], which are based on the *Shared Wireless Infostation Model* (SWIM): IMMUNE and IMMUNE-TX, using passive ACK, and VACCINE, using active ACK. In the first two strategies, nodes become "immune" after receiving a known obsolete message: when a node receives a message that has already been delivered to the destination, it rejects the new copy of this message, avoiding unnecessary retransmissions. The IMMUNE-TX mechanism also has the function of immunizing "the node" with an obsolete message, but it goes beyond, immunizing the neighbor that tried to pass the obsolete message. On the other hand, VACCINE tries to immunize "all nodes" once a message has been delivered.

In [6], authors present a DTN buffer management policy to deal with the message removal problem. When a node's buffer is full and needs space to store a new message, the longest message in the buffer is discarded. This strategy was called *droplargest (DLA)*. In [7] is presented a Reactive Weight Based Buffer Management Policy for DTN Routing Protocols. These mechanisms has however the drawback of affecting the operation of applications that generate large messages.

Additionally, in [8] a survey and comparison of various buffer management policies for DTNs are performed and a new policy is presented based on encounter rate of nodes and context information such as TTL, number of available replicas and maximum number of forwarded bundle replicas. These papers perform buffer management by removing messages (including messages not yet delivered) in order to avoid its overflow. So, these removal policies are not the subject of this paper, but can be investigated in future works.

Apart from the solutions dealing with the message removal problem, there is more and more a worry from the networking community in bringing a realistic consideration to simulation environments. This has been done by the use of real traces describing human mobility in network simulators. In [9], the authors conducted a study on the patterns of human mobility and its use in mobile network simulations. They found that the pattern of human mobility suffers from external influences and are not found as random because mobility is related to social contexts. Moreover, they observed the existence of different mobility patterns related to these contexts. Finally, it was emphasized the importance of using models that are able to provide more realistic results in the simulations such as the use of traces describing human mobility to increase the accuracy of simulations.

Authors in [10] perform a study on the use of real mobility traces in simulated environments. In this analysis, five real mobility traces were used together with the Epidemic protocol and compared with existent mobility models. Results clearly show the differences in the Epidemic protocol performance when applied to real or modeled mobility. Therefore, this outcomes strongly recommend, when possible, the use of real traces on the evaluation.

Based on the above described observations, the next sections describe a detailed evaluation of our proposed mechanism when compared to different message removal mechanisms and when real mobility traces are considered.

3 The Proposed Mechanism

This section introduces our mechanism named ReMO—*Removal Mechanism for Obsolete messages*. ReMO is based on the VACCINE strategy [5], that models the impact of the removal of redundant copies of already delivered messages that still occupy storage space in the intermediate nodes.

The operation of the ReMO can be described as follows. Each node has a list storing the identifier of each message that has been delivered to the destination. When a node comes in contact with another node, they then exchange/update their respective lists. If there is any record of a message already delivered to the destination, the node remove it from its buffer. After this phase (i.e., removal of locally stored obsolete messages), the nodes can then exchange messages as dictated by the used forwarding protocol.

A pseudo code of ReMO algorithm is described at Algorithm 1, where L is the delivered messages control lists, n_1 is the transmitter node and n_2 is the receiver node. The great advantage of the ReMO mechanism is that it is independent of the used routing protocol.

Algorithm 1 ReMO Algorithm of (n_1) transmitter node side

function REMO$(n_1)n_2$
    n_1 contact n_2
    if $L(n_1) <> 0$ **then**
        n_1 send $L(n_1)$
    if $(n_1$ no receive $L(n_2))\|(L(n_1) = L(n_2))$ **then**
        Start data message through the use routing protocol
    else
        n_1 updates $L(n_1)$
    while not at end of this $L(n_1)$ **do**
        read current
        if Regist $L(n_1)$ = buffer message **then**
            Delete message
    while Message destination **do**
        if (Message destination) \|\| (space in $L(n_1)\&L(n_2)$) **then**
            return $((L(n_1)\|L(n_2)) ==$ Id Message
        else
            if (Message destination \|\| (no space in $L(n_1)\&L(n_2)$) **then**
                repeat(delete old Id)
                until space in $L(n_1)\&L(n_2)$

4 The Influence of Removal Mechanisms on the Performance of DTN Routing Protocols

It is expected that, with the removal of obsolete messages in DTN, it will be possible to make better use of the nodes' buffers, thus allowing for a better network performance, especially by increasing the delivery rate and decreasing the delay of routing protocols that use messages replication techniques. The proposed mechanism and other evaluated mechanisms (IMMUNE, IMMUNE-TX, and TTL) were implemented in the ONE simulator.

The rest of this section will present the used routing protocols and also a description of the simulation parameters and of the performance metrics that were used.

4.1 Routing Protocols

According to [11], the routing protocols for DTN are classified as: (i) flooding based, (ii) replication (or controlled flooding) based and (iii) forwarding based. As the forwarding-based protocols do not create copies of a message, only the first two classes will be analyzed. Moreover, as there are many protocols for each class, the following well-known and most used ones were chosen: the Epidemic protocol, that is based on flooding, the Spray and Wait protocol, that is based on replication, and the BUBBLE Rap protocol, that is based on replication according to social context of nodes.

In [2] the Epidemic protocol is proposed and its operation is given as follows. Each node has a list of the messages in its buffer. When meeting neighbor nodes, their message lists are exchanged and, if there are different messages in their buffers (i.e., there are differences in their lists), nodes exchange the missing messages. We can observe two important factors in this protocol. First, it greatly increases the number of copies circulating in the network and, consequently, also increases the destination delivery probability. The second factor is that multiple copies spreading over network causes the filling of nodes' buffers more quickly. These factors should be mitigated by using a mechanism to remove messages.

In [12] is shown the Spray and Wait protocol, where the routing process is divided into two parts: (i) the spraying phase (*spray*), when $L - 1$ copies of a message are disseminated in the network, and (ii) the hold phase (*wait*), when a node only forward the message if the destination is met, i.e., direct transmission. To simulate this protocol, the L parameter was set to the default value of 6.

The BUBBLE Rap protocol [13] uses the social relations of nodes as a decision on forwarding messages. Its operation is based on the centrality and community metrics, where each node participates in a community and its centrality is proportional to its popularity (node connectivity degree). It also has a global centrality in the network for routing messages out of a community. It has two phases, *Bubble-up* in the global community and *Bubble-up* in the local community, always choosing the nodes with the highest centrality for routing messages. To simulate this protocol, the K parameter was set to the value of 5 and the *familiar Threshold* parameter to the value of 700, as [14].

4.2 Compared Mechanisms

The removal mechanisms aim to inform intermediate nodes that a message was delivered to the destination node, enabling the removal of message replies throughout the network. These are also known as *Anti-packet*, which prevent the node of receiving again a message already discarded. In this context, the mechanisms used for comparison reasons in our analysis are: the TTL set to 50 %[1] of the total simulation time (named TTL50 % hereafter), the IMMUNE, and the IMMUNE-TX. The mechanisms were implemented as mathematical models cited in [5] and can be combined with any routing protocol.

4.3 Description of Simulation Parameters

In the evaluation of different scenarios using a DTN, two real mobility traces were used: UCL1 [15] and RollerNet [16], both available in CRAWDAD,[2] which have significant differences in user mobility. The UCL1 mobility trace was obtained through the movement of people along the campus of the University College London. The trace of RollerNet refers to the movement of skaters through the streets of Paris. In this one, there is a peculiarity which is the "accordion effect" on movement of users over time, due to two existing mobility standards: one where the skaters agglomerate, waiting for the release of some junction, and the other when they are skating normally along a route [17]. This peculiarity periodically generates high connectivity between users, which may influence the operation of the DTN, unlike other scenarios as the UCL1.

For the representation of network and users of UCL1 scenario in the ONE simulator, each generated message file contains 8640 messages randomly distributed in time between 20 nodes over the six days of simulation. In RollerNet scenario, each generated message file contains 2160 messages randomly distributed in time between 62 nodes along the three hours of simulation.

In addition, 10 rounds of simulation were performed for each set of analyzed parameters. We then computed the confidence interval of 95 % from the obtained results, according to the *T-student* distribution.

4.4 Performance Metrics

We analyzed the following performance metrics: delivery ratio, average delay and messages overhead. *Delivery ratio* is given by the number of delivered messages

[1]The value of 50 % brings a good compromise between message delivery time and buffer occupancy of nodes.

[2]CRAWDAD is a project that provides data on real experiments of wireless networks.

divided by the number of messages created during a simulation. *Average delay* is defined as the average time it takes for a message to be delivered, since when it is generated until when it reaches its destination. *Message Overhead* is given by the number of forwarded messages divided by the number of messages delivered during a simulation.

5 Results

This section presents the obtained results and evaluates the mechanisms performance according to their delivery ratio, average delay, and messages overhead.

5.1 Delivery Ratio

Figure 1 analyzes the delivery ratio of UCL1 scenario. Figure 1a, related to Epidemic protocol, shows for different buffer sizes, an improvement of the protocol performance when jointly applied with removal mechanisms of obsolete messages: discards are lowered, increasing the delivery probability at destinations. The ReMO mechanism also obtained a better performance for this protocol, which is due to the removal of more obsolete messages, leaving more buffer space to store other messages and consequently, increasing the delivery probability. As the other mechanisms do not remove as many messages as ReMO, their performance were worst, but still, a bit better than the case where no removal mechanism is applied (cf. *Original*). Among the analyzed mechanisms, TTL50 % had the worst performance due to the fact that many messages were created in the beginning of the simulation and had not enough time to reach the destination, thus being discarded.

Figure 1b shows the results related to Spray and Wait protocol, which, contrarily to Epidemic, controls the number of copies of each message. As for Epidemic, ReMO had a better result due to its implementation, which enhances the removal of obsolete messages in the buffers. With TTL50 %, Spray and Wait protocol did not show a good result because messages are discarded after 50 % of the simulation time, preventing their delivery.

Figure 1c related to Bubble Rap shows that better performances were obtained when removal mechanisms were applied in nodes with small buffer sizes (i.e., between 1 and 10 MB). As the buffer sizes increases, we see a performance improvement of the Original case (where no removal mechanism is used), which gets better results than when removal mechanisms are applied, except for the ReMo protocol, which still presents better performance. This may be explained by the limited number of redundant copies generated by Bubble Rap protocol, which limits the benefits of removal mechanisms. For the mobility trace with few nodes, we observed that IMMUNE and IMMUNE-TX mechanisms present the worst performance for buffers greater than 12 MB. At this stage the buffer does not overflow anymore. As

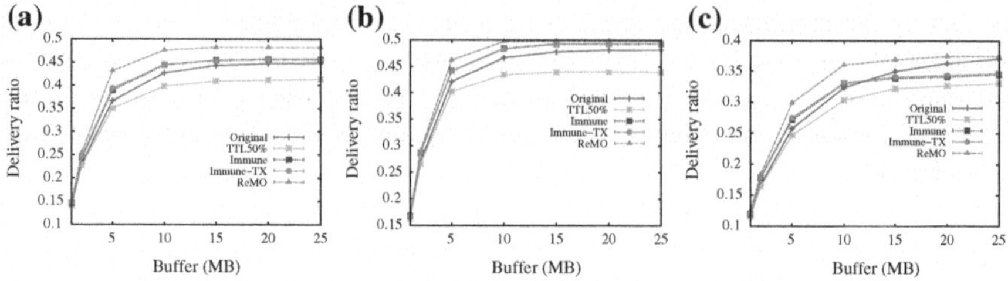

Fig. 1 Delivery ratio in the UCL1 scenario. **a** Epidemic. **b** Spray and Wait. **c** Bubble Rap

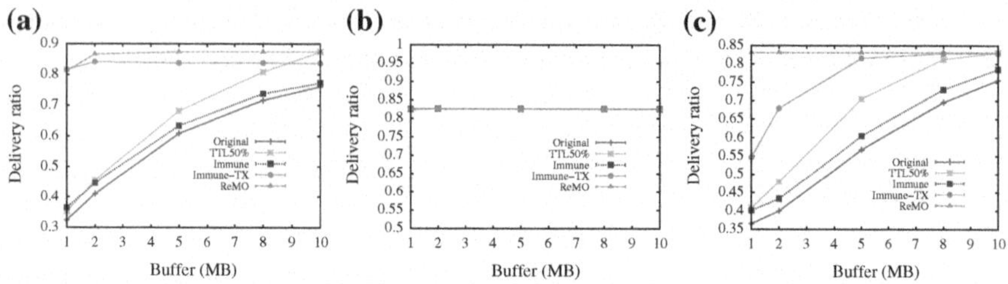

Fig. 2 Delivery ratio in the Rollernet scenario. **a** Epidemic. **b** Spray and Wait. **c** Bubble Rap

these mechanisms do not spread the list of delivered messages to all nodes, they exchange a greater number of obsolete messages, negatively impacting the results. In all cases, ReMO maintains a better performance up to 25 MB. The results for the TTL50 % mechanism is the worst. As it discards messages during the simulation time, delivery probability greatly reduces once Bubble Rap protocol uses social context to forward messages to the destination by selecting nodes with higher popularity. As the lifetime of messages reduces, it directly influences this mechanism worsening its performance.

Finally, IMMUNE and IMMUNE-TX mechanisms brought the same performance to the three routing protocols. Moreover, in all results of Fig. 1, the performance gains on protocols are usually bounded by the size of nodes buffer equal to 10 MB.

Delivery ratio of Rollernet scenario is shown in Fig. 2. In Fig. 2a, related to Epidemic protocol, it is observed an improvement of message delivery ratio when using ReMO. It maintains the buffer occupation below 80 % even when buffer size has only 1 MB, favoring the reception of new messages and thus, increasing the probability of delivering more messages to destinations. This improvement also occurs within IMMUNE-TX because node mobility increases its message removal mechanism, so performance gets close to ReMO. On the other side, IMMUNE does not exchange information with intermediate nodes, but still keeps a slightly better performance when compared to the case with no removal mechanism. If compared to TTL50 %, the performance of IMMUNE degrades as buffer size increases.

Figure 2b shows that message removal mechanisms were ineffective in the Rollernet when applied with the Spray and Wait protocol, which has the characteristic of low buffer occupancy. In Fig. 2c, related to Bubble Rap protocol, it is observed that the delivered ratio with ReMO mechanism is kept constant at approximately 83 % for any size of buffer, due to the removal of obsoleted messages, which increases the number of new received messages. The IMMUNE-TX mechanism also obtained a good result once it removed more messages than the other remaining mechanisms. TTL50 % surprisingly had a good performance. Having a closer look at existing contacts in the mobility trace, we observe that in the first half of the simulation, 1851 contacts happen between nodes, while 3080 contacts happen in the second half. Therefore, the delivery ratio at this second half of the simulation is improved by the higher number of contacts, what explain the obtained results.

5.2 Average Delay

Figure 3 shows the average delay obtained when the UCL1 scenario is used. In Fig. 3a, it can be observed that ReMO has the worst result. That is because the message removal affects the buffer space management policy, which is configured to remove oldest messages when buffer occupancy is above a certain limit. With less buffer occupancy, older messages are not removed and have a higher chance to be delivered to destination, which increases the average delay. The same situation occurred with IMMUNE and IMMUNE-TX mechanisms, but with lower average delays incurred since they remove less messages. TTL50 % provided the best result due to the metric definition. As its experiment time was divided in two half, time intervals between messages creations and their deliveries at the destinations are relatively small in each half part, leading to lower average delays.

The same observation is applied to the other two protocols shown in Fig. 3b, c. In Fig. 3b, it is observed that the removal mechanisms were quite ineffective once the Spray and Wait protocol limits message dissemination and that makes the results close to the case without message removal mechanism. In Fig. 3c related to Bubble Rap protocol, it is also observed that the average delays are worst when using message removals mechanisms. Again, this behavior is intrinsic related to the

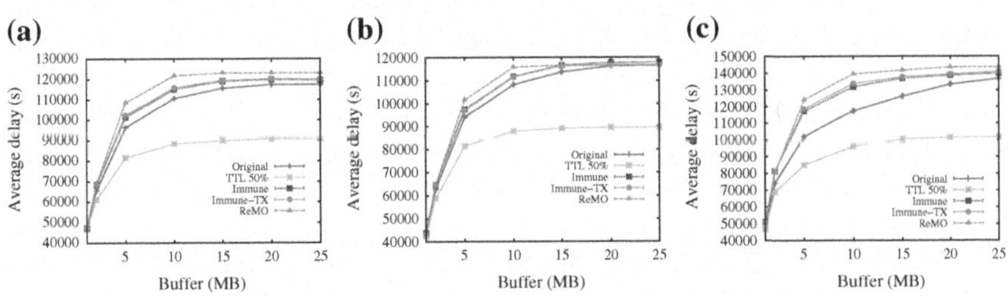

Fig. 3 Average delay in the UCL1 scenario. **a** Epidemic. **b** Spray and Wait. **c** Bubble Rap

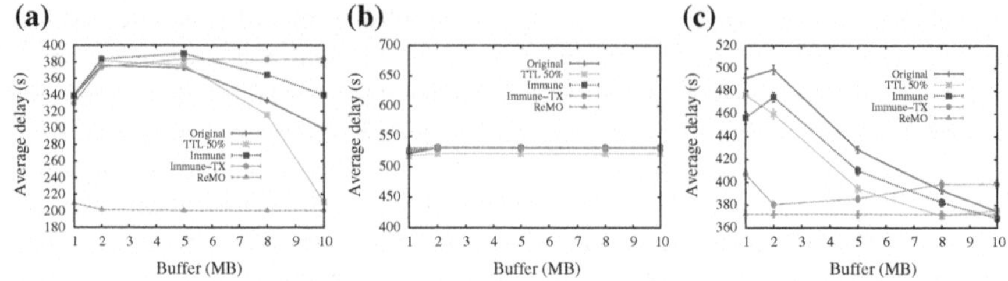

Fig. 4 Average delay in the Rollernet scenario. **a** Epidemic. **b** Spray and Wait. **c** Bubble Rap

metric definition and to the fact that removing messages already delivered helps to keep old messages in buffer, which would be discarded in situations of higher buffer occupancies.

Figure 4 shows the average delays obtained for the Rollernet scenario. In Fig. 4a, related to Epidemic protocol, it is observed that ReMO has significantly improved the average delay and kept it almost constant, compared to the others removal mechanisms and to the original case. In this scenario, node mobility is much higher, increasing contact frequency and consequently increasing message exchange. This favors message delivery (as shown in Fig. 2a) and reduces the incurred average delay. In Fig. 4b, related to Spray and Wait protocol, message removal mechanisms did not improved the average delay because of the limits imposed by this protocol at the message dissemination. Finally, in Fig. 4c, related to Bubble Rap protocol, it can be verified that ReMO kept average delay constant with buffer size increase and had the best performance for this metric in this scenario.

5.2.1 Messages Overhead

Figure 5 shows the message overhead generated by routing protocols in the UCL1 scenario. In Fig. 5a, related to the Epidemic protocol, it can be verified that ReMO reduces the message overhead generated by the protocol, which is due to the removal of obsolete messages that would be otherwise forwarded. The other mechanisms also reduced obsolete messages, but less than ReMO. With buffer size of 1 MB, ReMO removed an average of 1648 obsolete messages for this scenario, while IMMUNE-TX removed 1454 and IMMUNE removed 1419 obsolete messages. It can be observed that the number of removed messages is directly proportional to the result of message overhead. In Fig. 5b, related to the Spray and Wait protocol, it can be observed that even for this protocol, that forwards a lower number of messages, the removal of obsolete messages yet reduces message overheads, specially with the use of ReMO. In Fig. 5c, related to the Bubble Rap protocol, it can be observed that ReMO mechanism had the best result since it removed more obsolete messages than the other mechanisms.

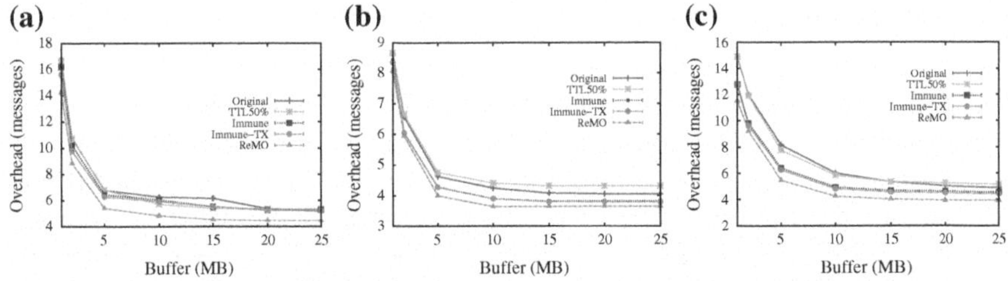

Fig. 5 Messages overhead in the UCL1 scenario. **a** Epidemic. **b** Spray and Wait. **c** Bubble Rap

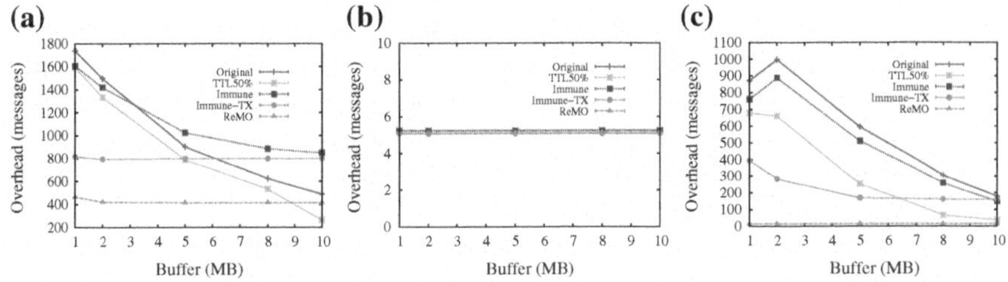

Fig. 6 Messages overhead in the Rollernet scenario. **a** Epidemic. **b** Spray and Wait. **c** Bubble Rap

Figure 6 shows the message overhead generated by routing protocols in the Rollernet scenario. In Fig. 6a, related to the Epidemic protocol, it can be observed a significative reduction of message overhead with the use of ReMO compared to the other mechanisms. The Rollernet scenario presents a great mobility of nodes, favoring the number of message forwarding by the Epidemic protocol and, consequently, message deliveries. The removal of obsolete messages lowers the number of copies of these messages that would be otherwise forwarded. The increase in buffer size also increases the average number of delivered messages, thus reducing message overhead. It can be observed that with ReMO the message overhead is almost the same despite the buffer size.

It is also possible to observe that the TTL50 % had less message overhead than ReMO for a buffer size of 10 MB. In Fig. 6b, related to the Spray and Wait protocol, once more it is observed that the obsolete message removal mechanisms did not affect message overhead incurred by this protocol. This protocol limits the number of copies forwarded by each node and the mobility characteristics of this scenario favors message deliveries at destinations, making obsolete message removal quite ineffective. In Fig. 6c, related to the Bubble Rap protocol, it can be observed that ReMO maintains a constant and extremely low message overhead over buffer variation, compared to the other solutions. This behavior can be explained by the characteristic of the Bubble Rap protocol of removing messages that has less chance to be delivered at the destinations, which is intensified by the ReMO mechanism and thus, drastically reduces the number of messages forwarded through the network.

6 Conclusions

This paper presented a mechanism called ReMO to remove obsolete messages already delivered to their destinations at Delay Tolerant Networks. ReMO was compared to other two mathematical mechanisms cited in [5], IMMUNE and IMMUNE-TX, and with the TTL of 50 % of total simulation time. All presented mechanisms, including ReMO, can be used independently of the DTN forwarding protocol. By analyzing the simulation results, it is possible to conclude that the ReMO mechanism presented the best overall performance. It was also possible to observe that, even when used with a forwarding protocol that controls the number of disseminated messages, like the Spray and Wait protocol, or when used with a social-based protocol that disseminates messages based on the social context of the node, like the Bubble Rap protocol, ReMO presented an improvement of the evaluated performance metrics. With the Epidemic protocol, that disseminates messages to every node with which a contact has been made (i.e., the worst case in terms of redundant message dissemination), ReMO also improved the results.

The use of real mobility traces enabled more realistic simulations and, from them, it was possible to observe that, depending on the mobility conditions, i.e., number of nodes and contact time intervals between nodes, the use of ReMO mechanism can better improve the overall performance of the DTN.

As future work, we intend to investigate the energy consumption incurred at nodes when ReMO is applied as well as apply energy-related improvements to the strategy, if necessary. Moreover, the study of how mobility affects the ReMO performance is also left to future works.

References

1. Fall, K.: A delay-tolerant network architecture for challenged internets. In: Proceedings of the ACM SIGCOMM '03, vol. 10, no. 863960, pp. 27–34. August 2003
2. Vahdat, A., Becker, D.: Epidemic routing for partially-connected ad hoc networks. Technical Report CS-200006, p. 18, Apr 2000
3. Yuen, W.H., Schulzrinne, H.: Message replication and deletion in delay tolerant networks under hop-based and time-based ttl schemes. Columbia University (2010)
4. Kaveevivitchai, S., Ochiai, H.E.: Message deletion and mobility patterns for efficient message delivery in dtns. In: IEEE PERCOM Workshops, pp. 760–763 (2010)
5. Haas, Z.J., Small, T.: A new networking model for biological applications of ad hoc sensor networks. IEEE/ACM Trans. Netw. **14**(1), 27–40 (2006)
6. Rashid, S., Ayub, Q.: Efficient buffer management policy dla for dtn routing protocols under congestion. (IJCNS) **2**(9), 118–121 (2010)
7. Rashid, S., Ayub, Q., Abdullah, A.H.: Reactive weight based buffer management policy for dtn routing protocols. Wireless Pers. Commun. **80**(3), 993–1010 (2015)
8. Iranmanesh, S.: A novel queue management policy for delay-tolerant networks. EURASIP J. Wireless Commun. Netw. **2016**(1) (2016)
9. Rhee, I., Shin, M., Hong, S., Lee, K., Kim, S.J., Chong, S.: On the levy-walk nature of human mobility. IEEE/ACM Trans. Netw. **19**(3), 630–643 (2011)

10. Thakur, G.S., Kumar, U., Hsu, W., Helmy, A.: Gauging human mobility characteristics and its impact on mobile routing performance. Int. J. Sensor Netw. **11**, 179–191 (2011)
11. Moreira, W., Mendes, P., Sargento, S.: Assessment model for opportunistic routing. IEEE Lat. Am. Trans. **10**(3), 1785–1790 (2012)
12. Spyropoulos, T., Psounis, K., Raghavendra, C.: Spray and wait: an efficient routing scheme for intermittently connected mobile networks. In: ACM WDTN, pp. 252–259 (2005)
13. Hui, P., Crowcroft, J., Yoneki, E.: Bubble rap: social-based forwarding in delay tolerant networks. IEEE Trans. Mob. Comput. **10**(11), 1576–1589 (2011)
14. Moreira, W., Mendes, P., Sargento, S.: Opportunistic routing based on daily routines. In: Proceedings of the IEEE WoWMoM, pp. 1–6, June 2012
15. Abdesslem, F.B., Henderson, T., Parris, I.: CRAWDAD trace st_andrews/locshare/2010/ucl1 (v. 2011-10-12) Oct 2011
16. Leguay, J., Benbadis, F.: CRAWDAD data set upmc/rollernet, Feb 2009
17. Tournoux, P., Leguay, J., Benbadis, F., Whitbeck, J., Conan, V.: Dias de Amorim, M.: Density-aware routing in highly dynamic dtns: the rollernet case. IEEE Trans. Mob. Comput. **10**(12), 1755–1768 (2011)

A Context-Aware Access Network Selection Based on Utility-Function for Handover in WLAN-LTE Environment

Maroua Drissi, Mohammed Oumsis and Driss Aboutajdine

Abstract Network selection plays an essential role in serving mobile users with the requisite Quality of Service (QoS) in the context of next generation networks (NGN). It boosts efficiently the use of radio resources in heterogeneous wireless networks environment. In order to be Always Best Connected in such environment that involve multiple networks with different access technologies, user's preferences and QoS requirements need to be considered during the Vertical Handover process. To address this issue, this paper proposes a context-aware access network selection based on utility-function that takes into consideration user's and QoS preferences. It aims at maximizing the user satisfaction while meeting application QoS when connecting to a target network. The proposed approach prioritizes networks with higher relevance to different types of applications and enables seamless connectivity to mobile user and applications. Thus, network resources are conveniently managed to support diverse services that might be considered by mobile users. Simulations results are provided to evaluate the performance of the proposed approach in low, medium and high mobility scenarios consisting in WLAN-LTE networks compared with the existing baseline scheme.

Keywords Heterogeneous networks · Network selection · Always best connected · Context-awareness · Utility function · ns-3

M. Drissi (✉) · M. Oumsis · D. Aboutajdine
LRIT, Associated Unit to CNRST URAC'29, Faculty of Sciences,
Mohammed V University, Rabat, Morocco
e-mail: drissimaroua@gmail.com

M. Oumsis
e-mail: oumsis@yahoo.com

D. Aboutajdine
e-mail: aboutaj@fsr.ac.ma

M. Oumsis
High School of Technology, Mohammed V University, Rabat, Sale, Morocco

R. El-Azouzi et al. (eds.), *Advances in Ubiquitous Networking 2*, Lecture Notes in Electrical Engineering 397, DOI 10.1007/978-981-10-1627-1_5

1 Introduction and Motivation

In recent years, various types of wireless access technologies have been deployed including 3G, WLAN, WiMAX, LTE and LTE Advanced. The most promising Next Generation Networks (NGN) are the heterogeneous networks. They are based on the coexistence and interoperability of the different types of Radio Access Technologies, and support existing and emerging networks, introducing by that the concept of Always Best Connected (ABC) Concept.

Authors of [1] assert that a terminal supports the ABC features means that it is not only always connected, but also connected through the best available network and access technology at all times. The ABC concept achieves a win-win partnership because it considers user's and operator's benefits. Indeed, Heterogeneous networks involve development of diverse paradigms of the concerned technologies, such as context-awareness of mobile devices. Communication in such environment has to cope with many provider's constraints (e.g., strong fluctuations of Real-time traffic and dynamic network topology) and also it has to meet user's application requirements (Fig. 1).

In addition to ABC functionalities, heterogeneous systems bring many promising paradigms aiming to deliver significantly higher capacity to meet the huge growth of mobile data traffic. A crucial challenge is that heterogeneous networks require strict QoS including better latency, reliability, higher spectral and energy efficiency, but also need an improved Quality of Experience (QoE) for users of wireless services in 4G and beyond networks. To meet these requirements, terminals have to select the suitable access network that fit for it QoS requirements of applications; escape

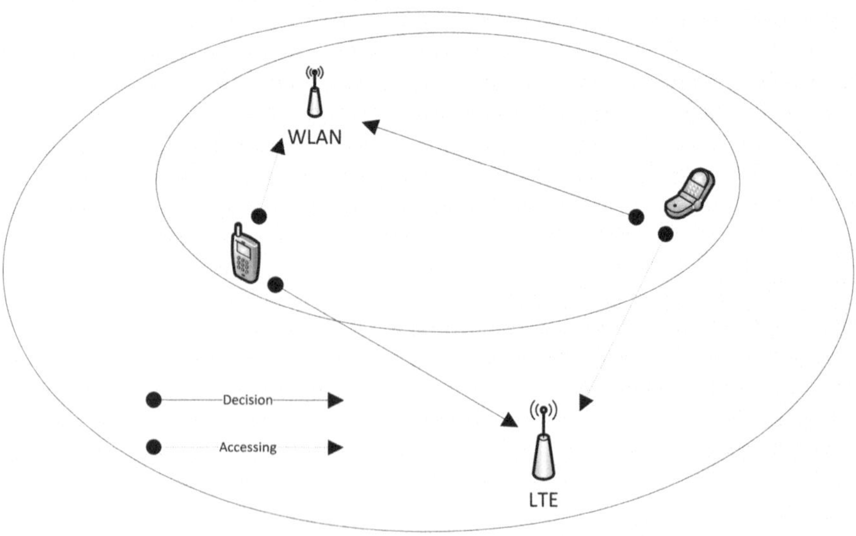

Fig. 1 Network topology for network selection

a network with high traffic load for avoiding congestion and also minimize costs by handling a context-aware network selection allowing mobile devices to make appropriate and timely decisions on behalf of users.

According to [2], context awareness is a capability to determine or influence a next action in telecommunication or process by referring to the status of relevant entities, which form a coherent environment as a context. Thus, this technology is a hot research topic in the field of communication. Context-aware network selection algorithms aim to provide users with satisfactory service quality. Not only the provider's constraints but also the user's preferences are considered during the process of vertical handover.

In this paper, we focus on the real time selection of the always best connected network in heterogeneous environment (WLAN and LTE), while maintaining QoS for multimedia services (Conversational, Streaming, Interactive and Background). We adopt, thereby, a utility-function based approach to enhance vertical handover decision; it enables a seamless real-time handover decision according to the network parameters and user's preferences. We adopt a scenario with low, medium and high speed. Implementation and simulation with Network Simulator NS3 are presented in order to validate our proposed goals. The results show that the proposed scheme achieves a significant improvement of the QoS Delay and Packets Loss metrics up to 7 % and 40 % respectively compared with baseline scheme.

The remainder of this paper is structured as follows. Section 2 gives a state of art about existing methods dealing with network selection. In Sect. 3 the utility-function based algorithm and the proposed system model are defined. Section 4 describes simulation parameters and results to illustrate the proposed scheme. Finally, Sect. 5 concludes the work.

2 Related Works

One of the challenges involved in the context of next generation networks is the development of solutions that consider several requirements in order to select the best network whenever it is needed to evaluate the transition of the mobile device between different networks.

There have been many works dealing with the Network Selection problem in different ways. Kassar [3] compared traditional handover decision strategies (RSS-based), and concluded that they are not good enough to make a vertical handover decision. They do not take into account the current context or user preferences. Therefore, vertical handover decision strategy involves complicated considerations and compromises. Authors of [4] studied the most important mathematical theories used for modelling the network selection problem in the literature. Authors compared the schemes of various mathematical theories in an unified scenario and discuss the ways to benefit from combining multiple algorithms together.

As a matter of fact, [5] proposed and optimized a common radio resource management techniques designed to efficiently distribute traffic among the available radio access technologies while providing adequate quality of service levels under heterogeneous traffic scenarios. Ma [6] also investigated in vertical handover in heterogeneous wireless networks. authors proposed a QoS-based vertical handover scheme for WLAN and WiMAX networks in order to provide always best service to users. Furthermore, authors of [7], as well, proposed an algorithm for network selection based on averaged received signal strength, outage probability and distance. Authors of [8] proposed a scheduling algorithm for the same cited goals, they proposed solution for scheduling packets while maintaining performance in wireless networks. The scheduling scheme is based on transmission link's condition from the media independent handover (MIH) protocol, type of call and classes of service.

In a similar context, Some researchers have been using utility function to select the best candidate network. Utility function refers to the satisfaction that a good or service provides to the decision maker. Indeed, [9] allocated terminals to the most appropriate network by jointly examining both user's and provider's preferences. Authors introduced three utility-based optimization functions based on the type of application that users request. In the same way, [10] presented a method that takes into account user preferences, network conditions, QoS and energy consumption requirements in order to select the optimal network which achieves the best balance between performance and energy consumption. The proposed network selection method incorporates the use of parametrized utility functions in order to model diverse QoS elasticities of different applications.

Lopez-Benitez [11] presented a selection mechanism that prioritizes networks with higher relevance to the application and lower energy consumption based on utility function. Wu [12] proposed a network selection scheme based on a utility function that take user's QoS demands, preferences and channel state information (CSI) into account.

These researches have moved us to combine utility-function with multi-attribute concept in real time application and varied speed mobile WLAN-LTE scenario for Network Selection presented in section below.

3 Utility-Function Based Network Selection

3.1 System Model

For network selection decision, utility function assigned to the satisfaction that a network provides to mobile users. Different available networks with different user preferences will have different utility values.

Thus, in this paper, we propose a context-aware scheme of network selection based on utility function and considers both mobile user's awareness and provider's constraints. The selection decision function is defined as a utility function consisting of four parameters Bandwidth, Delay, Jitter and Bit Error Rate. During the network selection procedure, we consider multiple attributes together, so the utilities of multi-

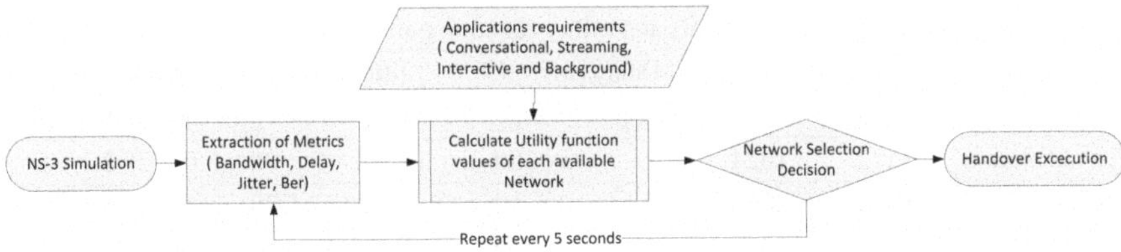

Fig. 2 The flowchart of the proposed scheme

ple attributes are combined as a total utility. We consider four real time applications. Simulation is conducted using NS-3 simulator.

As reported in the algorithm block in Fig. 2, simulation provides the system with the metrics in real time and a the utility function is applied according to each QoS application: Conversational, Streaming, Interactive and Background involving by that the context awareness of the users.

3.2 Utility Function

To capture the satisfaction level of mobile user when served by some network, we use utility function, which measures the normalized satisfaction of mobile user by taking into account Bandwidth, Delay, Jitter and Bit Error Rate of each available network. Hence, the utility should be high and the decision is made accordingly.

The utility function represents how mobile user satisfaction is varying from low to high values with respect to user's needs in terms of application. The applications that we study in this paper are Conversational, Streaming, Interactive and Background.

Network selection QoS metrics can be divided into two categories: Cost metrics and metric of performance. For metric of performance, the best utility value is the largest, like bandwidth, RSS, throughput, reliability degree, etc. Conversely, for a cost metric, the best utility value is the lowest, like delay, jitter, bit error rate, etc. However, in order to define an utility function that considers mobile user's needs, the network metrics are not enough. Thus, Another important factor that has been considered is the type of applications in terms of QoS requirements. Indeed, applications are classified as inelastic, partially elastic and perfectly elastic based on their sensitivity to QoS parameters. For example, real-time voice (Conversational) and video applications (Streaming) are inelastic in their demand for bandwidth and their delay requirements, whereas data transfer, e-mail or web browsing (Interactive and Background) applications are considered perfectly elastic, i.e. tolerant to variations in bandwidth and delay [13]. The equations below define the mathematical models of utility-function for both performance and cost metrics.

Table 1 Application QoS requirements and utility function parameters

	Bandwidth (kbps)		Delay (ms)		Jitter (ms)		Bit error rate (%)	
	x_{min}	x_{max}	x_{min}	x_{max}	x_{min}	x_{max}	x_{min}	x_{max}
Conversational	512	1024	5	100	2	30	0	2
Streaming	1024	2048	5	50	2	20	0	1
Interactive	512	1048	5	20	2	10	0	3
Background	256	512	5	120	2	40	0	5

For a metric of performance, the utility is calculated as follow:

$$f_{Performance}(x) = \frac{min(x, x_{max}) - x_{min}}{x_{max} - x_{min}} \tag{1}$$

For a cost metric, the utility is :

$$f_{Cost}(x) = \frac{x_{max} - max(x, x_{min})}{x_{max} - x_{min}} \tag{2}$$

x_{max} et x_{min} are the minimum and the maximum requirements of a metric in a specific type of application, the values of are provided in Table 1.

Finally, in order to determine the relevance R_{ij} of network i for an application j (Eq. (3)), we combine the utility values of all metrics using three calibration coefficients as following:

$$R_{ij} = \alpha \cdot f_{Bandwidth} + \beta \cdot f_{Delay} + \gamma \cdot f_{Jitter} + \delta \cdot f_{BitErrorRate} \tag{3}$$

where $\alpha + \beta + \gamma + \delta = 1$. Equation (3) $0 \leq R_{ij} \leq 1$ the closer R_{ij} is to 1, the more relevant network i is to application j.

In a scenario where bandwidth, delay, jitter and bit error rate are evenly set up, we take $\alpha = \beta = \gamma = \delta = 1/4$. However, many conceivable scenarios can be patterned by conveniently assessing α, β, γ and δ.

4 Performance Analysis

We have simulated the above algorithm with varying the velocity speed, 20 m/s, 30 m/s and 40 m/s and the type of traffic to analyse the performance in low, medium and high speed WLAN and LTE heterogeneous environment.

4.1 Simulation Parameters

In order to evaluate the Vertical Handover schemes, we conducted simulation experiments according to the algorithm block shown in Fig. 2.

Indeed, we extract the metrics data from simulation and utility function is applied according to application requirements (see Table 2). Eventually, we choose the network with the larger utility function in each application. We launch this process every 5 s. The obtained results are also compared with the baseline scheme. We implement the Simple Additive Weighting (SAW) method from [14] and use it as baseline scheme for comparison.

In all simulations, we use a network consisting of 10 mobile nodes. These nodes follow the same mobility model we managed to make users roam between WLAN and LTE conveniently. The simulations are performed with the Network Simulator NS3. Table 3 presents the detailed parameters considered in the scenario.

Table 2 Standardized QoS characteristics [15]

Resource type	Packet delay budget (ms)	Packet error loss rate	Example services	QoS classes
Guaranteed Bit Rate (GBR)	Up to 100	$\leq 10^{-2}$	Conversational voice	Conversational
	Up to 300	$\leq 10^{-6}$	Live streaming	Streaming
Non-GBR	Up to 100	$\leq 10^{-3}$	Interactive gaming	Interactive
	Up to 300	$\leq 10^{-6}$	e-mail, chat, ftp	Background

Table 3 Simulation parameters and settings

Simulation parameters and settings	
Size of the area (m^2)	$500 * 500$
Number of nodes	10
Available networks	WLAN and LTE
WLAN range (m)	100
Channel bandwidth of WLAN (MHz)	3
LTE range (m)	500
Channel bandwidth of LTE (MHz)	10
Node speed (m/s)	20–30–40
Mobility model	Adapted constant velocity mobility model
Application traffic	Conversational, streaming, interactive and background
Simulation time (s)	650

4.2 Evaluation Criteria

To compare the effectiveness of network selection during vertical handover, we handle the experiments with the network simulator NS3 in order to validate our proposed approach using utility-function, by analysing the impact of speed velocity on QoS. We analyse thereby the velocity speed by varying it: 20, 30 and 40 m/s.

Context awareness services are responsible for carrying a continuous uninterrupted stream to the user. Nonetheless, implementation always asks for the correct context information which leads to taking correct decisions. To this end, we analyse the Delay and Packet Loss over time, considering that those metrics change throughout the simulation. Delay is affected by the time taken by each algorithm to be executed which affects the time from the sending of a packet by the source until it is received by the destination and the packets dropped during the vertical handover execution which has an effect on the Packet Loss.

The following section details the development of the network throughout the simulation.

4.3 Simulation Results and Discussion

Figure 3 illustrates the behaviour of delay over time, it compares the performance of our scheme while varying the velocity speed. in terms of delay, its shows that the proposed scheme produces very good results at low 20 m/s, medium 30 m/s mobile user speed and allowable performance for high speed of 40 m/s still better by 7 % than the baseline scheme simulated with only 10 m/s (see Table 4).

In a similar way, Fig. 4 exposes the behaviour of packet loss over time, it contrasts the performance of our scheme while varying the velocity speed. The swiftness of the decisions made by the terminal to utility values influences also on the number of packets dropped all along the simulation. For packet loss, the proposed scheme provides an acceptable packet loss for all types of application according to the requirements shown in Table 1. Compared to the baseline, our proposed scheme reduces the packet by 40 %.

Table 4 Improvement of DELAY and LOSS PACKET by the proposed scheme

Traffic class	Delay (% ↓)	Packet loss ratio (% ↓)
Conversational	3.66	33.97
Streaming	3.64	39.41
Interactive	7.19	40.51
Background	4.68	36.24

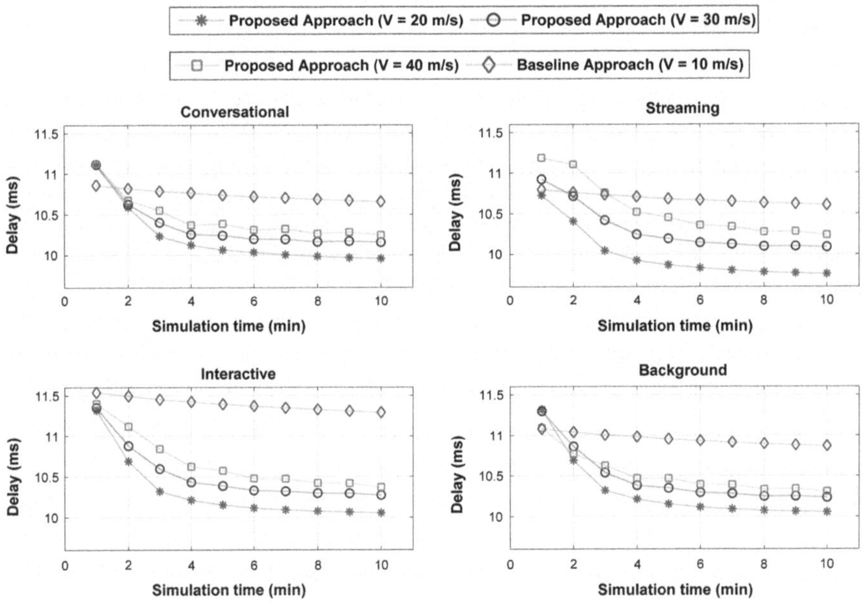

Fig. 3 Behaviour of delay over time

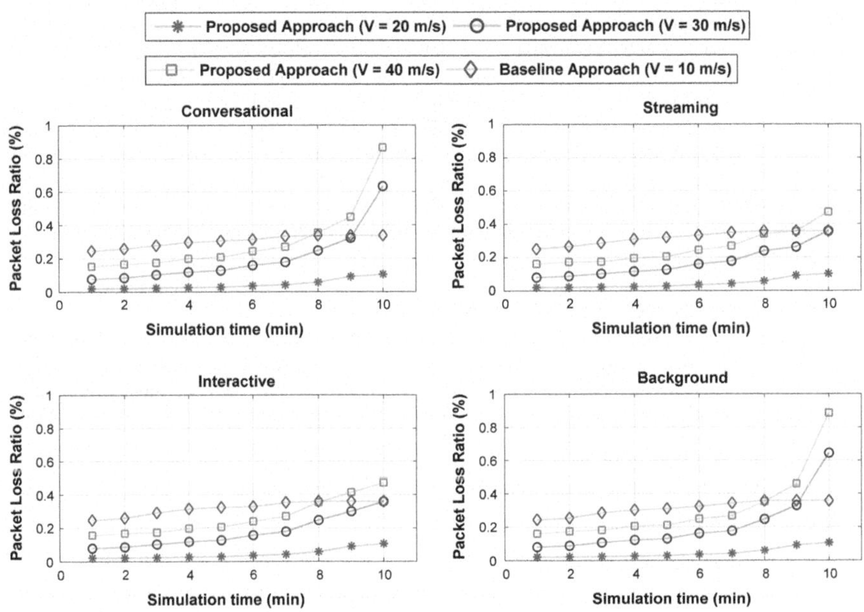

Fig. 4 Behaviour of packet loss over time

5 Conclusion

In heterogeneous networks, the required QoS can be achieved through an efficient vertical handover decision that combines the requirements of mobile users and networks. In this paper, we proposed a context-aware scheme of network selection based on utility function and considers both mobile user's needs and provider's constraints. The selection decision function is defined as a utility function consisting of four parameters bandwidth, delay, jitter and bit error rate. We have considered four real time applications. Simulation is conducted using NS-3 simulator consists of WLAN-LTE heterogeneous network. Simulation results show that the proposed scheme is advantageous for high data rate applications even if user move with the high speed of 40 m/s. It reduces the delay and packet loss ratio, and consequently improves QoS.

References

1. Gustafsson, E., Jonsson, A.: Always best connected. IEEE Wirel. Commun. **10**(1), 49–55 (2003). Feb
2. Itu, T.: Series y: global information infrastructure, internet protocol aspects and next-generation networks. Rec. ITU-T Y **2720** (2009)
3. Kassar, Meriem, Kervella, Brigitte, Pujolle, Guy: An overview of vertical handover decision strategies in heterogeneous wireless networks. Comput. Commun. **31**(10), 2607–2620 (2008)
4. Wang, L., Kuo, G.S.G.S.: Mathematical modeling for network selection in heterogeneous wireless networks—a tutorial. IEEE Commun. Surv. Tut. **15**(1), 271–292 (2013)
5. Lopez-Benitez, M., Gozalvez, J.: Common radio resource management algorithms for multimedia heterogeneous wireless networks. IEEE Trans. Mob. Comput. **10**(9), 1201–1213 (2011)
6. Ma, D., Ma, M.: A qos-based vertical handoff scheme for interworking of wlan and wimax. In: IEEE Global Telecommunications Conference, 2009. GLOBECOM 2009, pp. 1–6, Nov 2009
7. Ahuja, Kiran, Singh, Brahmjit, Khanna, Rajesh: Network selection algorithm based on link quality parameters for heterogeneous wireless networks. Optik—Int. J. Light Electron Opt. **125**(14), 3657–3662 (2014)
8. Mansouri, W., Mnif, K., Zarai, F., Obaidat, M.S., Kamoun, L.: A new multi-rat scheduling algorithm for heterogeneous wireless networks. J. Syst. Softw. (2015)
9. Kosmides, Pavlos, Rouskas, Angelos, Anagnostou, Miltiades: Utility-based RAT selection optimization in heterogeneous wireless networks. Pervasive Mob. Comput. **12**, 92–111 (2014)
10. Chamodrakas, Ioannis, Martakos, Drakoulis: A utility-based fuzzy TOPSIS method for energy efficient network selection in heterogeneous wireless networks. Appl. Soft Comput. **11**(4), 3734–3743 (2011)
11. Pirmez, L., Carvalho, J.C. Jr., Delicato, F.C., Protti, F., Carmo, L.F.R.C., Pires, P.F., Pirmez, M.: Sutil—network selection based on utility function and integer linear programming. Comput. Netw. **54**(13), 2117–2136 (2010)
12. Wu, X., Du, Q.: Utility-function-based radio-access-technology selection for heterogeneous wireless networks. Comput. Electr. Eng. (2015)
13. Shenker, S.: Fundamental design issues for the future internet. IEEE J. Sel. A. Commun. **13**(7), 1176–1188 (2006)
14. Drissi, Maroua, Oumsis, Mohammed: Multi-criteria vertical handover comparison between wimax and wifi. Information **6**(3), 399 (2015)
15. Alasti, Mehdi, Neekzad, Behnam, Hui, Jie, Vannithamby, Rath: Quality of service in wimax and lte networks. IEEE Commun. Mag. **48**(5), 104–111 (2010)

Study of the Impact of Designed Objective Function on the RPL-Based Routing Protocol

Hanane Lamaazi, Nabil Benamar and Antonio J. Jara

Abstract The routing over Low Power and Lossy network working group (ROLL) has specified RPL as an IPv6 routing protocol for Low Power and Lossy networks. RPL builds a Destination Oriented Directed Acyclic Graph (DODAG) based on a set of metrics and constraints trough a specific Objective Functions (OFs). This OF can specify the selection of the parent and the construction of the route. In this paper, the performances of RPL are analyzed based on two main objective functions: Minimum Rank with Hysteresis Objective Function (MRHOF) that uses number of expected transmission (ETX) as a condition to select the routes and the Objective Function Zero (OF0) based on the Minimum Hop Count to determine the best parent. The analysis of RPL performances with the two comparative Objective Function is made with different metrics such as ETX, Hop Count, lost packet, energy, and control traffic overhead. This comparison makes it possible to distinguish which objective function is the most optimal to guarantee good functioning of RPL especially in mobile environment and which one respond better to the requirements of its application.

Keywords RPL · MRHOF · OF0 · Sink · RWP · ETX · Hop count

H. Lamaazi (✉)
Faculty of Sciences, Moulay Ismail University, Meknes, Morocco
e-mail: Lamaazi.hanane@gmail.com

N. Benamar
High School of Technology, Moulay Ismail University, Meknes, Morocco
e-mail: benamar73@gmail.com

A.J. Jara
University of Applied Sciences Western Switzerland (HES-SO) Sierre, Vallais, Switzerland
e-mail: jara@ieee.org

R. El-Azouzi et al. (eds.), *Advances in Ubiquitous Networking 2*, Lecture Notes in Electrical Engineering 397, DOI 10.1007/978-981-10-1627-1_6

1 Introduction

The employment of the Internet of Things (IoT) [1] in a variety of domains imposes it to find solution for application that provides a constraint. One of these constraint applications is the Low power and Lossy networks (LLNs). This kind of networks is based on tiny and finer devices that have a limitation in terms of processing capability, transmission range and battery lifetime. Moreover, limited resources of constraint devices allow it to be restricted in its use in wide scale. Furthermore, the main parameters that allow overcoming these losses are to choose good routing protocol that can respond to these requirements. By considering the requirements mentioned above, IETF standardized a routing protocol for 6LoWPAN [2] networks called RPL (IPv6 routing protocol for Low Power and Lossy Networks). The specificity of RPL is that it can be used for LLNs Networks according to the application of an optimizing objective functions (OFs). Implementation of the OFs in separate way into the core protocol specification lets RPL to be easy adaptable with the change and optimization in its architecture. This allows it to be flexible in use and to satisfy network designs and application requirements.

The objective Function has, as main role, the selection and optimization of the route between nodes, it allows to decide which node can join the destination according to specific metrics/constraints as power consumption, delay, number of expected transmission count, hop count, latency, link quality level etc.

In this paper, RPL based on the default objective function (MRHOF) and the objective function zero (OF0) is implemented with different positions, number of nodes and mobility models. The performances of RPL are analyzed according to a set of parameters which, make to distinguish which Objective Function is better in these conditions.

The paper is organized as follows. Section 2 describes some research and studies related to RPL performances and Objective Functions. Section 3 describes an overview of RPL routing protocol and Objective Function used for this study. To analyze the performance of the Objective Functions, a set of metrics are explained in Sect. 4. In Sect. 5, an assessment of different results obtained is described. Finaly, we conclude this study by highlighting the suitable Objective function for the studied scenario for different metrics.

2 Related Works

In [3], authors propose a new Objective Function based on Scalable Context-Aware called SCAOF. This OF can adapt RPL to the environmental monitoring of Agricultural Low-power and Lossy Networks (A-LLNs). This adaptation is based on a combination of energy-aware, reliability-aware, robustness-aware and resource-aware according to the composite routing metrics approach. Performance evaluation of the proposed RPAL was verified on both simulation and field tests. Simulation results prove that in different simulation scenarios and hardware testbed, SCAOF can deliver the desired advantages on network lifetime extension, and high efficiency and reliability.

In [4], Authors propose a quantification of primary routing metrics based on specific formulas in order to capture impact related to the kind of network. They suggest a set of combination of routing metrics in additive and lexical way. The routing metrics used in the paper are primary or composite that conclude Hop Count (HC), Expected Transmission (ETX), Packet Forwarding Indicator (PFI) and Remaining Energy (RE). As a results for all combination, routing functions combining PFI with hop count offer better performances in terms of packet loss than HC used on own. Similarly, combination of HC and RE act better in terms of energy consumption than single HC in contrast with latency which HC provide better value than combined metrics.

In [5], Authors propose an assessment of RPL performances based on the Objective Function. They use the default objective function MRHOF and the Objective Function Zero (OF0) in order to distinguish which Objective Function provides better performances of RPL. The simulation environment uses a various radio models (Unit Disc Graph Model—Distance Loss, Unit Disc Graph Model—Constant Loss, Multi-Path Ray-Tracer Medium) and scaled network. To make comparison between the two objective functions a set of metrics are calculated: Packet Delivery Ratio, Control Traffic Overhead, Power Consumption and Network ETX. As a results, the objective function based on ETX metrics (MRHOF) provide better performances in all scenarios compared to the objective function based on hop count metric (OF0).

In [6], Authors study the performance of multiple RPL instances. They use a Cooja simulator to implement their proposed study using two types of traffic flows: regular and critical traffic. The two types of traffic can be generated by the forwarder nodes. Furthermore, comparison has been made between a multiple RPL instances and single instance by considering a set of metrics as routing tree convergence time, network latency and packet delivery rate. As results, the differentiating of traffic in different DAGs of multiple RPL instances gives a better performance in terms of PDR and Latency particularly in a substantial traffic network.

In [7], Authors propose an optimization objective of RPL that consider a context-awareness called Context-Aware Objective Function (CAOF). This optimization takes into consideration the limited resources of sensor nodes and their temporal changes. Moreover, this proposed CAOF has as main objective to make the routing decision to consider the battery level in order to optimize the power exploitation as a critical resource. The results show that the proposed objective function increases the lifetime compared to non-context-aware RPL OFs by up to 44 %. Additionally, CAOF improves the delivery ratio and guarantee more fairness in terms of battery exploitation for different nodes than non-context-aware OFs of RPL.

3 Preliminaries

In this section we present the necessary background discussion of both RPL and Objective Functions.

3.1 *RPL Routing Protocol*

The ROLL working group from IETF has specified a routing protocol designed for LLNs, called RPL. RPL can be supported by the lightweight devices. These devices are limited in terms of resources and memory and use a poor link. In RPL, nodes can use three traffics to communicate: Point-to-multipoint, Multipoint-to-point and point-to-point. The main idea of RPL is based on the concept of Directed Acyclic Graphs (DAGs). The DAG defines the default routes between nodes based on a tree structure where nodes can have more than one parent. Moreover, nodes are organized as Destination-Oriented DAGs (DODAGs) on which sink nodes can act as the roots of the DAGs. The default route for each node that it uses toward the roots is formed based on the best parent. RPL specify three control messages to exchange the traffic:

- DODAG Information Object (DIO) messages
- Destination Advertisement Object (DAO) messages
- DODAG Information Solicitation (DIS) messages

Node sent the DIO messages in multicast in order to build and maintain upwards (MP2P traffic) routes of the DODAGs. The DAO messages manage the downward (P2MP traffic) routes and it propagates the routing tables. Additionally, nodes send the DIS messages in order to solicit DIO messages from its neighborhood. This solicitation of DIO messages allow to update the routing information [3, 8] (Fig. 1).

3.2 *Objective Function*

One of the different specifications of RPL is that it can builds route toward the root based on objective function. For this reason, the OF is considered as the key facto to determine, in the network, the preferred parent of node from candidate neighbor. Node can have more than one parent especially in a network with huge density. For this, objective function try to choose which parent is suited for one node than other. In addition, the choice of parent by OF is based on a specific criteria as links, routing metrics or node metrics. These metrics can be specified by the designer according to its need [9]. Until now, the ROLL working group has specified two objective functions. The first one is the Objective Function Zero (OF0) [10], on which the criteria of selecting best parent is the minimum Hop Count that it

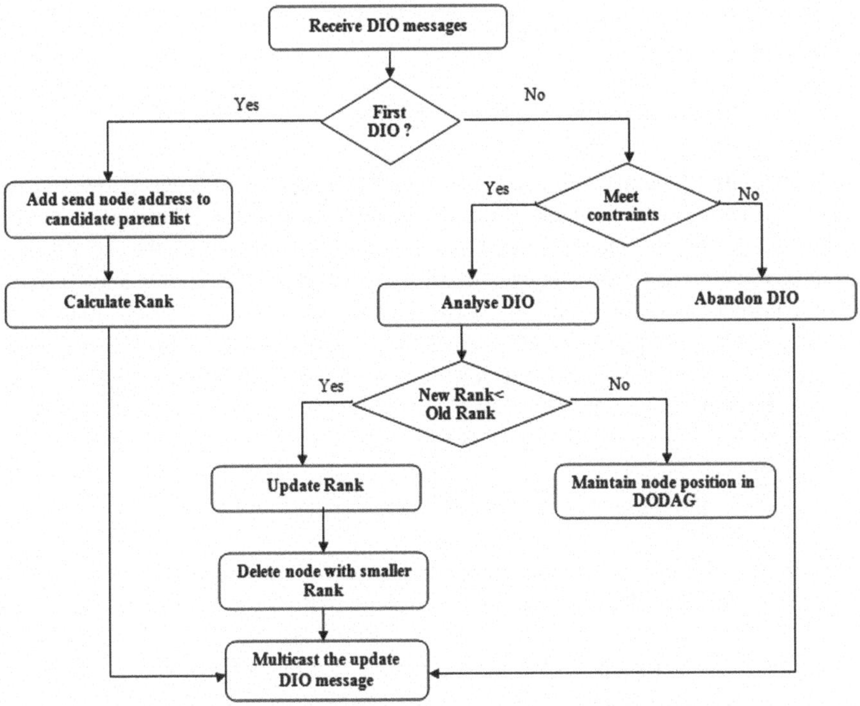

Fig. 1 Operation performed after receiving a DIO

provide. The second one is the Minimum Rank with Hysteresis Objective Function (MRHOF) [11]. In contrast to OF0, MRHOF selects routes based on Expected Transmission Count (ETX) metric. Minimum value of this metric means the optimal route to the sink node [12, 13].

4 Metrics Choice

The choice of the metrics plays a big role on the performance analysis of the objective function. In this paper, we focus on five metrics to investigate Objective Function of RPL. The results shows which objective function can perform better than others in different scenarios. In this section we describe all metrics that we have used in this study:

Control Traffic Overhead: it's the total number of control messages transmitted by nodes in order to build DODAG and to select the best parent between candidate neighbors. The control messages contain the DAO, DIS and DIO messages. It can be calculated as follow:

$$\text{Control Traffic Overhead} = \sum_{1}^{n} DIO + \sum_{1}^{n} DIS + \sum_{1}^{n} DAO \tag{1}$$

The stability of the network can be deduced by this metric, if it gets a lower value means that the network is stable if not the network is unstable. The control Traffic Overhead has a direct or indirect impact on the resource consumption of the network. For instance, when the network becomes denser it allows having congestion and collision between packets, and which makes the network very delayed. For this reason, nodes send more transmitted messages in order to check the availability of the network which, gets nodes to spend more power and consume more resources from the network [5].

Lost Packet: is the packet dropping during the transmission of messages between nodes. It can be calculated as follow:

$$\text{Packet Lost} = \sum_{1}^{n} Sent\ Packets - \sum_{1}^{n} Received\ Packets \tag{2}$$

Node Energy: it indicates the energy measured from nodes in the network over the network lifetime. The formula used to calculate the energy of nodes is:

$$\begin{aligned}\text{Energy (mJ)} = (&\text{Transmit} * 19.5\ \text{mA} + \text{Listen} * 21.5\ \text{mA} + \text{CPU\_time} * 1.8\ \text{mA} \\ &+ \text{LPM} * 0.0545\ \text{mA}) * 3\ \text{V}/(32768)\end{aligned} \tag{3}$$

Reducing energy is one of the important metric that objective function based-on. OF can select the route toward the sink node based on the low energy value which a candidate parent can provide. For this reason, application with energy-efficiency should take into consideration this metric [14].

Hop Count (HC): the Objective Function of the routing protocol can be based on this metrics to select the best parent. Low value of Hop Count means that node is suited to be considered as best parent. HC presents the number of hops between node and its neighbor towards root. Application that requires to be done in real-time consider the lower possible number of hops to join the destination [15].

Expected Transmission Count (ETX): it refers to the number of retransmission needed to a packet is successfully received to the destination. The ETX value can give information about stability network. Instead of the hop count, objective functions can based on the ETX value to select path to the root. Less value of ETX indicate that the network has good link stability which, allows to conclude that there are smallest retransmission of packets and nodes doesn't consume more resources [16]. The ETX value can be calculated using this formula as follow:

$$\text{ETX} = \frac{1}{\text{Df} * \text{Dr}} \tag{4}$$

The Df represents a forward delivery ratio which, is the measured probability that a packet is received by a neighbor. The Dr is a reverse delivery ratio which is the measured probability that an acknowledgment packet is successfully received [16].

5 Simulation Results

Needless to say that each metric we choose to analyze RPL behavior provides a particular side of this study. Performances investigation of RPL using different objective functions pushe to consider consumption of limited resources of nodes and network. For this reason, we have considered a set of metrics as ETX, Hop Count, Lost Packet, Energy and Control Traffic Overhead. These metrics show a set of information regarding to the network stability and resources consumption. In this paper, we propose three scenarios for our study. In the first one, we evaluate RPL based on the objective function by using scalable network. In the second, we use two mobility models with two different density of network. In the last one, we change the position of nodes by considering four positions: Random, Linear, Ellipse and Manual position.

To implement our proposed scenarios, we choose Cooja simulator. It is designed to simulate the IoT networks, as LowPAN and WSN. It is an open source simulator and can operate in different level and run over different platforms. For these reasons, the choice of Cooja is compatible for our study. In all simulations, we consider a multipoint-to-point topology, we use one sink and multi senders. All traffics are in upward i.e. all nodes send data to the root. The interference region and the transmission range are100 m for each one. The simulation time for each simulation in the scenarios is 21 min. The results are showing via graphs, they are plotted between the chosen metrics and the number of nodes, type of mobility models and nodes positions. In what follow, we show the graphical results that compare the two objective functions: MRHOF and OF0.

5.1 Analysis Based on Scalable Network

In this scenario, we increased the number of nodes in order to investigate the objective function behavior in a scale network.

Figures 2 and 3 show the main metrics that MRHOF and OF0 are based on to select route which, are ETX and Hop Count respectively. In Fig. 2, the ETX increases considerably for MRHOF when the network become huger while for OF0 it decreases. In Fig. 3, we show that both OFs diverge in a high value for MRHOF and low one for OF0. This can be explained by congestion and collision packets which, push nodes using MRHOF to keep sening packets to be received successfully. This requires more hops to reach destination. In contrast, nodes using OF0 can quickly find parents with minimum hop count when the network is denser

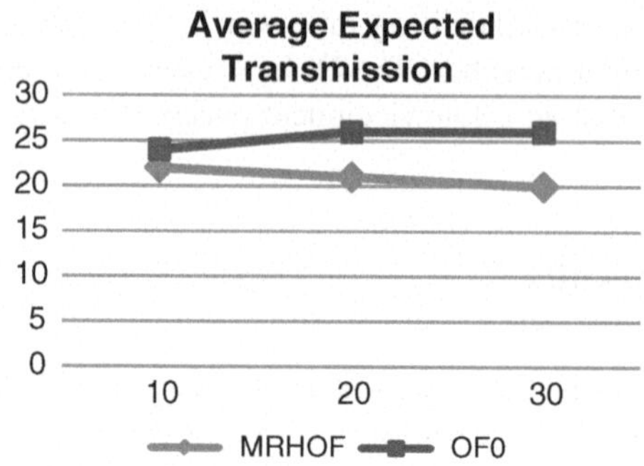

Fig. 2 Average ETX versus number of nodes

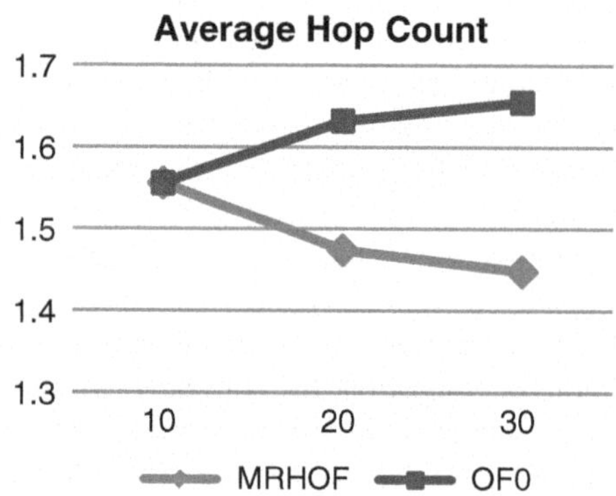

Fig. 3 Average hop count versus number of nodes

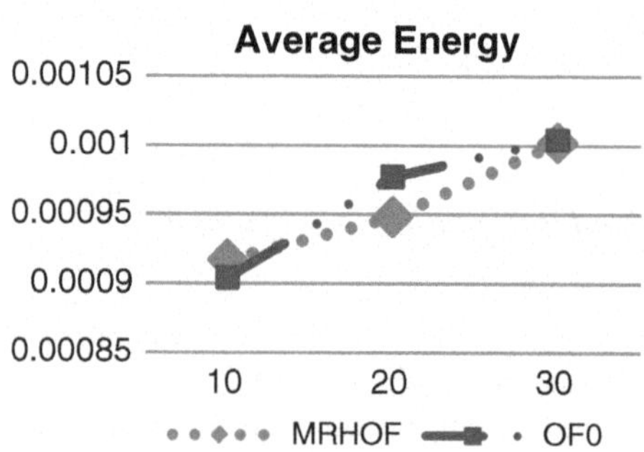

Fig. 4 Average energy versus number of nodes

regards to the number of candidate parents and instead of its packets are received or not. In Fig. 4, OF0 consumes slightly more energy than MRHOF, but both objective functions have an increase in energy when the network is dense. Indeed, the more nodes are used in the network the more the energy is consumed.

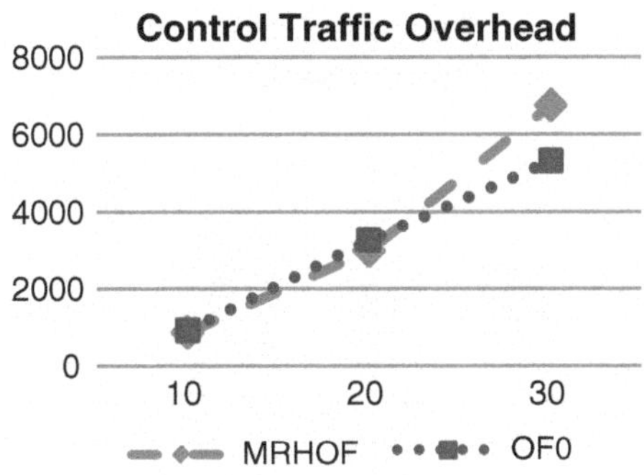

Fig. 5 Control traffic overhead versus number of nodes

The control traffic overhead is shown in Fig. 5. It is clear that nodes provide a low control traffic overhead value in the smallest network. In Contrast, in a huge network MRHOF increases considerably in comparison with OF0 which, can be explained by the needs of MRHOF nodes to retransmit more messages to select routes. Both metrics ETX and HC are complementary in the sense that each one of them impacts the other. Based on the ETX metric, MRHOF tries to choose path with the lower number of transmitted packet which, include the use of minimum HC (even if it is not considered by MRHOF as a metric). Similarly, OF0 use the minimum HC to reach destination which, allows node using less sending packets. In general, when network become dense, both HC and ETX metrics act in the same way. For MRHOF the ETX and HC metrics decrease in a huge network while in OF0 they increase.

5.2 Analysis Based on Mobility Models

In this scenario, we consider two mobility models with two different densities. This allows to show the impact of the mobility models on the OF operations. With regards to the mobility models that we choose in this paper, the behavior of OF change according to the movement of nodes. For Random Waypoint (RWP) model, it allows nodes to move in a random way. Each node into RWP travels separately compared to the other. In contrast, in Reference Point Group Mobility Model (RPG) nodes move as group with a dependently way which means that RPG realizes movement of nodes inside the group and of the entire group in arbitrary manner [17]. For these reasons, we apply this mobility models to demonstrate how objective functions react in both entity and group mobility models.

As shown in Fig. 6, MRHOF and OF0 have approximately same value of ETX with RPG model while it diverges with RWP model. In RPG, both objective functions provide similar number of ETX in a smallest network and a slight progress of OF0 in dense network. In the other side, MRHOF increases considerably with

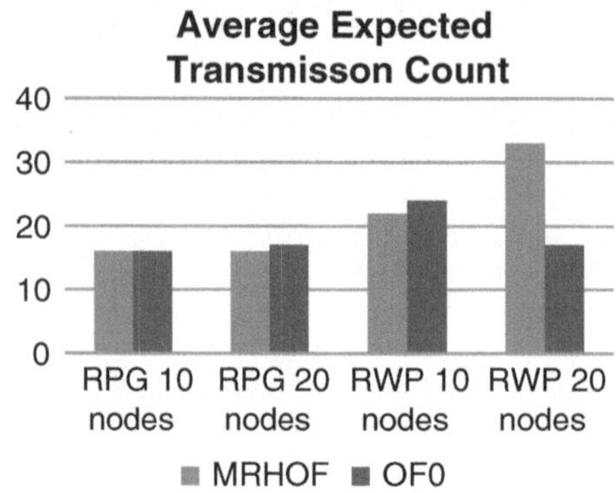

Fig. 6 Average ETX versus mobility models

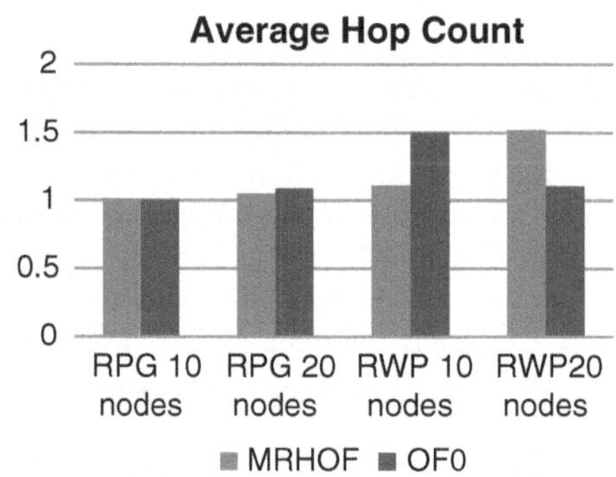

Fig. 7 Average hop count versus mobility models

RWP model and also in dense network while OF0 decrease. Similar to Fig. 6, in Fig. 7 both OFs provide nearest value of Hop Count with RPG while with RWP, MRHOF increase and OF0 decreases. In Fig. 8, routing protocol with MRHOF consumes less energy with RPG instead of the network is denser or not while with OF0 it growth slightly. However, MRHOF consumes more energy with RWP and increases with big density. In contrast, OF0 consumes less energy than MRHOF and increases very little with the growth of network. In Fig. 9, both MRHOF and OF0 have a less value of control traffic overhead in RPG model which, increase little with dense network. In contrast, OF0 provide high value compared to the MRHOF which, means that network is more stable in RPG with MRHOF than OF0. On the other side, OF0 operates better with RWP than MRHOF but both provide higher value of control traffic overhead in RWP compared to the case of RPG. This allows distinguishing that RPL is more stable in RPG than RWP. Figure 10 shows the packet lost during the movement of nodes. RPL looses fewer packets in RPG than RWP which, allows concluding that RWP is not suited for application that have a real-time

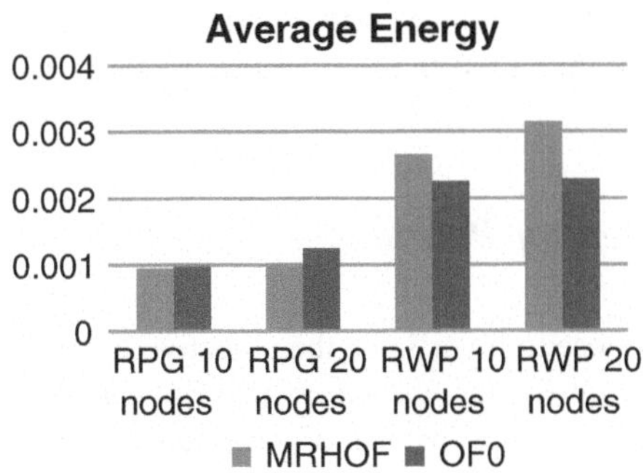

Fig. 8 Average energy versus mobility models

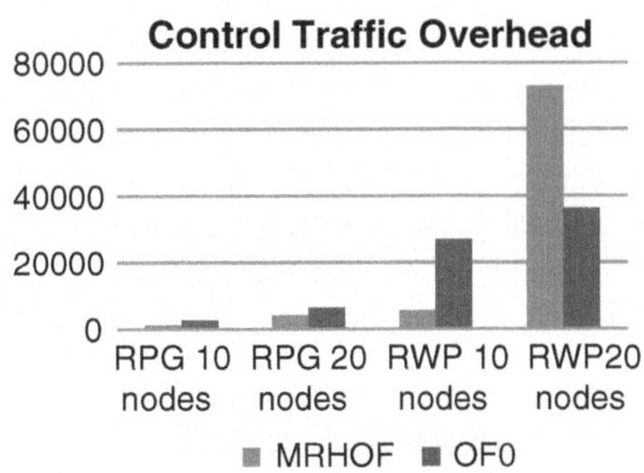

Fig. 9 Control traffic overhead versus mobility models

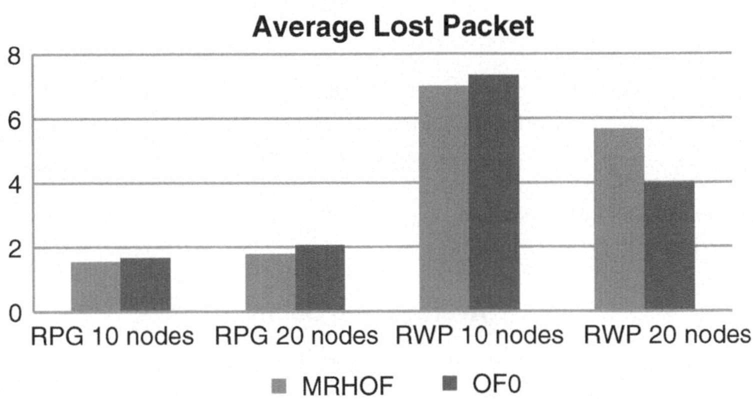

Fig. 10 Average lost packet versus mobility models

requirement regards to the number of lost packet that it causes. In Fig. 9 OF0 losts more packets than MRHOF and both increase similarly with the growth of network. Furthermore, OF0 provides more dropped packets than MRHOF but it decreases considerably more than MRHOF until it obtains a less value of lost packet when the network become denser. On the whole, RPL operates better with RPG than RWP and acts well with MRHOF than OF0 in terms of lost packets.

5.3 Analysis Based on Positions

In this scenario we propose to investigate the objective function reaction in different nodes distribution. For all figures, we notice that both objective functions provide similar results with all metrics instead of Average Energy and Control Traffic Overhead with linear position. In Fig. 11, MRHOF and OF0 have nearest value in terms of ETX metric in Random, Linear and Manual positions while they offer high value with Ellipse position. In agreement with this, nodes in Ellipse position need more Hop Count than other positions to reach destination as showing in Fig. 12. In the other case, nodes using MRHOF and distributed in linear position consume more power than nodes in random, ellipse and manual position. Besides, nodes using OF0 consume proximately same energy in all positions. Figure 13 show that nodes using MRHOF in linear position transmit more traffics than random, ellipse and manual positions. In contrast OF0 send less traffics in all positions compared to the MRHOF. Additionally, in random and manual position MRHOF still better than

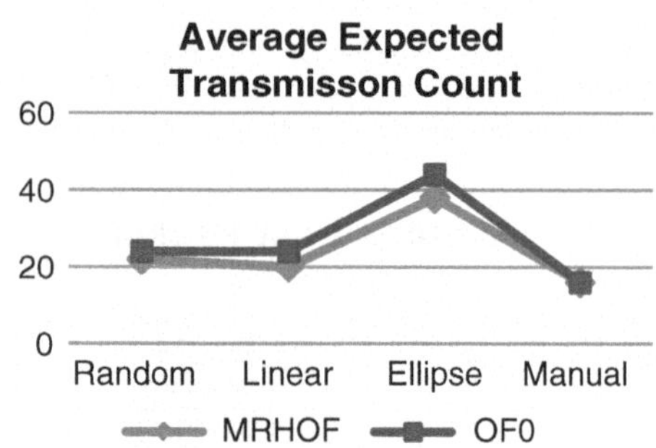

Fig. 11 Average ETX versus positions

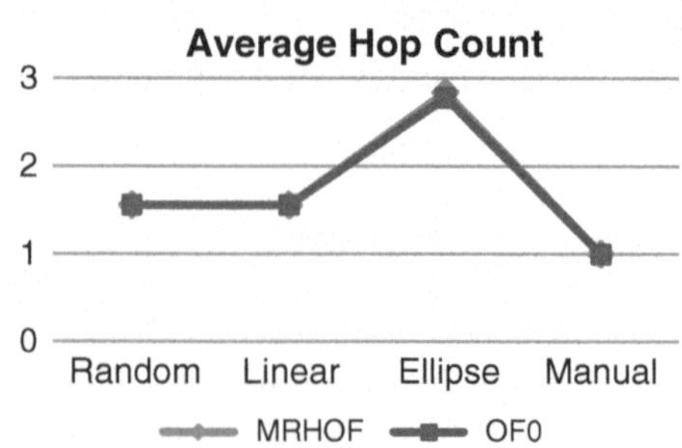

Fig. 12 Average hop count versus positions

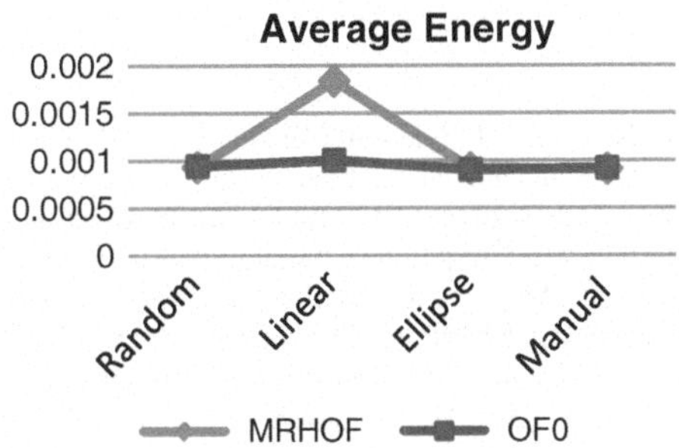

Fig. 13 Average energy versus positions

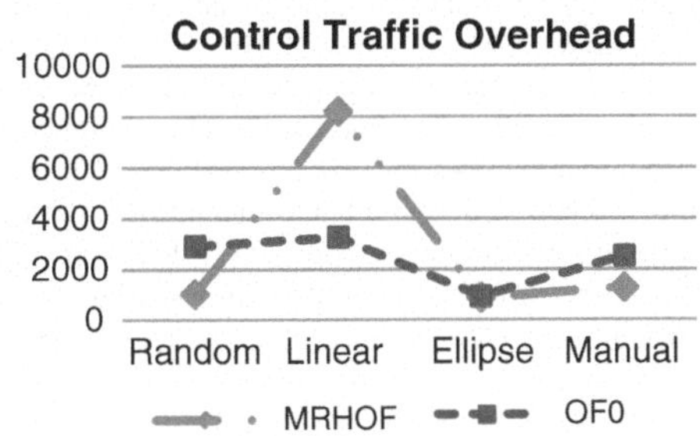

Fig. 14 Control traffic overhead versus positions

OF0 in contrast to the linear position where OF0 is better. Moreover, nodes in ellipse position provide nearest value for both MRHOF and OF0 (Fig. 14).

6 Conclusion

In this work, we have investigated the impact of the objective function on the RPL performances for different scenarios and for different metrics. Besides, the protocol performance is strictly influenced by the density of network, type of mobility models used and node positions. In this paper, comparison has been made using five metrics: ETX, Hop Count, lost packets, Energy and Control Traffic Overhead. The results show that MRHOF acts better with scalable network using a static nodes. In contrast, OF0 behaves better with mobile nodes than MRHOF. For different node distribution, both objective functions provide similar results for all metrics except for Linear position which consume more energy and provide high control traffic overhead. Briefly, one objective function cannot respond to all needs of the protocol. For that reason, as a future work, it would be interesting to designed a new objective function that consider mobility of nodes and provide better results that helps RPL to be more suitable for its applications.

References

1. Lamaazi, H., Benamar, N., Jara, A.J., Ladid, L., El Ouadghiri, D.: Challenges of the Internet of Things : IPv6 and network management. In: International Workshop on Extending Seamlessly to the Internet of Things (esIoT-2014), 2014 Eighth International Conference on Innovative Mobile and Internet Services in Ubiquitous Computing (IMIS), pp. 328–333 (2014)
2. Lamaazi, H., Benamar, N., Jara, A., Ladid, L., El Ouadghiri, D.: Internet of thing and networks' management: LNMP, SNMP, COMAN protocols. In: First International Workshop on Wireless Networks and Mobile Communications (WINCOM 2013), pp. 1–5 (2013)
3. Chen, Y., Chanet, J., Hou, K., Shi, H., De Sousa, G.: A scalable Context-Aware Objective Function (SCAOF) of Routing Protocol for Agricultural Low-Power and Lossy Networks (RPAL). Sensors 19507–19540 (2015)
4. Karkazis, P., Leligou, H.C., Sarakis, L.: Design of primary and composite routing metrics for RPL-compliant Wireless Sensor Networks, Feb 2016 (2012)
5. Sharma, R., Jayavignesh, T.: Quantitative Analysis and Evaluation of RPL with Various Objective Functions for 6LoWPAN, vol. 8, Aug 2015
6. Mai, B., Hieu, M., Nam, N., Kieu-Ha, P., Nugugen, T.H., Kris, S.: Performance evaluation of multiple RPL routing tree instances for Internet of Things applications. In: International Conference on Advanced Technologies for Communications (ATC), pp. 206–211 (2015)
7. Sharkawy, B., Khattab, A., Elsayed, K.M.F.: Fault-tolerant RPL through context awareness *. In: 2014 IEEE World Forum on Internet of Things (WF-IoT), pp. 1–5 (2014)
8. Farooq, M.O., Sreenan, C.J., Brown, K.N., Kunz, T.: RPL-based routing protocols for multi-sink wireless sensor networks. In: 11th International Conference on Wireless and Mobile Computing, Networking and Communications (WiMob) RPL-Based, IEEE, pp. 452–459 (2015)
9. Nurmio, J., Poellabauer, C.: Equalizing energy distribution in sensor nodes through optimization of RPL. In: Proceedings of the 15th IEEE International Conference on Computer and Information Technology (2015)
10. Thubert, P.: Objective function zero for the routing protocol for low-power and lossy networks (RPL). RFC 6552 (2012)
11. Gnawali, O., Levis, P.: The minimum rank with hysteresis objective function. RFC 6719 (2012)
12. Vasseur, D.B.J., Kim, M., Pister, K., Dejean, N.: Routing metrics used for path calculation in low-power and lossy networks. RFC 6551 (2012)
13. Gonizzi, P., Monica, R., Ferrari, G.: Design and Evaluation of a Delay-Efficient RPL Routing Metric, pp. 1573–1577 (2013)
14. Yunis, J.P., Dujovne, D.: Energy efficient routing performance evaluation for LLNs using combined metrics. In: IEEE Biennial Congress of Argentina (ARGENCON) Energy (2014)
15. Gaddour, O., Koubâa, A., Abid, M.: Quality-of-service aware routing for static and mobile IPv6-based low-power and lossy sensor networks using RPL. Ad Hoc Netw. Sci. Direct J. (2015)
16. Lamaazi, H., Benamar, N., Imaduddin, M.I., Jara, A.J.: Performance assessment of the routing protocol for low power and lossy networks. In: The International Workshop WINCOM (Wireless Networks and mobile COMmunications) in Marrakech, Morocco (2015)
17. Lamaazi, H., Benamar, N., Imaduddin, M.I., Habbal, A., Jara, A.J.: Mobility support for the routing protocol in low power and lossy networks. In: The 30th IEEE International Conference on Advanced Information Networking and Applications (AINA-2016) (2016)

Multi-channel Coordination Based MAC Protocols in Vehicular Ad Hoc Networks (VANETs): A Survey

Mina Ouyous, Ouadoudi Zytoune and Driss Aboutajdine

Abstract Multi-channel communications protocols in Vehicular Ad Hoc Networks (VANETs) is a very important topic that has been attracting the research community in the last decade. These communication protocols are based on the latest standard draft IEEE 802.11p and IEEE 1609.4, in which the channel is divided into multiple channels (i.e., control channel (CCH) and service channels (SCHs)) in order to improve open public road safety services, comfort and efficiency of driving. There are several survey papers that present and compare the Multi-channel communication protocols from various perspectives, but a survey on Multi-channel coordination based MAC protocols in Vehicular Ad Hoc Networks (VANETs) is still missing. In this work, we provide a comparative study of the existing literature on multi-channel coordination based MAC protocols in vehicular Ad Hoc Networks. We first define suitable criteria to classify existing solutions, and then describe them by separately addressing a set of protocols in order to compare them. We conclude the paper by addressing some open issues that need to be tackled in future studies.

Keywords VANET · Multi-channel MAC · IEEE 802.11p · IEEE 1609.4

M. Ouyous (✉) · O. Zytoune · D. Aboutajdine
LRIT, Associated Unit to CNRST (URAC 29) - Faculty of Sciences,
Mohammed V University in Rabat, B.P.1014, Rabat, Morocco
e-mail: ouyous.mina@gmail.com

O. Zytoune
e-mail: zytoune@univ-ibntofail.ac.ma

D. Aboutajdine
e-mail: aboutaj@hotmail.com

O. Zytoune
LGS, National School of Business and Management,
Ibn Tofail University, Kenitra, Morocco

R. El-Azouzi et al. (eds.), *Advances in Ubiquitous Networking 2*, Lecture Notes in Electrical Engineering 397, DOI 10.1007/978-981-10-1627-1_7

1 Introduction

Vehicular ad hoc network (VANET) is an application of mobile ad hoc network, it has been considered to be an important part of the Intelligent Transportation System (ITS). More precisely, a VANET is a self-organized network that can be formed by connecting vehicle, in order to offer to drivers a variety of safety applications (e.g., traffic coordination) and non-safety applications (e.g., Internet access). Two main types of communication are potentially provided in VANET, Vehicle to Vehicle communication (V2V) and Vehicle to Infrastructure communication (V2I) [1].

A VANET is mainly characterized by high mobility, rapidly changing network topology and time critical. All of these characteristics cause many challenging issues in the design of vehicle to vehicle (V2V) and vehicle to infrastructure (V2I) communication protocols [1].

IEEE 802.11p is an approved amendment to the IEEE 802.11 MAC/PHY standard to support wireless access in vehicular environments, this standard allows vehicular communications in Dedicated Short Range Communications (DSRC) spectrum [2]. The US Federal Communication Commission (FCC) allocated 75 MHz of the DSRC spectrum at 5.9 GHz to be exclusively used for vehicle to vehicle and infrastructure to vehicle communications. At the PHY layer, IEEE 802.11p is essentially based on the Orthogonal Frequency Division Multiplexing technique (OFDM) defined for IEEE 802.11a, with a 10 MHz wide channel instead of the 20 MHz which is usually used by 802.11a devices. At the MAC layer, IEEE 802.11p is basically based on the Enhanced Distributed Channel Access (EDCA) of IEEE 802.11e. EDCA provides a differentiated channel access to data traffic with different priorities [2].

In addition, the IEEE 1609 family of standards, is a higher layer standard on which IEEE 802.11p is based. This family of standards defines the architecture, communication model, management structure, security mechanisms and physical access for wireless communications in vehicular environment. Collectively, IEEE 802.11p and IEEE 1609.x are called Wireless Access in Vehicular Environments (WAVE) standards. In particular, the IEEE 802.11p and IEEE 1609.4 standards specify a multichannel Medium Access Contro (MAC) protocol draft for the Wireless Access in Vehicular Environment (WAVE) system. Channels used by the IEEE 1609.4 standard are allocated in the DSRC spectrum. As indicated in Fig. 1, the overall bandwidth in 5.9 GHz spectrum is divided into seven smaller frequency operation channels of 10 MHz for each one. One of these channels is allocated as a control channel (CCH), which is used as a public channel for safety-relevant applications on the road. The rest of channels are used as service channels (SCHs) for non safety service applications dedicated to the comfort of the driver [3].

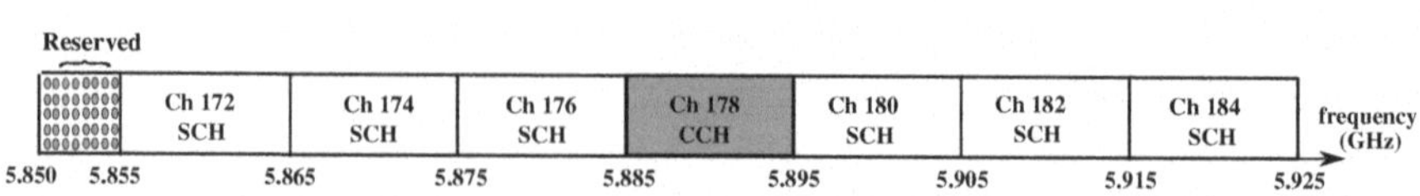

Fig. 1 US DSRC spectrum allocation

Nevertheless, as a contention-based mechanism, some requirements in the IEEE 1609.4 standard must be taken into consideration. For instance, safety related applications demand quick and reliable message delivery, while non-safety applications usually require high throughput and good fairness performance. However, the IEEE 1609.4 cannot meet the aforementioned requirements. Recently, several multi-channel coordination based MAC protocols have been proposed for VANETs, in order to achieve the above requirements.

The remainder of this paper is structured as follows. Section 2 introduces the IEEE 802.11p/1609.4 multi-channel operations. In Sect. 3, we summarize the operations of multi-channel coordination based MAC protocols in VANETs. Thereafter, we propose a qualitative comparison of these protocols, and point out some open issues that need to be tackled in future studies in Sect. 4. This paper is concluded in Sect. 5.

2 IEEE 802.11p/1609.4 Multi-channel Operations

The IEEE 802.11p MAC layer interacts with the IEEE 1609.4 to enable multi-channel operations of IEEE 802.11p-based radio interfaces. The WAVE system presents two types of channels, control channel (CCH) and service channel (SCH). As illustrated in Fig. 2, the available spectrum (in the frequency band of 5.85–5.925 GHz) is configured into one control channel (CCH) and six service channels (SCHs). The channel access time is divided into synchronization intervals (T_{Syn}) with a fixed length of 100 ms, which contain both CCH Interval (CCHI) and SCH Interval (SCHI), with 50 ms for each one. These synchronization intervals should follow an external time reference, which can be synchronized by the Coordinated Universal Time (UTC) that can be acquired from Global Positioning System (GPS) devices or from other vehicles [3].

During CCH Interval, all devices must monitor the CCH for safety and private service advertisements. When SCH Interval arrives, devices can optionally switch to SCHs, which are used for non-safety applications. Each of these intervals begins with guard interval to minimize the effect of timing inaccuracies. Typical values for the guard interval are between 4 and 6 ms. During the Guard interval, transmissions are not allowed.

To exchange the non-safety applications, two types of WAVE devices are defined: Provider device (called WAVE provider) and user device (called WAVE user) [4].

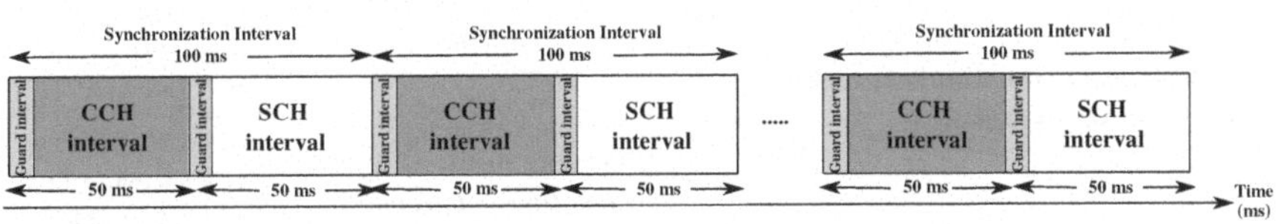

Fig. 2 Division of time into CCH intervals and SCH intervals in the IEEE 1609.4 standard

In WAVE communication services, the WAVE device may take the role of either a provider or a user on a given WAVE Basic Service Set (WBSS), this is determined by the role chosen by the application operating on the device. A WAVE provider could be either a roadside unit or a vehicle. During the CCH Interval, WAVE providers broadcast WAVE Service Advertisement (WSA) packets, to indicate there availability for data exchange on one or more SCHs. For the reliable delivery, each WSA packet will be broadcasted at least twice. WSA contains the information about the offered services and the network parameters necessary to join the advertised BSS. WAVE users interested in services offered by the WAVE providers can respond with an acknowledgment (ACK). After the CCH interval, WAVE users that successfully transmit the ACK in the CCH interval switch to SCHs, and start to exchange data with WAVE providers.

3 Multi-channel Coordination Based MAC Protocols

The multi-channel coordination based MAC protocols are compliant with the specified standard of IEEE 802.11p/WAVE. As shown in Fig. 3, we classify the current VANET Multi-channel coordination based MAC protocols into two main categories: dynamic interval division protocols and enhancing multi-channel operations protocols. In the former, CCH/SCH intervals are adjusted dynamically based on the traffic load in a distributed manner. In the latter, the length of CCH/SCH intervals is kept fixed while adding some mechanism to enhance the IEEE 1609.4 standard. In this section, we review existing works on multi-channel coordination based MAC protocols.

3.1 Dynamic Interval Division Protocols

Dynamic Interval Division multi-channel MAC (DID-MMAC) protocol: DID-MMAC presents the first protocol that proposes an adaptive MAC with dynamic interval division for WAVE system [5]. As explained in Fig. 4, in DID-MMAC

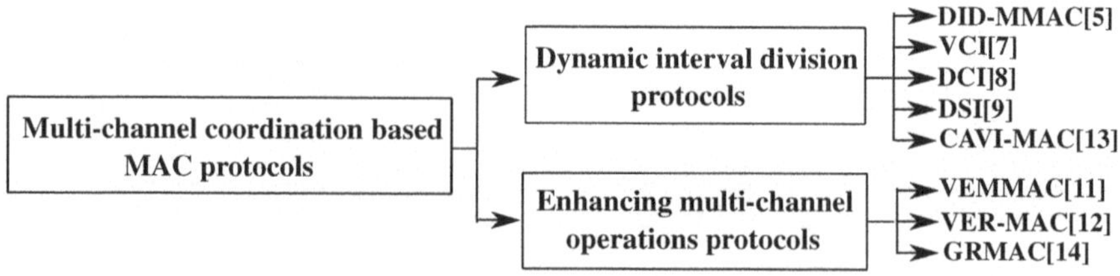

Fig. 3 Taxonomy of VANET Multi-channel coordination based MAC protocols

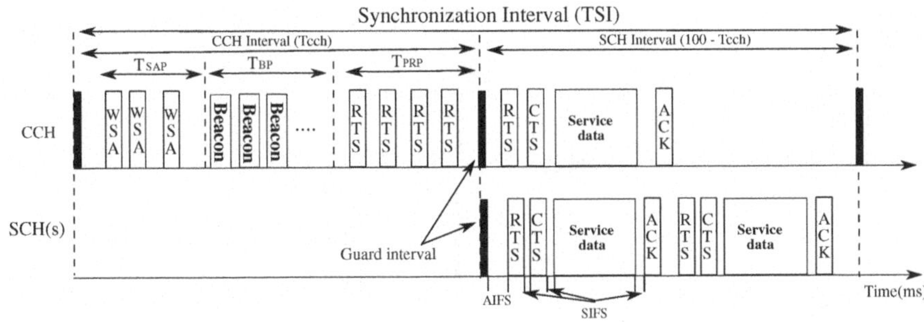

Fig. 4 DID-MMAC scheme

according to different types of frames, the CCH Interval is further splitted into Service Announce Phrase (SAP), Beacon Phrase (BP) and Peer-to-Peer Reservation Phrase (PRP). WSA frame and Beacon frame are transmitted in the SAP and BP, respectively. WSA frame is used to indicate the existing service in vehicle network, beacon frame is used to exchange vehicle status frame. While exchanges of control frame for SCH reservations are performed in PRP. DID-MMAC assigns different channel access priorities (i.e., contention window (CW) and the inter-frame space (AIFS)) for both frames SAP and BP, in order to adjust dynamically these frames in a distributed manner.

As shown in Fig. 4 PRP starts right after BP. In DID-MAC the duration of PRP is adjusted adaptively according to the estimated traffic load, based on two new fields called Service Indication (SI) and Traffic Indication (TI), added respectively into WSA and beacon frame. Since the SI field indicates the total data size of service, on the other hand, the TI field indicates the service status in vehicle side. However, this protocol did not provide channel selection algorithms, and did not take into consideration the collision from other vehicles that belong to the same contention domain.

Variable CCH Interval (VCI) protocol: VCI consists of well-known multi-channel MAC protocols for VANET, proposed by Wang et al. [6]. The VCI multi-channel MAC adjusts dynamically the CCH and the SCH intervals, to enhance the real-time delivery of safety/control packets, and to provide proper bandwidth for delivery of the application information. As shown in Fig. 5, VCI divides the CCH interval into safety interval and WSA interval. During the safety interval, vehicles transmit periodic beacons, while the vehicles exchange control messages for SCH reservations in the WSA interval. The WAVE providers broadcast WSA packets containing both information of service and identities of SCHs to be used. The WAVE users can either respond to the WSA packet with an acknowledgement (ACK) or initiatively send a Request For Service (RFS) packet to make an agreement with a WAVE provider.

In VCI scheme, the duration of safety interval is determined as:

$$T_{safety} = \frac{\alpha \times f \times N}{B_{cch}} \times 10^3$$

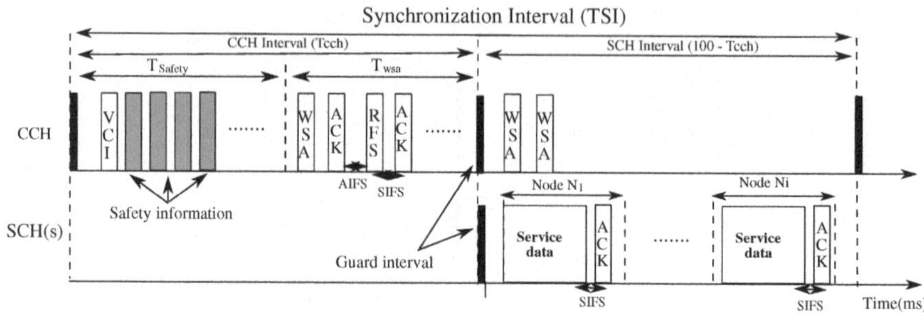

Fig. 5 VCI scheme

where N represents the total number of vehicles sending safety packets, f is the sending frequency of safety messages, B_{cch} is the data rate of CCH and α is a predefined factor.

The duration of T_{WSA} is estimated by the roadside unit (RSU), based on the average time consumed on the CCH for the negotiation of service packet transmission. Basically, both safety and WSA intervals are calculated by the RSU, then the RSU broadcasts a VCI packet containing the length of the CCH interval to the vehicles within their coverage range. Moreover, when no RSU can be detected, the VCI scheme selects one of vehicles within one hop to be a leader, in order to play the role of RSU. The main drawback of VCI protocol that it depends on RSU to calculate T_{safety} interval, and it does not address the hidden node problem.

Dynamic CCH Interval (DCI) MAC protocol: DCI works identically to VCI protocol except the calculation of the WSA interval, it was proposed by D. Zhu et al. [7]. In DCI the optimal WSA interval is calculated based on the probability distribution of the reservation time for service packet in the CCHI.

DCI defines k as the maximum number of service data packets which can be transmitted on a given SCH interval. Then, it derives the minimum WSA interval for reserving the transmission of service packets. Finally, DCI obtains the optimal WSA interval by finding the optimal k to minimize the difference between the sum of WSA and SCH interval and the residual synchronization interval except the safety interval.

Dynamic Safety Interval (DSI) protocol: DSI allows the safety interval to accommodate transmissions from vehicles within the same contention domain as well as from hidden vehicles [8]. The region within which hidden vehicles reside is affected by three types of ranges related to packet transmissions in the IEEE 802.11MAC: transmission range (r_t), carrier sensing range (r_c), and interference range (r_i). These ranges are tunable parameters, they affect significantly the MAC performance. Measurement studies such as [9] demonstrate that $r_t < r_i < r_c$.

Hidden nodes refer to the nodes located within r_i of the intended destination and out of r_c of the sender. When a receiver is receiving a packet, if a hidden node tries to start a concurrent transmission, collisions can happen at the receiver. DSI defines d_{sr} as the spatial reuse distance with $d_{sr} > r_t + r_i$. Then, it allows to each vehicle i to share the safety interval with vehicles whose distance from i is smaller than d_{sr}. In DSI, the extended contention domain (ECD) is defined as the region where vehicles sharing the safety interval reside. Given ECD, each vehicle calculates the number of vehicles within ECD, then derived the safety interval as:

$$T_{safety} = N \times (DIFS + \gamma + t_{trans})$$

where N represents the total number of vehicles, $DIFS$ is the inter frame space, γ is the average backoff time, and t_{trans} is the message transmission time. However, DSI calculates the length of safety interval in a deterministic way. Hence, the T_{safety} used to derive the safety interval is not optimal.

Context Aware Variable Interval MAC (CAVI-MAC) protocol: CAVI-MAC [10] reduces the interference for the propagation of critical event driven messages, by adjusting dynamically the control channel interval. In CAVI-MAC, the length of CCH is adjusted according to the neighborhood density and the type of safety message that is being disseminated currently in VANET. The safety message in VANET can be classified as periodic safety message and event-driven safety message. In periodic safety message, each vehicle automatically broadcasts safety messages at regular intervals (steady state). While in event-driven safety message, safety messages are broadcasted only in case of an unsafe situation (non-steady state).

Each vehicle listens to the exchanged information for one synchronization interval to gain insights about the context of the environment (i.e., steady or non-steady state). Based on the type of safety message, the CCH interval is given as: $T_{cch} = (L/D + AIFS) \times NBR\_CTR \times \alpha$, during steady state. Where L is the length of the packet, D is the data rate, AIFS is the Arbitration Inter Frame Space, NBR_CTR is a counter giving the number of beacon messages a vehicle receives during a given Synchronization Interval (SI) and α is a predefined scale up factor. Or as: $T_{cch} = (SI - 2 \times GI) \times LI/10$, during non steady state. Where LI is the Length Index which specifies the ratio between CCH and SCH, and GI represents the guard interval. CAVI-MAC protocol provides a virtual interference free environment, to tackle the hidden terminal problem. However, as the vehicle density increases, the CCHI increases and SCHI decreases in a given SI, which reduces the throughput over SCH.

3.2 Enhancing Multi-channel Operations Protocols

Enhanced Multi-channel MAC (VEMMAC) protocol: VEMMAC allows to vehicles to transmit non-safety messages during CCH interval and broadcast safety messages twice with each in the CCH and SCH interval [11]. To achieve this goal, the vehicles in VEMMAC have two transmission modes: Normal Transmission (noted N-Tx) which is the transmission performed within the SCH interval and Extended Transmission (noted E-Tx) which is the transmission performed in SCH interval and the upcoming CCH interval. Based on the traffic load, the sender decides which transmission mode is going to use.

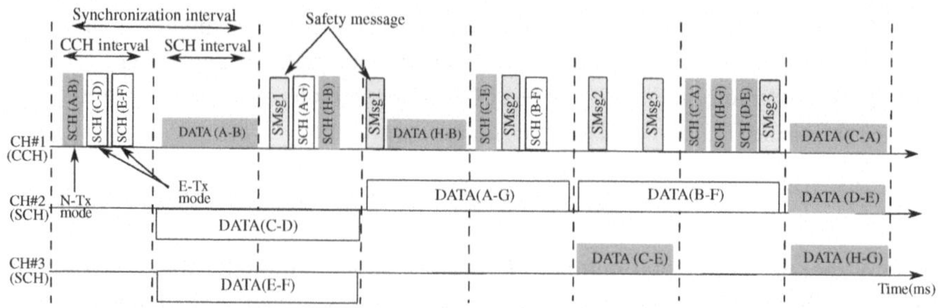

Fig. 6 VEMMAC scheme

As indicated in Fig. 6, the synchronized interval is divided into CCH and SCH intervals. To exchange non-safety messages, vehicles try to access the control channel to reserve one of the available service channels only during the CCH interval. When a node wants to broadcast the safety messages (SMsg), it has to switch to the CCH channel and contends the control channel to broadcast a safety message in the current interval (CCHI or SCHI). Then, it attempts to broadcast the safety message again if it is still valid in the next interval, in order to allow to all vehicles to receive the safety message.

To reserve the available service channel, vehicles in VEMMAC are based on two data structures called Neighbor Information List (NIL) and Channel Usage List (CUL). Since the NIL maintains the information of the neighbor vehicles, the CUL stores the information of channel. A serious drawback of VEMMAC is its incapability to oversee the high collisions at the beginning of the CCHI or SCHI, also vehicles might lose the emergency packets on the CCH due to the extended transmission mode.

Efficient and Reliable MAC (VER-MAC) protocol: VER-MAC allows vehicles to broadcast safety packets twice during both CCHI and SCHI to increase the safety broadcast reliability [12]. Moreover, the VER-MAC uses efficiently the SCH resources during the CCHI to enhance the service throughput. As indicated in Fig. 7, VER-MAC functions identically to VEM-MAC in transmitting the EMG packet (safety packet). In VER-MAC, On each SCH, the synchronized interval (SI) is divided into M transmission slots (TxSlots) which are used for the collision-free service data transmissions. Each vehicle pair performs a WSA handshake to select the available TxSlot of service channels (SCH), based on two data structures NIL and CUL as VEMMAC. The major drawback of VER-MAC is that it requires additional complex data structures and suffers from the large delay of emergency packets.

Group Reservation MAC (GRMAC) protocol: GRMAC allows vehicles to exchange their CCH bandwidth requirement and reserve the CCH bandwidth in SCHs during SCHI [13]. The aim of GRMAC is to minimize the number of collisions in CCH and regulate the ratio of the bandwidth consumed by both the WAVE Short Message (WSM) traffic and the WSA traffic. To achieve these goals, GRMAC divides vehicles into groups called RG_i, each of which comprises a set of vehicles using the same SCH_i.

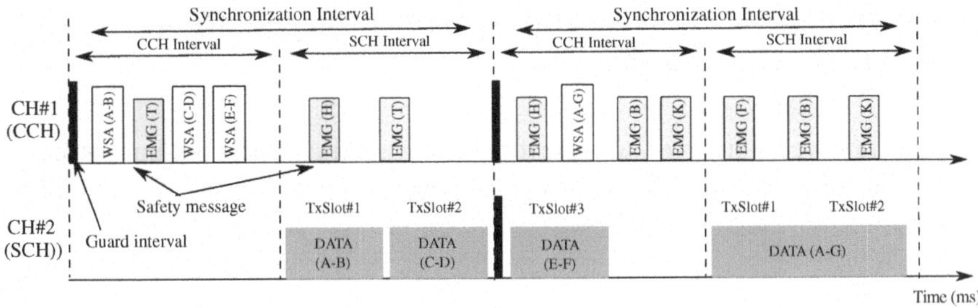

Fig. 7 VERMAC scheme

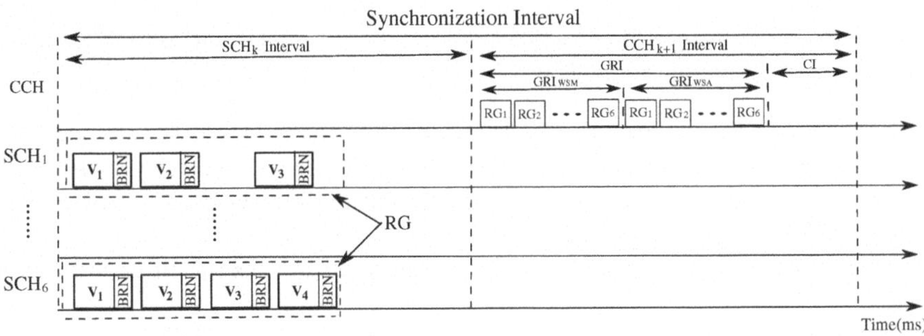

Fig. 8 GRMAC scheme

As displayed in Fig. 8, in GRMAC, the CCHI is further split into Group Reservation Interval (GRI) and Contention Interval (CI). The former is used for the reserved CCH bandwidth, and the latter is used by vehicles which did not reserve the CCH bandwidth in SCH during the previous SCHI. During the SCHI, each vehicle within the same SCH (RG) requiring CCH to transmit data can reserve CCH bandwidth, based on the bandwidth reservation notification (BRN). A BRN contains two one-bit fields: WSM and WSA fields, used by each vehicle to notify other vehicles of its demand to transmit WSM/WSA traffics in CCH in the following CCHI. The RGs sequentially (based on their SCH number) reserve the CCH bandwidth and transmit data over CCH during CCHI. The GRMAC divides the GRI into GRI_{WSM} and GRI_{WSA} for WSM and WSA traffics, respectively. Each RG sequentially transmits WSMs during GRI_{WSM} and then transmits WSAs during GRI_{WSA}. Both GRI_{WSM} and GRI_{WSA} have a maximum length (GRI_{WSMmax} and GRI_{WSAmax}) that prevents one of these two traffics from consuming all of the CCH bandwidth. Vehicles want to disseminate frames over CCH, which did not reserve CCH bandwidth in SCH during the preceding SCHI start transmission right after RGI. However, GRMAC does not discuss how to resolve the collision problem that occurs during CI.

4 Qualitative Analysis of Multi-channel Coordination Based MAC Protocols

In this section we qualitatively analyze the existing multi-channel coordination based MAC protocols, we also identify some open issues and possible directions of future research.

4.1 Qualitative Analysis

The qualitative analysis is done according to a few important metrics that we have deduced from the literature. Such as:

- SCH negotiation method: The SCH negotiation method used by protocols to achieve significant increase in channel reliability, throughput and delay constraints.
- Consideration of hidden nodes: Does the protocol consider the existence of hidden terminal?
- Dependency on RSU: Does the design of the protocol uses the Road Side Unit (RSU)?
- Control message overhead (bits/periodic message): Additional data sent along with the message through the network towards a destination uses a portion of the available bandwidth.
- Considered traffic condition: Refers to the type of traffic used by the protocol (saturated, unsaturated or both (none)).
- Considered application type: Indicates the type of application addressed by the protocol (safety, non safety application or both).

Table 1 gives a qualitative comparison of various existing multi-channel coordination based MAC protocols. DID-MMAC, VCI, DCI, VERMAC and GRMAC protocols are designed to work well in saturated traffic condition (i.e., the system always has a message to transmit), in a realistic scenario, it can exist cases where the traffic condition is unsaturated. Hence, in order to model a realistic scenario, DSI, VEMMAC and CAVI-MAC protocols are designed to work well under any traffic conditions. The fact that each node measures the traffic load in the dynamic interval division protocols, causes the message overhead by adding new fields into each beacon message. As shown in Table 1, each dynamic interval division protocol has its own message overhead. However, the overhead of a single message is insignificant compared to the size of control and beacon message ranging from 100 bytes to 300 bytes.

The deployment and maintenance of an RSU is a costly affair, hence MAC protocols must be autonomous, not depending on RSU. VCI and DCI protocols are dependent on the RSU for there normal working. On the other hand, the rest of protocols are functional without the help of RSU. Regarding the SCH reservation method, VCI, DCI, VEMMAC and VERMAC introduce a new SCH negotiation denoted by

Table 1 Qualitative comparisons

Protocols	Control message overhead (bits/ periodic message)	Considered application type	SCH negotiation method	Dependency on RSU	Considered traffic condition	Consideration of hidden nodes
DID-MMAC [5]	8	Safety, non-safety	Modified MMAC [14]	–	Saturated	–
VCI [6]	Size of VCI messages [6]	Safety, non-safety	WSA/RFS/ACK exchange	✓	Saturated	–
DCI [7]	Size of VCI messages [6]	Safety, non-safety	WSA/RFS/ACK exchange	✓	Saturated	–
DSI [8]	80	Safety	–	–	None	✓
VEMMAC [11]	–	Safety, non-safety	SCH-REQ/SCH-ACK/ SCH-RES handshake	–	None	–
VERMAC [12]	–	Safety, non-safety	WSA/ACK/RES handshake	–	Saturated	–
CAVI-MAC [10]	6	Safety	–	–	None	✓
GRMAC [13]	–	Safety, non-safety	–	–	Saturated	–

WSA/RFS/ACK exchange for VCI and DCI protocols, SCH-REQ/SCH-ACK/SCH-RES handshake and WSA/ACK/RES handshake for VEMMAC and VERMAC, respectively, while DID-MMAC exploits an existing protocol called the modified MMAC [14]. However, DSI, CAVI-MAC and GRMAC do not treat the SCH negotiation.

The aim of studying vehicular ad hoc networks (VANETs) is to ensure road safety (safety applications) and to maximize the throughput metric for non-safety applications. The DID-MMAC, VCI, DCI, VEMMAC, VERMAC and GRMAC protocols are studied and handled both types of application (safety and non-safety). However, DSI and CAVI-MAC protocols have just addressed the safety applications. Regarding interference from hidden nodes, DSI and CAVI-MAC protocols take into account this issue in the design of their protocols. DSI protocol handles this problem in a deterministic way, while CAVI-MAC protocol dynamically considers hidden terminals thereby minimizing the effect of interference in real time. DID-MMAC, VCI, DCI, VEMMAC, VERMAC and GRMAC protocols remain silent about this issue.

4.2 Open Issues and Future Research Directions

It is apparently seen so far that significant efforts have been made in designing effective MAC protocols for VANETs. However, there is a high potential to improve current MAC methods in the future. A few important open issues are summarized as follows:

(1) Safety applications require a low latency and high reliability communication service. In emergency situation, the safety-critical messages are limited lifetime due to mobility of nodes and fast topology changes. As a result, MAC protocols have to take into account an efficient broadcasting and reliable delivery for safety-critical messages with bounded latency requirement.
(2) Hidden terminal is a serious issue in the design of MAC protocols, because it deteriorate the MAC layer performance. Only a few researchers have discussed this problem for MAC protocols in VANETs. Hence, this issue needs to be studied and integrated in the design of MAC protocols in VANET.
(3) Mobility is one of the main characteristic of VANETs, it has an important impact on the performance of MAC protocols. Therefore, the design of this protocols should address mobility issues and estimate accurately the condition of the highly dynamic channel, in order to enhance the MAC performance and ensure better fairness by providing different priority levels to vehicles based on their mobility.
(4) Most of MAC protocols in VANET discus only the safety applications due to their importance, while VANET provides also the non-safety applications. Therefore, MAC protocols should allow a tradeoff between safety and non-safety applications in order to achieve better utilization of channel.

(5) Further research would be needed to design cross layer MAC protocols in VANETs. For instance, physical layer and network layer can achieve transmission range adjustment and efficient routing designing protocols, which can enhance and facilitate the designing of MAC protocols.

5 Conclusion

VANETs have attracted increasing attention in recent years for their extensive applications. Due to their characteristics, the design of MAC protocol is full of challenges in VANETs. In the past, much efforts have been made in designing effective MAC protocols for VANETs. In this paper, we proposed a survey of multi-channel coordination based MAC protocols which improve the performance of IEEE 802.11p- and IEEE 1609.4-based WAVE systems. Moreover, we summarized the operation of existing protocols and discussed the advantages and disadvantages of each one. Finally, we defined some open issues and possible directions for future research related to multi-channel coordination protocols for VANETs.

We hope that this survey provides a more expansive understanding of multi-channel coordination based MAC protocols in VANETs for readers, also gives an overview of the existing problems in the design of MAC protocols.

References

1. Karagiannis, G., Altintas, O., Ekici, E., et al.: Vehicular networking: a survey and tutorial on requirements, architectures, challenges, standards and solutions. IEEE Comm. Surv. Tut. (2011)
2. IEEE 802.11p: wireless LAN medium access control (MAC) and physical layer (PHY) specifications amendment 6: wireless access in vehicular environments (2010)
3. Booysen, M.J.: Survey of media access control protocols for vehicular ad hoc networks. IET Commun. (2011)
4. IEEE Std. 1609.4TM-2006: IEEE trial-use standard for wireless access in vehicular environments (WAVE)—multi-channel operation (2006)
5. Liu, L., Xia, W., Shen, L.: An adaptive multi-channel MAC protocol with dynamic interval division in vehicular environment. ICISE (2009)
6. Wang, Q., Leng, S., Fu, H., Zhang, Y.: An IEEE 802.11p-based multi-channel MAC scheme with channel coordination for vehicular ad hoc networks. IEEE Trans. Intell. Transp. Syst. (2012)
7. Zhu, D., Zhu, D.: Performance analysis of a multi-channel MAC with dynamic CCH interval in WAVE system. In: 2nd International Conference on System Engineering and Model (2013)
8. Hongseok, Y., Jinhong, K., Dongkyun, K.: A dynamic safety interval protocol for VANETs. In: 27th ACM Research in RACS (2012)
9. Anastasi, G., Borgia, E., Conti, M., Gregori, E.: Wi-Fi in ad hoc mode: a measurement study. In: 2nd IEEE Annual Conference on PERCOM 04 (2004)
10. Babu, S., Patra, M., Siva Ram Murthy, C.: A novel context-aware variable interval MAC protocol to enhance event-driven message delivery in IEEE 802.11p/WAVE vehicular networks. Veh. Commun. (2015)

11. Dang, D.N.M., Dang, H.N., Do, C.T., Hong, C.S.: An enhanced multi-channel MAC for vehicular ad hoc networks. In: IEEE WCNC (2013)
12. Dang, D., Hong, C., Lee, S., Huh, E.: An efficient and reliable MAC in VANETs. IEEE Commun. Lett. (2014)
13. Chen, Y., Lai, C., Lai, C., Li, Y.: A group bandwidth reservation scheme for the control channel in IEEE 802.11p/1609 networks. In: 10th International Conference (2015)
14. So, J., Vaidya, N.: Multi-channel MAC for adhoc networks: handling multi-channel hidden terminals using a single transceiver. In: 5th ACM International Symposium on MOBIHOC (2004)

Towards the Enhancement of QoS in 802.11e for Ad Hoc Networks

Fatima Lakrami, Mohamed El-Kamili and Najib Elkamoun

Abstract Multimedia applications running over Ad hoc networks must achieve some level of performance QoS guarantees. This topic is very broad and there is an extensive pool of solutions in the literature. For the last few years, works on enhancing Quality of service at MAC layer in wireless networks, have attracted tremendous research efforts. In this paper, we examine the QoS issue of MAC layer in Ad hoc networks. We present a state-of-the-art review and a comparison of typical enhancements of distributed versions of both 802.11 and 802.11e standards, designed first to work in infrastructure wireless networks. We start our work by highlighting problems of deploying the original algorithms in a distributed architecture such as MANETs. Then, we propose an enhancement of the 802.11e in order to improve the QoS and correct the reactivity of this standard when network conditions become highly disturbed.

Keywords QoS · MAC · EDCF · DCF · Wireless · Adhoc

1 Introduction

Wireless links are generally characterized by a high loss rate, large packet delay, jitter and many other transmission problems [1]. In fact, their characteristics are not constant and vary over time and place. The mobility of users for example may cause paths to change frequently and imperviously [2]. In infrastructure WLAN, the access point is in charge of scheduling communications and disseminating transmission parameters to each node using a centralized access function (Point Coordination Function

F. Lakrami (✉) · N. Elkamoun
STIC Laboratory, Chouaib Doukkali University, El Jadida, Morocco
e-mail: fatima.lakrami@gmail.com

N. Elkamoun
e-mail: elkamoun@gmail.com

M. El-Kamili
LIMS Laboratory, Sidi Mohammed Ben Abdallah University, Fez, Morocco
e-mail: mohamed.elkamili@usmba.ac.ma

R. El-Azouzi et al. (eds.), *Advances in Ubiquitous Networking 2*, Lecture Notes in Electrical Engineering 397, DOI 10.1007/978-981-10-1627-1_8

(PCF)/Hybrid Coordination Function (HCF)). In Ad hoc networks, a large number of nodes share the same channel. Nodes will compete to gain access to the link, using a CSMA/CA mechanism to avoid collisions. The default access function used is the distributed coordination function (DCF) [3], and it does not support transmission of real-time application, since it was designed for equal priorities. So, there is no differentiation between traffics.

To support QoS, enhancements of the legacy 802.11 [4] led to the development of the new 802.11e [5] standard that aims to resolve service prioritization issue. The new access function called the enhanced distributed channel access function (EDCA), enable a strict differentiation by granting each type of traffic specific transmission parameters, regarding its priority. However, this new EDCA still suffer from the common problems of wireless transmission such as hidden and exposed terminal, the fair sharing of the available bandwidth and many others. In this paper, we propose a survey of different solutions proposed to introduce the QoS to the wireless MAC layer in Mobile Ad hoc Networks (MANETs), which is revealed to be more difficult, due to the distributed nature of MANETs. We also propose a contribution for the enhancement of the 802.11e standard, through an adaptative algorithm that favors and enhances dynamically the transmission parameters of high priority traffics, while taking into account the residual state of channel.

The remainder of this paper is structured as follows. In Sect. 2, a brief presentation of the 802.11 standard is given, Sect. 3 describes the problem statement of QoS in the 802.11e standard. In Sect. 4, we present an enhancement of the 802.11e standard through an extension of the access channel function designed for distributed architectures. Our new version is compared with the original one through simulation using network simulator. Section 5 concludes the paper.

2 The QoS in 802.11

2.1 Presentation of the 802.11 standard

The IEEE 802.11 [4] is an international standard of physical and MAC layer specifications for WLAN. It provides mandatory support for 1 Mb/s data rate with optional support for 2 Mb/s [6]. These specifications have been altered to support higher data rates in succeeding versions, aiming to reach 54 Mb/s for future systems. The 802.11 standard can be applied to both infrastructure-based WLANs (that use fixed access points for wireless communication with mobile terminals) as well as infrastructure-less Ad hoc networks. In the first version of the IEEE 802.11 standard, there were two medium access functions: DCF and PCF. Next, QoS was introduced to the IEEE 802.11 MAC through the enhanced distributed coordination function (EDCF) [7] and the HCF controlled channel access (HCCA) [8] function. DCF is the fundamental access method of the IEEE 802.11 MAC and EDCA is an enhanced variant of

DCF. These functions were designed for both ad-hoc and infrastructure networks. PCF and HCCA are alternatives designed for infrastructure networks.

2.2 *QoS Limitations in 802.11*

DCF offers no QoS differentiation between traffic categories; in fact it was first developed to work as a best effort protocol. So, applications with QoS requirement will suffer the same treatment as best effort ones. DCF is also considerably affected by various overhead generated by physical layer, control frames backoff, etc. This problem becomes more consistent when data rate increases, in fact each time a station gain the access to the channel, the transmitted frame will carry on the same overhead, as the previous one, belonging to the same flow, which widely impact network performance.

2.2.1 Enhancements of DCF

802.11 is deprived of all kinds of QoS support. The standard 802.11e was proposed in order to ensure a certain level of QoS at MAC layer and overcome QoS limitations in 802.11 [9]. However works on improving DCF continue to emerge. Researchers defend their perceptions by the fact that most of the wireless equipment implement the DCF function, and so, it is wise to start first by adapting the existing access function, being widely deployed. Several approaches for improving the QoS in the DCF were presented. We mention here some examples of most interested works starting by [10], where a new protocol called MACA/PR (Multiple Access with Collision Avoidance Piggyback Reservation)is presented, which proposes to differentiate the policy of access to the medium according to the nature of flows. References [11, 12] suggest to design different backoff functions: by providing different values for the contention windows (CW): higher values for lower priority traffic, and vice versa, thereby giving a station a high chance to access the channel.

3 Enhanced Distributed Channel Function

3.1 *Presentation of the EDCF*

EDCF [13] is an extension of the DCF access method introducing service differentiation at MAC layer. It defines multiple Access Categories (AC), with AC-specific Contention Window (CW) sizes, Arbitration Inter-frame Space (AIFS) values, and Transmit Opportunity (TXOP). It differentiates access to the medium using the principle of priorities. The DCF algorithm has not been changed completely in EDCF, but the game of time interval on which it is based has been customized for each priority. Accordingly, these time intervals are adjusted to increase/decrease the probability of channel access and thus encourage/discourage the transmission of data flow high/low priority. We distinguish a total of eight levels of user priorities (UP or User Priority). Each priority is linked to what is called an access category (AC or Access Category). Each category is assigned an access logical function similar to DCF, called EDCAF (Enhanced Distributed Channel Access Function) which will participate in the contention phase in order to obtain a Transmission Opportunity (TXOP) for the benefit of the concerned AC and then enable it to access the medium.

3.2 *EDCF Limits*

The EDCF works as DCF, following a CSMA/CA scheme with BEB (Binary Exponential Backoff algorithm). Although the distributed EDCF is an important enhancement of the original 802.11, it is still insufficient for providing QoS guarantees. The no-deterministic nature of the EDCF, besides the absence of any distributed admission control algorithm, limits widely its performance, especially when the network conditions start to deteriorate (high density, mobility...). Another problem is related to the static values of the EDCF parameters (TXOP, Cwmin, Cwmax..), that remind fixed, regardless the fluctuation of both: environment constraints and traffic requirement. So, it appears that the EDCF parameters can require significant tuning to achieve better performance for the transmission of sensitive applications. Performance of the EDCF function were tested by simulation under mobility and traffic impact, in a network composed from 50 nodes exchanging 2 types of traffic (Voice and FTP) , the results are given in Figs. 1 and 2 only for high priority traffic which is in this case: voice.

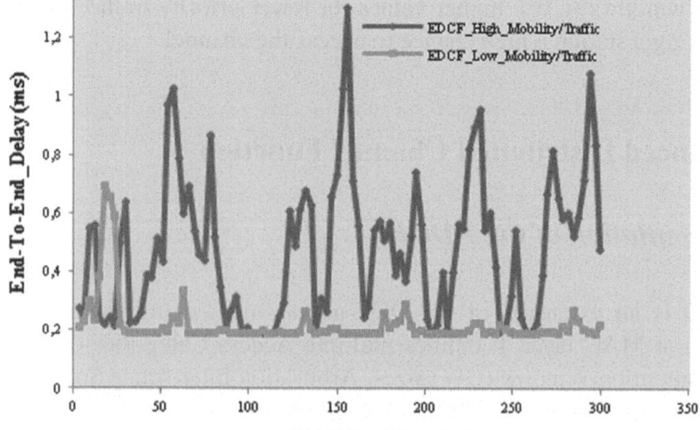

Fig. 1 End-to-End delay variation for EDCF performance under mobility and traffic impact

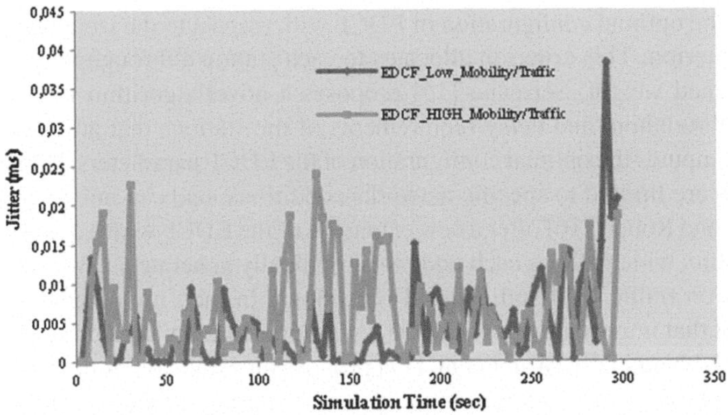

Fig. 2 The jitter variation for EDCF performance under mobility and traffic impact

As we can see, EDCF performance is vulnerable to the unpredictable change in the network status. Mobility of nodes can't be handled directly through a process adaptation of the EDCF, in fact there is no explicit manner to transmit mobility occurrence to higher layer. So, there is a huge need to find an implicit but adequate metric for mobility. For the traffic, every protocol suffers performance deterioration when the number of communicating nodes increases, this is happened principally because of the MAC latency, that occurs due to the size of local queues (if there is more than one traffic category competing to access the internal channel) and another latency due to the bandwidth congestion (high loss rate and high retransmission rate), then, and even if the EDCF grants favorable transmission settings to sensitive applications, it is still insufficient to guarantee a reliable end-to-end reliable transmission.

3.3 EDCF Enhancements

Managing quality of service in terms of service differentiation between multiple traffic categories is an important task regardless of the network architecture involved for this purpose. It allows granting a certain transmission privileges to real-time applications, in terms of access priority and channel occupation time. However, the static mechanism used by the EDCF loses its effectiveness when the competition between flows becomes more consistent, or the transmission conditions start to deteriorate.

Several research studies have been proposed to improve the quality of service in wireless networks, especially by focusing on the optimization of the calculation of a single EDCF parameter at the time, independently for each access category either the CW (contention Windows), AIFS (Arbitration Inter Frame Space or the TXOP (Transmission Opportunity). We cite the example of [14], where authors tried to search for the optimal configuration of EDCF with respect to the weighted max_min fairness criterion. This criterion allocates to each station a throughput proportional to its assigned weight. Serreano [15] proposes a novel algorithm for EDCF that, given the throughput and delay requirements of the stations that are present in the WLAN, computes the optimal configuration of the EDCF parameters. However, both approaches are limited to specific networks conditions, and can not be generalized. Gannoune and Robert [16] offer a new extension of the EDCF with a dynamic matching algorithm, which allows each node to periodically generate a new CWmin value depending on traffic load and channel conditions. In fact, improvement in packet delay is not that important since only the congestion window parameter CWmin can be adjusted dynamically. Wang et al. [17] propose an extension of the EDCF (ACT-EDCF) that enables mobile nodes to accommodate the size of the contention window and continuously transmit a number of high-priority packets while monitoring the state of the channel in terms of probability of failure of transmission or collision. The ACT-EDCF can starve the low-priority traffic for the sake of using the ACK burst mode in the EDCF, to support high-priority traffic. To share the channel capacity effectively between different traffic priorities and achieve improved QoS for real-time application, Romdhani et al. [18] propose an adaptive EDCF (AEDCF), which not only updates the contention window for each AC according to the ratio between the number of collisions and the total number of packets transmitted in a constant period, but also allows assigning a smaller value of PF to traffic with higher priority. Unlike MANETs, the use of the EDCF is very common in wireless networks with infrastructure, because the proposed approaches for improving the efficiency of the EDCF enable a more efficient management of the QoS in these type of networks. In fact the actual implementation of EDCF algorithm is still centralized at the Access point, which is responsible of coordinating and configuring transmission parameters for the stations communicating in the same network. Authors propose common approaches based on the combination of the EDCF with HCF, the two algorithms are modified at the access point, which has the major responsibility to organize communications according to the type of traffic. In Ad hoc networks, the distributed architecture complicates the implementation of the algorithm, the flow will experience a local competition but also across the network, each station has to probe the assessment of the conditions of its environment before transmitting its packages, evaluation that consists mainly on measuring the loss rate, traffic load and available bandwidth. We proposed in an earlier version [19] an extension of the EDCF that takes into account only the mobility occurrence, through measuring the channel error rate and updating Txoplimit using a specific formulas to increase/decrease transmission time of high priority traffics, to enhance QoS at Mac layer. We have chosen it to combine with other QoS mechanisms offered by routing and physical layer.

4 Our Contribution

4.1 The Proposed Approach

We present in this section an example of a modification of the original EDCF. We present an EDCF variant where various setting will not remind static but will dynamically depend on both the priorities of the applications and network constraints. In this context, we propose a new mechanism that allows in one hand to ensure a strict differentiation between AC and on the other hand to ensure differentiation intra AC (Access Category) based on the size of the flow. The mechanism proposed here is based on the reservation of the wireless channel for the priority traffic, through TXOPlimit parameter. In fact the TXOPlimit variable allows a station to send a sequence of packets (a packet burst), that contain a number of packets proportionally to the TXOPlimit value. Unlike EDCF that defines a static TXOPlimit value, the new algorithm allows each AC to dynamically adjust the value of its TXOPlimit. This value is calculated every time the AC wins the contention, and takes into account: the AC priority (to ensure inter-AC-QoS)and the flow rate (intra-AC-QoS). This adaptation is restricted to the higher classes AC2 and AC3 which represent Voice and video traffics, the calculation procedure is based on the error rate and the congestion rate (the channel busy time) based on the method explained in [20]. This method is developed primordially to design a rate adaptation scheme that provides high network performance in both congested networks and lightly-loaded networks. It has the advantage of being based on a passive estimation, so there is no need of adding an extra data or packets to probe the channel.

These two metrics enable approximatively to reveal the residual state of the channel, so, decision about transmitting traffic is taken through adjusting TxopLimit and CWmin values. However, and in order to limit the number of packets sent in a best-effort mode, we propose that the value of TXOPlimit for both AC1(Best effort) classes AC0(background) remains intact. We define for this purpose three parameters: channel_metric, threshold_min and Threshold_max. The channel is considered reliable, if the metric value is less than or comprised between Threshold_min and Threshold_max, so the Txoplimit for sensitive packet is increased (according to a specific formula). otherwise, if the measured metric exceeds Threshold_max then the TxopLimit is widely reduced as the same for Cwmin. the node continues to probe the channel while adapting the EDCF setting for each category. Choosing error and congestion rate as channel metrics is due to two principal factors, in fact we want to enhance EDCF reactivity to face the impact of both mobility and network overload. The error rate is considered as and indirect mobility metric, and once the system become highly mobile, the channel become very disturbed and the error rate increases significantly. As the same for traffic, once the network is congested, disseminating sensitive traffic while knowing the actual channel state for a long duration will, impact definitively the transmission quality of the voice/video stream. So it is wise to think about when a traffic should be privileged and for how long, regarding channel conditions. As explained, the proposed version adapts in addition to the

TXOP, the minimum contention window, depending on the threshold defined based on the metric resulted, from the combination of both channel metrics already cited. the decision is about decreasing simultaneously the TXOPlimit and the contention window, if the channel is very noisy, and vice versa in the contrary. The new function of the EDCF is defined as follows (Fig. 3).

The New_EDCF_Function function (channel_statut), updates the value of TXOPlimit and CWmin specifically for voice and video. The estimated duration of TXOPlimit affects directly the performance of the whole system. For an overestimation, the system is poorly exploited. However, for an underestimation, the transmitted stream can be delayed. In reality, it is quite important to correctly estimate the TXOP duration. The proposed mechanism provides an heuristic approach to increment, decrement or even leave unaltered the TXOP and CWmin. The goal is to save the voice packets from being transmitted for a long duration when the error rate of the channel is very high, however, the CWmin value is reduced to enable voice/video streams to rapidly access the channel via the backoff algorithm. The algorithm enable a better exploitation of network resources by multimedia traffic, when the transmission conditions are favorable. Otherwise, We choose the set the default mode of the original EDCF in order to protect sensitive packets from being mis-disseminated in the network.

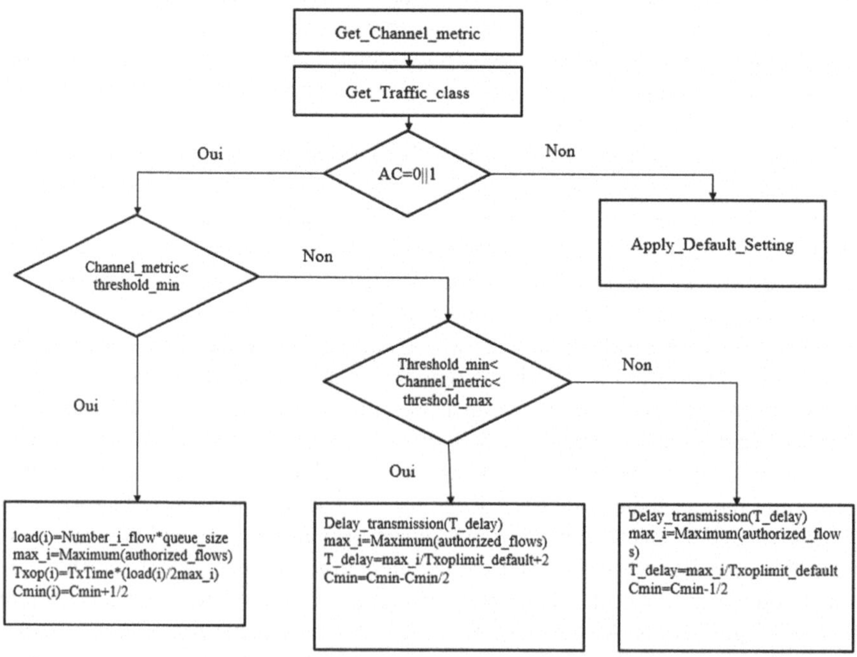

Fig. 3 The new EDCF adaptive-algorithm

4.2 Evaluation of the Approach

We use network simulator ns2.35 [21, 22] for the implementation and the evaluation of our proposition. We simulate a network of 200 nodes with an area of 2000 * 2000. We model a mobile Ad hoc network, under a TwoRayGround propagation model, We use 802.11e for MAC layer with a data rate of 11 Mb. We define two types of traffic: CBR (high priority traffic) and configure a best effort traffic to simulate the impact of traffic load on the performance of both original EDCF and the new_EDCF, we use as routing protocol the optimized link state routing (OLSR) [23]. Tables 1 and 2 resume different simulation parameters.

Table 1 Global simulation parameters

Simulation time	600 s
Simulated traffic	CBR (high priority) and Best_Effort_Traffic
Packet size	512 octets
Speed	Mob_H : Speed : 40 → 60 m/s Pause Time : 15,10 ms

Table 2 Simulated traffic parameters

	CBR (simulating voice traffic)	Best_Effort_Traffic
Throughput (Kbits/s)	10	10
Packet_size	256	512
Interarrival time (s)	0.025	0.025
Priority	0	2

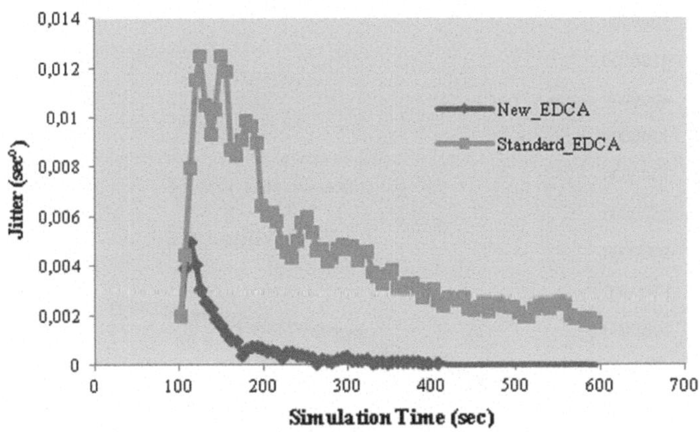

Fig. 4 Jitter variation of the original EDCF and the new enhanced EDCF under mobility and traffic impact

Figures 4, 5, 6 and 7 resume simulation results for the voice traffic specially for the following performance metrics: jitter, Packet_delay_variation, throughput, Traffic_Received.

By analyzing different results, we observe that the new system gives better performances, compared with the original EDCF. The new algorithm seems to react efficiently to channel fluctuation. The new modification enable the EDCF algorithm to take advantage from the dynamic adaptation according to specific metrics and with new calculation formulas, which enhance widely different performances of the transmission of the high priority traffic despite unfavorable conditions. In fact, and

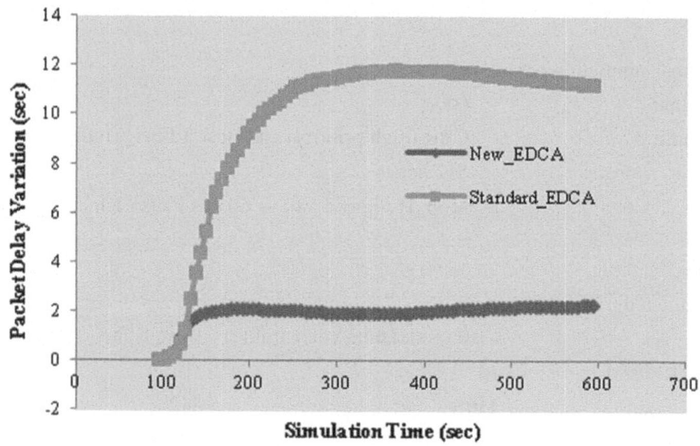

Fig. 5 Packet delay Variation of the original EDCF and the new enhanced EDCF under mobility and traffic impact

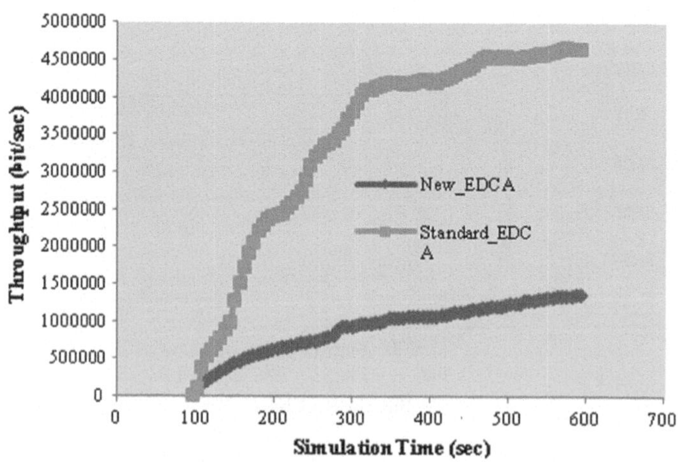

Fig. 6 Throughput of the original EDCF and the new enhanced EDCF under mobility and traffic impact

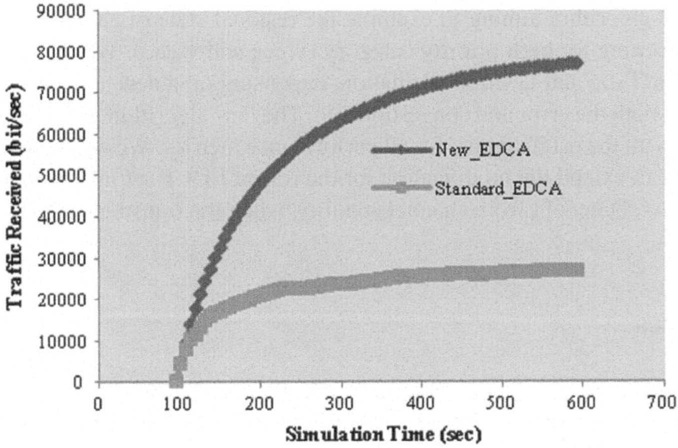

Fig. 7 Received traffic in the original EDCF an the new enhanced EDCF under mobility and traffic impact

specifically for high priority traffic, the system handle the internal congestion by allowing higher categories a rapid access to the channel if this one manifests a high stability. In the same context , the setting of the EDCF parameters is adapted according to the specific formulas shown in Fig. 4. In this new case, voice/video streams will take benefit from channel stability as long as possible, or will be delayed enough while probing channel process continues in order to estimate the actual state of transmission. For the best effort traffic, once the decision about not transmitting higher priority traffic is taken, best effort traffic gain the access, until the channel metric starts to increase, then the system concludes that the channel is about to get stable, and the variation between metric values is stored to keep a trace of the system evolution.

We conclude that the new algorithm enhances significantly the performance of the system for all performance metrics. And this is due to the new EDCF form. In fact the choice of using channel metrics to update EDCA configuration parameters is considered as the key factor for enhancing EDCF reactivity.

5 Conclusion

In this paper, we studied the evolution of enhancements of both 802.11 and 802.11e standards. We start our paper by highlighting different problems related to the access to wireless channel in Ad hoc networks, specially for sensitive applications with QoS requirements. We manage to aboard most relevant contributions for introducing a QoS support in both 802.11 and 802.11e, focusing on the distributed channel access functions: DCF and EDCF. For the enhancement of the EDCF through, we propose an adaptive algorithm aiming to examine the residual state of channel, and update the EDCF setting for high priority category (voice and video). We introduce a new formulas for Txop and Cwmin calculation, depensing on a new channel metric that depends on both the error and congestion rate. The new algorithm gives better results compared with the original one, for all performance metrics. We will consider in our future work to extend the modification for the rest of EDCF setting, to enable sensitive traffic to get benefit from channel stability, when the transmission still reliable.

References

1. Goldsmith, A.: Wireless communications. Cambridge University Press (2005)
2. Lakrami, F., Elkamoun, N.: Energy and mobility in OLSR routing protocol. J. Sel. Areas Telecommun. (JSAT) (2012)
3. Tardioli, D., Sicignano, D., Villarroel, J.: A wireless multi-hop protocol for real-time applications. Comput. Commun. **55**, 4–21 (2015)
4. Bianchi, G.: Performance analysis of the IEEE 802.11 distributed coordination function. IEEE J. Sel. Areas Commun. **18**(3), 535–547 (2000)
5. Kawata, T., Yamada, H.: Impact of multi-rate VoIP on quality of service in IEEE 802.11 e EDCA with link adaptation. In: IEEE International Conference on Communications, 2007, ICC'07, pp. 392–397. IEEE (2007)
6. Benedetto, S., Biglieri, E.: Principles of Digital Transmission: With Wireless Applications. Springer Science and Business Media (1999)
7. Xiao, Y., Li, H.: Evaluation of distributed admission control for the IEEE 802.11 e EDCA. IEEE Commun. Mag. **42**(9), S20–S24 (2004)
8. Qiang, N.: Performance analysis and enhancements for IEEE 802.11 e wireless networks. IEEE Netw. **19**(4), 21–27 (2005)
9. Perkins, C.E.: Ad hoc Networking. Addison-Wesley Professional (2008)
10. Karn, P.: MACA—a new channel access method for packet radio. In: ARRL/CRRL Amateur Radio 9th Computer Networking Conference, pp. 134–140 (1990)
11. Szott, S., Natkaniec, M., Canonico, R.: Detecting backoff misbehaviour in IEEE 802.11 EDCA. Eur. Trans. Telecommun. **22**(1), 31–34 (2011)
12. Xu, D., Sakurai, T., Vu, H.: An analysis of different backoff functions for an IEEE 802.11 WLAN. In : 68th Vehicular Technology Conference, 2008, VTC 2008-Fall, pp. 1–5. IEEE (2008)
13. Frederic, M.: WLAN QoS: 802.11 e. Tampere: Tampereen Teknillinen Korkeakoulu, Langattomat lhiverkot TLT-6556-kurssin luentokalvot (29 Mar 2007). Viitattu, vol. 6 (2008)
14. Banchs, A., Vollero, L.: Throughput analysis and optimal configuration of 802.11 e EDCA. Comput. Netw. **50**(11), 1749–1768 (2006)
15. Serrano, P., Banchs, A., Patras, P., et al.: Optimal configuration of 802.11 e EDCA for real-time and data traffic. Veh. Technol. IEEE Trans. 2010 **59**(5), 2511–2528 (2006)
16. Dong, Y., Wang, Y., Xia,Q.: A load adaptive IEEE 802.11 e EDCA backoff scheme with enhanced service differentiation. In: 2010 12th IEEE International Conference on Communication Technology (ICCT). IEEE (2010)
17. Wu, Y.-J., Chiu, J.-H., Sheu, T.-L.: A Modified EDCA with dynamic contention control for real-time traffic in multi-hop ad hoc networks. J. Inf. Sci. Eng. **24**(4), 1065–1079 (2008)
18. Wang, X., Zhang, Q., Li, X.: Protocol enhancement for IEEE 802.11 e EDCF. In: Proceedings of 12th IEEE International Conference on Networks, 2004 (ICON 2004), vol. 1. IEEE (2004)
19. Lakrami, F., El kamoun, N., El kamili, M.: An enforced QOS shceme for high Mobile Adhoc Networks. In: Proceedings of The International Conference on Wireless Networks and Mobile Communications (WINCOM15), Marrakech, Morocco, 20–23 Oct 2015. IEEE Explorer (2015)

20. LI, J., Blake, C., De couto, D.S.J., et al.: Capacity of ad hoc wireless networks. In: Proceedings of the 7th annual international conference on Mobile computing and networking, pp. 61–69. ACM (2001)
21. Mccanne, S., Floyd, S.: NS network simulator (1995)
22. SIMULATOR, Network. Network simulator (1998)
23. Lakrami, F., El kamoun, N., El kamili, M.: A survey on QoS protocol in MANETS. In: The International Symposium on Ubiquitous Networking (UNET2015), Casablanca Morocco, 08–10 Sept 2015. Published by Springer

DTN Network: Optimal Cluster Head in DTN Routing Hierarchical Topology (DRHT)

El Arbi Abdellaoui Alaoui, Said Agoujil, Moha Hajar and Youssef Qaraai

Abstract In this paper, we study the problem of data routing with an optimal delay in the bundle layer, by exploiting : the clustering, the messages ferries and the optimal election of cluster head (CH). We first introduce the DTN routing hierarchical topology (DRHT) which incorporates these four factors into the routing metric. We propose an optimal approach to elect a CH based on four criteria : the residual energy, the intra-cluster distance, the node degree and the head count of probable CHs. We proceed then to model a Markov decision process (MDP) to decide the optimal moment for sending data in order to ensure a higher delivery rate within a reasonable delay. At the end, we present the simulation results demonstrating the effectiveness of the DRHT. Our simulation shows that while using the DRHT which is based on the optimal election of CH, the traffic control during the TTL interval (Time To Live) is balanced, which greatly increases the delivery rate of bundles and decreases the loss rate.

Keywords Ad Hoc Networks · Delay Tolerant Networks DTN · Bundles · Hierarchical cluster · Cluster Head Election · Delivery success probability

E.A. Abdellaoui Alaoui (✉) · S. Agoujil · M. Hajar · Y. Qaraai
Faculty of Sciences and Technology, Department of Computer Science,
University Moulay Ismaïl, Errachidia, Morocco
e-mail: abdellaoui.e@gmail.com

S. Agoujil
e-mail: agoujil@gmail.com

M. Hajar
e-mail: moha_hajjar@yahoo.fr

Y. Qaraai
e-mail: qaraai_youssef@yahoo.fr

R. El-Azouzi et al. (eds.), *Advances in Ubiquitous Networking 2*, Lecture Notes in Electrical Engineering 397, DOI 10.1007/978-981-10-1627-1_9

1 Introduction

A DTN network is characterized by intermittent connectivity, asymmetric flow, high error rate, long or variable delivery delay, extensive networks and high mobility of nodes. These factors make the network spread on a large-scale, and therefore the delivery delay is very long and the delivery rate is potentially low [1, 2].

Thus, the need to develop optimal transmission systems to maximize the DTN network performance is then essential, in order to ensure a great autonomy to these networks which are typically deployed in hostile or inaccessible areas. The objective of this work is to solve effectively this problem of delivering information between different nodes of the network. Our approach to this problem is first based on the regrouping of nodes in clusters, then the selection of a cluster head (CH) in each cluster and finally the communication between CHs through ferries. The selected CH is responsible for coordinating the communication with mobile nodes in the same cluster (intra-cluster) and with nodes of the other clusters (inter-cluster). The elected CH must take into account the determined characteristics such as the battery lifetime and the minimum average distance between the nodes of a subset and the CH in a given cluster. For this, the problem of choosing a dynamic coordinator can be reduced to the problem of a CH election, which is a major problem of the mobile network. Furthermore, the solution proposed for the distributed system (e.g. WSN) cannot be applied in the DTN where the change in topology is frequent and links are intermittent [3–7]. Our work provides an effective intra-cluster communication due to an optimization on two levels: the optimal election of CHs and the communication between them via ferries in order to increase the QoS in the DTN networks. In other words, the proposed approach improves the delivery rate and the delivery delay compared to conventional approaches.

It is noteworthy to mention that the rest of this article is organized as the following: In Sect. 2, we present the system model and problem statement. Then, we describe in Sect. 3 the model of the DTN routing hierarchical topology (DRHT). In Sect. 4, we analyze our approach for obtaining the optimal cluster head (CH) in the DRHT. In Sect. 5 is devoted to the model of the success probability of delivery for a bundle with specific TTL and the average duration of inter-contact. In Sect. 6, we describe the environment and the simulation parameters. In Sect. 7, we will present the obtained results to assess the performance of the used topology control DRHT compared to Maxprop [8], Spray and wait [9] and Epidemic [10] protocols. Finally, in Sect. 8, we present a conclusion.

2 System Model and Problem Statement

2.1 Network Model

Let a DTN composed of $N + 1$ nodes, i.e. mobile nodes. Two nodes can communicate only when they enter the reciprocal communication range and we consider this

Table 1 Notations used for modeling

Notation	Definition
N	Total number of nodes of the shared network
C_k	Cluster of the network
N_k	Number of nodes in each area, with: $N = \sum_{k=1}^{K} N_k$
F	Message ferry
n_i	Node i in the network
v	Speed of the ferry
P	Ferry route
$\lvert P\rvert$	Length of a ferry route
l_{ij}^{P}	Distance between the node n_i and n_j
$t_{w_{ij}}$	Time of wait to n_i before being transmitted to the ferry
$t_{c_{ij}}$	Time of carrying to the ferry before being delivered to n_j
d_{ij}^{P}	Average delay to transmit a message of n_i to n_j
μ_i	Message size $(1 \leqslant i \leqslant M)$

as a "contact" between them in the network DTN. Let the interval of pairwise inter-contact between n_i and n_j denotes the time duration from the instant when they leave communication range of each other to the next instant when they enter it. To improve the performance of DTN in the existing analytical results, we use the same mobility model, in which the interval of pairwise encounter fulfills the exponential distribution with the same rate λ. This model has been widely supported in the literature [11] because it is considered as a good approximation for the interval of inter-contact in a significant number of realistic DTN networks [11].

2.2 *Notations*

For the rest of this work, we consider the following notations (Table 1):

2.3 *Hypotheses*

For the rest of this work, we consider the following hypotheses:

- (H1): The nodes have the same range of transmission;
- (H2): The regions of the network forming a cluster;
- (H3): The movement of nodes is random between K regions;
- (H4): The traffic in the network is unpredictable;
- (H5): The range length of each cluster is strictly lower than the ferry route length;
- (H6): The contact between the two regions C_k and $C_{k'}$ follows an exponential distribution of the parameter $\lambda = \lambda_{kk'} = \lambda_{k'k}$.

2.4 Problem Statement

A DTN network can be considered as a set of time-varying contacts (a contact is defined as an opportunity to send data). The maximum amount of data that can be transmitted on a contact is called the delivery rate, and is defined as the product of the contact duration and the number of messages received during this period. A path is defined as a sequence of contacts. The path volume is the minimum volume of contact of all contacts of the path. Messages are transferred along a path in storage and forwarding mode (store-and-forward) [12]. If the next contact is not available, messages are buffered until the contact becomes available or messages have expired.

2.5 Objective Function

The objective-function of the average delivery delay in the DRHT, for all the traffic in a given ferry route, is defined as:

$$\Delta_{DRHT} = \frac{\sum_{i=1}^{M} \mu_i d_{ij}^{P}}{\sum_{i=1}^{M} \mu_i} \tag{1}$$

3 DTN Routing Hierarchical Topology (DRHT)

3.1 Description of the Construction DRHT

The main idea is to build a topology of routing in a large-scale DTN network. The dominant character in the DRHT is the number of ferries that cross the diffusion paths to ensure connectivity between clusters. The choice of data carrying nodes is an important step in the construction of the set of clusters. In addition, each cluster is identified by three categories of nodes:

1. The cluster-head (CH) is a dominant node, it is the head of the cluster;
2. The center of the cluster (CC) is a point of exchange at which messages can exchange data between different CHs via ferries within each cluster;
3. The ordinary nodes (ON) are not dominating nodes.

3.2 Analysis of Delivery Delay in the DRHT

The delivery delay of this message is analyzed as follows [13, 14]:

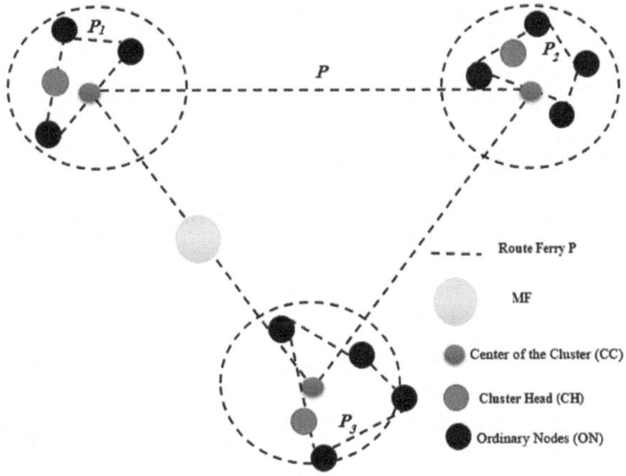

Fig. 1 Diagram of the DRHT

When nodes n_i and n_j are in the same cluster C_k The delay of single ferry routing is:

$$d^P_{ij_F} = \frac{|P|}{2v} + \frac{l^P_{ij_F}}{v} \tag{2}$$

In the DRHT, let P_k be the ferry route for cluster C_k and let $l^{P_k}_{ij}$ be the distance between nodes n_i and n_j in route P_k. The delay introduced by DRHT is $d^{P_k}_{ij}$:

$$d^{P_k}_{ij} = \frac{|P_k|}{2v} + \frac{l^{P_k}_{ij}}{v} \tag{3}$$

According to (2) and (3), we note that $d^{P_k}_{ij} < d^P_{ij_F}$ since $|P_k| < |P|$ and $l^{P_k}_{ij} < l^P_{ij_F}$. which means that when the node n_i and the node n_j belong to the same cluster, the delay of routing of DRHT is lower to the single message ferry.

When nodes n_i and n_j are situated in different clusters C_k and $C_{k'}$ Based on the Fig. 1, the delivery delay consists of three parts in the DRHT.

- Let $d^{P_1}_{ij}$ be the delivery delay in cluster C_1:

$$d^{P_1}_{ij} = \frac{|P_1|}{2v} + \frac{l^{P_1}_{ij}}{v} \tag{4}$$

- The delivery delay $d^{P_{CC}}_{ij}$ is the time of wait of the ferry and the time of carrying of the ferry in the point CC before delivering it to the cluster C_2:

$$d^{P_{CC}}_{ij} = \frac{|P_{CC}|}{2v} + \frac{l^{P_{CC}}_{ij}}{v} \tag{5}$$

- Let $d_{ij}^{P_2}$ be the delivery delay in cluster C_2:

$$d_{ij}^{P_2} = \frac{|P_2|}{2v} + \frac{l_{ij}^{P_2}}{v} \tag{6}$$

Therefore, the total delivery delay of the message is $d_{ij}^{P_{all}} = d_{ij}^{P_1} + d_{ij}^{P_{CC}} + d_{ij}^{P_2}$ and one writes:

$$\frac{|P_1|}{2v} + \frac{l_{ij}^{P_1}}{v} + \frac{|P_{CC}|}{2v} + \frac{l_{ij}^{P_{CC}}}{v} + \frac{|P_2|}{2v} + \frac{l_{ij}^{P_2}}{v} = \frac{(|P_1| + |P_{CC}| + |P_2|)}{2v} + \frac{l_{ij}^{P_1} + l_{ij}^{P_{CC}} + l_{ij}^{P_2}}{v} \tag{7}$$

From Fig. 1, we see that $|P_1| + |P_{CC}| + |P_2| < |P|$ and $l_{ij}^{P_1} + l_{ij}^{P_{CC}} + l_{ij}^{P_2} < l_{ij_F}^{P}$

4 Optimal Cluster Head Election in the DRHT

4.1 Objective Function for the Election of CH

In this section, we define our proposed objective function for effective execution of the election of the CH in the DRHT based on the Custody Transfer [15]. The main goal of the objective function is to optimize the combined effect of average distance between nodes in a cluster, residual energy, node degree and head count of probable cluster heads (i.e. the number of times a specific node served as a cluster head). The objective function, represented as $f(x_i(t))$ for the ith specific node is specified in the following equation:

$$f(x_i(t)) = optimize(\beta_1 \chi_1 + \beta_2 \chi_2 + \beta_3 \chi_3 + (1 - \beta_1 + \beta_2 + \beta_3)\chi_4) \tag{8}$$

Subject to:

$$\chi_1 = \sum_{\substack{\forall n_j \in C_k \\ x_i \in C_k}} \{\frac{||n_j, x_i||}{|C_k|}\} \tag{9}$$

$$\chi_2 = \frac{\sum\limits_{\substack{i=1 \\ x_i \in C_k}}^{N} E(p_i)}{\sum\limits_{\substack{j=1 \\ x_i \in C_k}}^{|C_k|} E(n_j)}, E_{min} \leqslant E(n_j) \leqslant E_{max} \tag{10}$$

$$\chi_3 = N_{deg}(p_i), 0 < \beta_1, \beta_2, \beta_3 < 1 \tag{11}$$

$$\chi_4 = \frac{1}{H(p_i)}, H(p_i) > 1, \beta_1 < \beta_2 < \beta_3 \tag{12}$$

As mentioned above $\beta_1, \beta_2, \beta_3$ are the weightage parameters. Moreover, $E(p_i)$, $N_{deg}(p_i)$ and $H(p_i)$ denote respectively the energy, the node degree and the head count of probable cluster heads, associated with the specific node p_i. Also, n_j is the le jth node of the kth cluster (C_k) and $|C_k|$ denotes the total number of nodes in the respective cluster. The Euclidean distance between the node n_j and the specific node p_i is represented by the notation $n_i; x_i$. It is clear, from the equation, that χ_1 is the average distance between the specific node p_i and all other nodes in the cluster and χ_2 is the energy measure of the specific node compared to the other nodes. The χ_3 parameter refers to the degree of the node associated to the specific node p_i. This criterion helps to select, around the specific node, the node with highest degree. The node that is connected to more number multiple of nodes reflects greater efficiency in receiving more bundles easily. χ_4 is the probability of choosing the specific node p_i on the basis of its head count the head count of probable CHs.

4.2 *Optimal Forwarding Instant of the CH*

In this section, we give a model the problem concerning the optimal instant for sending a message via a CH by a Markov decision process (MDP) which evolves in space and time. The objective of a CH is to maximize the percentage of the message reception by sending it while the bundle layer is free or little busy.

Markov Decision Process (MDP) Formulation The selected decision of the optimal moment of sending represents a compromise between the final gain; which is the reception rate, and the cost related to the potential delay. The resolution of this problem is obtained through a Markov decision process (MDP) modeling. The TTL lifetime of an information is limited and the delay tolerated for its forwarding too. Thus, we propose a set T of N periods in time included only in the TTL interval, of each t duration, during which CH can send its message or decide to delay it until the next period of time.

The set of time periods for the transmission of a message are: $T = \{T_1, \ldots, T_N\}$ with $N = \frac{V}{t}$, such as:

$T_{i+1} - T_i = t + \eta T_{Service} + \upsilon T_{garde}$; where $i < N$
where:

$$\eta = \begin{cases} 1 & \text{at the meeting of Service interval} \\ 0 & \text{if not} \end{cases}$$

$$\upsilon = \begin{cases} 1 & \text{at the meeting of the gard interval} \\ 0 & \text{if not} \end{cases}$$

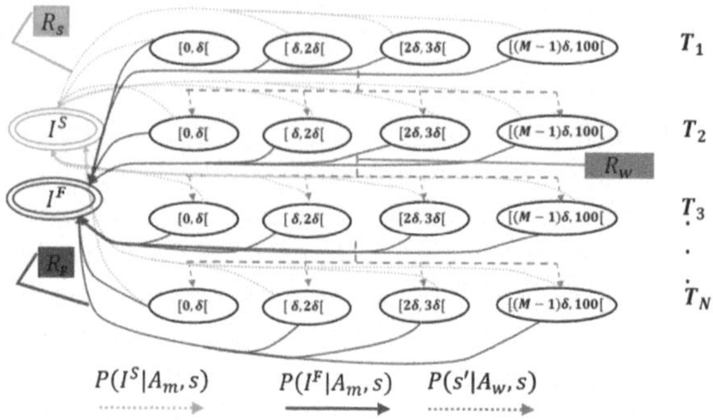

Fig. 2 Markov decision process of the transmission on Bundle layer

Our MDP model is composed of a set S of possible states for the system, actions $A_s^{T_i}$, rewards and costs $R(s'^{T_{(i+1)}}, s^{T_i})$ that depend on two process states, and finally transition probabilities $P(s'^{T_{i+1}}|a, s^{T_i})$ between the two states $s'^{T_{(i+1)}}$ and s^{T_i}, which are separated in time $(T_{i+1} - T_i)$, when the selected action is a.

(a) States: The set S of the process states includes two parts, the states C which relate to the occupation percentage of the bundle layer going from 0 % to 100 % for each period T_i of the information TTL. In addition to the two absorbing states I which represent the successful or failed forwarding status of a message. These states are achieved when a CH sends its message. This set of states is illustrated in Fig. 2. All these states are connected by transition probabilities, which result in costs R_f and R_w or rewards R_s.

$T = \{S_0^{T_i}, S_1^{T_i}, S_2^{T_i}, ..., S_j^{T_i}, ..., S_{M-1}^{T_i}\}, 0 < i \leqslant N$ Where $M = \frac{100}{\delta}$ and $S_j^{T_i} = [j\delta\%, (j+1)\delta\%]$, δ is the precision chosen for the intervals of the states C.
$I = \{I^S, I^F\}$

(b) Actions: Two actions $A_s^{T_i}$ can be chosen during a period of time T_i and for a state $s \in C$. The first action A_m consists in sending the message immediately; the second action A_w delays it for a period of time. A message is delayed until it meets a decision of immediate sending or until the expiry of its validity.

$$A_s^{T_i} = \begin{cases} \{A_w, A_m\} & \text{if } s^{T_i} \in C \text{ and } i < N \\ A_m & \text{if } s^{T_i} \in C \text{ and } i = N \end{cases}$$

(c) Rewards and costs: Each decision is taken in order to maximize a final gain; this latter represents the reception rate for a sent message. Its calculation depends on the $R(s'^{T_{i+1}}, s^{T_i})$ obtained during transitions between states; they can represent allotted rewards or deducted costs. A reward R_s is allotted when a message is sent successfully, whereas a cost R_f is inflicted when the sending fails. The cost induced by the adjournment of a message for a period of time,R_w, is the third parameter taken into account in the decision making.
The reward R_s is always positive, to motivate CHs to send their message. Whereas

the costs related to the sending failure and the additional time delay are either negative or equal to zero. The value of each of these parameters can be weighted to the access category (AC) of the message to send.

$$R(s'^{T_{(i+1)}}, s^{T_i}) = \begin{cases} R_s & \text{if } s'^{T_{i+1}} = I^S, s^{T_i} \in C, a = A_m \\ R_f & \text{if } s'^{T_{i+1}} = I^F, s^{T_i} \in C, a = A_m \\ R_w & \text{if } s'^{T_{i+1}}, s^{T_i} \in C, a = A_w, i < N \end{cases}$$

(d) Transition probabilities: Finally, a MDP model includes transition probabilities $P(s'^{T_{(i+1)}}|a, s^{T_i})$ for each action a chosen between two states of the process. When the chosen action is to delay the message A_w, the transition probabilities are the same as those concerning the occupation of the bundle layer at the time $T_{(i+1)}$. To have representative probabilities, each CH saves its local history of the occupation rate in the bundle layer during α TTL intervals. It then calculates the average percentage of occupation in time for each period of the TTL interval.
When the selected action is that of the sending; A_m, two probability are possible, either a successful forwarding or a failed one. One complements the other, they are calculated on the basis of the occupation percentage of the bundle layer at the sending time; let the state s, and the reception efficiency at this same period of time E^{T_i}. The efficiency is the ratio between the occupation time which was used for the successful reception of a number of messages NM^{T_i}, with an average size *Size* and a flow D, and the total occupation time of the bundle layer at this same period T_i, its calculation is given in the Eq. (13). These two parameters of occupation and efficiency of the bundle layer are weighted, in the sending probabilities with success or failure, by the variable $\rho \in [0, 1]$.

$$E^{T_i} = \frac{NM^{T_i} \times \frac{Size}{D}}{\frac{\delta \times s}{100} \times t} \tag{13}$$

$$P(s'^{T_{(i+1)}}|a, s^{T_i}) = \begin{cases} P(s'^{T_{(i+1)}}) & \text{if } s^{T_i}, s'^{T_{i+1}} \in C, a = A_w \\ P(s'|a, s^{T_i}) & \text{if } s^{T_i} \in C, s' \in I, a = A_m \\ 0 & \text{if not} \end{cases}$$

where

$$P(I^S|A_m, s^{T_i}) = \rho\frac{\delta \times s}{100} + (1 - \rho) \times E^{T_i}$$

$$P(I^F|A_m, s^{T_i}) = 1 - P(I^S|A_m, s^{T_i})$$

5 The Probability of Success to Deliver a Bunble with Specific TTL

5.1 Modeling of the Inter-contact Time

Proposition 1 *The punctual process* $\{T_n, n = 1, 2, ...\}$ *is a Poisson process with rate* λ *if and only if random variables* $\tau_n = T_n - T_{n-1}, n = 1, 2, ...$ *are independent and identically distributed according to an exponential law with parameter* λ*;* λ *is the intensity of inter-contact. We conclude this paragraph by calculating the average duration of inter-contact, which can be given using the following formula:*

$$E(\tau_n) = \frac{1}{\lambda} \tag{14}$$

5.2 Probability of Successful Delivery

We take again the Proposal 1, it is observed that the contact between nodes is distributed exponentially. We use these proposals to model the metrics of performance in the context of the DTN routing system. We use the delivery rate of bundles, the delivery delay and the buffer memory occupation, which are among the principal metrics of performance. Thereby, for a message entering the bundle layer at T_n time, let τ_n be the time of inter-contact between the nth and the $(n + 1)$th bundle. The probability that the bundle is delivered before the TTL expires is calculated using the following formula [16]:

$$P_r(T_n \leq t_{TTL}) = 1 - e^{(-\lambda t_{TTL})} \tag{15}$$

6 Simulation

Observing that they do not rely on analytical models, the exact evaluation of certain aspects of these protocols is very difficult. This is the reason that leads us to make simulations to study its performance. Our simulation is performed thanks to the ONE (Opportunistic Network Environment) simulator [17], which allows generating a classification of the different routing protocols studied using performance metrics.

Table 2 summarizes the simulation settings used to analyze the different DTN routing protocols in the simulated environment.

Table 2 Parameters of simulation

Parameter	Value
Total simulation time	12 h
World size	4500 × 3400 m^2
Routing protocol	Epidemic, Spray and wait and Maxprop
Node buffer size	5 M
No of nodes	10, 20, 30, …, 100
Interface transmit speed	2 Mbps
Interface transmit range	10 m
Message *TTL*	60 min
Node movement speed	Min = 0.5 m/s Max = 1.5 m/s
Message creation rate	One message per 25–35 s
Message size	50–150 KB

7 Results and Discussion

In the simulated environment, we focused on comparing the performance in terms of the two metrics: the delivery probability and the average duration of inter-contact.

7.1 The Delivery Probability

This metric characterizes how complete, correct and efficient a routing protocol is. It describes how many bundles were lost, as well as the maximum number of bundles that the network can support.

In Fig. 3 we show the ratio of delivery of bundles for each protocol by the number of nodes in the network. We noted that for a weak density (equal to 10 nodes) the three DTN protocols gave a low rate of bundles delivered. In fact, since the network's connectivity is weak because the density is weak, protocols do not find any path to reach some destinations, particularly after bundles' TTL expires. For medium density (between 20 and 25 nodes), the three protocols had a high ratio of bundles delivered. This is quite an interesting ratio and is much higher than 90 % of sent bundles. However, an observed drop with increasing density follows this ratio's increase. This drop is noticed for every protocol except Maxprop, which keeps a constant ratio for all values of density considered by all scenarios (until 100 nodes). In addition, for Epidemic and Spray and wait protocols, at high density, each node must be able to forward more traffic. This traffic increases the rate of collision, interferes with the data's traffic and therefore increases the loss of bundles. Because of its low traffic of control at high density, Maxprop keeps a constant ratio of delivered bundles. These results, which offer a fairly high delivery rate in the DRHT, can be explained by the use of multiple ferries and an optimal CH in each cluster.

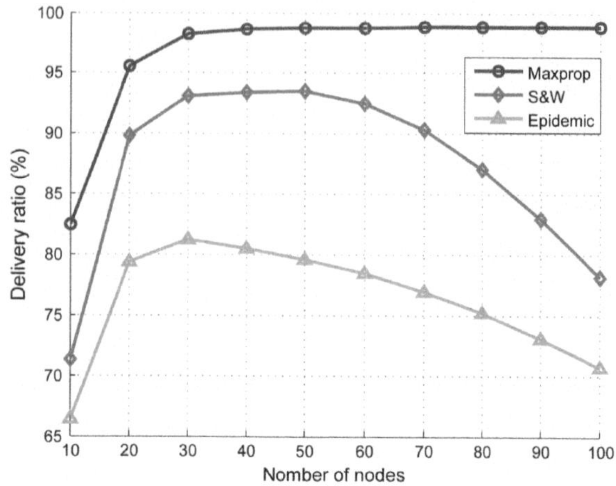

Fig. 3 Variation of the delivery probability depending upon the number of nodes

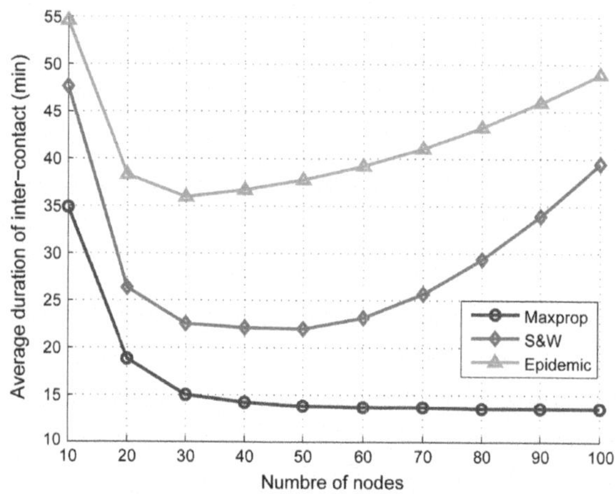

Fig. 4 Average duration of inter-contact of different protocols evaluated depending upon the number of nodes

7.2 *Average Duration of Inter-contact*

A shorter average life of contacts corresponds to a more dynamic topology of the network. In fact, great values of λ are reflected in shorter contact and inter-contact times and then an increase in contact opportunities. The bundles can benefit from it and their delivery probability increases when λ grows. Conversely, a very great instability of contacts and a lack of connection between nodes tend the delivery probability of bundles towards 0 because there are less contacts lasting in time.

The Fig. 4 shows clearly that, for a low density, the average duration of inter-contact of the three protocols is quite large because the distances separating nodes increase. We can also note that the average duration of inter-contact of the protocols decreases when nodes density increases. These results are explained by the increment of nodes average degree. Consequently, the end-to-end time becomes minimal. However, for Epidemic and Spray and Wait protocols with high density, each node is held to generate more traffic of control (overhead). This traffic of control increases the rate of collision and disturbs data traffic, and consequently the average duration of inter-contact increases. Thus, we note that Maxprop protocol remains the most efficient among the three routing protocols studied in terms of average duration of inter-contact, which will allow it to minimize the delay of delivery between the source and the destination.

8 Conclusion

In this research paper, we presented a proposition of a DTN hierarchical routing topology based on four fundamental notions: multiple ferries, ferry routes, the clustering and the election of a CH. In fact, the DRHT uses multiple ferries messages to make the whole network connected. Furthermore, the election of a dynamic CH in the DRHT has a major impact on the delivery rate and the delivery delay, by allowing the reduction of network resources. This election is based on specific criteria, among which we retain: the residual energy, the intra-cluster distance, the node degree and the headcount of probable CHs. The results show that Maxprop offers excellent performance in terms of the delivery rate and the delivery delay of bundles in the DRHT.

References

1. Fall, K.: A delay-tolerant network architecture for challenged internets. In: Proceedings of the 2003 Conference on Applications, Technologies, Architectures, and Protocols for Computer Communications, pp. 27–34. ACM, Karlsruhe, Germany (2003)
2. Abdellaoui Alaoui, E.A., Agoujil, S., Hajar, M., Qaraai, Y.: The performance of DTN routing protocols: A comparative study. WSEAS Trans. Commun. **14**, 121–130 (2015)
3. Maraiya, K., Kant, K., Gupta, N.: Efficient cluster head selection scheme for data aggregation in wireless sensor network. Int. J. Comput. Appl. (0975–8887), **23**(9), 10–18, June 2011
4. Sett, S., Thakurta, P.K.G.: Effect of optimal cluster head placement in MANET through multi objective GA. In: 2015 International Conference on Advances in IEEE Computer Engineering and Applications (ICACEA), pp. 832–837. 19–20 March 2015
5. Chen, J., Hong, Z., Wang, N.: Efficient cluster head selection methods for wireless sensor networks. J. Netw. **5**(8), 964–970 (2010)
6. Liu, X.: A survey on clustering routing protocols in wireless sensor networks. Sensors 2012, pp. 11113–11153 (2012). doi:10.3390/s120811113
7. Ferdous, R., Muthukkumarasamy, V., Sithirasenan, E.: Trust-based cluster head selection algorithm for mobile Ad Hoc Networks. In: 2011 IEEE 10th International Conference on Trust, Security and Privacy in Computing and Communications (TrustCom), pp. 589–596, 16–18 Nov 2011
8. Burgess, J., Gallagher, B., Jensen, D., Levine, B.N.: MaxProp: routing for vehicle-based disruption-tolerant networks. In: Proceedings of 25th IEEE International Conference on Computer Communications, pp. 1–11. Barcelona, Spain, April 2006

9. Spyropoulos, T., Psounis, K., Raghavendra, C.: Spray and wait: an efficient routing scheme for intermittently connected mobile networks. In: Proceedings of the 2005 ACM SIGCOMM Workshop on Delay-Tolerant Networking. ACM (2005)
10. Vahdat, A., Becker, D.: Epidemic routing for partially connected Ad Hoc Networks. Technical Report CS-200006, Duke University, Durham, April 2000
11. Zhu, H., Fu, L., Xue, G., Zhu, Y., Li, M., Ni, L.M.: Recognizing Exponential Inter-contact Time in VANETs. In: Proceedings of the 29th Conference on Information Communications (INFOCOM), pp. 101–105. IEEE (2010)
12. Chuah, M.C., Yang, P., Davison, B.D., Cheng, L.: Store-and-Forward performance in a DTN. In: Vehicular Technology Conference, Melbourne, Vic, vol.1, VTC 2006-Spring. IEEE 63rd, pp. 187–191, 7–10 May 2006
13. Zhao, W., Ammar, M.: Message ferrying: proactive routing in highly-partitioned wireless Ad Hoc Networks. In Proceedings of the Ninth IEEE Workshop on Future Trends of Distributed Computing Systems (FTDCS–03), pp. 308–314. Washington, DC, USA (2003)
14. Zhao, W., Ammar, M., Zegara, E.: A message ferrying approach for data delivery in sparse mobile Ad Hoc Networks. In: ACM MobiHoc (2004)
15. Fall, K., Hong, W., Madden, S.: Custody transfer for reliable Delivery in Delay Tolerant Networks. Technical Report, Intel Research, Berkeley (2003)
16. Abdellaoui Alaoui, E.A., Agoujil, S., Hajar, M.: Stochastic modeling and analysis of DTN Networks. In: Proceedings of The 2nd International Conference on Information Technology for Organizations Development (IT4OD). IEEE, 2016, March 30–April 1st 2016
17. Keränen, A., Ott, J., Kärkkäinen, T.: The ONE simulator for DTN protocol evaluation. In: SIMUTools'09: Proceedings of the 2nd International Conference on Simulation Tools and Techniques, ICST: New York, NY, USA, Article No. 55 (2009)
18. Zhao, W., Ammar, M., Zegara, E.: Controlling the mobility of multiple data transport ferries in a delay-tolerant network. In: IEEE INFOCOM, pp. 1407–1418 (2005)

Implementation of Bit Error Rate Model of 16-QAM in Aqua-Sim Simulator for Underwater Sensor Networks

Mohammed Jouhari, Khalil Ibrahimi and Mohammed Benattou

Abstract Aqua-Sim is a simulator built on top of NS-S, is designed for Underwater Sensor Networks(UWSNs). It can be used to simulate acoustic attenuation signal and packet collision in UWSNs. Aqua-Sim supports three-dimensional deployment that doesn't exist in the simulators developed for terrestrial wireless networks. Although this simulator doesn't consider any signal modulation scheme. In this paper, we introduce the Bite Error Rate(BER) model of 16-QAM modulation scheme in Aqua-Sim. Also we consider the ambient noise resulting from different environmental noise sources such as turbulence, shipping, waves and thermal noise in 16-QAM BER formula. The consideration of such BER model instead of random bit error improve the reliability of routing protocols simulated in Aqua-Sim. That is shown in this paper through the simulations of Vector Based Void Avoidance routing protocol (VBF).

Keywords BER · 16-QAM · Aqua-Sim · NS-2 · Underwater sensor networks · Acoustic channel · Thorp's equation · Ambient noise

1 Introduction

In several years ago Underwater Sensor Networks (UWSNs) has been used as a new type of Ad-hoc networks in underwater to collect data from 3D area of interest. Ad-hoc property allows the UWSNs to sense all depth not just the sea surface that was a huge challenge for scientific community working for understanding dynamics and ecosystems of ocean. In UWSNs sensor nodes are deployed in 3D area of

M. Jouhari (✉) · K. Ibrahimi
MISC Laboratory, Faculty of Sciences Kenitra, Ibn-Tofail University, Kenitra, Morocco
e-mail: jouhari4med@gmail.com

K. Ibrahimi
e-mail: khalil.ibrahimi@gmail.com
URL: https://sites.google.com/site/khalilibrahimishomepage/

M. Benattou
LASTID Laboratory, Faculty of Sciences Kenitra, Ibn-Tofail University, Kenitra, Morocco
e-mail: mbenattou@yahoo.fr

R. El-Azouzi et al. (eds.), *Advances in Ubiquitous Networking 2*, Lecture Notes in Electrical Engineering 397, DOI 10.1007/978-981-10-1627-1_10

interest, each sensor node use multi-hop forwarding strategy to deliver the sensing data from a source node into a destination. Sonobuoys are usually used as a destination of all source nodes (sensor nodes), and they're considered as a gateway, since they collect the data from a source nodes and forward it into the monitoring center. There are many different strategies to select the next-hop forwarder [1–4], some of them use the opportunistic routing where a packet is broadcasted to a forwarding set of nodes composed of several neighbor nodes, and others use the anycast greedy forwarding strategy that selects the best neighbor node having the smallest distance from the destination as next-hop forwarder. UWSNs are used for many applications such that oceanographic data collection, marine pollution monitoring, offshore exploration and disaster prevention and tactical surveillance [5].

The ocean is a dynamic and complex propagation environment, hence UWSNs can't use the electromagnetic and optical signals as a communications system, because of the high rate and high speed of absorption. These factors lead to the use of acoustic modems in communications between sensor nodes. Instead sonobuoys are equipped with both acoustic and radio frequency modems, which they use acoustic links to send command and receive data from sensor nodes and the radio links to forward data packets into base station. Each sensor node collects the data and sends it to a surface station (sonobuoys) through acoustic multihop communications. The main challenge in UWSN is the acoustic communication, such as acoustic channel often feature low bandwidths, long propagation delays due to the speed of acoustic signals in water which is about 1500 m/s and high and dynamic packet loss probabilities, which lead to lots of re-transmission, a high energy consumption and lower reliability [6, 7]. Thus the conventional re-active and pro-active routing protocols used in terrestrial wireless sensor networks are not suitable for UWSNs [8, 9].

Geographic routing protocols are the most used for UWSNs communications, they use the geographic position of sensor nodes to root the packet in the network. A source node is always aware of its position and only 1-hop neighbor locations, selects the node closest to destination as a next-hop node to forward the packets. This avoid the additional communication overhead and makes geographic routing protocol simple and scalable. However it suffers from a serious problem that is the disconnected nodes, this types of nodes can not reach any sonobuoys. The disconnected nodes refers to void and isolated nodes. The void nodes are the nodes that fail to locate next-hope node in its neighborhood. The isolated nodes are the nodes that can not reach any sonobuoys through multi-hope communication. Routing protocols should cope with this problem using some recovery methods or reduce the energy consumption to avoid the appearance of disconnected nodes [10].

Main contribution of this paper In our previous work untitled [10] we worked on the improvement of Greedy forwarding strategy by resolving the problem of the disconnected nodes through depth adjustment and transmission power control of underwater nodes, more details are given in the following section. In this work, we still working on the geographic routing protocols by improving the acoustic communication between two underwater nodes through the consideration of BER. Also, the ambient noise is introduced into signal-to-noise ratio SNR which consider different environment noise source such as turbulence, shipping, waves and thermal noise.

The consideration of BER is one of the problem facing the development of Aqua-Sim and the simulations preformed on it. The implementation works also for NS-2. Otherwise, the implementation of BER allows us to consider advanced propagation models (which take into account different path losses and many real parameters of underwater communication environment such as temperature, pressure and salinity.) in our simulation through the SNR included in BER formula presented below in this paper. Simulation of routing protocols using an advanced propagation model allows the authors in the fields to perform more realistic studies and get more credible results. The rest of the paper is organized as follows. Section 2 briefly review some related works. Section 3 presents the characteristics of the acoustic channel considered in the network model. In Sect. 4 16-QAM BER formulas are given. Section 5 explains the implementation of 16-QAM in Aqua-Sim. Section 6 evaluates the performance of VBR through simulations. Finally, Sect. 7 concludes the paper and discusses some future works.

2 Related Works

In this section we review some related works to our interested research field. [7] This paper is an overview of acoustic channel models for underwater wireless networks, where the author details the challenges facing the underwater wireless communication. Such as the presence of fading, multipath and refraction. A survey on ray-theory-based multipath Rayleigh underwater channel models for underwater wireless communication is presented. The strong aspect of this paper is the study of channel models is performed for both shallow and deep water, also many different path losses are introduced into the transmission loss function such as surface reflection and bottom bounce. In [1] authors proposed the DBR (Depth Based Routing) protocol where the receiver node decides to forward the packet based on its depth and the depth of its previous hop. Upon receiving a packet, a node decides to forward if its depth is smaller than the previous sender. Otherwise it discards the packet. DBR uses the greedy mechanism to advance the packet toward the destination. However it doesn't have any recovery strategy to cope with the communication void region problem.

Geographic routing protocol proposed in [2] use greedy forwarding strategy for the next-hop selection and topology control through depth adjustment of some nodes to better organize the network topology, this reduce the impact of communication void region problem in the network performance. In this work, authors consider the static network architecture where nodes can only move on the vertical axes, not on x y as occurs with ocean current. Two topology control are presented in this work. The first is centralized topology control (CTC) algorithm, where a monitoring center determines which nodes doesn't have a path to any sonobuoys (isolated nodes) and which ones are void nodes. Afterward move them to a new depths that allow them to communicate with one or more sonobuoys. The second, is a distributed topology control (DTC) where each node locally determines if it's a disconnected node and computes its new depth to overcome this problem. However, this protocol increases the energy consumption using the depth adjustment and control message during the process of topology control. The same authors as in [2] suggest an other geographic routing protocol called GEDAR [3], Geographic and opportunistic routing protocol with depth adjustment for mobile underwater sensor networks. In this work they consider a mobile network architecture, where nodes can move on the vertical axes to adjust their depth and on x y axes with ocean currents. GEDAR uses the greedy opportunistic mechanism to route the packet toward destination, and topology control through depth adjustment of some nodes to avoid the problem of disconnected nodes. In opportunistic routing each packet is broadcasted to a forwarding set composed of several neighbors then it will be re-transmitted only if none of the neighbors in the set receive it. In [10] we suggest a new topology control that uses the transmission power control to adjust the transmission range of sensor nodes. And we evaluate the network performance using this method and the topology control through depth adjustment described in [2]. This method resolve the problem of isolated and void nodes appeared in geographic routing protocols through the increase of transmission range.

3 Underwater Acoustic Channel Characteristics

Acoustic communication in underwater sensor networks has several challenge due to the presence of fading, multipath and refractive properties of the sound channel which necessitate the development of precise underwater channel models. In this paper, a simplified channel model that not consider multipath or multipath fading is used.

The attenuation called also path loss occurs in an underwater acoustic channel over a distance l for a signal of frequency f is calculated as follow

$$A(l,f) = l^k \alpha(f)^l \times 10^{-3} \tag{1}$$

where k is the spreading factor, which describes the geometry of propagation. If the spreading is spherical $k = 2, k = 1$ for cylindrical spreading and $k = 1.5$ for the so-called practical spreading, in this paper we consider the path loss in deep water corresponding to $k = 2$. $\alpha(f)$ is the absorption coefficient, it can be obtained by the use of *Thorp*'s equation for frequencies in kHz. *Thorp*'s equation for frequencies above a few hundred Hz:

$$\alpha(f) = 0.11\frac{f^2}{1+f^2} + 44\frac{f^2}{4100+f} + 2.75 \times 10^{-4} \times f^2 + 0.003 \tag{2}$$

For lower frequencies, *Thorp*'s equation can be used as follows:

$$\alpha(f) = 0.002 + 0.11\frac{f^2}{1+f^2} + 0.011 \times f^2 \tag{3}$$

In the current version of Aqua-Sim there is no implementation of ambient noise, this part is one of our own contributions is this work. It is well known that the ambient noise can be modeled using four sources: turbulence, shipping, waves and thermal noise [6]. The four noise component in dB re μ Pa per Hz are modeled as a functions of frequency in kHz as shown below:

$$10\log N_t(f) = 17\text{–}30\log f$$
$$10\log N_s(f) = 40 + 20(s + 0.5) + 26\log f - 60\log(f + 0.03)$$
$$10\log N_w(f) = 50 + 7.5w^{1/2} + 20\log f - 40\log(f + 0.4)$$
$$10\log N_{th}(f) = -15 + 20\log f$$

where N_t is the turbulence noise, it occurs only for the very low frequencies, $f <$ 10 Hz. $N_s(f)$ is the noise caused by distance shipping is influences in the frequency region 10–100 Hz, and it is modeled using the shipping activity factor s, whose value ranges between 0 and 1 for low and high activity, respectively. The noise caused by wind-driven waves, called surface motion noise and depicted by N_w, is the dominant factor contributing to the noise in the frequency region 100 Hz–100 kHz (this frequency region is the most used operating region by almost acoustic systems). This factor is obtained using the w wind speed in m/s. Finally, thermal noise depicted by N_{th} become dominant for frequencies above 100 kHz. The overall power spectral density on the ambient noise is given by:

$$N(f) = N_t(f) + N_s(f) + N_w(f) + N_{th}(f) \tag{4}$$

The signal-to-noise ratio is an important parameter in underwater acoustic communications, is used in the bit error probability also is used to determine whether the received signal is strong enough to be detected by the receiver or non. The *SNR* of a signal observed over a distance l with the frequency f and the power P is obtained using the attenuation $A(l,f)$ and the power spectral density of the ambient noise $N(f)$. Without considering the directivity gains and losses other than the path loss, the narrow-band *SNR* is given by

$$SNR(l,f) = \frac{P/A(l,f)}{N(f)B_N} \tag{5}$$

where B_N is the receiver noise bandwidth (Table 1).

4 16-Quadrature Amplitude Modulation (16-QAM) Bit Error Rate

The QAM is always considered as an independent PAM modulation on I arm and Q-arm respectively, let us take the example of 16-QAM scenario which is considered in our works. In this scenario each constellation point can be presented by $\log_2(16) = 4$

Table 1 Gray coded constellation mapping for 16-QAM

b0b1	I	b2b3	Q
00	−3	00	+3
01	−1	01	+1
11	+1	11	−1
10	+3	10	−3

bits, with two bits on the I-axis and two on Q-axis. For 16-QAM the I and Q axes take the value from the set $\{-3, -1, +1, +3\}$. The code Gray could be used to present the two bits on I and Q arm as shown in the following table The constellation diagram with the bits mapping is shown in the Fig. 1. We can see from the constellation diagram in Fig. 1 that with Gray codded mapping there is only one bit different for the adjacent constellation symbols. if the noise occurs on the constellation only one of k' bits will be in error, so the relation between bit error and symbol error is $P_b \approx \frac{P_s}{k'}$. It's more probable that the noise leads the constellation to fall near a diagonally located constellation for very low value of $\frac{E_s}{N_0}$. In this case each symbol error will leads to two bit errors. For important value of $\frac{E_s}{N_0}$ the chances of such events are negligible. As we know each symbol consists of k' bits, for 16-QAM modulation $k' = log_2(16) = 4$. The relation between the symbol energy and bit energy is given as follows:

$$\frac{E_s}{N_0} = k' \times \frac{E_b}{N_0} \tag{6}$$

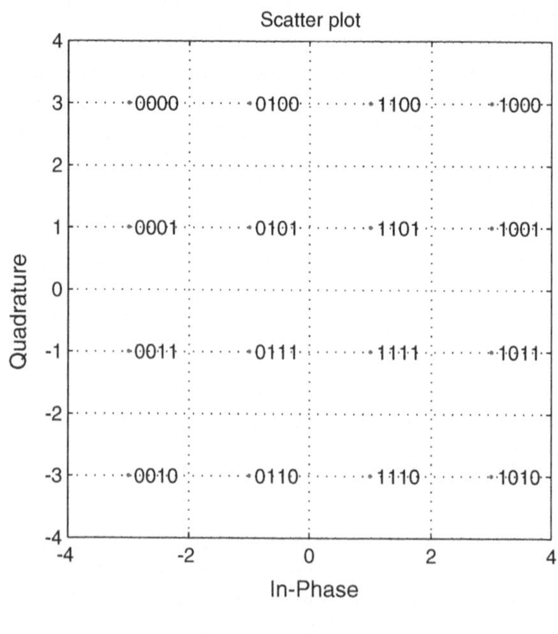

Fig. 1 16-QAM, Gray-coded Symbol Mapping

The symbol error rate is defined in [7] by

$$P_{s,16QAM} = \frac{3}{2} erfc\left(\sqrt{\frac{E_s}{10N_0}}\right) \tag{7}$$

Using the relation between symbol energy and bit energy in the Eq. (7) we can get the Bit Error Rate BER through the following equation. The BER can be considered as the probability of bit error.

$$P_{b,16QAM} = \frac{3}{2k'} erfc\left(\sqrt{\frac{k'E_b}{10N_0}}\right) \tag{8}$$

where erfc is the complimentary error function, k' is the number of bits per symbol. The BER of 16-QAM modulation corresponding to $k' = 4$ is obtained by replacing k' in the Eq. (8).

$$P_{b,16QAM} = \frac{3}{8} erfc\left(\sqrt{\frac{4E_b}{10N_0}}\right) \tag{9}$$

It's known that $E_b/N_0 = SNR\frac{B_N}{R}$, where $SNR = 10^{SNR_{dB}(l,f)/10}$, B_N is the noise bandwidth and R is the data rate. The received signal is given by $P_r = \frac{P}{A(l,f)}$. Combining these results with the Eqs. (5) and (9) we get the simple version of BER as follows:

$$P_{b,16QAM} = \frac{3}{8} erfc\left(\sqrt{\frac{4P_r}{10N(f)}}\right) \tag{10}$$

5 Implementation of 16-QAM in Aqua-Sim

In this section we explain the implementation of the modulation scheme in Aqua-Sim [11]. This simulator is designed to underwater sensor networks, developed on the top of NS-2 (Network Simulator for terrestrial sensor networks), and it considers the acoustic signal attenuation and packet collision in the underwater sensor networks. Moreover Aqua-Sim supports three-dimension deployment. The integration of Aqua-Sim in NS-2 is easy and any changes performed in Aqua-Sim do not affect the packages of NS-2.

Neither Aqua-Sim nor NS-2 consider a modulation scheme to calculate the Bit Error Rate in their classes respectively, underwaterphy.cc and wireless-phy.cc. Fig. 2 shows the interested classes of Aqua-Sim in which our work is focused. *uw_sink.cc* class presents the underwater node, it can be a routing agent (example: *uw_sink_vbva.cc* for Vector Based Void Avoidance routing protocol) or a source

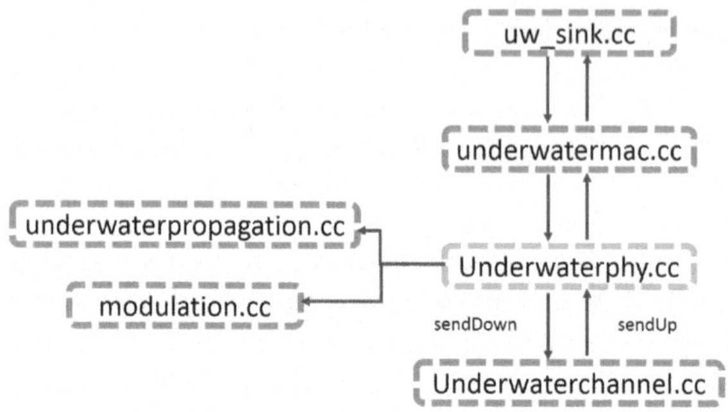

Fig. 2 Interested components and relations between them

```
54   int
55   Modulation::BitError(double Pr, double freq, double R)
56   {
57       double Pe; // probability of error
58       double x;
59       double N = Noise(freq);
60
61       Pe = ProbBitError(Pr, N, R);
62
63
64       Pe *= 1e3; //multiple assignment
65       x= (double)((Random::uniform()) * 1000);
66
67       if(x < Pe)
68           return 1; // bit error
69       else
70           return 0; // no bit errors
71   }
72
73
74   double
75   Modulation::ProbBitError(double Pr, double N, double R)
76   {
77       double Pe = (3/8)*erfc(sqrt((4/10)*Pr/(N*R))); // 16-QAM
78       return Pe;
79   }
```

Fig. 3 Main functions in modulation.cc

of traffic if it's attached to an application agent like as UDP agent. This component brings down the data rate information into the underwater physical interface through the mac layer. The underwater physical interface where the bit error rate is used to determine if the packet sends up from the underwater channel component have a bit error or non. Figure 3 shows the main functions of class Modulation where the Formula (10) is implemented in the function ProbBitError (line 75) where *Pr* is the power of received signal, *N* is the ambient noise and *R* is the data rate. In the function BitError where *freq* is the frequency of signal, we use the double x as the error probability threshold (line 67).

6 Performance Evaluation

This section focuses on the performance evaluation of routing protocol using BER model versus other using random bit error. This comparison is performed through three interesting metrics: Packet Delivery Ratio (PDR), Energy Consumption and Throughput. VBF: Vector-Based Forwarding Protocol for UWSNs is used for this study.

6.1 Simulation Settings

To evaluate the performance of VBF geographic routing protocol after the consideration of 16-QAM BER model we configure the TCL script of simulation following the main commands listed in the Table 2.

In this simulation we randomly deploy 480 underwater nodes in 3D region of size $\mathbf{500\,m \times 500\,m \times 500\,m}$, where only the nodes at the bottom $z = 500\,\text{m}$ can transmit their packets toward the nodes at the sea surface $z = 0\,\text{m}$. The nodes in the middle could move with speed 0.5 m/s. BroadcastMac protocol is used to transmit packets from Mac layer. For the Figs. 4 and 5 the simulation is performed with data rate 0.1 bps. The Simulation Time for the simulation performed in the Fig. 6

Table 2 Main parameters setting for TCL script

Parameter	Value
Set opt(txpower)	2.0
Set opt(rxpower)	0.75
Set opt(initialenergy)	10000
Set opt(idlepower)	0.008
Set opt(packet_size)	50
Set opt(nn)	480
Set opt(x)	500
Set opt(y)	500
Set opt(z)	500
Set opt(adhocRouting)	Vectorbasedforward
Phy/UnderwaterPhy set Pt_	0.5818
Phy/UnderwaterPhy set freq_	25
Phy/UnderwaterPhy set K_	2.0

Fig. 4 Packet delivery ratio versus simulation time

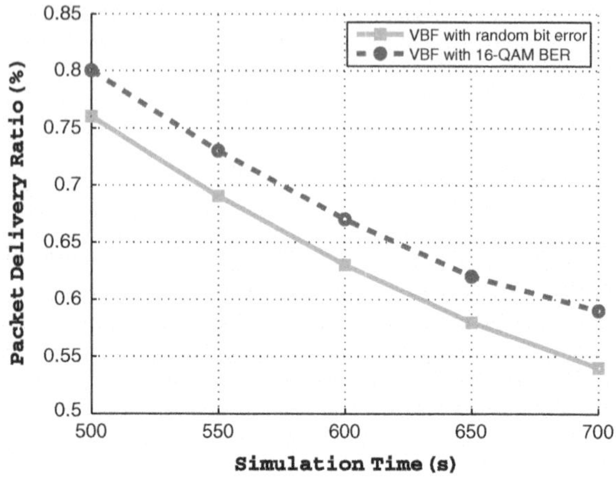

Fig. 5 Energy Consumption versus Simulation Time

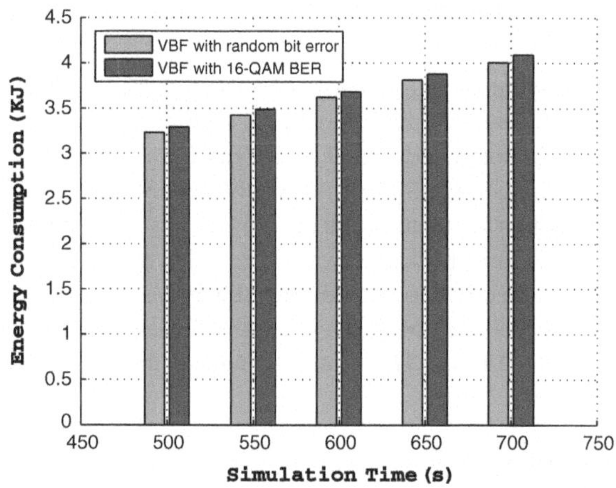

is set as 500 s. To control and compute the energy consumption the parameters, txpower, rxpower and idlpower are set as 2.0 W, 0.75 W and 0.008 W. Each node's initial energy is set as 10000 Joule. The parameters Pt_ and freq_ are respectively, the transmitted signal power and frequency, their values are respectively 0.5818 W and 25 kHz. The spreading factor K_ equal 2 corresponding to spherical spreading because the sea depth of 500 m is considered as deep water.

6.2 *Simulation Results*

In this section we show the results obtained from the simulation performed in Aqua-Sim. Our study is limited to three interesting parameters shown in the following graphs.

In Fig. 4 we mesure the PDR of the sink nodes to evaluate the reliability of VBF routing protocol using random bit error or 16-QAM BER model. In a simulation of 500 s, the source node sends 50 packets to the sink node in total. The proportion of the number of received packets by sink nodes is defined as packet delivery ratio. The result of this figure shows that when the simulation time increase the PDR decrease for both scenario. This may happen because sensor nodes including source, forwarding and sink nodes consume energy during the communication process, which result sensor nodes death that lead to void regions where the void nodes can't forward the received packet but just discard it. However, the PDR of VBF with 16-QAM BER is better than VBF with random bit error, Which mean that 16-QAM BER model improve the reliability of VBF routing protocol. This point is the mean aspect of this paper.

Figure 5 shows the evolution of energy consumption of all sensor nodes in the network topology during a simulation time. As we can see in the figure the energy consumption of the system increase for both scenario while increasing the simulation time. That happen because, while increasing the simulation time more packet will be sent from the source nodes and more node will participate in the forwarding task, even if all nodes become void nodes, there is a VBF beacon broadcast periodically from nodes using VBF to update their neighbors tables. However, the curve keep growing until the initial energy of sensor nodes is consumed, which the value is $480 \times 10 = 4800\,\text{kJ}$. Figure 6 examines the throughput of VBF considering random bit error and 16-QAM bit error using data rate ranging from $\{0, 1, \ldots, 20\,(\text{bps})\}$. Throughput is generally defined as the amount of data packet size delivered by a period of time. Until data rate contribute in BER formula of 16-QAM is used in the simulation in function of Throughput. Thus the influence of data rate changes on VBF routing protocol using both bit error probability is shown throughput. As it shown the figure the throughput of both scenario is not affected by the data rate changes. Otherwise, the consideration of BER model of 16-QAM do not improve the throughput of VBF.

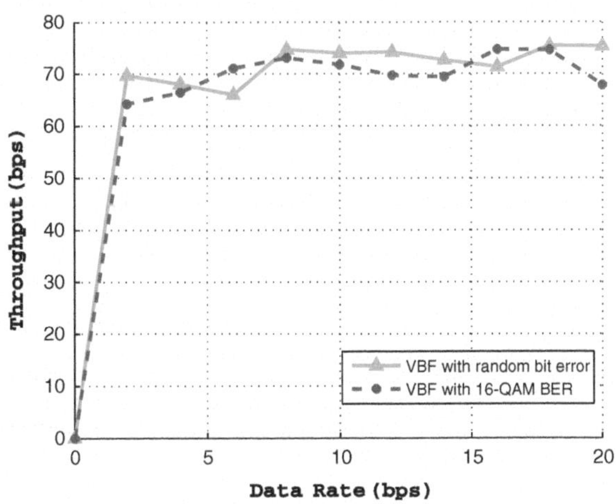

Fig. 6 Throughput versus data rate

7 Conclusion

In this paper we introduce the BER model of 16-QAM modulation scheme in Aqu-Sim instead of random bit error. This probability model is used to determine the packets containing bit error. Also the ambient noise resulting from different environmental noise sources such as turbulence, shipping, waves and thermal noise. Aqua-Sim is a good tool to evaluate the performance of underwater routing protocols And using the BER model suggested in this work can improve the reliability of these routing protocol as it shown in the simulation results. Figure 4 shows that clearly. In our future works we plan to consider more advanced acoustic channel and looking on the network coding scheme for UWSNs.

References

1. Yan, H., Shi, Z.J., Cui, J.-H.: Dbr: Depth-based routing for underwater sensor networks. In: Proceedings of the 7th International IFIP-TC6 Networking Conference on AdHoc and Sensor Networks, Wireless Networks, Next Generation Internet, NETWORKING'08, pp. 72–86. Springer-Verlag, Berlin, Heidelberg (2008)
2. Coutinho, R.W., Vieira, L.F., Loureiro, A.A.: Movement assisted-topology control and geographic routing protocol for underwater sensor networks. In: Proceedings of the 16th ACM International Conference on Modeling, Analysis; Simulation of Wireless and Mobile Systems, MSWiM'13, pp. 189–196. ACM, New York, NY, USA (2013)
3. Coutinho, R., Boukerche, A., Vieira, L., Loureiro, A.: Gedar: Geographic and opportunistic routing protocol with depth adjustment for mobile underwater sensor networks. In: 2014 IEEE International Conference onCommunications (ICC), pp. 251–256, June 2014
4. Noh, Y., Lee, U., Wang, P., Choi, B.S.C., Gerla, M.: Vapr: Void-aware pressure routing for underwater sensor networks. IEEE Trans. Mob. Comput. **12**, 895–908 (2013). May
5. Vasilescu, I., Kotay, K., Rus, D., Dunbabin, M., Corke, P.: Data collection, storage, and retrieval with an underwater sensor network. In: Proceedings of the 3rd International Conference on Embedded Networked Sensor Systems, SenSys'05, pp. 154–165. ACM, New York, NY, USA (2005)
6. Stojanovic, M.: On the relationship between capacity and distance in an underwater acoustic communication channel. In: Proceedings of the 1st ACM International Workshop on Underwater Networks, WUWNet'06, pp. 41–47. ACM, New York, NY, USA (2006)
7. Domingo, M.C.: Overview of channel models for underwater wireless communication networks. Phys. Commun. **1**, 163–182, Sept 2008
8. Akyildiz, I.F., Pompili, D., Melodia, T.: State of the art in protocol research for underwater acoustic sensor networks. SIGMOBILE Mob. Comput. Commun. Rev. **11**, 11–22 (2007). Oct.
9. Akyildiz, I.F., Pompili, D., Melodia, T.: Challenges for efficient communication in underwater acoustic sensor networks. SIGBED Rev. **1**, 3–8 (2004). July
10. Jouhari, M., Ibrahimi, K., Benattou., M.: Topology control through depth adjustment and transmission power control for uwsn routing protocols. In: The Proceedings of the International Conference on Wireless Networks and Mobile Communications (IEEE WINCOM15), Oct 2015
11. Xie, P., Zhou, Z., Peng, Z., Yan, H., Hu, T., Cui, J.H., Shi, Z., Fei, Y., Zhou, S.: Aqua-sim: An ns-2 based simulator for underwater sensor networks. In: OCEANS2009, MTS/IEEE Biloxi—Marine Technology for Our Future: Global and Local Challenges, pp. 1–7, Oct 2009

Energy Efficient In-Network Aggregation Algorithms in Wireless Sensor Networks: A Survey

Hafsa Ennajari, Yann Ben Maissa and Salma Mouline

Abstract Advancement in ubiquitous networking has led to the production of wireless sensor networks, consisting of many autonomous small devices called sensor nodes, able to observe and report various real world physical phenomena with no wired infrastructure. Onetheless, this feature precisely makes these nodes energy constrained, since most of the energy is consumed in data communication. In-network processing may be regarded as an efficient technique that reduces the amount of data to be transmitted in the network. We focus on data aggregation algorithms, whose fundamental idea is to gather, combine and compress data from different sources by applying simple functions in order to reduce the traffic load, thus enhancing the network's lifetime. However, it is difficult for developers to identify data aggregation algorithms strengths and weaknesses, nor to pinpoint current open research issues to be investigated. In this paper, we propose a survey of the most energy efficient data aggregation algorithms. After reviewing over 900 papers from which we selected 15 algorithms based on the energy efficiency criteria, we classify these protocols according to the network topology, then we describe each one in order to compare them. We conclude the paper with possible future research directions for aspiring researchers and algorithm developers.

Keywords Wireless Sensor Networks (WSNs) · In-network processing · Data aggregation · Routing · Energy consumption

H. Ennajari (✉) · S. Mouline
LRIT, Associated Unit to CNRST (URAC 29) - Faculty of Sciences,
Mohammed V University in Rabat, B.P.1014 RP, Rabat, Morocco
e-mail: ennajari.hafsa@gmail.com

S. Mouline
e-mail: mouline@fsr.ac.ma

Y. Ben Maissa
Telecommunication Systems, Networks and Services Lab, National Institute of Posts and Telecommunications, 2, Allal El Fassi Avenue, Rabat, Morocco
e-mail: benmaissa@inpt.ac.ma

R. El-Azouzi et al. (eds.), *Advances in Ubiquitous Networking 2*, Lecture Notes in Electrical Engineering 397, DOI 10.1007/978-981-10-1627-1_11

1 Introduction

Context. Recent advances in micro-electro-mechanical systems (MEMS) technology, wireless communications, and digital electronics have enabled a new generation of large-scale sensor networks suitable for a wide range of applications [1]. It consists of a spatially distributed autonomous devices called sensor nodes able to monitor physical or environmental conditions.

In-network processing is an approach where nodes of a network pre process collaboratively sensed data before transmission to the sink to avoid communicating a large amount of data and therefore extend lifetime of the network. Here the network's lifetime refers to the amount of time that a Wireless Sensor Network would be fully operative [2].

In-network processing can be classified into two main categories: data fusion and data aggregation. In the first one, nodes fuse the multiple received reports into a single one based on a decision criterion. For example, an explosion can result in a temperature above 80°C, a brightness above 500 lm and a sound level greater than 60 dB. In the second one, intermediate nodes usually compute an arithmetic function over a set of data (e.g., SUM, AVG, MAX) before transmitting the result to the sink. For example, calculating the average of several received temperature values.

We focus on data aggregation.

Problem. A wireless sensor network is limited in terms of resources: processing capabilities, storage and especially energy, due to usually non rechargeable batteries and hostile operational environment (e.g., near a volcano) unattainable by humans. It is widely recognized that the energy efficiency is the predominant performance criterion in wireless sensor networks even before the QoS criterion [3].

Under particular conditions, authors in [4] have found that the energy cost of transmitting a 1 kb packet over a distance of 100 m is approximately equal to executing 3 million instructions by a typical microprocessor. Therefore, domain experts tend to process as much as they can in the network, before transmission. In-network data aggregation can be considered as a candidate solution and a current trend [5]. There are multiple in-network data aggregation approaches. However, it is difficult for a researcher to identify strengths and weaknesses of each one as well as to pinpoint the open research issues.

Contribution. In this paper, we provide a comprehensive survey of in-network data aggregation algorithms according to a set of carefully selected criteria. We surveyed over 900 papers from which we selected 15 algorithms based on the energy efficiency criteria. These algorithms are fully studied, classified according to the network architecture and then compared regarding a set of selected criteria. Our ultimate goal is to identify open research issues for aspiring researchers and algorithm developers in order to enhance existing algorithms or to propose new ones.

Contents. This paper is organized as follows. In Sect. 2, we define fundamental concepts of wireless sensor networks and the data aggregation paradigm. In Sect. 3 we introduce the different approaches of data aggregation while addressing the main algorithms that belong to each approach. In Sect. 4 we propose a comparative synthesis of the already studied algorithms based on a set of criteria. Then, we provide directions and open issues for future research.

2 Background

In this section, we present an overview of wireless sensor networks and data aggregation concepts.

2.1 Wireless Sensor Networks

A Wireless Sensor Network is a set of autonomous nodes with sensors distributed in an environment. These nodes are able to detect a variety of physical phenomena deduced from a set of sensed variables such as temperature, humidity, sound, vibration, which are subsequently communicated via the wireless medium. Generally sensor nodes have three functions: sense physical quantities (e.g., light, temperature, sound), process all the collected data and communicate the result to a special device called sink or base station. These networks have many applications in several fields, such as military, health care, environmental and home monitoring.

Data aggregation in wireless sensor networks consists of replacing the individual readings of each node by a cooperative global overview on a given area. We can use for example a simple aggregation function such as MIN, MAX, AVG that allows from a set of n messages received by an intermediate node to send to the base station

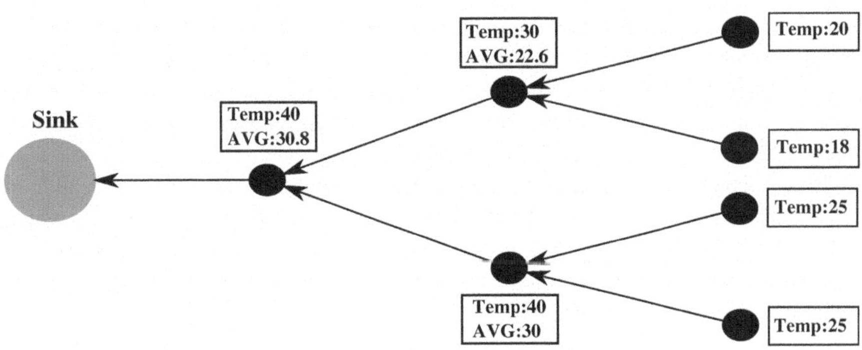

Fig. 1 Example of data aggregation concept using the average function

only a single message summarizing the information contained in these *n* messages. This reduces the number of messages sent and therefore saves energy [6] (Fig. 1).

Usually, we distinguish two data aggregation types:

Data aggregation with size reduction. Refers to the process that allows to combine and compress data received from different source nodes to reduce the information sent in the network. For instance, calculating the average of *n* received values, then sending the result into a single packet instead of forwarding the *n* received packets.

Data aggregation without size reduction. Refers to the process that merges received packets from multiple nodes into a single one without processing, assuming that they are two different physical quantities. For example, temperature and humidity can not be processed all together but can be sent in a single packet which reduces the header.

The choice of the approach depends on many factors, such as the application type, data rate, network characteristics, and so on. The two strategies above may include processing of data at different levels of the network.

Data aggregation consists generally in three basic components: (1) an efficient routing protocol (2) an efficient aggregation function (3) an efficient data representation.

Routing protocols. Data aggregation is based primarily on well designed routing protocols. In which nodes construct aggregation paths, share their routing decision and may change the transmission path of their data according to their network status. Indeed, aggregation points should be selected carefully so the information collected from these points is efficient and fast.

Aggregation functions. The choice of the aggregation function to use is closely related to the application for which it is dedicated. The aggregation functions are used to compress and merge data, according to one of these approaches: lossy or lossless data. In the former, the original values can not be restored after merging them by an aggregation function. Unlike the later, in which the original data can always be restored. Data aggregation functions can be as simple as statistics such as AVG, MIN, COUNT or simple operations such as removing duplicates. The aim of using simple aggregation functions is to decrease the complexity of aggregation, which leads to a lower load at the aggregation points.

Data storage. Because of its limited storage capacity, a node may not be able to store all the received and generated information in its memory. It must decide which data to store, discard, compress, or transmit. All these operations are necessary to represent the information in a memory efficient manner. Furthermore, the accuracy of the information must be preserved.

3 Classification of the Main Data Aggregation Algorithms Based on Network Architecture

In this section, we provide a comparative synthesis of the most energy efficient in-network aggregation algorithms in wireless sensor networks, since energy consumption is the major concern in these networks. We classify these algorithms according to the network's architecture, given that performance of data aggregation is strongly coupled with the architectural model of the network.

3.1 Algorithms Presentation

A variety of data processing techniques in wireless sensor networks are used to reduce the amount of data to transmit, therefore the energy consumption. To achieve this, many studies have proposed some solutions exploiting various routing structures to facilitate aggregation and data dissemination in the context of wireless sensor networks. We distinguish three major categories of data aggregation algorithms: (1) flat (2) hierarchical (3) location based data aggregation.

3.2 Flat Algorithms

In the flat approach, all sensor nodes have the same power level and play the same role in the network. In such networks, each node broadcasts its data to all its neighbours without taking into account whether they have already received the information or not. Data aggregation is performed at nodes along the multi-hop path.

Sensor Protocols for Information via Negotiation (SPIN) is a routing protocol based on a negotiation model in order to reduce redundant transmissions in the network. This negotiation is used via three types of messages: ADV, REQ, and DATA, to advertise presence of data, request data, and to tag data payloads respectively [7].

Rumor Routing (RR) presents a good alternative to event and query flooding. It uses the concept of agent which is a packet passing through the network node by node to establish relay tables. RR can be a good method for delivering queries to events in large networks as well as it handles node failure [8].

Directed diffusion (DD) is a routing protocol allowing data and interest aggregation. Its main idea is to disseminate data to nodes using a naming scheme for data called interest. DD selects empirically good paths and uses caching and processing data techniques in order to reduce the energy consumption [9].

3.3 Hierarchical Algorithms

The aim of hierarchical topologies is to maintain efficiently the energy consumption of sensor nodes by involving them in a multi-hop communication and performing data aggregation at special nodes. We distinguish three hierarchical data aggregation approaches: (1) tree based (2) cluster based (3) chain based aggregation.

Tree Based Aggregation

Tree based aggregation approach constructs an aggregation tree, where the sink is the tree's root and leaves are sensor nodes. In this technique, data is transferred from leaves towards the root and aggregation is performed at each parent node.

Tiny Aggregation (TAG) is an algorithm that uses the aggregation tree to route data from the sink to the rest of the network and vice versa. Queries in TAG are similar to SQL queries [10]. It uses the transmission slot mechanism in order to save energy since the sensor nodes can be set to idle state until the next transmission slot is programmed.

Energy-Aware Distributed heuristic to generate the Aggregation Tree (EADAT) is a tree based routing protocol which chooses nodes with higher residual energy to be non-leaf tree nodes [11]. Every node chooses the node with the higher residual energy and shorter path to the sink as its parent in order to enhance the network's lifetime.

Localized Power Efficient Data Aggregation (L-PEDAP) is a protocol that combines between Minimum Spanning Tree (MST) and shortest weighted path gathering algorithms to construct the aggregation tree [12]. Once it is done, each intermediate node aggregates the received packets from its children and transmits the result to its parent once in a round.

Cluster Based Aggregation

Cluster based approach consists of a hierarchical organization of the network. However, nodes here are subdivided into regular nodes and special nodes called cluster-heads (CH), who are elected to aggregate data locally and transmit the result to the sink.

Low-Energy Adaptive Clustering Hierarchy (LEACH) is a clustering algorithm where CHs are elected periodically based on their remaining energy. It proceeds in rounds in order to reduce the energy loss caused by a static clustering [13]. Every CH establishes a TDMA schedule for its members, aggregate the received data, then transmit the result directly to the sink.

Hybrid Energy-Efficient Distributed Clustering (HEED) is a distributed clustering algorithm, it selects CHs according to the remaining energy of nodes and the node proximity to its neighbor. CHs are probabilistically selected based on the average minimum reachability power and residual energy [14]. Communication between CHs and the sink is done using multi-hop routing in order to conserve energy.

Multi-hop Routing with Low Energy Adaptive Clustering Hierarchy (MR-LEACH) is an enhanced version of LEACH, it divides the network into different layers of clusters where each node in a given layer can reach the sink in equal number of hops which balances the energy consumption [15]. Here every node chooses the node with the highest residual energy as its CH.

Chain Based Aggregation

In the chain based approach, network nodes are arranged to form a large chain of close neighbours, where one node called leader is selected to transmit data to the sink [16].

Power-Efficient Gathering in Sensor Information Systems (PEGASIS) is an extension of LEACH which forms chains instead of clusters. The chain is constructed either by the sensor nodes or by the sink [17]. One node called leader is selected from this chain to transmit data to the sink. It is elected alternately according to the round-robin policy in order to enhance the network's lifetime.

CHIRON is a chain-based protocol, it divides the network into a number of smaller areas in order to create multiple shorter chains. The chain leader selection is based on the residual energy of nodes within each group. Chain leaders here, relay collaboratively their aggregated data to the sink, in a multi-hop, leader by leader transmission mode [18].

Energy-Efficient Chain-Based Routing Protocol (EECB) is a protocol based on the chain structure [19], it uses the distances between nodes and the sink and the energy level of nodes to decide which node will be qualified to be leader in the chain to take over data transmission to the sink. EECB adopts a distance threshold to avoid formation of long links.

3.4 Location Based Aggregation

In location based protocols sensor nodes are addressed according to their locations. The distance between neighboring nodes can be estimated based on the incoming signal strength, exchanging location information between neighbors or provided by GPS (global positioning system) [16].

Geographic and Energy Aware Routing (GEAR) consists in using geographical information during the queries broadcast to the target regions. GEAR limits the number of interests in Directed Diffusion by considering only a certain region rather than sending the interests to the whole network [20]. GEAR thus complements Directed Diffusion and conserves more energy.

Partial-partition Avoiding Geographic Routing-Mobile (PAGER-M) is a protocol that assigns a cost function to each sensor node using the geographical information of sensors and the sink [21]. It uses the greedy forwarding technique to transmit a packet to the sink. PAGER-M achieves energy efficiency thanks to its low control overhead and low path length.

Minimum Energy Relay Routing (MERR) is a protocol which distributes the energy consumption of sensor nodes uniformly in the network. MERR uses the concept of characteristic distance which refers generally to the distance between two sensor nodes that transmit data [22]. It chooses the nearest neighbor as a router in order to guarantee a reduced dissipation of energy.

Comparative Synthesis of In-Network Aggregation Algorithms

Here, we synthesize the main features of the aforementioned data aggregation algorithms and discuss their strengths and weaknesses. The comparison of their approaches is based on a set of criteria that we have deduced from the literature:

- **Classification**: The algorithm is based on which approach?
- **Mobility**: Does the protocol consider the nodes mobility?
- **Scalability**: Will the network implementing this algorithm continue operating with the same performance despite the addition of other nodes?
- **Network lifetime**: Does the protocol prolong the network lifetime?
- **Resource awareness**: Are the sensor nodes aware about their resource levels such as battery power or available memory?
- **Periodic message type**: What type of messages that nodes exchange periodically?

In Table 1, performances of the main protocols are compared.

4 Discussion

In this section, we propose a comparative synthesis of the aforementioned in-network aggregation algorithms based on a set of criteria. Thereafter, we provide open issues in this area and recommendations for inspiring algorithm developers and researchers.

4.1 General Analysis

In this sub-section, we discuss all the previously presented approaches and explain the strengths and weaknesses of each one.

Table 1 presents a comparative synthesis of various data aggregation algorithms based on network architecture. All of these algorithms focus mainly on improving the network's lifetime. Thus, RR protocol enhances more the network's lifetime compared to DD and SPIN wherein the whole network is flooded with query or event messages. Moreover, DD and SPIN can support limited mobility of nodes. Also RR is more scalable and robust than DD and SPIN due to the use of periodic hello messages to discover alive nodes in the network.

Table 1 Summary of the basic characteristics of the in-network aggregation algorithms

Algorithm	Classification	Advantages	Drawbacks	Mobility support	Scalability	Energy efficiency	Resource Awareness	Periodic message type
SPIN	Flat	Elimination of redundant data transfer	Not good for high-density distribution of nodes	Limited	Limited	Medium	No	ADV messages
RR	Flat	Reliable in delivering queries in large networks, handle the node failure	Send duplicated message to the same node	Low	Good	Good	No	Hello messages
DD	Flat	Energy efficiency achieved by selecting good paths, caching and processing data	Not suitable for application requiring continuous data delivery to the sink	Limited	Limited	Medium	No	Query messages
TAG	Hierarchical (Tree)	Improvement of the network's lifetime	does not support dynamic networks	Low	Low	Good	No	Query messages
EADAT	Hierarchical (Tree)	Network's lifetime increases linearly with the network density	The realistic path may be longer than the minimal path from the source to the sink	Low	Medium	Very good	Yes	Hello messages
L-PEDAP	Hierarchical (Tree)	Suitable for systems where all nodes does not communicate directly each other	High cost of setup and maintenance	Limited	Medium	Good	Yes	Hello messages
LEACH	Hierarchical (Cluster)	Distributed energy consumption	Dynamic clustering adds great overhead	Fixed BS	Good	Low	Yes	None

(continued)

Table 1 (continued)

Algorithm	Classification	Advantages	Drawbacks	Mobility support	Scalability	Energy efficiency	Resource Awareness	Periodic message type
HEED	Hierarchical (Cluster)	Uniform distribution of CHs, multi-hop communication between CHs and the sink	CHs near to the sink may die earlier because they have more workload	Fixed BS	Medium	Medium	Yes	CH advertisement messages
MR-LEACH	Hierarchical (Cluster)	Each node can reach the sink in equal number of hops	More overhead in CH selection and routing in multi-hop can fail	Fixed BS	Medium	Good	Yes	Hello and Head messages
PEGASIS	Hierarchical (Chain)	Reduced Transmission distance for most of nodes	Excessive delay for distant nodes	Fixed BS	Very low	Low	No	None
CHIRON	Hierarchical (Chain)	Chain length and redundant transmission paths are reduced	Imbalanced energy consumption with the increase of the network scale	Low	Medium	Very good	Yes	None
EECB	Hierarchical (Chain)	Distributes energy load evenly among the nodes	Long delay in case of a large network	Low	Medium	Medium	Yes	None
GEAR	Geographical	Balanced energy consumption	Table exchange periodically	Limited	Medium	Medium	Yes	Hello messages
PAGER-M	Geographical	Low routing overhead and low energy consumption are achieved	Does not require a node to memorize the past path	Low	Good	Very good	No	Hello messages
MERR	Geographical	Uniform distribution of energy consumption	Energy wasted in case of close nodes	Limited	Good	Good	No	None

In L-PEADAP, energy expenditure is decreased comparing to EADAT and TAG, by computing the minimum spanning tree. EADAT and L-PEDAP can contribute to load balancing to some extent, due to the consideration of the residual energy. However, TAG cannot realize real energy consumption balancing, because residual energy of nodes is not taken into account.

Energy efficiency is improved in MR-LEACH compared to LEACH and HEED, by role division among different nodes. Also MR-LEACH uses the multi-hop routing from cluster-heads to reach the sink in order to save energy, unlike LEACH where cluster heads reach directly the sink. Moreover, MR-LEACH and LEACH are more scalable than HEED.

CHIRON and EECB are more scalable than PEGASIS which uses only one chain. Furthermore, PEGASIS and EECB protocols suffer from low energy efficiency. However, CHIRON yields significant energy efficiency improvement when comparing to the EECB protocol which is an improvement over PEGASIS, it creates multiple shorter chains to reduce data propagation delay and redundant transmissions.

GEAR achieves energy balancing by taking an alternative path. PAGER-M and MERR are both efficient, because PAGER-M is more scalable than the others protocols of this scheme. It supports also mobility of nodes, while MEER conserves more energy thanks to the non use of periodic messages.

To summarize, in flat networks, data aggregation is performed by different nodes along the multi-hop path. Algorithms implementing this structure are very simple and not robust, because they suffer from the large amount of control packet overhead. Furthermore, the failure of the sink may result in the breakdown of the entire network, which introduces the lack of scalability.

However, in the tree based approach, data aggregation is performed by each parent node. Protocols belonging to this approach have many benefits: Simple topology structure, where data is not flooded over the network, and the energy consumption is much decreased compared with flat based data aggregation algorithms. However these algorithms suffer from large delay caused by too many communication hops, so this kind of topology shows its limitation of scalability.

In the cluster based approach, data aggregation is performed by each cluster-head in the network. Algorithms belonging to this approach are highly scalable, they perform very well in relatively static networks where the clustering structure remains unchanged during a long time, but they can be fragile when they are used in more dynamic environments. Often, the required cost to maintain the hierarchical structure is important.

In the chain based approach, each node in the network passes its aggregated data toward the designated leader, via its downstream neighbor, until the data reaches the sink. However large energy exhaustion is generated, due to long distance communication between the chain leaders to the sink.

Most of the routing protocols require geographical information of sensor nodes in the deployment field, in order to calculate the distance between two particular nodes on the basis of signal strength so the energy consumption can be estimated. But this may cause too much overhead and large energy consumption, especially in large networks. In other words, it exists the problem of scalability.

4.2 Open Research Issues

The main goal behind classifying data aggregation algorithms is to provide researchers with a diversified class of existing techniques which would be helpful to investigate challenges that confront effective solutions of the in-network aggregation problem in wireless sensor networks. We identify the following open research issues:

Cross layer aggregation. The majority of the reviewed papers focus on trying to merge routing with data aggregation using only simple aggregate functions but ignoring MAC, data representation or application issues. Since in-network aggregation concerns several layers of the protocol stack, further researches would be needed to address cross-layer in order to improve transmission performance, such as energy efficiency, data rate, QoS [23].

Security. Data aggregation obviously raises security issues in wireless sensor networks. These security issues are the same as in all information systems namely: data privacy, data integrity, authentication, etc. Further researches would be needed to design a secure system for data gathering and management in wireless sensor networks.

Aggregation points. The choice of efficient processing points in the network still an open research issue, because this decision must take into account the memory and the computational necessary resources of nodes to support data aggregation.

Quality of service (QoS). Data aggregation approaches in the aspect of QoS are also worth exploring, in order to offer high level of QoS to the end users, particularly in terms of packet loss rate, latency, throughput.

Data aggregation functions. Further researches must provide a deeper analysis of data aggregation functions, because the existing solutions rely on simple aggregation functions, which highly affects data precision.

Mobility. Much work still remains to be done to design new protocols supporting mobility of nodes in the network. In other words, it is the ability of maintaining the network functional under the topology changes.

Hybrid aggregation. Very few studies address the combination of the properties of different approaches, in order to design hybrid algorithms that benefit from the advantages of both structures for optimal performance (e.g., implementing cluster-based topology with tree-based topology in one network).

5 Conclusion

In this paper, we presented a comprehensive study of various energy efficient in-network aggregation protocols in wireless sensor networks. We have thoroughly reviewed 15 energy efficient data algorithms after surveying over 900 papers. We classified them according to the network structure into three approaches: flat networks, hierarchical networks (tree based, cluster based and chain based) and geographical networks. We have described main features, advantages and drawbacks of each protocol.

Our study revealed that flat algorithms are more suitable for small networks where nodes are stationary. Their performance decreases in case of the deployment of a large network. The hierarchical protocols are proposed to overcome this problem. In this case (i.e., cluster based), the network is divided into groups and data aggregation is allowed at the cluster heads in order to decrease the number of transmitted messages to the sink. However, the hierarchical approach may be fragile in case of more dynamic environments. Often, the required cost in maintaining the hierarchical structure is substantial. On the other hand, location based protocols can be applicable for high dynamic networks as they use the geographical information to calculate distance between nodes in order to extend the network's lifetime. Throughout our study, we pointed out some important open issues of in-network aggregation. We proposed 7 research directions and recommendations to aspire researchers and algorithm developers such as: cross-layer aggregation, data aggregation functions, aggregation points.

We hope that this survey may help researchers and algorithm designers to select appropriate logical topologies and data aggregation protocols for their specific applications. Moreover, we wish that the open research issues identified made it clearer for future data aggregation algorithm development.

References

1. Akyildiz, I.F., Can Vuran, M.: Wireless sensor networks. J. Wiley Sons (2010)
2. Ricardo, G., Loureiro, A.A.F., Mini, R.A.F.: QoS: requirements, design features, and challenges on wireless sensor networks. In: Handbook of Research on Developments and Trends in Wireless Sensor Networks (2010)
3. Akyildiz, I.F., et al.: A survey on sensor networks, IEEE Commun. Mag. **40**(8) (2002)
4. Pottie, G.J., Kaiser, W.J.: Wireless integrated network sensors. Commun. ACM **43**, (2000)
5. Krishnamachari, L., Estrin, D., Wicker, S.: The impact of data aggregation in wireless sensor networks. In: 22nd International Conference on Distributed Computing Systems Workshops (2002)
6. Patil, N.S., Patil, P.R.: Data aggregation in wireless sensor network. In: IEEE International Conference on Computational Intelligence and Computing Research (2010)
7. Kulik, J., Heinzelman, W., Balakrishnan, H.: Negotiation-based protocols for disseminating information in wireless sensor networks. Wirel. Netw. (2002)

8. Braginsky, D., Estrin, D.: Rumor routing algorithm for sensor networks. In: Proceeidngs 1st ACM International Workshop on Wireless Sensor Networks and Applications, USA, Atlanta (2002)
9. Intanagonwiwat, C., Govindan, R., Estrin, D., Heidemann, J.: Directed diffusion for wireless sensor networking. IEEE/ACM Trans. Netw. (2003)
10. Madden, S., Franklin, M.J., Hellerstein, J.M.: Tag : a tiny aggregation service for ad-hoc sensor networks. In: Appearing in 5th Annual Symposium on Operating Systems Design and Implementation (2002)
11. Cheng, X., Ding, M., Xue, G.: Aggregation Tree Construction in Sensor Networks. In: IEEE Vehicular Technology Conference, vol. 4, pp. 2168–2172 (2003)
12. Tan, H.O., Korpeoglu, I., Stojmenovi, I.: Computing localized power-efficient data aggregation trees for sensor networks. In: IEEE Trans. Parallel Distrib. Syst. (2011)
13. Rabiner Heinzelman, W., Chandrakasan, A., Balakrishnan, H.: Energy-efficient communication protocol for wireless microsensor networks. In: IEEE the Proceedings of the Hawaii International Conference on System Sciences (2000)
14. Younis, O., Fahmy, S.: HEED: a hybrid, energy-efficient, distributed clustering approach for adhoc sensor networks. IEEE Trans. Mob. Comput. (2004)
15. Farooq, M.O., Dogar, A.B., Shah, G.A.: MR-LEACH: multi-hop routing with low energy adaptive clustering hierarchy. In: 4th International Conference on Sensor Technologies and Applications (2010)
16. Akkaya, K., Younis, M.: A survey of routing protocols in wireless sensor networks. Elsevier Ad Hoc Netw. J. **3**(3), 325–349 (2005)
17. Lindsey, S., Raghavendra, S.: PEGASIS: Power-efficient gathering in sensor information systems. In: IEEE Aerospace Conference Proceedings (2002)
18. Chen, K., Huang, J., Hsiao, C.: CHIRON: an energy-efficient chain-based hierarchical routing protocol in wireless sensor networks. In: Wireless Telecommunications Symposium (2009)
19. Yu, Y., Song, Y.: An energy-efficient chain-based routing protocol in wireless sensor network. In: International Conference on Computer Application (2010)
20. Yu, Y., Estrin, D., Govindan, R.: Geographical and energy-aware routing: a recursive data dissemination protocol for wireless sensor networks. UCLA Computer Science Department Technical Report, UCLA-CSD TR-01-0023, May 2001
21. Zou, L., Lu, M., Xiong, Z.: PAGER-M: a novel location-based routing protocol for mobile sensor networks. In: International Conference on Management of Data (2007)
22. Zimmerling, M., Dargie, W., Reason, J.M.: Energy-efficient routing in linear wireless sensor networks. In: Proceedings 4th IEEE International Conference on Mobile Adhoc and Sensor Systems (2007)
23. Verd, J., et al.: The Impact of Traffic Aggregation on the Memory Performance of Networking Applications. ACM MEDEA, Antibes Juan-les-Pins, France (2004)

Towards an Autonomic Approach for Software Defined Networks: An Overview

Soukaina Bouzghiba, Hamza Dahmouni, Anouar Rachdi and Jean-Marie Garcia

Abstract Under the new paradigm Software Defined Networking (SDN), which involves decoupling control plane from data plane, and allowing control planes to be deployed on external servers, our main goal is to propose an overview of architecture that can effectively solve problems of network QoS caused by this separation. The overall objective is to study and evaluate the use of SDN networks as a cornerstone of a communication system that can effectively support distributed applications whose needs change over time. In this paper, we focus, in particular, on the controller placement problem in SDN, optimizing the latency, resilience, reliability, scalability and other network performance. The technical solutions to these problems will be studied to identify the components of SDN that can be improved.

Keywords SDN · Network slicing · 5G · NFV · Controller placement · QoS

1 Introduction

Recently, the unprecedented growth of cloud services and the explosion of mobile devices and content have a great impact on computer networks. It is also noted that today's applications have dynamic nature where traffic patterns have changed significantly and they will be continued in the future, especially with the emergence of new usage such as Big data and the Internet of Things. Such changes are among the trends

S. Bouzghiba (✉) · H. Dahmouni
INPT, 2, avenue Allal Al Fassi, Madinat Al Irfane, Rabat, Morocco
e-mail: bouzghiba@inpt.ac.ma

H. Dahmouni
e-mail: dahmouni@inpt.ac.ma

A. Rachdi
QoS Design, 6, avenue Marcel Doret, 31500 Toulouse, France
e-mail: rachdi@qosdesign.com

J.-M. Garcia
CNRS, LAAS, 7 avenue du colonel Roche, 31400 Toulouse, France
e-mail: jmg@laas.fr

R. El-Azouzi et al. (eds.), *Advances in Ubiquitous Networking 2*, Lecture Notes in Electrical Engineering 397, DOI 10.1007/978-981-10-1627-1_12

leading network operators review their infrastructures to make them more adaptable. A solution to this is to make current networks more flexible and programmable by adopting the software defined networking concept.

Indeed, Software Defined Networks (SDN) are based on a new paradigm, in contrast to traditional networks, which advocates the separation of control plane and data plane within network equipment. Thus, the control functions are centralized in a controller and the remote control of these functions are managed through Application Programming Interfaces (APIs) [1]. The control plane communicates with the forwarding layer to collect the information (via southbound APIs) and maintains the network topology; and it interacts with the applications and business logic (via northbound APIs) to implement various network functions. This allows building flexible networks that can adapt their operation on demand applications.

The control plane between service applications and network devices makes for more flexible and agile network service environments. The service providers can easily launch new applications, dynamically obtain network information from the control plane, and quickly control traffic flows for their needs in order to provide end-to-end QoS. Obviously, SDN is a key concept to bridge the gap between dynamic network resource management, on one hand, and the demand for connectivity and Quality of Service (QoS) type cloud applications, on the other hand.

Despite the technical benefits of SDN, especially for traffic engineering [2], cost-effective [3], and the support of distributed applications in a dynamic behavior [4], the separation of control plane and data plane can generate some performance issues related to network reliability and security, it could also generate new problems, such as the controller placement problem. This problem, initiated firstly by [5], focuses on seeking the best controller location to satisfy the optimal design for a given SDN topology. The authors of [5] show that when there is a small network, a single controller is enough. While for large scale networks, multiple domains are created and multiple controllers are necessary, thus how to deploy multiple controllers become among the fundamental research issues in SDN [6]. For the controller placement problem, it can be regarded as the task in network planing step which relies on the physical network topology to get the optimal number of controllers to be deployed and their best location in the network.

In fact, with the new SDN concepts, network performance of fixed and mobile networks can be affected [7]. So, the switch should be continuously controlled and assigned to controllers according to shortest path, controllers should be interconnected through an overlay network [8], and finally, controller failures or disconnections between control and data plane that may result packet loss or other network problems should be fixed [9, 10]. Thus, the formulation of controller placement problem to design and planning fixed and mobile networks must take into account these functional concepts of SDN networks.

The overall objective of our work is to study and evaluate the use of SDN networks as a cornerstone of a communication system that can effectively support distributed applications whose needs change over time. The idea is to deploy a virtual network dedicated exclusively to the application whose behavior can be reprogrammed dynamically based on application requirements. This network function, responsible for the deployment of the overlay virtual network, will be autonomous in the case of dynamic environments. In this paper, we focus on the controller placement problem, because, on the one hand, we believe that a formulation of the problem, taking into account all aspects of the SDN networks is the key to achieve our goal and to solving many other related problems in fixed and mobile networks, on the other hand, in the best of our knowledge this work can be considered as the first overview on this problem.

The rest of the paper is organized as follows. Next section give an overview on the problem statement with a description of controller placement problem. In Sect. 3, we present the different metrics used and discuss the different proposed solutions and algorithms. The last section draws a conclusion and some perspectives.

2 Problem Statement

2.1 SDN Concepts

According to the Open Networking Foundation (ONF),[1] Software Defined Networking is an emerging architecture decouples the network control and forwarding functions enabling the network control to become directly programmable and the underlying infrastructure to be abstracted for applications and network services. The control plane is becoming centralized in a controller node, and the "switch" is considered as a simple forwarding element executing the controller's rules on each traffic flow. The OpenFlow protocol [11] is considered as a foundational element for building SDN solutions.

Figure 1 shows the adapting changes of the traditional network Fig. 1a towards an SDN architecture Fig. 1b. In the current architecture, each network element "core" performs both forwarding and control operations. So, networking equipment manufacturer are offering their equipments as "closed box" with their own specific hardware and operating system, and they are the only ones who have access to the box.

Figure 1b shows the changes that should be occurred to the legacy network Fig. 1a. In large scale networks such Arpanet, multiple domains with probably different transmission technologies are created and multiple controllers should be placed [1]. Furthermore, each controller manages a significant number of switches in its domain, and have enough information for rules to be run on each switch. Otherwise, the controllers can be physically connected using various topologies such as a tree, ring, full mesh, etc. and they can have large capacities, as an example, the authors of [12] implement an SDN control serving up to 5000 switches and can achieve 14 million flows per second.

Another important concept of SDN is the network slicing, where a slice can be defined as a set of configured network functions, network applications, and under-

---

[1]ONF is a user-driven organization dedicated to open standards and SDN adoption. It was launched in 2011 by Deutsche Telekom, Facebook, Google, Microsoft, Verizon, and Yahoo!.

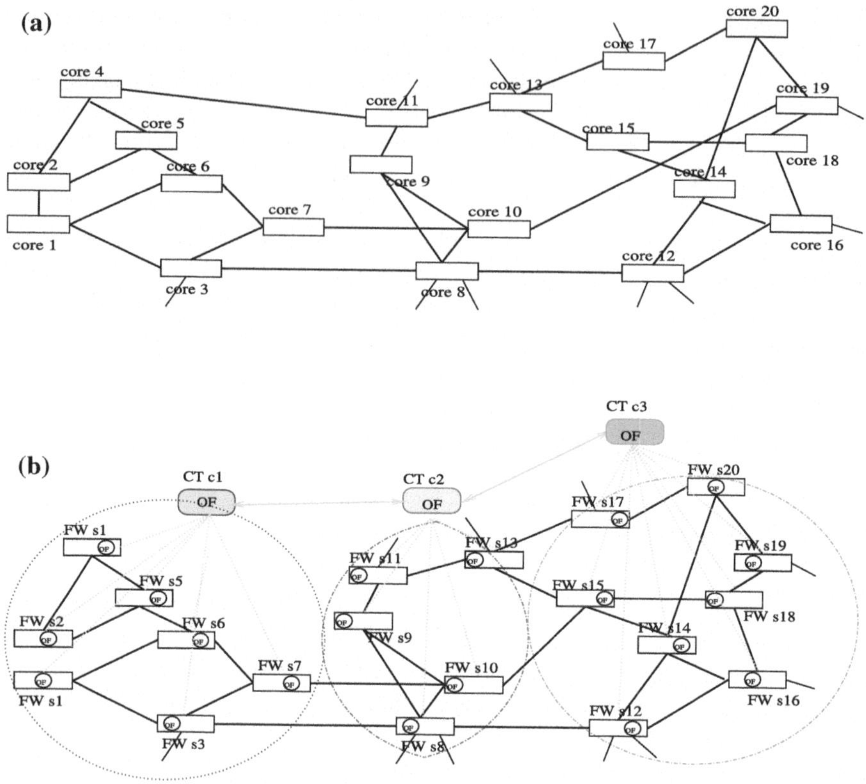

Fig. 1 Arpanet topology with and without SDN. **a** Traditional Network. **b** SDN Network

lying physical infrastructures to meet the requirement of a specific use case [13]. A detailed survey on SDN functionalities and concepts is given in [1].

2.2 SDN and Mobile Networks

The introduction of the SDN concept in mobile networks, especially in LTE and 5G, has solicited these last years the interest of the research community on networking. In fact, the current mobile network has several issues regarding its inherent design, furthermore, its implementation is very expensive and difficult to be modified or upgraded. In that regard, SDN is considered as a major trend in the evolution of mobile networks. In fact, applying SDN principle as shown in Fig. 2 leads to simplify core network nodes into pure forwarding elements and exporting the control plane functions to a centralized SDN controller node [14]. SDN principles can thus provide flexibility, openness, and programmability necessary to the mobile networks [15].

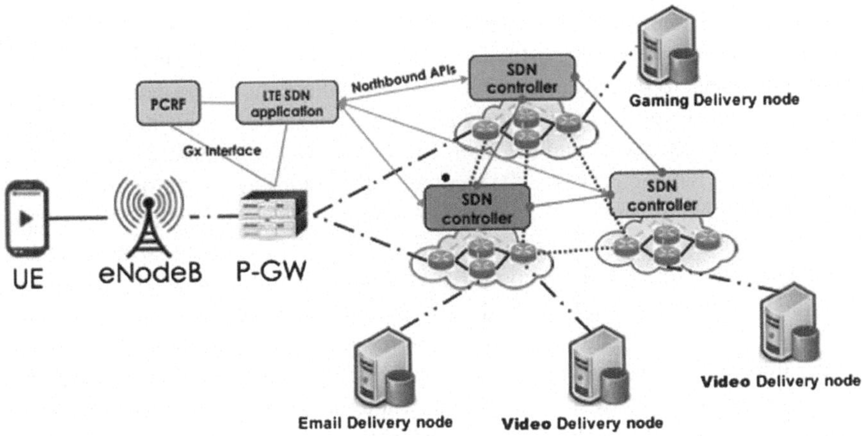

Fig. 2 LTE topology with SDN [14]

Otherwise, several works have shown that SDN can be simply applied to mobile networks, and elaborate the challenges of providing scalability and QoS-based dynamic flow control in these networks [7, 14]. The authors of [16] present an SDN-based architecture that implements MME and S-GW control functions in the controller. A detailed overview of the latest research works on SDN and virtualization in LTE mobile networks is given in [17].

For the 5G technology, there is up to date no formal specification or description of what this system will be, however, it is clear that 5G leads to provide a converged infrastructure including all new aspects of the network. This ecosystem will have to serve a variety of devices with different characteristics and needs such as Mobile Broadband, Massive IoT, and Mission-critical IoT [18]. Thus the exploitation of the network abstraction concept of SDN and virtualization (NFV), can provide the necessary tools as "network slicing" to delivery enhanced end-user quality of experience [19, 20].

2.3 *The Controller Placement Problem*

The SDN controller placement problem, for a given network topology, was initiated by Heller et al. [5]. Accordingly, several works have emerged on the topic in several contexts and use-cases. In this section, we present the controller placement problem, according to our point of view, for fixed and mobile networks using the topologies of Figs. 1 and 2 as examples. The problem will be thus addressed around three fundamental questions:

The first logical question to ask is, for a given network topology, "*How many SDN controllers are needed ?*". The answer of this essential question relies on the network characteristics among other user-composed requirements. Heller et al. [5] shows that a single controller would be mostly adequate, while for large scale networks, multiple domains have to be created and multiple controllers should be deployed. Obviously, the main constraints to be considered are the controller capacity, the number of domains in the network, and the inter- SDN controller communication approach.

Considering flat control plane [21], in which the controllers partition the network into disjoint domains, in Figs. 1 and 2, three domains are considered, and thus three OpenFlow controllers $\{c_1, c_2, c_2\}$ to oversee all OpenFlow-enabled switches should be deployed. Indeed, each SDN controller is able to control a part of the whole network, and computes a single logical view upon the network. So, interconnecting these controllers to share information and coordinate their decisions is important for routing information and providing end-to-end quality of service. Furthermore, network operators have interest to divide the whole network into multiple connected SDN domains for better scalability and reliability.

The second question to be asked is, "*Where SDN controllers should be placed?*" The answer of this question is fundamental because the location of the controller could affect the various aspects of network performance. Indeed, even a single controller is hard to place in a domain because we have to know, exactly or approximately, its optimal placement. Furthermore, the problem becomes even more complicated when we consider multiple controllers to place. While, controllers location has, directly, influences on communications between switches and a controller, the main performance functions that should be considered are latency, load-balancing, redundancy, connectivity, and survivability. These metrics will be defined and detailed in the next section.

Figure 1b illustrates a simple example of a network having 20 nodes and three controllers $\{c_1, c_2, c_3\}$ to be placed in any location among the 20 nodes $\{s_1, ...s_{20}\}$. So, firstly, the network should be partitioned on domains, according to network partition algorithm, and then each control should be placed within its domain (e.g., $c_1 \in \{s_1, ..., s_7\}$) according to the different performance constraints mentioned above. The physical topology connecting the controllers c_1, c_2 and c_3 should be considered. The authors in [22] provide the formulations for several topologies that can be used such as a tree, ring, and full mesh.

Regarding the mobile network, the controllers placement becomes even more complicated. In fact, the explosion of mobile devices and content have a great impact on network performance. It is also noted that today's mobile applications have dynamic nature where traffic patterns have changed significantly and fluctuate over time. This burstiness of the traffic can make the controller easily overloaded, hence, the placement of controllers should regard a dynamic readaptation of the number and the location of controllers [4]. Nevertheless, a synchronous vents such as controller failures or networks disconnections between controllers or switches may also lead to packet loss and performance degradation.

The last question is, *Which metrics should be considered?*. In SDN network, the switch should be continuously controlled and assigned to a controller according to the shortest path, in fact any flow unmatched by the switch should be forwarded to the controller to establish the applied rules. Thus, the most important performance metrics associated with an SDN controller that should be firstly considered is the flow setup time, called also switch-controller latency. Note that this metric can be strongly affected by the switch-controller distance, the link bandwidth, and the controller load.

Another issue to be considered is the network resilience, according to [9] the resilience can be expressed by four metrics: Controller failures, Network Disruption, Controller overload and Inter-Controller Latency. So, in case of controller failure, it should be possible to reassign all the switches of its its domain to neighboring controllers. Network Disruption is occurred by the failure of network links or nodes, thus some reassignment of nodes to other controllers is needed, and it would be preferable that a domain topology ensures redundancy to recover the failure within the domain. Controller overload consists to avoid that the controller might have too many switches to oversee, otherwise its average response time will increase, so it should distribute the load among controllers to avoid the controller becoming congested. Finally, inter-controller communications are necessary to synchronize their data bases and coordinate their decisions.

3 CPP Metrics and Algorithms

This section presents the different metrics and algorithms used in the literature, to solve the controller placement problem (CPP).

3.1 Placement Metrics

In wide area networks the latency is an important parameter that can affect controllers placement. Hence, it's considered as the commonly used metric for the CPP formulation. Recent studies have shown that an optimal placement using only delay is not sufficient. However, the CPP should also respect resilience constraints such as controller failures, network disruption, inter-controller latency and controller overload (load imbalance). Furthermore, in wide area networks as the size of the network rises a centralized architecture can not meet the need concerning scalability then distributed architectures with multiple domains and multiple controllers are used. Finally, a reliable control network improves the network availability but the problem is that the design of SDNs introduces some issues concerning the network reliability. According to [23] the reliability can be measured by the number of broken control paths due to network failures.

Define,

π_{prop}: the switch-to-controller propagation delay (can be considered as average or maximum) defined and used by [5].

π_{load}: represents the average load that can be handled by the controller, given in [24].
π_{fail}: is de controller failures metric which considers the distance to the backup controllers. Its expression is given in [25].
π_{imb}: is defined as the difference between the maximum and minimum number of nodes assigned to a controller. Its formulation is given in [25].
π_{disp}: is the network disruption metric related to nodes or links failures. The mathematical formulation is given in [25].
π_{con}: is the inter-controller latency calculated as the max latency between two controllers. This equation is given in [25].
π_{single}: single domain case.
π_{multi}: multiple domains case.

Table 1 summarizes the typical metrics used for the controller placement problem. Note that given the problem complexity, the set of metrics below is not representing an exhaustive view. Obviously an optimization algorithm could be applied to optimize one or several metrics.

Having defined metrics, in next subsections, we present the different solutions and algorithms proposed in different research works. To resolve the controller placement problem each authors have examined the problem taking in consideration different issues such as latency, resilience, etc.

3.2 Latency and Load Aware Algorithms

As mentioned above, the controller placement problem for SDN networks was first introduced in [5], in this paper, the authors examine the influence of controllers placement on the switch-to-controller propagation latency. They consider k-center (resp. k-median) algorithm to place controllers minimizing the average latency (resp. maximum latency) in the Internet2 OS3E network topology. They claim that the number of controllers depends on the network topology but in most cases one controller is enough. Their analysis focused only on latency without taking into consideration the controller load. However, the authors of [26, 28] present algorithms taking into account the load and capacity of controllers that minimizing the maximum latency. In [26] the placement problem is considered as a capacitated k-center problem, while in [28] the authors propose Network Clustering Particle Swarm Optimization Algorithm (NCPSO) to optimize the controller load under latency and load-balancing constraints.

The authors in [30] gave a solution from game theory to calculate the optimal number of controllers. The switches can dynamically be assigned to a controller to reduce the average latency between nodes and controllers and balancing the controllers' load. They have proved that the optimal number of controllers is random with Gaussian distribution and variable load condition. This approach ensures only the maximum utilization of controllers, without providing their locations. Whereas,

Table 1 Typical metrics used for controller placement problem

Metrics										Management	
	Latency		Resilience				Scalability		Reliability	Static	Dynamic
Papers	π_{prop}	π_{load}	π_{fail}	π_{imb}	π_{con}	π_{disp}	π_{single}	π_{multi}			
[5]	x							x		x	
[3, 26] [27, 28]	x	x					x			x	x
[28] [26]	x	x		x			x	x		x	
[9, 25] [29]	x		x	x	x	x		x		x	
[24] [23]	x	x					x	x	x	x	
[6, 30] [4, 31]	x	x		x			x	x			x
[32, 33] [34]	x	x				x		x	x	x	x

[27] introduces a dynamic controller placement that consists of determining the locations of controller modules to optimize latencies, and of determining the number of controllers per module to support the load. This approach helps to reduce the number of active controllers and enhances their utilizations.

Dixit et al. [31] on their side examine the problem of mapping between a switch and controller which is statically configured in distributed controllers architecture. To improve scalability they propose elastic distributed controller architecture, ElastiCon, in which the controller pool is grown according to traffic and load conditions and allows the node to migrate from one controller to others dynamically with minimal impact on response time. That is to say, allows balancing the load of controllers in real time.

Otherwise, according to [3] the problem can be solved using a mathematical model that aims to minimize the cost of the network planing considering the capacity of controllers, the path setup latency, the link cost and network traffic patterns. The cost function includes the cost of installing controllers, the cost of linking the controllers to the switches and the cost for linking the controllers together.

3.3 Resilience and Reliability Aware Solutions

No one can deny that minimizing latency is an important issue, but when nodes lose connection from controllers due to network failures the network lose its resilience. Therefore there are other metrics that require consideration in addition to propagation delay. Hu et al. [23, 24] propose placement algorithms to maximize the reliability of control network. These algorithms aim to automate placement decisions while minimizing the expected percentage of control path loss as metric. For solving the Reliable controller placement problem (RCP), which is considered as k-median problem. They have evaluated their work using real topologies. It shows that the reliability increases with the high number of controllers used and the placement quality depends on the algorithm used, moreover greedy algorithms provides optimal results.

Moreover, reliability enhances when number of controllers to place is higher. In contrast, the tradeoffs between reliability and latency performance depend on the number of controllers, it is necessary to choose between optimizing for reliability or latencies but the corresponding latencies when optimizing for reliability is sufficient to achieve the goal.

The authors in [34] proposes a controller placement strategy, Survivor, which aims to avoid controller overload by adding capacity-awareness in the controller placement, it consider also path diversity to enhance connectivity between nodes and controllers to reduce the chance of connectivity loss and developing failover mechanisms for composing controllers list of backup.

So the survivable controller placement problem is finding the controller placement while optimizing the survivability of the network: connectivity, capacity, and recovery. The optimization problem consists of: firstly, the definition of controllers locations such that connectivity is maximized and capacity constraints is satisfied. Secondly, the definition of list of backup controllers for each node using a given heuristic. To maximize connectivity, the placement that yields the highest number of node-disjoint paths between the nodes and controller is selected. For capacity aspect, each controller has a percentage of capacity reserved as backup to resist for overload. The algorithm used is implemented using Integer Linear Programming (ILP).

The resilience issue is discussed in [9, 25] using controller failures, network disruption, controller overload and inter-controller latency as metrics. They present their framework, Pareto-based Optimal COntroller placement (POCO), that provides optimal placements. The framework considers load balancing between controllers and inter-controller latency. They propose a heuristic approach to find out the optimal number and locations of controllers for the large scale SDN. Their analysis shows that in most topologies where a single controller would be enough from a latency point of view as proved in [5] many more controllers are necessary for network resilience.

The authors show also that more than 20 % of all nodes need to be controllers to face any failure scenario assuring connectivity. Furthermore, the optimization placement taking account both inter-controller and node-to-controller latencies metrics cannot be achieved. When higher importance is given to node-to-controller latency, the controllers are more distributed in the network. Otherwise, when inter-controller-latency is given higher priority, the controllers are placed closer together. In the same direction, [33] focuses on fault tolerant controller placement, which means each node must be continuously controlled. To this end, a heuristic algorithm is developed which take into account the fact that at least an operational path between node and controller should exist.

4 Conclusion

In this paper, we have presented an overview on one of the main problems of SDN deployment in fixed and mobile networks. While SDN comes with many advantages, this new paradigm have some issues affecting network performance that should be fixed. The controller placement problem (CPP), defining how many controllers are needed and where should be placed in the network, is widely introduced and discussed in this paper. We have then presented different solutions and algorithms proposed in some researches where each work examined the CPP using one or more metrics. This study allows as to evaluate the use of SDN networks as a cornerstone of a communication system that can effectively support distributed applications whose needs change over time. Our perspective is to deploy a virtual network dedicated exclusively to the application whose behavior can be reprogrammed dynamically based on application requirements. This network function, responsible for the deployment of the overlay virtual network, will be autonomous in the case of dynamic environments.

References

1. Nunes, B., Mendonca, M., Nguyen, X.N., Obraczka, K., Turletti, T.: A survey of software-defined networking: past, present, and future of programmable networks. IEEE Commun. Surv. Tutor. **16**(3), 1617–1634 (2014)
2. Jain, S., Kumar, A., Mandal, S., Ong, J., Poutievski, L., Singh, A., Vahdat, A.: B4: Experience with a globally-deployed software defined WAN. Proc. ACM SIGCOMM **12**, 3–14 (2013)
3. Sallahi, A., St-Hilaire, M.: Optimal model for the controller placement problem in software defined networks. In IEEE Commun. Lett. **19**(1), 3033 (2015)
4. Bari, M.F., Roy, A.R., Chowdhury, S.R., Zhang, Q., Zhani, M.F., Ahmed, R., Boutaba, R.: Dynamic controller provisioning in software defined networks. In: International Conference on Network and Service Management, p. 188, May 2013
5. Heller, B., Sherwood, R., McKeown, N.: The controller placement problem. In: Proceedings of HotSDN'12, p. 712 (2012)
6. Yao, L., Hong, P., Zhang, W., Li, J., Ni, D.: Controller placement and flow based dynamic management problem towards SDN. In: IEEE ICC (2015)
7. Li, L., Mao, Z., Rexford, J.: Toward software-defined cellular networks. In: European Workshop on Software Defined Networking (EWSDN), p. 712 (2012)
8. Yan-nan, H., Wen-dong, W., Xiang-yang, G., Xi-rong, Q., Shi-duan, C.: On the placement of controllers in software-defined networks. ELSEVIER Sci. Direct **19**, 9297 (2012)
9. Lange, Stanislav, Gebert, Steffen, Zinner, Thomas, Tran-Gia, Phuoc, Hock, David, Jarschel, Michael, Hoffmann, Marco: Heuristic approaches to the controller placement problem in large scale SDN networks. IEEE Trans. Netw. Serv. Manage. **12**(1), 4–17 (2015)
10. Yeganeh, S.H., Tootoonchian, A., Ganjali, Y.: On scalability of software-defined networking. In: IEEE Communication. Magazine, pp. 16–141, Feb 2013
11. McKeown, N., Anderson, T., et al.: OpenFlow: enabling innovation in campus networks. In: ACM SIGCOMM Computer Communication Review, vol. 38, no. 2, Mar 2008
12. Voellmy, A., Wang, J.: Scalable software defined network controllers. In: Proceedings of the ACM SIGCOMM, pp. 289–290 (2012)
13. Gutz, S., Stor, A., Schlesinger, C., Foster, N.: Splendid isolation: a slice abstraction for software-defined networks. In: HotSDN12, Helsinki, 13 Aug 2012
14. Nam, H., Calin, D., Schulzrinne, H.: Intelligent content delivery over wireless via SDN. In: IEEE WCNC (2015)
15. Pentikousis, K., Wang, Y., Hu, W.: MobileFlow: toward software-defined mobile networks. IEEE Commun. Mag. **51**(7), 44–53 (2013)
16. Ben Hadj Said, S., Sama, M., Guillouard, K., Suciu, L., Simon, G., Lagrange, X., Bonnin, J.-M.: New control plane in 3GPP LTE/EPC architecture for on-demand connectivity service. In: IEEE Cloud Networking (CloudNet). USA, Nov 2013
17. Nguyen, V., Do, T., Kim, Y.: SDN and virtualization-based LTE mobile network architectures: a comprehensive survey. Wirel. Pers. Commun. **86**, 1401–1438 (2016)
18. NGMN 5G White Paper. https://www.ngmn.org/uploads/media/NGMN_5G_White_Paper_V1_0.pdf (2005). Accessed Feb 2015
19. Nikaein, N., et al.: Network store: exploring slicing in future 5G networks. In: ACM MobiArch'15, pp. 8–13. USA (2015)
20. Szabo, N.D., Nemeth, F., Sonkoly, B., Gulyas, A., Fitzek, F.H.P.: Towards the 5G revolution: a software defined network architecture exploiting network coding as a service. In: ACM Conference on Special Interest Group on Data Communication, Aug 2015
21. Schmid, S., Suomela, J.: Exploiting locality in distributed SDN control. In: HotSDN, vol. 13, Aug 2013
22. S. Chamberland, M. St-Hilaire, and S. Pierre, An analysis of different co-located router network topologies within a POP in IP networks. In: IEEE CCECE 2003, vol. 2, pp. 733–736 (2003)
23. Hu, Y.N., Wang, W.D., Gong, X.Y., Que, X.R., Cheng, S.D.: Reliability-aware controller placement for software-defined networks. In: Proceedings IFIP/IEEE International Symposium IM, p. 675 (2013)

24. Hu, Y.N., Wang, W.D., Gong, X.Y., Que, X.R., Cheng, S.D.: On the placement of controllers in software-defined networks. J. China Univ. Posts Telecommun. **19**, 9297 (2012)
25. Hock, D., et al:, Pareto-optimal resilient controller placement in SDN-based core networks. In: Proceedings of 25th ITC, p. 19 (2013)
26. Yao, G., Bi, J., Li, Y., Guo, L.: On the capacitated controller placement problem in software defined networks. IEEE Commun. Lett. **18**(8), 1339–1342 (2014)
27. Huque, M., Jourjon, G., Gramoli, V.: Revisiting the controller placement problem. In: 40th Annual IEEE Conference on Local Computer Networks (LCN), pp. 450–453. Florida, USA (2015)
28. Liu, S., Wang, H., Yi, S., Zhu, F.: NCPSO: a solution of the controller placement problem in software defined networks. In: 15th International Conference, ICA3PP, pp. 213–225
29. Guo, M. Bhattacharya, P.: Controller placement for improving resilience of software-defined networks. In: Proceedings of ICNDC2013, pp. 23–27. Los Angeles (2013)
30. Rath, H.K., Revoori, V., et al.: Optimal controller placement in Software Defined Networks (SDN) using a non-zero-sum game. In: Proceedings of IEEE WoWMoM, p. 16. Sydney,NSW (2014)
31. Dixit, A., Hao, F., Mukherjee, S., Lakshman, T.V., Kompella, R.: Towards an elastic distributed SDN controller. HotSDN **13**, 7–12 (2013)
32. Aoki, H., Shinomiya, N.: Controller placement problem to enhance performance in multi-domain SDN networks. In: The Fifteenth International Conference on Networks (ICN) (2016)
33. Ros, F.J., Ruiz, P.M.: Five-nines of southbound reliability in software-defined networks. In: Proceedings of 3rd Workshop Hot Topics Softw. Defined Net., p. 3136 (2014)
34. Muller, L., Oliveira, R., Luizelli, M., Gaspary, L., Barcellos, M.: Survivor: an enhanced controller placement strategy for improving SDN survivability. In: IEEE Global Communication Conference (GLOBECOM), Dec 2014

Privacy Preservation in the Internet of Things

Fatima Zahra Berrehili and Abdelhamid Belmekki

Abstract The Internet of Things (IoT) is the future of Internet where users, machines and everyday Things have the ability to sense, communicate and interact with their environment. IoT promises a wide variety of applications to make the human's life more comfortable, safe and improve quality of life. It's also considered as big business opportunity for enterprises based on huge quantity of data gathered by connected Things. In the IoT applications such as smart home, smart city, heath and so on, Things cohabit with humans and deal with their personal data even the most private ones. When this data is collected massively and exposed to Internet without an explicit person's agreement, Things constitute by this way a threat for privacy, which is a universal human right. Thus privacy preservation is one of the most prominent issues in IoT. This paper analyzes the privacy in the context of IoT based on case study, and proposes mechanisms to improve security and preserve privacy. This analysis considers also the economic advantages related to the use of IoT as new business opportunity without personal private data disclosure. The proposed solutions are based on data anonymity technologies with recommendation for users and for developers of IoT application.

Keywords Internet of Things · Privacy · Data anonymization · IoT security

1 Introduction

Internet of Things (IoT) as described in an IEEE special report [1] as being "A network of items each embedded with sensor which are connected to the Internet". It has a wide range of applications in industry and life, such as vehicle with sensors

F.Z. Berrehili (✉) · A. Belmekki (✉)
STRS Lab, National Institute of Posts and Telecommunications,
2, Av.Allal El Fassi madinat al Irfane, Rabat, Morocco
e-mail: berrehili@inpt.ac.ma

A. Belmekki
e-mail: belmekki@inpt.ac.ma

R. El-Azouzi et al. (eds.), *Advances in Ubiquitous Networking 2*, Lecture Notes in Electrical Engineering 397, DOI 10.1007/978-981-10-1627-1_13

[2], biochip on people and animal's body, heart monitoring, devices that assist disabled etc. Deploying IoT suffer from different major challenges such as power management [3], Things identification [4], standardization [5] and security [6]. These Things contain according to Gartner [7] embedded technology to communicate and sense or/and interact with their internal states or with their external environment.

In the IoT literature, different terms are used to designate the « Thing » [8], which is component that has communication capability over Internet, exchange data and can interact with its internal and external environment. This paper will use such term Things or smart device indifferently. But in literature the same component can be designated by other terms such as Objects [9], Smart Things [10], Smart Objects [11], Smart Devices [12] and in some case Blogjects [13] (which are Things able to communicate their status by blogging) or Spimes [14] (which is a combination of 'SP'ace and t'IME', it represents objects that can be tracked in the time).

Things can have access to personal data particularly in IoT applications related to human life such as, personal device, smart home, smart health, smart city and so on. In such applications Things are used to gather huge amount of personal data used by these application's providers for different aims, such as profiling for advertising, digital marketing, statistics etc. The use of this personal data, collected in many case without the explicit agreement (or forced agreement), is considered as violation a universal human right which is the privacy of users. This can undermine the privacy because the Things facilitate the collection, storage, processing and combination of this personal data.

This situation will be worst in future if we consider the fact that It is predicted that IoT will interconnect more than 50 billion devices by 2020 [15] to form a gigantic and unprecedented network of devices, which will be pervasively deployed and will enable new applications. The constrained nature of Things require the design and the adoption of standardized communication models and adapted security and privacy mechanisms in order to build inter operable, scalable and safe IoT environment.

The aims of this work are to analyze and evaluate the risks for privacy over the IoT network and propose solutions to deal with this problem. In the first section, we present an IoT overview and its architecture for security concerns. The second section focuses on privacy preservation as a security issue, and present how Things threat the user's privacy. The last section analyzes and suggests solutions presented as two categories: the best practices and the technical solutions.

2 Motivation and Related Work

Even if the IoT has many advantages, the use of smart Things in personal life presents weaknesses that can facilitate the violation of privacy. The dilemma is that in one hand, these Things help to improve the quality of life, better autonomy and environment control. But, on the other hand, these same Things and their applications gather a huge amount of personal data that can be used in malicious way. Indeed, while we, as consumers in the context of the IoT, are using these connected things much more extensively, they are in turn using us (and our data) as consumers.

To make the analysis more concrete, let's consider a real case study in which two users, Alice and Bob, have an appointment in a location X and want to keep this as a private event, but the applications lunched in their smart phone can disclose their identities, location and time (D1 = identity, D2 = localization and D3 = Date) that can be processed and sent by some of these applications.

At this stage and according to the hierarchy cited below and in [16], it will be easy to conclude that Bob met Alice for an exact duration in place X. This can be easily done particularly because this data are available on Internet. So the privacy of Bob and Alice is violated because third part can know about their private meeting.

This work aims to make the use of IoT in personal life less intrusive regarding the privacy. After critical analyses of the classical solutions already used in the computer network, we deduce that they are not suitable to be used in the IoT environment because of its constrained features in term of CPU, memory, and energy consumption, for that, we propose in this work mechanisms based on best practices and anonymization of data.

As the benefits of the IoT can't overshadow privacy concerns, there is a significant amount of research [17, 18, 19] in the area of privacy preservation in this context proposing a several mechanisms with the goal of how the user can benefit the maximum from the connected Things without affecting his private life.

In [20] authors present the use of secure multi-party computations (SMC) within the context of IoT and prove that the SMC creates a new development opportunities of privacy preserving in ubiquitous applications. Another work [21] presents a performance evaluation of Attribute-Based Encryption(ABE) approach that focuses on execution time, data and network overhead, energy consumption, CPU and memory usage according to the connected Things constrained features.

3 IoT Architecture and Data Exploitation

3.1 IoT Architecture

Numerous projects interested in studying the requirements of IoT architecture such as IEEE P2413 [22], ETSI [23] (even if it doesn't mention the word 'Internet of Things' the concept discussed such machine-to machine is similar) and OneM2M [24] (developing standards for machine to machine enabling large-scale implementation of IoT with the aim to provide an architecture reference). In this paper, the three layered architecture [23] is used as a reference one. The advantage of this architecture is that it includes all component involved in the process of data collection and exploitation. It is also a generic one which fit with different application of IoT.

In this architecture, the block « Domain Things » represents the things in their environment. The paper considers two categories of Things, those with the capabilities to communicate directly via Internet infrastructure. The second category represents Things that need special gateway to communication with Internet. This second category communicates basically in their environment, and with the gateway, by using limited scope wireless technologies such as bluetooth. The « Network Domain » represents any existing ICT IP network infrastructure that can permit communication between Things and application servers on the Internet. The « Application Domain » designates different IoT application and aims to use data gathered by Things.

3.2 Data Exploitation in IoT Context

Following the hierarchy proposed by Bernstein, J. H in the paper referenced [16], data collected from the lower level (Things domain) can be coupled, filtrated, reduced and refined to finally compose useful information for different third part. It explains how wisdom is derived from the lower phenomena.

As we move from the lower level of this hierarchy to the higher one, the meaning is created. At the data level an unmeaning data are generated such as numbers, names, identities, locals and other data according to the Things capabilities and application. Then in the level above elements composed by piece of data are linked together to form information. At the knowledge level information are organized in accordance with the application field then at the very higher layer we have the final state of data collected which is a synthesize of user's behavior and which represent in our work the asset to protect against any privacy violation.

If we consider as example, the smart phone of someone, where different installed application can send to application server on the cloud three basic data: D1 = identity, D2 = GPS position (x, y) and D3 = Date. Taken separately, these data don't have meaning. But if we suppose that these three pieces of data are issued from the same Things owned by one person "Bob" we can deduce that "Bob was in location (x, y) at instant indicated by Date". If we add to these data, from other source, that location (x, y) is the university, we can deduce that "Bob was in the university at the instant indicated by Date".

3.3 Privacy Concerns in IoT

In the universal declaration of human rights [25] the third article specifies that "Everyone has the right to life, liberty and security of person". In the twelfth article: "No one shall be subjected to arbitrary interference with his privacy, family, home or correspondence, nor to attacks upon his honor and reputation. Everyone has the right to the protection of the law against such interference or attacks", also, private data is defined as being any information pertaining to a "concerned individual that reveals racial and ethnic origin, political, philosophical, religious opinions or trade union affiliation, or that concern life or health, or that concern sex life or health, including the genetic data".

While using the connected Things, many private data related to person are sent and collected for different aims by different applications running on Things. Users know that they are sending their data but they aren't aware to whom exactly and for which purpose. For example, while installing Facebook application on smart phone, we should absolutely accept the conditions in which there are a several implicit use conditions as follow:

Allow the application to access to these following data. Information related to the activity on the device list of current executed applications, history and favorites navigation.

Identity of the user. The application is allowed to use: accounts on the device, profile data.

Agenda. Information from the agenda.

Contacts. The application is allowed to use the contact information.

Position. Use the user's position.

SMS. The installed application can use SMS, MMS, supplement fees.

Photos/multimedia/files. The application can use recorded files on the device (images, video, audio or external storage).

Camera and Micro. The application uses cameras and microphones of the device.

Device ID. The application is allowed to determine the phone number and the ID of the device, if the user is calling and who is calling.

Information about the WiFi connection. Application is able to access the WiFi to which the device is connected and then know information about other devices on the same network.

We consider in this paper privacy as a security feature, because the idea behind the privacy is to ensure the confidentiality of data related to private life of a person. By this way, this feature ensures firstly that personal data are neither readable over the network nor unattainable except by the owner or explicit mandated entities. And secondly the process and the use of the user's gathered data should be regulated and absolutely not be sold [26] to interested parties for marketing purposes, targeted advertising or used as way to pressure and blackmail.

Depending on application fields, the privacy can concern different categories of data, basically [27]:

Location. To avoid being tracked, the person may prefer not to disclose his location to others.

Identity. The gathered data should not refer to his owner, so it will be useless to any malicious part able to intercept it.

Profiling. This is the threat of compiling information about individuals and inferring interests by correlation with other profiles and data.

Linkage. The service provider or the architecture component cannot link different end user's accesses.

Personal data. The data exchanged in IoT networks should be unattainable and unreadable, and the user should be the only who can decide which data, when, and to whom it could be sent.

Privacy can also be viewed in various ways, e.g. as a right to confidentiality of communications, a right to be left alone, a right to control one's own life or a right to the protection of one's personal data [28].

4 Privacy Risk Analysis in IoT

The reference architecture model [23] illustrated in Fig. 1, will be used as fundamental architecture in this paper to analysis the privacy risk in the context of IoT. According to this architecture and focusing on data that can be source of privacy violation, we can deduce that the threat are in each component of this architecture that held this data and help to exchange it. Basically, the Thing itself, the network component used to exchange this data, and the application server hosting the IoT application. The perimeter of this study of risk analysis is composed by (Things, Gateway, Public Infrastructure, Application server).

4.1 Risk on Things Domain

At this level, Things can communicate with application even directly or via special gateway. In the first case, the Things can hold data and send it to related application

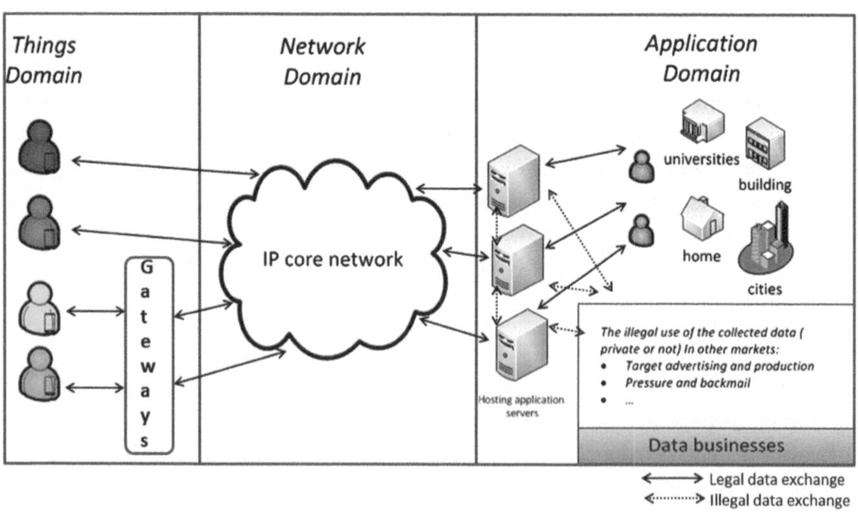

Fig. 1 The three layered architecture

on the cloud via Internet. This category of Things is more vulnerable to attack because they are directly accessible on Internet. In the second case, Things has low communication capability, so that it can only communicate with closest other Things or with the gateway responsible to forward data to application on the cloud. It is more difficult to attack this second category, because they are not visible directly on Internet, so we need to be so closer to them or we need to attack and gain control of the gateway. In these two cases, the data available on Things can be disclosed and used in privacy violation.

4.2 Risk on Network Domain

The network domain is basically composed by the network infrastructure owned by different providers and used to exchange data between Things and server application. As in the classic computer networks, the data exchange via network are exposed to sniffing attack that can help to disclose precious data. In this domain, it is difficult to implement solution to protect data. Indeed, the security policies of each provider are different and independent from those of others, and the user doesn't have any ability to implement any security mechanism on the components of these networks. So any security mechanism proposed in this paper will be out of this domain.

4.3 Risk on the Application Domain

The application domain is composed by different applications that are used in the IoT context. These applications are hosted on dedicated server. All data gathered by client application lunched within the Things are processed and stored in these servers. So we are in dilemma situation in which the user is the owner of the Things, but not the owner of the content even his personal data, especially when this data is forwarded to application servers in the cloud. Once this personal data are in the cloud, the risk for privacy become important and the user is not the only one who has access and control to it. Third parties have also access to this data.

4.4 Summary of Risk Related to the Confidentiality of Private Data

The preservation of confidentiality must be ensured throughout the lifetime of data, to prevent unauthorized or abused access during its process of capture, transmission, aggregation, storage, and processing. To, presents considered risk in this

Table 1 Privacy risk over the IoT network

	Data location			
Risk	Things domain		Network domain	Application domain
	Things	Gateways		
Sniffing, traffic analysis, black or grey hole	Intruder Things can sniff data and intercept communications	Intercept data from Things to application or the inverse	Out of scope[a]	Application servers are hosting all data gathered from the lower level. A rich spot of sniffing attack.
Data replication	Things can be cloned and recover data from environment, other Thing or from application	Gateways can be cloned and intercept data crossing them	Out of scope[a]	Application servers can be cloned and recover data from other Thing or from application
Destruction, theft or taking control of the entity	Things can be physically destroyed, attacked, and then data on it are recoverable	Gateways can be physically destroyed, attacked, and then data on it are recoverable	Out of scope[a]	The hosting servers are generally protected physically against the malicious attacks
Malicious node	Compromised Things can replace Things and interfere exchanges to cause a malfunction	Compromised gateways can replace the legitimate one and read data.	Out of scope[a]	Fake application behaving as a legitimate one and gather data from Things without permission
Wormhole attack	Communication route between Things can be diverted to a malicious Thing.	Malicious gateways can enforce the data flow to transit over it.	Out of scope[a]	Data in the application domain can be forged to transit on a specific routes
Sybil attack	A fake Things can recover data as a legitimate one	A fake gateway can recover data as a legitimate one	Out of scope[a]	Fake application can change identities and gather data from Things as the true one

[a]The Network Domain is out of scope of this analysis because the risk to capture private data at one node of this domain is minimal. Indeed, the difficulty is in fact that it is not easy to know on which node traffic issued from the Things is forwarded to application server, and have control of this node in order to sniff this traffic

paper. Based on some critical threats identified in [29], and by using risk analysis methodology such as EBIOS [30], we can summarize the risk related to the privacy, in Table 1, according to the location of personal data in the architecture.

For that, we need to have mechanism to implement in the Things itself, to protect data when transmitted by Things to the application, and to secure application that is responsible for data collection.

5 Proposed Solutions

According to risk analysis presented in previous section, each solution to preserve privacy must be able to ensure the confidentiality of data related to the users and its private life. Whereas no, preserving privacy means that the recipient of information can have access to the data. While access control and authentication are deployed against direct disclosures and alone fail to ensure privacy preservation.

For example, in the situation when Things generate data (D1, D2, D3...) about their environment via different applications (A1, A2, A3...) and D1 = identity, D2 = position(x, y), D3 = Date, when these data are massively available on the cloud, there is a risk to combine between different part and constitute a useful information:

$$D1 = \text{Alice},\ D2 = \text{Position}\,(x1, y1),\ D3 = 12 - 02 - 2016 - 14{:}30{:}00$$
$$D1 = \text{Bob},\ D2 = \text{Position}\,(x1, y1),\ D3 = 12 - 02 - 2016 - 14{:}30{:}00$$

We deduce from this set of data that Alice and Bob were meeting for a specific duration in location identified by its GPS position (x1, y1).

To preserve privacy we propose to act in two levels. The first one is recommendation for users and application developers. The second one is technical solution based on anonymization of data.

5.1 Recommendations for IoT Users and Developers

Different actions and best practices can reduce the risk of privacy violation in IoT application related to human life. We classify them to respond to different risks and threats sources identified in the previous section.

An awareness strategy at user level. Users should be aware of the disclosure of their personal data, and should be able to limit the type of data collected, communicated or stored, also they have to be able to secure their Things themselves.

At Things level. Things (sensor, gateway, robot, camera, RFID reader, phone etc.) must authenticate the source of update files and periodically check it using cryptography mechanism.

At Application level. Application must authenticate each Thing responsible for gathering data. Application should also verify the consistency of this data and use cryptography mechanism for that aims and for hardening communication with Things level, especially if we consider the fact that the network domain can be vulnerable but we don't have any control of its components to implement mechanisms on it. Sharing data between applications must be controlled to make it difficult to exploit private data gathered from different application.

5.2 Anonymization for Privacy Preservation

Data anonymization is a promising category of approaches to achieve the privacy preservation goal [31], at the same time it keeps opportunities for the economic business to use the data gathered from Things to customize the services for users. Anonymity is used in other context to protect data [32, 33]. The idea behind the anonymization is to hide or vary part of data in order to make deduction of information from directly accessible data more difficult. For example masque the identity if not need, or give interval instead of exact value, etc. The anonymization process has three objectives [34]:

- Protect the privacy of monitored users.
- Hide any information about the internal infrastructure of the network.
- The anonymized traffic traces have to be as realistic as possible, that means as close as possible to the non-anonymized packet stream.

And thus by different way, the most used are: 'The encryption personally identifiable information' [35], which ensure the integrity of data and encrypt just the data related to the user identity, 'k-anonymity' [36], an exchanged data is k-anonymized means that it cannot be distinguished from at least k-1 individuals whose information are the same, there is also the 'Generalization' [37] and 'perturbation' which means that specificities are removed from data and this is not suitable for the context of IoT because of the inaccuracy of data after the anonymization, and data need to be exact to offer the best services. In the proposed case study of Alice and Bob, we can for example use one of anonymization technique such as generalization and masking part of data. The result can be:

$$D1 = Al^{***}, D2 = \text{Position}\,(x1 + A,\ y1 + B),\ D3 = 12 - 02 - 2016 - 14{:}30{:}00$$
$$D1 = {}^{***}b,\ D2 = \text{Position}\,(x1 + C,\ y1 + D),\ D3 = 12 - 02 - 2016 - 14{:}30{:}00$$

In this case we use masking technique (character *) in part of data field (D1) containing identification. And the second technique use generalization of data for the data field (D2) containing the position by varying the exact information by random values (A, B, C and D) chosen by applications according to the aims of gathering these data.

5.3 Anonymization Over IoT Network

In the application of anonymity mechanisms, in order to prevent privacy attacks, data should be anonymized properly before crossing the threat zone. The anonymization in this case is based on the re-identification of data which means that data is real and is exactly what was gathered to the predefined service but with a new identity, the new identities can be deterministic; the same value every time, or not, these are dependent on the technique used for anonymization, that is to allow businesses to benefit from generated data for the economic needs.

For this, the entity which re-identifies data; anonymizes, should be able to support the anonymization needs in terms of processing, memory, power and so on.

When re-identifying data, we should obviously think about the reversibly way to return services on the application layer or on the Things layer in the case of actuating Things.

Applying the anonymization mechanism on Things is the best way to preserve the user privacy but it is challenging because of the constrained features of the connected Things in term of computing capacities, memory, and the autonomy which are necessary for the process of the re attribution of the new identities and the inverse operation. Then, we push the application toward the data destination, on the gateways but with more risks from the internal areas which are the constrained devices which are then able to track and recover data by using the side of the constrained communication unit.

The more the anonymization is deployed toward the applications level the more the risk of privacy violation grow, till the application level then the data are protected just against the illegal use of the service provider when application are exchanging user's data with benefice and use them with third part for business matters.

6 Conclusion

IoT is nowadays one of hottest research domain and is related to different industrial and human life fields. The huge amount of gathered data is a big opportunity for enterprise and economy. But, part of this concerns individual and an inappropriate use of them is a violation of privacy which is a human right. The paper contributes in the field of user's privacy in IoT. It presents a deep risk analysis of the privacy threat and proposed two approaches to preserve privacy. The first one is recommendations for user and IoT application developers. The second is the use of anonymization technique to hide data recorded that can be used as threat to violate privacy. As future work, we focus on implementing such mechanism in typical application and the choice of which set of data must be anonymized. We also focus on evaluating the impact of such mechanism on business related to data gathered by things.

References

1. T. I. IEEE. Special Report: The Internet of Things
2. Vinayaga Sundaram, B., Ramnath, M., Prasanth, M., Varsha Sundaram, J.: Encryption and hash based security in internet of things. In: (ICSCN) 3rd International Conference on Signal Processing, Communication and Networking, pp. 1–6. Chennai (2015)

3. Lee, J., Dong, M., Sun, Y.: A preliminary study of low power wireless technologies : ZigBee and Bluetooth low energy. In: (ICIEA) 10th IEEE Conference on Industrial Electronics and Applications, pp. 135–139. Auckland (2015)
4. Friese, I., Heuer, J., Kong, N.: Challenges from the identities of things: introduction of the identities of things discussion group within Kantara initiative. In: WF-IoT IEEE Worlds Forum on Internet of Things, pp. 1–4 (2014)
5. Miorandi, D., Sicari, S., De Pellegrini, F., Chlamtac, I.: Internet of things: vision, applications and research challenges. Ad Hoc Netw. J. **10** 1497–1516 (2012)
6. Sundmaeker, H., Guillemin, P., Friess, P., Woelfflé, S., Sundmaeker, H.: Vision and Challenges for Realising the Internet of Things (2010)
7. Internet of Things: http://www.gartner.com/it-glossary/internet-of-things
8. Oriwoh, E., Conrad, M.: "Things" in the internet of things: towards a definition. Int. J. Internet Things 1–5 (2015)
9. Bohn, J., Coroama, V., Langheinrich, M., Mattern, F., Rohs, M.: Living in a world of smart everyday objects- social, economic, and ethical implications. J. Hum. Ecol. Risk Assess. **10**, 763–786 (2004)
10. Pintus, A., Carboni, D., Piras, A., Giordano, A.: Connecting smart things through web services orchestrations. Curr. Trends Web Eng. **6385**, 431–441 (2010)
11. Kortuem, G., Vasughi, F., Sundramoorthy, V., Fitton, D.: Smart objects as building blocks for the internet of things. IEEE Internet Comput. **14**, 44–51 (2010)
12. The Internet of things : Networked objects and smart devices. www.theinternetofthings.eu/sites/default/files/Rob%20van%20Kranenburg/networked_objects.pdf (2010)
13. Bleecker, J., Nova, N.: Blogjects and the new ecology of things. http://blog.nearfuturelaboratory.com/2006/03/15/report-from-blogject-workshop-lift06/ (2006)
14. McFedries, P.: The Age of Spimes, IEEE Spectrum. http://spectrum.ieee.org/at-work/innovation/the-age-of-spimes (2010)
15. Evans, D.: The Internet of Things how the next evolution of the Internet is Changing everything. White paper. http://share.cisco.com/internet-of-things.html (2011)
16. Bernstein, J.H.: The data-information-knowledge-wisdom hierarchy and its antithesis. Nasko **2**, 68–75 (1989)
17. Feng, H., Fu, W.: Study of recent development about privacy and security of the internet of things. In: WISM International Conference on Web Information Systems and Mining, pp. 91–95. Sanya. http://ieeexplore.ieee.org/search/searchresult.jsp?searchWithin=%22Authors%22:.QT.Wenxiu%20Fu.QT.&newsearch=true (2010)
18. Roman, R., Zhou, J., Lopez, J.: On the features and challenges of security and privacy in distributed internet of things. Comput. Netw. **57**(10), 2266–2279 (2013)
19. Zhou, B., Pei, J., Luk, W.: A brief survey on anonymization techniques for privacy preserving publishing of social network data. ACM SIGKDD Explor. Newslett. **10**(2), 12–22 (2008)
20. Oleshchuk, V.: Internet of things and privacy preserving technologies. In: Wireless VITAE 1st International Conference on Wireless Communication, Vehicular Technology, Information Theory and Aerospace & Electronic Systems Technology, pp. 336–340 (2009)
21. Wang, X., Zhang, J., Schooler, E., Ion, M.: Performance evaluation of attribute-based encryption : toward data privacy in the IoT. In: ICC IEEE International Conference on Communications, pp. 725–730 (2014)
22. IEEE Project, Standard for an Architectural Framework for the Internet of Things. https://standards.ieee.org/develop/project/2413.html
23. Lin, F., Ren, Y., Cerritos, E.: A feasibility study on developing IoT/M2 M applications over ETSI M2 M architecture. In: ICPADS International Conference Parallel and Distributed Systems, pp. 558–563 (2013)
24. OneM2 M Technical Specification: oneM2 M Functional Architecture Baseline. http://www.onem2m.org/# (2014)
25. The Universal Declaration of Human Rights. http://www.un.org/en/universal-declaration-human-rights/

26. Garcia-Morchon, O., Keoh, S., Kumar, S., Hummen, R., Struik, R.: Security Consideration in the IP-based Internet of Things. https://tools.ietf.org/html/draft-garcia-core-security-06 (2013)
27. Ziegeldorf, J.H., Morchon, O.G., Wehrle, K.: Privacy in the Internet of Things : threats and challenges. Secur. Commun. Netw. J. (2013)
28. Friedewald, M., Wright, D., Gutwirth, S., Mordini, E.: Privacy, data protection and emerging sciences and technologies: towards a common framework. Innovation- Eur. J. Soc. Sci. Res. **23**, 61–67 (2010)
29. Abomhara, M., Køien, G.M.: Security and privacy in the internet of things: current status and open issues. In: (PRISMS) International Conference on Privacy and Security in Mobile Systems, pp. 1–8 (2014)
30. EBIOS: (In French: Expression des Besoins et Identification des Objectifs de sécurité). http://www.ssi.gouv.fr/guide/ebios-2010-expression-des-besoins-et-identification-des-objectifs-de-securite/
31. Fung, B.C.M., Wang, K., Chen, R., Yu, P.S.: Privacy-preserving data publishing: a survey of recent developments. ACM Comput. Surv. **42** (2010)
32. Reiter, M.K., Aviel, D.: Crowds: anonymity for web transactions. Trans. Inf. Syst. Secur. J. **1**, 66–92 (1998)
33. Deibel, K., Petersen, A., Schwerin, A.: Cone of silence: a layered approach for network-level protocol anonymization. http://www.cs.washington.edu/homes/deibel/papers/cse561-cos/cse561-cos%.pdf
34. Koukis, D., Antonatos, S., Antoniades, D., Markatos, E.P., Trimintzios, P.: A generic anonymization framework for network traffic. In: IEEE International Conference on Communication, pp. 2302–2309. Istanbul (2006)
35. Stringer, J.: Protecting personally identifiable information : What data is at risk and what you can do about it. A Sophos White Paper, 1–6 (2011)
36. Machanavajjhala, A., Kifer, D., Gehrke, J., Venkitasubramaniam M.: L-Diversity: privacy beyond k-Anonymity. ACM Trans. Knowl. Discov. Data (2007)
37. Zhang, L., Zhang, W.: Generalization-based privacy-preserving data collection. Data Warehousing Knowl. Discov. **5182**, 115–124 (2008)

Packet Delay Analysis in Wireless Sensor Networks Using Fountain Code Enabled-DCF

Rachid Aouami, Mohamed Hanaoui, Mounir Rifi and Mohammed Ouzzif

Abstract Many applications currently exploit wireless sensor networks for long term data gathering, ranging from environmental sensing, etc., and many more are under development. This paper introduces an analytical model to investigate the delay analysis for the Fountain-Code-Enabled–Distribution Coordination Function (FCE-DCF) for the IEEE 802.11 protocol in the Wireless Sensor Networks (WSN). The state transition of buffering queue data in node is described by a two-dimensional Markov chain model. The delay is developed by extending the throughput analysis introduced by our model. This study is validate by comparison with the result obtained by Bianchi's model. The packet delay analysis results present as a function of a number of station, packet size and the effects of contention windows are obtained using DCF for asynchronous packets transmission with four-way handshaking technique.

1 Introduction

A key difficulty in the mathematical modeling performance the 802.11 MAC layer has been study extensively in the scaling exponentially with the number of node. When the range of single-hop wireless communication limited by distance or harsh radio propagation conditions. Bianchi [1] addressed this difficulty by assuming that every node has a data to be transmitted and the packets collision probability is

R. Aouami (✉) · M. Hanaoui · M. Rifi · M. Ouzzif
RITM Laboratory, CED Engineering Sciences, EST, ENSEM,
Hassan II University of Casablanca, Casablanca, Morocco
e-mail: aouami@est-uh2c.ac.ma

M. Hanaoui
e-mail: hanaoui.mohamed@gmail.com

M. Rifi
e-mail: rifi@gmail.com

M. Ouzzif
e-mail: ouzzif@gmail.com

R. El-Azouzi et al. (eds.), *Advances in Ubiquitous Networking 2*, Lecture Notes in Electrical Engineering 397, DOI 10.1007/978-981-10-1627-1_14

constant regardless of the station. The authors of [2] inserted an additional state to the Bianchi chain, called "idle" to take into account the situation where no frame is either in the queue at the MAC layer or being transmitted. Many main approaches were used to account for the saturation and non–saturation load assumption.

The fountain code-enabled IEEE 802.11 DCF standard is based on the Carrier Sense Multiple Access with Collision Avoidance (CSMA/CA) protocol supporting the packet collision transmission. Since no hidden nodes are considered, collisions take place because two or more contending stations choose the same backoff slot to transmit. The time needed for a frame transmission is considered to start when a frame becomes head of the station's queue and is finalized when a positive acknowledgement is received. Assuming that the frame drop probability is very low and can be neglected.

In wireless sensor network, the access delay is also an important metric. Generally, total delay in a communication network includes processing delay, queuing delay, access delay, and propagation delay. In [3] Vukovic reduce the delay by improving the model used in the literature, and calculate the delay from the finite and infinite packet retransmission. In this paper, we only focus on average access delay since our main purpose is to evaluate performance of CSMA/CA [4] scheme and a random backoff (windows, stage) following the channel condition. Time needed to transmit a packet is define as the instant from the moment packet is at the head of its MAC queue and ready to be transmitted to the moment when coordinator receives packet. This time is also denote as packet service time.

The remainder of the paper is structured as follows: In Sect. 2 of this document, we give a briefly description of the analytical model based on a discrete time Markov chain. This model used in Sect. 3 to calculate the delay analysis. Finally, Sect. 4 presents the numerical solution of our model.

2 The Analytical Modeling

2.1 Markov Chain Approach

The development of the mathematical model, has enabled us to design a fairly comprehensive analytical model to represent/evaluate the performance of the wireless sensor networks working with the IEEE 802.1DCF/FCE-DCF. Many researchers have been studied the performance of 802.11 DCF based on two dimensional Markov chain analytical model. The 802.11 standard for wireless networks incorporates two kinds of medium access methods, which are the distributed coordination function (DCF). We consider the corresponding discrete time Markov chain as follows: We assume that packet frame size is equal to a single MAC frame size and that the whole is transmitted as a single MAC frame. The state space of the Markov chain describing the decoding process using fountain code. However, needs only include the states that are irreducible in the sense that they cannot be reduced by the receiver. The number of possible distinct packets is 2^{n-1}. The number of different sets of received distinct packets is then 2^{2n-1} including the initial state. The node has a data moves into the idle state when the MAC buffer becomes empty either after a successful transmission or after the limited number of retransmissions to access in the channel.

2.2 FCE-DCF Approach

The model also calculates the probability of a packet transmission failure due to collision using a fountain code in two part first in the sender and at the receiver. It assumes that the channel is in ideal conditions, there is no hidden terminal and capture effect (Fig. 1).

We assume that the network consist of n contenting nodes, each node has a packet available for transmission, the backoff timer is uniformly chosen in the range [0, Wi−1] and the probability of collision p of the transmission packet has constant and independent of the retransmission. At the first transmission attempt of a packet equal CW_{min} and after unsuccessful transmission CW is doubled up to the maximum value CW_{max}.

Let b(t) be the stochastic process representing the back-off time counter for o given station. The counter k, is initially chosen between [0, Wi−1]. The counter is decremented when the medium is sensed idle and the transmission is attempted when k = 0.

Let s(t) be the stochastic process representing the back-off stage (0….m) of the station at the time t, the back-off stage i, starts at 0 at the first transmission attempt and is increased by 1 every time a transmission attempts results a collision, the maximum value is m.

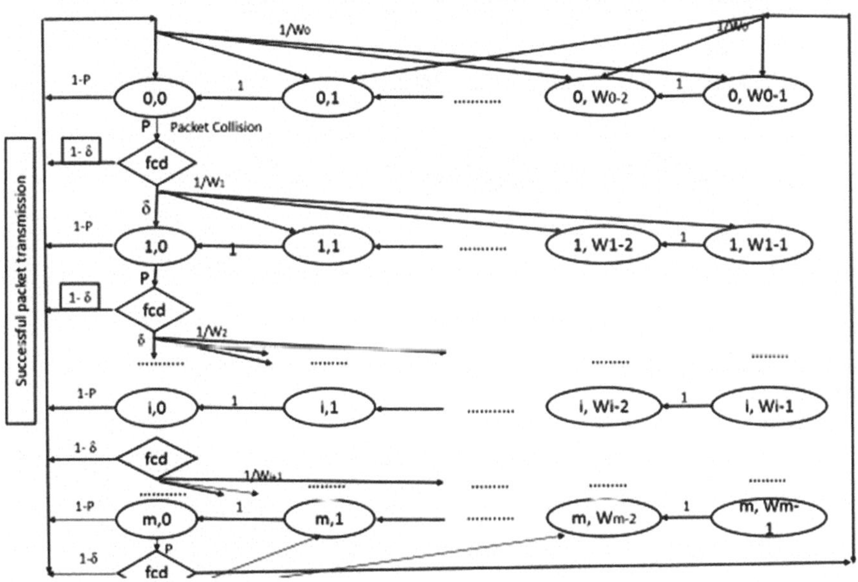

Fig. 1 Markov chain model for the backoff window size using the fountain decoding algorithm

2.3 Backoff Window Procedure

When the de station has a data packet to transmit and senses the channel to be idle for a period of Distribution inter frame Spacing (DISF) then the station proceeds with this transmission/If the channel is busy, the station defers until an idle DIFS is detected and then generates a random backoff interval before transmitting in order to minimize collision. The backoff time counter is decreased in terms of slot time as long as the channel is sensed idle. The counter is stopped when the channel is busy and resumed when the channel is sensed idle again for more than DIFS. A station transmits a packet when its backoff timer reaches zero. If the destination station successfully receivers the packet, its waits for a Short Inter-Frame Space (SIFS) time interval, the data packet is assumed to have been lost and the station schedules a retransmission. Each station holds a retry counter that is increased by one each time a data packet is unsuccessfully transmitted. If the counter reaches the retransmission limit m, the packet is discarded (Fig. 2).

The RTS/CTS access scheme follows the same backoff rules as basic access. When the backoff timer reaches zero, the station sends a short RTS packet first instead of the data packet. The receiving station responds with a CTS packet after a SIFS time interval. The send is allowed to transmit the data packet only if it receives valid CTS. Upon the successful reception of the data packet, the receiver transmit an ACK frame (Fig. 3).

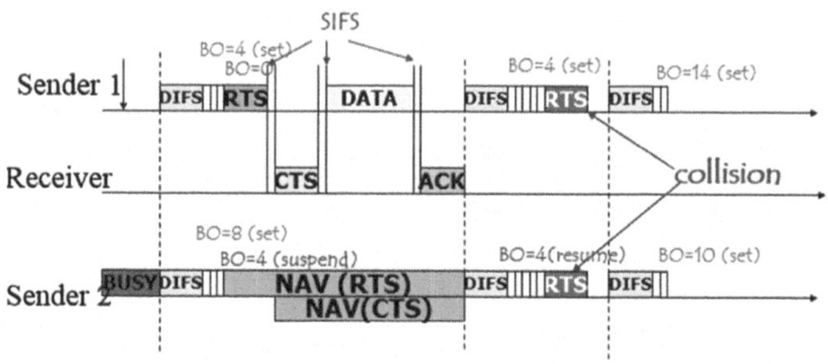

Fig. 2 Mechanism of random backoff time between two consecutive DIFS

Fig. 3 Mechanism of the RTS/CTS for four-way handshaking

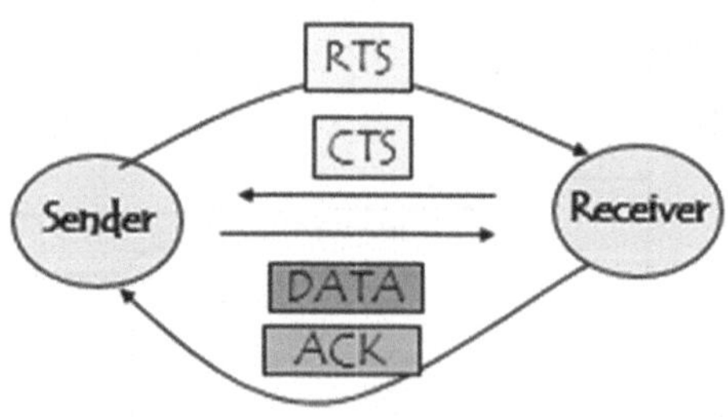

2.4 Transmission Probability Per Station τ

Proceeding with the traditional computation of the stationary distribution of the Markov chain like in [5, 6], a transmission occurs when the back-off time counter is equal to zero (k, 0), so tau should be the sum of the frequency probabilities of all states (k, 0). Finally, considering Eqs. (2), (3) and (5) we can write the probability of a station transmit in randomly chosen slot time by a probability of successful coding delta δ at the sender and decoding (1 − δ) at the receiver:

$$\tau = \frac{(2-4p\delta)}{(1-2p(1-\delta))(W+1)+p(1-\delta)W(1-(2p(1-\delta))^{m})} \tag{1}$$

2.5 Probability of Collision P

We describe our analytical model of the FCE-DCF. In each transmission attempt, regardless of the number of retransmission suffered, each data collides with constant and independent probability: p is the probability that, when one station is sending a packet, collisions occur if there is at least one of the n − 1 remaining stations transmits as well. If at sate ach remaining station transmits a packet with probability τ.

$$P = 1-(1-tau)^{n-1} \tag{2}$$

2.6 Throughput Analysis

From the system of two nonlinear equations that has a unique solution and can be solved numerically for the values of p and τ. Once these probability are obtained, the saturation throughput, which is the average information payload transmitted in a slot time over the aver duration of o slot time, can be computed as follows:

$$\mathrm{S} = fracP_{tr}P_sL^2(1-P_{tr})\sigma + P_{tr}(1-P_s)T_c + P_{tr}P_sT_s \tag{3}$$

where $P_{tr} = 1-(1-\tau)^n$ is the probability that here is at least a transmission in the considered slot time, L is the average length packet payload size and T_{id} is the duration of the idle period.

Let Ps be the probability that one station transmits by a successful transmission in the channel, or when packets encounters a collision which the successful fountain decoding, it is conditioned by there is only one station sending in a time slot.

The probability that there is a successful packet in the slot time is:

$$P_s = \frac{(n\tau - (1 - \delta))(1 - \tau)^{n-1} + (1 - \delta)}{1 - (1 - \tau)^n} \tag{4}$$

Let Ts, Tc the time that the channel is sensed by a successful transmission, and missed transmission.

$$\begin{cases} Ts = RTS + SIFS + 4\sigma + CTS + SIFS + H + L + SIFS + ACK + DIFS \\ Tc = RTS + DIFS + \sigma \end{cases} \tag{5}$$

where H the transmission times needed to send the packet header, L the payload, ACK the acknowledgment, and σ is the propagation delay. Their value are independent of system parameters. They are listed in Table 1 as defined in [6]. For calculate the throughput it be must find the solution of nonlinear equation.

2.7 Analytical Model of Delay Analysis for Ideal Channel

We can now define the delay for a successfully transmission packet as the time elapsed between the generation of a frame and its successful reception, until an acknowledgement for this packet is received. If a packet reaches the specified retry limit then this packet dropped and its time delay is not included in the calculation of the average packet delay. The transmission delay occurs later as the transmission delay to send a piece of information and takes into consideration the time of issue, the propagation time, the computation time and the time of reception of a packet.

Let D be the random variable representing the frame delay and E[D] its the means-value for a successfully transmitted packet. Packet delay is defined to be the

Table 1 Describes the items that are kept for different simulations

Parameters	Values
Packet payload	8184 bits
MAC header	272 bits
PHY header	128 nits
ACK	112 bits + PHY header
RTS	160 bits + PHY header
CTS	112 bits + PHY header
Channel bit rate	1 Mbits/s
Slot time	50 μs
Propagation delay	1 μs
SIFS	28 μs
DIFS	128 μs
ACK_Timeout	300 μs
CTS_Timeout	300 μs

time interval from the time a packet is at the head of its MAC queue ready for transmission, until its successful reception in the destination. E[D] is given by:

$$\mathrm{E[D]} = \mathrm{E[X]} \cdot \mathrm{E[length\ of\ a\ slot\ time]} \tag{6}$$

where E[length of a slot time] is the average length of slot time is given by:

$$E[slot] = (1 - p_{tr})\sigma + p_{tr}p_sT_s + p_{tr}(1 - p_s)T_c \tag{7}$$

where E[X] the average number delay depends on the value of its counter and the duration the counter freezes when the station detect transmissions from others stations. Considering that the counter of station is at state $b_{i,k}$ then a time interval of k slot times is needed for the counter to reach state 0. This time interval denote by a random variable X and the average as given by:

$$E[X] = \sum_{i=0}^{m} \sum_{k=1}^{w_i - 1} k \times b_{i,k} \tag{8}$$

After some algebra, E[X] is given by:

$$E[X] = \frac{(1 - 2p) + (w + 1) + pw(1 - (2p)^m)}{2(1 - 2p)(1 - p)} \tag{9}$$

We are now able to give the delay of FCE-DCF model

$$D[X] = \frac{((1 - 2p) + (w + 1) + pw(1 - (2p)^m)) \cdot ((1 - p_{tr})\sigma + p_{tr}p_sT_s + p_{tr}(1 - p_s)T_c)}{2(1 - 2p)(1 - p)} \tag{10}$$

2.8 Mathematical Model of Delay Analysis for Ideal Channel

In this part we use an approximation mathematical to give a new value of the delay for the ideal channel using the equations cited in [16]. The packet transmission probability tau_{ap} and the probability of collision of the FCE-DCF for P_{ap} written as:

$$\begin{cases} \tau_{ap} = \dfrac{2(1 - 2\delta p)}{W(1 - \delta p(1 + (2\delta p)^m)} \\ p_{ap} = \dfrac{2\delta(n - 1)}{W + 2\delta p(n - 1)} \end{cases} \tag{11}$$

Si n $\gg 10$ P_{ap} rewritten as follows.

$$p_{ap} = \frac{2\delta n}{\mathrm{W} + 2\delta n} \tag{12}$$

where $P_{tr-ap} = 1 - (1 - \tau_{ap})^n$ is the probability that here is at least a transmission in the considered slot time,

Let P_{s-ap} be the probability that one station transmits one the channel, which is conditioned by the fact at least one station transmits, and when packets encounters a collision which the successful fountain decoding.

$$p_{s-ap} = \frac{(n\tau_{ap} - (1-\delta))(1-\tau_{ap})^{n-1} + (1-\delta)}{p_{tr-ap}} \tag{13}$$

From the system of two nonlinear equations that has a unique solution and can be solved numerically for the approximation values of pap and tau_{ap}. The average delay t, which is the average information payload transmitted in a slot time over the average duration of o slot time, can be computed as follows:

$$\mathrm{E_{ap}}[\mathrm{D}] = \mathrm{E_{ap}}[\mathrm{X}] \cdot \mathrm{E_{ap}}[\text{length of a slot time}] \tag{14}$$

where E[length of a slot time] is the average length of slot time.is giving by:

$$E_{ap}[slot] = (1 - p_{tr-ap})\sigma + p_{tr-ap}p_{s-ap}T_s + p_{tr-ap}(1 - p_{s-ap})T_c \tag{15}$$

where $E_{ap}[X]$ the average number delay depends on the value of its counter and the duration the counter freezes when the station detect transmissions from others stations. Considering that the counter of station is at state $b_{i,k}$ then a time interval of k slot times is needed for the counter to reach state 0. This time interval denote by a random variable X and the average as given by:

$$E_{ap}[X] = \sum_{i=0}^{m} \sum_{k=1}^{w_i - 1} k \times b_{i,k} \tag{16}$$

After some algebra, $E_{ap}[X]$ is given by:

$$E_{ap}[X] = \frac{(1 - 2p_{ap}) + (w+1) + p_{ap}w(1 - (2p_{ap})^m)}{2(1 - 2p_{ap})(1 - p_{ap})} \tag{17}$$

We are now able to give de delay of FCE-DCF model

$$D_{ap}[X] = \frac{((1 - 2p_{ap}) + (w+1) + p_{ap}w(1 - (2p_{ap})^m)) \cdot ((1 - p_{tr-ap})\sigma + p_{tr-ap}p_{s-ap}T_s + p_{tr-ap}(1 - p_{s-ap})T_c)}{2(1 - 2p_{ap})(1 - p_{ap})} \tag{18}$$

3 Numerical Results

In this section, we present numerical results for our linear approximation analytical model and mathematical model. Solving the Markov Chain, one can find the probability tau that a transmitter transmits in a randomly chosen time.

The analysis is carried out for the relationship between packet transmission probability tau, condition collision probability P and the probability of fountain decoding $(1 - \delta)$. The two non-linear equation can be solve firstly by using Matlab tool to find the values of tau and p, and secondly by approximation mathematical based on equation used in [16], to calculate the performance of the analysis delay.

The idea here is to use a probability of fountain code to improve the success probability for the ideal channel conditions. The figures illustrate this operation, using the parameters reported in Table 1.

Simulation results are shown the average packet delay versus number of node. In Fig. 4 we fixed backoff stage and we make changing the value of the backoff window. We easily find that the delay analysis significantly increases under condition of collision avoidance, and with the number of contending stations increase, the delay decreases. We conclude that as large nodes having a data attempts to access a channel, more collision has occurs between data and increasing number of retransmission data. On the other hand, more packet lost due to time out because the

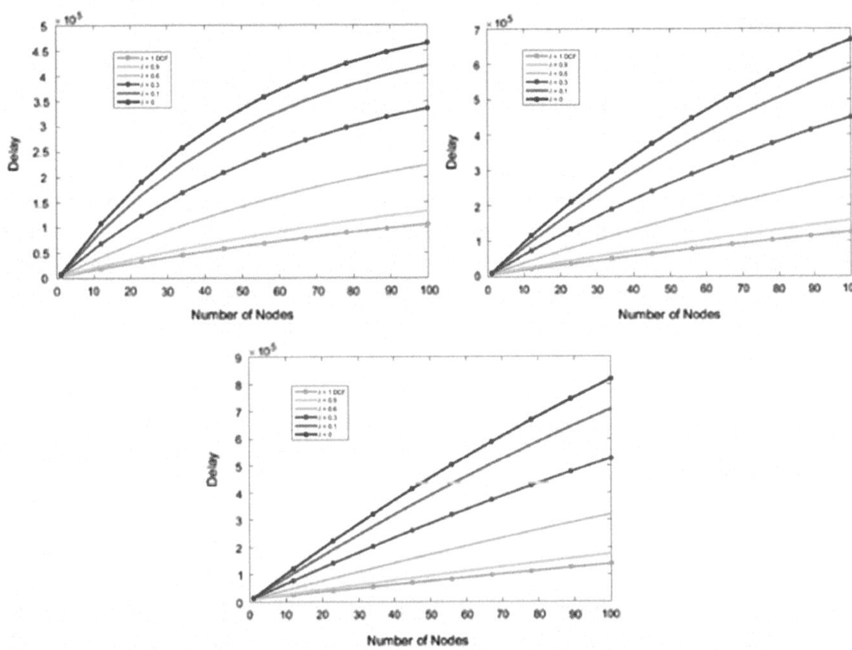

Fig. 4 The average packet delay versus number of node at m = 6 for the different value of contention windows: **a** for w = 256, **b** for w = 512 and **c** for m = 1024

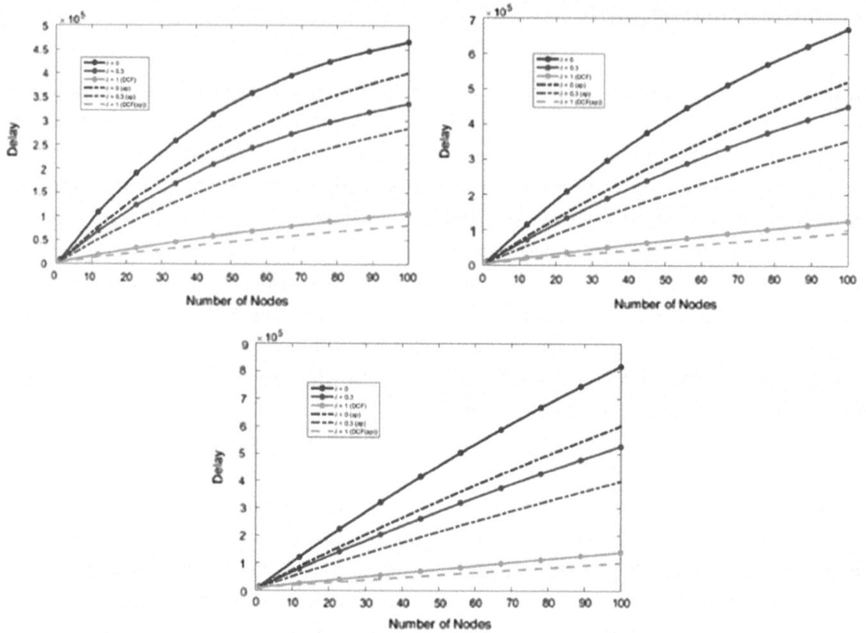

Fig. 5 Comparison between the analytical and mathematical model results for performance of the analysis delay various number of nodes for the RTS/CTS mechanism. At m = 6 for the different value of contention window: **a** for w = 256, **b** for w = 512 and **c** for m = 1024

station suffer a larger delay. However, by using the FCE-DCF we are able to minimizing this time (Fig. 5).

This figure demonstrates comparison between our linear approximation mathematical model and the analytical model for a system delay analysis versus number of nodes at m = 6 and different values of backoff window (CW). It's an interesting study because his give a same result when we increase the backoff window and number of nodes as like a greater network application. However, we can find that this is a small difference between two ways when we have increase the number of nodes.

4 Conclusion

In this paper, we presented an analytical model and approximation mathematical for the ideal channel based on a Markov chain and generating functions to compute the average delay of the IEEE802.11 in the new analytical model of the FCE-DCF. According to the comparison by simulation results presented in previous figure, we conclude that our model present a smaller delay than a Bianchi's approach. So we need to tread carefully and provide opportunities for all, not some, even to succumb to the delights of exciting technologies. This work may have some interest for the future planning of wireless sensor networks in presence of parameters inter-operating wireless technologies. And completed by simulation which the simulator Opnet to evaluate our approach in the future works.

References

1. Raptis, P., Vitsas, V., Paparrizos, K., Chatzimisios, P., Boucouvalas, A.C.: Packet delay distribution of the IEEE 802.11 distributed coordination function. In: Proceedings of IEEE International Symposium on a World of Wireless, Mobile and Multimedia Networks (WOWMOM 2005). Taormina, Italy, June 2005
2. Lee, W., Wang, C., Sohraby, K.: On use of traditional M/G/1 model for IEEE 802.11 DCF in unsaturated traffic conditions. In: IEEE Wireless Communications and Networking Conference, 2006. WCNC 2006, vol. 4, pp. 1933–1937 (2006)
3. Bisnik, N., Abouzeid, A.: Queuing network models for delay analysis of multihop wireless ad hoc networks. Ad Hoc Netw. **7**(1), 79–97 (2009)
4. Jacobs, I.S., Bean, C.P.: Fine particles, thin films and exchange anisotropy. In: Rado, G.T., Suhl, H. (eds.) Magnetism, vol. III, pp. 271–350. Academic, New York (1963)
5. Bianchi: Performance analysis of the IEEE 802.11 distributed coordination function. IEEE J. Sel. Areas Commun. **18**(3), 535–547 (2000)
6. Clerk Maxwell, J.: A Treatise on Electricity and Magnetism, vol. 2, pp. 68–73, 3rd edn. Clarendon, Oxford (1892)
7. Shokrollahi: Raptor codes. IEEE Trans. Inf. Theory **52**(6), 2551–2567 (2006)
8. Aouami, R., Said, E., Rifi, M., Ouzzif, M.: Fountain code enabled of IEEE 802.11DCF for optimization throughput in wireless sensors. In: The 10th International Conference for Internet Technology and Secured Transactions (ICITST-2015). London, UK. Accessed 14–16 Dec 2015
9. Vukovic, I., Smavatkul, N.: Delay analysis of different backoff algorithms in IEEE 802.11. In: Proceedings of IEEE Vehicular Technology Conference (VTC). Los Angeles CA, Sept 2004
10. Alabady, S.A., Salleh, M.F.M.: Analysis and Throughput Performance of IEEE 802.11 DCF in Multi-hop Wireless Networks. Springer Science + Business Media, New York (2014)
11. Bianchi: Performance analysis of the IEEE 802.11 distributed coordination function. IEEE J. Sel. Areas Commun. **18**(3), 535–547 (2000)
12. Singh, K., Awasthi, A.K., Mishra, R. (eds.): QSHINE 2013, LNICST 115. © Institute for Computer Sciences, Social Informatics and Télécommunications Engineering 2013, pp. 86–103 (2013)
13. Aouami, R., Rifi, M., Ouzzif, M.: Comparative analysis of contention oriented power saving based medium access control Protocols for wireless sensor networks. In: Proceedings IEEE of the 2nd World Conference on Complex Systems (WCCS). Agadir, Morocco. Accessed 10–13 Nov (2014)
14. Alabady, S.A., Salleh, M.F.M.: Analysis and Throughput Performance of IEEE 802.11 DCF in Multi-hop Wireless Networks. Springer Science + Business Media, New York (2014)
15. Yue, J., Lin, Z., Vucetic, B., Xiao, P.: The design of degree distribution for distributed fountain codes in wireless sensor networks. In: IEEE ICC 2014—Wireless Communications Symposium

Part II
Main Track 2: Mobile Edge Networking and Virtualization

Cost-Precision Tradeoffs in 3D Air Pollution Mapping Using WSN

Ahmed Boubrima, Walid Bechkit, Hervé Rivano and Lionel Soulhac

Abstract Air pollution has become a major issue of modern megalopolis, where the majority of world population lives. Measuring air pollution levels is an important step in designing and assessing air quality related public policies. Unfortunately, existing solutions are inadequate to get insights on the real exposition of citizens. In particular, high quality sensors deployed today are too large and too costly to envision a three dimensional deployment at the scale of a street. In this paper, we investigate the deployment of wireless sensor networks (WSN) used for building a three-dimensional mapping of pollution concentrations. We consider in our simulations a 3D model of air pollution dispersion based on real experiments performed in wind tunnels emulating the pollution emitted by a steady state traffic flow in a typical street canyon. Our contribution is to analyze the performances of different 3D WSN topologies in terms of the trade-off between the economical cost of the infrastructure and the quality of the reconstructed air pollution mapping.

1 Introduction

Air pollution affects human health dramatically. According to the World Health Organization (WHO), exposure to air pollution is accountable to seven million casualties in 2012 [15]. In 2013, the International Agency for Research on Cancer (IARC) classified particulate matter, the main component of outdoor pollution, as carcino-

A. Boubrima · W. Bechkit · H. Rivano (✉)
Univ Lyon, Inria, INSA Lyon, CITI, 69621 Villeurbanne, France
e-mail: herve.rivano@inria.fr

A. Boubrima
e-mail: ahmed.boubrima@insa-lyon.fr

W. Bechkit
e-mail: walid.bechkit@insa-lyon.fr

L. Soulhac
LMFA, Univ Lyon, CNRS UMR 5509 ECL, INSA Lyon, Univ Claude Bernard, 69134 Ecully, France
e-mail: lionel.soulhac@ec-lyon.fr

R. El-Azouzi et al. (eds.), *Advances in Ubiquitous Networking 2*, Lecture Notes in Electrical Engineering 397, DOI 10.1007/978-981-10-1627-1_15

genic for humans [14]. Air pollution has become a major issue of modern megalopolis, where the majority of world population lives, adding industrial emissions to the consequences of an ever denser urbanization with traffic jams and heating/cooling of buildings. As a consequence, the reduction of pollutant emissions is at the heart of many sustainable development efforts, in particular those of smart cities. Monitoring urban air pollution is therefore required by both municipalities and the civil society to develop pollution mitigation public policies.

Current air quality monitoring is mostly operated by independent authorities. Conventional measuring stations are equipped with multiple lab quality sensors [13]. These systems are however massive, inflexible and expensive. An alternative—or complementary—solution would be to use wireless sensor networks (WSN) [7] which consist of a set of lower cost nodes that can measure information from the environment, process and relay them to some base stations, denoted as sinks [17]. The progress of electrochemical sensors, that are smaller and cheaper while keeping a reasonable measurement quality, makes the use of WSN for air quality monitoring viable [10]. The main advantage of the use of WSN for air pollution monitoring is to obtain a finer spatiotemporal granularity of measurements, thanks to the resulting lighter installation and operational costs [11]. Although some WSN-based air quality monitoring systems are already operating [3, 4, 8], the deployment issue of these tiny nodes while taking into account the precision of the resulting network has not yet been investigated.

Minimizing the deployment cost is a major challenge in WSN design. The problem consists in determining the optimal positions of sensors and sinks so as to cover the environment and ensure network connectivity while minimizing the deployment cost [19]. The deployment is constrained by the cost of the nodes and sinks, but also by operational costs such as the energy spent by the nodes. The network is said connected if each sensor can communicate information to at least one sink [18]. The coverage issue has often been modeled as a k-coverage problem in which at least k sensors should monitor each point of interest. Most research work on coverage uses a simple detection model which assumes that a sensor is able to cover a point in the environment if the distance between them is less than a radius called the detection range [2]. This can be true for some applications like presence sensors but is not suitable for pollution monitoring. Indeed, a pollution sensor detects pollutants that are brought in contact by the wind. The notion of detection range is thus irrelevant in this context. Therefore, a deployment model is still needed for the air quality monitoring application.

In this paper, we investigate the deployment of wireless sensor networks (WSN) used for building a three-dimensional mapping of pollution concentrations. We base on interpolation methods to evaluate the accuracy of a given wireless sensor networks topology. Then, we present an optimization model for optimal air pollution mapping. We consider in our simulations a 3D model of air pollution dispersion based on real experiments that we have performed in wind tunnels. Our contribution is to analyze the performances of different 3D WSN topologies in terms of the trade-off between the economical cost of the infrastructure and the quality of the reconstructed air pollution mapping in terms of precision.

The remainder of this paper is organized as follows. First, we review in Sect. 2 the most used methods in the estimation of air pollution concentrations. Then, we present in Sect. 3 the formulation used for assessing the accuracy of a given WSN topology. After that, we present the simulation data set and the obtained results in Sect. 4. Finally, we conclude and give some perspectives in Sect. 5.

2 Air Pollution Mapping

As claimed in the introduction, our goal is to evaluate how much the estimation of pollution concentrations by a given WSN topology is good. Air quality estimation allows to determine pollution concentrations of locations where no sensor is deployed, and this based on pollution concentrations gathered by the deployed sensors [9]. Three major methods are used to do so: atmospheric dispersion, interpolation and land-use regression [6].

Atmospheric dispersion models take as input locations of pollution sources, the pollutant emission rate of each pollution source and meteorological data in order to measure the pollutant concentration at a given location [6]. The obtained concentrations can then be calibrated using the measurements of sensors.

Interpolation methods formulate the estimated concentration $\hat{\mathcal{Z}}_p$ at a given location $p \in \mathcal{P}$ as a weighted combination of the measured concentrations $\mathcal{Z}_q, q \in \mathcal{P} - \{p\}$ [16]. The weights of the measured concentrations $\mathcal{W}_{pq}$ can be evaluated in a deterministic way based on the distance between the location of the measured concentration and the location of the estimated concentration. In this case, which is called the Inverse Distance Weighting interpolation, $\hat{\mathcal{Z}}_p$ is evaluated using formula (1). The concentration weights can also be evaluated in a stochastic way, the most used method doing so is called kriging.

$$\hat{\mathcal{Z}}_p = \frac{\sum_{q \in \mathcal{P}-\{p\}} \mathcal{W}_{pq} * \mathcal{Z}_q}{\sum_{q \in \mathcal{P}-\{p\}} \mathcal{W}_{pq}} \tag{1}$$

The last method is land-use regression models, which are a kind of stochastic regression models [5]. The idea behind these models is to evaluate the pollution concentration at a given location based on the concentrations of locations that are similar in terms of land-use parameters such as the elevation and the distance to the closest busy road.

In the next section, we present the placement model allowing to determine sensor optimal positions in such a way that the estimation error is minimized, and hence evaluate the trade-off between the number of sensors and the accuracy of the reconstructed pollution map. In order to design our air quality coverage formulation, we use the so-called inverse distance weighting interpolation as interpolation method. Our choice is motivated by the fact that in this latter, weights are given in a deterministic way, which allows to integrate them into the ILP deployment model.

3 Optimization Model of Pollution Mapping

3.1 Inputs and Objective Function

We consider as input of our model the map of a given urban area that we call the deployment region. We start by discretizing the deployment region in order to get a set of points $\mathcal{P}$ approximating the urban area at a high-scale ($|\mathcal{P}| = \mathcal{N}$). Our goal is to be able to determine with a high precision the concentration value at each point $p \in P$. We ensure that for each point $p \in P$, either a sensor is deployed or the pollution concentration can be estimated with a high precision based on the data gathered by the neighboring deployed sensors.

In general case, the set $\mathcal{P}$ is thus considered as the set of potential positions of WSN nodes. However, in smart cities applications, some restrictions on node positions may apply because of authorization or practical issues. When this is the case, we do not consider as potential positions the points $p \in P$ where sensors cannot be deployed. We use decision variables x_p to specify if a sensor is deployed at point p or not. All the potential positions of sensors are supposed linked to the base station, thus we focus only on the constraint of pollution mapping. The objective function to minimize is thus given as follows.

$$\mathcal{F} = \sum_{p \in \mathcal{P}} x_p \tag{2}$$

3.2 Constraints

Using numerical atmospheric dispersion models, we first get simulated pollution concentrations that may be considered as reference pollution concentrations. This does not mean that these reference concentrations are real but they reflect the best today's pollution knowledge. Let $\mathcal{Z}_p$ denote the reference concentration value at point p. Given the set of selected points where sensors will be deployed $\{p \textit{ where } x_p = 1\}$, we evaluate the estimated pollution concentrations $\hat{\mathcal{Z}}_p$ at points $\{p \textit{ where } x_p = 0\}$ based on reference values corresponding to the selected points, i.e. based on $\mathcal{Z}_p \textit{ where } p \in \{p \textit{ where } x_p = 1\}$, as follows.

$$\begin{cases} \hat{\mathcal{Z}}_p = \dfrac{\sum_{q \in \mathcal{P}-\{p\}} \mathcal{W}_{pq} * \mathcal{Z}_q * x_q}{\sum_{q \in \mathcal{P}-\{p\}} \mathcal{W}_{pq} * x_q}, p \in \mathcal{P} \,\&\, x_p = 0 \\ \displaystyle\sum_{q \in \mathcal{P}-\{p\}} \mathcal{W}_{pq} * x_q > 0, p \in \mathcal{P} \,\&\, x_p = 0 \end{cases} \tag{3}$$

We ensure that the denominator of $\hat{\mathcal{Z}}_p$ is never equal to zero using the second part of (3). The $\mathcal{W}_{pq}$ parameter is the correlation coefficient between points p and

q and is calculated using (4) based on the distance between the two points. $\mathcal{D}(p, q)$ is the distance function. α is the attenuation coefficient of the correlation distance, this means that for greater values of α, very low correlation coefficients are assigned to far points. The last parameter of (4) is the maximum correlation distance, which defines the range of correlated neighboring points of a given point.

In order to take into account the impact of the urban topography on the dispersion of pollutants, let $\mathcal{D}$ be the shortest distance along the roads network. This allows to assign small correlation values to points that are separated by buildings, even if they are close.

$$\mathcal{W}_{pq} = \begin{cases} \frac{1}{\mathcal{D}(p,q)^{\alpha}} \text{ if } q \in Disc(p,d) - \{p\} \\ 0 \text{ if } q \notin Disc(p,d) \end{cases} \quad (4)$$

In order to ensure that the concentration is estimated with high precision at points where no sensor is deployed, we introduce the constraint (5). The $\mathcal{E}_p$ parameter corresponds to the estimation error that is tolerated at point p. The choice of different values of $\mathcal{E}_p$ in function of p allows to assign low tolerated estimation errors to locations that are sensitive to air quality such as hospitals, primary schools, etc.

$$\left|\hat{\mathcal{Z}}_p - \mathcal{Z}_p\right| \le \mathcal{E}_p, \quad p \in \mathcal{P} \;\&\; x_p = 0 \quad (5)$$

By replacing $\hat{\mathcal{Z}}_p$ by its expression given in (3), we obtain the coverage constraints (6) and (7).

$$\left|\frac{\sum_{q\in\mathcal{P}-\{p\}} \mathcal{W}_{pq} * \mathcal{Z}_q * x_q}{\sum_{q\in\mathcal{P}-\{p\}} \mathcal{W}_{pq} * x_q} - \mathcal{Z}_p\right| \le \mathcal{E}_p, p \in \mathcal{P} \;\&\; x_p = 0 \quad (6)$$

$$\sum_{q\in\mathcal{P}-\{p\}} \mathcal{W}_{pq} * x_q > 0, p \in \mathcal{P} \;\&\; x_p = 0 \quad (7)$$

4 Simulation Results

The constraints introduced in the previous section can be linearized in order to obtain an Integer Linear Program. The details of the linearization are presented in [1]. The resulting ILP takes as an input a set of potential positions for the deployment of sensors, a set of points where the error of estimation has to be bounded and the ground truth of pollution concentrations. The output is the topology of the minimum cost wireless sensor network respecting the bound on pollution estimation error.

In the following, we first describe the ground truth taken as input, generated in an experimental wind tunnel emulating an actual street canyon. We then study the cost-precision trade-off in three different ways. We focus on a vertical plan and show the impact of the targeted precision on the cost of the infrastructure. We also investigate the impact of the quality of sensors, which is an important cost-factor of the devices. We then constrain the deployment to achievable positions in a urban area and limit the evaluation of the precision at positions where people may be exposed to the pollutants. We finish by investigating the impact of a longitudinal variation of the pollution concentration on the cost of a full 3-dimensional deployment of sensors.

4.1 Ground Truth Pollution Concentration in a Street Canyon

When studying city scale, 2-dimensional deployments of sensors, one can take as input historical data of pollution dispersion over the map of the city [1]. In order to evaluate the cost-precision trade-off for a 3 dimensional mapping of the pollution concentrations in a street, a more detailed dataset is required. In particular, the vertical dispersion of the pollutants has to be known.

The ground truth that are used in this paper are measurements generated in an instrumented wind tunnel test bed. The experimental set up emulates a street canyon described in Fig. 1. The emulated street is 100 m long (Y axis, coordinates in $[-50, 50]$), 20 m large (X axis, $[-10, 10]$) and 20 m high (Z axis, $[0, 20]$). The details on the wind and pollutant emissions as well as the physics involved for scaling the measurements on the wind tunnels into the pollution concentrations on the emulated street are found in [12]. The pollution emissions of a steady state urban vehicular traffic is emulated along the longitudinal $(0, Y, 0)$ axis, the wind being perpendicular to it. The pollution concentrations are constant along this dimension. Sensors are deployed in the vertical $(X, 0, Z)$ plan.

The ground truth concentrations that are considered in our experiment are in the zone of interest depicted in Fig. 1: the square of the width of the street and 10 m high that corresponds to the zone where people can be exposed and where pollution can get into the first two to three floors in apartments.

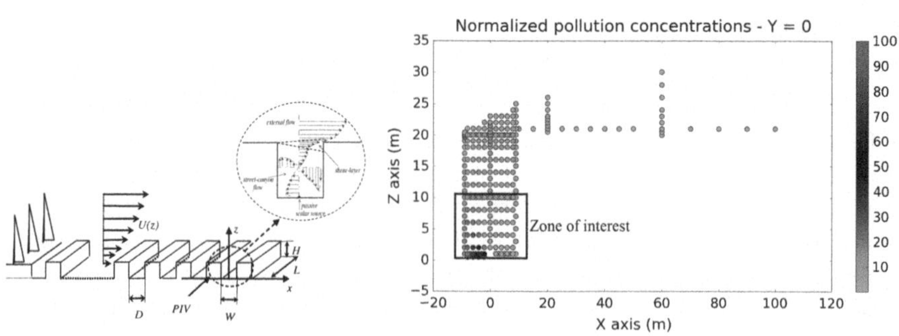

Fig. 1 Wind tunnel set up and measurements in the (X, O, Z) plan [12]

4.2 Precision Cost on a Vertical Plan

In this first scenario, we evaluate the cost-precision trade-off in the mapping of the (X, O, Z) vertical plan. The cost of the infrastructure is mainly the number of sensors to be deployed. The precision of the mapping is evaluated by the maximum difference between a ground truth and the result of the linear interpolation of the deployed sensors measurements. Here, the set of potential positions for sensors and the set of points where the precision of the mapping is evaluated are the same. They are the points of a 2-D regular grid over the zone of interest of the wind tunnel data set, as depicted in Fig. 2, together with the cost-precision trade-off obtained by our model.

As expected, the cost of the infrastructure decreases when the precision of the interpolation is more tolerated. Interestingly, increasing the attenuation of the correlation improves on the linear interpolation when a high precision is required. As a matter of fact, when looked at a small scale, the diffusion pattern of the pollution is quite different from a linear field. With a weak attenuation in the linear interpolation, the values of the distant sensors have a too strong impact on the estimation, introducing errors, hence requiring a higher density of sensors. When a lower precision is required, the small variations on the lower-left side of the street (because of the wind

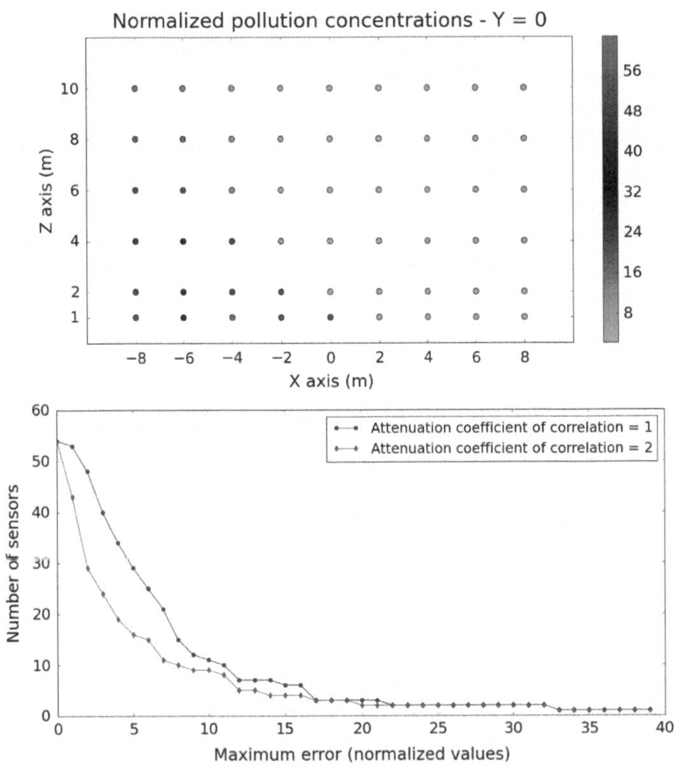

Fig. 2 Ground truth at potential sensor positions and cost-precision trade-off

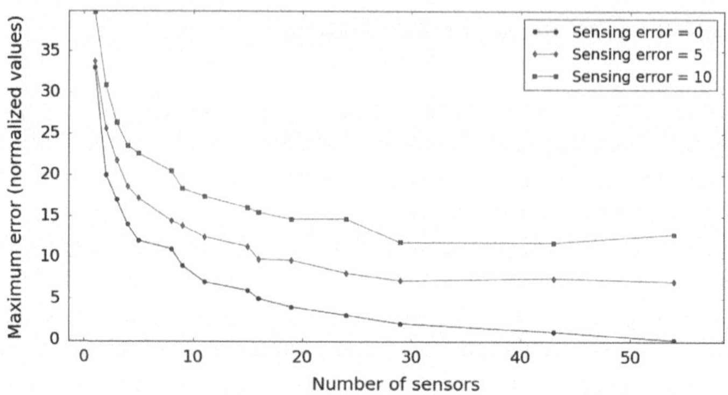

Fig. 3 Quality of sensor versus precision

direction in this scenario) fall within the error margin and the field of concentration values is closer to a linear one. Hence, the lesser impact of the attenuation.

We now investigate the impact of sensing quality. If we assume that the sensors are accurately calibrated, one of the most important cost factors is the quantity of random errors that is added to the sensor's readings and which depends on the quality of the power source and the electronic components of the sensor. In this simulation, we consider that these errors are a Gaussian noise of a mean $\langle\langle sensing\_error \rangle\rangle$. The impact of the sensing errors on the precision of estimation is depicted in Fig. 3. The impact of the sensing varies with the density of deployed sensors. When a sensor is deployed at each potential position, the maximum error of the estimation is the the raw sensing error. When the deployment is sparse, the maximum error combines the errors induced by the linear interpolation without considering sensing errors, plus the sum of the noises on the sensors contributing to the estimation. These noises being in this case less important than the interpolation errors, their impact on the maximum error is less significant.

4.3 Realistic Deployment and Citizen's Exposure

In practical situations, the sensors cannot be deployed at the potential positions used in the previous scenario. In order to deploy a sensor, one need a urban furniture or a wall. In the following, we restrict the potential positions to be at the vertical of sidewalks: no sensors are allowed on the roadway. Obviously, this will decrease the precision of the estimation.

On the other hand, the positions at which the precision matters are those that have an impact on the exposure of people. We therefore focus on the precision of the estimation on points lower than 2 m (where someone is directly exposed), and points close to walls (since these are in interaction with buildings). The resulting set of potential positions is depicted in Fig. 4.

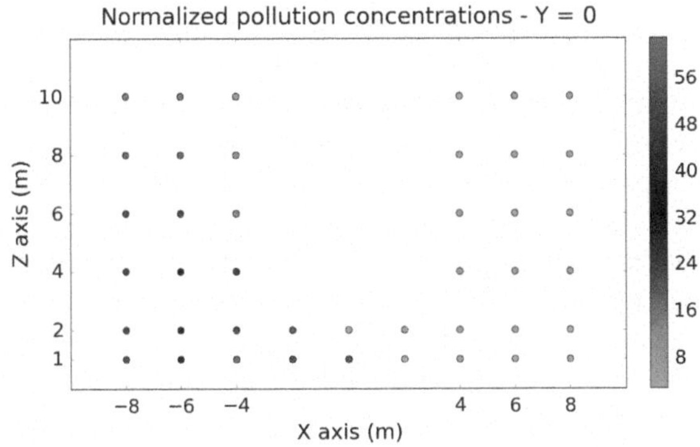

Fig. 4 Meaningful positions for citizen's exposure

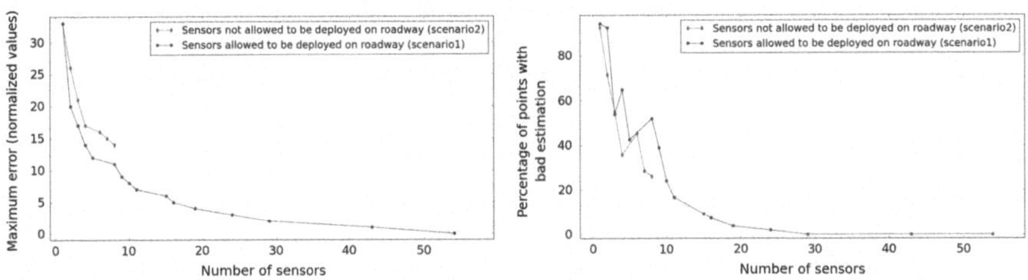

Fig. 5 Cost-precision tradeoff—focus on citizen's exposure and reading

We depict the obtained results in Fig. 5, left hand side. As expected, the fact that no sensor can be deployed on the roadway increases the maximum error. The plot stops when the best achievable precision is reached: additional sensors do not improve the result. The maximum error obtained is high in particular because there is a peak of concentration on the roadway.

Another viewpoint on the quality of estimation is the proportion of the mapping that is accurate. When communication about air quality toward citizens is at stake, actual values of pollution concentration are not given. Authorities prefer more readable "Air Quality Indicators" (AQI) which are some kind of discretization of the pollution concentration into classes. The results in Fig. 5 right hand side depict the percentage of points on which the estimation gives a wrong AQI. As expected, few exceptions apart, when the maximum absolute error decreases, the proportion of wrong AQI decreases also. More surprisingly, the restricted deployments gives less wrong AQIs. Indeed, if the positions on the roadway are prone to higher errors, there is only a small number of them.

4.4 3D Mapping

In the following, we consider the impact of the variability of the pollution concentration on the cost of the infrastructure required for producing a three dimensional mapping.

We extend the 2D grid used in the previous scenario into a 3D one by considering different planes along the Y axis. We consider 9 possible values for Y $\in [-40, 40]$, i.e. a plane each 10m, in order to avoid the two ends of the street, where very different phenomenon may occur. Let $S(x, y, z)$ be the concentration value at point (x, y, z). The 2D grid used in the previous simulation case corresponds to the concentrations at the center of the street, $S(x, 0, z)$.

The dataset that is taken as input has been produced with homogeneous traffic assumptions. Consequently, pollution concentrations are constant along the longitudinal axis (Y). In order to generate longitudinally varying concentrations, we use the sinusoidal model given in (8). This model is theoretical and does not claim to represent a real situation.

$$S(x, y, z) = \left(\frac{cos(\frac{y}{40} * \phi)}{m} + \frac{m-1}{m} \right) * S(x, 0, z) \tag{8}$$

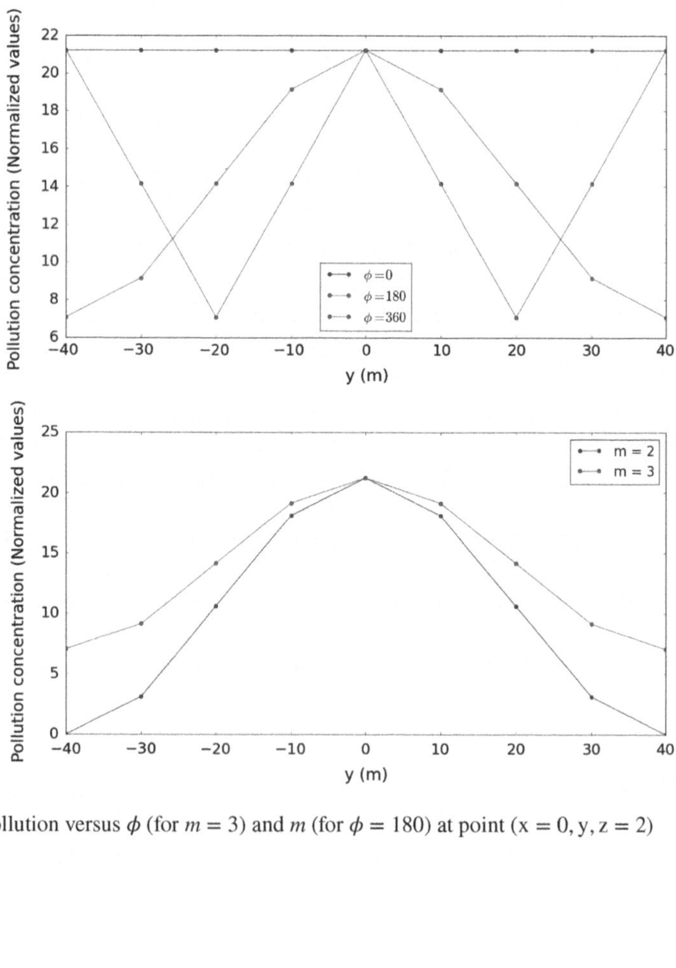

Fig. 6 Pollution versus ϕ (for $m = 3$) and m (for $\phi = 180$) at point (x = 0, y, z = 2)

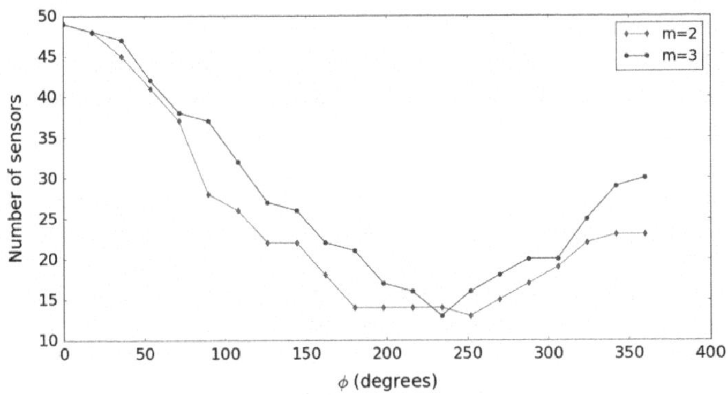

Fig. 7 Cost of a 3D deployment versus longitudinal variability

Two parameters characterize the variation of pollution concentrations along the Y-axis, ϕ and m. The m parameter allows to define the range of the concentrations: $S \in [\frac{m-2}{m}, 1], m \geq 2$.

The ϕ parameter defines the variation of concentration from a plane y to a neighboring one. It somehow captures the presence of peaks in the production of pollutants where (8) takes maximal values as depicted in Fig. 6.

Figure 7 depicts the number of optimally deployed sensors depending on parameters ϕ and m for a given precision. Surprisingly, the situations requiring the highest number of sensors to deploy are those when the pollution concentration varies the less: the scenarios with $m = 3$ cost more than those with $m = 2$, and when $\phi = 0$, the concentration is constant along the Y axis. When $\phi = 360$, the increase of the cost of infrastructure may be an artifact of the combination of the periodicity of the variation and the discretization of the space by the potential positions. Indeed, the values of the concentrations are constant on the positions in Fig. 6.

The explanation of this phenomenon is yet to be confirmed and investigated. We conjecture that the reason comes from the isotropy of the correlation function in the interpolation. Indeed, in our case the Y axis is a very particular direction and the isotropy of the interpolation does not take that into account. In particular, the correlations in the X Z plans are different than the one in the Y axis. In a future work, we will investigate the integration of the bias induced by the urban environment in the interpolation function to improve on the cost-precision tradeoff.

5 Conclusion and Future Work

In this paper, we investigate the cost-precision tradeoffs in 3D air pollution mapping using wireless sensor networks. Our main contribution is the analysis of different WSN topologies, confronting their deployment cost—mainly the number of deployed sensors—and the accuracy of a linear interpolation of their readings to build a maps of pollution concentrations. We present and apply our deployment model on pollution concentrations obtained from a testbed made of wind tunnels, emulating the diffusion of the pollution emitted by a steady state traffic flow in a typical urban canyon. We show how the deployment cost evolve with the estimation error that is tolerated.

As a future work, we plan to improve our correlation function in order to take into account the bias induced by urban topography and weather conditions.

Acknowledgments This work has been supported by the "LABEX IMU" (ANR-10-LABX-0088) and the "Programme Avenir Lyon Saint-Etienne" of Université de Lyon, within the program "Investissements d' Avenir" (ANR-11-IDEX-0007) operated by the French National Research Agency (ANR).

References

1. Boubrima, A., Bechkit, W., Rivano, H.: Optimal deployment of dense wsn for error bounded air pollution mapping. In: International Conference on Distributed Computing in Sensor Systems (DCOSS 2016). IEEE (2016)
2. Chakrabarty, K., Iyengar, S.S., Qi, H., Cho, E.: Grid coverage for surveillance and target location in distributed sensor networks. IEEE Trans. Comput. **51**(12), 1448–1453 (2002)
3. Devarakonda, S., Sevusu, P., Liu, H., Liu, R., Iftode, L., Nath, B.: Real-time air quality monitoring through mobile sensing in metropolitan areas. In: Proceedings of the 2nd ACM SIGKDD International Workshop on Urban Computing. p. 15. ACM (2013)
4. Hasenfratz, D., Saukh, O., Walser, C., Hueglin, C., Fierz, M., Thiele, L.: Pushing the spatio-temporal resolution limit of urban air pollution maps. In: 2014 IEEE International Conference on Pervasive Computing and Communications (PerCom), pp. 69–77. IEEE (2014)
5. Hoek, G., Beelen, R., De Hoogh, K., Vienneau, D., Gulliver, J., Fischer, P., Briggs, D.: A review of land-use regression models to assess spatial variation of outdoor air pollution. Atmos. Environ. **42**(33), 7561–7578 (2008)
6. Jerrett, M., Arain, A., Kanaroglou, P., Beckerman, B., Potoglou, D., Sahsuvaroglu, T., Morrison, J., Giovis, C.: A review and evaluation of intraurban air pollution exposure models. J. Exposure Sci. Environ. Epidemiol. **15**(2), 185–204 (2005)
7. Kumar, A., Kim, H., Hancke, G.P.: Environmental monitoring systems: a review. Sens. J. IEEE **13**(4), 1329–1339 (2013)
8. Marjovi, A., Arfire, A., Martinoli, A.: High resolution air pollution maps in urban environments using mobile sensor networks. In: 2015 International Conference on Distributed Computing in Sensor Systems (DCOSS), pp. 11–20. IEEE (2015)
9. Marshall, J.D., Nethery, E., Brauer, M.: Within-urban variability in ambient air pollution: comparison of estimation methods. Atmos. Environ. **42**(6), 1359–1369 (2008)
10. Mead, M., Popoola, O., Stewart, G., Landshoff, P., Calleja, M., Hayes, M., Baldovi, J., McLeod, M., Hodgson, T., Dicks, J., et al.: The use of electrochemical sensors for monitoring urban air quality in low-cost, high-density networks. Atmos. Environ. **70**, 186–203 (2013)
11. Rajasegarar, S., Havens, T.C., Karunasekera, S., Leckie, C., Bezdek, J.C., Jamriska, M., Gunatilaka, A., Skvortsov, A., Palaniswami, M.: High-resolution monitoring of atmospheric pollutants using a system of low-cost sensors. IEEE Trans. Geosci. Remote Sens. **52**(7), 3823–3832 (2014)
12. Salizzoni, P., Soulhac, L., Mejean, P.: Street canyon ventilation and atmospheric turbulence. Atmos. Environ. **43** (2009)
13. Air Rhône-Alpes: The air quality monitoring organization of the lyon agglomeration. http://www.air-rhonealpes.fr (2016). Accessed 27 Jan 2016

14. International Agency for Research on Cancer: Iarc: Outdoor air pollution a leading environmental cause of cancer deaths, on. http://www.iarc.fr/en/media-centre/iarcnews/pdf/pr221_E.pdf (2016). Accessed 27 Jan 2016
15. World Health Organization: Burden of disease from household air pollution for 2012, on http://www.who.int/phe/health_topics/outdoorair/databases/FINAL_HAP_AAP_BoD_24March2014.pdf (2016). Accessed 27 Jan 2016
16. Wong, D.W., Yuan, L., Perlin, S.A.: Comparison of spatial interpolation methods for the estimation of air quality data. J. Exposure Sci. Environ. Epidemiol. **14**(5), 404–415 (2004)
17. Yick, J., Mukherjee, B., Ghosal, D.: Wireless sensor network survey. Comput. Netw. **52**(12), 2292–2330 (2008)
18. Younis, M., Akkaya, K.: Strategies and techniques for node placement in wireless sensor networks: a survey. Ad Hoc Netw. **6**(4), 621–655 (2008)
19. Zhu, C., Zheng, C., Shu, L., Han, G.: A survey on coverage and connectivity issues in wireless sensor networks. J. Netw. Comput. Appl. **35**(2), 619–632 (2012)

A Novel Architecture with Dynamic Queues Based on Fuzzy Logic and Particle Swarm Optimization Algorithm for Task Scheduling in Cloud Computing

Hicham Ben Alla, Said Ben Alla, Abdellah Ezzati and Ahmed Mouhsen

Abstract Cloud computing is an emerging high performance computing paradigm for managing and delivering services using a large collection of heterogeneous autonomous systems with flexible computational architecture. Task scheduling is one of the most challenging aspects to improve the overall performance of the cloud computing such as response time, cost, makespan, throughput etc. Task scheduling is also essential to reduce power consumption, processing time and improve the profit of service providers by decreasing operating costs and improving the system reliability. This paper focuses on Task Scheduling using a novel architecture with Dynamic Queues based on hybrid algorithm using Fuzzy Logic and Particle Swarm Optimization algorithm (TSDQ-FLPSO) to optimize makespan and waiting time. The experimental result based on an open source simulator (CloudSim) show that the proposed TSDQ-FLPSO provides an optimal balance results, minimizing the waiting time, reducing the makespan and improving the resource utilization compared to existing scheduling algorithms.

Keywords Task scheduling · Cloud computing · TSDQ-FLPSO · Fuzzy logic · PSO algorithm · CloudSim

H. Ben Alla (✉) · S. Ben Alla · A. Ezzati · A. Mouhsen
LAVETE laboratory, Science and Technical Faculty,
Mathematics and Computer Science Department,
Hassan 1 University, 26000 Settat, Morocco
e-mail: hich.benalla@gmail.com

S. Ben Alla
e-mail: saidb_05@hotmail.com

A. Ezzati
e-mail: abdezzati@gmail.com

A. Mouhsen
e-mail: mouhsen.ahmed@gmail.com

R. El-Azouzi et al. (eds.), *Advances in Ubiquitous Networking 2*, Lecture Notes in Electrical Engineering 397, DOI 10.1007/978-981-10-1627-1_16

1 Introduction

Recently, with the rapid evolution of information technology, the Cloud Computing has been one of the most important computing paradigms, as an emerging technology that describe the concept of providing services over networks on demand. According to the NIST [1] the Cloud Computing refers to the concept of allowing for users to request a variety of services like storage, computing power, applications at anytime, anywhere and in any quantity, on the basis that pay only what they use. However, the consumers use the services through the cloud service delivery model, and do not know where the services are located in the infrastructure [2]. The cloud architecture has three levels of services: SaaS (Software as a service), PaaS (Platform as a Service) and IaaS (infrastructure as a Service).

Cloud task scheduling is a NP complete problem which consists of m machines and n tasks with different characteristics such as processing times, priority, etc. In the process of scheduling the tasks, the cloud scheduler receives the tasks from the users and maps them to available resources, taking into consideration the characteristics, attributes, and requirements of tasks, as well as the resource parameters and properties. So, an efficient/optimal task scheduling algorithm considers the load balancing of the system by achieving a good and efficiency utilization of resource with maximum profit and a high performance computing. To achieve these goals, this paper presents a novel architecture for task scheduling based on Dynamic Queues algorithm, Fuzzy logic and Particle Swarm Optimization algorithm (TSDQ-FLPSO). The rest of the paper is organized as follows: Sect. 2 presents related works. In Sect. 3, the proposed work is described. Section 4 discusses the experiment setup and simulation results of the proposed work. The paper gives a conclusion in Sect. 5.

2 Related Works

Recently, most of works discuss the concept of task scheduling in cloud computing, and aim to achieve and ensure a good performance and maximum utilization of resources on the basis of requirements of users and the Cloud Provider. So some single objective can influences the tasks scheduling process such as: Cost, QoS, energy consumption, waiting time, deadline, and makespan that means the overall execution time of all the tasks.

In the paper [3], authors proposed a novel dynamic task scheduling algorithm based on improved genetic algorithm (IGATS). The proposed algorithm works take into consideration the scalability of the cloud, and can effectively improve the throughput and reduce the execution time of task scheduling. In the paper [4], the key role of algorithm proposed is the QOS-driven based on the priority of task which in turn is computed using many task attributes such as user privilege, expectation, and the length of task, next, this task scheduled onto the service which has a minimum completion time. The results show that the proposed algorithm achieves good performance and load balancing by the priority and the completion time. The authors in paper [5], propose an algorithm to optimize the bi-objective makespan and cost using meta heuristic search techniques for scheduling independent tasks. The paper proposes a new variant of continuous Particle Swarm Optimization (PSO) algorithm, named Integer-PSO to solve two objective task scheduling problem in cloud.

In the paper [6], a novel task scheduling algorithm MQoS-GAAC with multi-QoS constraints is proposed, considering the time-consuming, expenditure, security and reliability in the scheduling process. The algorithm integrates ant colony optimization algorithm with genetic algorithm (GA). The results experiments show that this algorithm has preferable performance both in balancing resources and guaranteeing QoS. In the paper [7], the authors focus on a Cost-deadline Based Task Scheduling Algorithm which takes care of deadline and cost based on the concept of space shared policy. The experimental results illustrated that the proposed approach is more effective in defined parameters as task profit, task penalty, throughput, provider profit and user loss. The paper [8] present a task scheduling using a multi-objective nested Particle Swarm Optimization algorithm (TSPSO) that can solve the task scheduling problem. The aim of the work is to optimize energy and processing time. The experimental results show that the proposed algorithm provides an optimal balance results for multiple objectives. The authors in paper [9] design and evaluate an efficient dynamic fuzzy load balancing algorithm based on fuzzy system, which use different parameters such as memory, bandwidth and disk-space. The fuzzy algorithm proposed could efficiently predict the virtual machine where the next job will be scheduled, and the simulation results shows good performances with respecting the response time and data center processing time.

3 Proposed Architecture

3.1 Scheduling Problem Description

As one of the most interesting aspects of the Cloud Computing, The process of tasks scheduling has become an important issue to be solved, due the big overlap between the users and the cloud provider requirements, such as the quality of service (QoS), the cost of service, the priority of users, etc. Whereas the cloud service providers search to gain maximum profit and satisfying users when schedule their tasks with optimal way and minimum execution time as well as minimum waiting time.

In a cloud computing environment as shown in the Fig. 1, a large set of independent tasks of different size have submitted by different users to be handled by the Cloud Provider. The choice of the tasks to be served is determined by multiple factors and QoS requirements assured by broker that play a key role in this process.

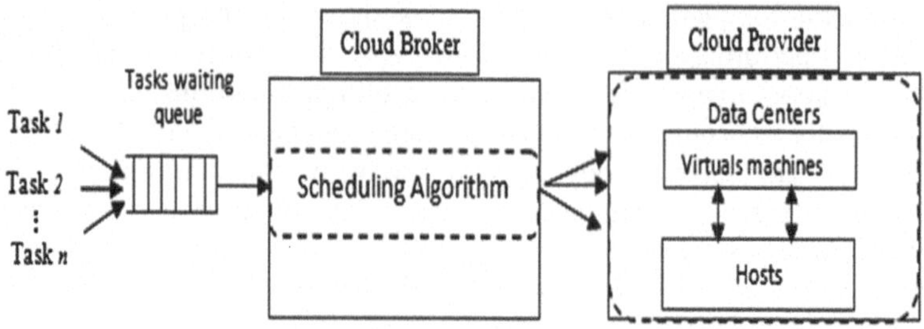

Fig. 1 Cloud scheduling environment

The cloud broker is the main component of tasks scheduling process, which is responsible for making scheduling decisions of task to specific and particular resource. The user and provider satisfaction in the Cloud Computing can be attained by assuring a good QoS and maximum profit for user and Cloud Provider respectively. However, there are some issues to be taken into account. Firstly, when users submit their tasks, they join the queue of entire system, and have to wait while the resources are not yet available, what effectively extending the queue length of system, and increasing the waiting time, so this queue have to be managed with a better method rather than FCFS policy. Secondly, in the process of handling the tasks, a most of parameters can be considered as single objective or multi-objective simultaneously. Where the makespan is one, it refers to the time spend for executing all tasks. In fact, the makespan has a direct effect on utilization of resources. In other words, when the resources utilization increases, the makespan surely decrease. Therefore, an optimal tasks scheduling algorithm should be designed and implemented in the Cloud broker to satisfy the QoS constraints imposed by cloud users in one hand, on other hand perform load balancing among virtual machines which effectively improve the resource utilization and maximize the profit earned by the cloud provider. On the basis of issues mentioned above, the main objectives of the proposed architecture are: (a) To find the right order that can reduce the waiting time of tasks, and minimize the length of queue, by applying a Particle Swarm Optimization algorithm (PSO). (b) To dispatch the tasks among dynamic queues based on Decision Algorithm and scheduling the tasks using an algorithm based on Fuzzy Logic and PSO. These algorithms aim to optimize specific performance metrics, particularly; we focus to minimize the makespan, maximize utilization of resources, and reduce the cost of services demanded.

3.2 *Particle Swarm Optimization (PSO)*

Particle swarm optimization (PSO) is a population based stochastic optimization technique was first introduced in 1995 [10], inspired by social behavior of bird flocking or fish schooling. The PSO algorithm consists of a set of potential solutions evolves to approach a convenient solution (or set of solutions) for a problem. It is used to explore the search space of a given problem to find the settings or parameters required to maximize a particular objective. It can be implemented and applied easily to solve various function optimization problems. PSO is initialized by creating a group of random particles (solutions) and then searches for optimum solution in the problem space by updating generations. We consider that the search space is d-dimensional. In every iteration, each particle is updated by following two position values, p_i called personal best (pbest), is the best position achieved so long by particle i and p_g called global best (gbest), is the best position found by the neighbors of particle i. After finding the two best values, the particle updates its velocity and positions with following Eqs. (1) and (2):

$$v_i^{t+1} = \omega . v_{id}^t + c_1 . r_1 . \left(p_i^t - x_i^t\right) + c_2 . r_2 . \left(p_g^t - x_i^t\right) \tag{1}$$

$$x_i^{t+1} = x_i^t + v_i^{t+1} \tag{2}$$

where v_i^t and x_i^t are the component in dimension d of the ith particle velocity and position in iteration t respectively. c_1,c_2 are constant weight factors.r_1, r_2 are random factors in the [0, 1] interval. ω is the inertia weight. For calculating p_g, the particle used is depends on the type of neighborhood selected such as global (gbest) or local (lbest). To calculate p_g, in the case of global neighborhood, all the particles are considered. However, in the local case, neighborhood is only composed by a certain number of particles among the whole population. The parameters ω, c_1 and c_2 must be selected properly for increasing the capabilities of PSO algorithm [11]. The inertia weight is an important parameter for providing balance between exploration and exploitation process, and can affects the overall performance of the algorithm in finding a potential optimal solution in less computing time. For this, many strategies have been proposed to choose the proper value of ω, such as Chaotic Inertia Weight [12], The Linearly Decreasing Inertia Weight strategy (LDIW) [13], Random Inertia Weight (RIW) [14] and fuzzy particle swarm Optimization (FPSO) [15]. However, the original PSO version is designed for continuous function optimization problems, not for discrete function optimization problems. So, a binary version of PSO (BPSO) algorithm was developed to solve discrete function optimization problems [16]. In the binary version, the particle position is not a real value, but an integers in {0, 1}. The logistic sigmoid function of the particle velocity is used as the probability distribution for the position, that is, the particle position in a dimension is randomly generated using that distribution. The logistic sigmoid function shown in (3), can be used for the needs that the probability stay in the range of [0, 1], in other word, it used to limit the speed of the particle:

$$S\left(v_i^{t+1}\right) = \frac{1}{1 + e^{-v_i^{t+1}}} \tag{3}$$

The equation that updates the particle position becomes the following:

$$x_i^{t+1} = \begin{cases} 1 & \text{if } r_3 < S\left(v_i^{t+1}\right) \\ 0 & \text{otherwise} \end{cases} \quad (4)$$

where r_3 is a random factor in the [0, 1] interval.

3.3 Fuzzy Logic Controller

The Fuzzy logic is a mathematical logic that attempts to solve problems by assigning values to an imprecise data in order to calculate the most accurate decision possible. The fuzzy controller uses a form of quantification of imprecise information (input fuzzy sets) to generate by an inference scheme, which is based on a knowledge base of control, a precise control force to be applied on the system.

The logical controller has three main components as shown in Fig. 2: (a) a Fuzzifier component, where the information is quantified by means of fuzzy sets; (b) a fuzzy inference engine component, which convert the input fuzzy sets into control force fuzzy sets, through rules collected in the knowledge base, and aggregate the resulting fuzzy sets, and (c) a Defuzzifier component, where the output fuzzy information (aggregated fuzzy set) is converted into a precise value.

3.4 Proposed Architecture Description

In the Fig. 3, a novel architecture based on TSDQ-FLPSO is proposed. The tasks arrive in the queue, once all tasks are queued, the optimization waiting time algorithm is applied to calculate the waiting times of all possible tasks sequences, then, return the minimum value, which refers to the right order of the tasks that can minimize the waiting time as well as the queue length.

The function fitness used to calculate the solution of each particle, and examine the solutions to find an optimal solution for the problem under consideration is given in (5) and (6).

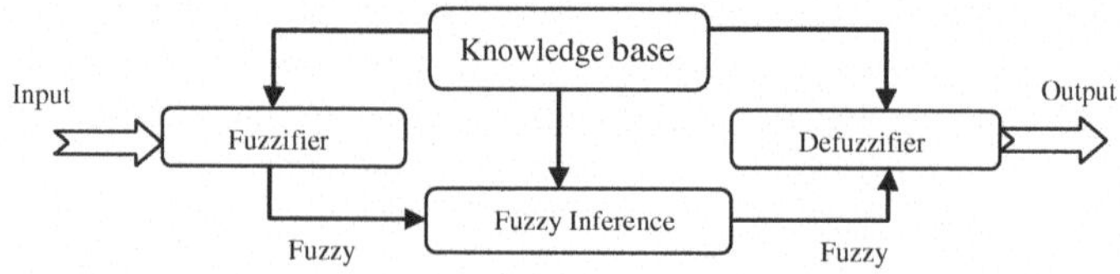

Fig. 2 Architecture of the fuzzy logic controller

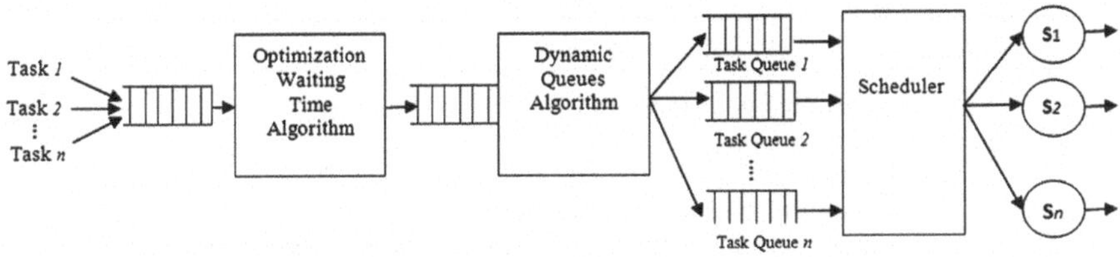

Fig. 3 Scheduling architecture proposed

$$\text{Fitness} = \text{Min}T_{WT} \tag{5}$$

where T_{WT} is the total waiting time of tasks given by:

$$T_{WT} = \frac{1}{n}\sum_{i=1}^{n} WT_{Task\ i} \tag{6}$$

where $WT_{Task\ i}$ is the waiting time of task *i*, and *n* is the number of tasks in the queue. Next, when we get the best/optimal orders of tasks, a Task Scheduling based on Dynamic Queues Algorithm (TSDQ) that manages the tasks automatically is applied. The TSDQ Algorithm start to calculate the sum of task length until arrive a threshold, then make decision to create a queue and dispatch the appropriate tasks to this queue, and execute the algorithm again until to dispatch all tasks arrived. Applying the TSDQ will generate a dynamic queue on the basis of a decision threshold. The TSDQ-FLPSO algorithm steps are presented as follows:

(a) Create list of tasks received.
(b) Keep the Burst time of each task.
(c) Calculate the total waiting time of each combination possible of tasks.
(d) Applying PSO to get the minimum waiting time.
(e) Return the best sequence with minimum waiting time founded.
(f) Calculate the Decision Threshold Function, and keep the best value.
(g) Create an empty queue Q. Next, Put tasks in the queue while the accumulate sum of task length is less than the best Decision Threshold. Else create the next queue.
(h) For each queue generated, calculate the execution time of tasks, then Apply FLPSO algorithm to get best fitness solution.
(i) Scheduling tasks among VMs on the basis of the optimal allocation scheme.

After creating dynamic queue on the basis of the TSDQ, the scheduler receives all queues generated, selects each queue and scheduling tasks to appropriate resource (servers) using an hybrid algorithm based on Fuzzy Logic and PSO algorithm. The pseudo-code of FLPSO is described in Algorithm 1.

Algorithm 1. Pseudo-code of FLPSO

```
Initialize particles with random position and velocity
For all iterations do
    For all particles do
        Calculate the positions and velocity
        Calculate Fuzzy Inertia weight value based parameter controller
        Calculate Fuzzy fitness value based parameter controller
        Evaluate fitness solution
        Update personal and global best
        Update velocity and position
    end for
end for
```

The main of objective of TSDQ-FLPSO algorithm is to assign the most suitable resources to tasks based on the computational capabilities of the resources and the tasks' characteristics. In fact, According to the relationship between the users and providers requirements, an efficient process of tasks scheduling require an optimal/efficient algorithm that take into consideration the characters of tasks and resources configuration. So, the proposed algorithm using Fuzzy Logic and PSO algorithm considers these requirements. The performance metric here is to minimize processing time of tasks as well as the makespan, achieve a high utilization of resources which mean keep the resources as busy as possible while scheduling tasks, and reduce the cost of utilization of service demanded. The Fuzzy theory is used in two steps in the proposed algorithm. First, it is used to calculate the inertia weight. Second, it is used to calculate the fitness value of each PSO particle. Accordingly, using the fuzzy logic controller assure that the PSO converge in each iteration to the optimum solution value based of the fitness values.

The Fuzzy Logic used in these two steps focus to assigns tasks to the most suitable resources with optimizing the performance metrics under consideration. In the first step, as the inertia weight w is one of the most important adjustable parameters in PSO. So, it was adaptively adjusted using the fuzzy theory based on iteration parameter as input and the inertia weight as output. In the second step, the input parameters of Fuzzy Logic are the task length, CPU speed, RAM memory and the status of the resources (i.e. the occupancy rate of the VM), and the output is the fitness value of the particle. However, the objective function of the PSO algorithm is to get the maximum value of all fitness values founded by Fuzzy logic for all possible tasks sequences on cloud resources, in other words, to get the most suitable resources to process the tasks. The scheduling of tasks using the FLPSO algorithm can effectively maximize the resource utilization and reduce the makespan as well as the cost of using resources. To get the output value of Fuzzy Logic, the fuzzy inference convert the input fuzzy sets into control force fuzzy sets, through rules collected in the knowledge base. In this work, we use Mamdani [17] inference system. So, to calculate inertia weight value, the input variables is the iteration, and the output variable is the inertia value. However to calculate the fitness value, we

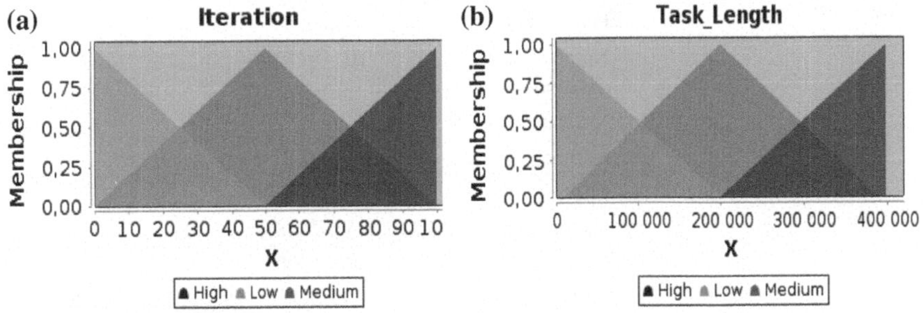

Fig. 4 **a** Fuzzy sets for iteration parameter **b** Fuzzy sets for task length parameter

Fig. 5 Fuzzy sets for status of VM parameter

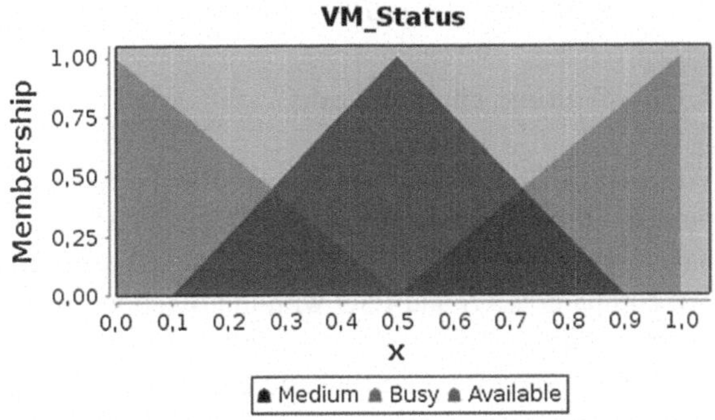

use the task length, CPU speed, RAM memory and the status of VM as parameters, the output determines the fitness value of each particle. The Figs. 4 and 5 show the fuzzy sets for the iteration, task length and the status of VM parameters, which are created by using The JFuzzyLogic [18, 19] library for the Fuzzification, Defuzzification and to define the rule blocks.

4 Testing and Analysis of the Experimental Results

In order to evaluate the proposed architecture based on TSDQ-FLPSO comparing with other algorithms, a simulation is implemented using Cloudsim 3.0.3 simulator [20] that allow modeling and simulating extensible Clouds, and testing the performance of developed application service in a controlled environment. The CloudSim toolkit supports modeling of Cloud system components such as data centers, virtual machines (VMs) and resource provisioning policies. Cloudsim give several functionalities such as generate a different workload, with different scenarios and perform robust tests based on the custom configurations. The simulation is done under the following conditions described in Table 1. In the simulation, the NASA Ames iPSC/860 log was used to generate different Workload [21].

Table 1 Resource parameters

Parameters	Values
Number of datacenter	10
Number of hosts	2–6
Number of VMs	5–30
MIPS	10000–30000
VM memory(RAM)	256–2048
Bandwidth	500–1000
Tasks source	Workload NASA Ames iPSC/860 log

4.1 Measures of Effectiveness

4.1.1 Waiting Time of Tasks

We assume that 3 tasks are received to be handled by the Cloud Provider. The number of possible combination of tasks is 3! = 6. In this paper, 10 experiments have done where the burst time has changed for different tasks in each experiment. We compare FCFS and PSO with the function fitness to get the minimum value of waiting time.

Table 2 shows the results of comparison in term of waiting time of each sequence and total average waiting time. For example, in the first serial in Table 2, There are three tasks (T1, T2, T3) which requires processing time (15, 10, 29) respectively. Using FCFS algorithm, the waiting time in this case is 13,33. However, by using PSO algorithm, the sequence founded as best solution that give the minimum time is T2, T1, T3, and the result is 11,67. The total average waiting time using PSO is lesser than using FCFS algorithm. The PSO can effectively give an optimized solution, which increases speed and efficiency of managing the tasks queue by minimizing the waiting time of tasks, and reduce the queue length.

Table 2 Comparison of results

Serial no.	Burst time of tasks			FCFS	PSO
	T1	T2	T3		
1	15	10	29	13,33	11,67
2	20	55	49	31,67	29,67
3	33	22	11	29,33	14,67
4	38	8	73	28,00	18,00
5	23	50	39	32,00	28,33
6	59	34	44	50,67	37,33
7	12	20	3	14,67	6,00
8	9	24	13	14,00	10,33
9	40	23	18	34,33	19,67
10	19	10	45	16,00	13,00
Total average waiting time				264,00	188,67

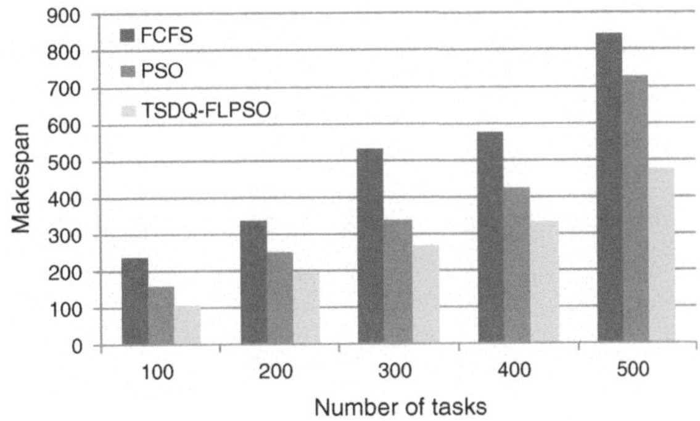

Fig. 6 Average makespan with different number of tasks

4.1.2 Execution Time of Tasks

In the simulation experiments, we compare the proposed algorithm TSDQ-FLPSO with FCFS, and PSO. Several experiments with different parameter setting were performed to evaluate the efficiency of proposed algorithm. The evaluation is done using independent tasks from workload data, and by executing the simulation 20 times, which represents the number of independents experiments done. The performance evaluation is compared in term of the average makespan with different number of tasks and average Resource Utilization. Figure 6 shows that the makespan is increasing when the number of tasks increases. But the makespan of our proposed algorithm TSDQ-FLPSO is better and is reduced when compared with the FCFS algorithm and PSO algorithm.

4.1.3 Resource Utilization

The Resource Utilization is an important metric in the scheduling task process, the objective is to maximize and achieve a high utilization of resources. To calculate the Resource utilization, the formula (7) is used where T_{VM_i} is the time taken by the VM_i to finish all tasks, and N is number of resources [22]:

$$\text{Average Resource Utilization} = \frac{\sum_{i=1}^{N} T_{VM_i}}{\text{Makespan} \times N} \tag{7}$$

Figure 7 shows that the Average Resource Utilization of our proposed algorithm TSDQ-FLPSO outperforms FCFS and PSO algorithms. This is because that TSDQ-FLPSO keeps the resources as busy as possible while scheduling tasks. This metric is gaining significance as service providers want to earn maximum profit by renting limited number of resources.

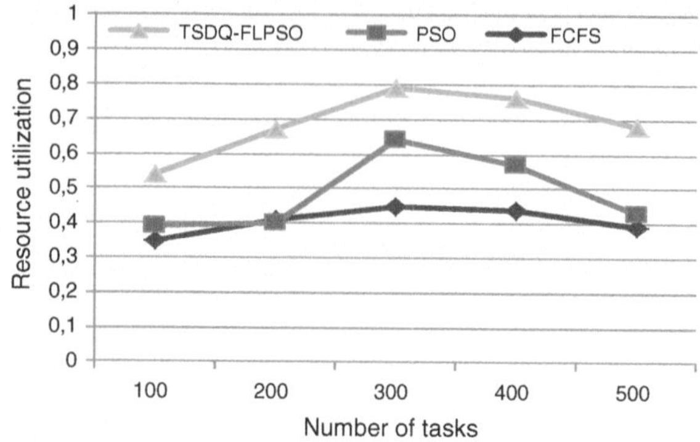

Fig. 7 Average resource utilization

5 Conclusion

This paper proposes a novel architecture with Dynamic Queues for task scheduling based on hybrid algorithm using Fuzzy Logic and Particle Swarm Optimization algorithm (TSDQ-FLPSO). The experimental results demonstrated the effectiveness of the proposed algorithm and the average performance is better than other two algorithms FCFS and PSO in term of minimizing the waiting time as well as the length of queue, reducing the makespan, achieving a high utilization of resources and considering the load balancing when distributing tasks to resources in the cloud. As a future work, we intend to enhance our architecture by considering and adding more QoS parameters.

References

1. Mell, P., Grance, T.: The NIST Definition of Cloud Computing. National Institute of Standards and Technology, the NIST Special Publication 800-145. ACM (2011)
2. Armbrust, M., Fox, A., Griffith, R., Joseph, A.D., Katz, R., Konwinski, A., Lee, G., Patterson, D., Rabkin, A., Stoica, I., Zaharia, M.: A view of cloud computing. Commun. ACM **53**(4), 50–58 (2010). ACM
3. Ma, J., Li, W., Fu, T., Yan, L., Hu, G.: A novel dynamic task scheduling algorithm based on improved genetic algorithm in cloud computing. In: Wireless Communications, Networking and Applications, pp. 829–835. Springer (2015)
4. Wu, X., Deng, M., Zhang, R., Zeng, B., Zhou, S.: A task scheduling algorithm based on QoS-driven in cloud computing. In: Procedia Computer Science, vol. 17, pp. 1162–1169. Elsevier (2013)
5. Beegom, A., Rajasree, M.: A particle swarm optimization based pareto optimal task scheduling in cloud computing. Lecture Notes in Computer Science, pp. 79–86. Springer (2014)
6. Dai, Y., Lou, Y., Lu, X.: A task scheduling algorithm based on genetic algorithm and ant colony optimization algorithm with multi-QoS constraints in cloud computing. In: 7th International Conference on Intelligent Human-Machine Systems and Cybernetics, pp. 428–431. IEEE (2015)

7. Himani, Sidhu, H.: Cost-deadline based task scheduling in cloud computing. In: Second International Conference on Advances in Computing and Communication Engineering, pp. 273–279. IEEE (2015)
8. Jena, R.: Multi objective task scheduling in cloud environment using nested PSO framework. In: Procedia Computer Science, vol. 57, pp. 1219–1227. Elsevier (2015)
9. Zulkar Nine, M., Azad, M., Abdullah, S., Rahman, R.: Fuzzy logic based dynamic load balancing in virtualized data centers. In: International Conference on Fuzzy Systems (FUZZ-IEEE), pp. 1–7. IEEE (2013)
10. Kennedy, J., Eberhart, R.: Particle swarm optimization. In: International Conference on Neural Networks, vol. 4, pp. 1942–1948. IEEE (1995)
11. Clerc, M., Kennedy, J.: The particle swarm—explosion, stability, and convergence in a multidimensional complex space. IEEE Trans. Evol. Comput. **6**(1), 58–73 (2002). IEEE
12. Feng, Y., Teng, G., Wang, A., Yao, Y.: Chaotic inertia weight in particle swarm optimization. In: Second International Conference on Innovative Computing, Information and Control (ICICIC 2007), p. 475. IEEE (2007)
13. Xin, J., Chen, G., Hai, Y.: A particle swarm optimizer with multi-stage linearly-decreasing inertia weight. In: International Joint Conference on Computational Sciences and Optimization, pp. 505–508. IEEE (2009)
14. Yue-lin, G., Yu-hong, D.: A new particle swarm optimization algorithm with random inertia weight and evolution strategy. In: International Conference on Computational Intelligence and Security (CISW 2007), pp. 199–203. IEEE (2007)
15. Kumar, S., Chaturvedi, D.: Tuning of particle swarm optimization parameter using fuzzy logic. In: International Conference on Communication Systems and Network Technologies, pp. 174–179. IEEE (2011)
16. Kennedy, J., Eberhart, R.: A discrete binary version of the particle swarm algorithm. In: International Conference on Systems, Man, and Cybernetics. Computational Cybernetics and Simulation, pp. 4104–4108. IEEE (1997)
17. Mamdani, E.: Application of fuzzy algorithms for control of simple dynamic plant. Proc. Inst. Electr. Eng. UK **121**(12), 1585 (1974). IEEE
18. Cingolani, P., Alcala-Fdez, J.: jFuzzyLogic: a java library to design fuzzy logic controllers according to the standard for fuzzy control programming. In: International Journal of Computational Intelligence Systems, pp. 61–75. IEEE (2013)
19. Cingolani, P., Alcala-Fdez, J.: jFuzzyLogic: a robust and flexible Fuzzy-Logic inference system language implementation. In: International Conference on Fuzzy Systems (FUZZIEEE), pp. 1–8. IEEE (2012)
20. Calheiros, R., Ranjan, R., Beloglazov, A., De Rose, C., Buyya, R.: CloudSim: a toolkit for modeling and simulation of cloud computing environments and evaluation of resource provisioning algorithms. J. Softw. Pract. Experience **41**(1), 23–50 (2011). ACM
21. Parallel Workloads Archive: NASA Ames iPSC/860. http://www.cs.huji.ac.il/labs/parallel/workload/l_nasa_ipsc/
22. Kalra, M., Singh, S.: A review of metaheuristic scheduling techniques in cloud computing. Egypt. Inf. J. **16**(3), 275–295 (2015). Elsevier

A Vehicular Cloud for Secure and QoS Aware Service Provision

Mouna Garai, Slim Rekhis and Noureddine Boudriga

Abstract The Vehicular Cloud (V-Cloud) is a new technology that will have impact on traffic management and road safety. It enables the release of various reliable and low cost services, including on board advertisement, multimedia, and traffic management. This paper presents a secure and QoS aware three-layer Vehicular Cloud architecture enabling a tree-based connection of vehicles to the network. We propose a certificate based authentication and privacy preservation protocol for secure vehicular communication. The proposed protocol allows generating and using public key certificate with zone based temporal vehicles identities to prevent Sybil and Tracking attacks.

Keywords Vehicular Cloud · Security · Sybil attacks · Qos

1 Introduction

The development of Vehicular Ad-hoc Networks (VANET) has brought the concept of Vehicular Cloud (V-Cloud) which made it possible to release different scalable and reliable services. However, the delivery of these services raises several challenges. In fact, seamless connectivity, capable of handling the Quality of Service (QoS) requirements and value added services constraints (e.g., reducing delays and jitter), is needed. Moreover, Security and privacy problems need to be addressed if V-Clouds will be widely adopted. Many forms of security attacks against V-Cloud have emerged which could affect the privacy, the resource

M. Garai · S. Rekhis (✉) · N. Boudriga
Communication Networks and Security Research Lab (CNAS),
University of Carthage, Tunis, Tunisia
e-mail: slim.rekhis@gmail.com

M. Garai
e-mail: mouna.garai@gmail.com

N. Boudriga
e-mail: noure.boudriga2@gmail.com

R. El-Azouzi et al. (eds.), *Advances in Ubiquitous Networking 2*, Lecture Notes in Electrical Engineering 397, DOI 10.1007/978-981-10-1627-1_17

availability, and the anonymity of users. Among these attacks, we distinguish Sybil attacks, by which malicious vehicles could forge and claim multiple identities at the same time to disturb the traffic routing and prepare for denial of service or selfish attacks.

The security challenges for VANET clouds have not yet been addressed widely in the literature and the main challenges are expected to be the same faced by VANET and cloud computing. However, authors in [1] identified and analyzed a number of security and privacy challenges in vehicular cloud. Several challenges resulting by vehicular clouds' features, e.g., authentication of highly mobile nodes and complex trust relationships between vehicles in multi-hop connections, are discussed. Moreover, a directional security scheme is described to provide an appropriate security architecture and overcome some security and privacy challenges in vehicular clouds. The authors recommended that future works should provide feasible security and privacy solutions in vehicular cloud. In [2], authors were interested in analyzing the security challenges and the potential privacy threats in Vehicular Cloud Computing (VCC). They addressed some major design issues that will affect the future implementation of VCC, and developed two types of clouds. The first one is an infrastructure based cloud (IVC) and the second cloud is called autonomous vehicular cloud (AVC). In [3], a comprehensive cloud framework was developed to provide a reliable and secure vehicular communication network. This framework proposed a novel interconnection of vehicular cloud architecture that provides several cloud services such as Authentication as a service (AaaS) to enforce security in Vehicular Networks. The interconnection of cloud services rises several issues such as privacy and interoperability, but the proposed framework does not provide solutions to solve these issues. In [4], Road safety applications are predicted to offer a wide range of services such as Computing as Service (CompaaS), Storage as a Service (STaaS), Network as a Service (NaaS), Cooperation as a Service (CaaS), Entertainment as a Service (ENaaS), Information as Service (INaaS) and Traffic-Information as Service (TIaaS). The authors also proposed another service called WaaS, which provides the vehicles with timely warning messages in cases of hazardous situations along with the necessary security measures that must be taken by the vehicles in the AoI (Area of Interest). Privacy-enhancing solutions and data-centric misbehavior solutions can be found in [5]. The proposed scheme enables the law enforcement authorities to trace the route of a particular node within an expected time span to revoke it. However, the proposed plan is computationally costly and many other system attacks are not reported. The problems of privacy, anonymous withdrawal, and route tracing via VANET have been addressed in [6] using Clouds (VuC). The authors proposed a lightweight privacy-aware revocation and route tracing mechanism for VuC where beacons are stored in the cloud infrastructure to be used for route tracing and anonymous withdrawal. In [7], a hierarchical Vehicular Cloud architecture integrating Cloud Computing and VANET networks was proposed. In this paper, a resource management and vehicle connectivity solution is proposed, but it does not address the security issues. In [8] a study of the limitations of the previously proposed vehicular ad hoc network-based secure navigation protocols, is performed. Based the findings, a new model for secure navigation systems based on the concept of vehicular cloud, is proposed, and a secure navigation protocol based on single-use anonymous credentials, is designed. Through an analytical model, the proposed protocol is shown to provide an efficient protection mechanism against insider threats and data leakage. A VCC Service-oriented Security Framework (VCC-SSF) is proposed in [9]. The proposed Framework provides two services: (a) security services guaranteeing Confidentiality, Integrity, Availability, and Privacy Protection for users, and (b) new user-oriented payment management and active accident management services. However, issues related to key distribution and management in VCC environments, are not handled.

In this paper, we propose a secure and QoS-aware Cloud architecture. We propose a Connection as Service to the cloud while taking into consideration the different QoS requirements of the provisioned services. To ensure a secure communication between mobile vehicles moving throughout the vehicular cloud, a certificate-based vehicle authentication protocol is provided, allowing to protect against Sybil attacks, provide vehicles authentication, and prevent tracking attacks.

The contribution of the paper is three-fold. First, we propose a tree-based connection scheme providing rapid and simple access of vehicles to the cloud. The mobile vehicles are behaving as mobile brokers; they authenticate other vehicles and attach them to a tree topology network, allowing to maximize the network coverage beyond the area covered by Road Side Units (RSUs). Second, we propose a solution to prevent and detect Sybil attacks in the vehicular Cloud. This solution provides a certificate-based vehicle authentication protocol and controls access control to the network. Third, we propose a solution based on a secure and periodic generation of temporal vehicles' identities to preserve vehicles' anonymity and prevent tracking attacks.

2 Network Architecture

The proposed architecture aims to enforce QoS guarantees for cloud services. We consider the Vehicular Cloud Network (VCNet) architecture as illustrated in Fig. 1, which is divided into three layers. The first layer is the Central Cloud (CC), where cloud services are deployed. This cloud is established among a group of servers in the Internet to support complex computations and massive data storage. The CC often uses virtualization to encapsulate the requested services in virtual machines (VMs) such that they can be configured, deployed, migrated, suspended, and consolidated in multi core servers. The second layer is the R-Cloudlet. The R-Cloudlet is a local cloud which is composed of a set of neighboring RSUs that are attached to a set of dedicated local cloud servers which are connected to the internet through the Central Cloud. Since access to the central cloud incurs long latency due to wide area network delays and networks contention, the use of R-Cloudlets promotes the provision of QoS. In fact, R-Cloudlets aim to bring cloud services closer to mobile vehicles. Each R-Cloudlet includes a broker, say R- Broker, which acts as an intermediary between the consumer (i.e., mobile vehicle) of a cloud computing service and the service providers (Central Cloud or Internet). The R-Cloudlet is accessible by vehicles that pass through the radio coverage of the set of RSUs and activate their 4G interface. In each R-Cloudlet, servers aggregate their resources to run Virtual Machines (VMs) and bring services closer to consumers. The third layer is the V-Cloudlet which is proposed in order to extend the network coverage beyond the area covered by the RSUs and to provide connectivity to the CC through the R- Cloudlet. The V-Cloudlet is a local cloud established among a set of vehicles that cooperate with each other to provide an access service to any vehicle that is far from an RSU. The networks of vehicles inside a V-Cloudlet are organized in a tree topology where each intermediate vehicle, named also Vehicular Broker (V-Broker), shares its allocated resources with the child nodes in that tree. Neighboring vehicles parts of the tree are connected together using IEEE 802.11p network interfaces. The vehicle root of the tree plays the role of a gateway. It is connected to an RSU using a 4G access network, and is connected to the child vehicles using 802.11p. In the proposed vehicular cloud architecture, resources are inter-networked, that is, each V-Broker shares a part of resources with the vehicles that are directly connected to it. For efficiency, one vehicle in a V-Cloudlet can be a V-broker based on some selected metrics (e.g., connectivity to vehicles, connection lifetime, the residence time in the RSU coverage). The use of mobile brokers in the V-Cloudlet is used to help cloud customers obtaining highly efficient services even if they are located in uncovered areas. To provide a secure connection to the vehicular cloud, we propose a certificate based authentication protocol. A hierarchical certificate authority is proposed as shown in Fig. 1.

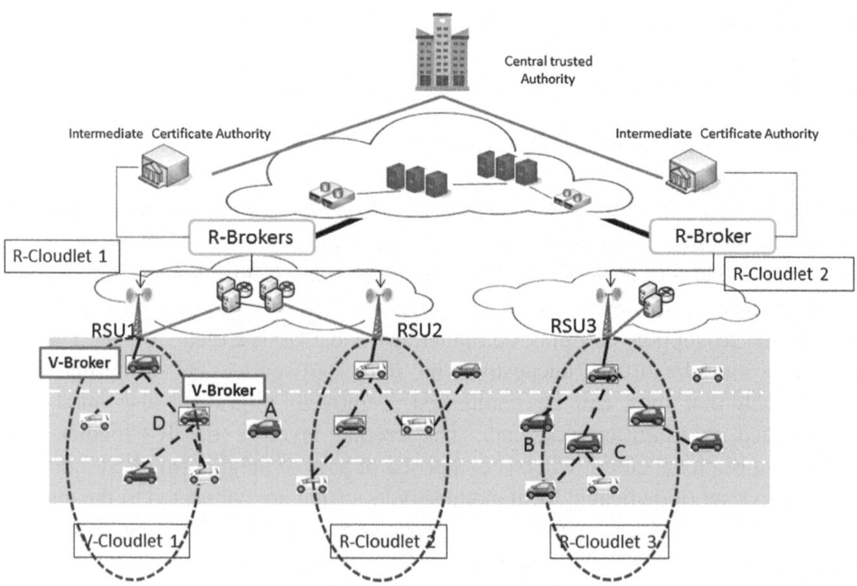

Fig. 1 Network architecture

The first level, which is the Central Certificate Authority (CCA), is the root of the system. It is located at the Central Cloud layer and is in charge of generating public key certificates for Intermediate Certification Authorities (ICA). An ICA is attached to every R- Cloudlet, and is in charge of generating certificates to vehicles connected over RSUs of the same R-Cloudlet. Only a vehicle owning a certificate related to a temporary identity generated by an ICA can connect to the related R-Cloudlet and benefit from services. An Intermediary Registration Authority (IRA) is attached to each ICA. It manages vehicles' certification requests and delivers them to the ICA to issue certificates.

3 QoS Aware Attachment Scheme

We propose a QoS aware network attachment scheme. The proposed scheme allows vehicles connected over a tree topology to control throughput, delay, loss rate and jitter. These QoS parameters are stored in a QoS vector that is periodically transmitted to the R-Cloudlet layer to monitor the connection established and prevent the drop in QoS level below the level promised during the establishment of the connection. If an unconnected vehicle requests a service with a given QoS level, in terms of throughput, loss, and delays, our system must provide a connection that can satisfy the needed QoS level. The connection offer is described with a vector collecting the QoS parameters that are guaranteed, in terms of: throughput, delay, average packet loss, and route lifetime. The QoS vector is generated and updated by each connected vehicle (whether it is V-Broker or not). To look for the suitable offer, the vehicle receives connection offers broadcasted by neighbor brokers. Let B_0 be a fixed R-Cloudlet broker (Typically an RSU), and B_i be a mobile broker vehicle. The route connecting B_i to B_0 is denoted by α_i where B_1 can be directly connected to B_0 through a 4G link, and connected to B_2 in the tree through an 802.11 connection. The remaining set of vehicles $\{B_i, B_{i-1}, \ldots, B_2\}$ are connected in multi-hop manner (using local 802.11 connections) through a set of intermediate mobile broker nodes $\mathrm{B}_{i-1}, \ldots, \mathrm{B}_1$. We denote by $\mathrm{Q}_{\alpha_i} = \langle \mathrm{T}_{\alpha_i}, \mathrm{D}_{\alpha_i}, \mathrm{L}_{\alpha_i}, \mathrm{R}_{\alpha_i} \rangle$ the QoS vector computed by the V-Broker B_i to characterize the dynamic QoS metrics of the route α_i in the tree S connecting it to the R-Cloudlet. We describe each metric in Q_{α_i} as follows:

1. T_{α_i} is the available throughput computed by the fixed broker B_0. It is equal to the difference between the bandwidth of the wireless link connecting B_0 and the vehicular broker B_1, and the total bandwidth reserved by all vehicles in the tree. Note that the reserved bandwidth can be split in two parts, a fixed sub-band guaranteed by the network, plus a second one where the network has no QoS mechanisms to guarantee the promised QoS: $T_{\alpha_i} = T(B_1, B_0) - \sum_{v \in S} T_v$, where: (a) $T(B_1, B_0)$ is the available throughput announced by the B_1; (b) T_v is the bandwidth reserved by the vehicle v connected to the tree S.
2. D_{α_i} is the transport delay computed iteratively as follows. Having received the delay $D_{\alpha_{i-1}}$ from the vehicle B_{i-1}, the broker B_i computes $D(B_i, B_{i-1})$ and sets:

$D_{\alpha_i} = \max\{D_{\alpha_{i-1}}, D(B_i, B_{i-1})\}$, where: $D(B_i, B_{i-1})$ is the delay between the B_i and its broker B_{i-1}, which is equal to the sum of the transmission delay, the decoding delay, and the queuing delay in the vehicle B_{i-1}.

3. L_{α_i}, is the average packet loss computed iteratively as follows. The vehicle B_{i-1} computes the average packet loss $L_{\alpha_{i-1}}$, of the path α_{i-1}, and forwards it to B_i. The broker B_i computes the packet loss of the link connecting the broker B_i and B_{i-1}, $L(B_i, B_{i-1})$ and sets $L_{\alpha_i} = \max\{L_{B_{i-1}}, L(B_i, B_{i-1})\}$, where: $L(B_i, B_{i-1})$ represents the percentage of frames that are dropped by the decoder in B_i if the packet arrival time exceeds the playback deadline.
4. R_{α_i} is the route lifetime observed at B_i. It is iteratively computed as follows. The vehicle B_{i-1} computes the route lifetime $R_{\alpha_{i-1}}$ and forwards it to B_i. The broker B_i computes $r(B_i, B_{i-1})$ and sets $R_{\alpha_i} = \min\{r(B_i, B_{i-1}), R_{\alpha_{i-1}}\}$, where $r(B_i, B_{i-1})$ is the lifetime of the link connecting B_i to B_{i-1}; and $R_{\alpha_{i-1}}$ is the lifetime of the route connecting the B_{i-1} to the R-Cloudlet. We denote by $r(B_i, B_{i-1})$ the lifetime of the link (B_i, B_{i-1}), i.e., the remaining time before B_i becomes out of coverage of B_{i-1}. It is equal to: $r(B_i, B_{i-1}) = (TR_{B_i} - \sigma_{i,i-1}) / |s_{B_i} - s_{B_{i-1}}|$, where TR_{B_i} is the transmission range of the vehicle B_i, $\sigma_{i,i-1}$ is the distance between the vehicle B_i and B_{i-1}, and s_{B_i} is the speed of the vehicle B_i.

4 Secure Access to the Tree

We introduce a set of security analyses that are associated to the VCNet architecture presented above. In VCNet, an external vehicle connects to the Cloud, using an access scheme based on unsecure and vulnerable communication protocols. These vulnerabilities are mainly related to lack authentication, and integrity verification.

4.1 Security Attacks in the Proposed System

In this context, we describe a set of attacks that affect the particular tree topology of the provided VCNet.

DoS Attacks: In vehicular networks, the most common destructive attack regarding communication networks is the DoS attack. Such an attack denies all services provided by the network, preventing legitimate users from using the services of the victim node. DoS attacks can be carried out in many ways; for example, a malicious vehicle forges and broadcasts a message to its neighbors that announces a good QoS offer in terms of delay and throughput. Therefore, each vehicle that wants to apply for other services and to improve its QoS, executes a handover and attaches itself to the attacker. The latter will disconnect from the old tree after a period of time, as there is a timeout triggered by the previous attachment point to keep the connection alive. Moreover, an attacker prevents the tree to be extended by not broadcasting the CAD messages received from the RSU or its father vehicle in the tree. Therefore, the vehicle, which has already executed the handover, will be unable to receive the service.

Selfish attack: In vehicular networks, vehicles cooperate to forward packets from one vehicle to another. In the case where the vehicle is not willing to cooperate with other nodes and refuse to forward packets in order to preserve its limited resources, the latter is said to be a selfish node. In VCNet, a selfish attack can be executed following several scenarios. For example, when the attacker sees that the connection requests it should forward, will allow other vehicles to benefit from additional resources, it just simply drop them, by not forwarding them to its parent node in the tree. The prevented requests will allow the selfish vehicle to spare the network resources and increase the likelihood of succeeding to get more network resources (e.g., bandwidth) later.

Sybil Attacks: In a Sybil attack, a malicious vehicle generates multiple identities, either by forging or stealing them from neighboring vehicles. It can steal identities by overhearing broadcast messages within its communication range. In such a situation, the data received from Sybil node attackers may seem as if it was received from many distinct vehicles. In VCNet network, the Sybil attacker could perform several connections using different identities. These connections can be with nodes in the same tree or to nodes located in different trees, to receive a high percentage of shared resources. This behavior prevents the network from guaranteeing the fairness of resources and can prepare for denial of service attacks if the vehicles would be prevented from getting the resources they required in the future. This attack can also impact the tree construction process, by emulating an exchange between virtual (forged identities) nodes, which lead to the construction of a false tree. Moreover, the attacker takes advantage of the network resources and prevents the tree expanding in depth (i.e., number of nodes in the route) and breadth (i.e., number of child nodes).

4.2 Security Requirements

In this section we present the requirements to be fulfilled and the solutions proposed to mitigate the aforementioned potential security threats. For our proposed system, the security requirements include authentication, integrity, and Sybil attacks detection. First, authentication ensures that the message is generated by a legitimate user. In VCNets, participants need to verify the sender authenticity and message integrity. In addition, the recipient should be able to verify that the message has not been tampered with in transit. Sender authenticity and message integrity prevents malicious outsiders from injecting rogue messages that might disrupt the normal operation of the VCNet. Second, we should have a solution toward the prevention of Sybil attacks in the VCNet. If we choose to use certificates and digital signatures to authenticate vehicles and preserve the message integrity, the certificates should not show the real identity of the vehicles to preserve their privacy by using a temporary identity of the vehicle instead of its real identity. Third, a basic privacy requirement is that an attacker cannot link messages sent by a vehicle to the real vehicle's identity, and cannot also track the vehicle's mobility by linking or correlating temporal identities together.

4.3 Authentication Protocol

To ensure an anonymous authenticity and to keep the vehicles' global identity secret, the vehicles should avoid showing their global identities to the V-Brokers. Therefore, the authentication is done by R-Broker which is able to verify the vehicle's identity and generate an available certificate using a new temporary identity. That generated certificate will be useful to perform authentication with the V-Broker. Figure 2 shows the proposed authentication protocol before starting the generation of the route requests (i.e., the protocol described in the previous section), we assume that: (a) all the vehicles in VCNet are pre-registered with a CA before they are assigned to a network; (b) the CA is the root authority that has an ICA under every RC. The ICA is in charge of distributing the keys and certificates to the RC and the vehicles, (c) each vehicle has a tamper proof device to store its private keys, and (d) a vehicle will obtain short lifetime certificate from the ICA to be used in the new zone.

When a vehicle asks for a connection that satisfies its QoS requirements, it proceeds based on the attachment process described in Sect. 3 and selects a neighbor V-Broker that offers the best QoS in the broadcasted message. The Broker (either V-broker or R-Broker) sends together with the CAD message a random number N, its certificate $Certif_B$ (if the Broker is a vehicular Broker) and the $Certif_{RB}$ of the R-Broker. All these information are signed by the Broker using its private key. If the vehicle selects the Broker's offer, it responds by a message that contains its global identity and the Random number 'N' collected from the received offer. All these information are encrypted by the key shared with the IRA ($K_{V,IRA}$). In turn, the Broker sends a signed message M to the IRA. The latter verifies the global identity of the vehicle and computes its temporary identity, ID_T, in order to be sent to the Broker. To establish a mutual authentication, the IRA generates a random number X and sends it in a message encrypted by the shared key $K_{V,IRA}$ to the Broker, together with the new vehicle temporary identity, ID_T. Upon receiving the ACK message, the Broker decrypts the received message extracts the number X, verifies the number N + 1, and sends a signed message M′ to the vehicle. The vehicle saves ID_T and extracts the number X in order to compute the output of HMAC function calculation over X and message M″. After verifying the received number X, the mutual authentication is established and the Broker sends the encrypted CSR to the IRA. The latter forwards the CSR related to the temporary identity to the ICA in order to generate the vehicle certificate. Finally, the generated certificate is forwarded to the vehicle.

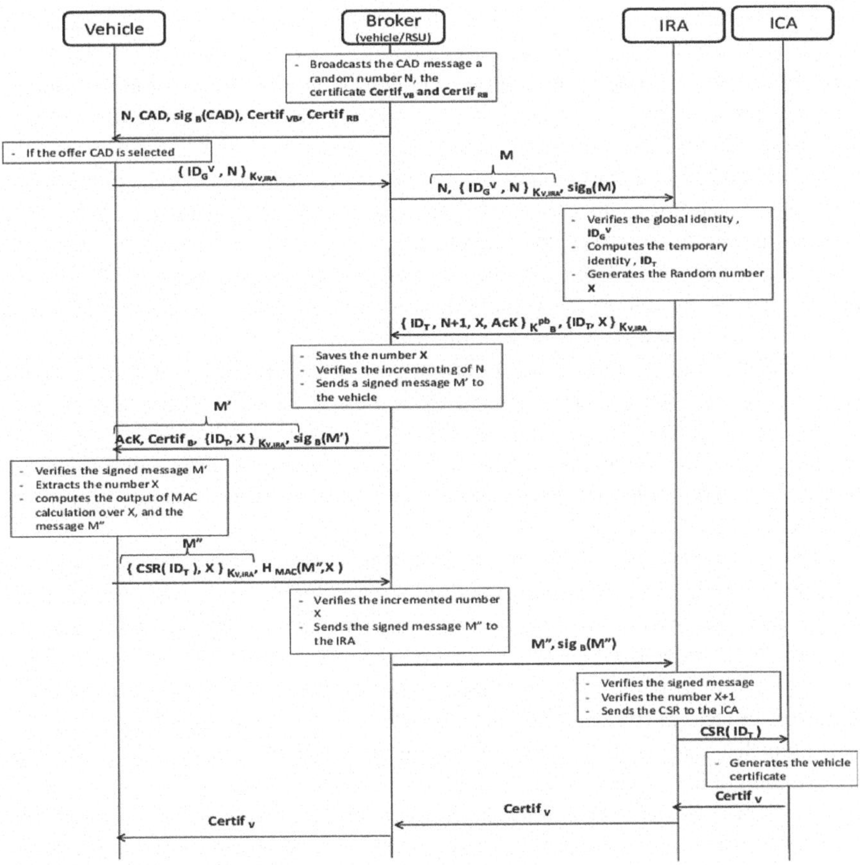

Fig. 2 Authentication protocol

5 Mobility Management

To fulfill the mobile vehicles' QoS and security requirements, and cope with the high mobility of vehicles, a Mobility management scheme is developed. Two types of hangovers are considered: (a) Inter-R-Cloudlet Handover; and (b) Intra-R-Cloudlet Handover. Let us consider the vehicle B illustrated in Fig. 1 as a node which needs to establish a handover. It listens to broadcast CAD and saves the routes that satisfy its QoS requirements in a route list, say θ. Then, it sends an Alert message, containing the Route List θ, to its current father Broker which will in turn forward it to the R-Broker. The R-Broker selects the best offers from θ. If the new point of attachment is located in another neighbor R-Cloudlet, an Inter-R-Cloudlet Handover is established. Before the HO is completely achieved, the vehicle B remains connected to the old R-Broker for a short period of time in order to estimate the delay variation. To avoid the handover delay and the delay variation caused by a modification in the node level, we use a de-jitter buffer at the R-Broker side [7]. The latter allows compensating the delay variation due to potential modifications on the used routes. In fact, it tries to maximize the end-user perception quality, by integrating parameters related to the estimation of the duration of HO and the variation of the level of attachment between the new and the old route. In addition, an estimate of the delay and delay variance is made to calculate the buffer size (i.e., the playout time). Then, the R-Broker calculates the new provisioned size of the De-jitter Buffer to estimate the quantity of supplementary data to be sent to the vehicle. It consequently increases the throughput in order to avoid a potential delay variation. Moreover, in order to keep the VM close to the mobile vehicle, the virtual machine, which serves to the execution of the requested service, needs to be migrated to the new R-Cloudlet. If the new point of attachment is located in the same R-Cloudlet, an Intra-R-Cloudlet Handover is triggered. Thus, the use and recalculation of the size of the De-jitter buffer is needed in order to avoid a potential QoS degradation. However, there is no need for a VM migration.

In the case where the vehicle, which is in a state of handover, plays the role of a V-Broker, a tricky process is performed in order to find a new point of attachment and to keep its promise to provide a connection service to the nodes that are attached to it. Indeed, when the V-Broker, say "Y", detects a decrease in QoS parameters, it first sends a warning message to its child nodes to inform them that a Handover process is triggered and they should keep their connection to it. Then, it remains in listening state for a period of time in order to receive a CAD message from its neighbors. If the V-Broker "Y" does not find any neighbor V-Broker, it tries to connect to the closest RSU. In the other case, the vehicle Y selects the best offer that satisfies its needs and sends an attachment request to the source node. In the case where none of the received offer satisfies the QoS requirement, we propose to split the tree in order to decrease the QoS requirements. The objective of Tree splitting is to keep child vehicles connected as a V-Broker and detach others vehicles. To achieve this purpose, child vehicles are ordered by the V-Broker based on the lowest required Throughput and nodes that use low throughput are detached starting with the non-broker nodes until reaching the Throughput offered by future attachment point.

5.1 Certificate Renewal

The certificate renewal consists in the generation of another certificate which has the same serial number and the same vehicle identity, but has a new validity period. In the case where a connected vehicle remains in the same R-Cloudlet for a long period of time exceeding the certificate lifetime, its certificate should be renewed in the zone to which it is connected. The certificate lifetime is pre-defined depending on the R-Cloudlet width. If the current certificate is expired, the vehicle requests a new certificate over the R-Broker to which it is authenticated. The certificate lifetime is pre-defined with respect to the R-Cloudlet width and the mean period of vehicles stay under a same R-Cloudlet coverage. Noted that the choice of certificate lifetime should also come to a compromise between: (a) a long lifetime period, which allows to avoid renewing the certificate frequently but may show the need to revoke the certificate if the vehicle leaves the actual R-Cloudlet while its certificate is still valid; and (b) a short lifetime period which allow avoiding to revoke the certificate, but leads the mobile vehicle to generate several renewal requests for the same certificate.

5.2 Certificate Revocation

As the vehicle moves from one R-Cloudlet to another, it should obtain a new certificate. If the certificate that it obtained in the old R-Cloudlet is still valid, the vehicle becomes a holder of two valid certificates. Consequently, it becomes able to conduct a Sybil attack as it holds the two certificates generated with two different identities. Thus, to protect the network against such attacks, the old valid certificate should be revoked before generating a new one. In order to inform other R-Broker that the vehicle obtains a new certificate, the new R-Cloudlet generates and broadcasts a message containing the vehicle global identity and the starting date of the certificate validity. Once this message is received by other R-Brokers, they verify if the vehicle has a valid certificate to revoke it.

6 Simulation

In this section, the system performance will be analyzed by assessing the average number of certificate renewal, the average number of certificate revocation per vehicle/R-Cloud, and the average tree-route length. The simulation setup is conducted considering a one-directional highway segment, whose length is 26 km, and which contains three lanes. The vehicle's speed ranges from 60 to 100 km/h (60 km/h in the first lane, 80 km/h in the second lane, and 100 km/h in the third lane). We set the vehicle and the RSU coverage radius equal to 200 and 1000 m, respectively. Table 1 lists the simulation parameters used, unless a change is mentioned explicitly. In our simulation, the arriving vehicle rate is defined by the injection in each timeslot of a vehicle with a probability λ in each lane of the highway. We assume that the speed of vehicles in each lane is constant and we respect the safety distance between the injected vehicles. Each R-Cloudlet is composed of a set of three neighboring RSUs .

In the first simulation, both of the Average Number of Certificate Renewal per vehicle/R-Cloud, say RnR, and the Average Number of Certificate Revocation per vehicle/R-Cloud, say RvR, with respect to the Certificate Lifetime were evaluated. As depicted in Fig. 3, the RnR decreases as long as the certificate lifetime increases.

Table 1 Value of parameters used in simulation

Parameters/Description	Values
Simulation duration	7500 time-slot (set to 2 s)
Probability of service request/vehicle	10 %
The minimum required bandwidth: T^{min}	$T^{min} = 5$ Mbps
The maximum variation beyond of T^{min}	T^{Δ}
Duration of a service	200 s
Security distance between two vehicles in each lane	5/9 of the vehicle speed in km/h

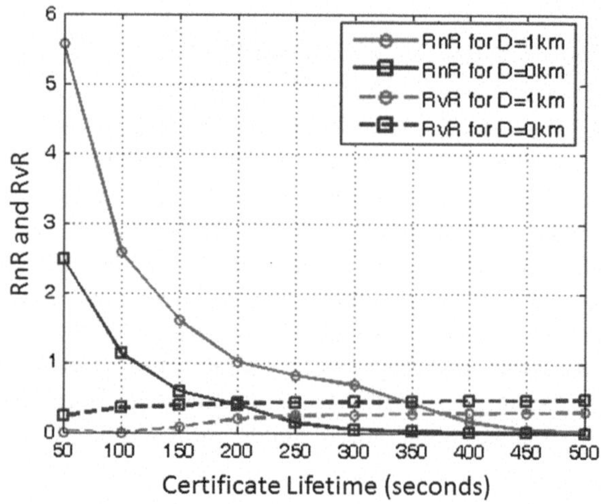

Fig. 3 RnR and RvR w.r.t. certificate lifetime

However, the longer is the certificate lifetime, the higher will be RvR. In addition, both Certificate Renewal and Revocation ratios decrease as long as the uncovered distance between RSU, say D, decreases. In fact, as the certificate lifetime increases, the probability that vehicles cross the R-Cloudlet while the generated certificate lifetime does not expired, will increase. On the other hand, the probability that vehicles having valid certificates, reach the next R-Cloudlet will increases. Consequently, the RnR will decrease and the RvR will increase. Besides, as long as the width of R-cloudlet increases the RnR increases and the RvR decreases. In fact, while the width of the R-Cloudlet increases, the certificate will be renewed more often, since the period the vehicle stays in the R-Cloudlet increases. However, when the distance between two neighbor R-Cloudlets increases, the probability that vehicles having valid certificates reach the next R-Cloulet, will decreases. Thus, the renewal process will be more executed when the distance between RSUs increases. Beside, we detect that the value of Certificate lifetime where RnR and RvR are equal, are respectively, 200 s for D = 0 km and 380 s for D = 1 km. These values represent the most convenient values of certificate lifetime for the considered architecture (i.e., length of uncovered area, uncovered distance between RSUs). In fact, when varying the certificate lifetime, we should come to a compromise between a long lifetime period, which allows to avoid renewing the certificate frequently, but may show the need to revoke the certificate if the distance between neighbor R-Cloudlet is short; and a short lifetime period which avoids revoking the certificate, but leads the mobile to generate several renewal requests for the same certificate.

Figure 4 shows a 3D graph of the Average number of Certificate Renewal per vehicle per R-Cloudlet, RnR, with respect to the Certificate lifetime (in seconds) and the arrival rate of vehicles λ. The graph shows that the RnR decreases as long as the Certificate lifetime increases. In fact, since the width of a R-Cloudlet is fixed, long lifetime Certificates took longer to be renewed during the travel over the same R-Cloudlet than short-lifetime certificates. In addition, the graph shows that while increasing the arrival Rate, λ, the RnR tends to increase at a decreasing rate, especially for long certificate lifetime. However, this increase is more important for a short certificate lifetime when the arrival rate λ increases. In fact, as long as the network becomes dense, mobile nodes are likely to obtain adjacent vehicles that meet their needs. Since, the connectivity increases, the RnR decreases as long as the certificate lifetime increases. That is why when the certificate lifetime is short, the RnR increases more strongly.

In Fig. 5, the 3D graph shows an increase of the average number of certificate revocation per vehicle, RvR with the increase of the Certificate lifetime and the arrival rate (i.e., λ). In fact, as long as the certificate lifetime increases, the probability that the connected vehicle reaches the next R-Cloudlet with a valid certificate increases, requiring consequently revoking their certificates. Moreover, since the network becomes more dense if the arrival rate increases the likelihood of successful connections increases due to the increase of the neighborhood rate in the network. Therefore, the revocation rate increases, and this increase is more important for long certificates lifetime.

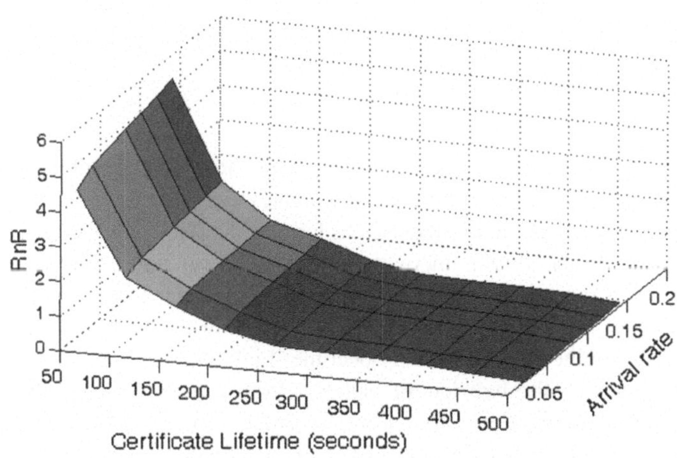

Fig. 4 Average number of certificate renewal per vehicle/R-Cloudlet, RnR

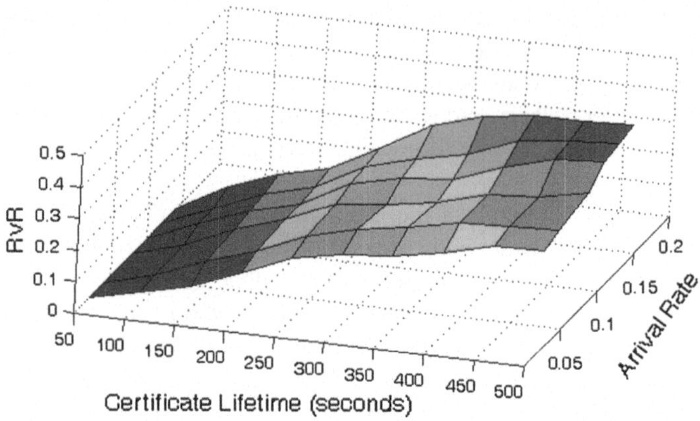

Fig. 5 Average number of certificate revocation per vehicle/R-Cloudlet, RvR

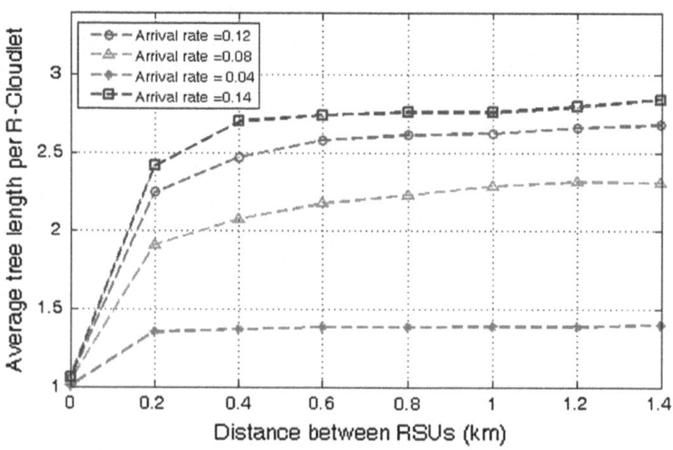

Fig. 6 Average tree length per R-Cloudlet

Figure 6 represents the Average tree length per R-Cloudlet with respect to the distance between RSUs for different arrival rates (i.e., $\lambda = 0.14$, 0.12, 0.08 and 0.04). The Average tree length increases as long as the uncovered distance between RSUs increases. In fact, in the uncovered area, in order to solve the problem of the absence of RSUs coverage, mobile nodes try to be connected to a neighbor vehicle having sufficient remaining bandwidth. Thus, as long as the uncovered distance becomes large and the network density becomes high, the tree length becomes high. Beside, we note that the tree length plateaus from a given distance between the RSUs. This plateau effect appears earlier for small λ (i.e., $\lambda = 0.04$) due the low network density which lowers the likelihood of finding a connected neighbor vehicle in the uncovered area. Also, we note that the tree length increases decreasingly while the arrival rate λ increases and the distance between RSUs increases due to the decrease of the network density. In fact, as long as the uncovered area between RSUs increases, the ability to cover this area using IEEE 802.11 connection is limited and mobile nodes are not able to find adjacent connected vehicles that meet their QoS requirements.

7 Conclusion

In this paper we provided a secure Vehicular Cloud architecture integrating Cloud Computing with Vehicular Ad-hoc Networks. We proposed a secure and QoS aware network access scheme based on a tree topology. To protect from Sybil attacks in the vehicular Cloud, we proposed a certificate-based vehicle authentication protocol to strengthen and secure access to the tree-based network topology. Moreover, in order to preserve users' anonymity and protect them from tracking attacks, we proposed a solution for the secure and periodic generation of vehicles' temporal, and an anonymous authentication protocol based on the use of these identities. In a future work, we propose to develop a secure billing model for the proposed Vehicular Cloud, and extend the simulation work to provide a comparison with the different existing Vehicular Cloud architectures.

References

1. Yan, G., Wen, D., Olariu, S., Weigle, M.C.: Security challenges in vehicular cloud computing. IEEE Trans. Intell. Transp. Syst. **14**, 284–294 (2013)
2. Yan, G., Rawat, D.B., Bista, B.B.: Towards secure vehicular clouds. In: Proceedings of International Conference, Complex, Intelligent and Software Intensive Systems, (Palermo), July 2012
3. Danquah, W.M., Altilar, D.T.: Mobile and wireless technology 2015, Ch. V. In: Cloud: A Security Framework for VANET, pp. 1–13. Springer Berlin, Heidelberg (2015)
4. Hussain, R., Oh, H.: Cooperation-aware vanet clouds: providing secure cloud services to vehicular ad hoc networks. J. Inf. Process. Syst. **10**, 103–118 (2014)
5. Hussain, R., Abbas, F., Son, J., Eun, H., Oh, H.: Privacy-aware route tracing and revocation games in vanet-based clouds. In: Proceedings of IEEE International Conference, Wireless and Mobile Computing, Networking and Communications, (Lyon), Oct 2013
6. Hussain, R., Oh, H.: A secure and privacy-aware route tracing and revocation mechanism in vanet-based clouds. J. Korea Inst. Inf. Secur. Cryptol. **24**, 795–807 (2014)
7. Garai, M.: Communication as a service for cloud vanets. In: Proceedings of the 20th IEEE Symposium on Computers and Communications (ISSC'2015), (Larnaca, Cyprus), July 2
8. Sur, C., Park, Y., Rhee, K.H.: An efficient and secure navigation protocol based on vehicular cloud. Int. J. Comput. Math. **12**, 1–20 (2014)
9. Kang, W.M., Lee, J.D., Jeong, Y.S., Park, J.H.: Vcc-ssf: service-oriented security framework for vehicular cloud computing. Sustainability **7**(2), 2028–2044 (2015)

A Conceptual Architecture for a Cloud-Based Context-Aware Service Composition

Soufiane Faieq, Rajaa Saidi, Hamid Elghazi and Moulay Driss Rahmani

Abstract In today's ubiquitous environments, more and more companies use cloud computing services to achieve their everyday operations and processes. However, the development of these services is a tedious and time consuming task. Moreover, existing solutions rarely take into account the personalization and the adaptation of services to the context of their use. In this paper, we propose an architecture for context-aware service composition in cloud environments to address the challenges mentioned above. The architecture takes advantage of service composition as a way to create new composite services from a set of atomic or composite ones, causing the reduction of development efforts and time to market, and also, the introduction of context-awareness to manage the dynamics of ubiquitous environments.

Keywords Context-awareness · Service composition · Cloud computing · Ubiquitous computing · Business processes

1 Introduction

The rapid growth in the number of mobile devices (e.g. smart phones, tablets, PDAs, wearables, etc.) and the advances achieved in wireless communications, made it

S. Faieq (✉) · R. Saidi · M.D. Rahmani
LRIT, Research Unit Associated to the CNRST (URAC 29), Faculty of Sciences,
Mohammed V University in Rabat, Rabat, Morocco
e-mail: soufiane.faieq@gmail.com

M.D. Rahmani
e-mail: mrahmani@fsr.ac.ma

R. Saidi
SI2M Laboratory, National Institute of Statistics and Applied Economics,
Rabat, Morocco
e-mail: r.saidi@insea.ac.ma

H. Elghazi
International University of Rabat, Rabat, Morocco
e-mail: hamid.elghazi@uir.ac.ma

R. El-Azouzi et al. (eds.), *Advances in Ubiquitous Networking 2*, Lecture Notes in Electrical Engineering 397, DOI 10.1007/978-981-10-1627-1_18

easier to access information from anywhere and anytime, driving us a step closer to Mark Weiser's vision of the ubiquitous computing paradigm [1]. Organizations saw this evolution as an opportunity to provide mobile support to their workers and clients, to increase their productivity, improve the organization's business processes and bring more competitive edges for their business [2, 3].

However, ubiquitous environments are described as being highly heterogeneous, dynamic and mobile, which in turn makes them highly complex. To deal with this complexity, context-awareness has been explored to provide contextual information about the involved entities. Context has been defined as "any information that can be used to characterise the situation of an entity. An entity is a person, place or object that is considered relevant or the interaction between a user and an application, including the user and applications themselves" [4].

Meanwhile, cloud computing received great attention as a way to deliver ICT services over the Internet. It is defined by the National Institute of Standards and Technology (NIST)[1] as "a model for enabling ubiquitous, convenient, on-demand network access to a shared pool of configurable computing resources (e.g., networks, servers, storage, applications, and services) that can be rapidly provisioned and released with minimal management effort or service provider interaction. This cloud model is composed of five essential characteristics, three service models, and four deployment models" [5]. The five essential characteristics are: *on-demand self-service*, for on-demand provision of resources with minimal interaction with the provider. *Broad network access*, for ubiquitous access through different types of clients (e.g. mobile phones, tablets, desktops, wearables, etc.). *Resource pooling*, to serve multiple consumers demands using a multi-tenant model. *Rapid elasticity*, enables automatic scaling of the resources according to the user's need. *Measured service*, allows the user to pay in pay-per-use model.

The three service models are: *Software as a Service* (SaaS), *Platform as a Service* (PaaS) and *Infrastructure as a Service* (IaaS). In the SaaS model, the provider offers software functionality as a service, that can accessed either through a web browser or an API. In the PaaS model, the service vendor provides the tools required to manage, develop, test, deploy applications. In the IaaS service model, the cloud provider offers resources (e.g., processing, network and storage) as services according to the consumer's needs. While the four deployment models include: *Private cloud*, where the cloud resources are used by a single organization. *Community cloud*, in which a community of organizations with similar activities share the cloud resources. *Public cloud*, where the resources are open for public use and *Hybrid cloud*, which is a combination of two or more of the deployment models mentioned above.

From a business perspective, moving ICT services to the cloud, allows organizations to benefit from the high performance and availability of cloud services in a pay-per-use model. Thus, reducing upfront costs and adding elasticity to support the execution of their IT enabled, complex business processes. In this regard, SaaS is the most adopted service model in industry. It allows consumers to use software functionality; running on a cloud infrastructure; as a service over the Internet, usually

[1] http://www.nist.gov/.

using web-based clients (navigator, web services) or APIs. However, the development of said software is usually tedious and time consuming, impeding the rapid development and interoperability of this kind of services. On an other hand, SOA (Service Oriented Architecture) introduced service composition, and has since then been widely explored and adopted, as a way to create new composite services from a set of atomic or composite ones. This had the benefits of reducing development time, cost and risks.

The goal of this paper is to propose a novel architecture for a service composition system, running on cloud environments, that is capable of: adapting to the changes in the ubiquitous environments while considering the user's preferences and situation.

The rest of the paper is organized as follows: Sect. 2 provides a summary of related works proposed in literature. In Sect. 3, we present our motivating scenario. Then, we propose our conceptual architecture in Sect. 4. Finally, we conclude and outline our future works.

2 Related Work

In literature, the service composition problem is mainly studied as a planning and optimization problems. On one hand, the planning problem consists of generating an automatic composition plan to achieve the composition goal. According to [6], most researches conducted are usually derived from workflow composition or AI planning. A recent trend is the use of MDE (Model Driven Engineering) to model services and services composition, and automate code generation [7].

On the other hand, the optimization problem deals with the optimization of QoS attributes of each participant service in the service composition, to satisfy the user's requirements. Jula et al. [8] cover the different approaches used to address QoS optimization problem; considered an NP-hard problem; in cloud service composition.

In their efforts to identify issues related to dependability, ubiquity, personalization in Web services composition, authors of [9] investigated the life cycle of Web services compositions and surveyed the main standards, research prototypes, and platforms. Only eight out of the thirty research works selected and studied in [9], include context-awareness in the service composition process. Though mostly, the context is only considered to improve user's experience, the contextual information used in the selected works is limited to a fixed set. They have also identified *services composition in ubiquitous computing* as one of the main open issues needing extensive research efforts.

In short, many research works have covered the problem of cloud and Web service compositions, but only very few of them explored context-awareness as a mean to improve the quality of the composite services, through personalization to increase the user's satisfaction and situational awareness for the composition system. Authors of [10] partially cover context-aware service composition in cloud environments based on Microsoft Azure Platform.[2] However, context information used in their approach is not formally defined in their work.

Through our conceptual architecture, we aim to address the service composition problem in ubiquitous computing via the leveraging of contextual information in the different phases of the composition life cycle and providing cloud support for the composition process.

3 Motivating Scenario

To cover all the aspects that we are trying to deal with in this paper, we suggest making travel arrangements as our motivating scenario. Though is seems trivial, but making travel arrangements covers a lot of concepts in different areas.

In service computing and SOA, travel booking is one of the most explored scenarios for Web service composition.[3, 4] Case studies include being able to book several travel related services, via Web services (e.g. booking flights, taxi cabs or rental cars, hotel rooms, restaurants, museums, etc.). These Web services are usually provided by the entities hosting the services or third parties.

In Business Process Management, businesses try to model their employees' booking arrangements for business trips or meetings, taking place abroad. An example of such processes is provided by the OMG in [11]. From this perspective, the booking process may interact with other IT-enabled services, involving other business processes (e.g. Billing, Human Resources Management, Customer Relation Management, etc.).

In context-aware computing, several works covered travel and tourism scenarios in their efforts to make travel and tourism applications more user-centered than service-centered [12], usually by means of personalization and recommendation systems. In [13], authors consider context in travel and tourism to be the bridge between a specific need and the necessary information to fulfill that requirement.

Figure 1 shows some of the services involved in the tourism industry. Transport services can be airlines, cruise lines, car rentals, cab companies and rail companies. Accommodations can be hotels and motels, apartments, guest houses, bed and breakfast establishments and cabins, while catering facilities offer a variety of outlets for food and refreshments including local restaurants, roadside joints, cafeterias, and retail outlets serving food and beverages. Attractions such as theme parks and natural attractions including scenic locations, cultural and educational attractions, monuments, events, and medical, social or professional causes are also a major component of the tourism industry. The tourist information and guidance providers include a number of services such as those offering insurance, communication, and

[2]https://azure.microsoft.com/.

[3]http://www.keller.com/xml/travelscenario.html.

[4]https://www.w3.org/2002/04/17-ws-usecase.html.

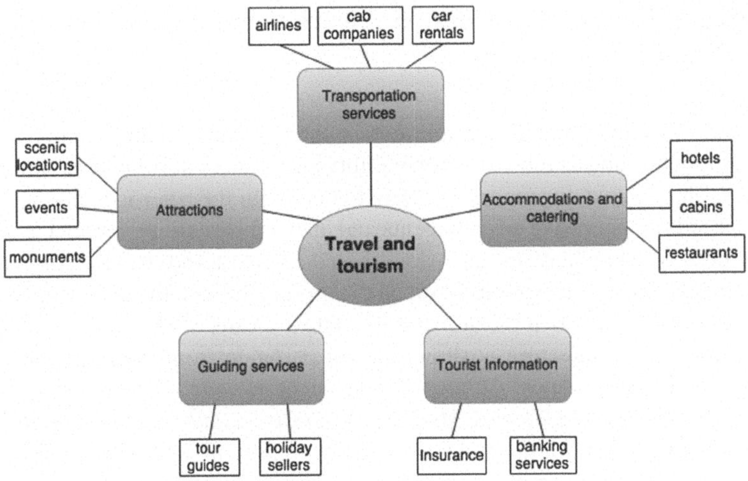

Fig. 1 Services from the tourism domain

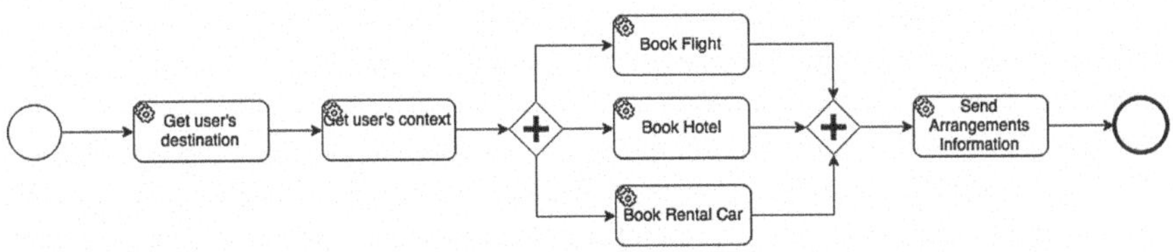

Fig. 2 Abstract process for travel planning

banking services; government agencies; tour guides; industry associations and holiday sellers.[5]

We envision that our architecture would help users deal with the complex and time consuming tasks such as travel planning, as well as provide them with better travel experience, through the use of context information (e.g. profile, preferences, interests, location, weather, etc.). Figure 2 shows the example of an abstract business process for travel planning based on service composition, specified in BPMN (Business Process Modeling Notation). In this example, the context would give us information about the user's preferences (e.g. Travel class, favorite car or airline, hotel types, budget, etc.). Another scenario, could be the use of user's interests, location or the use of other travelers' experiences to plan a tour guide for the current user. The applications of our architecture can be as rich and divers as context information can be. Also, the use of SaaS as a delivery model enables ubiquitous access to the system's services and enables the later to benefit from the rapid elasticity and measurability that characterise the cloud.

[5]https://www.dnb.co.in/Travel_Tourism/Indian_Travel_and_Tourism_Industry.asp.

4 An Architecture for Context-Aware Service Composition in Cloud Environments

In our work, we aim to provide a framework capable of merging the benefits of SOA, as a paradigm that allows reuse, interoperability and reduces development time, costs and risks; and context-awareness as a way to deal with the heterogeneity and complexity of today's pervasive environments and also improve user experience. To this end, we propose an architecture of a cloud-based context-aware service composition system, that leverages the benefits of cloud computing through the SaaS (Software as a Service) service model, in the delivery of context-aware, SOA based applications.

The proposed architecture is shown in Fig. 3. Aside from the user, the model is made up of five modules: *Context Management Module, Design and Planning Module, Discovery and Selection Module, Composition and Execution Module* and *Request Management Module*. The databases represent additional information required to perform the functions of each module.

- The *Context Management Module* provides context information to the above mentioned modules to improve their decision making processes.
- The *Request Management Module* is responsible for analysing the user's requirements and converting them to a set of intermediary goals and one final goal to be achieved, through domain ontologies.
- The *Design and Planning Module* takes the set of goals provided by the *Request Management Module* and constructs an abstract model describing the process to be performed to achieve those goals and if no process can be automatically generated the user is asked to complete it.

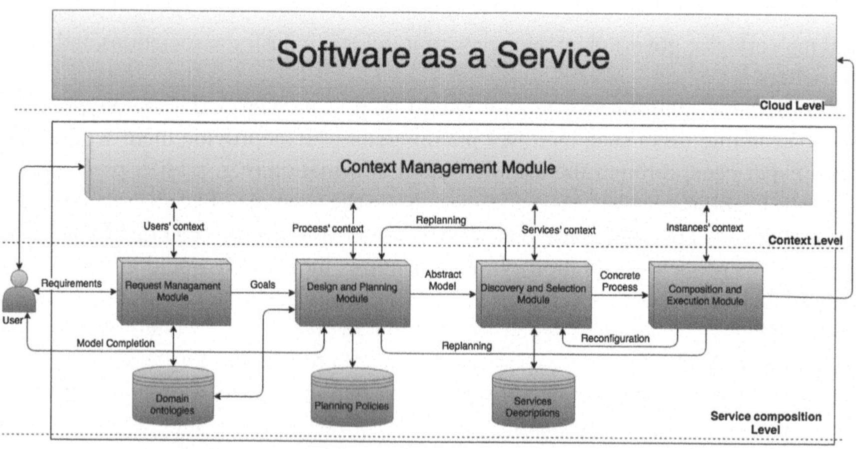

Fig. 3 Cloud-based context-aware service composition

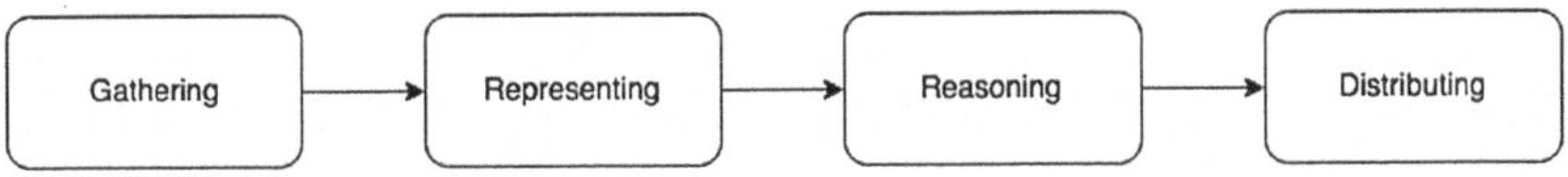

Fig. 4 Context Management life cycle

- The *Discovery and Selection Module* receives the abstract model and looks for the concrete services capable of accomplishing the models tasks and selects the best service for the task, through QoS attributes and user preferences, and builds a concrete process.
- The *Composition and Execution Module* takes as input the concrete process and transforms it into an executable code and then delivers the application in SaaS manner.

4.1 Context Management Module

In literature, numerous works have studied the impact of context in different areas such as mobile computing and Internet of things. In [14], context is seen as a way of describing or characterizing the current status or situation of an entity or object. Context-aware computing can be seen as the collecting, representing, reasoning and distributing of context information (see Fig. 4). The collecting phase concerns the ways of acquiring the context data around an entity. The representing phase deals with the modeling and storage of the data acquired through the previous phase. The reasoning phase is responsible for synthesising and aggregating the data stored to get better knowledge and more meaningful information out of them. The distributing phase specifies the way the context information gathered is provided to the interested parties. Authors in [15], proposed an architecture to provide a more complete model of the information relevant to a mobile user and making this data available to interested applications.

In order to represent and reason about context, a high level abstraction of the concepts involved needs to be specified. Model Driven Engineering (MDE), provides us with such ability in metamodelling. Figure 5 presents our context metamodel, which is mainly inspired by the concepts in the definition of context given in [4] (Situation, Entity, Association) and enriched by other concepts from some literature works [16, 17]. The metamodel is generic and MOF[6] (Meta Object Facility) compliant. The **C_Situation** class describes the conditions of the involved entities and their relationships at a given time. The **C_Entity** represents an entity in which context we are interested (e.g. Person, mobile, environment, etc.). The **C_Association** is used to describe the relations between entities (e.g. user owns mobile phone). Each **C_Entity** has a set of **C_Property**, which is a property that can be simple or composed. A **C_Property** has a provider/**C_Provider** and can be either static/**Static** (e.g. name, birthday, etc.), profiled (e.g. work experience), user defined/**UserDefined** (e.g. through dialog interface), derived/**Derived** (e.g. derive age from birthday or

[6]http://www.omg.org/spec/MOF/.

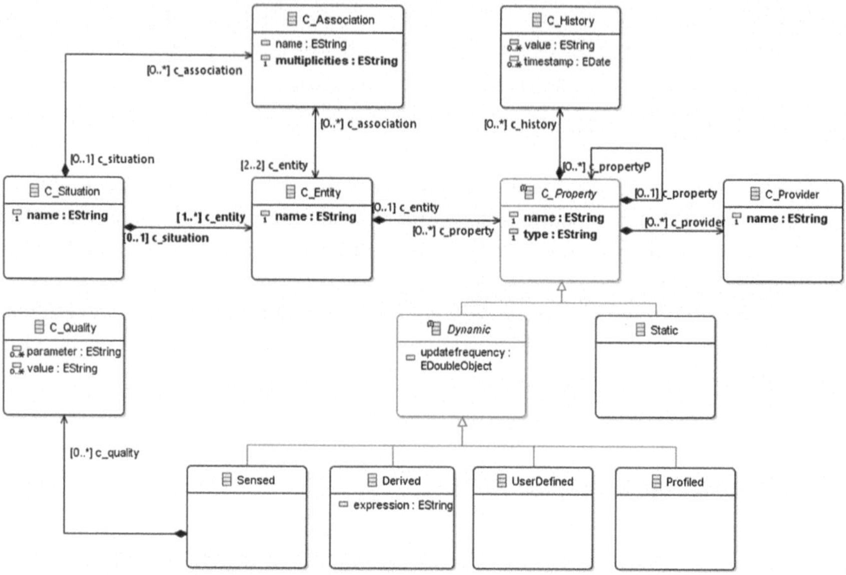

Fig. 5 MOF context metamodel

night/day from time) or it can be sensed/**Sensed** (e.g. temperature, speed, position, etc.) and has a quality/**C_Quality** (e.g. precision of the GPS position). Each **C_Property** has a history of use/**C_History** that stores its previous values.

In our architecture, the *Context Management Module* interacts with the user and the other modules to provide additional contextual information, in order to refine and enrich the input data resulting in improving the performances of each module. It interacts with the *Request Management Module* to provide additional information about the user (e.g. profile, preferences, location, etc.), that can be useful in the query analysis. The *Design and Planning Module* can also benefit from context information by taking into account other people's histories of use and expert knowledge [15] (e.g. tour planning, payment methods, etc.). The *Context Management Module* can also work with the *Discovery and Selection Module* to make a context-aware service discovery and selection system [18]. The *Context Management Module* can also be used to monitor the execution of the composition at the *Composition and Execution Module* level, so that it can request the reconfiguration (e.g. if a service's quality degrades and needs to be replaced) or re-planning (e.g. if the user's context changes).

4.2 Request Management Module

Though service composition is usually realised by individuals with high technical expertise (software designers, engineers, etc.), we envision that our architecture would simplify the composition process enough so that users with minimal technical skill could benefit from it in their daily lives.

To that end, the *Request Management Module* would allow end users to specify their requirements in a natural language. These requirements would then be converted through semantic means (ontologies) and context information to a set of intermediary goals or tasks that would result in the achievement of the user's request. For example, the user can enter the query "arrange travel to Paris" and the module is responsible for analysing the query and converting it to a formal one, which will be executed against the semantic repository [19], resulting in the acquiring of the semantic concepts related to the travel domain. In the example above (Fig. 2), that would mean the goals is the booking of a hotel, a flight and a rental car and sending the arrangements information.

The module also interacts with the context Management module to refine and enrich the queries through capturing user intents behind the queries (e.g. current time, weather, user's location, schedule and preferences, etc.). Examples of such systems can be found in [14, 20]. The user can also interact with the module at any moment to further specify or modify details about his requirements (e.g. enter the date, budget, etc.).

4.3 Design and Planning Module

This module is responsible for the generation of a high level process or model for the composition, through which the composition goal can be achieved. Though as mentioned in Sect. 1 the planning problem in service composition is widely explored, there was little efforts to integrate context information in the composition planning to the best of our knowledge. We argue that it is important for our planning module to benefit from context, as we believe context information (e.g. Other people's histories of use, expert knowledge) can provide the planning module with current and valuable information that could improve the planning process and the user's satisfaction.

To automate the planning process, works have investigated the use of techniques from the AI (Artificial Intelligence) domain [21]. For our module, we are considering the use of M.A.Ss (Multi-Agent Systems), to provide the autonomy and automation needed to perform the planning policies. The M.A.S takes as input the composition goals expressed in semantics and the contextual information, analyses them and then builds an abstract model based on the planning policies provided. In our example (Fig. 2), the booking of the services could be parallel or sequential, and several context information as the process level can influence the planning (e.g. the user could prefer walking the streets instead of booking a rental car if the weather is nice or prefer other transport means, book another type of accommodations based on other travelers' recommendations or expert ratings, etc.).

The user can also complete the model if the M.A.S is unable to build it, or modify the model if he wants a different plan. This is done in a simple design manner using a graphical language to ease the task for non-IT users. The model is then transformed to the planning language before being sent to the next module. The work in [22], also considered multi-agent systems, semantics and cloud integration for the automatic construction of business processes using Web services composition. However, they did not consider context in their system.

4.4 Discovery and Selection Module

In their work [9], authors covered service discovery and selection and explained the mechanisms to implement them. They defined service discovery as the process of finding suitable service(s) for a given activity based on functional properties, which is the prerequisite of service selection, in which the best service is selected based on non-functional properties of the services (e.g. price and other QoS attributes). This module encapsulates the two concepts, as it discovers and selects the best service to achieve each task in the abstract model (process) and then generates a concrete composition process (mostly expressed in BPMN or a UML activity diagram).

First, the *Discovery and Selection Module* takes the abstract model as input from the previous module and tries to discover the available services to perform each task of the model. Mostly, this is done through matchmaking techniques (semantic or synthetic). Based on [9], semantic matchmaking increases the level of automation in services discovery. After that, each candidate service is evaluated through his non-functional properties (e.g. performance, reliability, scalability, capacity, etc.) and the service with the best score is selected. The selection can be done based on the task level (maximize the score of the services for each task) or on the model level (maximize the score of the services for the whole model).

The module can also interact with the *Context Management Module* to consider the user's preferences in service selection process (e.g. travel class and airline, hotel type, car type, etc.), thus improving user's satisfaction. However, if the module isn't capable of finding a service for some task. It sends a re-planning request to the *Design and Planning Module* to further decompose the service at hand or find another plan to achieve the composition goal.

4.5 Composition and Execution Module

The *Composition and Execution Module* represents the last component of our architecture, it is responsible for the deployment and execution of the composite service on the cloud. In [23] authors, explained the respective deployment and execution options. We argue that cloud is the best option for deployment due to its advantages discussed in Sect. 1. And as such, we are considering the use of SaaS delivery model. As for the execution Engine, we consider the use of business process engines fitting, due to the process-like nature of the composition and the fact that most BPM solution vendors are taking their solutions to the cloud more and more. Thus, the module will use the SaaS delivery model to provide the applications to the user using BPEL (Business Process Execution Language) or BPMN.

However, to broaden the choices of techniques to be used in the previous modules, which will result in different output formats. We introduce the composition component; as a mean to leverage adaptability; as the transformation of the concrete process (output of the previous module) to a machine comprehensible code to deliver the application according to the different delivery models.

Also, as we discussed in Sect. 4.1, this module can also interact with the *Context Management Module* to be informed about the context of the user and the composition instance(i.e. each and every participant service in the composition execution), so that it can request the reconfiguration (e.g. if a service's performance degrades and so needs to be replaced) or re-planning (e.g. if the user's context changes).

5 Conclusion and Future Work

Across the studies described above, the goal was to speed up the development time of cloud software services through service composition. We presented in this paper a novel conceptual architecture for a context-aware service composition in cloud environments. The idea was to improve the service composition process on three levels. First, enabling the composition process to deal with the dynamics of today's ubiquitous environments, through the integration of contextual information in its various phases and the integration with the cloud technology to allow ubiquitous access and elastic execution. Second, exposing the composition process to non-IT users through the use of a Request Management system. Three, improving the user's experience through the consideration of his preferences and situation in the process.

Each module in our architecture can be considered as a separate problem that has issues to be addressed. We've started our work on the *Context Management Module* and presented the metamodel in Fig. 5 that would enable us to represent the possible context data. This work is introductory and serves to get an overview of our vision to offer businesses a shorter time to market for their services, and a richer experience for end-users while dealing with the environment's changes. To implement our architecture, we intend on focusing on each module, one at a time while taking into account the whole composition picture, starting with the context management module. Our motivating scenario will serve us as a case study to evaluate the significance of our propositions.

We also believe our architecture could benefit greatly from recommender systems, as they rely heavily on contextual information. The integration of such systems could further enrich our architecture by allowing users to include and react to events in the composition process in real-time.

We are also considering Smart cities as a potential environment that could benefit from our system, as the true potential of the smart objects; as constituents of the smart environment; lies in their capability to interact and share their services to remedy the problems related to smart cities [24].

References

1. Weiser, M.: The computer for the 21st century. Sci. Am. **265**, 94–104 (2011)
2. Yufei, Y., Wuping, Z.: Mobile task characteristics and the needs for mobile work support: a comparison between mobile knowledge workers and field workers. In: Eighth International Conference on Mobile Business, pp. 7–11 (2009)
3. Yousfi, A., de Freitas, A., Dey, A., Saidi, R.: The use of ubiquitous computing for business process improvement. In: IEEE Transactions on Services Computing, vol. PP(99), pp. 1–1 (2015)
4. Abowd, G.D., Dey, A.K., Brown, P.J., Davies, N., Smith, M., Steggles, P.: Towards a better understanding of context and context-awareness. Lect. Notes Comput. Sci. **1707**, 304–307 (1999)
5. Mell, P., Grance, T.: The NIST definition of cloud computing. In: Computer Security Division, Information Technology Laboratory, NIST (2011)
6. Rao, J., Su, X.: A survey of automated web service composition methods. Lect. Notes Comput. Sci. **3387**, 43–54 (2005)
7. Bezerra, E., Lopes, D., Abdelouahab, Z.: Dynamic Web service composition with MDE approaches and ontologies. Lect. Notes Electr. Eng. **151** (2013)
8. Jula, A., Sundararajan, E., Othman, Z.: Cloud computing service composition: a systematic literature review. Expert Syst. Appl. **41**, 3809–3824 (2014)
9. Sheng, Q.Z., Qiao, X., Vasilakos, A.V., Szabo, C., Bourne, S., Xu, X.: Web services composition: a decades overview. Inf. Sci. **280**, 218–238 (2014)
10. Zhou, J., Athukorala, K., Gilman, E., Riekki, J., Ylianttila, M.: Cloud architecture for dynamic service composition. Int. J. Grid High Perform. Comput. **4**, 17–31 (2012)
11. Object Management Group, Inc.: BPMN 2.0 by Example. Version 1.0 (2010)
12. Meehan, K., Lunney, T., Curran, K., McCaughey, A.: Context-aware intelligent recommendation system for tourism. In: Pervasive Computing and Communications Workshops, pp. 328–331 (2013)
13. Lamsfus, C., Xiang, Z., Alzua-Sorzabal, A., Martin, D.: Conceptualizing context in an intelligent mobile environment in travel and tourism. Inf. Commun. Technol. Tour. 1–11 (2013)
14. Subbaraj, R., Venkatraman, N.: A systematic literature review on ontology based context management system. Adv. Intell. Syst. Comput. **338**, 609–619 (2015)
15. Williams, C., Mathew, J.: An architecture for mobile context services. Lect. Notes Electr. Eng. **313**, 61–68 (2015)
16. de Farias, C., Leite, M., Calvi, C., Pessoa, R., Filho, J.: A MOF metamodel for the development of context-aware mobile applications. ACM Symp. Appl. Comput. **22**, 947–952 (2007)
17. Jaouadi, I., Djemaa, R., Abdallah, H.: A generic metamodel for context-aware applications. Adv. Intell. Syst. Comput. **330**, 587–594 (2015)
18. Fenza, G., Furno, D., Loia, V.: Hybrid approach for context-aware service discovery in healthcare domain. J. Comput. Syst. Sci. **78**, 1232–1247 (2012)
19. Tablan, V., Damljanovic, D., Bontcheva, K.: A natural language query interface to structured information. Lect. Notes Comput. Sci. **5021**, 361–375 (2008)
20. Yao, Y., Yi, J., Liu, Y., Zhao, X., Sun, C.: Query processing based on associated semantic context inference. Inf. Sci. Control Eng. **2**, 395–399 (2015)
21. Cugola, G., Ghezzi, C., Pinto, L.S.: DSOL: a declarative approach to self-adaptive service orchestrations. Computing **94**, 579–617 (2012)
22. Coria, J.A.G., Castellanos-Garzón, J.A., Corchado, J.M.: Intelligent business processes composition based on multi-agent systems. Expert Syst. Appl. **41**, 1189–1205 (2014)
23. Lemos, A.L., Daniel, F., Benatallah, B.: Web service composition: a survey of techniques and tools. ACM Comput. Surv. **48**(33) (2015)
24. Han, S.N., Khan, I., Lee, G.M., Crespi, N., Glitho, R.H.: Service composition for IP smart object using realtime Web protocols: concept and research challenges. Comput. Stand. Interfaces **43**, 79–90 (2016)

Knowledge Flows Within Open Source Software Projects: A Social Network Perspective

Noureddine Kerzazi and Ikram El Asri

Abstract Developing software is knowledge-intensive activity, requiring extensive technical knowledge and awareness. The abstract part of development is the social interactions that drive knowledge flows between contributors, especially for Open Source Software (OSS). This study investigated knowledge sharing and propagation from social perspective using social network analysis (SNA). We mined and analyzed the issue and review histories of three OSS from GitHub. Particular attention has been paid to the socio-interactions through comments from contributors on reviews. We aim at explaining the propagation and density of knowledge flows within contributor networks. The results show that review requests flow from the core contributors toward peripheral contributors and comments on reviews are in a continuous loop from the core teams to the peripherals and back; and the core contributors leverage on their awareness and technical knowledge to increase their notoriety by playing the role of communication brokers supported by comments on work items.

Keywords Knowledge flows · Expertise · SNA · Open source

1 Introduction

Open source communities can be perceived as knowledge-sharing ecosystems in which contributors learn from the community and from each other [1]. They share both domain and technical knowledge through contributions to the source code repositories or by reviewing source code from one another. Interactions between contributors, which can be materialized by looking to co-edited files [2], constitutes

N. Kerzazi · I. El Asri (✉)
National Higher School for Computer Science and System Analysis (ENSIAS), Rabat, Morocco
e-mail: ikram.asri@um5s.net.ma

N. Kerzazi
e-mail: n.kerzazi@um5s.net.ma

R. El-Azouzi et al. (eds.), *Advances in Ubiquitous Networking 2*, Lecture Notes in Electrical Engineering 397, DOI 10.1007/978-981-10-1627-1_19

the backbone of socio-technical perspective which has gained increased attention over the past decade [3–5]. Social Network Analysis (SNA) has been used to capture and understand such information about relations among people [6] with the aim at enhancing team performance and software product quality.

Previous research has shown that there are expert reviewing technical contributions involved in most OSS projects [7]. However, this past research does not explain how developers identify experienced contributors to review their code and how awareness and knowledge are spread through the contributors' community. Many open source (OSS) projects adopt the practice of code reviews to increase the quality of their software products [8]. Collaboration on code review aims not only to improve the quality of code changes made by contributors [9], but also for the purpose of knowledge transfer and awareness [10, 11]. If we could explain the propagation of knowledge flows within contributor networks throw code source reviews, we can enhance the quality of the code and improve the signal to noise ratio of comments on commits which decrease teams' performance. One way of locating reputed domain expert, to ask for reviewing a piece of code, is to build contributors networks and analyze it.

Historically, SNA has been known to be effective in many areas [12]. In this paper, we examine the socio-technical interactions for three OSS. Using histories of version control data, we constructed contributors' networks based upon which files are commonly modified by contributors. Using network analysis, we can uncover details of knowledge sharing and the circulation of knowledge flows between the core and peripheral contributors. Our research questions can be summarized as follows:

RQ1. Is there a Relationship between Contributors' Network and Knowledge Sharing?
RQ2. Does the network position of contributors affect the review process and the number of comments on GitHub projects?
RQ3. Does the Socio-Technical analysis make knowledge transfer an actionable concept?
RQ4. What Kind of Knowledge is transferred?

The main contributions of the paper are as follows:

- A thorough Social Network Analysis of three OSS projects that provides insights into socio-interaction of contributors and their knowledge sharing;
- A view of knowledge circulation through code review practice along with the kind of knowledge that is transferred;
- An exploration of how SNA metrics can inform to answer whether or to what extent an open source community has a good underpin knowledge and awareness sharing mechanisms;
- An understanding of whom are requesting code review; whom are commenting on reviews; and whom are performing reviews according to their network position and degree.

Paper organization. The remainder of the paper is organized as follows. Section 2 presents related work and background. Section 3 introduces our SNA-based method for identifying knowledge flows. Section 4 describes the selected projects from GitHub and data collection process. Section 5 provides our study results. Section 6 discusses our finding and points out practical implications. Section 7 discloses the threats of validity. Section 8 concludes and outlines future work.

2 Related Work and Background

Considering people at the heart of OSS projects, SNA in software development teams shows that social networking contains tremendous information that can be leveraged for purposes such as: defects prediction [3, 13], teams' organization and coordination [14], team productivity [15], and tools or techniques for the purpose of studying developer communities [1, 4]. We first summarize related work according to these three different perspectives. Then we introduce previous work on knowledge sharing and propagation. Finally, we present what we know about SNA measures.

Defects Prediction—Rigby and Storey [16] examined manually hundreds of code reviews across five high-profile OSS projects aiming to investigate the mechanisms and behaviours that developers use to find code changes they are competent to review. They found that the Apache project adopted a broadcast-based style of code review, meaning increasing the awareness of new changes, but annoying the community with a high amount of irrelevant notification. Baysal et al. [9] studied the factors that influence the outcome of the review process and found that review positivity (i.e., the proportion of accepted patches) can be influenced by non-technical factors such as organization.

Furthermore, a recent qualitative study at Microsoft [10] showed that identification of defects is not the only motivation for code review, but **sharing knowledge among team members** is also considered as a very important motivation of modern code review. This related work indicates that our findings are not specific to the open source community but can be applied within commercial organizations.

Organization and Coordination—Recently, there has been considerable interests and work on improving the coordination between software team's members [17]. Knowledge dependencies drive the need to coordinate software process activities. Saying that an SNA approach can support identification of coordination needs by identifying previous collaboration and communications. Social Network metrics arise as a response to those questions such as who should do what, when it is required.

Productivity—It has been reported that higher socio-technical congruence usually correlates with higher developer productivity [15] and reduces integration failures [17]. Both researchers and OSS projects leads could use STC to diagnose project members' collaboration and improve team coordination [18].

Social Knowledge Sharing—Prior work has shown that social networking contains plenty of information that can be leveraged for other purposes [14]. For instance, the socio-cultural learning theories state that people learn from each other through observation, interaction and communication [19]. Seeing learning through its social aspect emphasizes the fact that OSS projects are increasingly growing. Contributors are part of a community of practice, organization, and belong to a group of people where there is competence knowledge already established. Source code review practice and comments are seen as ideal vehicles for leveraging tacit knowledge and learning.

3 SNA-Based Knowledge Flows

According to SNA [20], a Network consists of a set of nodes and a set of edges. Thus, we represent contributors as **nodes** as shown in Fig. 1. **Connections**, between those nodes, are weighted and represented based on the number of files the pair has collaborated on.

When two contributors are directly connected by an edge they are *adjacent*. The number of adjacent connections for a given contributor is called the *Degree* of that contributor. As illustrated in Fig. 1, C_3 has a degree of 3 and C_4 has a degree of 1.

Geodesic path refers to the shortest social distance between two contributors represented such as adjacent and unique connections. While networks' **diameter** refers to the longest path between two contributors.

3.1 Contributor Network Metrics

Connectivity metric measuring direct connections between nodes. SNA has come up with three distinct structural properties to measure the centrality of a given node.

Centrality metrics measure how closely contributors are indirectly connected to each other in the network. SNA measures the centrality based on two metrics: *closeness* and *betweenness*.

Closeness refers to the average distance from a node to any other node in the network. For example, Closeness for $C_1 = (1 + 1 + 2)/3 = 4/3$ noticing that the

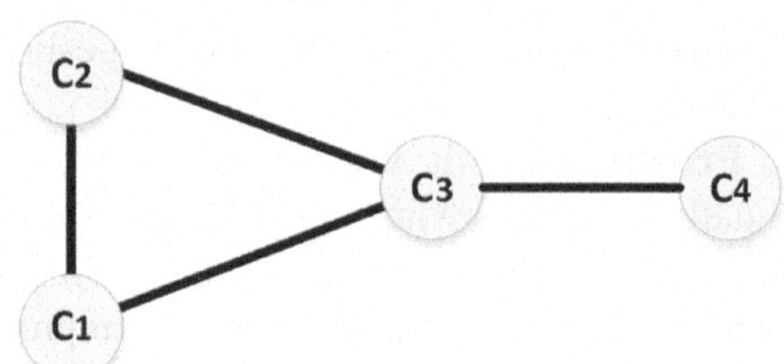

Fig. 1 An example of Contributors' Network with four nodes

shortest paths from C_1 to (C_2 and C_3) are each 1 and the shortest path from C_1 to C_4 is 2. For instance, Fig. 1 shows that C_3 has the maximum possible degree (3) meaning that it is central in this network. While C_1 and C_2 have a degree equal to 2; and C_4 has a degree of 1 meaning that this contributor is peripheral in this network.

Betweenness is another centrality metric calculated for a given node as the number of shortest paths that include this node divided by the total number of shortest paths in the network. In the example of Fig. 1, we have a total of 6 shortest paths. Saying that the betweenness of C_1 and C_2 is 3/6, while the betweenness of C_3 is 5/6.

4 DataSets

We focus our study on three large and rapidly evolving open-source systems which are highly stared projects from GitHub. Our choice of projects was based on the following criteria: (i) project should be among the 100 most stared projects; (ii) should be still under active development; and (iii) involving at least 250 contributors. Table 1 summarizes the characteristics of our selected projects including the programming language, the total number of developers, number of releases; number of lines of code; number of requested reviews; and the total of commits. Table 2 shows the characteristics of each network.

For each project we queried the GitHub API with the query https://api.github.com/repos/<owner>/<repo>/commits?page=<n>, where <owner> is a GitHub

Table 1 Overview of the studied systems

	Overview				Commits	
	Language	Contributors	Releases	LOC	Request review	Total
Angular. Js	JavaScript	1403	161	369.574	349	7534
Docker	Go	1314	129	670.722	2399	22318
JQuery	JavaScript	250	134	62.566	10	6050

Table 2 Metric of the networks

	Clustering coefficient	Network centralization	Avg. # of neighbours	#Nodes	Network density	Network heterogeneity
Angular. Js	0.897	0.918	66.428	1393	0.048	1.674
Docker	0.874	0.794	81.385	1314	0.062	1.582
JQuery	0.834	0.725	61.179	244	0.252	0.837

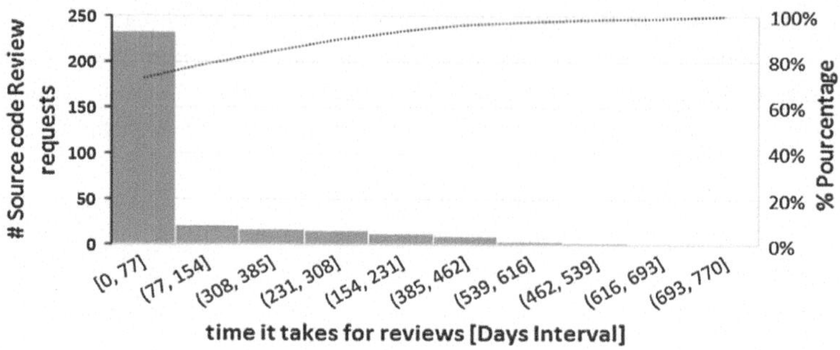

Fig. 2 Time required to close a request for code reviews

user account and <repo> is the name of the repository. Hence, we extracted the commits data for each project including details such as the programming language used, the time period covered, the number of commits and developers, information about releases as well as the number of edited files.

Once the commits and edited files were linked, we were interested by all requests of reviews for each project. Since our study is focused on knowledge sharing between contributors, we also extracted all comments on each code review. Our query retrieves open and closed issues (i.e., state = all), along with labels tagged as 'need review'. Figure 2 summary interval times needed to close code review requests.

5 Study Results

RQ1. *Is there a Relationship between Contributors' Networks and knowledge sharing?*

We were interested to find the position of contributors within the contributors' network, that are asking for code reviews, commenting on code reviews, and carrying out reviewing activity. Figure 3 shows a comparative between Angular and Docker projects. We represent in red colour contributors' network on top of which we map sub-networks. For instance, Fig. 3a1 illustrates the social network of *Angular* (red colour) and a sub network of contributors that requested code review (blue colour). One can observe the differences between the two projects in terms of density and position of requesters of code reviews.

Figure $3b_{1-2}$ show the mapping of the contributors who have commented on code reviews (green). And finally, Fig. 3c. compares the relative network position of contributors that carried out the code reviews.

We found that core contributors act such as knowledge brokers and boundary spanners across comments loops not only from the periphery to the

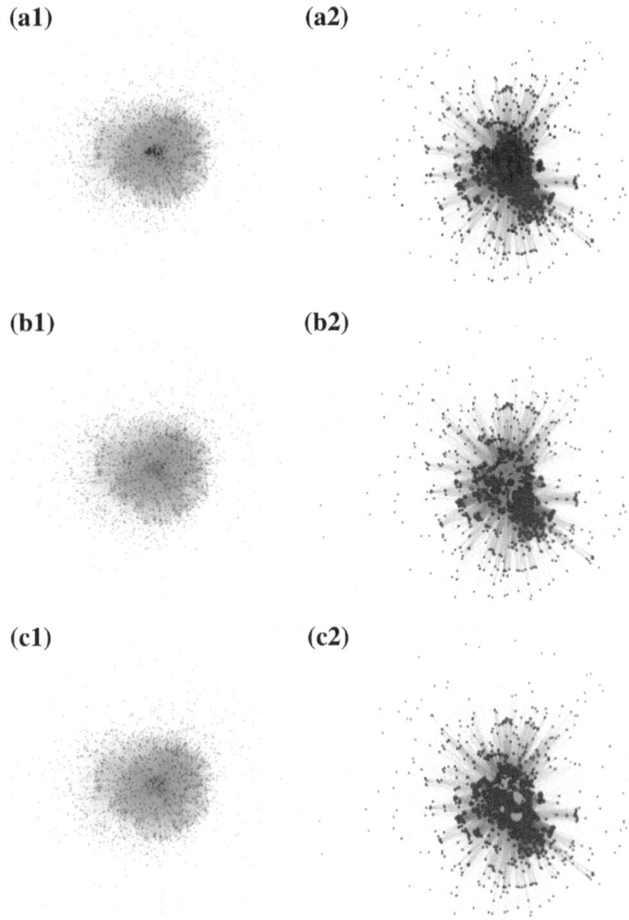

Fig. 3 Comparing different networks of Angular and Docker Projects. **a1 Angular** project Contributors Network which requested a code review, **a2 Docker** project Contributors Network which requested a code review, **b1 Angular** Network of commenters on code review, **b2 Docker** Network of Commenters on code review, **c1 Angular** Network of Contributors that Resolve and close the code review, **c2 Docker** Network of Contributors that Resolve and close the code review

core, but also from the core to the periphery. We pay a close attention to how core contributors (experts) influence communication patterns through comments on code reviews and issues in OSS projects as well as transferring and spreading knowledge to peripherals.

RQ2. *Does the network position of contributors affect the review process and the number of comments on GitHub projects?*

We found a strong correlation between the position of contributors in the Network (*Degree*) and the number of comments on code reviews. Figure 4

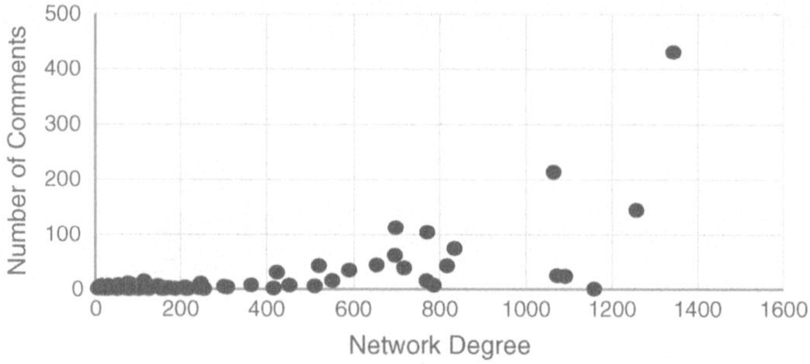

Fig. 4 Mapping the Contributors Centrality Degree Metric with the Number of Comments

illustrates the trend of communication in the *Angular* project. One can also notice that more the Degree is highly likely the contributors are active in transferring their knowledge and awareness to other contributors. For instance, surprisingly in *Angular* project we have identified 16.7 % of contributors such as core[1] developers that generate 81.6 % of communication against 83.3 % of peripheral developers generating only 18.4 % of the comments flow.

RQ3. Does the socio-technical analysis make Knowledge Transfer an actionable concept?

Core contributors are communication brokers that have awareness and both technical and domain knowledge. SNA allows us to identify central core contributors. We segregate contributors according to the degree of centrality they have. Our analysis shows that we can go further with SNA metrics and patterns that can support studying knowledge flows in OSS communities similar to previous studies that attempt to predict software failures based on SNA metrics [13].

Figure 5 shows a comparative of the betweenness centrality metric as well as the distribution of degree for contributors.

Table 3 summarizes SNA metrics for each network. Those metrics help to understand the nature of the project as well as the architecture. For instance, we observed a high density for Docker project probably meaning cohesive architecture of this project.

Table 4 shows intrinsic SNA metrics emphasizing characteristics of interactions between contributors such as degree and centrality metrics (betweenness and closeness).

[1]We define a Degree threshold > 500 to filter on core developers which are marked as central in our SNA analysis.

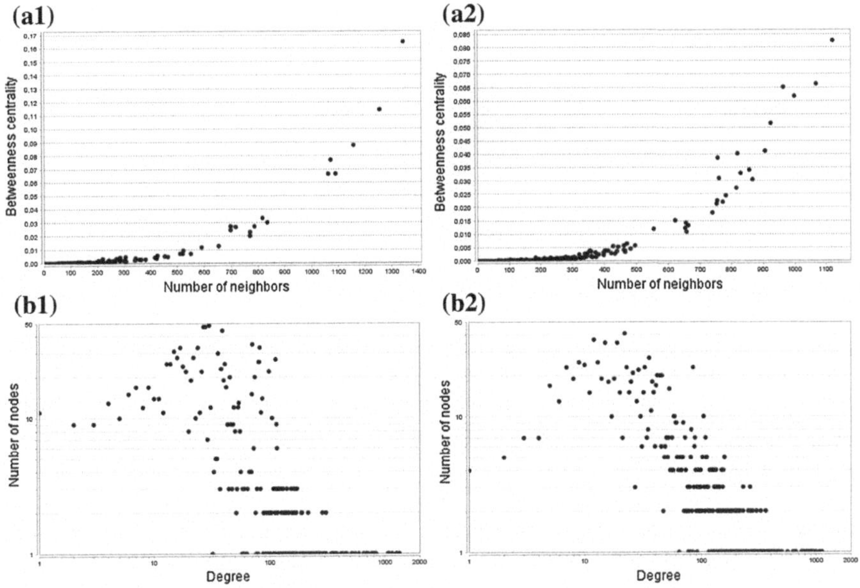

Fig. 5 Comparing networks metrics of Angular and Docker Projects. **a1 Angular** Betweenness metric, **a2 Docker** Betweenness metric, **b1 Angular** Degree Metric, **b2 Docker** Degree Metric

Table 3 Network SNA mertics

Projects	Density	Centrality	Avg. neighbours	Clustering coef.
JQuery	0.252	0.725	61.197	0.834
Angular	0.048	0.918	66.428	0.897
Docker	0.062	0.794	81.385	0.874

Table 4 Contributors SNA metrics

Projects	Degree	Betweenness	Closeness	Clustering coef.
JQuery	[1–236] Median = 48	[0–0.8]	[0.46–0.97]	[0–1]
Angular	[1–1343] Median = 38	[0–0.16]	[0.35–0.96]	[0–1]
Docker	[1–1122] Median = 38	[0–0.08]	[0.41–0.87]	[0–1]

6 Discussion

Review assignments are not sufficient to explain comments flows in a project. Figure 6 shows the distribution of contributors pre-assigned for code reviews (yellow) within the overall contributors' network. Nerveless, we have seen quite lot of code reviews carried out by other contributors with different degrees of centrality, which is not necessarily problematic but may indicate areas where the

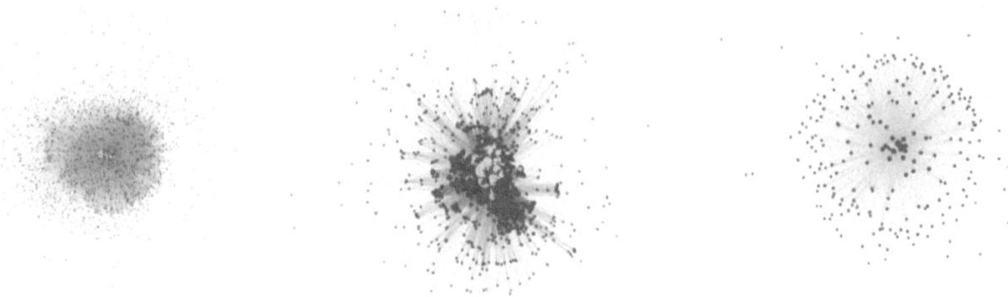

Fig. 6 Assigned contributors for reviewing source code

review assignment was not supported by either awareness or cross-functional knowledge or the distribution of domain knowledge in the core team. OSS Team leads can optimize the team configuration when forming new teams, especially for the code review activity.

The core contributors are assumed to be structurally more central, in the contributors' networks, than other contributors. They have enough either awareness or knowledge about the product to manage other developers' contributions.

RQ4. *What kind of Knowledge is transferred?*

We manually classify comments according to technical or domain knowledge. Another category emerged throughout our classification process: Awareness. The majority of knowledge transfer is about Awareness (46.3 %), then technical (34.5 %) generating large contributors' debates and domain knowledge (19.1 %). For example, contributor 13286 in Angular project commented on an implementing approach that he perceived as an anti-pattern:

> [I'm not quite sure why you're against this. The job of 'inject' is to inject a function, as its name implies. Not inject a function and eliminate its return value. I would argue, instead, that is an anti-pattern of function decorators. It's confusing and unnecessary....] 13286.

7 Threats to Validity

Construct Validity—In this paper, we adopt co-edited files as a heuristic to build the graph of contributors' networks. We do not consider the time frame such as co-edition within one month or under releases. In fact, we could rely on comments for SNA instead of co-edited files. However, focusing only on comments will hide the big analysis of all socio-interactions. Furthermore, our heuristic based on file co-edition does not consider the amount of LOC the contributors make. However, file editing is considered by many studies as a fine-grained enough indicator of developers' collaboration [2]. Furthermore, we assume that all communications occur with either review requests or comments within the review process. We cannot assume that developers on GitHub are not using an external social media or mailing list to communicate.

Internal Validity—We are aware that we might miss transitive dependencies between technical elements. For instance, changing the framework on which depends many files is unseen such as a technical interaction. Moreover, software development is dynamic, and as contributions are made over time, the nature of the socio-interaction changes. We mitigated this threat by studying multiple open source projects, using different languages, within the GitHub community. Furthermore, our analysis is time-agnostic. Since contributors are changing over time, the number of core developers may vary as well. We plan to conduct a temporal analysis of core contributors in future work to get more insights on how those contributors rich their actual position in the Network.

External Validity—In this study, we choose three projects which therefore might limit the generation of our results. However, we choose carefully mature and long-lived projects running in different languages and with an amount of contributors ranging from 250 to 1403. We filtered away projects that have fewer than 250 contributors or fewer than 1,000 edited files to remove projects that are immature or without an underpinning socio-technical interaction, and thus alleviate potential bias.

8 Conclusion

In this paper, we have performed Social Network Analysis on three open source projects. We showed how knowledge is transferred between core contributors and peripherals when using code review activity. We build contributors' networks based on co-edited files and then we build sub-networks for contributors requesting code reviews, commenting on, and those performing the code reviews. SNA visualization makes the identification of the structural interactions analysis of those networks possible. We found that there is a strong correlation relationship between the degree centrality of contributors and their implication on knowledge and awareness transfer.

By understanding the knowledge flows between OSS collaborators, socio-technical interactions structure, OSS communities gain an increased ability to facilitate code reviews in their projects. We hope this will lead to software projects with more efficient knowledge transfer, less overhead of review assignment, and increased leverage of the software quality and teams' performance.

References

1. VonHippel, E., VonKrogh, G.: Open source software and the "Private-Collective" innovation model: issues for organization science. Organ. Sci. **14**(2), 209–223 (2003)
2. Dabbish, L., et al.: Social coding in GitHub: transparency and collaboration in an open software repository. In: The Conference on Computer Supported Cooperative Work. Seattle, WA, USA (2012)

3. Begel, A., DeLine, R., Zimmermann, T.: Social media for software engineering. In: FSE/SDP Workshop on Future of Software Engineering Research, pp. 33–38. Santa Fe, New Mexico, USA (2010)
4. Yang, X.: Social Network Analysis in Open Source Software Peer Review, pp. 820–822 (2014)
5. Yang, X., et al.: Understanding OSS Peer Review Roles in Peer Review Social Network (PeRSoN), pp. 709–712 (2012)
6. Bird, C., et al.: Latent social structure in open source projects. In: Proceedings of the 16th International Symposium on Foundations of Software Engineering (FSE'08). Atlanta, Georgia (2008)
7. Asundi, J., Jayant, R.: Patch review processes in open source software development communities: a comparative case study. In: The 40th Annual Hawaii International Conference on System Sciences (2007)
8. Bissyande, T.F., et al.: Got issues? Who cares about it? A large scale investigation of issue trackers from GitHub. In: 24th International Symposium on Software Reliability Engineering (ISSRE) (2013)
9. Baysal, O., et al.: The influence of non-technical factors on code review. In: Proceedings of the 20th Working Conference on Reverse Engineering. Koblenz, Germany (2013)
10. Bacchelli, A., Bird, C.: Expectations, outcomes, and challenges of modern code review. In: Proceedings of the 35th International Conference on Software Engineering (ICSE'13). San Francisco, CA, USA (2013)
11. Kilamo, T., et al.: Knowledge transfer in collaborative teams: experiences from a two-week code camp. In: 36th International Conference on Software Engineering (ICSE'13), pp. 264–271. Hyderabad, India (2014)
12. Yarosh, S., et al.: I need someone to help!: a taxonomy of helper-finding activities in the enterprise. In: Proceedings of the 27th International Conference on Computer Supported Cooperative Work (CSCW'13), pp. 1375–1386. Texas, USA (2013)
13. Meneely, A., et al.: Predicting failures with developer networks and social network analysis. In: International Symposium on Foundations of Software Engineering (FSE'11). Atlanta, Georgia (2011)
14. Hossaina, L., Zhub, D.: Social networks and coordination performance of distributed software development teams. J. High Technol. Manage. Res. **20**(1), 52–61 (2009)
15. Cataldo, M., Herbsleb, J.D.: Coordination breakdowns and their impact on development productivity and software failures. Trans. Softw. Eng. **39**(3), 343–360 (2013)
16. Rigby, P.C., Storey, M.-A.: Understanding broadcast based peer review on open source software projects. In: Proceedings of the 33rd International Conference on Software Engineering (ICSE'11). 2011. Waikiki, Honolulu, USA
17. Kwan, I., Schroter, A., Damian, D.: Does socio-technical congruence have an effect on software build success? a study of coordination in a software project. Trans. Softw. Eng. **37**(3), 307–324 (2011)
18. Cataldo, et al.: Identification of coordination requirements: implications for the design of collaboration and awareness tools. In: Proceedings of the 20th International Conference on Computer Supported Cooperative Work. Banff, Alberta, Canada (2006)
19. Nam, K.K., Ackerman, M.S., Adamic, L.A.: Questions in, knowledge in?: a study of Naver's question answering community. In: Proceedings of the SIGCHI Conference on Human Factors in Computing Systems. Boston, MA, USA (2009)
20. Kadushin, C.: Understanding Social Networks: Theories, Concepts, and Findings. Oxford University Press (2011)

A Pub-Sub Based Architecture for IDS as Service

Maïssa Mbaye and Cheikh Ba

Abstract Cloud services have become increasingly popular and massively deployed over the past years. However, providing security as cloud service, accessible from outside the cloud, remains one of the most challenging research problems in this topic. The main problem comes from the fact that it is hard to maintain scalability with client growth while ensuring an efficient intrusion detection requests management inside cloud services. In this paper, we propose the use of Pub-Sub communication mechanism to provide a highly available and distributed IDS cloud service. Our aim is to reduce intrusion detection time and increase accuracy by specializing IDS nodes according to various taxonomies based attack categories. Our cloud IDS service is available from outside via web service interfaces, and is appropriate for limited-capacity devices such as smartphones or tablets.

Keywords IDS · Pub-Sub · Computer attack taxonomies · Cloud computing · Web services

1 Introduction

Intrusion detection is the process of monitoring the events occurring in a computer system or network and analyzing them for signs of intrusions, defined as attempts to compromise the confidentiality, integrity, availability, or to bypass the security mechanisms of a computer or network [1].

Based on the location in a network, IDS can be categorized into two groups: Host-based IDS (HIDS) and Network-based IDS (NIDS). HIDS refers to the class of intrusion detection systems that reside on and monitor an individual host machine, while NIDS monitors network link.

M. Mbaye · C. Ba (✉)
LANI (Laboratoire d'Analyse Numérique et d'Informatique),
Université Gaston Berger, BP 234, Saint-Louis, Senegal
e-mail: cheikh2.ba@ugb.edu.sn

M. Mbaye
e-mail: maissa.mbaye@ugb.edu.sn

R. El-Azouzi et al. (eds.), *Advances in Ubiquitous Networking 2*, Lecture Notes in Electrical Engineering 397, DOI 10.1007/978-981-10-1627-1_20

Nowadays, intrusion detection tasks become more complex because of large amount of data generated by users and the low technical level needed to reproduce attacks. Modern IDS need more computing resources to be efficient and to respond with an acceptable delay. Cloud Computing can be a suitable solution to this challenging task. Cloud Computing provides a cheaper and reliable infrastructure in which service providers may deploy applications or store data to be used by their customers.

However, having more computing resources doing the same task of intrusion detection in parallel does not necessarily accelerate detection delay. IDS in the cloud should rely on an extensible and efficient communication architecture. In this paper we address this challenge of distributing intrusion detection requests amongst IDS nodes inside a distributed cloud service. To achieve this goal we use Pub-Sub communication model inside cloud IDS. Service IDS nodes are specialized according to an attack taxonomy so that all existing attacks are covered and each node has lesser search space.

Our first step goal is to provide a flexible and distributed cloud IDS architecture, based on Pub-Sub paradigm. This service can then be used inside a cloud service or accessible from outside via web services. To the best of our knowledge, this work is the first attempt that aims to provide an IDS cloud service, based on the Pub-Sub communication mechanism and partners interoperability using web services. We also expect to reduce the intrusion detection time by specializing IDS nodes using a computer attack taxonomy. Beyond the classical personal computer, we want our solution to be particularly appropriate for limited-capacity devices such as smartphones, tablets or classical network.

The paper organisation is as follow: Sect. 2 presents some background concepts. Section 3 discusses related works. Section 4 details our proposition. Section 5 concludes the paper and presents future works.

2 Background Concepts

2.1 Cloud Computing and Intrusion Detection Systems

Cloud Computing is an internet based computing where virtual shared servers provide software, infrastructure, platform, devices and other resources [2]. Access to the resources is granted to users by offering a client-server infrastructure that can be used to perform tasks at three different levels of abstraction [3]: Cloud *Infrastructure as a Service* (IaaS), Cloud *Platform as a Service* (PaaS) and Cloud *Software as a Service* (SaaS). Services may be used to query data or to perform computations.

Since our IDS is available as an on-demand service hosted in the cloud, we place ourselves on the SaaS level of abstraction in relation to the cloud infrastructure. We would like to point out that the monitoring for our IDS is processed by the client. The reason is that this system is an on-demand service handling host-based intrusion detection as well as network based intrusion detection.

2.2 Pub-Sub Paradigm

Pub-Sub (*Publish-Subscribe*) is a mechanism for disseminating information (also called events) through distributed systems [4]. Participants to the communication can act as *publishers* or *subscribers*. *Publishers* submit information (or messages) to the Pub-Sub system, whereas *Subscribers* express their interest in specific types of information, within the system, in the form of a subscription. The system matches published information to subscriptions and delivers messages to interested subscribers using a notification mechanism. There are various model of subscription based on how subscribers express their interest for a certain information and how the matching is done by the notification system so that the subscribers only receive information that he is interested in [4, 5]. These models are topic-based, attribute-based and content-based, even if some works consider attribute-based Pub-Sub as one kind of content-based Pub-Sub [5].

Our proposal focuses on the topic-based system which is the simplest one and delivers us from the intensive computation of content-based systems. Moreover, it is well-suited to our work since messages routing is simple through multicast group to nodes that match subscription topics.

3 Related Works

Various research projects [2, 6–9] combining the two concepts "IDS" and "cloud" exist in the literature. However, many of them address different IDS or cloud models, and often focus on a different goal. For instance, some researchers focus on adapting or extending [10, 11] classic IDS techniques to the cloud context. Some others propose distributed IDS in which several IDS nodes cooperate for a specific task, mostly against distributed attacks like DDoS [12].

The work in [6] aims to discover coordinated attacks on local sites. To this end, they developed an IDS based on Cloud Computing architecture, a two-parts architecture (the global and local sites) to achieve a global monitoring view of the network resources. The goal of the global site is to collect and process the alerts from the local sites. Along the same lines, the work in [8] discusses the use of a distributed strategy to detect and block attacks originated by misbehaving customers of a Cloud Computing provider, by testing different deployments of existing IDS.

Works that are closer to ours are those that provide IDS as a service in the cloud. Authors in [7] introduce *IDSaaS* (Intrusion Detection System as a Service) framework, which is a network and signature based IDS for the cloud model that targets the Infrastructure as a Service (IaaS) level of the cloud. This framework, which is an IDS adoption for the cloud consumers contains a centralized Intrusion Engine, the brain of the system. It preprocesses the incoming packets and examines their payload section looking for any matching pattern of a threat defined in the loaded attacking rules. A similar work is proposed in [9], apart from the fact that they provide an IDS as a service in which data privacy concerns have been discussed.

Most IDSs use centralized architecture and detect intrusions that occur in a single monitored system. Nevertheless, there is a recent increasing trend towards distributed and coordinated attacks, where multiple machines are involved, either as attackers (e.g. DDoS) or as victims (e.g. large volume worms). Moreover, there is another weakness of centralized IDSs due to the fact that a unique agent is responsible for the whole system security.

The main contributions of our work are the specialization of IDS nodes based on computer attack taxonomy and the use of Pub-Sub communication mechanism to provide a flexible, highly available and distributed IDS cloud service architecture. The flexibility comes with a transparent deployment of new IDS in the system without heavy configurations. Load-balancing process is implicit and does not need central knowledge about all the system. Each intrusion detection request is routed to specialized nodes through Pub-Sub brokering system. We expect to reduce the intrusion detection time by specializing IDS nodes in various taxonomy based attack categories.

4 Pub-Sub Based Cloud IDS Service

4.1 Basic Architecture

The overall system works as illustrated in Fig. 1. On one side, we have the clients of the IDS cloud service which submit data in the form of an intrusion detection request. On the other side, we have the IDS service composed of specialized IDS (sIDS), a brokering system and a web service interface. The client and the IDS service communicate via a secured VPN.

Clients are considered as publishers while sIDS are subscribers. With this aim in mind, a client may apply to the service throughout a web service interface.

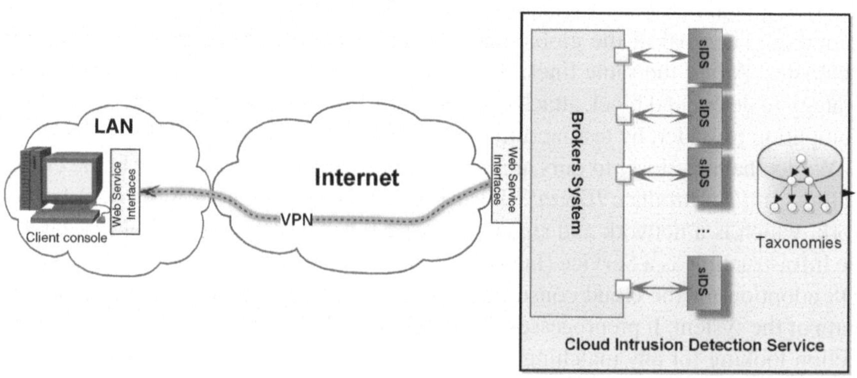

Fig. 1 Basic architecture

The brokering system (a set of brokers) is responsible for routing intrusion detection requests to sIDS nodes (Definition 1). These nodes can detect faster a specific kind of network attack because they are specialized in specific categories of attacks. The set of sIDS covers all kinds of known attacks. The response to a request may be formatted with respect to an extended version of IDMEF.[1]

Definition 1 (*sIDS distribution*) Let B be the set of brokers in our brokering system. The distribution (dist) of all sIDS (S) over the brokers $B_i \in B$ is a bijective application that maps B to a partition of S.

$$dist: \quad B \rightarrow Part(S)$$
$$B_i \mapsto p_i \subseteq S$$

We remind that a partition of the set S of sIDS is a set of non-empty subsets of S such that every element S_i in S is in exactly one of these subsets, i.e., S is a disjoint union of the subsets. Thus, the three following conditions hold:

$$(B_i, \emptyset) \notin dist \tag{1}$$

$$\bigcup_{(B_i,p_i)\in dist} p_i = S \tag{2}$$

$$(B_i, p_i) \in dist \wedge (B_j, p_j) \in dist \wedge p_i \neq p_j \Rightarrow p_i \cap p_j = \emptyset \tag{3}$$

These conditions, coupled with the definition of the sIDS distribution, mean that:

1. The partition does not contain $\emptyset$, so each broker handles at least one sIDS.
2. The union of the subsets in the partition is equal to S. Thus each sIDS has to be assigned to a broker.
3. The partition's elements are pairwise disjoint. So the same sIDS cannot be assigned to different brokers. □

The main advantage of this architecture is that attacks can be detected faster by distributing intrusion detection to all nodes. This distribution eliminates the problem of single-points-of-failure. Furthermore, assigning sIDS to brokers increase the system flexibility: a single entry point that makes the overall system management transparent to the client.

4.2 Taxonomy Based Topics

Topics are one of the core concepts in our architecture. In topic based Pub-Sub, topics are keyword patterns that determine recipient subscribers. Taxonomy based topics are keyword patterns extracted from an attack taxonomy.

[1] Intrusion Detection Message Exchange Format, https://www.ietf.org/rfc/rfc4765.txt.

A taxonomy is a classification scheme that partitions a body of knowledge and defines the relationship of the pieces [13]. A network and computer attacks taxonomy is a classification of individual known attacks into categories according to their relationship [14]. Many taxonomies have been developed according to different criteria or dimensions [15, 16]. Authors in [14] proposed several dimensions for computer attack classification: *attack vector*, *targets of attack*, *vulnerabilities and exploits* that the attack uses. Taxonomies can be very important for security evaluation and help to develop defence techniques by extending the ones from well-known attacks in the same attack category.

In this work we focus on *target* dimension and *attack vector* dimension proposed in [14] to create the two attack taxonomy trees. The target oriented taxonomy organizes computer attacks according to their targets. A target of an attack can be an operating system, a software application, a network protocol, a specific device, etc. This taxonomy is used to create topics in which sIDS nodes subscribe to specialize themselves. An IDS client requesting intrusion detection is supposed to gather information about target context in the form of keywords, and send it as metadata with the request. This kind of information is always available and simple to retrieve. The gateway brokers are responsible for making correspondences between these keywords and topics.

The target oriented taxonomy is organized in a tree structure with nodes that represent keywords identifying classes of attack. Each attack class in the taxonomy has a parent class (except root class representing the dimension) and might have subclasses. Each attack class is more specific than its parent and more general than its sub-classes. For instance, considering the partial target based taxonomy in Fig. 2, attacks that target Windows Family Operating System are less general than the ones that target all OSs.

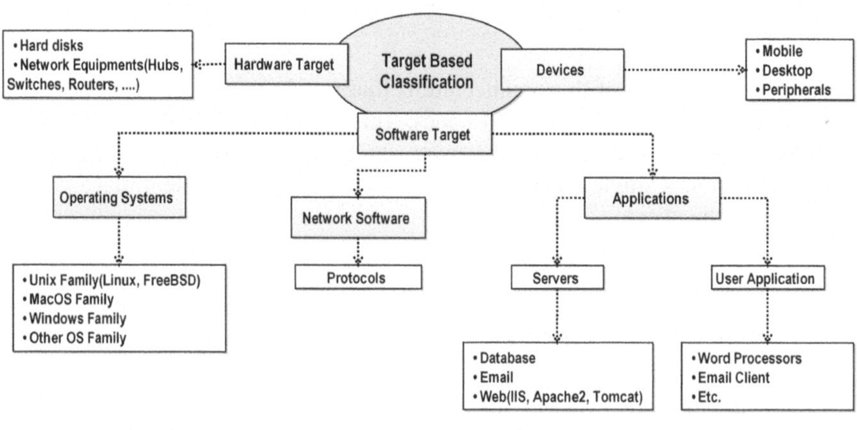

Fig. 2 A target based attack taxonomy $\mathbb{T}$

More formally, our target based taxonomy $\mathbb{T}$ is a tree of attack classes such that for any two nodes u and v, if u is parent of v then u is more general than v. In other words, v is a sub-tree of u. This generality relation ($>$) is a partial relation order in $\mathbb{T}$. For all $u, v \in \mathbb{T}$, and $u \neq v$, we have:

$$(u > v) \quad \Rightarrow \quad \mathcal{A}(u) \supset \mathcal{A}(v) \; and \; (u \cup v) = u$$

where $\mathcal{A}(u)$ is the set of all attacks targeting node u. A topic θ, based on attack taxonomy $\mathbb{T}$, is a sub-tree identified by the path of the highest attack category involving θ. For instance, the path */TargetBased/OS/MS_Windows/SMB* identifies a topic that corresponds to attacks that target SMB Service on MS Windows operating systems. As said earlier, a sIDS subscribes to the most general topics that covers all attacks it can handle. On the other hand, clients send intrusion requests $R = \langle \mathbb{K}, d \rangle$ composed of data d to be analyzed and a set $\mathbb{K}$ of keywords that describe the context. This set of keywords $\mathbb{K} = \{k_1, k_2, \ldots, k_n\}$ has a semantic equivalence with a set of topics $\varphi(\mathbb{K})$ in $\mathbb{T}$:

$$\varphi(\mathbb{K}) = \bigcup_{i=1}^{n} \{\theta \in \mathbb{T} \mid k_i \cong \theta\}$$

where θ is a topic, k_i is a keyword and the operator $\cong$ is a semantic equivalence. This semantic equivalence extends simple string matching. For instance "*Windows 7*" and "*Windows Seven*" are semantically equivalent even if they are syntactically different. The function $\varphi(\mathbb{K})$ is a semantic projection of $\mathbb{K}$ into $\mathbb{T}$.

The Fig. 3 illustrates a semantic projection where $\varphi(\{k_1, k_2, k_3\}) = \{\theta_1, \theta_2, \theta_3\}$. Also, since topics are sets, union and intersection can be applied to them.

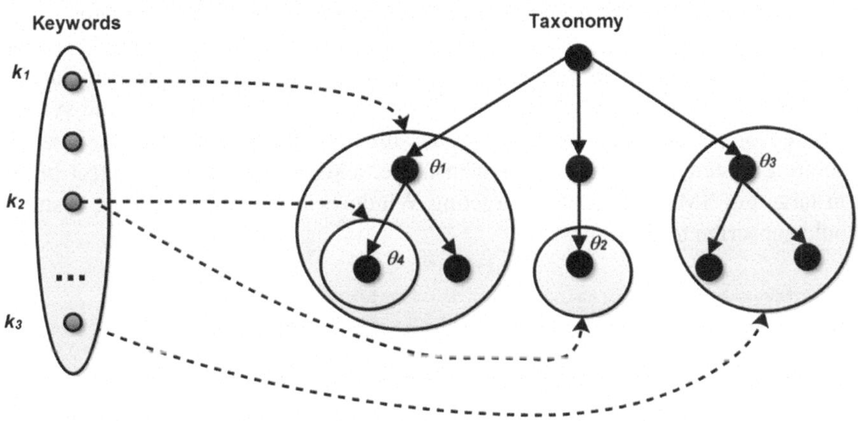

Fig. 3 A keyword projection into the taxonomy tree

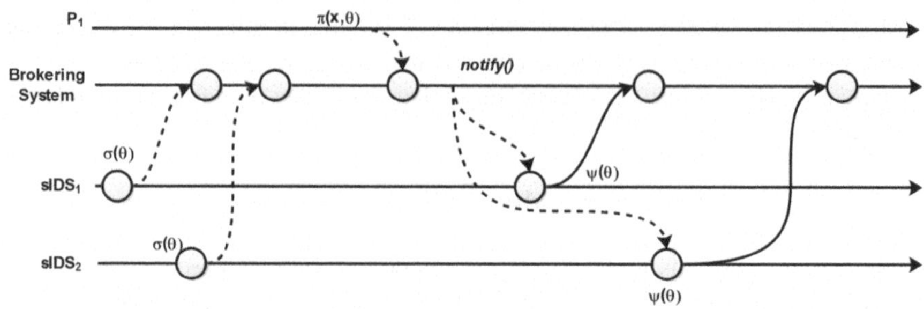

Fig. 4 A Simplified representation of Pub-Sub processes

4.3 Publish-Subscribe IDS Interfaces

In our system, Pub-Sub Interface enables one-to-many communications with a limited API namely *Publish*, *Subscribe*, *Unsubscribe*, *Notify* and *Consume*.

Subscribers select information they want to receive using *subscribe*(θ) or $\sigma(\theta)$ routine. Subscriptions are processed by *brokers* that have a database of nodes and topics in which they subscribed. A subscriber can also leave a topic with the *unsubscribe*(θ) or $\overline{\sigma}(\theta)$ routine. A process that wants to submit data x in topic θ uses *publish*(x, θ) or $\pi(x, \theta)$ routine.

Brokering system stores published data and notifies subscribers that data is available for them with the *notify*() routine. Then, the subscribers receive data with *consume*(θ) or $\psi(\theta)$ routine. Figure 4 illustrates a simplified process.

More formally a process P_i that subscribe to topic θ receives published data x with $\pi(x, \theta')$ if and only if $\theta \geq \theta'$, where $>$ is the generality relation operator.

In our context, sIDS are subscribers and client publisher by the mean of gateway brokers. sIDS subscribe to topics that corresponds to the keyword patterns identifying the kind of intrusion of attack they can handle. These topics in the system are based on an attack taxonomy that classifies attacks according to their target (operating system, software, ...). For instance, a sIDS node that can detect intrusion targeting SMB Services or targeting Windows XP family operating systems should subscribe to the topics:

```
/SoftwareAttack/{operatingsystems/windowsfamily/xp*
                                 + application/server/SMB}
```

The corresponding subscribe operation can be illustrated by the following listing:

```
OPERATION : Subscribe
TOPICS    : /SoftwareAttack/operatingsystems/windowsfamily/xp*
            /SoftwareAttack/application/server/SMB
```

In other words, sIDS use target information to declare what corresponding security topics they can handle based on taxonomy. The taxonomy simplifies the specialization process of sIDS and limits the number of topics.

We recall that a topic θ, based on attack taxonomy $\mathbb{T}$, is a sub-tree identified by the path of the highest attack category involving θ. A sIDS can subscribe to some topics by using σ routine with a kind of regular expression on paths (topics) in $\mathbb{T}$. This regular expression on paths denotes a set of topics. To have a *non-formal* view of this option, let's have a look at Example 1.

Example 1 We assume the three subscriptions below.

1. σ `(/target/software/network/services)`
2. σ `(/target/software/network/os/windows/xp*)`
3. σ `(/target/software/network/{services + os/windows/xp*})`

A sIDS that uses the routine (1) subscribes to topic $\theta = /target/software/network/services$. The one which uses the routine (2) subscribes to all the topics in the taxonomy that have the same prefix "*/target/software/network/os/windows/xp*". In the same manner, the routine (3) is equivalent to the routine (1) followed by the routine (2). The semantics of a subscription $\sigma(\theta)$ is that a subscriber (a sIDS) can handle all attacks whose target is in the space defined by the path θ. □

Clients submit requests containing data that they want to be analyzed accompanied by a set of keywords describing the potential target context. These keywords may not syntactically match the ones in taxonomy but they can have semantic equivalent items in it. For instance, a client that wants an intrusion detection intended to SMB Services on Wind. XP could submit something like:

```
OPERATION: Submit Intrusion Detection
KEYWORDS : Windows XP;SMB Server; DST_IP: 150.142.56.78
DATA      : MYME-TYPE : application/vnd.tcpdump.pcap;
<binary data>   ...
```

Actual publishing is done by gateway brokers that find suitable topics. Gateway brokers are special ones that receive intrusion requests from clients and map (semantic projection) keywords to the taxonomy based topics. The following listing illustrates the publication of the previous request:

```
OPERATION: Publish
TOPICS    : /SoftwareAttack/operatingsystems/windowsfamily/xp*
            /SoftwareAttack/application/server/SMB
DATA      : MYME-TYPE : application/vnd.tcpdump.pcap;
<binary data>   ...
```

All intrusion detection requests are intended to sIDS that subscribed to the corresponding topics. If a sIDS detects an attack, an intrusion description is extracted from the taxonomy that classifies attacks according to their vector, i.e. our **vector** based taxonomy $\mathbb{V}$. Figure 5 illustrates this detection process.

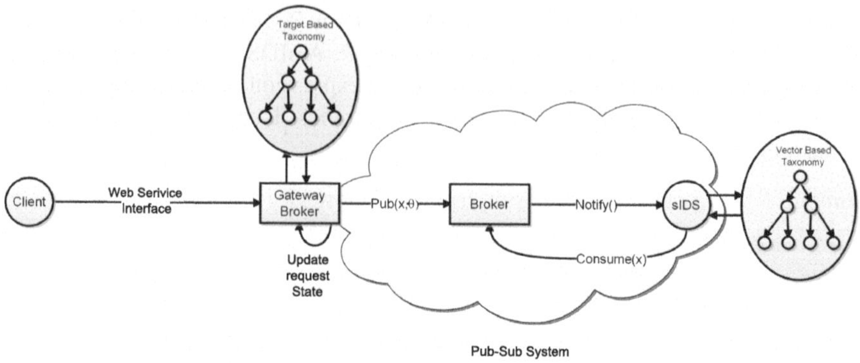

Fig. 5 Intrusion detection process

4.4 Service Supplying

This section describes the cloud IDS service supplying, from a client request to the response it receives. The clients of our IDS Cloud service see the system as a black box that receives intrusion detection requests and produces responses that indicate whether or not an attack is detected. We recall that a client sends a two-part intrusion request $R = \langle \mathbb{K}, d \rangle$ composed of the request context and the request data. The context is a set of keywords $\mathbb{K} = \{k_1, k_2, \dots, k_n\}$ that defines the potential target while information to be processed is in the request data d. We assume that this client is connected to the cloud throughout a virtual private network (VPN), and sends a continuous flow of packets to the entry point of the system. This entry point is an asynchronous web service.

Once the set $\mathbb{K}$ of keywords is at the getaway broker level, the system uses its semantic equivalence $\varphi(\mathbb{K})$, a set of topics, for the request dispatching. In fact, the brokering system, which is a set of brokers, is responsible for routing intrusion detection requests to sIDS nodes. Each sIDS is assigned to a unique broker, and each broker covers a subset of sIDS. The set of all sIDS in the brokering system covers all kinds of known attacks defined in the target based taxonomy. More formally, Let S be the set of all sIDS in the brokering system, and let $S_i \in S$ be a sIDS. The set $\{\theta_i^1, \dots, \theta_i^n\}$ of n topics handled by S_i is given by the function Θ below.

$$\begin{aligned} \Theta : S &\rightarrow 2^{\Sigma} \\ S_i &\mapsto \{\theta_i^1, \dots, \theta_i^n\} \end{aligned} \qquad \Sigma = \bigcup_i \Theta(S_i)$$

Since S handles all topics, the set of all attacks defined in the target-based taxonomy is Σ. This condition ensures that every incoming request can be handled by at least one sIDS. In the brokering system, each broker that receives a request with a topic θ checks among its assigned sIDS the ones that have subscribed to this topic, i.e. the sIDS in the set $\{S_i \mid \theta \in \Theta(S_i)\}$. In others words, when a client sends an intrusion request with a set $\mathbb{K}$ of keywords, the sIDS S_i receives this request if and only if

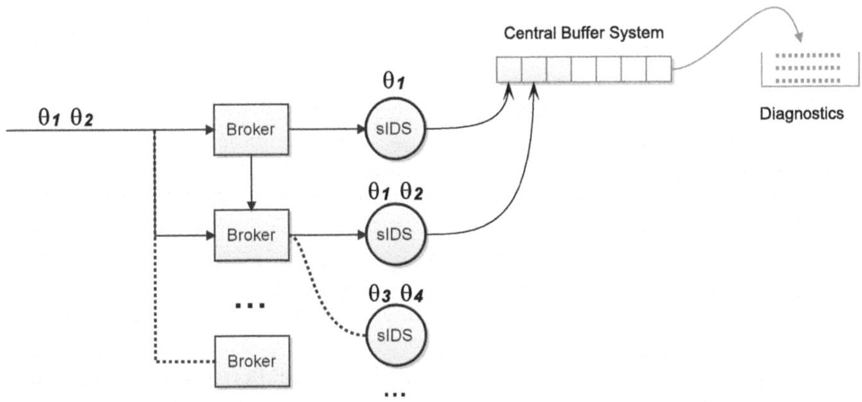

Fig. 6 Intrusion request dispatching

$$\Theta(S_i) \cap \varphi(\mathbb{K}) \neq \emptyset$$

The brokers notify only the matched sIDS of the availability of a request for the intrusion detection process. For instance, in Fig. 6 where a request comes with the set of keywords $\mathbb{K}$ such that $\varphi(\mathbb{K}) = \{\theta_1, \theta_2\}$, the two matched sIDS are the ones that subscribed to topics $\{\theta_1\}$ and $\{\theta_1, \theta_2\}$. The node that subscribed to topics $\{\theta_3, \theta_4\}$ will not receive the request.

A sIDS that receives a request with topic θ means that it can detect attacks that target this kind of related software or service. A *Central Buffer System* (*CBS*) is available for collecting, from the matched sIDS, the results of different intrusion detection processes. Each sIDS that detects an attack will put into the *CBS* the complete attack description, using the vector based taxonomy $\mathbb{V}$.

The final result of the intrusion detection process will be sent to the client. This result is in the form of a *diagnostic* built by the *CBS*. A diagnostic is simply a list of the individual responses weighted with the following norm:

$$\|\Theta(S_i), \varphi(\mathbb{K})\| = \frac{|\Theta(S_i) \cap \varphi(\mathbb{K})|}{|\varphi(\mathbb{K})|}, \quad \varphi(\mathbb{K}) \neq \emptyset$$

This norm means that the more a sIDS covers the request topics, the more its response is significant. The weight is therefore a mean to indicate the relevance of individual sIDS responses. The following listing is an example of a diagnostic obtained from sIDS implemented with *snort*.[2]

```
Diagnostic
Alert 1, 70%,
09/22-21:03:36.305795 [**] [1:2013976:10] ET TROJAN
Zeus POST Request to CnC - URL agnostic [**]
```

[2]https://www.snort.org/.

```
[Classification: A Network Trojan was detected]
[Priority: 1] {TCP} 192.168.3.65:1033 -> 188.72.243.72:80

Alert 2, 50%
09/22-21:03:36.316004 [**] [1:16435:6] FILE-IDENTIFY
Ultimate Packer for Executables/UPX v0.62-v1.22
packed file magic detection [**] [Classification:
Misc activity] [Priority: 3] {TCP} 188.72.243.72:80 -> 192.168.3.65:1035

Alert 3, 50%
09/22-21:03:36.341625 [**] [1:12798:4] SHELLCODE
base64 x86 NOOP [**] [Classification: Executable code
was detected] [Priority: 1] {TCP} 188.72.243.72:80 -> 192.168.3.65:1035
```

This example shows a diagnostic composed of three alert messages, weighted respectively by 70 %, 50 % and 50 %.

The final response intended to the client is then a diagnostic that shows different responses gathered from sIDS. Since the diagnostic is built by the *CBS*, a timer is used to determine the moment the diagnostic has to be sent to the client. The timer is initialised by the first broker that submits an intrusion detection request with a unique request ID. The *CBS* send the diagnostic every time the timer expires and all responses coming after are ignored.

5 Conclusion and Future Works

In this work we propose an architecture of cloud IDS with a topic based Pub-Sub paradigm. Our proposition uses a target based taxonomy to create topics. Intrusion detections are handled by sIDS that specialized themselves to intrusion categories according to the taxonomy. This simplifies requests forwarding to suitable sIDS nodes. The clients of such an architecture can be low power devices that are connected to the Internet (tablets, smartphones, ...) and that cannot embed powerful IDS or antiviruses. The main outcomes of our proposition are: a flexible and distributed IDS based on Pub-Sub communication model and a highly available IDS service with specialized IDS nodes. These specialized IDS nodes have smaller signatures database and can decide faster for a specific kind of security attack.

We are implementing our proposed architecture in OpenStack Cloud platform, NS3 and Snorts. In our evaluations, we use the following performance metrics: detection speed in the cloud IDS, false positives and negatives, CPU computing charge during detection, efficiency of publishing and subscribing, load balancing among sIDS.

Acknowledgments This work has been supported by the African-Excellence-Center in Mathematics, Computer Science and ICT(CEA-MITIC) at Gaston Berger University of Saint-Louis/Senegal.

References

1. Mell, P., Bace, R.: NIST Special Publication on Intrusion Detection Systems (2001)
2. Ms. Shelke, P.K., Ms. Sontakke, S., Dr. Gawande, A.D.: Intrusion detection system for cloud computing. Int. J. Sci. Technol. Res. **1**, 67–71 (2012)
3. Robert, P.-C., Voas Mark, J.M., Badger, L., Grance, T.: Cloud computing synopsis and recommendations. Technical report, Gaithersburg, MD, United States (2012)
4. Yusuf, S.: Survey of publish subscribe communication system. Adv. Internet Appl. Syst. Des. **24** (2004)
5. Li, M., Ye, F., Kim, M., Chen, H., Lei, H.: Bluedove: a scalable and elastic publish/subscribe service. Int. Parallel Distrib. Process. Symp. (2011)
6. Xiao-yu, L., Wang, X., Ting-lei, H.: Research on the intrusion detection mechanism based on cloud computing. In: International Conference on Intelligent Computing and Integrated Systems (ICISS), pp. 125–128 (2010)
7. Alharkan, T., Martin, P.: Idsaas: Intrusion detection system as a service in public clouds. In: CCGRID, pp. 686–687 (2012)
8. Mazzariello, C., Bifulco, R., Canonico, R.: Integrating a network IDS into an open source cloud computing environment. In: IAS, pp. 265–270 (2010)
9. Meng, Y., Li, W., for Kwok, L., Xiang, Y.: Towards designing privacy-preserving signature-based IDS as a service: a study and practice. In: INCoS, pp. 181–188 (2013)
10. Vaid, C., Verma, H.K.: Anomaly-based IDS implementation in cloud environment using BOAT algorithm. In: IEEE-ICRITO'14, pp. 1–6 (2014)
11. Alqahtani, S.M., Balushi, M.A., John, R.: An intelligent intrusion detection system for cloud computing (SIDSCC). In: IEEE-CSCI'14, pp. 135–141 (2014)
12. Mohamed, H., Adil, L., Saida, T., Hicham, M.: A collaborative intrusion detection and prevention system in cloud computing. In: IEEE-AFRICON 2013, pp. 1–5 (2013)
13. Radatz, J.: The IEEE Standard Dictionary of Electrical and Electronics Terms, 6th edn. IEEE Standards Office, New York, NY, USA (1997)
14. Hansman, S., Hunt, R.: A taxonomy of network and computer attacks. Comput. Secur. **24**(1), 31–43 (2005)
15. Zheng, W., Yang, O., Liu, Y.: A taxonomy of network and computer attacks based on responses. Int. Conf. Inf. Technol. Comput. Eng. Manage. Sci. **1**, 26–29 (2011)
16. Zafar, B., Khan, F.A., ud Din, F., Ahmad, W., Hayat, Z., Shah, I.: A survey on taxonomies of attacks and vulnerabilities in computer systems. Int. J. Comput. Sci. Telecommun. **3**, 93–97 (2012)

Towards a Service Broker for Telecom Service Provision and Negociation in IMS Network

Imane Haddar, Brahim Raouyane and Mostafa Bellafkih

Abstract The world of telecommunications is undergoing a rapid change in designing and developing applications. Telecom operators are becoming increasingly forced to diversify the portfolio of available services to ensure customer loyalty. Indeed the operator's business has far exceeded the supply of traditional telephony to open up to the most demanding services. For this purpose IP Multimedia Subsystem (IMS) has been a major work effort of the 3GPP (3rd Generation Partnership Project) standards bodies for several years now. However, the acquisition of services by operators slows down the development of services offered because of the integration difficulties, and the negotiation issues. In this paper we propose a new negotiation approach. The idea is to set up a Service Broker able to look for the best offer of the service and execute the Service Level Agreement (SLA) between Service Provider, network operator, and customer. To ensure a transparent communication at all levels, the NGOSS (New Generation Operations Systems and Software) Framework is used in a consistent manner.

Keywords IP Multimedia Subsystem (IMS) · Service Broker (SB) · enhanced Telecom Operations Map (eTOM) · Service Level Agreement (SLA) · Service Provider

I. Haddar · M. Bellafkih (✉)
Department of Telecommunications Systems, Networks and Services,
STRS, National Institute of Posts and Telecommunications, Rabat, Morocco
e-mail: mbella@inpt.ac.ma

I. Haddar
e-mail: haddar@inpt.ac.ma

B. Raouyane
Hassan II University, Casablanca, Morocco
e-mail: raouyane_brahim@yahoo.fr

R. El-Azouzi et al. (eds.), *Advances in Ubiquitous Networking 2*, Lecture Notes in Electrical Engineering 397, DOI 10.1007/978-981-10-1627-1_21

1 Introduction

With the advent of the Internet, IP has become the most interesting and used protocol in the world. Globally 3.2 billion people used the Internet by 2015. Indeed, IP has become increasingly useful due to the appearance of real-time applications such as voice over IP and voice/video conferences. All these aspects have motivated the research of a network architecture that is all IP, and which is called later, NGN (Next Generation Networks). However, this innovation is unable to ensure by itself an easy and rapid deployment of new services based on IP or control access to these services. That is why an innovation in the control plan was necessary. Hence the creation of IMS approach (IP Multimedia Subsystem). Indeed, IMS is designed to allow full control over the offered IP services. This architecture is the path to a unified framework to provide services for fixed and mobile networks. IMS services can be classified into on demand content, live content, managed service, data services, and etc. The IMS services can be provided at anytime and anywhere.

Even though there are many advantages to use the IMS network, there are also some challenges. Due to the high quality IMS services and to support of massive data traffic over IMS network, it is not possible to guarantee the sufficient amount of all parameters for IMS services every time. A Service Level Agreement (SLA), managed by a Service Broker with the integration of eTOM (enhanced Telecom Operations Map) processes, where eTOM is a standard framework maintained by the TM Forum, an association for service providers and their suppliers in the telecommunications and entertainment industries, will guarantee the sufficient of the important parameters for any service requested by a customer. Until now no researcher proposed the SLA architecture for IMS network. As a result of our SLA negotiation and Service Broker integration, all parameters are well distributed to enable to customers to get the best service offer.

This paper is organized as follows. In Sect. 2 only a related work is being briefly described. Based on vital standards and mechanisms, IMS and enhanced Telecom Operations Map (eTOM) are outlined in Sect. 3. The proposed SLA architecture with the Service Broker integration is presented in Sect. 3.3. Section 4 concludes this paper, sums up and explains next steps.

2 Related Work and Motivation

2.1 Related Work

Recently, considerable attention has been paid to IMS network and service delivery in an efficient manner. Managing the interactions of service capability between any types of IMS application servers appears important to deliver requested services from customers. However an SLA contract is needed to define all conditions of supply, which include the type of the service, the quality of service (QoS), the price, service modification, the responsibilities and obligations of the supplier and the customer, and etc. Almost all SLA contracts deal only with the service provider and neglect the customer side. However for a transparency and credibility operation, all parties should have a guarantee SLA on their side.

Several authors [1, 2] have specified the function of the Service Broker on being an entity that manages the interactions of service capabilities, they discussed the issues and considerations without embarking on the implementation phase of the Service Broker whether alone or with the SLA negotiation procedure. The work on the future Service Broker had started and several approaches have been proposed. However until now, these propositions are not deployed. In [3], according to the authors, Service Broker can be part of AS, or takes a part of the IMS entity S-CSCF (Serving-Call Session Control function), or reside in between AS and HSS (Home Subscriber Server). The integration and deployment of services is described in [4] as complicated and expensive operation; authors specified that there is no common integration of the different services from the point of view of the end user.

Existing Service Broker approaches have addressed the implementation propositions, Nguyen Tai Hung et al. in [5] adopted a rules based strategy in which Service Broker will invoke a specific service (from AS) based on static and dynamic rules, however some issues from an architectural aspect exist, like Service Broker needs to have depth details about all the services to be invoked, and interface between the Service Broker and application servers to provide the identification of individual services.

2.2 Motivation

Moreover, the service provider and the IMS network operator may not be the same. Hence, different operators are involved for the IMS services over the networks. Due to these diverse operators, there should have Service Broker agreements to provide adequate parameters (like QoS, price, etc.) for the IMS users.

To guarantee these service parameters, there should be SLA between the IMS service provider and network provider. A SLA between the service provider and IMS network provider can be developed so both operators benefit and the different parameters of IMS users are ensured. The author in [6] explains that a typical SLA contains basic information about the service, and a set of appropriate procedures and objectives between the involved parties in order to specify the normal information of a commercial contract, such as the technical terms of the QoS, the period of application, responsibilities, penalties, and so on. The technical part of SLA is SLS (Service Level Specification) [7]; it usually contains definitions of QoS measures for the service in question, and the declaration and implementation procedures.

Our approach differs from others by providing a Service Broker who enables to operators and service providers to interconnect via SLA (Service Level Agreement), so the billing process for charging services is well defined, SLA scenarios and other processes from eTOM Framework are integrated with the Service Broker to afford the best service to the customer. SLA can directly react with customers and service providers or via Service Broker.

3 Overview and New Approach

3.1 IMS Concept

The IMS standard is proposed by 3GPP [8]. In a formal way, IMS is defined as a new core network domain. The IMS architecture can be structured in layers as identified in the Fig. 1.

The user can connect to IMS in various ways through the access layer and register his devices with the higher layers (Control and Application layer). The control layer controls the authentication, routing, and distribution of IMS traffic between the transport layer and the application layer. Most of the traffic in this layer is based on the session initiation protocol (SIP). In addition to routing SIP messages to their appropriate services, the control layer also provides the capability to interface the application layer with other services. The main entity is the CSCF (Call Session

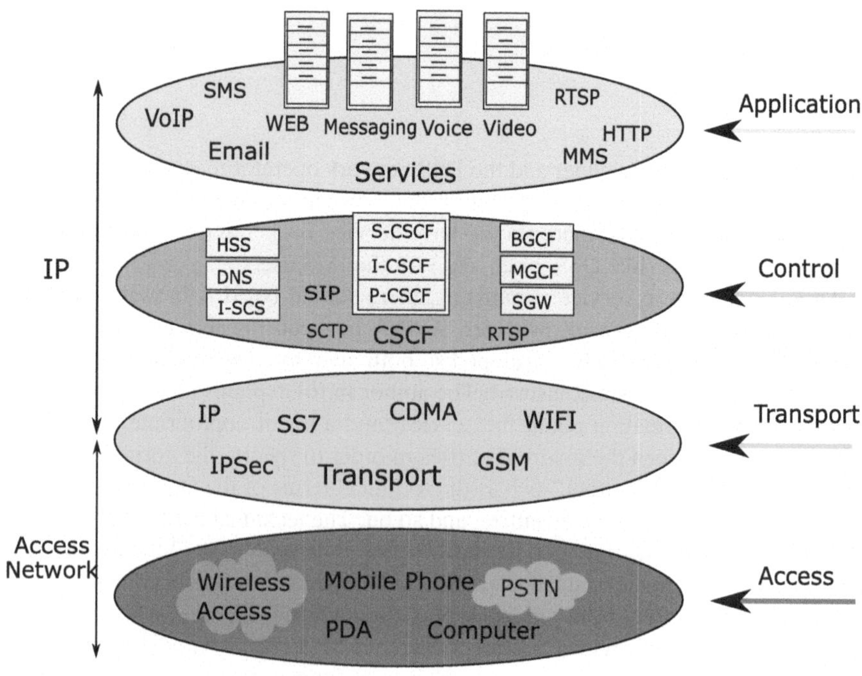

Fig. 1 IMS network architecture

Control Function) which facilitates the correct interaction between the application servers, media servers, and the HSS (Home Subscriber Server), which is the centralized repository for all subscriber account information [9]. The application layer includes traditional voice services (like voicemail, announcements, interactive voice response, and so on) as well as new applications built on the IMS architecture. This is the final layer of abstraction that gives IMS architecture the power and flexibility to rapidly deploy new services.

Application servers are characterized by implementing a SIP interface toward the S-CSCF. There are three types of AS; the SIP application Server (SIP-AS) is the native AS in the IMS [10]. The Open Service Access—Service Capability Server (OSA-SCS) provides the gateway functionality to execute OSA services in the IMS [10], and last, the IP Multimedia Service Switching Function (IM-SSF), which provides a gateway to legacy service networks that implement CAMEL (customized Applications for Mobile network Enhanced logic) services [10].

3.2 Enhanced Telecom Operation Map Framework

eTOM (enhanced Telecom Operation Map) Framework is the functional analysis viewpoint of the NGOSS (New Generation Operations Systems and Software) Framework. eTOM is considered as a repository of process elements at various levels of detail that can be combined and applied in specific applications. It provides a common language to describe business processes carried out in telecom activities. Flow diagrams used in eTOM illustrate end to end processes like Fulfillment process. Its technical content is more mature lately, with an increasing emphasis on guidelines for its application and usage. It is characterized by a hierarchy of process definitions and a common language for business processes. eTOM Framework provides a standardized telecom-oriented Business Process map covering all functions of an operator, including service integration and supply. Two operators communicate with each other in a Customer/Provider way. eTOM processes manage the most important tasks of an operator, however, to our knowledge, there is no real implementation for IMS network using eTOM specifications with Service Broker task, these still standard for all type of networks [11].

In the Framework of eTOM level 1, fulfillment, assurance, billing (FAB) is the core processes of operation processes as shown in the Fig. 2. The horizontal processes represent functional view points and vertical processes represent business view points.

As we highlighted and for efficient and agile management, we need to make projection of eTOM in IMS Network, it requires the integration of specific processes with Service Broker to identify the principal steps for conception and implementation as seen in the Fig. 3. At the first glance, we distinguish two views, one is business, and the other network.

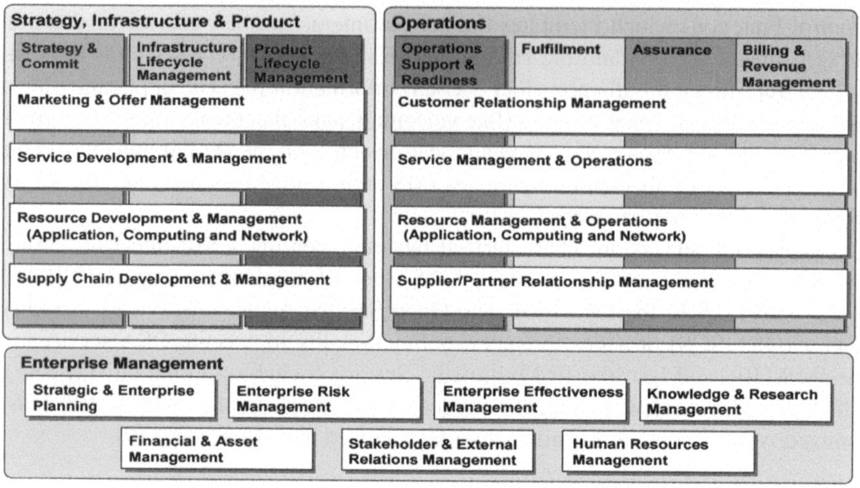

Fig. 2 eTOM business process framework—Level 1 [12]

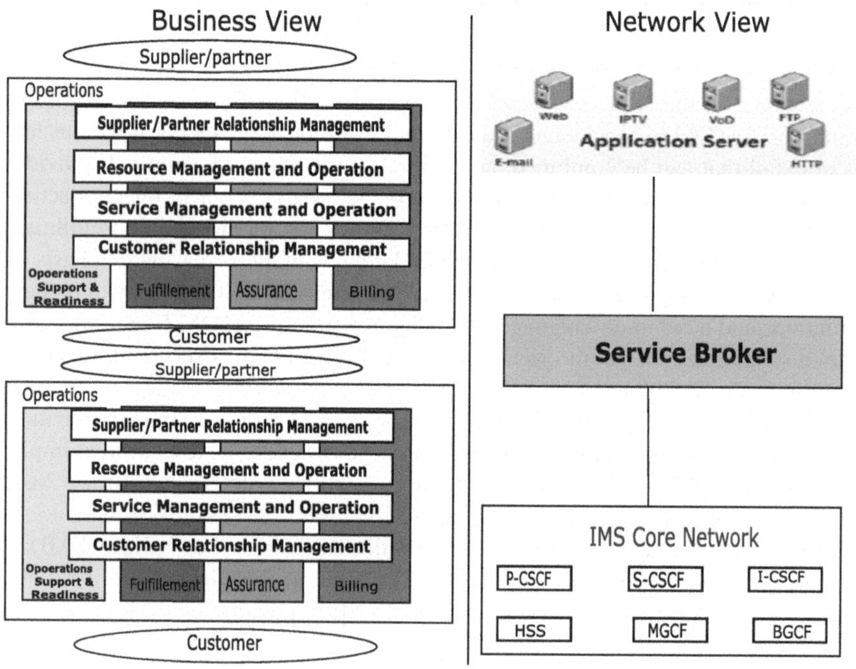

Fig. 3 Projection of eTOM on IMS architecture with Service Broker integration

eTOM takes two positions in the business view, however, in the network view; Service Broker is associated with the application layer in one side, and with the core of IMS in the other. Service Broker will react in situation-dependent, he can take several roles, including, service provider, customer, and even operator. When Service Broker took the role of a service provider, he will interface and interact with a customer to agree the provisioning of a service, however, in the customer role, SB interacts with the service provider to procure a service, he makes a deal and a contract with the service provider in order to provide the service requested. In another case SB take the operator's role and manage the service and the network resources. For these purposes, Service Broke must:

- Retrieve service requests.
- Search the best offer possible available.
- Calculate an estimation for the service provider (SLA client).
- Deliver the service to the customer.

To identify the customer's request, the Service Broker must verify the customer's profile and his class category (Platinum, Gold, Silver or Bronze), the Service Broker must also detect the type of the service asked (IPTV, presence, VoIP (Voice over IP) and so on), also verify the Quality of Service (QoS) and security, right after, the Service Broker will look for the best deal available respecting the criteria mentioned before, and then calculate an estimation vis-à-vis service providers through the entity service-level agreement Client (SLA client) to finally send the service to the customer.

3.3 Proposed Network Architecture

Our proposed network architecture for Service Broker and SLA deployment over IMS network is shown in Fig. 4. The IMS service provider is responsible for the type of service that will be provided to users. There is a Service Level Agreement

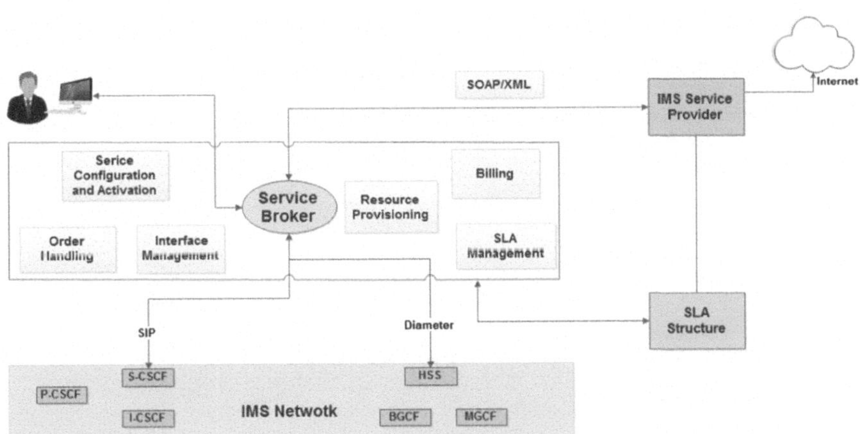

Fig. 4 Network architecture for Service Broker and SLA based IMS services in IMS network

between the IMS service provider and the IMS network operator. The Service Broker executes the SLA between the IMS service provider and the network operator. The Service Broker efficiently distributes all needed parameters among the IMS services.

The Service Broker makes a policy access, resource reservation, service configuration, and service activation for the SLA framework. Service Broker reserves the needed parameters for IMS services. The Service Broker calculates the average requested of the previous service history for a certain period of time. Then the Service Broker allocates this amount of parameters (QoS, bandwidth, and so on) for IMS users from the network capacity. Thus parameters used of IMS network satisfy IMS user's significantly. Service Broker is responsible for allocating preferred service to users as requested. Service Broker has a module database to keep the information on, with a method of using that database to authenticate requesters. The requests include the service type, the customer profile, and the time period when service is required. The Service Broker verifies all components whether checked to meet the request or not.

Figure 5 illustrates the SLA between IMS network and other entities. There is a SLA negotiation between the IMS network operator, the customer, and the IMS service provider to ensure the best offer suited the IMS users. The Service Broker is the central entity that is responsible for the execution of SLA negotiation and resource reservation. The Service Broker manages the services and their parameters, requested QoS, type of services, and etc. of the customer networks (IMS network

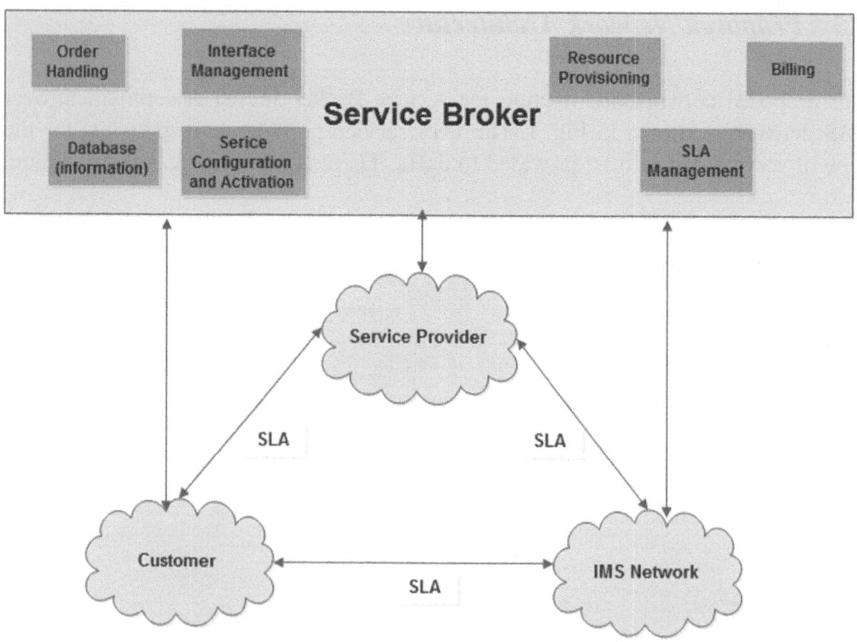

Fig. 5 SLA between IMS, customer, and service provider to guarantee the IMS service delivery

operator, IMS service provider). The Database stores the customers' information, SLA parameters, and services history. The Service Broker can configure the parameters of the SLA according to the feedback from the SLA management process. The amount of IMS service is calculated according to the Billing process, Order Handling result, and collected information from Database.

The satisfaction of the customer depends on the availability of requested service. Available service with acceptable quality of service equal to or greater than requested service promotes a high satisfaction. When the quality of the service is less than the customer expectations, there is the possibility of disappointed customers. Hence, the SLA between service provider and IMS network operator ensures the reservation of the adequate bandwidth and QoS for the IMS services.

Service Broker is implemented using the Service-Oriented Architecture (SOA) which is an architectural approach that facilitates the creation of loosely coupled, and encourages the reuse of applications [13]. SOA is about integration and software reuse, the common thread for planning SOA operations usage is the eTOM, because it illustrates the points where service features need to be activated or provisioned. Operators plan out this operations dimension in order to ensure compatibility with their business processes and also to ensure that any other service interactions are compatible with established operations principles. SOA respects the layers compositional structure of eTOM and their presentation needs business logic with protocol communication based XML (SOAP (Simple Object Access Protocol)). While there may be different implementations, typical SOA makes use of Web Services Description Language (WSDL), and Business Process Execution Language (BPEL) and web services. Interacting with SOA allow Service Broker to connect with web services. New applications can be built with less effort and existing applications can be efficiently adapted to changing requirements, reducing maintenance and development cost.

To describe the interactions between the different items, we took for instance service provisioning as a basic scenario in IMS network.

The approach starts starts when the Customer Interface Management (CIM) system from eTOM Framework receives a new IMS service order, and completes the customer's request and the related service by adding some information (SLA, profile, customer's profile). The Order Handling process structures the service request with customer information, validate the order, and verify that the required information for processing the order is available. Thus Service Broker redirects this information to Service Configuration and Activation (SC&A) process, this step creates a new instance of the service using the service configuration, Service Broker checks the resource capacity and configure resource elements involved to provision a service.

After the success of resource reservation on both sides by internal Resource Provisioning (RP) and by external business processes, Service Broker notifies the customer that the reservation and the design compilation are made perfectly, and the service proceeds to the activation step, details about the service and the customer are added by Service Broker to the billing system, at the end, Service Broker sends provision requested to the service provider, update service and SLA (Service Level Agreement) with Billing and Collection Management process.

Service Broker can take several roles including Service Provider by interfacing and interacting with customer to agree the provisioning of a service, however, in the customer role, he procure a service, makes a deal in order to provide the service requested. Service Broker can also manage the service and the network resources. Service Broker can interact between two users invoking one service, or manage the conflict between different services invoked at the same time.

Regards the negotiation process can be done on behalf of any entity, proposals and alternative proposals are exchanged between concerned parties. The offer based on the service, the price, the class of the customer, the QoS, is the first step that the service provider presents. The Service Broker chooses the offer that maximizes the service provider's payoff by taking all features into consideration and after calculating the payoff related with other offers possible. A customer sends an acceptance message to the Service Broker if he is convinced about the offer; the SLA is finalized at this stage. However, the customer can call off the negotiation or send an alternative proposal if he does not accept the initial offer. In this case, the Service Broker must update the customer needs and contact the adequate service provider to figure out an offer for him. If both sides agree, and the offer is acceptable, an SLA is created, otherwise Service Broker sends another proposition. This operation remains until one of the concerned parties quit or accepts the offer, or simply, the time allowed for this operation expires.

4 Conclusion

In this work, we proposed a new approach to Service Broker implementation in IMS environment, which bring insights to the NGN service provisioning. Service Level Agreement are difficult to specify in a clear and simple manner. Based on the architectural issues, in ongoing work, the Service Broker will negotiate the quality of service between the customer and the service provider, from one side, and between the customer and the network, from the other side.

References

1. Chua, H.-N., Tan, C.-M.: Service broker function in IMS architecture—issues and considerations In: 12th WSEAS International Conference on COMPUTERS. Heraklion, Greece, 23–25 July 2008
2. Goveas, R.P.S., Packard, H., Sunku, R., Das, D.: Centralized Service Capability Interaction Manager (SCIM) architecture to support dynamic blended services in IMS network. In: 2nd International Conference on Internet Multimedia Services Architecture and Applications, IMSAA 2008 (2008)
3. Chua, H.-N., Tan, C.-M.: Malaysian Research Centre, British Telecommunications Group, Kuala Lumpur, Malaysia: Service Broker Function in IMS Architecture—Issues and Considerations, Sept 2008

4. OMA service environment Approved version 1.0.4, 01 Feb 2007, MA-AD-Service-Environment
5. Hung, N.T., Thanh, N.H., Magedanz, T., Mueller, J.: Toward a full implementation of SCIM functional block in IMS framework. In: 2013 Fifth International Conference on Ubiquitous and Future Networks (ICUFN)
6. Blokidijk, G.: Service Level Agreement—Simple Steps to Win, Insights and Opportunities for Maxing Out Success (2014)
7. Goderis, D., Van Den Bosch, S., Poupel, O., Jacquenet, C.: Service level specification semantics and parameters In: Internet draft, Jan 2002
8. Ip Multimedia Subsystem (IMS); Stage 2, 3GPP, TS 23,228, Release 9, 2010
9. Navarro, M., Donoso, Y., Rodríguez, V.: An IMS architecture with QoS parameters for flexible convergent services. In: 2010 IEEE Symposium on Computers and Communications (ISCC)
10. Camarillo, G., Garcia-Martin, M.A.: The 3G IP Multimedia Subsystem (IMS) merging the internet and the cellular worlds. In: Third edition of this Best-Selling Guide to IMS: Fully Revised, and Updated with Brand New Material
11. Enhanced Telecom Operation Map (eTOM), The Business Process Framework. For The Information and Communications Services Industry. Process Decompositions and Descriptions, Release 6.0, TMF GB921V, Nov 2005
12. Chang, B.-Y.: Business process management of telecommunication companies: fulfillment and operations support and readiness cases. Int. J. Future Gener. Commun. Networking **4**(3) (2011)
13. Zhang, X., Song, J., Qu, H.: Research on the architecture of telecom services platform based on IMS and SOA. In: Conferences on Pervasive Computing (JCPC) (2009)

A New Approach for Modeling Strategic IT Governance Workflow

Meriem Chergui, Aziza Chakir, Hicham Medromi and Mostafa Radoui

Abstract In this article we propose a new approach to govern an information system (IS) through The Control Objectives for Information and related Technology Business (COBIT) in an intelligent way. The purpose of this approach is monitoring IS good governance and enabling other frameworks to apply specific processing, if needed. In fact, Information Technologies Management (ITM) is a recent discipline aiming at guiding and controlling organization resources to achieve business goals, by relating human resources, financial resources, and technical resources to IT tools. ITM needs Information Technologies Governance (ITG) methodologies and frameworks to achieve competitive edge in the marketplace. COBIT, a base of ITG, offers a generic framework of structured IT control activities. It is designed to ensure harmonization of terms and principles to facilitate its integration with other frameworks. To benefit from this generic aspect, we propose an IT Governance kernel based on COBIT with an intelligent learning layer for Enterprise knowledge. We appealed to the Loose Inter-Organizational Workflow to address the constraints of heterogeneity and difference between IS components. We use both the multi-agent technology to insure the issues of autonomy, cooperation and coordination and the semantic web to understand business stakeholders' languages to express their needs. An implementation of this solution was done in J2EE technology to ensure its performance.

Keywords Matchmaking · Expert system · Inference engine · Web semantic · IT governance · Business strategy · IT services

M. Chergui (✉) · A. Chakir · H. Medromi · M. Radoui
LISER Lab, ENSEM, Hassan II University, Casablanca, Morocco
e mail: chergui.meriyem@gmail.com

A. Chakir
e-mail: aziza1chakir@gmail.com

H. Medromi
e-mail: hmedromi@yahoo.fr

M. Radoui
e-mail: m.radoui@ensem.ac.ma

R. El-Azouzi et al. (eds.), *Advances in Ubiquitous Networking 2*, Lecture Notes in Electrical Engineering 397, DOI 10.1007/978-981-10-1627-1_22

1 Introduction

The structure of information technology (IT) could have an impact on business performance. Today, it plays an important role in modern social and economic life. It has not only changed the traditional methods by which people obtain information, but also broke the old production management schemes and profoundly changed the business organization structure in space and time. It plays an important role in inter-organizational transactions and relationships. Thus, it has become a valuable asset and resource in the literature of contemporary business strategy.

The business world is changing rapidly and requires flexibility and responsiveness. Consequently, Information Technologies and information systems (IS) are closely linked and agility is a challenge for many business.

Information technologies have become essential to support the sustainability and growth of the company. Excessive use of technology has created a critical dependency on IT that requires particular interest to IT governance (ITG). Some studies have shown that companies that have good models ITG generate superior returns on their IT investments than their competitors With IT investments that are an important part of companies' budgets and increasing external pressure to control and monitor costs, effective ITG is considered as an essential means to ensure a return on IT investments and improving organizational performance.

ITG requires the development of good practices proposed by the market leaders and experts. Repositories and guides have been proposed serving companies to allow their information technologies to deliver value, manage risk and resources, to align business strategy to IT matters and also to measure performance of business processes. Quite sensitive and delicate operation for the multitude of existing solutions, this research work proposes a new automation approach to govern an information system based on COBIT framework.

The article is structured as following; after the introduction, Sect. 2 presents IT Governance state of art; Sect. 3 talks about the proposed approach in a functional and an organizational dimension Sect. 4 gives an overview of intelligent agents and Inter-organizational Workflows to understand the technical architecture proposed in Sect. 5. In Sect. 6 we present the implemented solution and give after that a conclusion and perspectives.

2 An Overview on IT Governance

2.1 IT Governance Definitions

Governance is a concept that can be used in many contexts; there are different types of governance: "Enterprise Governance Corporate Governance and IT governance.

In fact, Enterprise Governance is balanced between conformity and performance. For performance we talk about business governance and it concerns strategic decision making, plans, and scorecards for value creation, as for conformity it's mainly about control to insure liability: internal audit, human and materials resources, and it's what we call Corporate Governance [1]. IT Governance (ITG) supports both Corporate and Business Governance inside the Enterprise Governance by managing processes to enable the business to drives IT correctly.

These types of governance are correlated and we should treat them as "Global Governance" with dependencies between them and an order to go with.

Governance of Information Technology (ITG) can't exist in isolation, but must be a subset of corporate governance and is also commonly called the subset of corporate governance [2], we conclude that ITG is the lowest level of the three types of governance and more specific and targeted.

Since information technology (IT) and information systems (IS) are strongly connected, the lack of clarity on the concept of ITG is not surprising given that SI is a relatively new discipline that has emerged in variety of disciplines, but certainly not limited to the social sciences and computer science Many studies continue to focus on the definition of the ITG [3] we quote for example:

- In 2005 [4] define ITG as the process by which decisions are made about IT investments. How decisions are made, who makes the decisions, who is responsible, and how the results of the measured decisions are monitored by all parts of the ITG.
- In 2006 [5] define ITG as strategic alignment of IT with business so that the maximum value of companies is achieved through the development and maintenance of an effective information control and accountability, performance management and management risks.
- In 2010 [6] define ITG as the process to ensures the efficient and effective use of information to enable an organization to achieve its objectives.

2.2 Information Technologies Governance Frameworks

In the literature of the IT governance, many frameworks are presented, with a variety of strengths, advantages and limitations. Thus, each framework provides a specific level of detail in its field. One of the concerns of IT governance is frameworks coexistence to exhibit their complementarily. We will present in the order the main ITG repositories for:

- Overall management,
- Service and projects control,
- IT security.

As for corporate Governance:

1. **COSO** [7] (Committee of Sponsoring Organizations of the Tread way Commission) published in 1992 base on internal control to help companies assess

and improve their internal control systems. Internal control is a process described as the liability established in order to achieve the objectives grouped in the following areas:

- Efficiency and operation
- Reliability of financial information;
- Compliance with laws and regulations.

2. **Balanced Scorecard** (BSC) [8] is a performance to explain the vision and strategy of the company, and translate them into action plans. This gives a return on internal processes and external constraints, in a continuous improvement strategy. Its authors, Robert Kaplan and David Norton, describe it as follows: "The BSC enables traditional financial results, but these results highlight the past, which was normal in the industrial age, with long-term investment and this little customer relationships. These financial elements are insufficient, however, to control the companies in the information age, which should build their future value through investment in customers, suppliers, employees, processes, technology and innovation."
3. In terms of security management, several standards, methods and safety standards of the information systems are set up. These methodological guides to ensure consistent safety approach. ISO has implements the ISO/IEC standard 27000;
   This number corresponds to a series of safety standards, namely 27000, 27001, 27002 and 27006 are published. These standards are either for certification or good practice guides:

   - ISO/IEC 27000: Vocabulary and definitions of the security, applicable to each of the standards;
   - ISO/IEC 27001: the management policy of the computer security for the company as a certification;
   - ISO/IEC 27002: good practice of IT security guide;
   - ISO/IEC 27003: Implementation Guide;
   - ISO/IEC 27005 requirements for information security, low on the PDCA (Plan, Do, Check, Act), complementing the ISO 27001.

4. **ITIL** [9] "IT Infrastructure Library" is a study launched by the British government in 1990 to identify best practice IT management services, the results have given rise to a library named "IT Infrastructure Library" or ITIL, documenting an IT management approach to support users' business organizations. In 2007, the V3 version of the ITIL framework is based on five pounds of good practice, offering complements sector or market and generic models (process maps, etc.) including:

   - Service Strategy describes the general strategy and value delivery service, all in dealing with business alignment and IT governance.
   - Service Design provides procedures, architectures and documents to create the service management process.

- Service Transition provides practical guides integration of service management processes between businesses and operations.
- Service Operation provides guides to achieve the QoS objectives in the interests of efficiency and effectiveness.
- Continuous Development Service provides guides to identify and improve processes. It combines the methods of quality management and improvement of PDCA loop.

2.3 *COBIT a Federator Framework for ITG*

COBIT (Control Objectives for Information and related Technology Business), developed in 1994 (published in 1996) by ISACA (The Information System Audit and Control Association) is an IT governance tool that was developed initially for the control by declining COSO guidelines on the objectives of information technology [10].

COBIT in its 4.1 version is a repository of information systems governance that divides any Information System into 34 processes, which are divided into four functional areas:

- Planning and organization (Planning and Organization) (10 processes).
- Acquisition and installation (Acquire and Implement) (7 processes).
- Providing service and support (Deliver and Support) (13 processes).
- Monitoring (Monitor) (4 processes).

Using COBIT, we identifies three stages:

- Goal Management,
- Determination and Resource Management,
- Processes Management.
- Control and measuring performance

COBIT consists of several components in the service internal and external stakeholders of the SI. These components are interconnected and aimed at meeting the needs of IT governance, management, and control of different players, as shown in the figure below (Fig. 1):

The major contribution of COBIT is that mobilizes over the Information Systems Department (ISD) business managers and the Board of Directors, with the IT support of the company's business objectives. To do so, COBIT deploys five axes of IT governance namely:

- **Strategic alignment**: IT plans are aligned to business plan
- **Value creation**: the added value of IT to the enterprise sales business
- **Resource management**: the optimization of infrastructure and knowledge
- **Risk management**: awareness of potential risks, compliance and treatment

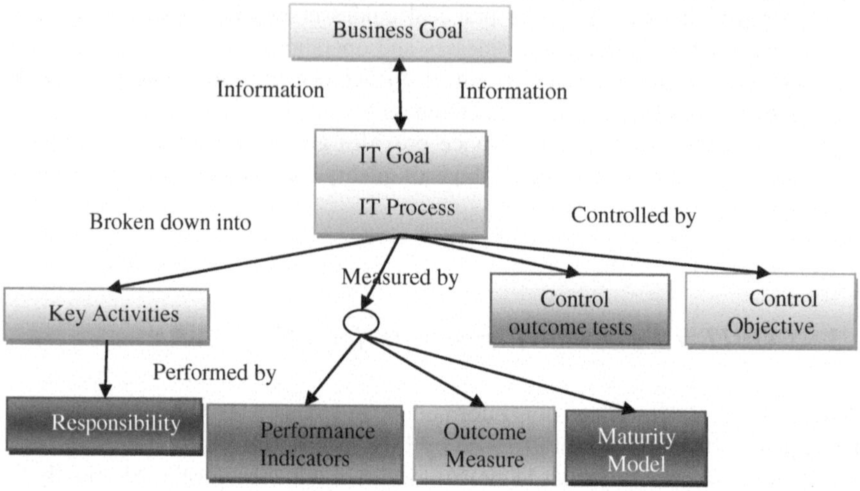

Fig. 1 Relationship between COBIT Components

- **Performance measuring**: monitoring the implementation of the strategy.

COBIT was designed with a strategic vision and a control vision, related both to the operational activities through the process. Indeed, it is a repository based on established frameworks such as Capability Maturity Model for Software Engineering Institute, ISO 9001, ITIL and ISO 17799 (standard security framework, now ISO 27001). Due to its cross-coverage of the areas of the company and because it is based on many existing practices, COBIT can act as an integrator which includes several practices in a single framework, also linking these practices to strategic business objectives [11].

3 IT Governance Workflow New Approach

First, the governance of information systems within the company is not just a set of guidelines to implement by the Information System Management (ISM) and/or top management, but it is an exchange between each other, or even rebound information from the business departments. These exchanges can be both demanding new features and controls for computer services feedback. Indeed, for better governance of the IS, the mobilized actors are as follows:

a. The internal stakeholders interested in the business by IT investments generating value such as:

   - Investment Policymakers,
   - Officials who define the requirements,
   - Users of IT services.

b. Internal and external stakeholders who provide IT services such as:

- The managers of the organization and IT process
- The daily users of the information system.

c. Internal and external stakeholders who have responsibilities in the control and the risk such as:

- The security officials, for privacy and risk,
- The Responsible for compliance with laws and regulations,
- The general Auditor

So this is indeed a set of information flow (Workflow) moved from end to end between the ISM and potential users on the one hand and the ISM and decision makers on the other hand.

Our modeling approach is therefore to design a workflow that manages the IS governance flows between all these actors for:

- Better alignment of information technology with business strategy.
- Business Value creation through well governed IT.
- IS and business risk management
- Better Management of human resources and equipment.
- A measure of the Enterprise IT performance.

The type of workflow depends on the constraints of the ITG context. Faced most often with heterogeneous information systems with loosely coupled components, we opted for a loose inter-organizational Workflow [12].

Second, in the previous section, we highlighted the unifying aspect of COBIT as a global repository of the ITG and its ability to trigger other repositories, so we will base this approach on COBIT to capitalize these strengths and especially the diversity of its components for a successfully ITG. However to not limit this approach to a single repository or fall into the trap of computerization repository that has not been a real success, we will use as COBIT core and we will try to combine the company specific results for increasingly efficient. Especially, COBIT is oriented repository process, something that opens the programming perspectives and fairly extensive computerization.

4 Intelligent Agents and Inter-Organizational Workflows

4.1 Workflow and Inter-Organizational Workflow

A workflow in general is the total or partial automation of business process execution, execution during which documents, information tasks from one participant to another to perform specific activities according to predefined rules.

There are many kinds of workflows namely [13]:

- **Administration Workflow**: devoted to manage administrative procedures whose rules of conducts are established and known by everyone in the company.
- **Production Workflow**: devoted to manage the production process in the company.
- **Collaboration Workflow**: devoted to manage awareness and group collaboration in a project of creative work
- **Ad hoc Workflow**: is a class of workflows for specific situations where the flow logic to be followed is set during execution. It forms a hybrid solution collecting characteristics administration, production, and collaboration

The interested on these kinds of Workflow will find in the references more details about them the advantages and drawbacks of every one.

Inter-organizational Workflow (IOW): is an extension of the classical Workflow aiming at cooperating between heterogeneous and autonomous organizations. The reason why it was chosen as a workflow model for this Audit solution.

4.2 Multi-agent System

Multi-agent systems (MAS) are widely used for modeling coordination system [14]. It seems to be appropriate to describe the coordination of IOW as a dynamic system aiming at finding "supply and demand service" and adopting the negotiation between partners. In fact, agent technology is a custom frame for IOW abstraction: it resolves its constraint of distribution, heterogeneity, autonomy and flexibility:

- Autonomy: every organization of the IOW can be encapsulated in an Agent as autonomous entity having its intentions goals and resources and able to be executed aloe or in an environment, depending on the context.
- Distribution: IOW is a distributed context and MAS includes specific architecture, communication protocols and languages to support this constraint.
- Heterogeneity: Agent technology allows communication and interaction between heterogonous agents through Agent-Communication-Languages (ACL). It also provides synchronous and asynchronous ways of communication depending on the agent localization and constraints.

MAS offer many Meta-Models to cover the organizational aspect of Workflow. It also covers the scalability and security worries in loose IOW context.

5 Proposed Architecture

Faced to a competitive market continuously changing IT solutions, and information systems are made of heterogeneous components with various information flows and processes increasingly complex. The decision of top management in the field of IT

Governance became increasingly sensitive (poor visibility) Hence the need of adequate IT governance tools.

In this perspective, this research focuses on modeling IT governance solution for enterprise with different business flow and heterogeneous partners assisting the Information system process orchestration.

In fact after a benchmarking done in previous works [15], the existing solutions of ITG, are mainly Enterprise Governance platform dealing with a specific subject (health/finance/public services…), or a part of an ERP difficult to deploy in any IS.

The objective of this work is to propose a workflow model that encompasses IT Governance support on good practices (we opted for COBIT) and adaptability to the complexity and changes with agile appearance, distributed and cooperative through a workflow-based on Multi-Agent Systems (MAS).

The proposed architecture is a process oriented solution that enables:

1. Strategic analysis of an information system through the WIO
2. Exploit the strengths of COBIT for Information Systems Governance namely:
   - List the computer activities to implement
   - Propose any previous optimization and control
   - Deduce the different levels of maturity, measures and performance
   - Define the responsibility matrix.
   - Provide adequate control tests
3. Distribution, autonomy and learning through MAS.
4. Semantic efficiency and portability on the web through the IT Governance Ontology. In addition, this solution is intended for all users of the IS for a self-audit in real time by combining the raw material of the COBIT framework and know-how of the company.

The architecture of the loose Inter-Organizational Workflow (IOW) of IT Governance is a solution that allows to govern each IS component without consideration of its technical characteristics and its interconnection with the rest of the components. It is based on multi-agent systems and COBIT framework.

The proposed architecture is based on the Inter-organization Workflow reference architecture, intelligent agents and the bond between the components of COBIT 4.1 framework, it contains:

1. IS Workflow Agents: each agent represents an IS business application not necessarily communicating with each other.
2. Business goal Agents: manages a set of IT goals that appeal in its turn to IT processes to measure IS business goals.
3. Multi-Agent System Framework Manager: three agents the first one called business goal manager agent who managed business goal agents (creation/suspension/resource sharing), the second one is update agent who's responsible for framework update, the third one is a learning agent who persists enterprise IT processes in the knowledge base to reuse them.

4. Multi-Agent System IS Manager: It contains the interface agent responsible for the dynamic configuration and IS Manager Agent (creation/suspension/resource sharing), who manage IS Workflow agents.
5. Connection Server Agents: Yellow Page for the publication of responses and requests respectively Framework agents and IS Workflow Agents.
6. Mediation Expert system: an intelligent system who establishes the correspondence between demand of IS Workflow Agents and supply of Business goal Agent.

6 Demonstration

6.1 General Presentation

As described before, the proposed architecture aims at ensuring IT Governance of a Complex Information System. It is based on three essential components:

1. Loose Inter-organizational Workflow of ITG
2. Matchmaking Expert system with semantic inference engine
3. IT Governance Framework Multi-Agent system

To implement this architecture we proposed a web solution multi-users linked to a knowledge base, intelligent agents are deployed to:

- Capture uses needs
- Interpret their requests to ITG understood goals
- Propose convenient IT Processes from ITG framework to users' requests.
- Update the used framework

To evaluate the platform results we compare them to ITG expert ones for the same request, since one of the main objectives of this research work is to computerize ITG audit mission.

6.2 Technical Presentation

The proposed platform is a web solution developed in Java using the J2EE Technology with Frameworks JPA, EJB, JSF2.2 and MySQL database Management system.

As for multi-agent systems we used Madkit 5 API.

As for semantic analysis we used the Solr server version 5.1.1.

As for ontology we used OWL-S language in the editor Protégé 4.3 and Fact++ compiler.

6.3 Functional Presentation

As presented before, the platform main functionalities' are:

- Static configuration
- Dynamic request creation
- Results visualization with details
- Report edition
- System logging
- EAS IT-GRC launching

The screens bellows represent the platform main functionalities:

The Fig. 2 presents the login interface; there are essentially 3 users, super root, Information System Manager (ISM) and Business Manager (BM). As for Fig. 3 it presents the main menu of ISM about static configuration of the platform done in many steps.

The static configuration consists on defining the organization, its department, the IS actives and RH the organization. Also, introduced informations are used to define both enterprise specificities for better governance and concerned profiles.

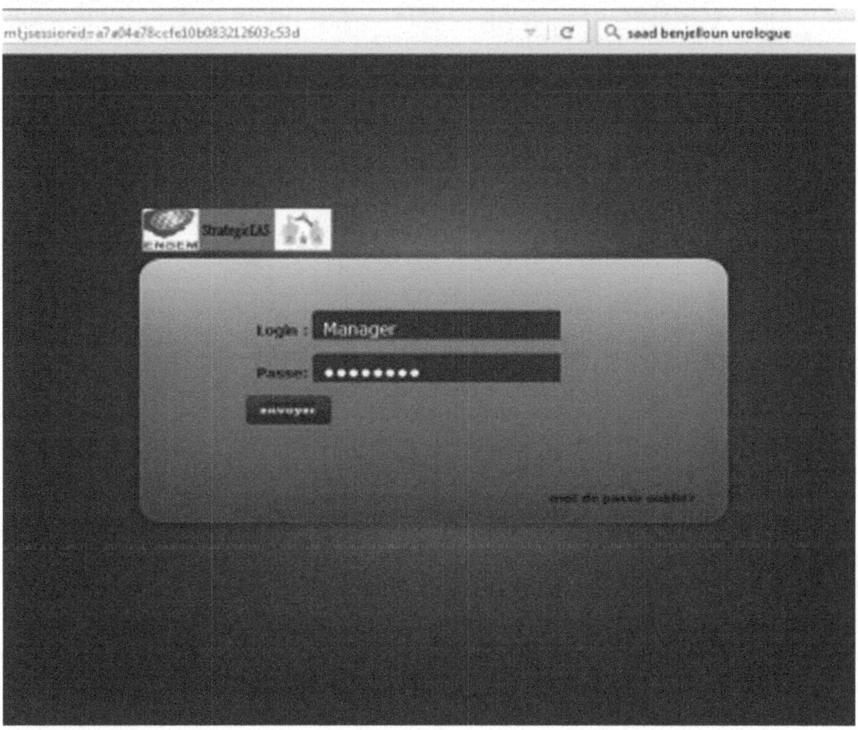

Fig. 2 Strategy evaluation

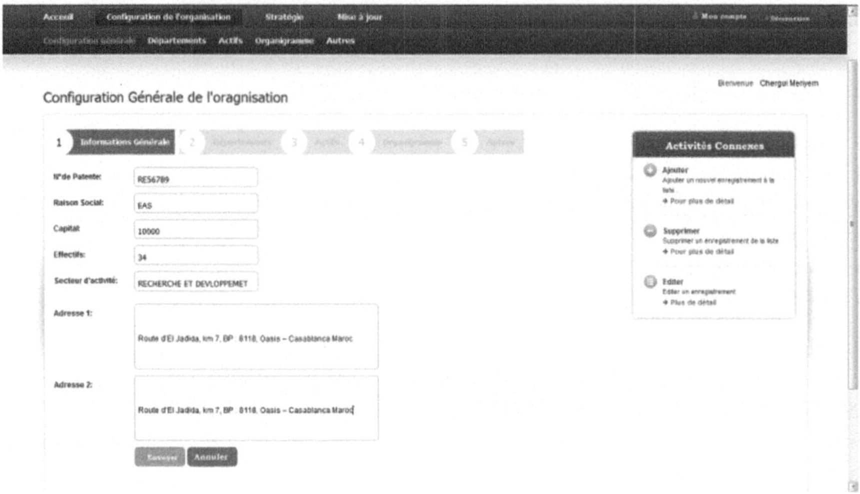

Fig. 3 Authentification

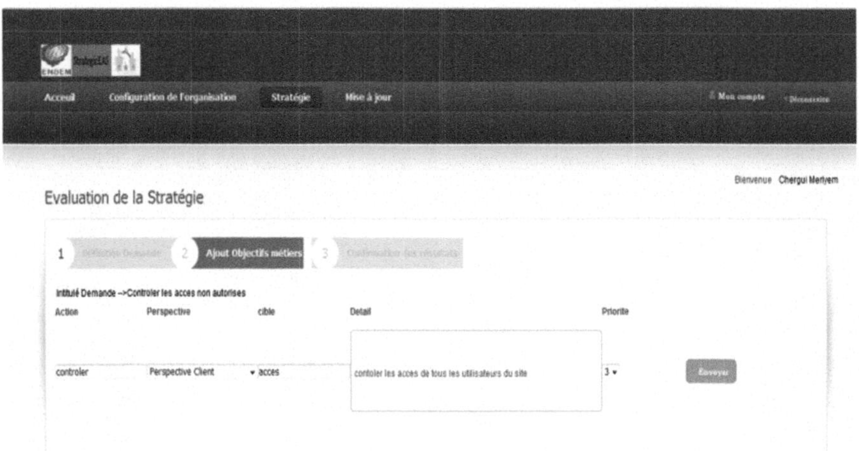

Fig. 4 Static configuration advanced step

The results concerns both Business Manager and ISM in which they can express their request about an application in the information system. a priority is define for the request to be treated by the system.

After launching the treatment, the results are detailed as shown in Fig. 4, about convenient Business Goals, IT goals and IT processes. A report is generated with other details such as metrics, Maturity models, key activities…etc.

The component responsible for these results is the mediation layer which is based on a Matchmaker expert system that provides:

- Requested and editing services
  Explanation of services through "AuditOntology" [16]
- Semantic matchmaking

These three tasks in practical terms exceed the capabilities of a cognitive agent. Especially in matchmaking knowledge base consulting is required to match the best offers on demand.

The matchmaking Expert system contains three components:

- **Persistence unit:** to edit demand and offers in the Knowledge Base, in fact through connection servers real time demands are sent to be understood and answered, to extend the knowledge base for future uses.
- **Semantic unit** based on ITG ontology to understand users' requests in an IT Governance way, it's a dynamic layer where Audit ontology in OWL-s format is created and saved. In fact it's the hierarchical description of demand services and supply ones. This layer communicates with an ontology Data-base, in this paper, Protégé save Ontologies by default in web localization; so data-base could be replaced with an XML file containing Ontologies URL.
- **Matching unit** with comparison algorithm to join business goals to user request. It's the comparison and link between a demand and convenient offers; it's a return of convenient Business Agent Addresses to IS Workflow Agent. The comparison is based on the Audit Ontology defined in Processing layer and need an algorithm to filter offers (not yet done). This is the intelligent layer of the Matchmaker agent and it sends results to the Knowledge Base through the persistence unit.

7 Conclusion

As conclusion the purpose of this paper is to propose a new approach to provide permanent and interactive Governance of Information systems.

Many literature issues were invoked namely:

- Inter-Organizational Workflows
- Multi-agent System and artificial intelligence
- Mediation Expert system
- Semantic Web and Ontologies

The choice of every issue has an added value for this solution; in fact, Inter-organization Workflows provide the orchestration of heterogeneous components of an IS in an autonomic way.

Multi-agent system insures the intelligent dimension of the solution with high level communication protocol and modeling architecture.

Mediation in MAS gives a theoretical model of matching services among intelligent entities.

Ontologies offer the semantic alignment of stakeholders with IT Governance vocabulary.

In fact, the IT Governance IOW role is not only to find the convenient Business Objectives for user demands but to find the best IT processes to launch with efficient priority order.

References

1. Harford, J., Mansi, S.A., Maxwell, W.F.: Corporate governance and firm cash holdings in the US. In: Corporate Governance, pp. 107–138. Springer Berlin Heidelberg (2012)
2. Van Grembergen, W.: Introduction to the Minitrack IT Governance and its Mechanisms-HICSS 2013. In: 2013 46th Hawaii International Conference on System Sciences (HICSS), pp. 4394–4394. IEEE, Jan 2013
3. Peterson, R.: Crafting information technology governance. Inf. Syst. Manag. **21**(4), 7–22 (2004)
4. Doidge, C., Karolyi, G.A., Stulz, R.M.: Why do countries matter so much for corporate governance? J. Financ. Econ. **86**(1), 1–39 (2007)
5. Webb, P., Pollard, C., Ridley, G.: Attempting to define IT governance: wisdom or folly?. In: Proceedings of the 39th Annual Hawaii International Conference on System Sciences, 2006. HICSS'06, vol. 8, pp. 194a–194a. IEEE, Jan 2006
6. Gerrard, M.: Defining IT Governance: The Gartner IT Governance Demand/Supply Model. Gartner ID G, 140091 (2010)
7. De Haes, S., Van Grembergen, W., Debreceny, R.S.: COBIT 5 and enterprise governance of information technology: building blocks and research opportunities. J. Inf. Syst. **27**(1), 307–324 (2013)
8. Gibbons, R., Kaplan, R.S.: Formal Measures in Informal Management: Can a Balanced Scorecard Change a Culture? (2015)
9. Sahibudin, S., Sharifi, M., Ayat, M.: Combining ITIL, COBIT and ISO/IEC 27002 in order to design a comprehensive IT framework in organizations. In: Second Asia International Conference on Modeling & Simulation, 2008. AICMS 08, pp. 749–753. IEEE May 2008
10. Ridley, G., Young, J., Carroll, P.: COBIT and its Utilization: A framework from the literature. In: Proceedings of the 37th Annual Hawaii International Conference on System Sciences, 2004, pp. 8–pp. IEEE, Jan 2004
11. Hardy, G.: Using IT governance and COBIT to deliver value with IT and respond to legal, regulatory and compliance challenges. Inf. Secur. Techn. Rep. **11**(1), 55–61 (2006)
12. Chebbi, I., Dustdar, S., Tata, S.: The view-based approach to dynamic inter-organizational workflow cooperation. Data Knowl. Eng. **56**(2), 139–173 (2006)
13. van Der Aalst, W.M., Ter Hofstede, A.H., Kiepuszewski, B., Barros, A.P.: Workflow patterns. Distrib. Parallel Databases **14**(1), 5–51 (2003)
14. Khamphanchai, W., Pipattanasomporn, M., Kuzlu, M., Zhang, J., Rahman, S.: An Approach for Distribution Transformer Management With a Multiagent System. IEEE Trans. Smart Grid **6**(3), 1208–1218 (2015)
15. Chergui, M., Sayouti, A., Medromi, H.: IT Governance through an Inter-Organizational Workflow based on Multi-Agent System. Int. J. Appl. Inf. Syst. Found. Comput. Sci. FCS **6** (6), 10–16 (2013). New York, USA
16. Chergui, M., Adil, S., Hicham, M.: IT Governance ontology building process: example of developing audit ontology. Int. J. Comput. Tech. (IJCT) **V2**(1), 134–141, Jan–Feb 2015. Published by International Research Group-IR (9). www.ijctjournal.org. ISSN: 2394-2231

Decentralized Control of Substations in Smart Cities

Mohamed Nouh Dazahra, Faycel Elmariami, Aziz Belfqih, Jamal Boukhrouaa, Lakbich Anass and Cherkaoui Nazha

Abstract The role of smart electric substation in smart cities becomes more important compared with the traditional one. Some benefits of smart substation and the concept of a decentralized control with an example of simulation are presented in this paper.

Keywords Smart city · Smart substation · Smart grid

1 Introduction

Cities all over the world are experiencing a real increase in population, economy and industry, to meet the challenges of this growth many governments are starting to adopt the concept of smart city. The concept of smart city includes smart infrastructure, smart operation, smart service and smart industry by utilizing next generation of ICT (Information and communication technology), such as internet of things, cloud computing, the city can fulfill high intelligent management of the city resources and facilitate people's life.

However the use of new technologies will increase the necessity of a more stable electrical power for the city, most cities contain many distributed substations which offer power to different services of the city, Moreover many cities have started using green energy to cover their power needs by installing different solar and wind farms which require upgrading of electrical networks and also building electrical sub-

M.N. Dazahra (✉)
Laboratory of Electric Systems and Energy, Team of Electrical Networks and Static Converters, Casablanca, Morocco
e-mail: m.n.dazahra@gmail.com

F. Elmariami
Superior National School of Electricity and Mechanics (ENSEM), Casablanca, Morocco

A. Belfqih · J. Boukhrouaa · L. Anass · C. Nazha
University Hassan II of Casablanca, PO Box 8118 Oasis, Casablanca, Morocco

R. El-Azouzi et al. (eds.), *Advances in Ubiquitous Networking 2*, Lecture Notes in Electrical Engineering 397, DOI 10.1007/978-981-10-1627-1_23

stations that are more intelligent, efficient and secure. In this paper we propose an innovative solution to update the infrastructure of existing substations into new smart substations which can respond to the needs of a smart city.

2 Smart Substations and Smart Grid

A Smart Grid provides electricity from suppliers to consumers using digital technology to save energy, reduce costs and increase reliability. It links everyone to abundant, affordable, clean, efficient and reliable electric power anytime, anywhere, providing means to treat energy independence and global warming issues [1–3].

Smart Grid is a concept and may look different to the different actors however, the concept of the smart substation for smart city will consider:

- Motivate and include customers
- Resistance to attack
- Provide quality energy for the 21st century
- Gather all storage and generation technologies
- Buying and selling of energy
- Optimize energy and operate efficiently
- Be self-healing

The implementation of Smart Grid provides complete solutions which will help improve the reliability of supply, operational performance and productivity for utilities. By making the network intelligent, the energy consumption is managed effectively, and customers will be able to save money without compromising life. The optimal integration of renewable energy into the grid is a major advantage of the implementation of smart grid, and there will be a significant penetration of renewable energy in a Smart Grid scenario. Smart Grid will provide significant, measurable and sustainable benefits to all stakeholders by increasing energy efficiency.

3 Smart Substation for Smart City

3.1 Smart Substation in Smart City

The smart substation should have the characteristics and specifications listed below [4–6].

- Reliability: the smart substation must ensure the stability of the voltage and current during abrupt load changes.
- Security: the smart substation communication network must be protected against cyber attacks, access control, port security and encryption should be secured.

- Measurability: the smart substation must be able to monitor in real time all the measures in substation.
- The controllability: the ability to act on all organs (breaker; switch) of the substation.
- Flexibility: the control must be flexible and can be executed from several points.
- Scalability: the ability to evolve and adapt to different hazards.
- Availability: the availability of communications and energy, the smart substation must quickly return in normal state after occurrence of electrical or communication faults.
- Resistance: the ability of recovery and restoration after any destruction or failure that may occur as a result of natural disasters or malicious activity.
- Maintainability: maintenance of smart substation should be quick and simple.
- Durability: the smart substation must consume less energy and ensure that energy is properly used.
- Interoperability: technologies and protocols should be interoperable to facilitate communication and connection between different communication technologies.
- Optimization: the cost of installing and operating the smart substation must be optimal.

3.2 Limitations of Existing Substations

Although, the use of digital command control systems in substations has started to grow since 2000, which allowed an evolution in the way of control and management of the electrical grid, the traditional concept of substations remains limited to new challenges facing the electric power sector.

The limits of substations are shown in the following Table 1.

4 Proposed Solution

In order to meet the need of a smart substation in smart city and facilitate their integration into the existing grid we propose in this article a solution that includes the use of the 61850 protocol and substation-substation communication in order to reply to the necessary security, Flexibility, Scalability, Availability and Interoperability [1, 7–9].

The common Substation Automation Systems SAS architecture used in substations are presented in Fig. 1, where the SAS is divided in 3 levels. Level 1 is for communication of intelligent equipment device IED, such as protection relay and measurement unit, which communicate using legacy protocol DNP3, MODBUS and IEC 60870-5-103 with the Bay Controller Unit BCU. The level 2 contains BCU, gateway GTW, Switch and the Human Machine Interface HMI which

Table 1 Limitations of traditional substations

Limits	Impact on the integration of the Smart substation
Communication does not exist between substations	Inability to implement a buy and sell system between customers and energy producer
Centralized control of substations for the regional dispatching center RDC	The inability to develop a decentralized control system of substations
The communication protocol of the various digital control systems used is not standard	Limitation in interoperability which causes equipment integration issues
	Limits in the expansion of existing stations to the same Substation Automation System SAS
Meter not communicating with SAS	The impossibility of real-time power management
SAS engineering is done locally and not online	Less optimization of resource costs
The protections adjusting is fixed and not dynamic	The inability to adapt and evolve with network changes
Basic architecture of SAS	Availability problem, controllability and measurability
Restrictions of optimum use of transmission lines due to the low use of FACTS	No optimization of electrical power and the quality provided
Manual locking procedure	Safety issue when working people

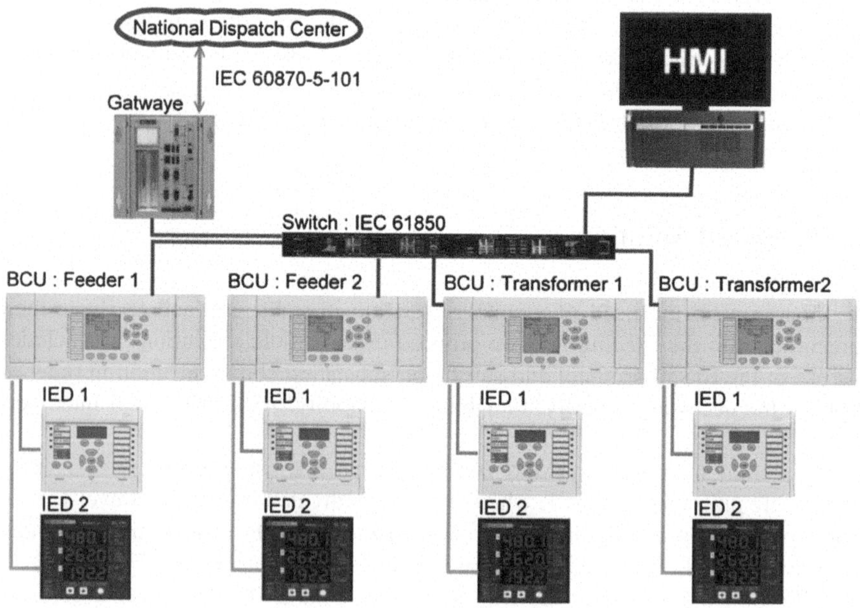

Fig. 1 SAS architecture

communicate between each other using 61850. The level 3 is dedicating for the communication between Substations and the Regional Dispatch Center RDC using IEC 60870-5-101 protocol.

The disadvantage of using IEC60870-5-101 for communication between Substations and RDC is that all the information are centralized in the RDC so the loss of communication or problem in the RDC will lead to a loss of control of substations which can cause dangerous problems.

We propose to use the 61850 protocol between Substation and RDC and also between Substations Fig. 2, which will allow benefiting from many advantages.

As an upgrade of the existent substation we suggest the following applications which will help improving the SAS.

- Decentralized HMI
- Constraints monitoring
- Voltage stability prediction
- Automatic locking
- Strengthening of automation systems

Each of the proposed application is discussed next.

1. Decentralized HMI

Substation will be communicating using 61850, so we suggest building small HMI in the gateway of the substation using circuit breakers, disconnectors positions and measurements of the connected substations. This will allow taking control of other substation in case of fault in RDC.

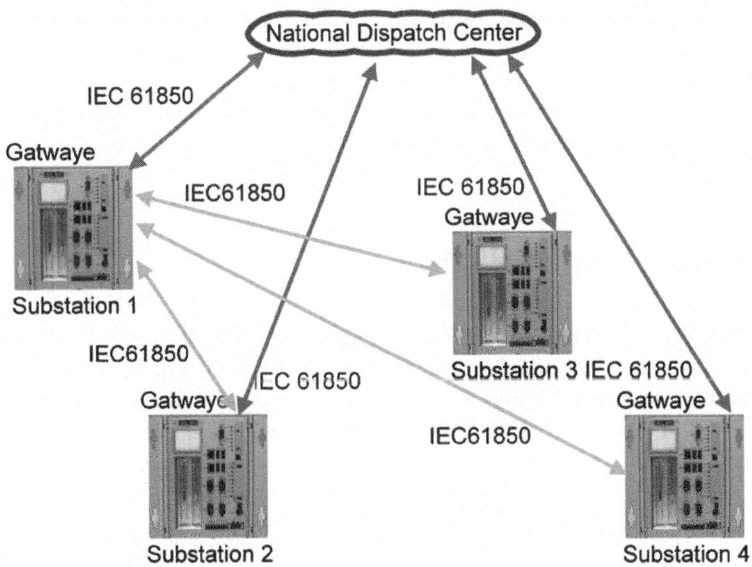

Fig. 2 Proposed architecture

2. Constraints monitoring

We suggest implementing algorithm in the gateway for monitoring the violations of voltage and transmitted power and also build automated scenarios to protect the grid in such case; this will allow taking quick action and self healing.

3. Voltage stability prediction

We suggest using algorithm for calculation of voltage stability index using measurements of the connected substations (active power, reactive power, voltage, current, power factor, frequency …). Also include power flow calculation, which will allow to predict voltage instability, voltage collapse and load flow.

The prediction will allow taking quick action and preventing collapse and contingencies of the grid also scenarios of self healing can be implemented.

4. Automatic locking

In the existent substations during maintenance of lines or feeder the lock of disconnectors and circuit breakers is done manually following a procedure of communication between electrical workers, the misunderstanding of message or the locking of inappropriate organs can lead to catastrophic accidents on people and materials.

To avoid this problem we suggest using automatic locking between substations using GOOSE as they provide secure and quick response time. So the automatism can be activated in substation which will send orders of opening or closing to organs until the locking is done.

5. Strengthening of automation systems

In the case of a breaker faille 50BF the protective relay send order of tripping to other frame circuit breaker to isolate the failed breaker in substitution, and send also order to breaker of the feeder of the other substations. We suggest using Generic Object Oriented Substation Events GOOSE for tripping which will strength and ensure the tripping control.

5 Simulation and Application

To validate our suggested solution we will apply it to 6 buses network every bus is considered as a substation and its gateway. Every gateway will be simulated by an IEC61850 server simulator Fig. 3. For example we choose bus 5 to be the control substation. We developed in MATLAB program the cited applications in previous chapter.

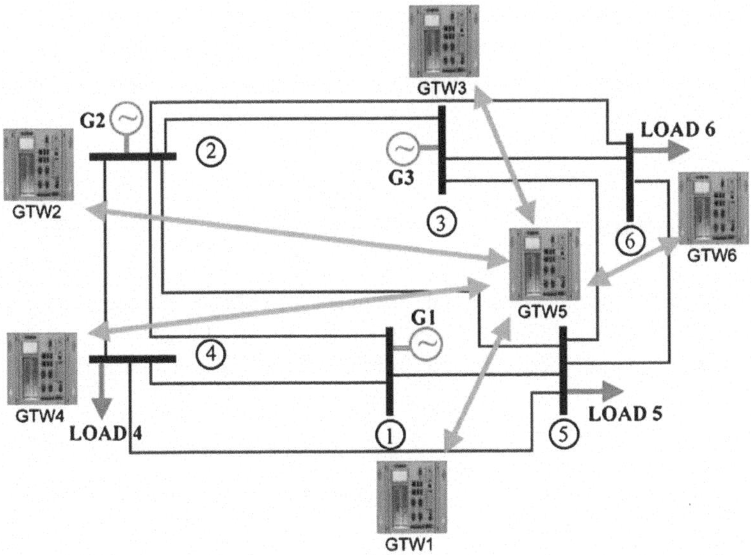

Fig. 3 Simulation architecture

5.1 Application 1

The first application is used for monitoring the maximum power allowed for transmission in line 2–5. If the Power is above the limit power for 30 s the program in GTW5 will send GOOSE to GTW2 and RDC indicating that the power transient is critical this GOOSE is mapped in booth GTW2 and RDC as a "VIOLATION OF POWER".

Example of the executing program:

```
START P2-5>0.7*Plimit
February 26, 2016 10:00:00.433 PM
----------------------------
CONFIRMATION P2-5>0.7*PlimitFOR 30s
February 26, 2016 10:00:30.379 PM
----------------------------
SEND GOOSE 01 TO GTW5
February 26, 2016 10:00:30.850 PM
----------------------------
SEND GOOSE 02 TO RDC
February 26, 2016 10:00:30.851 PM

-------------------------
```

The advantage of this application is that information will be sent to RDC and GTW2 so if there is any problem in the RDC the decentralized control substation 2 will be informed and can take control of network, which was impossible in case of traditional substation.

5.2 Application 2

The second application is voltage stability index for voltage prediction; the program will calculate the Fast Voltage Stability index FVSI of each line connected to GTW5 using provided measurements from other gateways. The algorithm is given in Fig. 4.

$$FVSI_{ij} = \frac{4 * Z^2 * Q_j}{V_i^2 * X} \tag{1}$$

Z = line impedance, X = line reactance, Q = the reactive power flow at the receiving end, Vj = sending end voltage.

Example: instability in line 2–5

```
FVSI2-5>0.5
February 26, 2016 11:02:00.521 PM
---------------------------
SEND GOOSE 03 TO RDC
February 26, 2016 11:02:00.952 PM
---------------------------
```

In traditional substation it was impossible to change measurement between substations; However with the use of the proposed solution substations can communicate between them in a way to allow calculating indexes to monitor the electrical network.

Fig. 4 FVSI algorithm

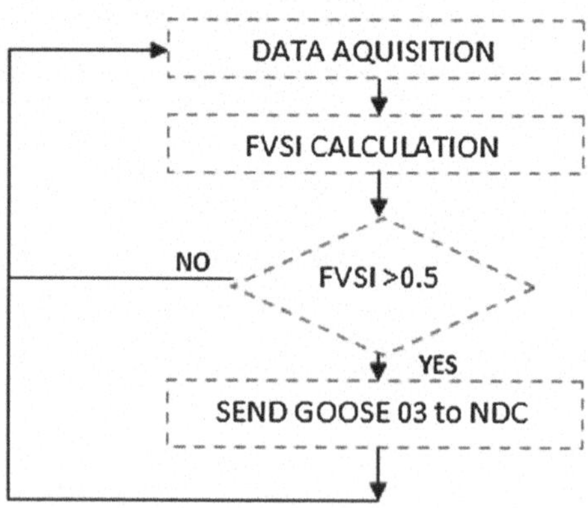

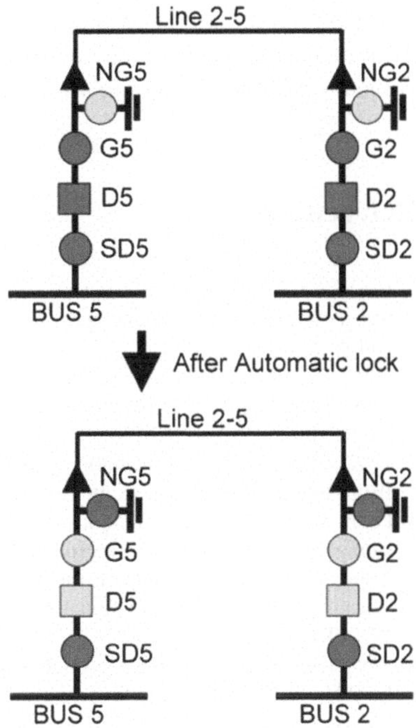

Fig. 5 Automatic locking

5.3 *Application 3*

In the application of automatic lock we take an example of locking line 2–5. The locking consist of opening the general disconnectors (G2 and G5) and breaker (D5 and D2) and closing earth switch (NG5, NG2), the status of feeders before and after the lock is given in Fig. 5.

After initializing the locking automatism, GTW2 and GTW5 will exchange GOOSE necessary for the locking automatism, if any problem occurs wail executing the automatism, the automatism will be blocked and RDC will be informed about the situation. Such as application will help improving time of maintenance and avoid any dangers on personnel or materials causing by maneuvering mistakes.

6 Conclusion

The construction of Smart Substation and traditional substation migration is a long-term process and must be implemented step by step. It is obvious that the implementation of the IEC 61850 protocol for communication between substations will open doors to many applications that can ensure security; Reliability and flexibility of the substation to ensure a better power for a smart city. The development of Smart substation will become the mainstream in the future.

References

1. Bin, S., et al.: Applied research of supervision and control system in 110 kV smart substation based on three layers of three networks. In: 2015 IEEE International Conference on Information and Automation (2015)
2. Feng, Z., et al.: Research of digital metering system and calibration technology of smart substation. In: 2014 IEEE Workshop on Advanced Research and Technology in Industry Applications (WARTIA) (2014)
3. Ji, Z., et al.: Analysis of technology and economy of new generation smart substation. In: 2015 IEEE Power & Energy Society General Meeting (2015)
4. Peiqi, F., et al.: Key techniques in smart substation. In: 2014 International Symposium on Computer, Consumer and Control (IS3C) (2014)
5. Yang, G., et al.: Research on the key technology of process-level network switch of smart substation. In: 2014 International Conference on Power Engineering and Renewable Energy (ICPERE) (2014)
6. Zeynal, H., et al.: Intelligent substation automation systems for robust operation of smart grids. In: 2014 IEEE Innovative Smart Grid Technologies—Asia (ISGT Asia) (2014)
7. Elgargouri, A., et al.: IEC 61850 based smart grid security. In: 2015 IEEE International Conference on Industrial Technology (ICIT) (2015)
8. Hyung, L., Sidhu, T.S.: Design of a backup IED for IEC 61850-based substation. IEEE Trans. Power Deliv. **28**(4), 2048–2055 (2013)
9. Xueyang, C., et al.: Electrical substation automation system modernization through the adoption of IEC61850. In: 2015 IEEE/IAS 51st Industrial & Commercial Power Systems Technical Conference (I&CPS) (2015)

Part III
Main Track 3: Enablers, Challenges and Applications

Comparative Analysis of Different Excitation Techniques for Cylindrical Dielectric Resonator Antenna

Kaoutar Allabouche, Fabien Ferrero, Najiba El Amrani El Idrissi, Mohammed Jorio, Jean Marc Ribero, Leonardo Lizzi and Abdellatif Slimani

Abstract Direct Microstrip line, Microstrip slot-coupled feed and hybrid coupler techniques are investigated for Cylindrical Dielectric Resonator Antenna (CDRA). The Dielectric resonator has a dielectric constant of 30, and etched on Arlan dielectric substrate having a relative permittivity of 3.58 and dimensions of $150 \times 150\ \text{mm}^2$. The structures are numerically analyzed using the numerical software HFSS. Radiation characteristics including return loss, gain, directivity and VSWR versus frequency characteristic are presented and compared based on the excitation method employed for the studied CDRA. The simulation results proved that the 90° hybrid coupler provides good performances particularly: a wider impedance bandwidth of 1.15 GHz and a maximum coupling of −45 dB.

Keywords Cylindrical dielectric resonator antenna · Excitation techniques · HFSS · Microstrip line · Microstrip slot-coupled line · 90° hybrid coupler

1 Introduction

Dielectric Resonator (DRs) made of low loss and high dielectric constant materials have been widely used as an energy storage component or resonant cavities for various applications like microwave filters, tuners, amplifiers and oscillators. However, the remarkable study of dielectric material as an Antenna element by Long et al. in 1983 has completely changed its scope of application [1]. The dielectric materials with permittivity values in the range 10–100, when properly fed will act as excellent radiators. Different DRAs commonly used shapes have been

K. Allabouche (✉) · F. Ferrero · J.M. Ribero · L. Lizzi
LEAT, CNRS, UNICE, Nice, France
e-mail: Kaoutar.allabouche@usmba.ac.ma; Kaoutar.ALLABOUCHE@unice.fr

K. Allabouche · N. El Amrani El Idrissi
LSSC, FST Fez, USMBA, Fez, Morocco

K. Allabouche · M. Jorio · A. Slimani
LERSI, FST Fez, USMBA, Fez, Morocco

R. El-Azouzi et al. (eds.), *Advances in Ubiquitous Networking 2*, Lecture Notes in Electrical Engineering 397, DOI 10.1007/978-981-10-1627-1_24

investigated such as cylindrical [2, 3], Rectangular [4], Spherical and Hemispherical [5, 6], Half-split Cylindrical [7], Equilateral triangular [8].

Dielectric Antennas have proved themselves to be perfect candidates for application by offering many advantages [9], that describes their capability and applicability such as: Design & shape flexibility, wider impedance bandwidth, ease of integration with other antennas, negligible dielectric losses and High temperature tolerance [10, 11]. DRs are small in size, light in weight, having low cost and high temperature stability [9–12]. The next generation communication systems trends are shifting to higher frequencies limiting the performances of metallic antennas. In these circumstances, the performance of DRAs offers best results compared to other families of antennas, among of them the Microstrip patch antennas, which are the most commonly used [13].

Flexible excitation techniques are also another advantage of DRAs. In fact, numerous types of feeding techniques have been used to feed a linear array of DRAs, such as Microstrip lines [14], dielectric image line [15], waveguide [16], coplanar slotted waveguide [17], slot coupled DRA [18, 19] and hybrid coupler technique [20].

One major DRA disadvantage is their limited bandwidth, since for a single-mode excitation; the bandwidth of DRA is always below 10 % [21], which is not sufficient for numerous applications. To overcome this limitation, various bandwidth enhancement techniques have been investigated such as optimizing the feeding mechanisms [22]: A. Baba has performed a study [23] on cylindrical dielectric resonator antenna (CDRA) with an aperture coupled dielectric image line (DIL) around 2.45 GH and found a bandwidth that doesn't exceed 80 MHz. An other excitation techniques has been proposed and examined by Raggad [24] that consists of a stair slot, this proposed structure enhanced the gain but the bandwidth didn't surpass a 200 MHz.

Three excitation techniques are the subject of this paper. The first one, is the direct Microstrip feed line method which consists of a conducting strip, connected directly to the antenna. The second one, is an indirect feeding mechanism which is the aperture coupled line, that excites the antenna via coupling the energy from the feed line through an opening in the ground plane [25]. The third technique is the hybrid coupler, which is a four-port device for equally splitting input signals with a 90° phase shift between output ports.

The three techniques are applied in this work, for a Cylindrical DRA and are investigated numerically. The structures are studied and compared in terms of their return loss, impedance bandwidth and VSWR characteristics versus frequency. Far field Radiation Pattern at 2 GHz, directivity and gain are also investigated through simulation studies that was performed by using HFSS package which is a commercial 3-D electromagnetic field solver based on the Finite Element Method (FEM) [26].

2 Antenna Structures & Design

In this section, we describe the geometry of the proposed cylindrical dielectric resonator antenna. Simulation results and discussion will be provided in the next section. The views of the geometry of the CDRA with the different excitation techniques are illustrated in Fig. 1a, b, c.

The Cylindrical Dielectric Resonator Antenna in all cases consists of a bloc of dielectric material with a relative permittivity $\varepsilon_r = 30$. Height h and radius R of the cylinder are computed from equation to excite the HEM mode (1) [27], and gives respectively values of 12 mm and 15 mm.

$$K_0 R = \frac{6.324}{\sqrt{\varepsilon_r + 1}}\left[0.27 + 0.36\left(\frac{R}{2H}\right) + 0.02\left(\frac{R}{2H}\right)^2\right] \quad \text{For } 0.4 \le \frac{R}{H} \le 6 \tag{1}$$

The overall structure is placed on Arlan dielectric substrate with the following parameters ($\varepsilon_r = 3.58$; $\tan\delta = 0.0035$; $h = 0.8$ mm) and a ground plane with dimensions of 150 × 150 mm^2 and a thickness of 0.035 mm. In Fig. 1a, direct 50 Ω Microstrip feed Line as a transmission media, consist of a rectangular metallic

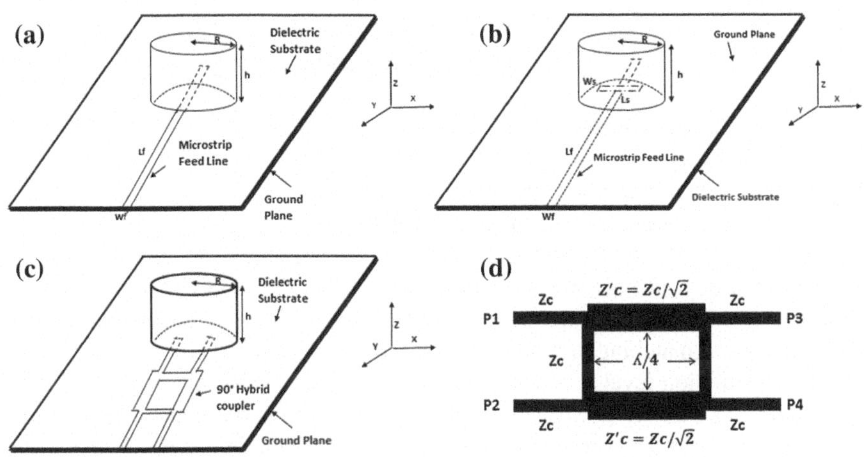

Fig. 1 The geometry of the CDRA fed by: **a** Direct Microstrip feed line technique, **b** Microstrip aperture-coupled line technique, **c** 90°–3 dB Hybrid coupler technique, **d** Configuration of the narrow band 3 dB 90° hybrid coupler

line of Wf = 2.2 mm and Lf = 100 mm dimensions, placed under the radiating element. The CDRA and the feed line are placed on the Arlan dielectric substrate, which is metalized at its underside.

The dimensions of the 50 Ω Microstrip feed line have been calculated through the tool Linecalc of ADS software [28].

The second structure consists of a Microstrip-Slot coupled CDRA as it is shown in Fig. 1b. A 50 Ω Microstrip feed line is etched on the bottom side of the dielectric substrate. The narrow coupling slot of dimensions Ws = 1.2 mm and Ls = 24 mm, is printed at the center of the ground plane and it's used to excite the DR. The most important feature of this structure is to place the feed network of the antenna at the backside of the ground plane.

The third structure consists of a linearly polarized CDRA fed through 90° hybrid 3-dB coupler, as exposed in Fig. 1c, d. The coupler used here, achieves not only power division but also phase shifting. The coupler delivers to the CDRA balanced power division and consistant 90° phase difference. The quarter-wavelength microstrip branch along the x-axis, has a characteristic impedance $Z_c' = 35.36\ \Omega$, a length of 22 mm and a width of 2.6 mm. The branch along the Y-axis, has a characteristic impedance $Z_c = 50\ \Omega$, and width of 1.7 mm. For 90° phase shifting, the lengths of microstrip branches, were set equals to $\lambda_g/4$ (22 mm), where λ_g refers to the guided wavelength at the design center operating frequency of 2 GHz.

The width of the different branch-lines, has been calculated by using the following formula [29]:

$$Z_c = \frac{87}{\sqrt{\varepsilon_r + 1.41}} \ln\left(\frac{5.98 \times h}{0.8 \times W + t}\right) \tag{2}$$

Where ε_r dielectric constant of the dielectric substrate
h Substrate's height
W Width of the microstrip center conductor
t Thickness of microstrip center conductor

The coupling and transmission coefficients, corresponding to the 3 dB-hybrid coupler are shown in Fig. 2a. The return losses S11, S22, S33 & S44 are about -45 dB, around the resonant frequency 2 GHz. Hence, it can be concluded that we obtained a good matching around the operating frequency 2 GHz.

The transmission level is about 3 dB throughout the operating frequency band, which confirms that the power is halved on both output ports 3 and 4. In terms of phase, as it can be seen from Fig. 2b. The output signals on ports 3 and 4 are approximatly in phase quadrature (around 90, 96°).

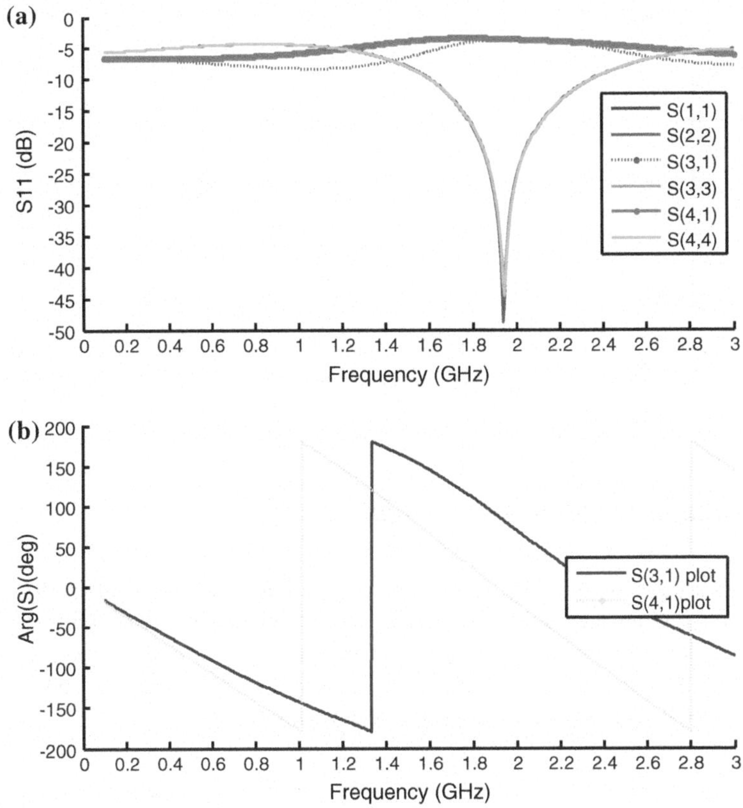

Fig. 2 Simulated amplitude (**a**) and phase responses (**b**) of the hybrid coupler

3 Results and Discussion

3.1 Return Loss Versus Frequency

Figure 3 shows simulation results for return loss characteristics of the Microstrip line-fed, Microstrip slot-coupled and hybrid coupler fed CDRA versus frequency using HFSS.The lowest value of return loss represents the maximum coupling from the excitation line to the CDRA. It can be seen from the curves illustrated in Fig. 2, that the coupling is greater in the hybrid coupler technique than in microstrip and slot coupled line fed with a value of reflection coefficient up to –45 dB, while for the other cases it doesn't exceed –30 dB. Additionally, the impedance bandwidth is much wider compared to the two other coupling techniques with a value of 1.11 GHz.

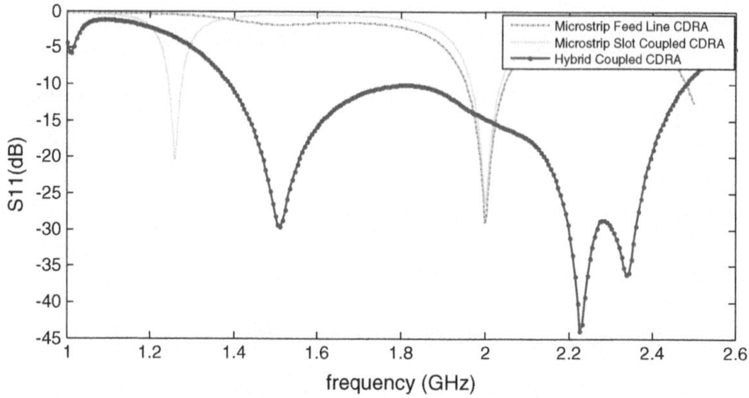

Fig. 3 Return loss *curves* for the different excitation techniques CDRA

3.2 Voltage Standing Wave Ratio

From the curves shown in Fig. 4, it can be deduced that the bandwidth of the hybrid coupler excitation method offers much wider impedance bandwidth then the microstrip slot-coupled and microstrip feel line techniques. Furthermore, it is well matched to the source since the minimum value of VSWR is equal to 1.04 which is less than the VSWR values of the other techniques (1.09).

3.3 Fairfield Simulation

Figure 5 shows the radiation patterns of the CDRA, for the used coupling mechanisms at 2 GHz. Simulations were conducted by using HFSS. The radiation

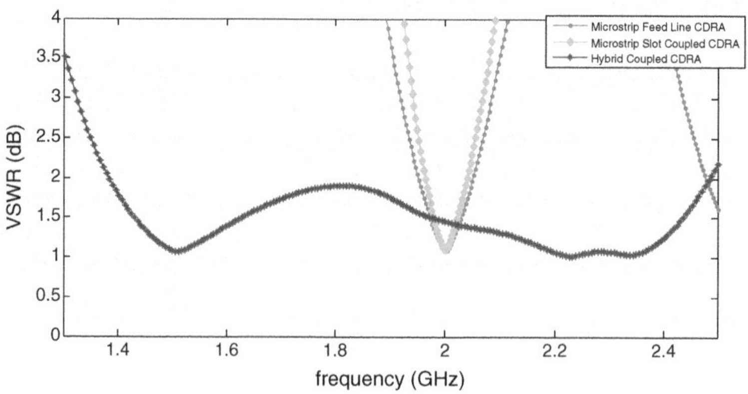

Fig. 4 VSWR versus frequency

parameters extracted from this figures show that hybrid coupler method provides higher gain and improved directivity, in comparison to those of the other coupling methods. Hence, a better radiation efficiency is obtained in this case.

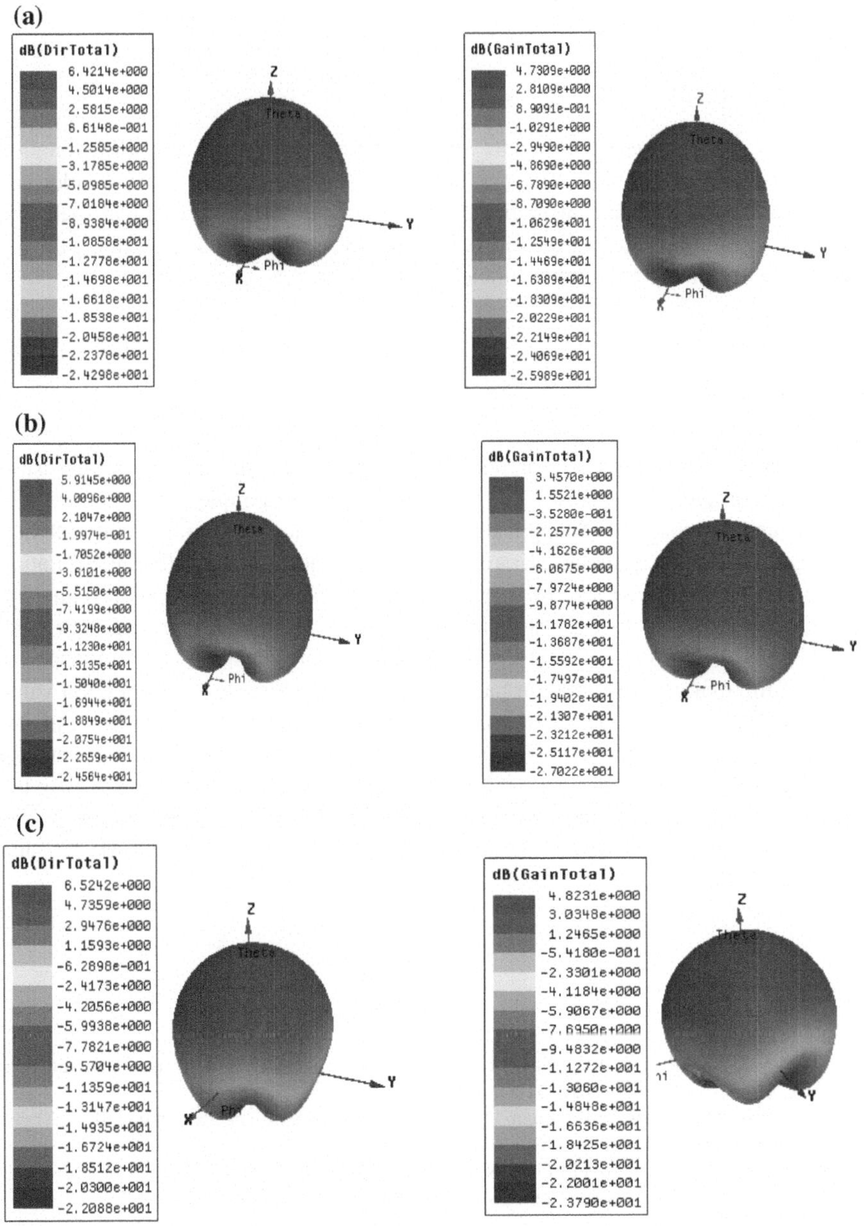

Fig. 5 3D far field pattern of gain and directivity at 2 GHz for: **a** Microstrip Feed line **b** Microstrip Aperture-coupled line; **c** 3 dB-Hybrid coupler

Table 1 Comparative results of CDRA for the different coupling mechanisms

Antenna parameters	Microstrip Feed line	Microstrip slot coupled line	3 dB-Hybrid coupler feed
Return Loss(dB)	–29.14	–26.61	Up to –45
Bandwidth	93.2 MHz	69.2 MHz	1.11 GHz
VSWR	1.08	1.09	1.04
Gain(dB)	4.7309	3.4570	4.8050
Directivity(dB)	6.4214	5.9145	6.4749

Table 1 summarizes the results of simulation performed for all excitation mechanisms around 2 GHz. We can affirm that the 90° hybrid 3 dB coupler can improves significantly all the radiation characteristics of the cylindrical dielectric resonator antenna, including the return loss, the impedance bandwidth, the gain and the directivity and the radiation efficiency.

4 Conclusion

In this paper cylindrical dielectric resonator antenna fed respectively, by Microstrip fed line, slot-aperture line and 3 dB-hybrid coupler techniques, was numerically investigated for the UMTS application at 2 GHz, using the commercial software package HFSS. The analysis of radiation performances of the different excitation techniques, including return loss, bandwidth and VSWR as a function of frequency, have been carried out as well as radiation variation in terms of gain and directivity at 2 GHz. The results show that the hybrid coupler presents good performances in terms of the majority of radiation parameters such as return loss with a good value of –45 dB and a good adaptation, in comparison to the simple microstrip and slot coupled feed line techniques. Also this excitation technique enhances the gain and the directivity of the CDRA. In addition, this technique increases significantly the cylindrical dielectric resonator antenna bandwidth (1.11 GHz), which allows to the antenna to cover wide range of applications like the GSM1800, GSM1900 and UMTS2000 bands.

References

1. Long, S.A, McAllister, M.W., Shen, L.C.: The resonant cylindrical dielectric cavity antenna. IEEE Trans. Antennas Propag. **31**(3), 406–412 (1983)
2. Long, S.A., McAllister, M.W., Shen, L.C.: The resonant dielectric cavity antenna. IEEE Trans. Antennas Propag. **31**(3), 406–412 (1983)
3. Allabouche, K., El Amrani el Idrissi, N., Jorio, M., Mazri, T.: Cylindrical Dielectric Resonator Antenna for Multiband Communication Systems. IEEE Conference 978-1-4799-7054-4/14/ $31.00 ©2014 (2014)

4. McAllister, M.W., Long, S.A.: Rectangular dielectric resonator antenna. IEEE Electron. Lett. **19**, 218–219 (1983)
5. McAllister, M., Long, S.A.: Resonant hemispherical dielectric antenna. IEEE Electron. Lett. **20**, 657–659 (1984)
6. Allabouche, K., El Amrani el Idrissi, N., Jorio, M., Mazri, T.: Design of new UMTS hemispherical dielectric resonator antenna. Int. J. Eng. Res. Technol. (IJERT), **3**(2) (2014)
7. Mongia, R.K.: Half-split dielectric resonator placed on metallic plane for antenna applications. IEEE Electron. Lett. **25**, 462–464 (1989)
8. Lo, H.Y., Leung, K.W., Luk, K.M., Yung, E.K.N.: Electron. Lett. **35**, 2164–2166 (1999)
9. Luk, K.M., Leung, K.W.: Dielectric resonator antennas. Research Studies Press Ltd, Baldock, Hertfordshire, England (2003)
10. Luk, K.M., Leung, K.W.: Dielectric Resonator antennas (2002)
11. Eshrah, I.A., et al.: Theory and implementation of dielectric resonator antenna excited by a waveguide slot. IEEE Trans. Antennas Propag. **53**(1), 483–494 (2005)
12. Petosa, A.: Dielectric Resonator Antennas Handbook. Artech House Inc (2007)
13. Allabouche, K., et al.: Comparative analysis of microstrip and dielectric resonator antennas for UMTS application. Int. J. Commun. Antenna Propag. (IRECAP) **5**(1) (2015)
14. Wee, F.H., et al.: Investigation of the characteristics of barium strontium titanate (BST) dielectric resonator ceramic loaded on array antennas. Prog. Electromagn. Res. **121**, 181–213 (2011)
15. Al-Zoubi, A.S., Kishk, A.A., Glisson, A.W.: Analysis and design of a rectangular dielectric resonator antenna fed by dielectric image line through narrow slots. Prog. Electromagn. Res. **77**, 379–390 (2007)
16. Lee, R.Q., Simons, R.N.: Bandwidth enhancement of dielectric resonator antennas. In: Antennas and Propagation Society International Symposium, AP-S Digest (1993)
17. Eshrah, I.A., et al.: Theory and implementation of dielectric resonator antenna excited by a waveguide slot. IEEE Trans. Antennas Propag. **53**(1), 483–494 (2005)
18. Ee, L., Ong, M.L.C.: Aperture coupled, differentially fed DRAs. In: Asia Pacific Microwave Conference, pp. 2781–2784 (2009)
19. Buerkle, A., Sarabandi, K., Mosallaei, H: Compact slot and dielectric resonator antenna with dual resonance, broadband characteristics. IEEE Trans. Antennas Propag. **53**(3),1020–1024 (2005)
20. Khoo, K.-W.: Wideband circularly polarized dielectric resonator antenna. IEEE Trans. Antennas Propag. **55**(7), (2007)
21. Luk, K.M., Leung, K.W.: Dielectric Resonator Antennas. Research Studies Press Ltd, Baldock, Hertfordshire, England (2003)
22. Kishk, A., Yin, Y., Glisson, A.W.: Conical dielectric resonator antenna for wideband applications. IEEE Trans. Antennas Propag. **50**, 469–474 (2002)
23. Baba, A.A., Zakariya, M.A., Baharudin, Z., Khir, M.H.M., Ali, S.M.: 2.45 GHz cylindrical dielectric resonator antenna fed by dielectric image line. In: 2013 IEEE Business Engineering and Industrial Applications Colloquium (BEIAC). IEEE (2013). 978-1-4673-5968-9/13/ $31.00 ©2013
24. Raggad, H., Latrach, M., Razban, T., Gharsallah, A.: Cylindrical dielectric resonator antenna fed by a stair slot in the ground plane of a microstripline. In: General Assembly and Scientific Symposium, 2011 XXXth URSI, 13–20 Aug. 2011. 2009
25. Cormos, D., Laisne, A., Gillard, R., Le Bolzer, E., Nicolas, C.: Compact dielectric resonator antenna for WLAN applications. Electron. Lett. **39**(7) (2003)
26. Ansoft High Frequency Structure Simulator (HFSS), ver. 11.0, Ansoft Corp (2007)
27. Ragad, H.: Etude et conception de nouvelles topologies d'antennes à résonateur diélectrique dans les bandes UHF et SHF. Université de Tunis El Manar et Université de Nantes, Thèse (2013)
28. http://www.keysight.com/en/pc-1297113/advanced-design-system-ads?cc=IT&lc=italnstitute
29. Balanis, C.A.: Antenna Theory. Analysis and Design, 3rd edn., pp. 865–866. Wiley (2005)

Recognition of OFDM and SCLD Signals Based on Second-Order Statistics

Mohamed Firdaoussi, Hicham Ghennioui and Mohamed El-Kamili

Abstract This work addresses the problem of the modulation recognition signal in the context of cognitive radio. In particular, the discrimination between OFDM (Orthogonal Frequency Division Multiplexing) and SCLD (Single Carrier Linear Digitally) signals. We present a new method based on second order statistics. The main advantages of this method are fast, robust in a context of AWGN (Additive White Gaussian Noise) channel, frequency and timing offsets. Computer simulations are provided in order to illustrate the behavior of the proposed algorithm.

Keywords Cognitive radio · Modulation recognition · OFDM · SCLD · AWGN · Second order statistics

1 Introduction

In recent years, Modulation Recognition (MR) plays an important role in various civilian and military applications, such as electronic warfare, surveillance and control of broadcasting activities, spectrum monitoring, management and cognitive radio. To respond to the needs of users in terms of voice and high rate data multimedia applications, expanding the allocated band is an adequate solution to increase without much difficulty; also the rate of a system while the band is scarce must be used sparingly, because it appears that the spectral resource by regulatory agencies already seems almost completely occupied. The first path explored to increase the

M. Firdaoussi (✉) · M. El-Kamili
Faculté des Sciences Dhar EL Mehraz, LIMS, USMBA, Fes, Morocco
e-mail: mohamed.firdaoussi@usmba.ac.ma

M. El-Kamili
e-mail: mohamed.elkamili@usmba.ac.ma

H. Ghennioui
Faculté des Sciences Et Technique, LSSC, USMBA, Fes, Morocco
e-mail: hicham.ghennioui@usmba.ac.ma

R. El-Azouzi et al. (eds.), *Advances in Ubiquitous Networking 2*, Lecture Notes in Electrical Engineering 397, DOI 10.1007/978-981-10-1627-1_25

effective capacity of wireless systems has been to conceive techniques of digital communications namely cognitive radio, [1, 2], which is a form of wireless communication in which a transmitter/receiver can detect intelligently communication channels that are in use and those that are not, and can move in the unused channels. In practice, this means knowing the type of modulation before to know and classify the present system on each frequency band over which it could communicate the receiver. However, most current systems is based on OFDM, that the wireless industry has shown great interest in OFDM technology, due to several advantages of OFDM, such as high capacity data transmission, immunity to multipath fading and impulsive noise and simplicity in equalization [3, 4]. OFDM has been adopted in a variety of applications, such as IEEE 802.11 a [5] and IEEE 802.16a [6]. Furthermore, the up-and-coming OFDM wireless communication technology lead to a new challenges for the designers of intelligent radios, such as discrimination of OFDM alongside Single-Carrier (SC) modulations. Solutions to tackle such new signal recognition problems need to be required [7]. Modulation recognition (MR) for single carrier signals has been studied for at least a decade [7]. So, algorithms recognize OFDM against single carrier linear digital (SCLD) modulations have been reported in [8–12]. The algorithms anticipated in [8–10] enforced an estimation of signal-to-noise ratio, carrier frequency recovery, and both carrier frequency and timing recovery, respectively. Besides, the algorithm wished-for in [11, 12] does not involve such preprocessing tasks. Most of the proposed methods are cyclostationarity based [13–20]. Some of these used the cyclic prefix (CP) induced cyclic statistics and obtained by the autocorrelation function (AF) [13–15]. Others require the detection of cyclostationary signatures intentionally embedded in the OFDM signals [16] by the redundant transmission of message symbols on multiple subcarriers, on the other hand the pilot-induced cyclic statistics were employed in [21]. Another approach, based on the cyclostationarity properties associated with the block transmitted-single carrier linearly digitally modulated (BT-SCLD) signals in order to discriminate between BT-SCLD, OFDM, and SCLD signals, was described in [19]. After all, the proposed algorithm is applicable to the recognition of OFDM against SCLD modulations in Additive White Gaussian noise (AWGN) channel. Here, we enlarge the applicability of this algorithm to time dispersive channels. Generally, in this paper one approach is proposed to tackle the MR problem, this approach is based on the correlation function of the received signal and the correlation ratio test is used for decision making. This can provide an optimal solution, in the sense that it distinguishes the OFDM against the SCLD modulations. Both recognition performance and complexity of the proposed method are investigated for AWGN and time dispersive channels into numerical computations. The rest of this paper is organized as follows. The SCLD and OFDM signals models are presented in Sect. 2. The proposed recognition approach based on the correlation ratio of the received SCLD and OFDM signals are presented in Sect. 3. Simulation results are discussed in Sect. 4, and conclusions are drawn in Sect. 5.

2 Signal Model

In this section, we present the OFDM and SCLD transmitted and received signal models that will be used throughout the paper.

2.1 OFDM Signal Model

The continuous-time baseband equivalent of a transmitted OFDM signal is written as follows [13]:

$$s_{OFDM}(t) = \frac{1}{\sqrt{N}} \sum_{k=0}^{K-1} \sum_{n=0}^{N-1} a_{k,n} e^{-2i\pi \frac{n(t-DT_c-kT_s)}{NT_c}} g^{tr}\left(t - kT_s\right), \tag{1}$$

where the data symbols $a_{k,n}$ represents the unknown information data of the subcarrier n and the OFDM block k, $a_{k,n}$ are assumed to be zero-mean and be independent and identically distributed (i.i.d) random variables. N is the number of subcarriers, K is the number of OFDM symbols and T_c is the chip duration such as $1/T_c$ is the information symbol rate in the absence of guard interval. The intercarrier spacing is then equal to $1/(NT_c)$. DT_c represents the length of the cyclic prefix. The T_s is the OFDM symbol period given by $T_s = (N + D)T_c$. The transmit pulse shaping window $g^{tr}(t)$ [4] is assumed to be equal to 1 if $0 < t < T_s$ and 0 otherwise.

At the receive-side, the continuous-time baseband equivalent is given by:

$$y_{OFDM}(t) = \frac{e^{2i\pi\delta ft}}{\sqrt{N}} \sum_{l=1}^{L} \sum_{k=0}^{K-1} \sum_{n=0}^{N-1} \lambda_l e^{2i\pi n \frac{\tau_l}{NT_c}} a_{k,n} e^{-2i\pi \frac{n(t-DT_c-kT_s)}{NT_c}} g\left(t - \tau_l - kT_s\right) + b\left(t\right), \tag{2}$$

with L represents the number of paths performed when the transmitted signal passes through a multipath fading channel. The amplitude and the delay of the $l-th$ path are respectively denoted by λ_l and τ_l. The $b(t)$ is a zero-mean white Gaussian noise with variance σ^2 and where δf is the offset frequency due to local oscillator drift or Doppler effect [13]. A discrete-time baseband received OFDM signal, $y[m]$, is obtained by oversampling $y_{OFDM}(t)$ at the sampling frequency $1/T_e$, where T_e is the sampling period, and with $M = \lfloor T_0/T_e \rfloor$ be the number of available samples at reception where $\lfloor X \rfloor$ represents the integer part operator, and T_0 be the observation window duration. The expression for the discrete time baseband OFDM signal can be easily written as in [13]:

$$y_{OFDM}[m] = \frac{1}{\sqrt{N}} \sum_{l=1}^{L} \sum_{k=0}^{K-1} \sum_{n=0}^{N-1} \lambda_l e^{2i\pi n \frac{\tau_l}{NT_c}} a_{k,n} e^{-2i\pi nm \frac{T_e}{NT_c}} e^{2i\pi(k+1)\frac{DT_c}{NT_c}}$$
$$\times g\left(mT_e - \tau_l - k(N+D)T_c\right) e^{2i\pi \Delta f m} + b[m], \quad (3)$$

where $b[m] = b(mT_e)$, and $\Delta f = \delta f T_e$ the normalized carrier frequency offset.

2.2 SCLD Signal Model

For the SCLD signals there is a single carrier ($N = 1$), therefore the transmitted continuous-time signal has the general form

$$s_{SCLD}(t) = \sum_{n=0}^{N-1} a_n g^{tr}(t - nT), \quad (4)$$

where $g^{tr}(t)$ represents a basic pulse shape and a_n is the binary data sequence of $\{\pm 1\}$ transmitted at a rate of $1/T$ bits/s, drawn either from a quadrature amplitude modulation (QAM) or phase shift keying (PSK) constellation.

Referring to [22], the continuous time received SCLD signal can be expressed as:

$$y_{SCLD}(t) = a e^{i\theta} e^{2i\pi \delta f_c t} \sum_{n=0}^{N-1} a_n g(t - nT - \varepsilon T) + w(t), \quad (5)$$

where a is the amplitude factor, θ is the phase, δf_c is the carrier frequency offset, T is the symbol period, ε is the normalized timing offset ($0 < \varepsilon < 1$), $g(t)$ is the overall impulse response of the transmit and receive filters, and $w(t)$ is zero-mean complex Gaussian noise.

The discrete-time baseband signal, $y_{SCLD}(m)$, obtained by oversampling $y_{SCLD}(t)$ at a rate $1/T_e$, with m as the number of samples per symbol, is given by,

$$y_{SCLD}[m] = a e^{i\theta} e^{2i\pi \Delta f_c m} \sum_{n=0}^{N-1} a_n g\left(mT_e - nT - \varepsilon T\right) + w[m], \quad (6)$$

where $w[m]$ is stationary zero-mean complex Gaussian noise.

The proposed method in Sect. 3 rely on signal model provided by Eqs. (3) and (6). while considering an AWGN channel, i.e., $L = 1$, $\lambda_l = 1$, and $\tau_l = 0$. However, impact and robustness of this method are addressed in Sect. 5 devoted to numerical computations.

3 Proposed Recognition Method

In this section, we present a method for recognition of OFDM against SCLD signals by using the correlation function in order to define the decision criteria that allows us to discriminate the OFDM against SCLD signals. In other words, our method uses the second-order statistics of the received OFDM or SCLD signals to select the right type modulation under the assumption of Gaussian channel AWGN and perfect time and frequency synchronization ($L = 1; \lambda_l = 1; \tau_l = 0; \Delta f = 0; \theta = 0; \delta f_c = 0$; and $\varepsilon = 0$). We assume that the symbols, $a_{k,n}$, a_n, are zero mean, independent and identically-distributed random variables. Let $R_y^{OFDM}(m,p) = \mathbb{E}\left[y_{OFDM}(m+p)\, y^*_{OFDM}(m)\right]$, $R_y^{SCLD}(m,p) = \mathbb{E}\left[y_{SCLD}(m+p)\, y^*_{SCLD}(m)\right]$ be the correlation function of the received OFDM and SCLD signals given by Eq. (3) and Eq. (6), respectively, are written in the following form [13, 23],

$$R_y^{OFDM}(m,p) = \frac{\mathbb{E}|a_{k,n}|^2}{N} \sum_{k=0}^{K-1} g\left(m+p-k\frac{T_s}{T_e}\right) g^*\left(m-k\frac{T_s}{T_e}\right) \sum_{n=0}^{N-1} e^{-2i\pi np\alpha} + \sigma^2 \delta(p), \tag{7}$$

$$R_y^{SCLD}(m,p) = \mathbb{E}|a_n|^2 \sum_{n=0}^{N-1} g\left(m+p-n\frac{T}{T_e}\right) g^*\left(m-n\frac{T}{T_e}\right) + \sigma^2\delta(p), \tag{8}$$

where $\alpha = \frac{T_e}{NT_c}$ and the superscript $(.)^*$ stands for the complex conjugate.

The correlation function of the received OFDM signal can be expressed in this form, [13],

$$R_y^{OFDM}(m,p) = R_y^{OFDM}(m,0)\,\delta_{p,0} + R_y^{OFDM}\left(m,\alpha^{-1}\right)\delta_{p,\alpha^{-1}} + R_y^{OFDM}\left(m,-\alpha^{-1}\right)\delta_{p,\alpha^{-1}}. \tag{9}$$

The recognition of OFDM and SCLD signals can be formulated into C_d (defined as the ratio of the correlation function of the α^{-1} and the sum of the same function on all points except the origin point).

$$C_d = \frac{R_y^{\bullet}\left(m,\alpha^{-1}\right)}{\sum_{p>0} R_y^{\bullet}(m,p)}. \tag{10}$$

where $(.)^{\bullet}$ stands for the OFDM or SCLD signals.
Replacing $R_y^{\bullet}\left(m,\alpha^{-1}\right)$ into Eq. (7) or Eq. (8), and $\sum_{p>0} R_y^{\bullet}(m,p)$ into Eq. (9), respectively, lead to the following models:

$$R_y^{\bullet}\left(m,\alpha^{-1}\right) = \mathbb{E}|a_{k,n}|^2 \sum_{k=0}^{K-1} g\left(m+\alpha^{-1}-k\frac{T_s}{T_e}\right) g^*\left(m-k\frac{T_s}{T_e}\right) + \sigma^2\delta\left(\alpha^{-1}\right).$$

$$\sum_{p>0} R_y^{\bullet}(m,p) = 2R_y^{\bullet}\left(m,\alpha^{-1}\right)\delta_{p,\alpha^{-1}}.$$

The Decision criteria can be expressed as,

$$\begin{aligned} |C_d| > 1-\xi = \eta \ \ for \quad OFDM, \\ |C_d| < 1-\xi = \eta \ \ for \quad SCLD, \end{aligned} \tag{11}$$

where ξ, η are a low value threshold and the decision threshold, respectively, that helps us to distinguish OFDM against the class of single carrier linear digital (SCLD) modulations.

4 Simulations

In this section, we evaluate the proposed technique by means of numerical simulations. Thc displayed results are averaged over 1000 Monte-Carlo trials. We consider the lengths of the observed symbols are 0.083, 0.416, 1.66, and 4.166 ms for single carrier signals. For OFDM signals, there are the same of single carrier in terms of the lengths observed symbols, except each with 64 subcarriers and *CP* length was 4. All subcarriers are modulated either using QPSK or 16-QAM.

We have generated a signal among OFDM (IEEE802.16.e, as known the useful time duration NT_c of each standard systems is stocked in our database) and SCLD modulations. The threshold experimental η value correspond to a 0.5.

In Fig. 1, the decision criteria of the OFDM and SCLD signals, is plotted versus SNR, for several short observation intervals. As we can see, the OFDM signals are superior of the value of decision threshold, On the other hand, the SCLD signals are inferior of the value of decision threshold, which is demonstrated Theoretically by the Eq. (11). As we can see in Fig. 1, as the observation interval increases, the performance to select easily the single Carrier from OFDM signals in lower values of SNR increases.

In Fig. 2, we plot the correct detection rate of our proposed method versus SNR, while increasing the observation interval of OFDM signals we get a higher results even we have a lower values of SNR.

In Fig. 3, the performance of our algorithm is studied based on the number of OFDM symbols, i.e., depending on the received signal size, as we can notice if we increase the number of symbols we always get the maximum of detection's performance.

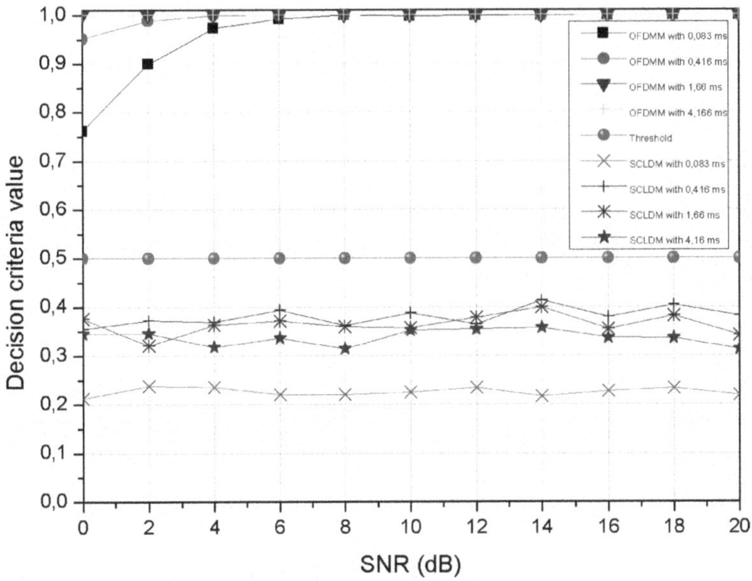

Fig. 1 Detection criteria versus SNR for SCLD and OFDM signals

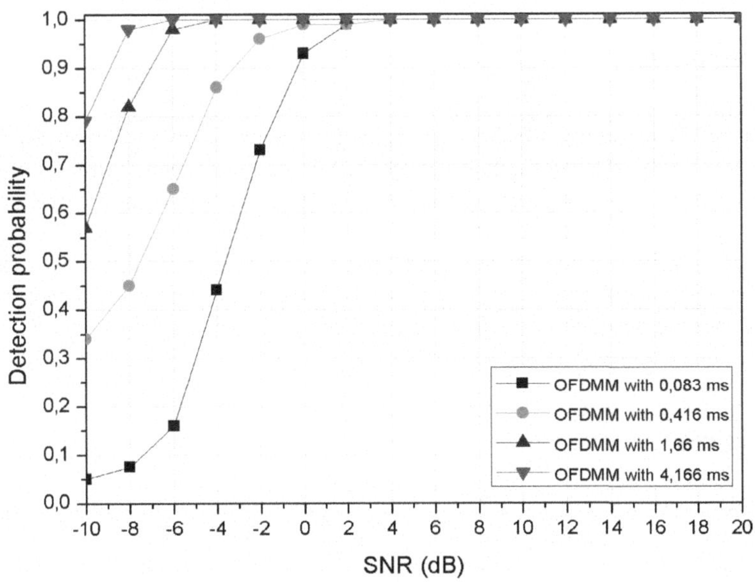

Fig. 2 Detection probability of OFDM modulations versus SNR

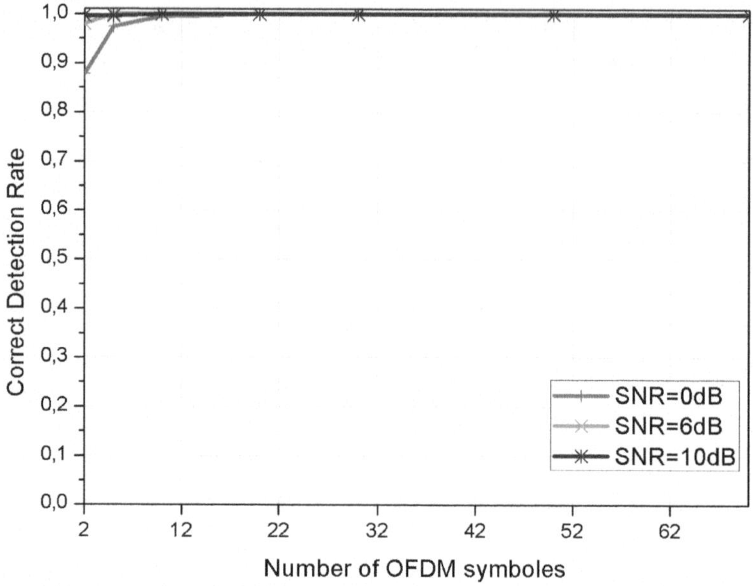

Fig. 3 Correct detection rate versus number of available OFDM symbols

5 Conclusion

To sum-up, the method has been described in this paper allows via a received signal to distinguish OFDM from single carrier signals by using second order statistics. Monte Carlo simulations show that OFDM and SCLD classification based on this method is generally quite robust and more effective than different existing scenarios [24]. The performance of the method and its simplicity were illustrated with an example. Theoretical extension of the temporal and frequency missynchronization scenario will be presented in a future paper.

References

1. Mitola, J.: Cognitive Radio: an Integrated Agent architecture for Software Defined Radio. Ph.D. thesis, Royal Institute of Technology, Stockholm, Sweden (2000)
2. Haykin, S.: Cognitive radio: brain-empowered wireless communications. IEEE J. Sel. Areas Commun. Spec. Issue Cogn. Netw. **23** (2005)
3. Bingham, J.A.C.: Multicarrier modulation for data transmission: an idea for whose time has come. IEEE Commun. Mag. **28**, 5–14 (1990)
4. Nee, R.V., Prasad, R.: OFDM for Wireless Multimedia Communications. Artcch House (2000)
5. Part11: Wireless LAN Mediwn Access Control (MAC) and Physical Layer (pHY) Specifications: High Speed Physical Layer in the 5GHz, IEEE Standanl802.11a-1999
6. Local and Metropolitan Area Networks-Part 16, Air Interface for Fixed Broadband Wireless Access Systems, IEEE Standard IEEE 802.160-2001

7. Dobre, O.A., Abdi, A., Bar-Nees, Y., Suo, W.: A survey of automatic modulation classification techniques: classical approaches and new trends. IET Commun. **1**, 137–156 (2007)
8. Wang, B., Ge, L.: A novel algorithm for identification of OFDM signal. In: Proceedings of IEEE WCNC, pp. 261–264 (2005)
9. Grimaldi, D., Rapunao, S., Truglia, G.: An automatic digital modulation classifier for measurement on telecommunication networks. In: Proceedings of IEEE IMT, pp. 957–962 (2002)
10. Akmouche, W.: Detection of multicarrier modulations using 4th order cumulants. In: Proceedings of IEEE MILCOM, pp. 432–436 (1999)
11. Punchihewa, A., Dobre, O.A.: Cyclostationarity-based algorithm for blind recognition of OFDM and single carrier linear digital modulations. In: Proceedings of IEEE PIMRC (2007)
12. Punchihewa, A., Dobre, O.A.: Cyclostationarity-Based Recognition of Orthogonal Frequency Division Multiplexing and Single Carrier Linear Digital Modulations, report submitted to DRDC, Mar 2007
13. Bouzegzi, A., Ciblat, P., Jallon, P.: New algorithms for blind recognition of OFDM based systems. Signal Process. **90**(3), 900–913 (2010)
14. Punchihewa, A., Zhang, Q., Dobre, O.A., Spooner, C., Rajan, S., Inkol, R.: On the cyclostationarity of OFDM and single carrier linearly digitally modulated signals in time dispersive channels: theoretical developments and application. IEEE Trans. Wireless Commun. **9**, 2588–2599 (2010)
15. Oner, M., Jondral, F.: On the extraction of the channel allocation information in spectrum pooling systems. IEEE J. Sel. Areas Commun. **25**, 558–565 (2007)
16. Sutton, P.D., Nolan, K.E., Doyle, L.E.: Cyclostationary signatures in practical cognitive radio applications. IEEE J. Sel. Areas Commun. **26**, 13–24 (2008)
17. Socheleau, F.-X., Ciblat, P., Houcke, S.: OFDM system identification for cognitive radio based on pilot-induced cyclostationarity. In: Proceedings of 2009 IEEE WCNC, pp. 1–6
18. Al-Habashna, A., Dobre, O.A., Venkatesan, R., Popescu, D.C.: Second-order cyclostationarity of mobile WiMAX and LTE OFDM signals and application to spectrum awareness in cognitive radio systems. IEEE J. Sel. Topics Signal Process. **6**, 26–42 (2012)
19. Zhang, Q., Dobre, O.A., Eldemerdash, Y.A., Rajan, S., Inkol, R.: Second-order cyclostationarity of BT-SCLD signals: theoretical developments and applications to signal classification and blind parameter estimation. IEEE Trans. Wireless Commun. **12**, 1501–1511 (2013)
20. Jerjawi, W., Eldemerdash, Y.A., Dobre, O.A.: Second-order cyclostationarity-based detection of LTE SC-FDMA signals for cognitive radio systems. IEEE Trans. Instrum. Meas. **64**, 823–833 (2015)
21. Socheleau, F.-X., Houcke, S., Ciblat, P., Assa-El-Bey, A.: Cognitive OFDM system detection using pilot tones second and third-order cyclostationarity. Signal Process **91**(2), 252–268 (2011)
22. Proakis, J.G.: Digital Communications, 4th edn. McGraw Hill, New York (2000)
23. Firdaoussi, M., Ghennioui, H., El Kamili, M.: New algorithm for blind recognition of OFDM based systems using second-order statistics. In: Proceedings of 2015 IEEE WINCOM, pp. 1–4
24. Shi, M., Laufer, A., Bar-Ness, Y., Su, W.: Fourth order cumulants in distinguishing single carrier from OFDM signals. In: Proceedings of IEEE MILCOM (2008)

MDE-Based Languages for Wireless Sensor Networks Modeling: A Systematic Mapping Study

Fatima Essaadi, Yann Ben Maissa and Mohammed Dahchour

Abstract Wireless Sensor Networks (WSNs) are ubiquitous systems of small devices equipped with sensors that collaborate to sense physical quantities in an area. However, the design constraints, the behavior requirements and the error prone nature, make the development of WSNs and their deployment an extremely challenging task. The Model Driven Engineering (MDE) approach helps tackling these issues by using models and automatic transformation to generate code or analyze WSNs against their requirements. In this paper, we propose a systematic mapping study which presents the existing WSNs MDE-based modeling languages. We surveyed a total of 1852 papers from which we selected 21 languages satisfying 7 selection criteria. We analyze these languages according to 5 rigorous research questions and 12 comparative criteria. Then we provide a precise view on the existing languages and their weaknesses mainly regarding mobility and data fusion. Finally, we propose research directions and recommendations for aspiring languages developers.

Keywords Wireless sensor networks · Model driven engineering · Domain specific modeling language · Model transformation · Systematic mapping study

1 Introduction

Context and Problem. Wireless Sensor Networks (WSNs) have emerged as one of the most promising technologies for the future [1]. WSNs have achieved great results in several domains such as health monitoring. Nevertheless, the application development in such a domain is a complicated process for the following reasons:

F. Essaadi (✉) · Y. Ben Maissa · M. Dahchour
Telecommunication Systems, Networks and Services Lab,
National Institute of Posts and Telecommunications, Rabat, Morocco
e-mail: essaadi@inpt.ac.ma

Y. Ben Maissa
e-mail: benmaissa@inpt.ac.ma

M. Dahchour
e-mail: dahchour@inpt.ac.ma

R. El-Azouzi et al. (eds.), *Advances in Ubiquitous Networking 2*, Lecture Notes in Electrical Engineering 397, DOI 10.1007/978-981-10-1627-1_26

First, once the network deployment is carried out, it will be very costly and time-consuming to make modifications if the network goes down.

Second, the sensor node, as an embedded system, is characterized by the hard coupling between the software and the hardware [2]. As a result, expert developers must deal with low-level implementation details in their applications development which results in highly platform-dependent designs. This makes the designs a hard task to maintain, modify and reuse.

Third, the deployment of WSNs is strongly influenced by several constraints which can be categorized in application and design constraints. While the application constraints are application dependent, the limited energy is the most important design constraint [3]. The energy consumption has the higher priority even more than quality of service [3]. All these constraints make WSNs application development complicated and error prone.

In order to tackle these difficulties, the current proposals insist on adopting a Software Engineering Paradigm to support the application life cycle development. The Model Driven Engineering (MDE) approach is proposed for this aim. MDE is a software engineering approach which is based on the principle of *Everything is model* [4]. The most relevant advantages presented by MDE are:

First, it raises the abstraction level and hides low-level implementation details. So, the domain experts can focus on the domain problems instead of focusing on technological aspects (e.g., target platforms).

Second, MDE is less error-prone, because it allows analysis in the early stage of life cycle development which reveal errors (e.g., missed performance requirements) before network deployment. Then the developers can customize their designs in regard to detected errors.

Third, MDE introduces the concept of Model Transformation which incorporates a set of rules to apply sequentially over models usually in order to produce the system code or the analysis model. Thus the code generation becomes an easy task regarding its complexity in traditional programming languages.

Contribution. In this paper, we propose a Systematic Mapping Study (SMS) to study the existing MDE-based languages. A SMS differs from a Systematic Literature Review (SLR). While in SLR, we deeply review existing primary studies and summarize their methodologies and results, a SMS is a defined method to provide a *precise* classification scheme and structure an area of interest [5]. In this SMS, we explored 1852 papers regarding 7 selection criteria, 1831 were discarded and 21 languages were finally selected. These languages are thoroughly studied and classified according to 5 mapping questions and a comparison framework of 12 criteria. Finally we propose some research directions to help developers and researchers in their future works.

Contents. The remainder of this paper is organized as follows. Section 2 shortly introduces the concepts of wireless sensor networks and model driven engineering. Section 3 is devoted for the research mechanism while in Sect. 4, we report the achieved results. Finally, we discuss the principal findings and their implications for researchers and developers before concluding the paper.

2 Background

Wireless Sensor Networks consist of small, cheap, limited battery-powered and spatially distributed sensor nodes which communicate through wireless links. These devices are able of sensing, computing and transmitting real-time information autonomously from the physical area.

Model Driven Engineering is a software engineering approach based on creating and exploiting models in order to address software systems complexity. This approach allows to raise the abstraction level and to discard the low-level details. The model is the core concept in MDE. It is an abstraction of the system under study. The MDE variant proposed by OMG (Object Management Group) is known as MDA (Model Driven Architecture). MDA is based on three models: computation-independent model (CIM), platform-independent model (PIM) and platform-specific model (PSM). MDA transforms a CIM to a PIM and a PIM to a multiple PSMs using Model transformations. A model describes the system using either a general-purpose modeling language such as UML or specific modeling language called Domain Specific Modeling Language (DSML). A DSML is defined using Metamodels or Grammars which underline both the relations between concepts and their static semantics. Thus, the developers can use a DSML to model the system at a higher abstraction level. After that and using Model Transformations, the model can be transformed into another model usually for analysis or code generation purposes.

3 Research Process

To build our research, we decided to conduct a systematic mapping study in order to provide a detailed view of the existing WSNs MDE-based languages. In this study, we follow several steps. The first is the choice of research questions, upon which our future analysis will be carried out. The second step consists in identifying the primary studies that present MDE-based languages for modeling WSNs. Finally, we describe the framework used to compare different languages in the fifth mapping question.

3.1 Mapping Questions

To develop a complete view on the MDE-based languages which are used to model the WSNs area, this study answers five mapping questions. These questions are presented in Table 1, with their corresponding main motivations.

- **MQ1** identifies the publication source and channel for each modeling language.
- **MQ2** aims to classify the languages per year of publication in order to have a clear idea about the research publications trend over time.

Table 1 Mapping questions

	Mapping question	Rationale
MQ1	In which sources and channels are WSNs MDE-based languages published?	To highlight where existing languages can be found
MQ2	What is the publications frequency of WSNs MDE-based languages?	To indicate the publication trends over time of this research area
MQ3	What are the existing MDE-based languages for modeling WSNs?	To identify the existing MDE-based languages that are used to model WSNs
MQ4	What are the main motivations for developing MDE-based languages?	To determine if the language is implemented to facilitate code generation or to allow analysis
MQ5	What are the main characteristics of existing MDE-based modeling languages?	To establish a comparison between the identified languages with respect to a framework of comparison criteria

- **MQ3** provides an overview of the existing MDE-based languages for modeling WSNs. The papers selection criteria are described in the Sect. 3.2.
- **MQ4** delineates the relevant motivation for the language implementation and design. The motivation can be classified as:
  - Code Generation: MDE fosters the design of a platform-independent model that can be used to automatically generate a source code, alleviating the expert developers from the complexity of traditional programming languages.
  - Analysis: MDE allows to transform a system model to another model for analysis purposes (e.g., simulation, model checking, theorem proving).
- **MQ5** establishes a comparison between the different languages. This question allows to identify the features and the weaknesses of each language. So the future studies can be aimed to address these gaps in their languages. The comparison is based on a set of criteria explained in detail in the Sect. 3.3.

3.2 Paper Selection Criteria

To identify the existing MDE-based languages, we explore several search engines and databases, namely, Google scholar, IEEE Digital Library, Science Direct and ACM Digital Library. The following search string is used to conduct the automatic search: ("*Wireless Sensor* Network**" ***OR*** "*Wireless Sensor* and Actuator* Network**" ***OR*** *WSN* ***OR*** WSAN) ***AND*** (("*Model Driven*" ***AND*** (*Engineering* ***OR*** *Architecture* ***OR*** *Development* ***OR*** *Approach**)) ***OR*** "*Domain Specific Language**" ***OR*** "*Domain Specific Modeling Language**" ***OR*** *MDE* ***OR*** *MDD* ***OR*** *MDA* ***OR*** *DSL* ***OR*** *DSML*).

Any paper that satisfies all the inclusion criteria and no one of the exclusion criteria is considered in our investigation.

Inclusion Criteria

- The paper shall be published in English.
- The paper shall be published after 2000 and before 2016.
- The paper's main contribution shall be the introduction to a new MDE-based language for modeling WSNs.
- If a study is conducted on the same modeling language, the latest published paper shall be selected with the aim of staying up-to-date.

Exclusion Criteria

- The paper targets Embedded Systems, Cyber-Physical Systems or Internet of Things (WSN is a part of these systems).
- The paper is built around a modeling language which was previously considered in our investigation.
- The paper involves the state of the art, the summary and the review of existing researches.

3.3 Comparison Framework

In the fifth mapping question, a comparison between the languages is established according to the following criteria:

MDE Instance: What is the MDE instance used in the modeling language? Possible values are MDA or Other. The most used value is MDA.

Modeling Language Origin: The presented language is a new DSML or it is based on a generic existing one. Possible values are DSML or GENERIC.

Modeling Scope: What is the modeling capacity of the selected language? It can model: node level (N) or group of nodes level (G) or network level (Net) or environment (Env). A Group of nodes combines all the nodes that are intended to have the same task in one entity.

Modeling Sensors: Is the language capable of modeling more than one sensor per node? Possible values are Yes or No.

Physical Location: Does the language support the physical location of nodes?

Mobility: Does the modeling language support the mobility of nodes?

Topology: Does the language model the WSNs topology?

Data Aggregation and Fusion: Is the selected language able to model the data aggregation and fusion? Possible values are Yes or No.

Presentation: Does the language model WSNs graphically (GRAPH) or textually (TEXT) or using both of them (MIX)?

Target Language: In case the modeling language is designed to facilitate the code generation, what is the target language?

Evaluation Method: What is the method used to evaluate the language?

4 Research Results

In this section, we report the answers to the mapping questions identified in Table 1. Figure 1 shows the number of produced papers after the execution of the search string in different search engines and databases. Considering the inclusion and exclusion criteria, we select 8 papers from IEEE Digital Library, 6 from ACM Digital Library and 7 from Google Scholar. The 21 selected languages are thoroughly analyzed and discussed.

4.1 MQ1. Publication Sources and Channels

Table 2 showcases the publication sources and channels for the selected MDE-based languages. Most of the sources targeted by selected studies are Conferences and Workshops. They represent 43 % and 29 % respectively. Around 14 % of studies are published in Journals and the same percentage is presented at Symposia. Different channels are identified, but the SESENA Workshop is the unique channel where more than one study are presented. Recall that the SESENA workshop is concerned with topics related to the Software Engineering for Sensor Networks.

4.2 MQ2. Publication Trends

The papers are selected from 2000 to 2016 because WSNs have gained an interest in the twenty first century [1, 3]. Figure 2 shows the number of published languages per year. Generally, we notice that all the languages are published after 2007 and in an increasing way, which can be interpreted by the rise of the third commercial generation of sensor networks technology [6]. This generation is characterized by some properties that attract a plethora of applications, such as small sensor nodes with a long life span, standard wireless protocols and low power-consumption based processors. The demand on such WSNs invites developers to rely on software development, especially the MDE approach, to alleviate their applications design and deployment.

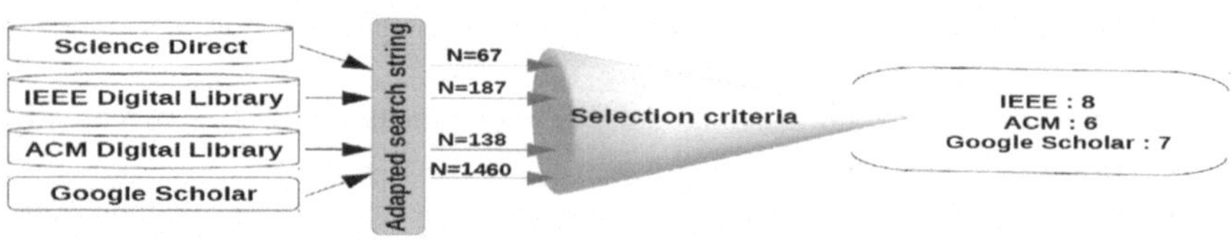

Fig. 1 Selection process

Table 2 MQ1, MQ2, MQ3 and MQ4 results

Language[a]	Publication year	Publication source	Publication channel	Modeling purpose
MEDWSA [10]	2007	Journal	CIS	Code generation
Wada et al. [11]	2007	Conference	SEA	Code generation
ScatterClipse [12]	2008	Symposium	ISPA	Code generation
Escolar et al. [13]	2008	Conference	DCOSS	Code generation
Baobab [14]	2009	Conference	UNISCON	Code generation
Flow [15]	2009	Conference	SENSYS	Code generation
SM4RCD [16]	2010	Workshop	SESENA	Code generation
Moppet [17]	2011	Workshop	BADS	Code generation
Xuan Thang et al. [18]	2011	Conference	MOMM	Code generation
BPMN4WSN [19]	2012	Conference	BPM	Code generation
VeriSensor [7]	2012	Workshop	PNSE	Analysis
Doddapaneni et al. [9]	2012	Workshop	SESENA	Code generation—analysis
Harbouche et al. [20]	2013	Workshop	WETICE	Code generation
SyVad [8]	2013	Symposium	ISPS	Analysis
Lwissy [21]	2013	Workshop	SESENA	Code generation
Vujovic et al. [22]	2014	Symposium	SACI	Code generation
WiSeN [23]	2014	Conference	INDIN	Code generation
Tei et al. [24]	2014	Journal	SMCS	Code generation
WML [25]	2014	Conference	IDCS	Code generation
AMF [26]	2015	Conference	ECMFA	Code generation
ArchWiSeN [27]	2015	Journal	SSM	Code generation

[a]By convention, for unnamed languages, the authors are referenced instead

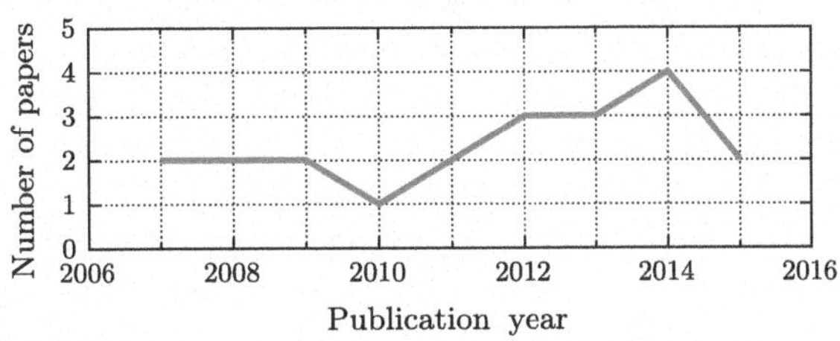

Fig. 2 Number of papers published per year from 2000 to 2016

4.3 *MQ3 and MQ4. Modeling Motivations*

Table 2 shows 21 selected languages. They are sorted by year of publication for the sake of clarity. Around 90 % of selected languages model the WSNs applications in order to facilitate the code generation. The other percentage represents three languages, VeriSensor [7], SyVad [8] and [9], which are proposed to model applications for analysis purposes. In VeriSensor, Ben Maissa et al. introduce a DSML and its transformation to a formal model (Instantiable Transition Systems) for reliability model checking purposes. In SyVad, Berrani et al. propose a SysML model and its mapping to a Modelica model for simulation and verification purposes, while in [9], Doddapaneni et al. present a DSML and its transformation to Castalia scripts for simulation purposes.

4.4 MQ5. Languages Features

A summary of the main characteristics of the selected languages is presented in Table 3. Around 29 % of selected MDE-based languages adopt the MDA approach, this is a significant value that confirms the idea which presents the MDA as the popular MDE instance. Most languages (62 %) introduce one or more DSMLs to model WSNs applications. This percentage points out that the majority of developers prefer introducing their own languages. As shown in Fig. 3, most of the DSMLs have the option to model WSNs graphically which facilitate the programming task from the cognitive viewpoint. The other 38 % are built on generic languages for designing WSNs. BPMN4WSN [19] extends Business Process Modeling Notation (BPMN) to graphically model WSNs and integrate them with Business Processes. While the 7 other studies propose to add UML stereotypes and profiles. We notice that in the last two years, there is a common trend to use UML.

Around 67 %, 9 % and 5 % of the studies model respectively the node-level, the group-level and the network-level of WSNs applications as shown in Fig. 4. In other studies like [21], several DSMLs are proposed to model the application at different levels which increase the learning cost. The use of group-level or network-level models is more suitable for average developers because they don't require as much details as node-level models. The language which handles the environment aspect is presented by Doddapaneni et al. [9]. They model the physical area where sensor nodes are deployed and present some obstacles that attenuate the transmitting power.

Several papers don't express all the information about their modeling languages. So, no definitive conclusions could be made about the rest of comparison criteria. Around 12 selected languages allow to model several sensors per node. This worthwhile option allows to reduce the number of deployed sensor nodes as well as the hardware cost. Regarding the physical location, it is important to determine the space coordinates of a sensor node. In this way, we can estimate where an event has happened or where an important data is collected. But half of the languages don't model this criterion.

In our mapping study, Harbouche et al. introduce the only language [20] which supports mobility. Indeed, they present a health monitoring system where the collector node was mobile in order to collect the different sensory data from the physiological sensors. Around 11 languages support the modeling of topology whereas 6 languages don't. The WSNs topology allows to understand the nodes positions and their neighborly relations.

Table 3 Languages features

Language[a]	MDE instance	Language origin	MScope	MS	PL	Mob	Top	DA	DF	Presentation	TL	Evaluation method
MEDWSA [10]	MDA	DSML	G	Yes	No	No	Yes	No	No	MIX	NesC	Case study
Wada et al. [11]	Other	GENERIC (UML)	N	No	Yes	No	Yes	No	No	GRAPH	NesC	Case study
ScatterClipse [12]	Other	DSML	N, Net	Yes	No	No	Yes	No	No	MIX	C	Case study
Escolar et al. [13]	MDA	DSML	N	Yes	–	–	–	–	–	GRAPH	NesC	–
Baobab [14]	Other	DSML	N	No	No	No	Yes	Yes	No	TEXT	NesC	Case study
Flow [15]	Other	DSML	N	–	–	–	–	–	–	GRAPH	C	–
SM4RCD [16]	Other	DSML	Net	–	–	–	–	–	–	GRAPH	C++	Experiments
Moppet [17]	Other	DSML	N	No	Yes	No	No	Yes	No	TEXT	NesC	Case study
												Simulation
												EE
Xuan Thang et al. [18]	MDA	DSML	N	Yes	No	No	No	Yes	No	MIX	NesC	Case study
BPMN4WSN [19]	Other	GENERIC (BPMN)	N	Yes	No	No	No	Yes	No	GRAPH	–	Case study
VeriSensor [7]	Other	DSML	N	Yes	Yes	No	Yes	No	No	TEXT	–	Case study
Doddapaneni et al. [9]	Other	DSML	N, Env	Yes	Yes	No	Yes	No	No	MIX	–	Case study
Harbouche et al. [20]	Other	GENERIC (UML)	N	–	–	Yes	Yes	Yes	No	GRAPH	NesC	Case study
SyVad [8]	MDA	GENERIC (UML)	N	No	No	No	No	No	No	GRAPH	–	Case Study
												Simulation

(continued)

Table 3 (continued)

Language[a]	MDE instance	Language origin	MScope	MS	PL	Mob	Top	DA	DF	Presentation	TL	Evaluation method
Lwissy [21]	MDA	DSML	N,G,Net	Yes	No	No	Yes	Yes	No	MIX	NesC	SE
Vujovic et al. [22]	Other	DSML	N	Yes	No	No	No	No	No	MIX	–	–
WiSeN [23]	Other	GENERIC (UML)	N	Yes	Yes	No	Yes	Yes	No	GRAPH	C++	Case study
Tei et al. [24]	Other	DSML	N,G,Net	No	Yes	–	Yes	Yes	Yes	MIX	NesC	Case study
												User study
WML [25]	Other	GENERIC (UML)	N	Yes	No	No	No	No	No	GRAPH	–	Case study
AMF [26]	Other	GENERIC (UML)	N	–	–	–	–	–	–	GRAPH	NesC	Case study
ArchWiSeN [27]	MDA	GENERIC (UML)	G	Yes	No	No	Yes	Yes	No	GRAPH	NesC	Case study
												User study

[a]By convention, for unnamed languages, the authors are referenced instead

Acronyms Modeling Scope (MScope), Modeling Sensors (MS), Physical Location (PL), Mobility (Mob), Topology (Top), Data Aggregation (DA), Data Fusion (DF), Target Language (TL), Empirical Experiments (EE), Statistical Experiments (SE)

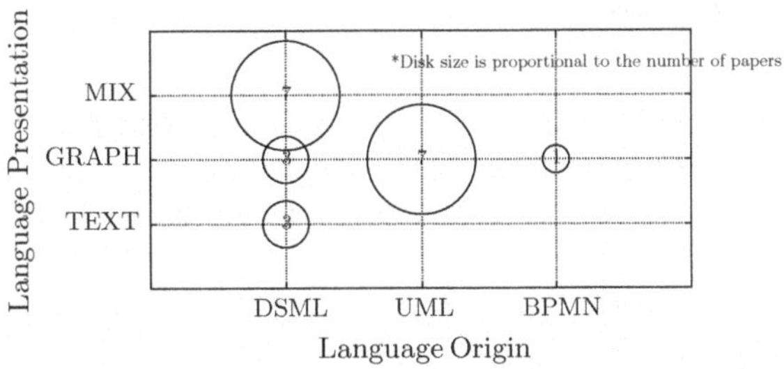

Fig. 3 Languages origins and presentations

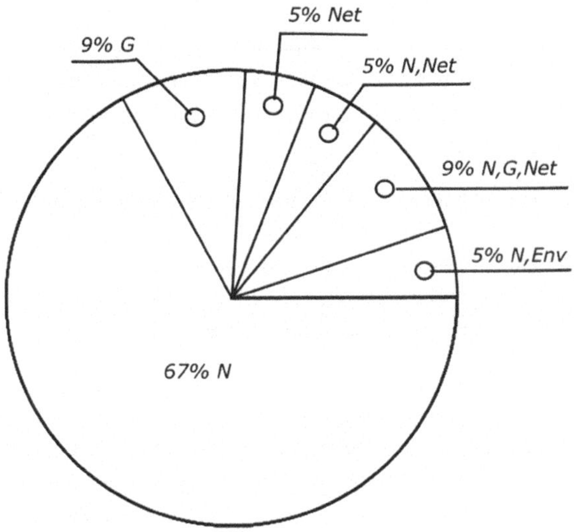

Fig. 4 Languages modeling scope. *Acronyms* Node level (N), Group of nodes level (G), Network level (Net), Environment (Env)

Data aggregation is a process that collects data from multiple sensor nodes, eliminates redundancy using special techniques (e.g., average, maximum) and transmits the reduced data in one copy [28]. Data fusion processes the gathered data in order to achieve relevant, precise and complete inferences [29] (e.g., sensor nodes transmit their own locations and their event detection times which can be fused to obtain the event position). In our study, 9 languages allow to model the data aggregation while just [24] allows the data fusion.

Most languages (58 %) for code generation support nesC as the target code language. nesC is a component-oriented extension to the C language used to design networked embedded systems. All the languages target one language but in [20], Harbouche et al. use two languages, nesC for the sensor nodes and Java for the mobile collector. Around 76 % of the selected studies are validated and evaluated through case studies. In other languages, [24, 27], particular studies are also conducted to test the understandability and the ease of use of the language for the average developers.

5 Discussion

In this section, we analyze the principal findings and the conclusions which can be drawn from the mapping results and present some implications and recommendations for researchers and aspiring languages developers.

5.1 Principal Findings

Analysis. In critical systems where a failure can cause dangerous economic, human or environmental damages, modeling for analysis purposes is very important to realize rigorous and exhaustive verification. However, few studies handle this issue given the following reasons:

1. The exhaustive verification is based on formal models which require a strong experience and skills in mathematics.
2. The process is complicated, because developers must transform a model to another model and conduct analysis. After that, developers must translate the analysis results in the original model language.

Physical Environment. The environment affects strongly the WSNs behavior. But we notice this aspect is considered in only one language [9].

Mobility. The only language which supports the mobility is introduced by Harbouche et al. [20]. In fact, the language models mobile collector which moves throughout the sensing area in order to collect data from sensor nodes. In this context, the mobility allows to reduce the number of transmission hops and thus to minimize the required transmission energy. This result reflects the modeling difficulty of this feature. But, with the growing demand on Mobile WSNs (MWSNs), MDE-based languages must deal with the mobility aspect.

Data Fusion. In energy-constrained sensor networks, sensor nodes can process the data collected from the physical environment. Processing data allows to reduce traffic load and energy consumption and to overcome sensor failures. Note that the data processing requires less energy compared to data communication. Indeed, assuming a 1 GHz carrier frequency, an antenna elevation of 1/2 wavelength, an efficient digital modulation, Rayleigh fading, fourth-power distance loss, 10^{-6} error probability, an ideal receiver and a general-purpose processor with 100 MIPS/W power, transmitting 1kb over a distance of 100 m and executing 3 million instructions consume approximately the same energy [30]. But, according to our study, developers are more attracted to the data aggregation given its modeling ease, than data fusion which is relatively hard to model.

Language Extensibility. In some studies, developers propose instruments to facilitate the language extensibility which encourage other developers to adopt the language and to tailor it to their needs.

5.2 Recommendations for Developers and Researchers

Analysis. Analysis is recommended especially in critical systems. Considering the complicated mathematical aspect of formal verification (e.g., model checking and theorem proving), developers should at least propose more languages for simulation purposes.

Physical Environment. The environment affects strongly the application behavior. For example, the presence of obstacles attenuates the transmitting power, the environment changes (e.g., earthquake, fire) can trigger the nodes mobility or destruction, also the environmental noises as temperature or pressure can affect the data precision. These situations must be considered by developers.

Mobility. Mobility is an important challenge and shall be more explored by developers. Developers can model mobile nodes in several scenarios:

- Sensor nodes which are programmed to move using the mobilizer unit. So, developers can propose a model to the mobilizer unit.
- Sensor nodes which move due to external forces, such as when attached to a vehicle.
- Sinks or collectors which move throughout the sensing region to collect data from sensor nodes. In this case, developers can follow the example proposed by Harbouche et al. in [20].
- Actuator nodes can be mobile to perform actions upon the maximum of the sensing area. These nodes are important in several domains such as environmental monitoring. Note that, in some applications, integrated sensor/actuator nodes are used, which means that the node contains both the sensing and actuating units.

Developers can use some specific mobility models presented by Rezazadeh et al. [31]. These models try to simply present the real behavior of mobile sensor nodes.

Data Fusion. There is no comprehensive theoretical framework to unify the various algorithms proposed in the literature for data fusion [29]. The lack of such unified framework raises the issue of data fusion modeling. The researchers are thus invited to elaborate such framework to alleviate the task of developers. Otherwise, these latter are constrained to model the existing specific algorithms.

Language Unification. Based on our study, we select 21 different MDE-based languages in the WSNs area. Most of them propose instruments to extend the languages to help developers in their future works. However, language unification should receive more attention from researchers to resolve the languages diversity problems and proliferation. This concern has been raised by [32]. Researchers can get inspiration from the UML unification process. UML is an unified modeling language which can be extended by profiles mechanisms.

6 Conclusion

In this paper, we proposed a systematic mapping study on the existing MDE-based languages in the WSNs area. From 1852 surveyed papers, 21 languages were selected according to 7 inclusion and exclusion criteria. These languages were deeply reviewed and analyzed according to 5 research questions and 12 rigorous comparative criteria. Our main results showed that MDE had gained an increasing interest in the WSNs area since 2007 and MDA was the most used MDE instance. The study revealed that 19 languages were designed to facilitate the code generation and just 3 were implemented for analysis purposes. Another interesting result was that nesC was the main targeted programming language. Our study pointed out the lack of some important modeling features such as mobility, environment and data fusion. Finally, we proposed 5 research directions and recommendations to help developers address the languages weaknesses: analysis support, physical environment, mobility, data fusion modeling and language unification.

In the future, along with the existing MDE-based WSNs languages, more studies shall be conducted to provide a better insight into the modeling weaknesses and to propose more solutions according to the proposed recommendations.

References

1. Chong, C., Kumar, S.P.: Sensor networks: evolution, opportunities, and challenges. J. Proc. IEEE **91**, 1247–1256 (2003)
2. Akyildiz, I.F., Can Vuran, M.: Wireless Sensor Networks. Wiley (2010)
3. Akyildiz, I.F., Su, W., Sankarasubramaniam, Y., Cayirci, E.: Wireless sensor network: a survey. J. Comput. Netw. **38**, 393–422 (2002)
4. Bzivin, J.: On the unification power of models. J. Softw. Syst. Model. **4**, 171–188 (2005)
5. Petersen, K., Feldt, R., Mujtaba, S., Mattsson, M.: Systematic mapping studies in software engineering. In: 12th International Conference on Evaluation and Assessment in Software Engineering, pp. 68–77 (2008)
6. Sohraby, K., Minoli, D., Znati, T.: Wireless Sensor Networks: Technology, Protocols, and Applications. Wiley-Interscience (2007)
7. Ben Maissa, Y., Kordon, F., Mouline, S., Thierry-Mieg, Y.: Modeling and analyzing wireless sensor networks with VeriSensor. In: International Workshop on Petri Nets and Software Engineering (PNSE), pp. 60–76 (2012)
8. Berrani, S., Hammad, A., Mountassir, H.: Mapping sysml to modelica to validate wireless sensor networks non-functional requirements. In: 11th International Symposium on Programming and Systems (ISPS), pp. 177–186 (2013)
9. Doddapaneni, K., Ever, E., Gemikonakli, O., Malavolta, I., Mostarda, L., Muccini, H.: A model-driven engineering framework for architecting and analysing wireless sensor networks. In: The 3th International Workshop on Software Engineering for Sensor Network Applications (SESENA), pp. 1–7 (2012)
10. Vicente-Chicote, C., Losilla, F., Alvarez, B., Iborra, A., Sanchez, P.: Applying MDE to the development of flexible and reusable wireless sensor networks. J. Coop. Inf. Syst. **16**, 393–412 (2007)
11. Wada, H., Boonma, P., Suzuki, J., Oba, K.: Modeling and executing adaptive sensor network applications with the MATILDA UML virtual machine. In: 11th IASTED International Conference on Software Engineering and Applications, pp. 216–225 (2007)

12. Al Saad, M., Fehr, E., Kamenzky, N., Schiller, J.: ScatterClipse: a model-driven tool-chain for developing, testing, and prototyping wireless sensor networks. In: International Symposium on Parallel and Distributed Processing with Applications, pp. 871–885 (2008)
13. Escolar, S., Carretero, J., Isaila, F., Tartari, G.: A MDA-based development framework for sensor networks applications. In: 4th IEEE/ACM International Conference on Distributed Computing in Sensor Systems (DCOSS) (2008)
14. Akbal-Delibas, B., Boonma, P., Suzuki, J.: Extensible and precise modeling for wireless sensor networks. In: 3rd International United Information Systems Conference (UNISCON), pp. 551–562 (2009)
15. Naumowicz, T., Schroter, B., Schiller, J.: Prototyping a software factory for wireless sensor networks. In: 7th ACM Conference on Embedded Networked Sensor Systems (SenSys), pp. 369–370 (2009)
16. Glombitza, N., Pfisterer, D., Fischer, S.: Using state machines for a model driven development of web service-based sensor network applications. In: Workshop on Software Engineering for Sensor Network Applications (SESENA), pp. 2–7 (2010)
17. Boonma, P., Suzuki, J.: Model-driven performance engineering for wireless sensor networks with feature modeling and event calculus. In: 3rd Workshop on Biologically inspired algorithms for distributed systems (BADS), pp. 17–24 (2011)
18. Xuan Thang, N., Zapf, M., Geihs, K.: Model driven development for data-centric sensor network applications. In: 9th International Conference on Advances in Mobile Computing and Multimedia (MoMM), pp. 194–197 (2011)
19. Tranquillini, S., Spieß, P., Daniel, F., Karnouskos, S., Casati, F., Oertel, N., Mottola, L., Oppermann, F.J., Picco, G.P., Römer, K., Voigt, T.: Process-based design and integration of wireless sensor network applications. In: 10th International Conference on Business Process Management (BPM), pp. 134–149 (2012)
20. Harbouche, A., Erradi, M., Kobbane, A.: A flexible wireless body sensor network system for health monitoring. In: 22nd Workshop on Enabling Technologies: Infrastructure for Collaborative Enterprises (WETICE), pp. 44–49 (2013)
21. Dantas, P., Rodrigues, T., Batista, T., Delicato, F.C., Pires, P.F., Li, W., Zomaya, A.Y.: LWiSSy: a domain specific language to model wireless sensor and actuators network systems. In: 4th International Workshop on Software Engineering for Sensor Network Applications (SESENA), pp. 7–12 (2013)
22. Vujovic, V., Maksimovic, M., Perisic, B., Milosevic, V.: A Graphical editor for RESTful sensor web networks modeling. In: 9th IEEE International Symposium on Applied Computational Intelligence and Informatics (SACI), pp. 61–66 (2014)
23. Paulon, A.R., Frohlich, A.A., Becker, L.B., Basso, F.P.: Wireless sensor network UML profile to support model-driven development. In: 12th IEEE International Conference on Industrial Informatics (INDIN), pp. 227–232 (2014)
24. Tei, K., Shimizu, R., Fukazawa, Y., Honiden, S.: Model-driven-development-based stepwise software development process for wireless sensor networks. J. IEEE Trans. Syst. Man Cybern. Syst. **45**, 675–687 (2014)
25. Ruiz-Zafra, A., Noguera, M., Benghazi, K.: Towards a model-driven approach for sensor management in wireless body area networks. In: 7th International Conference on Internet and Distributed Computing Systems (IDCS), pp. 335–347 (2014)
26. Berardinelli, L., Di Marco, A., Pace, S., Pomante, L., Tiberti, W.: Energy consumption analysis and design of energy-aware WSN agents in fUML. In: 11th European Conference on Modelling Foundations and Applications (ECMFA), pp. 1–17 (2015)
27. Rodrigues, T., Delicato, F.C., Batista, T., Pires, P.F., Pirmez, L.: An approach based on the domain perspective to develop WSAN applications. J. Softw. Syst. Model. 1–29 (2015)
28. Maraiya, K., Kant, K., Gupta, N.: Wireless sensor network: a review on data aggregation. J. Sci. Eng. Res. **2** (2011)
29. Abdelgawad, A., Bayoumi, M.: Resource-Aware Data Fusion Algorithms for Wireless Sensor Networks. Springer (2012)

30. Pottie, G.J., Kaiser, W.J.: Wireless integrated network sensors. J. Commun. ACM **43**, 51–58 (2000)
31. Rezazadeh, J., Moradi, M., Ismail, A.S.: Mobile wireless sensor networks overview. J. Comput. Commun. Netw. **2** (2012)
32. Malavolta, I., Muccini, H.: A study on MDE approaches for engineering wireless sensor networks. In: 40th Euromicro Conference on Software Engineering and Advanced Applications (2014)

Multi-homing as an Enabler for 5G Networks: Survey and Open Challenges

Salma Ibnalfakih, Essaïd Sabir and Mohammed Sadik

Abstract Driven by the unprecedented growth in the number of connected devices and mobile data traffic, 5G wireless network is expected to afford a full scale Internet of Things. Various wireless access networks will continue to coexist; one concept is allowing a device to virtually and simultaneously be connected to other devices using all available network resources at its location which is multi-homing. In this paper, we draw an overall picture of multi-homing based solution for provisioning Quality of Service and Quality of Experience taking into consideration the economics issues and the challenging problems that must be solved in the future.

Keywords 5G · IoT · Single-homing · Multi-homing · Ultra-dense small cells · QoE · QoS · Energy consumption

1 Introduction

Mobile Internet and Internet of Things (IoT) are the two main drivers of the 5th new radio standard [1]. Academia, industries and governments are looking for another mobile network to cope with the expected future growth in mobile data traffic (the mobile data traffic would increase to 24.3 Exabytes per month [2], while the number of connected IoTs is estimated to reach 50 Billion by 2020 [3]). Both 5G technological and standardization aspects consider IoTs issue [4]. 5G is a promising revolution in wireless communication and wearable devices with

S. Ibnalfakih (✉) · E. Sabir · M. Sadik
NEST Research Group, ENSEM, Hassan II University of Casablanca, Casablanca, Morocco
e-mail: ibnalfakih.salma@gmail.com

E. Sabir
e-mail: e.sabir@ensem.ac.ma

M. Sadik
e-mail: m.sadik@ensem.ac.ma

R. El-Azouzi et al. (eds.), *Advances in Ubiquitous Networking 2*, Lecture Notes in Electrical Engineering 397, DOI 10.1007/978-981-10-1627-1_27

artificial intelligence capabilities [5]. 5G technologies are foreseen to change the way we use Internet services and introduces the world of pervasive and always-connected mobile services. This perspective includes supporting an Integrated Smart Home and Smart City (ISHSC) [6]. In the context of smart cities, user-to-machine interactions are most needed (smart building, health care, and so forth) [7] but many governments are challenged by poor financial capabilities to invest in [8].

Research activities on 5G communication technology involve the worldwide deployment of the existing wireless standards. All the ongoing access technologies will continue to coexist; mobile devices need to improve their reliability and performance by efficiently benefiting from the network resources available at their location. Simultaneously integrating many access technologies in a network transmission is referred to as multi-homing. In an IP based system, Multi-homing allows at a giving time to benefit from a set of IP address (many connections) instead of only one wireless access network to reach all destinations (single-homing).

The existing wireless communication medium is a heterogeneous network (HetNet) paradigm with multiple access networks that offer many of access options in terms of bandwidth, coverage area and costs. HetNets will offer the required seamless connectivity for promising IoT through many mechanisms for coordination and management [9, 10]. Using multi-homing in HetNets (WIFI, 3G, 4G, B4G, 5G, etc.) can reinforce ubiquitous access.

However, the main envisioned characteristics of 5G paradigm are reaching 10 Gbps peak data rate, using ultra dense small cells, allowing a variety of Machine to Machine (M2M) services, maintaining network connectivity between devices regardless of time and location, adopting Device to Device (D2D) communications to reduce end-to-end communication latency [11]. To address these 5G requirements, research on both the network side, as well as on the device side is needed. The network has to be able to integrate and manage heterogeneous network elements in a seamless way to provide an optimum service experience. Devices have to be able to transmit and receive higher data rates at lower or equal cost as the devices of today.

2 Small Cells in Ultra-Dense Networks

From the network perspective, 5G will encompass basically new wireless network designs such as deploying a large number of small cells to carry the majority of the traffic in a given area and to manage them virtually. Small cells are low-powered radio access nodes which operate in licensed and unlicensed spectrum that have a range of 10 m to 1 or 2 km. On the contrary, macro-cells are the wide area high powered base stations covering areas up to few tens of kilometers (depending on the propagation environment). Small cells are vital elements to 3G data offloading and many mobile network operators see small cells as vital to managing

LTE Advanced spectrum more efficiently compared to using just macro-cells (statically planned).

The fifth generation network is expected to be much denser compared to the fourth one. Therefore, users can enjoy uniform QoE. In urban areas, a separate layer of small cells is required to provide the capacity and in-building penetration. They are dynamically planned and commissioned quickly in small office buildings, private homes and shopping malls, where an operating microcells network is the most needed to take the load off the macro cellular network. Moreover the switching doesn't make any interruption to the user's conversation. By configuring each small cell with neighbor lists, any possible interruption is avoided. The large bandwidth in the millimeter wave (mmWave) bands using those higher frequencies is one of 5G key enabling technologies. Since, the notably high propagation loss of mmWave makes it convenient for dense small cells to take advantage from the higher spatial frequency reuse [12].

Furthermore, conventional static network topologies with a central controller have an "edge", the reach of the central controller. However, the concept of a user-centric virtual cell that contains a group of cooperating BSs is constantly reformed so that any user will always be at the "center" of the cell [12]. To deploy many small Base Stations (BSs), it is necessary to use distributed and self-configured network technologies. For BSs cooperative communication, an In-band wireless backhaul can be used to minimize the network complexity and cost. Thus, using a network topology aware holistic smart backhaul solution for 5G small cells named Self-optimizing Wireless Mesh Network (SWMN) [13]. SWMN is a transparent 5G small cell transport network allowing ease of deployment, network resilience and flexible QoS scheme.

3 Multi-homing-Based Solution for Provisioning QoS and QoE

Managing both volume of devices and amount of data is no longer feasible by traditional network architectures. The 5G technologies should manage the load of traffic and the network resources efficiently to allow the coexistence of different services with different QoS requirements [14]. Multi-homing capability allows each MT to obtain its required QoS from all available wireless access networks in its location. This capability includes supporting applications with high data rate, handling mobility issue and balancing the traffic load across different wireless access networks [15].

Given the increasing degree of heterogeneity, the traditional notion of a device belonging to a specific cell is changing toward ubiquitous connectivity with uniform QoE to provide the user with an immersive experience even while he is on the move. A device would choose the most convenient connection from the various connections available around. In such a setting, new concept had recently been introduced in the context of the IoTs, referred to as Downlink and Uplink Decoupling (DUDe) [16, 17] concept. It's a situation when a wireless device that sees multiple BSs may access the infrastructure in a way to the DownLink (DL) traffic from one BS and sends UpLink (UL) traffic through another BS, given that transmission powers differ significantly between DL and UL [4].

Multi-homing is necessary to integrate in different layers of the network protocol stack: network layer, link layer or transport layer [18]. At transport and network level, studies have been made on some protocols related to multi-homing in IPv4 and multi-homing in IPv6 [19]. There are two network classifications depending on the number of upstream ISP that the multi-homing networks connect:

A multi-homed network: a network that is connected to more than one ISP to avoid connection failure.
A multi-attached network: a network that is related to a single ISP with multiple connections.

A multi-homed stub is an enterprise that has Internet access links to multiple providers [31] to cope with the increasing trend in network traffic load. One of the challenging issues for multi-homed stub networks is non aggregation problem: a stub network is multi-homed to multiple Autonomous System (AS) and thus may generate many IP prefixes from various upstream Ass [18, 20, 21], while in traditionally stub network, the routes advertised by the stub networks with the same upstream AS can be aggregated to one route by this AS. However, there are protocols that can solve the non-aggregation problem in multi-homed stub networks.

3.1 Deploying Multi-homing Technology in IPv4

Two main solutions are available to deploy multi-homing technology in a stub network: multi-homing with Border Gateway Protocol (BGP) only, referred to as BGP multi-homing and multi-homing with a network address translation (NAT) mechanism, referred to as NAT multi-homing [18].

BGP Multi-homing is the default inter-AS routing protocol. It determines how to exchange the network reachability information via different ASs. BGP runs on the edge routers of the AS, called external BGP (eBGP). By default, BGP chooses the shortest route according the AS hop counts as well as the preference level designed by each AS.

NAT Multi-homing is like a translator between the public Internet address and the internal local network address. Local networks adopt NAT to reuse the IP addresses inside their networks to help implement multi-homing. Moreover, NAT multi-homing has no requirement regarding the size of the network nor it has a requirement for the AS number.

BGP Multi-homing VS. NAT Multi-homing: BGP multi-homing and NAT multi-homing are different in address management, routing process control and failure handling process. In addition, BGP is the standard Internet inter-AS protocol; whereas NAT is introduced as a functionality to map local IP addresses inside a network with the global IP addresses outside the network. Therefore BGP and NAT themselves also create differences between BGP multi-homing and NAT multi-homing.

Being the standard Internet inter-AS protocol, BGP and provides the largest support for the upper level applications. On the other hand, NAT removes the uniqueness of the IP and does not support all of the upper level applications. The BGP is recommended for large organizations as it guarantees the uniqueness of the host IP address: supporting the upper level applications and controlling the routing process beyond the local networks; Whereas NAT is recommended for small to mid-size organizations not needing to be involved in the routing process beyond the local network: It aggregates the network traffic [21].

3.2 Deploying Multi-homing Technology in IPv6

The overall issue of IPv6 site multi-homing is to provide two main functionalities: full fault-tolerance (IPv6 multi-homing solution should insure transport-layer survivability across failure connections) and traffic engineering capabilities (IPv6 multi-homing solution should fulfill the load-sharing, performance and policy motivations for multi-homing [20]. In particular, the use of several IPv6 addresses per end host introduces many architectural change compared with today's IPv4 Internet, where hosts are only identified by a unique IPv4 address.

At the network level, recent protocols such as MTCP and mSCTP allow session continuity and offer the possibility of seamlessly adapting IP routing during a transmission session. These protocols are mainly based on the use of a set of IP addresses that can be associated to the same terminal and user session. But at the access level, efficient packet scheduling mechanisms on multiple interfaces are still required.

Thanks to multi-path based protocols such as mobile Stream Control Transmission Protocol (mSCTP) [22] or MultiPath Transmission Control Protocol (MPTCP) [23], mobile terminals are expected to be able to use simultaneous connections and/or to seamlessly switch between different access technologies during a communication sessions. With some cost, these rising protocols allow the use of multiple IP (Internet Protocol) paths for TCP sessions and can enable seamless session mobility [24].

Any multi-homing solution for IPv6 must respect several technical and non-technical constraints. The main constraint is to preserve the size of the BGP routing tables in the Internet. A second constraint is that a multi-homing solution should not preclude filtering procedures, for security reasons. The filtering consists in dropping the customers' packets entering in the ISP network that is coming from a source address not legitimately in use by the customer network. A third constraint is that a solution for IPv6 multi-homing must not require cooperation between the providers of the multi-homed site.

4 Multi-homing: Energy Efficiency Considerations

A MT with Multi-homing capabilities can benefit from the simultaneous use of multiple wireless network interfaces ranging from wireless hotspots to high speed cellular networks. However, the coexistence of several applications and wireless technologies emerge the challenge of energy consumption. An Energy Aware cross layer Scheduling algorithm (EAS) based on 802.11n and LTE energy models it uses both MAC and PHY parameters as well as network loads through the knowledge of the number of active stations [24]. The EAS algorithm aims to maintain high QoS levels while reducing mobile power consumption. Its utilization starts by selecting the available new networks with an acceptable Received Signal Strength (RSS) then checking if they can afford the application quality requirements. Secondly, the available networks classification is calculated according to the power cost and the time duration needed to send or receive a packet. Thirdly, the most energy efficient network is selected for each technology type to finally select the most efficient technology for the packet transmission [24]. This new energy scheduling and network selection algorithm needs to be studied for all the existing wireless networks to correctly select the most appropriate interfaces for each data transfer.

MTs with multi-homing capability can communicate with Access Points (APs) of the multiple Radio Access Technologies (RATs) simultaneously. However, meeting the increasing demands of multimedia applications also leads to increasing the power consumption of the MTs. Then, the Energy Efficiency (EE) has become very crucial in resource allocation [25]. Authors in [26, 27], have studied downlink energy-efficient communication in heterogeneous wireless networks. In [28], the authors design a resource allocation framework that maximizes the minimum energy efficiency among all MTs but without QoS guarantees.

However, some other studies focus on energy efficiency in heterogeneous wireless networks to formulate an energy-efficient resource allocation problem and minimize the sum of energy efficiency of the MTs respecting a satisfying quality of service QoS [29]. Another optimal algorithm that allocates bandwidth and power to each connection between a user and a BS has been proposed to investigate the network energy efficiency in resource allocation for uplink heterogeneous multi-homing. The proposed design has been developed based on the Dinkelbach method, which obtains the solution of the fractional program by solving series of convex problems [30]. As studies have shown that is the main source of energy consumption in wireless networks is the BS, many radio resource allocation mechanisms have been suggested to save the BS transmission power while affording the end users with an acceptable QoS performance [31].

Taking into account the fact that BSs of various networks are generally operated by different service providers, studies have proposed a decentralized downlink power allocation strategy that minimizes the total power consumption and insures mutual benefit among different service providers. The associate mechanism is referred to a win-win cooperative resource allocation strategy [32]. Power saving in a multi-operator heterogeneous wireless medium based on a decentralized framework using a Nash Bargaining Solution (NBS) among multiple operators is encouraging service providers to cooperate [32], unlike the sum minimization solution. The sum minimization solution based on networks cooperation to minimize the total power consumption in the geographical region reduces power consumption of one service provider and increases power consumption of another one.

5 Simultaneous Presence of Both unihomed Networks and Multi-homed Networks

Both single-network and multi-homing services are expected to still coexist after the 5th Generation standardization. An MT with multi-homing capabilities can benefit from various multi-homing services, but the network selection should depend on the residual energy at the MT. When there is no sufficient energy available at the MT, the MT should switch to the single-network instead of multi-homing service. Therefore, interesting energy saving at the MT level is acquired since it uses only the best radio interface available whereas all other interfaces are turned off.

For this reason, some researchers work is about developing a radio resource allocation algorithm that can support both single-network and multi-homing services. They have created a Centralized Optimal Resource Allocation (CORA) algorithm which takes account of both single network and multi-homing services [33]. The objective of the CORA algorithm is to find the optimal network assignment for MTs with single-network services and to investigate the corresponding optimal bandwidth allocation for MTs with both single-network and multi-homing services. Moreover, to support MTs with single-network and multi-homing services, a central resource manager can't be practical in a case that these HetNets are operated by different service providers which raises some issues: (1) choosing the network in charge of the operation and maintenance of the central resource manager; (2) making the changes required, (3) dealing with network failure limited in the central resource manager [33]. Hence, in such HetNets, it is desirable to have a decentralized solution enabling each BS/AP to perform its own resource allocation and admission control while simultaneously cooperating with available BSs/APs of other networks. Towards this end, a Decentralized

Sub-optimal Resource Allocation algorithm (DSRA algorithm) can be implemented in heterogeneous wireless access mediums to support both single-network and multi-homing services [33]. The DSRA algorithm accounts for the system dynamics, in terms of call arrivals and departures to achieve an acceptable call blocking probability and provide a sufficient amount of allocated resources to each call [33].

6 Multi-homing: Economics Considerations

Multi-homing techniques have been considered as a promising solution for both user and service provider. From user perspective, user tries to maximize its utility including QoS and cost to pay. While from service provider perspective, it is necessary to set the cost of each access networks to promote maximum utility function (profit and energy efficiency) over user reaction. To solve both pricing and load distribution problem of multi-homing in heterogeneous wireless access networks taking into account the users and service provider satisfaction, the problem can be formulated as a Stackelberg game [34]. In this Stackelberg model, service provider takes a role of the leader (makes a pricing strategy p to impose cost on the users), while users take role of followers (decide the rate-allocation for each interface based on provider's strategy to maximize their utility). By varying different parameters (e.g. rate allocation, cost, energy efficiency) and analyzing the relationship between users and service provider in terms of best response to finally reach an equilibrium point [34].

The Content Delivery Networks (CDN) serves as an essential element in providing content delivery services on the Internet; however, large content providers often use multiple CDNs to adapt to variations of network conditions and user demands and support the large-scale content delivery over the Internet. This approach is referred to as content multi-homing. Nevertheless, using multiple CDNs to provide high quality services is economically inefficient, in particular for small content providers [35]. Content multi-homing enables a content provider to afford ubiquitous QoE, whereas it could be an unprofitable business for small or medium content providers. This is due to the CDN charging policy based on two principals: (i) you pay as you go, and (ii) the more you use, the cheaper the average price you get.

The authors of [35] have defined a particular Content Delivery Networks (CDN-0). CDN-0 referred to a situation when a content provider only signs one contract with an authoritative CDN To make an optimal operating plan for CDN-0 operator [35], there is some recent research that has developed an algorithm for solving the problem of Multi-homing Content Delivery Networks-Concave Minimization (MCDN-CM) [36]. MCDN-CM modeling is taking advantage from the special form of the practical linear concave CDN-pricing function through an iterative procedure to help the operator of CDN-0 make an optimal operating plan under realistic settings.

7 Conclusion

5G is a promising technology that likely would revolutionize our daily life. It indeed allows and enables several new networking paradigms such as IoT, D2D, MTC and highly flexible and infrastructure-less communications. Even a huge real-time autonomous interconnection between devices would be possible, providing ubiquitous multi-Gbps data rate regardless of the user's location. Hence, in such a networking environment with overlapped coverage from different complementary service networks in terms of bandwidth, coverage area, and cost; Multi-homed mobile networks and terminals will make it worthwhile for operators and service providers to promote productivity and efficiency. Their interests will include performance analysis of communication networks and network security.

References

1. Chih-Lin, I., Han, S., Xu, Z., Wang, S., Sun, Q., Chen, Y.: New Paradigm of 5G Wireless Internet (2016)
2. Cisco: Cisco Visual Networking Index: Global Mobile Data Traffic Forecast Update: 2013-2018. Cisco, Feb 2014
3. UMTS: Mobile Traffic Forecasts 2010-2020 Report, UMTS Forum, Jan 2011
4. Palattella, M.R.: Internet of Things in the 5G Era: Enablers, Architecture and Business Models (2016)
5. Gohil, A., Modi, H., Patel, S.K.: 5G Technology of Mobile Communication: A Survey, Apr 2013
6. Skouby, K.E., Lynggaard, P.: Smart Home and Smart City Solutions enabled by 5G, IoT, AAI and CoT Services. CMI, Aalborg University Copenhagen, Denmark (2014)
7. Want, R.: The physical web. In: Proceedings of the 2015 Workshop on IoT Challenges in Mobile and Industrial Systems, ser. IoT-Sys'15, pp. 1–1. ACM, New York, NY, USA (2015)
8. Frost, A.: Smart city as a service. White paper (2015)
9. Andrews, J.: Seven ways that hetnets are a cellular paradigm shift. IEEE Commun. Mag. **51** (3), 136–144 (2013)
10. Condoluci, M., Dohler, M., Araniti, G., Molinaro, A., Zheng, K.: Toward 5G densenets: architectural advances for effective machine-type communications over femtocells. IEEE Commun. Mag. **53**(1), 134–141 (2015)
11. Talwar, S., Choudhury, D., Dimou, K., Aryafar, E., Bangerter, B., Kenneth Stewart. Enabling Technologies and Architectures for 5G Wireless, 2014
12. Samsung Electronics Co., Ltd. Samsung 5G vision, White paper, Feb 2015
13. Chen, D.T., Schuler, J., Wainio, P., Salmelin, J.: Technology & Innovation Research-Nokia Networks Arlington Heights, Illinois, USA5G Self-Optimizing Wireless Mesh Backhaul (2015)
14. Amani, M., Mahmoodi, T., Tatipamula, M., Aghvami, H.: SDN-based data offloading for 5G mobile networks. ZTE Commun. Mag. 2 (2014)
15. Kim, S.: Energy-per-bit minimized radio resource allocation in heterogeneous networks. IEEE Trans. Wirel. Commun. **13**(4) (2014)
16. Elshaer, H., Boccardi, F., Dohler, M., Irmer, R.: Downlink and uplink decoupling: a disruptive architectural design for 5G networks. In: Global Communications Conference (GLOBECOM), 2014 IEEE, pp. 1798–1803. IEEE (2014)

17. Boccardi, F., Andrews, J., Elshaer, H., Dohler, M., Parkvall, S., Popovski, P., Singh, S.: Why to decouple the uplink and downlink in cellular networks and how to do it (2015). arXiv:1503.06746
18. Liu, X., Xiao, L.: A Survey of Multihoming Technology in Stub Networks: Current Research and Open Issues. Michigan State University (2007)
19. Li, J., Veeraraghavan, M.: A stub multi-homing solution for IPv6 networks. IEEE ICC 2014-Next Generation Networking Symposium (2014)
20. de Launois, C., Bagnulo, M.: The paths toward IPv6 multihoming. 2nd Quarter 2006, vol. 8, no. 2 (2006)
21. Guo, F., Chen, J., Li, W., Chiueh, T.: Experiences in Building A Multihoming Load Balancing System (2004)
22. Stewart, R., et al.: Stream control transmission protocol. In: IETF RFC 2960, Oct 2000
23. Ford, A., Raiciu, C., Handley, M.: TCP Extensions for Multipath Operation with Multiple Addresses, draft-ford-mptcp-multiaddressed-02 (2009)
24. Ghariani, T., Jouaber, B.: Energy aware cross layer uplink scheduling for multihomed environments. Globecom Worshop (2013)
25. Lorincz, J., Matijevic, T.: Energy-efficency analyses of heterogeneous macro and micro base station sites. Comput. Electr. Eng. **20**(2), 330–349 (2014)
26. Kim, S., Lee, B.G., Park, D.: Energy-per-bit minimized radio resource allocation in heterogeneous networks. IEEE Trans. Wirel. Commun. **13**(4), 1862–1873 (2014)
27. Lim, G., Xiong, C., et al.: Energy-efficient resource allocation for OFDMA-based multi-RAT networks. IEEE Trans. Wirel. Commun. **13**(5), 2696–2705 (2014)
28. Ismail, M., Gamage, A.T., Zhuang, W., Shen, X.: Energy efficient uplink resource allocation in a heterogeneous wireless medium. In: Proceedings of 2014 IEEE ICC, pp. 5275–5280
29. Zou, J., Xi, Q., Zhang, Q., He, C., Jiang, L., Ding, J.: QoS-Aware Energy-Efficient Radio Resource Allocation in Heterogeneous Wireless Networks (2015)
30. Vu, Q.-D., Tran, L.-N., Juntti, M., Hong, E.-K.: Optimal Energy Efficient Resource Allocation for Heterogeneous Multi-homing Networks (2014)
31. Ismail, M., Gamage, A.T., Zhuang, W., (Sherman) Shen, X.: Energy Efficient Uplink Resource Allocation in a Heterogeneous Wireless Medium. Canada (2014)
32. Ismail, M., Serpedin, E., Qaraqe, K.: A Win-Win Cooperative Downlink Resource Allocation for Green Communications in a Heterogeneous Wireless Medium (2014)
33. Ismail, M.: Decentralized radio resource allocation for single-network and multi-homing services in cooperative heterogeneous wireless access medium. IEEE Trans. Wirel. Commun. **11**(11) (2012)
34. Yun, S., Lee, J., Shah Newaz, S.H., Choi, J.K.: Energy efficient pricing scheme for multi-homing in heterogeneous wireless access networks: a game theoretic model and its analysis. In: 2015 IEEE Wireless Communications and Networking Conference
35. Wang, J.M., Zhang, J., Bensaou, B.: Content multi-homing: an alternative approach. In: IEEE ICC 2014 - Next-Generation Networking Symposium
36. Wang, J.M., Zhang, J., Bensaou, B.: Content multi-homing: an alternative approach. The Hong Kong University of Science and Technology, Technical Report HKUST-CS14-01 (2014)

Fast Algorithm for 3D Local Feature Extraction Using Hahn and Charlier Moments

Abderrahim Mesbah, Aissam Berrahou, Mostafa El Mallahi and Hassan Qjidaa

Abstract In this paper, we propose a fast algorithm to extract 3D local features from an object by using Hahn and Charlier moments. These moments have the property to compute local descriptors from a region of interest in an image. This can be realized by varying parameters of Hahn and Charlier polynomials. An algorithm based on matrix multiplication is used to speed up the computational time of 3D moments. The experiment results have illustrated the ability of Hahn and Charlier moments to extract the features from any region of 3D object. However, we have observed the superiority of Hahn moments in terms of reconstruction accuracy. In addition, the proposed algorithm produces a drastic reduction in the computational time as compared with straightforward method.

Keywords Hahn moments · Charlier moments · Matrix multiplication · Region of interest · 3D local reconstruction

1 Introduction

Moments and moment invariants have been widely applied in the field of image analysis and pattern recognition [1–5]. Geometric moments and their translation, scaling and rotation invariants were introduced by Hu [6]. Teague in [7] proposed

A. Mesbah (✉) · M. El Mallahi · H. Qjidaa
Faculty of Sciences Dhar el Mehraz, Sidi Mohamed Ben Abdellah University, Fez, Morocco
e-mail: abderrahim.mesbah@usmba.ac.ma

M. El Mallahi
e-mail: mostafa.elmallahi@usmba.ac.ma

H. Qjidaa
e-mail: qjidah@yahoo.fr

A. Berrahou
Mohammedia School of Engineering, Mohamed V University, Rabat, Morocco
e-mail: aissamberrahou@research.emi.ac.ma

R. El-Azouzi et al. (eds.), *Advances in Ubiquitous Networking 2*, Lecture Notes in Electrical Engineering 397, DOI 10.1007/978-981-10-1627-1_28

the concept of orthogonal continuous moments such as Legendre and Zernike moments to represent image with minimum amount of information redundancy. The major problem associated with these moments is the discretization error, which increases by increasing the moment order [1]. To solve the above problem of the continuous orthogonal moments, Mukundan et al. in [8] introduced the notion of discrete orthogonal moments and proposed the set of Tchebichef moments. The use of Tchebichef polynomials as basis function for image moments eliminates the discrete approximation associated with the continuous moments. Moreover, it represents an image with the minimum amount of information redundancy. Recently, another set of discrete orthogonal moments have been introduced in the domain of image processing, such as Hahn moments [9] and Charlier moments [10]. Their experimental results make them superior to Zernike, Legendre and Tchebichef moments in terms of global image reconstruction.

Local descriptors have become more interest tools in image analysis and used in many scientific fields such as object recognition [11, 12], image retrieval [13], and shape matching [14]. They can be computed more efficiently than global descriptors, are robust to occlusion, generally less sensitive to viewpoint changes, and do not require segmentation [11, 15]. Some orthogonal moments such as Krawtchouk [16] and Hahn moments have the ability to extract the features of any selected region-of-interest (ROI) in an image. This is can be achieved by varying the parameters given in the definition of their polynomials. The locality property of these moments was discussed only for 2D case.

3D imaging gains more interest due to its precise descriptions of 3D objects. In fact, the representation and the reconstruction of 3D object are very important in different scientific areas such as medical imaging [17], multimedia [18] and molecular biology [19]. The direct computation of 3D discrete moments is time consuming process and the computational complexity increased by increasing the moment order. Therefore, some algorithms have been developed to accelerate the computational time of discrete moments for the case of 2D images [20, 21]. Only few works were devoted to compute 3D discrete moments.

In this paper, we have focused our study on 3D local feature extraction of Hahn and Charlier moments. Then, we compared the performance between these moments in terms of local reconstruction of 3D image. We also proposed an algorithm based on matrix multiplication to compute 3D discrete orthogonal moments in fast way. The conducted experiments studied the accuracy of our descriptors in terms of 3D local features extraction and fast computation of 3D moments.

The rest of this paper is organized as follows: in Sect. 2, an overview of Hahn and Charlier orthogonal polynomials is given. 3D orthogonal moments and the proposed fast algorithm are presented in Sect. 3. Section 4 discuss the capacity of Hahn and Charlier moments to extract 3D local features. Section 5 is devoted to the simulation results. Finally, concluding remarks are presented in Sect. 6.

2 Orthogonal Polynomials

In this section we will briefly present the mathematical foundations behind the moment theory including Hahn and Charlier polynomials.

2.1 Hahn Polynomials

Hahn polynomials of one variable x, with the order n, defined in the region of $[0, N-1]$ have the following representation [9]:

$$h_n(\alpha,\beta,N|x) = {}_3F_2\left(\begin{matrix} -n, n+\alpha+\beta, -x \\ \alpha+1, -N \end{matrix} \middle| 1\right) \quad (1)$$
$$n, x = 0, 1, \ldots, N-1$$

where α, β are free parameters, and ${}_3F_2$ is the generalized hypergeometric function given by

$${}_3F_2\left(\begin{matrix} a_1, a_2, a_3 \\ b_1, b_2 \end{matrix} \middle| z\right) = \sum_{k=0}^{\infty} \frac{(a_1)_k (a_2)_k (a_3)_k}{(b_1)_k (b_2)_k k!} z^k \quad (2)$$

The Hahn polynomials satisfy the orthogonal property

$$\sum_{x=0}^{N-1} h_n(\alpha,\beta,N|x)\, h_m(\alpha,\beta,N|x) w_h(x) = \rho_h(n)\delta_{mn} \quad (3)$$

where $w_h(x)$ is the weighting function giving by

$$w_h(x) = \frac{(\alpha+1)_x (\beta+1)_{N-x}}{(N-x)!x!} \quad (4)$$

while ρ_h is the squared-norm expressed by

$$\rho_h(n) = \frac{(-1)^n n! (\beta+1)_n (\alpha+\beta+n+1)_{N+1}}{(-N)_n (2n+\alpha+\beta+1) N! (\alpha+1)_n} \quad (5)$$

The set of the weighted Hahn polynomials is defined as

$$\tilde{h}_n(\alpha,\beta,N|x) = h_n(\alpha,\beta,N|x)\sqrt{\frac{w_h(x)}{\rho_h(n)}} \quad (6)$$

The set of Hahn polynomials obeys the three term recurrence relation:

$$\tilde{h}_n(\alpha,\beta,N|x) = A\sqrt{\frac{\rho_h(n-1)}{\rho_h(n)}}\tilde{h}_{n-1}(\alpha,\beta,N|x) - B\sqrt{\frac{\rho_h(n-2)}{\rho_h(n)}}\tilde{h}_{n-2}(\alpha,\beta,N|x) \tag{7}$$

$$n = 2, 3, \ldots, N-1.$$

where

$$A = 1 + B - x\frac{(2n+\alpha+\beta+1)(2n+\alpha+\beta+2)}{(n+\alpha+\beta+1)(\alpha+n+1)(N-n)}$$

$$B = \frac{n(n+\beta)(\alpha+\beta+n+N+1)(2n+\alpha+\beta+2)}{(2n+\alpha+\beta)(\alpha+\beta+n+1)(\alpha+n+1)(N-n)}$$

The initial values for the above recursion can be obtained from

$$\tilde{h}_0(\alpha,\beta,N|x) = \sqrt{\frac{w_h(x)}{\rho_h(0)}}$$

$$\tilde{h}_1(\alpha,\beta,N|x) = \left(1 - \frac{x(\alpha+\beta+2)}{(\alpha+1)N}\right)\sqrt{\frac{w_h(x)}{\rho_h(1)}}$$

2.2 Charlier Polynomials

The Charlier polynomials is defined by using hypergeometric function as:

$$C_n^{a_1}(x) = {}_2F_0(-n, -x;; 1/a_1) \tag{8}$$

The normalized Charlier polynomials are given by:

$$\check{C}_n^{a_1}(x) = C_n^{a_1}(x)\sqrt{\frac{w_c(x)}{d_n^2}} \tag{9}$$

with $w_c(x) = \frac{e^{-a_1}a_1^x}{x!}$ and $d_n^2 = \frac{n!}{a_1^n}$

The discrete Charlier polynomials satisfy the following three-term recurrence relation:

$$\check{C}_n^{a_1}(x) = \frac{a_1 - x + n - 1}{a_1}\sqrt{\frac{a_1}{n}}\check{C}_{n-1}^{a_1}(x) - \sqrt{\frac{n-1}{n}}\check{C}_{n-2}^{a_1}(x) \quad (10)$$

with $\check{C}_n^{a_1}(x) = \sqrt{\frac{w(x)}{d_0^2}}$ and $\check{C}_n^{a_1}(x) = \frac{a_1 - x}{a_1}\sqrt{\frac{w(x)}{d_1^2}}$ the orthogonality condition becomes:

$$\sum_{x=0}^{N} \check{C}_n^{a_1}(x)\check{C}_m^{a_1}(x) = \delta_{nm} \quad (11)$$

3 Fast Computation of 3D Discrete Orthogonal Moments

3.1 3D Discrete Orthogonal Moments

The 3D discrete orthogonal moments of order $m+n+l$ of an image intensity function $f(x,y,z)$ are defined over the cube $[0, M-1] \times [0, N-1] \times [0, L-1]$ as:

$$M_{mnl} = \sum_{x=0}^{M-1}\sum_{y=0}^{N-1}\sum_{z=0}^{L-1} f(x,y,z)\tilde{p}_m(x)\tilde{p}_n(y)\tilde{p}_l(z) \quad (12)$$

where $\tilde{p}_m$, $\tilde{p}_n(y)$ and $\tilde{p}_l(z)$ denote the normalized polynomials.

Due to the orthogonal property of the normalized polynomials, the 3D image/object intensity function $f(x, y, z)$ can be expressed over cube $[0, M-1] \times [0, N-1] \times [0, L-1]$ as:

$$f(x,y,z) = \sum_{m=0}^{M-1}\sum_{n=0}^{N-1}\sum_{l=0}^{L-1} M_{mnl}\tilde{p}_m(x)\tilde{p}_n(y)\tilde{p}_l(z) \quad (13)$$

3.2 Fast Algorithm

The direct method of 3D orthogonal moments, using Eq. (12), is time consuming process and the computational complexity increased by increasing the moment order. For this reason, we have used in this paper an algorithm based on matrix multiplication to reduce the computation cost of moment and inverse transformations especially for high order moments and large size objects cases. Therefore, the expression of orthogonal moments in Eq. (12) can be written as follow:

$$M_{mnl} = \sum_{z=0}^{N-1} \tilde{p}_l(z) \left\{ \sum_{x=0}^{N-1} \sum_{y=0}^{N-1} \tilde{p}_m(x) \tilde{p}_n(y) f(x,y,z) \right\} \tag{14}$$

In the first step, we calculate L temporary matrices along z-direction by using the matrix form of discrete moments for 2D image:

$$M = p_1 A p_2^T \tag{15}$$

where M is an $m \times n$ matrix of moments $M = \{M_{ij}\}, 0 \le i,j \le N-1, p_1 = \{\tilde{p}_i(x)\}$, with $0 \le i \le m-1$ and $0 \le x \le N-1$, $p_2 = \{\tilde{p}_j(y)\}$, with $0 \le j \le n-1$ and $0 \le y \le N-1$, and $A = \{f(x,y)\}, 0 \le x,y \le N-1$

The L temporary matrices are given as follow:

$$z \nearrow \quad \begin{bmatrix} M'_{00}(N-1) & M'_{01}(N-1) & \cdots & M'_{0n}(N-1) \\ M'_{10}(N-1) & M'_{11}(N-1) & \cdots & M'_{1n}(N-1) \\ \vdots & \vdots & \ddots & \vdots \\ M'_{m0}(N-1) & M'_{m1}(N-1) & \cdots & M'_{mn}(N-1) \end{bmatrix}$$

$$\begin{bmatrix} M'_{00}(1) & M'_{01}(1) & \cdots & M'_{0n}(1) \\ M'_{10}(1) & M'_{11}(N-1) & \cdots & M'_{1n}(1) \\ \vdots & \vdots & \ddots & \vdots \\ M'_{m0}(1) & M'_{m1}(1) & \cdots & M'_{mn}(1) \end{bmatrix}$$

$$\begin{bmatrix} M'_{00}(0) & M'_{01}(0) & \cdots & M'_{0n}(0) \\ M'_{10}(0) & M'_{11}(0) & \cdots & M'_{1n}(0) \\ \vdots & \vdots & \ddots & \vdots \\ M'_{m0}(0) & M'_{m1}(0) & \cdots & M'_{mn}(0) \end{bmatrix}$$

These matrices will be rearranged, in the second step, and they are multiplied by the polynomial matrices $p_3 = \{\tilde{p}_k(z)\}$, with $0 \le k \le l-1$ and $0 \le z \le N-1$. Then we obtained the matrices of 3D orthogonal moments.

$$\begin{bmatrix} M_{000} & \cdots & M_{0n0} \\ M_{001} & \cdots & M_{0n1} \\ \vdots & \ddots & \vdots \\ M_{00l} & \cdots & M_{0nl} \end{bmatrix} = \begin{bmatrix} \tilde{p}_0(0) & \tilde{p}_0(1) & \cdots & \tilde{p}_0(N-1) \\ \tilde{p}_1(0) & \tilde{p}_1(1) & \cdots & \tilde{p}_1(N-1) \\ \vdots & \vdots & \ddots & \vdots \\ \tilde{p}_l(0) & \tilde{p}_l(1) & \cdots & \tilde{p}_l(N-1) \end{bmatrix} \times \begin{bmatrix} M'_{00}(0) & M'_{01}(0) & \cdots & M'_{0n}(0) \\ M'_{00}(1) & M'_{01}(1) & \cdots & M'_{0n}(1) \\ \vdots & \vdots & \ddots & \vdots \\ M'_{00}(N-1) & M'_{01}(N-1) & \cdots & M'_{0n}(N-1) \end{bmatrix}$$

$$\begin{bmatrix} M_{100} & \cdots & M_{1n0} \\ M_{101} & \cdots & M_{1n1} \\ \vdots & \ddots & \vdots \\ M_{10l} & \cdots & M_{1nl} \end{bmatrix} = \begin{bmatrix} \tilde{p}_0(0) & \tilde{p}_0(1) & \cdots & \tilde{p}_0(N-1) \\ \tilde{p}_1(0) & \tilde{p}_1(1) & \cdots & \tilde{p}_1(N-1) \\ \vdots & \vdots & \ddots & \vdots \\ \tilde{p}_l(0) & \tilde{p}_l(1) & \cdots & \tilde{p}_l(N-1) \end{bmatrix} \times \begin{bmatrix} M'_{10}(0) & M'_{11}(0) & \cdots & M'_{1n}(0) \\ M'_{10}(1) & M'_{11}(1) & \cdots & M'_{1n}(1) \\ \vdots & \vdots & \ddots & \vdots \\ M'_{10}(N-1) & M'_{11}(N-1) & \cdots & M'_{1n}(N-1) \end{bmatrix}$$

$$\begin{bmatrix} M_{m00} & \cdots & M_{mn0} \\ M_{m01} & \cdots & M_{mn1} \\ \vdots & \ddots & \vdots \\ M_{m0l} & \cdots & M_{mnl} \end{bmatrix} = \begin{bmatrix} \tilde{p}_0(0) & \tilde{p}_0(1) & \cdots & \tilde{p}_0(N-1) \\ \tilde{p}_1(0) & \tilde{p}_1(1) & \cdots & \tilde{p}_1(N-1) \\ \vdots & \vdots & \ddots & \vdots \\ \tilde{p}_l(0) & \tilde{p}_l(1) & \cdots & \tilde{p}_l(N-1) \end{bmatrix} \times \begin{bmatrix} M'_{m0}(0) & M'_{m1}(0) & \cdots & M'_{mn}(0) \\ M'_{m0}(1) & M'_{m1}(1) & \cdots & M'_{mn}(1) \\ \vdots & \vdots & \ddots & \vdots \\ M'_{m0}(N-1) & M'_{m1}(N-1) & \cdots & M'_{mn}(N-1) \end{bmatrix}$$

By using the same algorithm, we can calculate the intensity of the object by the matrix method. The image intensity function can be written in the matrix form for 2D case as follow:

$$A = p_1^T M p_2 \tag{16}$$

where A, p_1, M and p_2 are defined above.

4 Extraction of Local Moments

In this section, we will discuss the capability of 3D weighted Hahn and Charlier moments to capture local features from a region-of-interest in an object.

4.1 Local Hahn-Based Moments

Hahn moments can be used to extract features from different locations of an object by setting the polynomial parameters. With the parameters α = β the values of

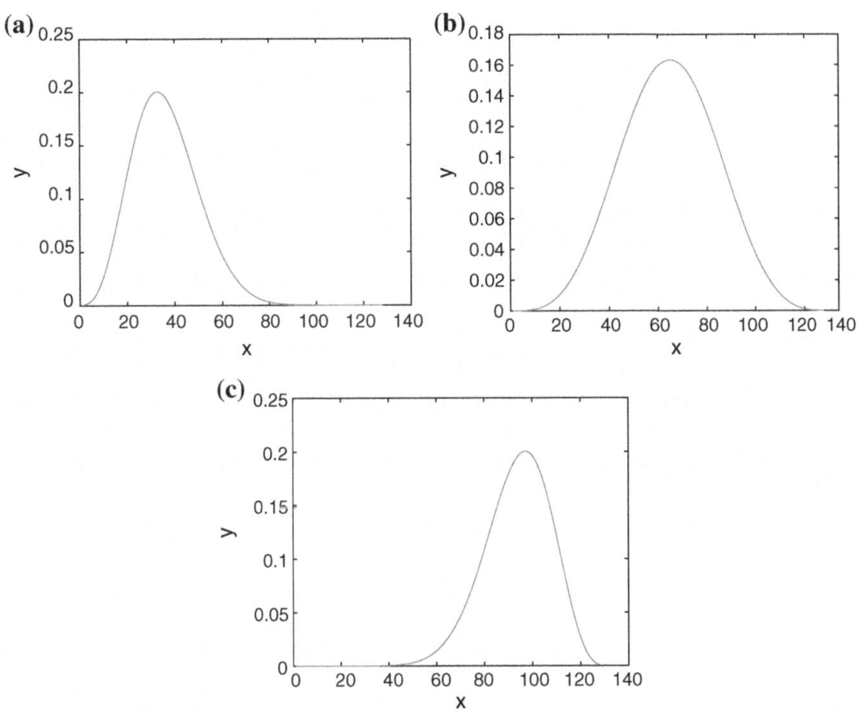

Fig. 1 Plots of weighted Hahn polynomials of one variable (N = 128), order = 0. **a** (α = 10, β = 30), **b** (α = 10, β = 10), **c** (α = **30**, β = 10)

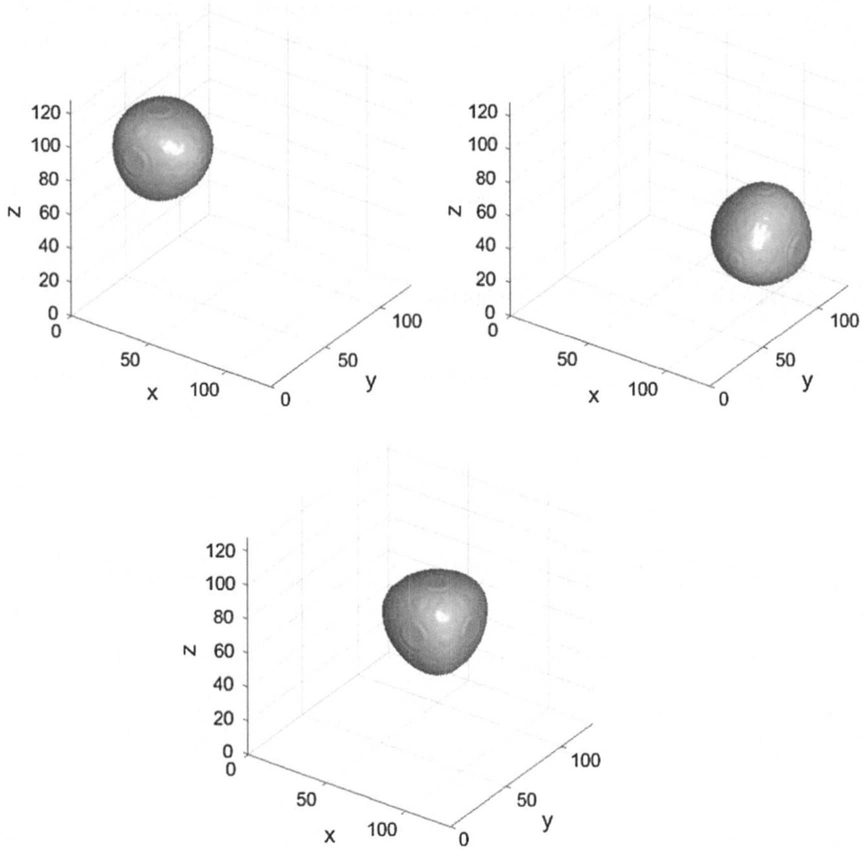

Fig. 2 Weighted Hahn polynomials in three dimensional space. ($\alpha_x = 10$, $\beta_x = 30$, $\alpha_y = 10$, $\beta_y = 30$, $\alpha_z = 30$, $\beta_z = 10$), ($\alpha_x = 30$, $\beta_x = 10$, $\alpha_y = 30$, $\beta_y = 10$, $\alpha_z = 10$, $\beta_z = 30$) $\alpha_x = 30$, $\beta_x = 10$, $\alpha_y = 10$), ($\beta_y = 30$, $\alpha_z = 10$, $\beta_z = 10$

polynomials are symmetrically distributed about the center of the x-axis as shown in Fig. 1b. In the Fig. 1a, c we observed that when $\alpha < \beta$, the ROI tend to left of the definition of the domain and vice versa. Figure 2 shows the plot of 3D weighted Hahn polynomials at different positions as determined by the parameters α and β.

4.2 Local Charlier-Based Moments

Charlier moments can be set into local feature extraction mode by adjusting the parameter a_1. Figure 2b shows that when a_1 is equal to around $N/2$, the Charlier polynomials' ROI resides in the center of the x axis. If the value of a_1 tends to N, ROI is shifted horizontally to the right of the central x value as shown in Fig. 3c, while for $a_1 < N/2$, ROI is shifted to the left as illustrate in Fig. 3a. Figure 4 shows

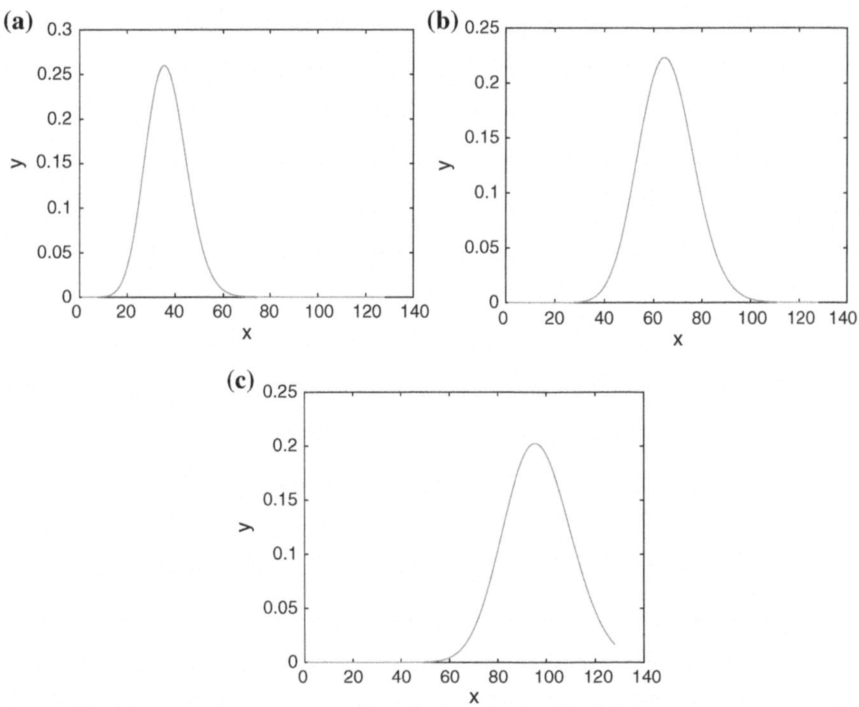

Fig. 3 Plots of weighted Charlier polynomials of one variable (N = 128), order = 0. **a** (a_1 = 30), **b** (a_1 = **64**), **c** (a_1 = 95)

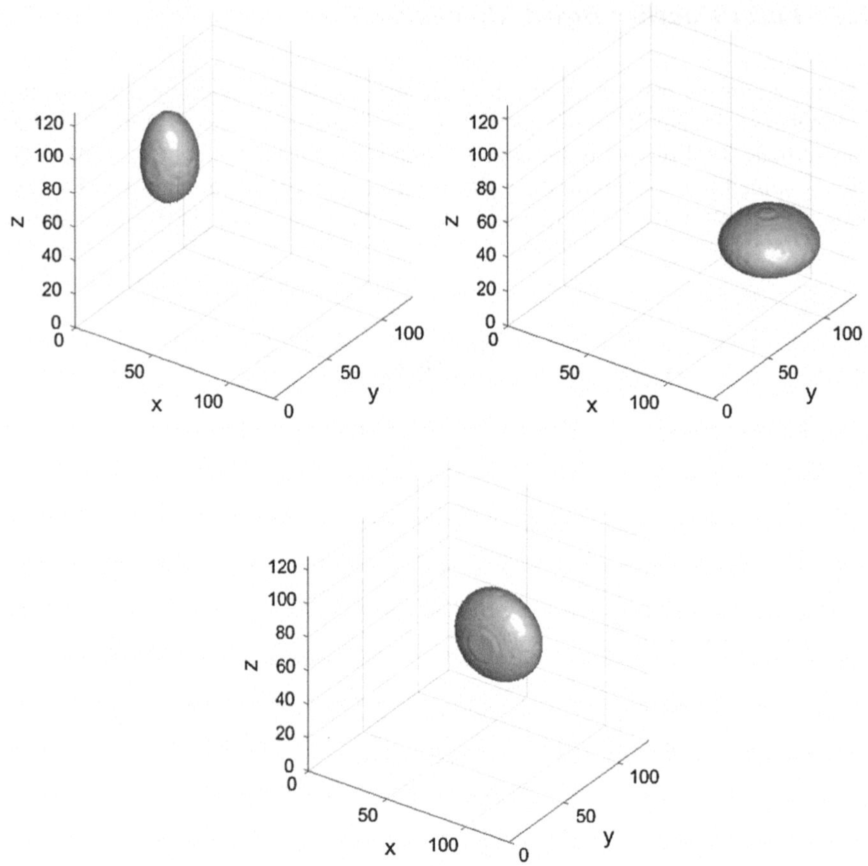

Fig. 4 weighted Charlier polynomials in three dimensional space for different values of (a_x, a_y, a_z), (35, 35, 95), (95, 95, 35), (95, 35, 95)

the plot of 3D weighted Charlier polynomials for different value of (a_x, a_y, a_z). It can be observed that the density plot moves to the region of interest corresponding to the selected value of (a_x, a_y, a_z).

5 Numerical Simulations

In this section, the theoretical framework presented in the previous sections is validated by experiment results. The ability of Hahn and Charlier moments to accurately extract local features is discussed. We also conducted a comparison between them in terms of reconstruction error. Figure 5 shows the original model of the size $128 \times 128 \times 128$ composed by eight binary volumetric images from Princeton database [22]. Each object was resized at $64 \times 64 \times 64$ voxels. Figures 2 and 4 show that the proposed moments can be used to capture the local information of an object when the parameters α, β and a_1 are set to the local feature extraction mode. By varying these parameters, the corresponding moments emphasize different regions of an object. From Fig. 6 we can observe that the extracted objects using Hahn moments show more resemblance to the original ones. However, Charlier moments performs better in some locations and lesser in the other.

Fig. 5 Original model composed by 8 objects

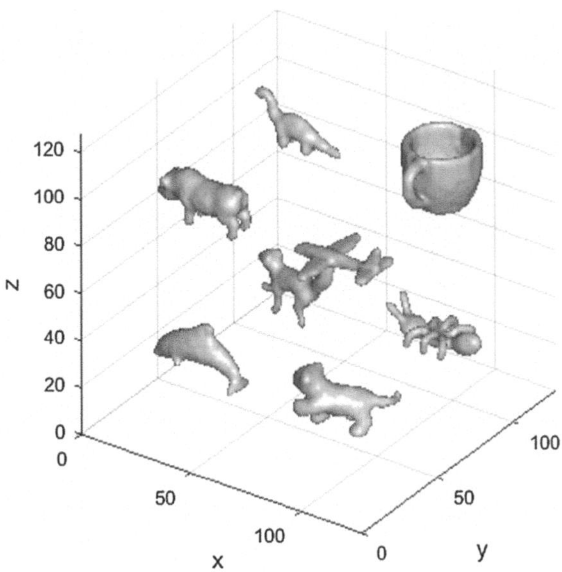

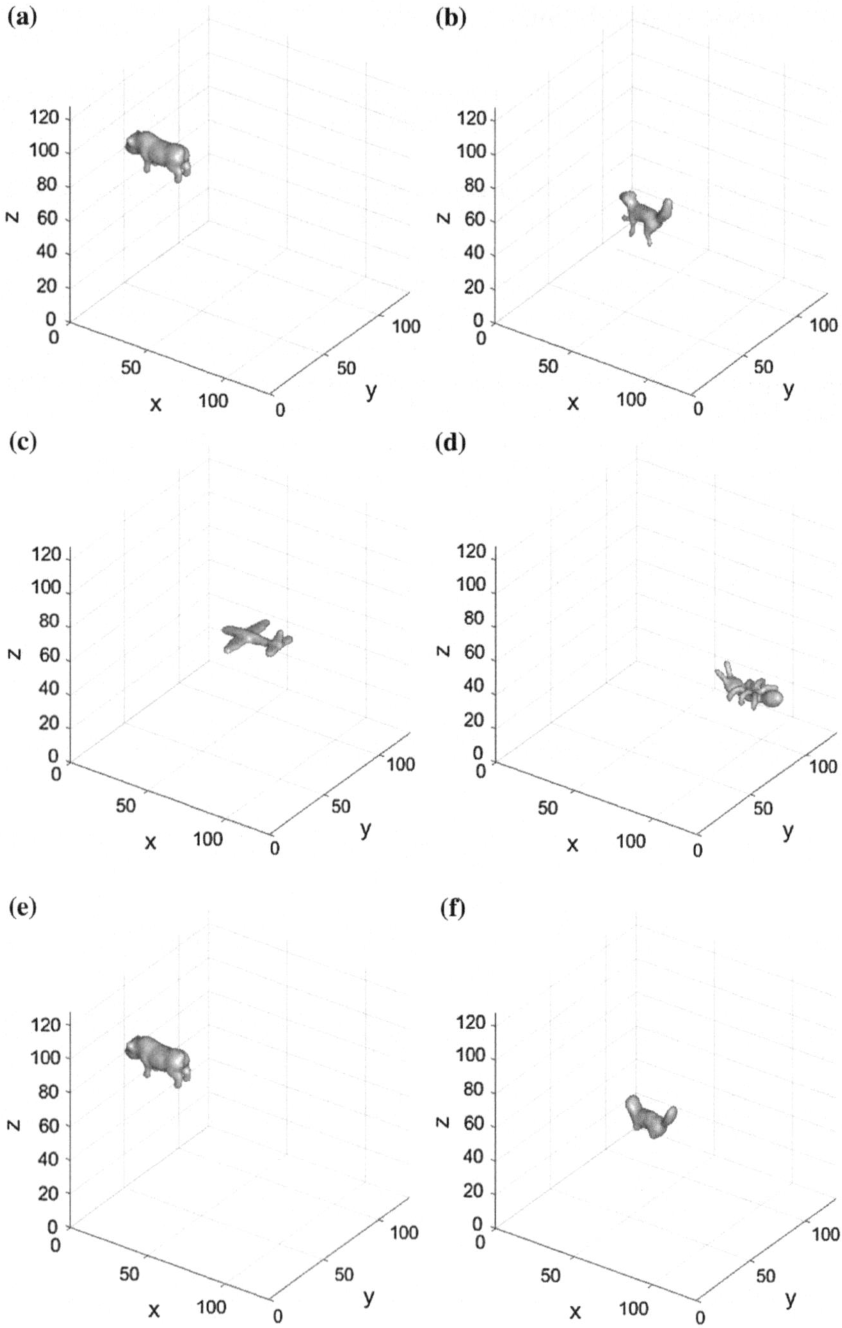
(a)
(b)
(c)
(d)
(e)
(f)
120
100
80
60
40
20
0
z
x
y
0
50
100

◀ **Fig. 6** Reconstructions of the original object using the weighted Hahn moments **a** (($\alpha_x = 50$, $\beta_x = 1000$), ($\alpha_y = 50$, $\beta_y = 1000$), ($\alpha_z = 1000$, $\beta_z = 50$)), **b** (($\alpha_x = 50$, $\beta_x = 1000$), ($\alpha_y = 1000$, $\beta_y = 50$), ($\alpha_z = 50$, $\beta_z = 1000$)), **c** (($\alpha_x = 1000$, $\beta_x = 50$), ($\alpha_y = 50$, $\beta_y = 1000$), ($\alpha_z = 1000$, $\beta_z = 50$)), **d** (($\alpha_x = 1000$, $\beta_x = 50$), ($\alpha_y = 1000$, $\beta_y = 50$), ($\alpha_z = 50$, $\beta_z = 1000$)), **i** (($\alpha_x = 1000$, $\beta_x = 50$), ($\alpha_y = 1000$, $\beta_y = 50$), ($\alpha_z = 1000$, $\beta_z = 50$)), **j** $\alpha_x = 50$, $\beta_x = 1000$), ($\alpha_y = 1000$, $\beta_y = 50$), ($\alpha_z = 1000$, $\beta_z = 50$)), **k** (($\alpha_x = 1000$, $\beta_x = 50$), ($\alpha_y = 50$, $\beta_y = 1000$), ($\alpha_z = 50$, $\beta_z = 1000$)) **l** (($\alpha_x = 50$, $\beta_x = 1000$), ($\alpha_y = 50$, $\beta_y = 1000$), ($\alpha_z = 50$, $\beta_z = 1000$)) and the Charlier moments **e** ($a_x = 50$, $a_y = 50$, $a_z = 125$), **f** ($a_x = 50$, $a_y = 125$, $a_z = 50$), **g** ($a_x = 125$, $a_y = 50$, $a_z = 125$), **h** ($a_x = 125$, $a_y = 125$, $a_z = 50$), **m** ($a_x = 125$, $a_y = 125$, $a_z = 125$), **n** ($a_x = 50$, $a_y = 125$, $a_z = 125$), **o** ($a_x = 125$, $a_y = 50$, $a_z = 50$), **p** ($a_x = 50$, $a_y = 50$, $a_z = 50$) up to order 40

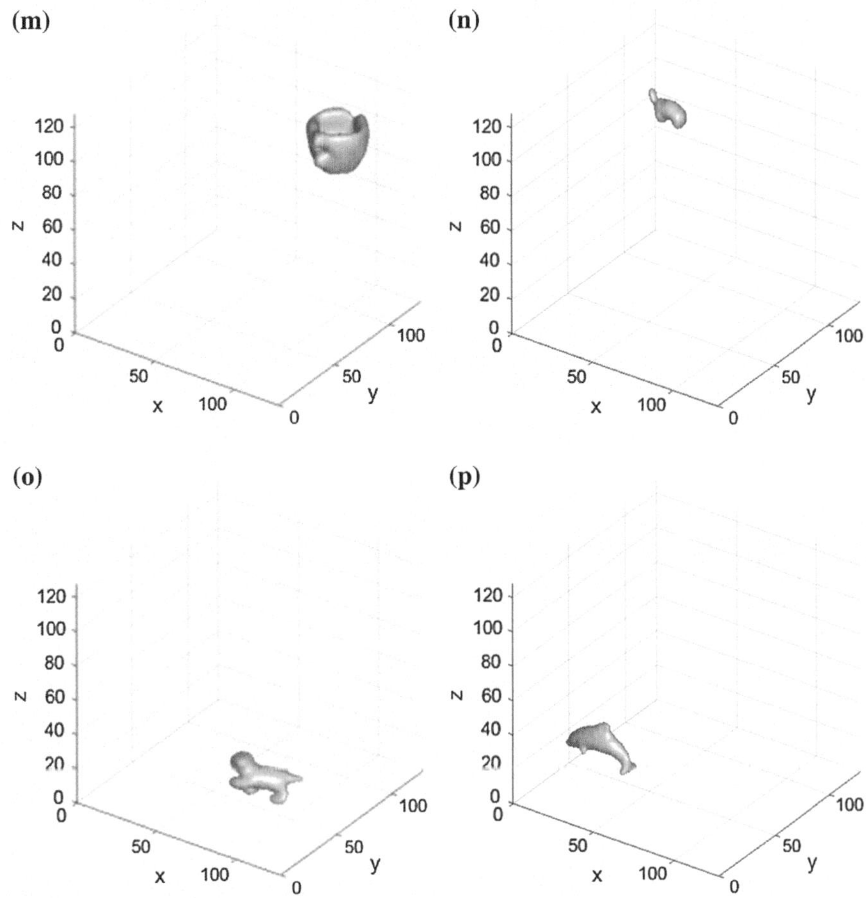

Fig. 6 (continued)

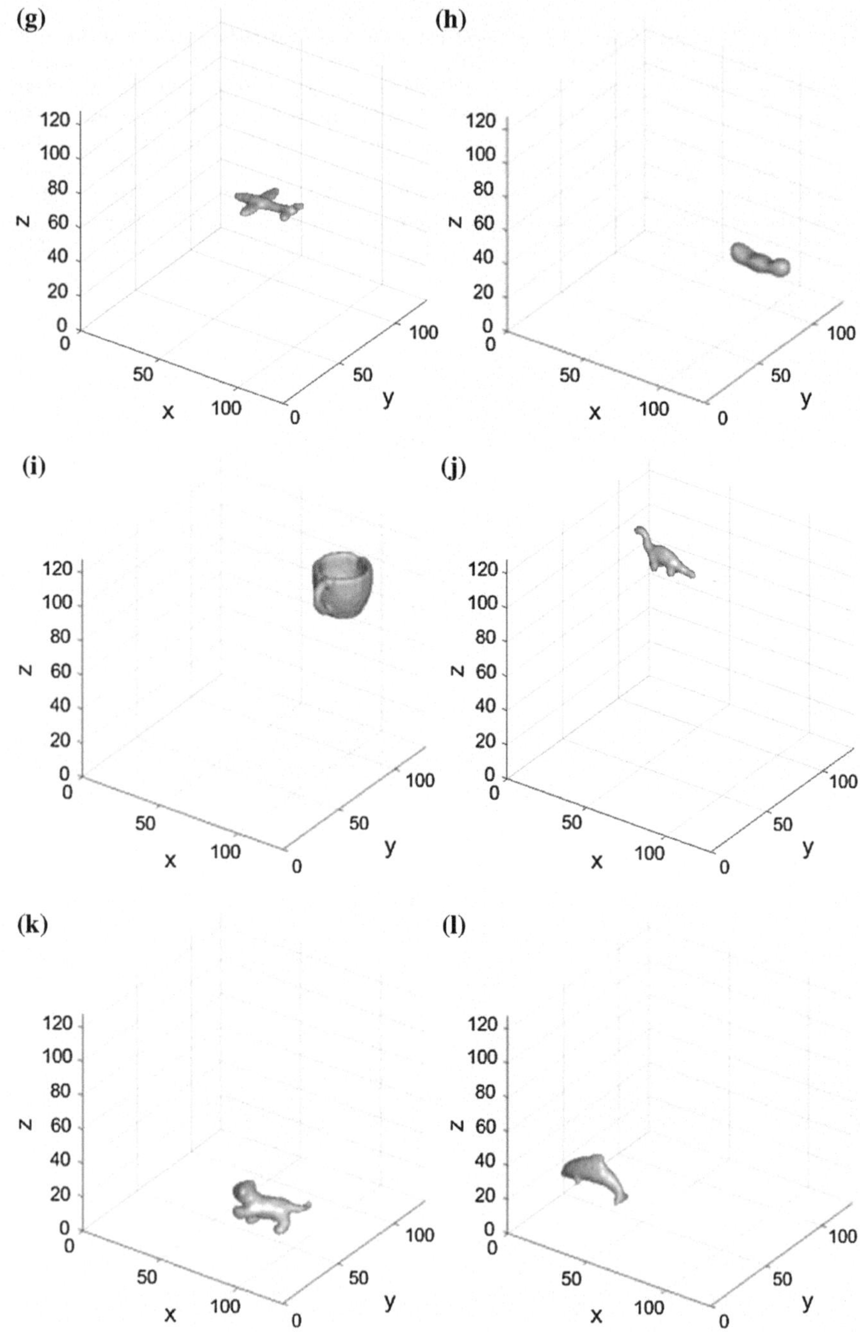

Fig. 6 (continued)

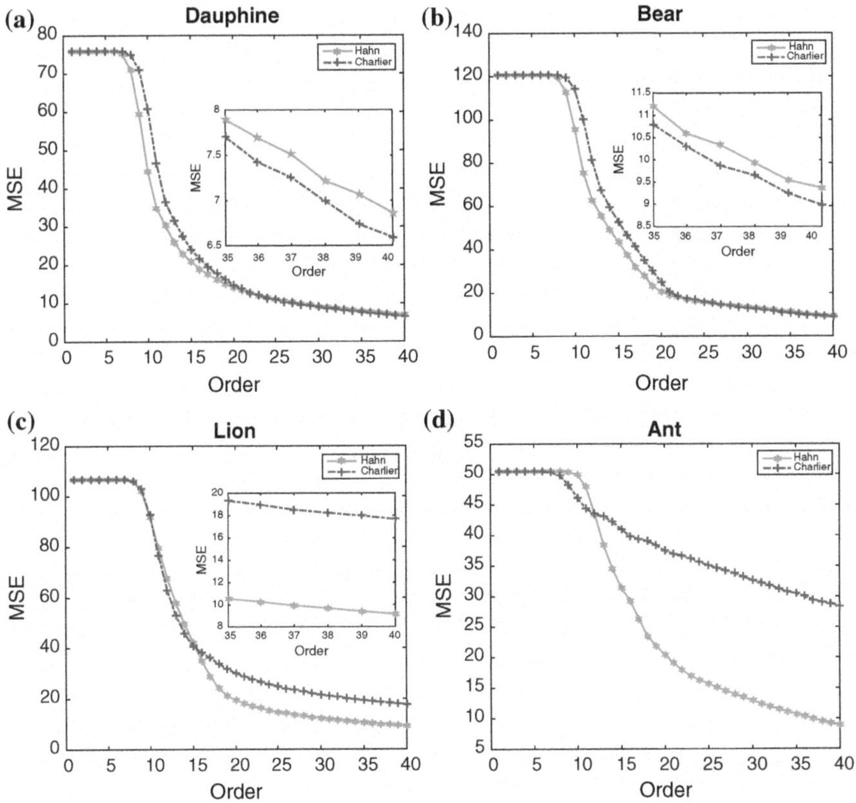

Fig. 7 Comparative study of reconstruction errors by using Hahn and Charlier moments

Figure 6h shows that it is difficult to recognize the extracted object. In order to evaluate the performance of different moments to extract local features, we use mean square error (MSE) which is defined as follows

$$MSE = \frac{1}{N^3} \sum_{x=0}^{N-1} \sum_{y=0}^{N-1} \sum_{z=0}^{N-1} \left[f(x,y,z) - \hat{f}(x,y,z)\right]^2 \tag{17}$$

where $f(x,y,z)$ and $\hat{f}(x,y,z)$ denote the original object and the reconstructed object, respectively. Figure 7 shows the detailed plots of the reconstruction error for Hahn and Charlier moments of maximum order up to 40, on different locations. As can be observed, for Hahn moments, the reconstruction error decreases monotonically with the increase of the order as predicted from Fig. 6. Figure 7a, b show that Charlie moments perform slightly better than Hahn moment in terms of reconstruction error, which is consistent with Fig. 6e, l. However, Fig. 7c, d illustrate that Charlier moments performs poorly. Figure 8 shows the graphical representation of elapsed CPU time in seconds for 3D Hahn moments computation. The experiments were

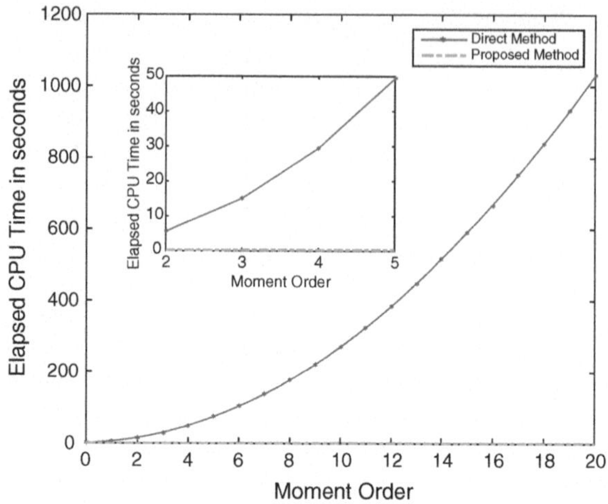

Fig. 8 Elapsed CPU time in second for 3D Hahn moments computation

performed on a laptop equipped with 2.5 GHz Intel Core i5 and 8 GB RAM. The results show that the proposed algorithm performs by far better than the classical method.

6 Conclusion

This paper investigated the capability of the Hahn and Charlier moment to extract local features from 3D object and proposes a new method for 3D moments computation by using an algorithm based on matrix multiplication. Simulated result clearly showed that the Hahn moments perform better than Charlier moments in terms of local features extraction. The elapsed CPU time is clearly reduced as compared with the classical algorithm.

References

1. Khotanzad, A., Hong, Y.: Invariant image recognition by Zernike moments. IEEE Trans. Pattern Anal. Mach. Intel. 12 **5**, 489–497 (1990)
2. Belkasim, S., Shridhar, M., Ahmadi, M.: Pattern recognition with moment invariants: a comparative study and new results. Pattern Recogn. **24**(12), 1117–1138 (1991)
3. Flusser, J., Suk, T.: Pattern recognition by affine moment invariants. Pattern Recogn. **26**(1), 167–174 (1993)
4. Hsu, H.S.: Moment preserving edge detection and its application to image data compression. Optim. Eng. **32**(7), 1596–1608 (1993)

5. Zhu, H., Shu, H., Zhou, J., Luo, L., Coatrieux, J.L.: Image Analysis by discrete orthogonal dual Hahn moments. Pattern Recogn. Lett. **28**(13), 1688–1704 (2007)
6. Hu, M.K.: Visual pattern recognition by moment invariants. IRE Trans. Inform. Theory **8**(2), 179–187 (1962)
7. Teague, M.R.: Image analysis via the general theory of moments. J. Opt. Soc. Am. **70**(8), 920–930 (1980)
8. Mukundan, R., Ong, S.H., Lee, P.A.: Image analysis by Tchebichef moments. IEEE Trans. Image Process. **10**(9), 1357–1364 (2001)
9. Yap, P.-T., Paramesran, R., Ong, S.-H.: Image analysis using hahn moments. IEEE Trans. Pattern Anal. Mach. Intell. **29**(11), 2057–2062 (2007)
10. Zhu, H., Liu, M., Shu, H., H. Zhang, H., Luo, L.: General form for obtaining discret orthogonal moments. IET Image Process. **4**(5) 335–352 (2010)
11. Fergus, R., Perona, P., Zisserman, A.: Object class recognition by unsupervised scale-invariant learning. IEEE Comput. Soc. Conf. Comput. Vis. Pattern Recogn. **2**, II-264–II-271 (2003)
12. Ke, Y., Sukthankar, R.: PCA-SIFT: a more distinctive representation for local image descriptors. In: IEEE Computer Society Conference CVPR, vol. 2, pp. II-506–II-513 (2004)
13. Mikolajczyk, K., Schmid, C.: Indexing based on scale invariant interest points. In: 8th IEEE ICCV, vol. 1. Pp. 525–531 (2001)
14. Chen, L., Feris, R., M. Turk, M: Efficient partial shape matching using Smith–Waterman algorithm. In: IEEE Comput. Soc. Conf. CVPRW, pp. 1–6 (2008)
15. Mikolajczyk, K., Schmid, C.: A performance evaluation of local descriptors. IEEE Trans. Pattern Anal. Mach. Intell. **27**(10), 1615–1630 (2005)
16. Yap, P.-T., Paramesran, R.: Image analysis by krawtcouk moments. IEEE Trans. Image Process. **12**(11), 1367–1377 (2003)
17. Broggioa, D., et al.: Comparison of organs' shapes with geometric and Zernike 3D moments. Comput. Methods Programs Biomed. **111**(3), 740–754 (2013)
18. Lin, Y-H.: 3D multimedia signal processing. In: Proceedings of the 20th ACM international conference on Multimedia., pp. 1445–1448 (2012)
19. Jiang. Y et al.: Gold nanoflowers for 3D volumetric molecular imaging of tumors by photoacoustic tomography. Nano Research **8**(7), 2152–2161 (2015)
20. Venkataramana, A., Ananth Raj, P.: Recursive computation of forward krawtchouk moment transform using clenshaw's recurrence formula. In: Third National Conference on Computer Vision, Pattern Recognition, Image Processing and Graphics (2011)
21. Ananth Raj, P., Venkataramana, A.: Fast computation of inverse krawtchouk moment transform using clenshaw's recurrence formula. In: Third National Conference on Computer Vision, Pattern Recognition, Image Processing and Graphics (2011)
22. Princeton, Princeton Shape Benchmark, http://shape.cs.princeton.edu/benchmark/ (2013)

Automatic Detection of Suspicious Lesions in Digital X-ray Mammograms

Abdelali Elmoufidi, Khalid El Fahssi, Said Jai-Andaloussi, Abderrahim Sekkaki, Gwenole Quellec, Mathieu Lamard and Guy Cazuguel

Abstract Mammography remains the most effective tool for the early detection of breast cancer, as well as the systems of computer-aided detection/diagnosis (CAD) is typically used as a second opinion by the radiologists. So, the main goal of our method is to introduce a new approach for automatic detecting the suspicious lesions in mammograms (regions of interest) for early diagnosis of breast cancer. This study has two phases: The first one is the preprocessing step and the second one is the detection of Regions of Interest (ROIs). Our method has tested with the well-known Mammography Image Analysis Society (MIAS) database and we've used Free-Receiver Operating Characteristics (FROC) to measure methods performance. The obtained experimental results show that our algorithm's performance has sensitivity of 94.75 % at 0.54 false positive per image.

Keywords Computer aided detection · FROC analysis · Image processing · Mammography · Segmentation

1 Introduction

Breast cancer is one of the most typical sorts of cancer in worldwide, and the most frequent cancer in women. From where, It contributing to the rise in mortality among women worldwide. Recent statistics have shown that one women in eight within the

A. Elmoufidi (✉) · K. El Fahssi · S. Jai-Andaloussi · A. Sekkaki
Faculty of Sciences, Department of Mathematics and Computer Sciences,
Hassan II University of Casablanca, Casablanca, Morocco
e-mail: Abdelali.Elmoufidi09@univcasa.ma

G. Quellec · M. Lamard · G. Cazuguel
Inserm, UMR 1101, 29200 Brest, France

M. Lamard
Univ Bretagne Occidentale, 29200 Brest, France

G. Cazuguel
Institut Mines-Telecom, Telecom Bretagne, UEB, Dpt ITI, 29200 Brest, France

R. El-Azouzi et al. (eds.), *Advances in Ubiquitous Networking 2*, Lecture Notes in Electrical Engineering 397, DOI 10.1007/978-981-10-1627-1_29

United States and one in ten women in Europe develop breast cancer throughout their lifetime [1, 2]. For the fight against this serious disease, Early detection of breast cancer is the most important factor affecting possibility of recovery from the disease. For it, Mammography is an X-ray technique has been developed specifically for soft tissue radiography of breast, and remains the best and most exact tool for early detection of breast cancer [2, 3]. It based on the differential absorption of X-rays between the various tissues components of the breast such as fat tissue, fibroglandular tissue; extremely dense tissue (can be tumor), This appearance described "Breast density". Breast density contains lobular elements, ducts and fibrous connective tissue of the breast [4]. The sturdy relationship between breast density and also the risk of developing carcinoma was planned, first off by Wolfe [5]. It was confirmed by other researchers, like Karssemeijer [6], Boyd et al. [7]. Breast Imaging-Reporting and Data System of American College of Radiology (ACR BI-RADS), aims at providing a standardized classification system for reporting mammographic breast densities [8]. ACR BI-RADS 4th Edition classifies breast density into four major categories: (1) The breasts are almost entirely fatty; (2) the breasts scattered areas of fibroglandular density; (3) The breasts are heterogeneously dense; (4) The breasts are extremely dense.

In order to improve the accuracy of interpreting mammogram images, a variety of CAD systems that perform computerized mammogram analysis have been proposed. These CAD systems used as a supplement to the radiologists' assessment and their role in modern medical practice considered to be important and significant in the early detection and diagnosis of breast cancer.

Generally, the procedure to develop a CAD system for detection of suspicious regions within mammograms takes place in two phases: The first one is a Computer-aided detection (CADe) contains two steps: (1) preprocessing step, (2) Image Analysis. And the second one is a Computer-aided diagnosis (CADx) also contains two steps: (1) Extraction and selection of features of ROIs, (2) the Classification of ROIs detected in the first phase [9, 10].

In this paper, we've proposed fully automatic and robust algorithm for detecting suspicious lesions in a mammogram. Firstly, we've started by preprocessing step. Secondly, we've detected the regions of interest (ROIs). The proposed algorithm is a very accurate technique for detecting of breast cancer by using mammogram images. The obtained qualitative and quantitative results prove the efficiency of this method and confirm possibility of using it in improving the CAD system.

Paper organization: The setup of the paper is organized as follows: An introduction is given in Sect. 1; Sect. 2 discusses related work; Sect. 3 presents materials and method Sect. 4 describes our proposed research; The results and performance are presented in Sect. 5; Sect. 6 includes a conclusion; References are given at the end.

2 Related Work

Many methods have been proposed for detection/diagnosis of abnormalities in mammogram, Such as: Statistical methods [11], methods based wavelets [12, 13], Markov models [14] and fuzzy set theory. In addition, Several researches have published about computer aided breast cancer detection and diagnosis. i.e., Ganesan et al. [15] provided an overview about recent developments and advances in the field of Computer-Aided Diagnosis (CAD) of breast cancer using mammograms. Veta et al. [16] presented an overview of methods that have been proposed for the analysis of breast cancer histopathology images. Jalalian et al. [9] presented the approaches which they applied a method to develop a CAD systems on mammography and ultrasound images. Detection of regions of interest (ROIs) is a capital step in a development CADe system. Hence, a number of researchers have published on segmentation of breast tissue regions according to differences in density and texture. For example, Elmoufidi et al. [1] proposed a method for dynamically and automatically segmented the different breast tissue in mammogram based on the breast density. Elmoufidi et al. [17, 18] proposed a method for Detection of Regions of Interest in Mammograms by Using Local Binary Pattern, Dynamic K-Means Algorithm and Gray Level Co-occurrence Matrix.

3 Materials and Method

To develop and check the proposed method we've used the Mammographic Image Analysis Society (MIAS) database [19]. In addition, for implementing our method, we've used the Seed Based Region Growing (SBRG) techniques and Local Binary Pattern (LBP): SBRG algorithm to remove the pectoral muscle and LBP to detect the regions of interest. They're two simples methods of segmentation and better choice for easy implementation and applying them on a larger dataset.

3.1 Database

In this paper, to develop and test the proposed method we've used the Mammographic Image Analysis Society (Mini-MIAS) database [19]. Mammograms have a size of 1024 × 1024 pixels in Portable Greymap (PGM) format, and resolution of 200 micron. Each pixel in the mammograms represented as an 8-bit word, where the images are in grayscale format with a pixel intensity of range [0, 255]. This database composed of 322 mammograms of right and left breast, from 161 patients, where 52 mammograms have diagnosed as malignant, 63 benign and 207 normal. In addition, MIAS database provides appropriate details, for example, character of background tissue, severity of the abnormality, image-coordinates of center of abnormality, and approximate radius (in pixels) of a circle enclosing the abnormality.

3.2 Seed Region Growing (SBRG)

Seed region growing (SBRG) algorithm for segmentation introduced by Adams et al. [20] is a simple method of segmentation which is rapid and free of tuning parameters. SBRG algorithm is a better choice for easy implementation and applying it on a larger dataset.

Seeded region growing approach to image segmentation is to segment an image into regions with respect to a set of n seeds as presented in [21] discussed here.

3.3 Local Binary Pattern (LBP)

Local Binary Pattern (LBP) operator combines the characteristics of statistical and structural texture analysis. The LBP operator is used to perform gray-scale invariant two-dimensional texture analysis. The LPB operator labels the pixel of an image by Thresholding the neighborhood (i.e. 3 × 3) of each pixel with the center value and considering the result of this Thresholding as a binary number [15, 22]. When all the pixels have been labeled with the corresponding LBP codes, histogram of the labels are computed and used as a texture descriptor. Formally, given a pixel at (x_c, y_c), the resulting LBP can be expressed in decimal form as follows:

$$LBP_{P,R}(x_c, y_c) = \sum_{P=0}^{P=1} S(i_p - i_c)2^P \tag{1}$$

where i_c and i_p are, respectively, gray-level values of the central pixel and P surrounding pixels in the circle neighborhood with a radius R, and function s(x) is defined as:

$$S(x) = \{^{1,x\geq 0}_{0,x<0} \tag{2}$$

4 Our Proposed Research

In this paper, we've implemented a new method for automatic detecting suspicious lesions in mammograms. Our method planned in two major blocks, namely: (1) pre-processing, (2) detection of regions of interest (ROIs). In the first block, we've applied a preprocessing of the mammogram. This block performs three steps are: (1) Labels removal, (2) artifact and digital noise suppressed, (3) remove pectoral muscle and additional background, contrast enhancement. In the second block, we've detected ROIs by using Local Binary Pattern (LBP). The details about (LBP) discussed above.

One among the novelties of our algorithm, that we've proposed a new technique to detect all suspected areas (not just lesions) in mammograms and considered as regions of interest (ROIs) thereafter. In addition, our algorithm is able to detect the different objects within mammogram: the masses, the classifications and the micro-classifications. The obtained quantitative and qualitative results demonstrate the efficiency of this method and confirm possibility of its use in improving the CAD system.

4.1 Preprocessing

Preprocessing: the purpose of this stage is to prepare images for the next stage of operations, such as detection of regions of interest and classification. Images as acquired by the mammography could have deficiencies such as artifact, noise, blurs, etc. [10]. So, computer image preprocessing techniques have applied to enhance quality of mammograms. In this stage, the aim is to extract only the breast profile region without labels, artifact, noise, pectoral muscle and additional background. Firstly, a threshold value used to remove the labels. Secondly, we've used median filtering two-dimensional (2D) in a 3-by-3 neighborhood connection to remove artifact and noise. Thirdly, we've used region growing technique to remove the pectoral muscle. In addition, the mammogram is usually basically low contrast [1], therefore a step of enhancement of contrast has applied (see Figs. 1 and 2).

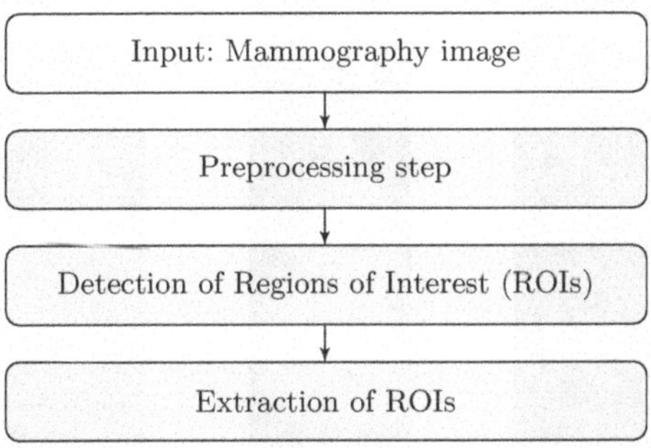

Fig. 1 Organization chart of our proposed method

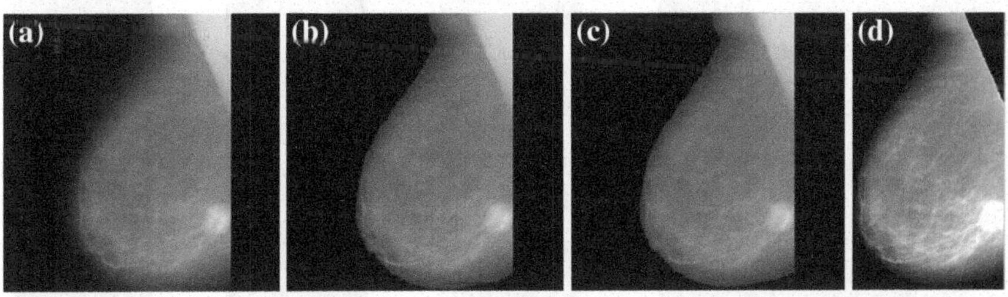

Fig. 2 Mammogram preprocessing step: **a** Original mammogram, **b** Label suppressed, **c** Artifact and noise removal, **d** Remove of pectoral muscle and additional background, contrast enhancement

4.2 Detection of Regions of Interest (ROIs)

Detection of regions of interest (ROIs) is a capital step in developing a CAD system, and detecting several false positive causes a weak system. For that, we've considered a suspicious lesion correctly detected if its area overlapped by at least of 75 % from ground truth. To perform this task, we've implemented the Local Binary Pattern (LBP) algorithm.

Experimental results of detection phase

Example 1 The normals mammograms. Figure 3.

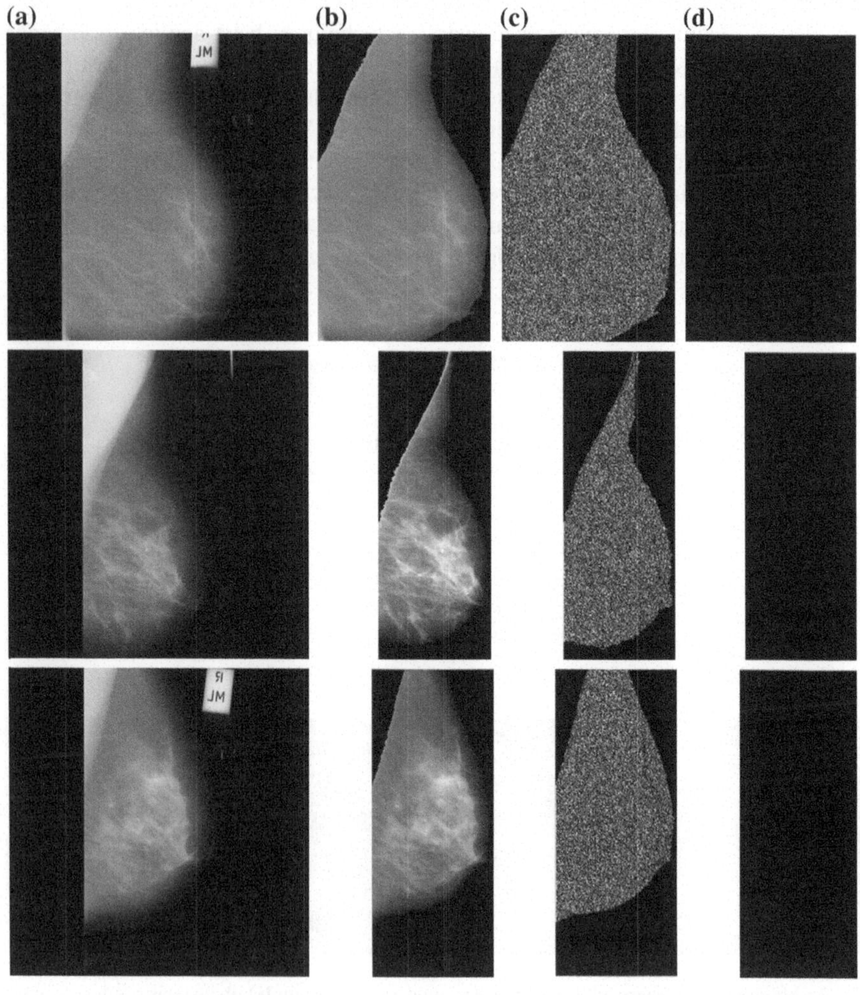

Fig. 3 Detection of ROIs: **a** Original mammograms, **b** Mammograms after preprocessing step, **c** LBP has applied, **d** Any regions of interest have detected

Example 2 Mammograms, which contains a single suspicious lesion has been correctly detected without false positive. Figure 4.

Example 3 Mammograms, which contains a single suspicious lesion has been correctly detected with another ROI as false positive. Figure 5.

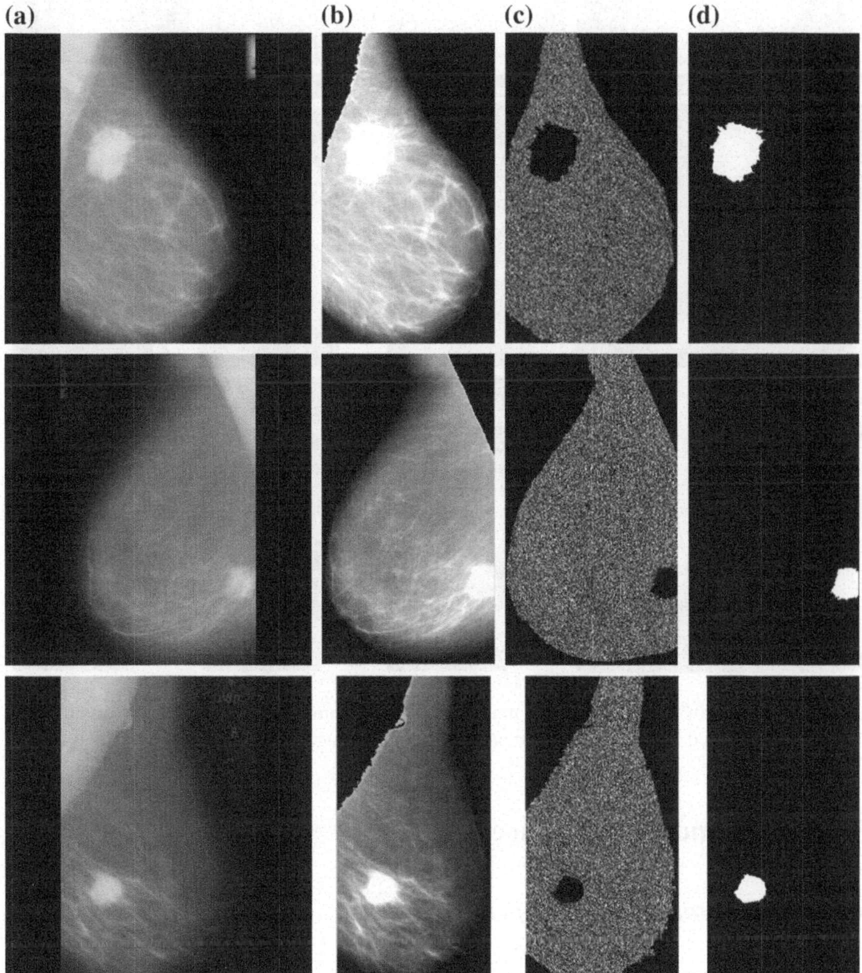

Fig. 4 Detection of ROIs: **a** Original mammograms, **b** Mammograms after preprocessing step, **c** LBP has applied, **d** Regions of interest have correctly detected without any false positive

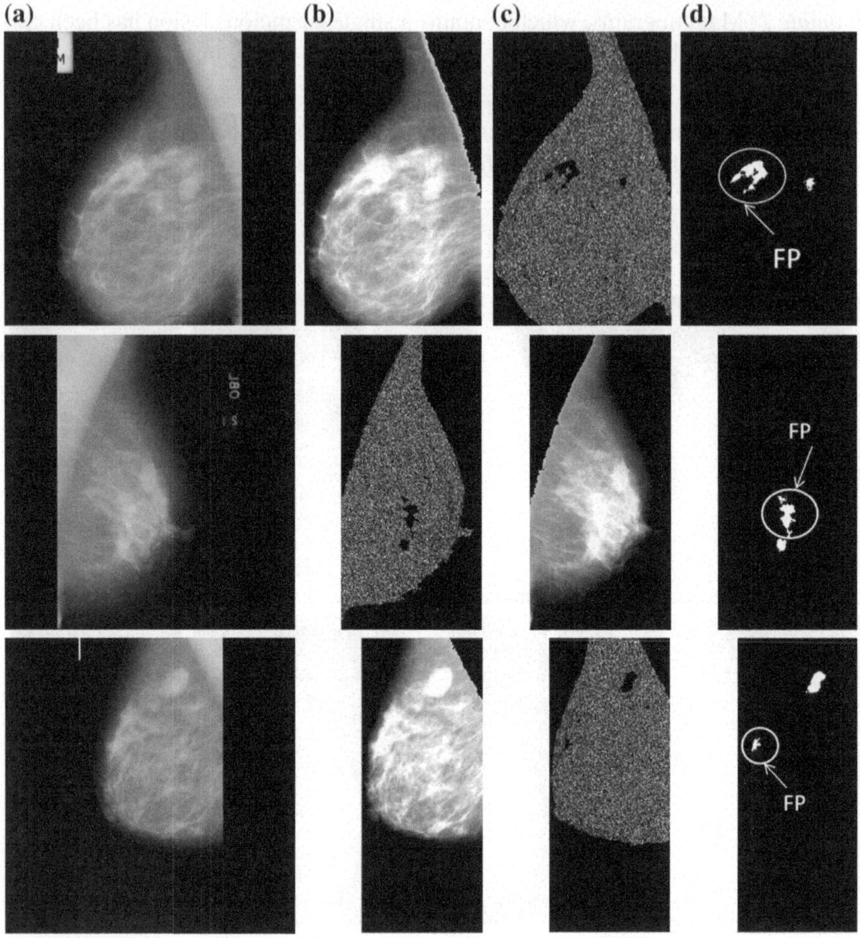

Fig. 5 Detection of ROIs: **a** Original mammograms, **b** Mammograms after preprocessing step, **c** LBP has applied, **d** Regions of interest have detected with another ROIs as false positive

5 Results and Performance

5.1 Performance Detection Evaluation

Detected of suspicious regions in mammograms is a crucial step in a CADe system and detection of a maximum portion of the true lesion is necessary because geometric features are very important for further true lesion detection, for that we've considered a correct lesion detected if its area overlapped by at least of 75 %. We've obtained a good detection result, i.e., 100 %, for MISC and 95.45 %, for CIRC. The obtained detection result of SPIC (89.47 %) is relatively reliable, because just not all

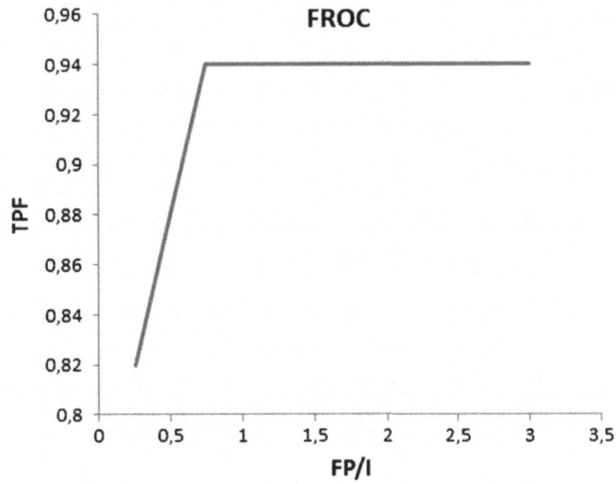

Fig. 6 Plot illustrating FROC *curve*

the overlapping of SPIC obtained with high accuracy, hence, the detection results are not in high accuracy. Generally, we've obtained sensitivity of 94.75 % at 0.54 False Positive per Image in the detection stage. To evaluate our method, we've used Free-Receiver Operating Characteristics (FROC) for drawing the FROC curve, representing the True positive Fraction (TPF) according False Positive per Image (FP/I) see Fig. 6 (Table 1):

$$False\ Positive\ per\ Image = \frac{Number\ of\ false\ positive}{Number\ of\ image} \tag{3}$$

Table 1 The details of obtaining results regrouped by average in percentage of each classes of anomaly

Class of abnormality present	Number of images	Sensitivity (%)
Normal	207	96.2
CIRC—Circumscribed masses	23	95.45
SPIC—Spiculated masses	19	89.47
ARCH—Architectural distortion	19	94.73
ASYM—Asymmetry	15	93.3
MISC—Other, ill-defined masses	14	100
CALC—Calcification	25	94.15
Total	322	94.75

Table 2 The performance obtained comparison with papers published recently

Authors	Method used	Accuracy (%)
Hu et al. [2]	Detection of suspicious lesions by adaptive thresholding based on multiresolution analysis in mammograms	91.3
Veena et al. [23]	CAD Based System for Automatic Detection and Classification of Suspicious Lesions in Mammograms	92.13
Our method	Automatic Detection of Suspicious Lesions in Digital X-ray Mammograms	94.75

5.2 The Comparison of Performance of Our Method with Papers Recently Published

We've considered as a correct lesion detected if its area overlapped al last 75 %, and we've obtained sensitivity of 94.75 in the detection stage of our algorithm at 0.54 false positive per image (Table 2).

6 Conclusion and Future Work

In this paper, an algorithm for breast mass detection implemented under the MATLAB environment for automatic detecting of suspicious lesions in mammograms. The obtained results demonstrate the efficiency of this method and comparable to other solutions. Our proposed algorithm can contribute to solving the main problem in mammography image processing such as: detection of the masses and the calcifications. Performance of our algorithm has evaluated by using Free-Receiver Operating Characteristics (FROC). The experimental results show that the detection phase has sensitivity of 94.75 % at 0.54 false positive per image. In the future work, we are going to generate and extract the features of ROIs and classifying them to Normal/Abnormal and Benign/Malignant.

References

1. Elmoufidi, A., et al.: Automatically density based breast segmentation for mammograms by using dynamic K-means algorithm and seed based region growing. In: I2MTC, et al.: International Instrumentation and Measurement Technology Conference, PISA, ITALY, 11–14 May 2015 (2015)
2. Hu, K., et al.: Detection of suspicious lesions by adaptive thresholding based on multiresolution analysis in mammograms. IEEE Trans. Instrum. Meas. **60**(2), 462–472 (2010)
3. Ferrero, A., et al.: Uncertainty evaluation in a fuzzy classifier for microcalcifications in digital mammography. In: I2MTC 2010—International Instrumentation and Measurement Technology Conference Austin, TX, 3–6 May 2010

4. Saidin, N., et al.: Density based breast segmentation for mammograms using graph cut and seed based region growing techniques. In: 2nd International Conference on Computer Research and Development (ICCRD'10), Kuala Lumpur, Malaysia, pp. 246–250, May 2010
5. Wolfe, J.N.: Risk for breast cancer development determined by mammographic parenchymal pattern. Cancer **37**(5), 2486–2492 (1976)
6. Karssemeijer, N.: Automated classification of parenchymal patterns in mammograms. Phys. Med. Biol. **43**, 365–378 (1998)
7. Boyd, N.F., et al.: Quantitative classification of mammographic densities and breast cancer risk: results from the Canadian National Breast Screening Study. J. Natl. Cancer Inst. **87**(9), 670–675 (1995)
8. American College of Radiology. American College of Radiology Breast Imaging Reporting and Data System (BIRADS), 4th edn. American College of Radiology, Reston, VA (2003)
9. Jalalian, A., et al.: Computer-aided detection/diagnosis of breast cancer in mammography and ultrasound. Clin. Imaging **37**, 420426 (2013)
10. Shirmohammadi, S., Ferrero, A.: Camera as the instrument: the rising trend of vision based measurement. IEEE Instrum. Meas. Mag. **17**(3), 41–47 (2014)
11. Chan, H., et al.: Computerized analysis of mammographic micro calcifications in morphological and feature spaces. Med. Phys. **25**(10), 2007–2019 (1998)
12. Mencattini, A., et al.: Mammographies images enhancement and denoising for breast cancer detection using dyadic wavelet processing. IEEE Trans. Instrum. Meas. **57**, 1422–1430 (2007). doi:10.1109/TIM.915470
13. Wang, T., Karayiannis, N.: Detection of microcalcification in digital mammograms using wavelets. IEEE Trans. Med. Imaging **17**(4), 498–509 (1998)
14. Li, H., et al.: Marcov random field for tumor detection in digital mammography. IEEE Trans. Med. Imaging **14**(3), 565–576 (1995)
15. Ganesan, K., et al.: Computer-aided breast cancer detection using mammograms. IEEE Rev. Biomed. Eng. **6** (2013)
16. Veta, M., et al.: Breast cancer histopathology image analysis: a review. IEEE Trans. Bio-med. Eng. **61**(5), 140011 (2014)
17. Elmoufidi, A., et al.: Detection of regions of interest in mammograms by using local binary pattern, dynamic k-means algorithm and gray level co-occurrence matrix. In: Fifth International Conference on Next Generation Networks and Services (NGNS'14) 28–30 May 2014, Casablanca, Morocco (2014)
18. Elmoufidi, A., et al.: Detection of regions of interest in mammograms by using local binary pattern and dynamic K-means algorithm. Int. J. Image Video Process. Theor. Appl. **1**(1), 30 Apr 2014. ISSN: 2336-0992
19. Suckling, J., et al.: The Mammographic Image Analysis Society digital mammogram database. In: Exerpta Medica, International Congress Series, vol. 1069, pp. 375–378 (1994)
20. Adams, R., Bischof, L.: Seeded region growing. IEEE Trans. Pattern Anal. Mach. Intell. **16**(6), 641–647 (1994)
21. Harikrishna Rai, G.N., et al.: Gradient based seeded region grow method for CT angiographic image segmentation. Int. JRI Comput. Sci. Netw. 1(1) (2009)
22. Huang, D., et al.: Local binary patterns and its application to facial image analysis: a survey. IEEE Trans. Syst. Man Cybern. Part C Appl. Rev. **41**(6) (2011)
23. Veena, et al.: CAD based system for automatic detection et classification of suspicious lesions in mammograms. Int. J. Emerg. Trends et Technol. Comput. Sci. (IJETTCS), **3**(4) (2014). ISSN: 2278-6856

Smart Antenna System Using Butler Matrix Based Beamforming Network for X Band Applications

Hayat Errifi, Abdennaceur Baghdad, Abdelmajid Badri and Aicha Sahel

Abstract This paper presents the optimum design of a 4 × 4 planar Butler matrix array as a key component of a switched beam smart antenna system operating at 10 GHz for X band applications. In 4 × 4 Butler matrix, four input ports are used for the input signal connections and four output ports can be connected to an array of four micro-strip edge feed patch antennas to form the beamforming network. Such a network is capable of production of four orthogonal uniform beams (at −50°, −10°, 10° and 50°) with effective coverage over 120°, when feed with electromagnetic signals. Conception details, and simulation results are also given for the components (hybrid coupler, crossover, phase shifter) used to implement the matrix. Finally, the simulation results using Ansoft HFSS show a great improvement of Gain and HPBW of the four generated orthogonal beams.

Keywords Beamforming network · Butler matrix · Microstrip patch antenna array

1 Introduction

In a high-speed Wireless communication, it becomes a necessary to separate desired signal from delay or interference signal. Thus to overcome these problems, smart antenna systems have been developed to be able to offer the solution [1–4]. Now Smart antenna [5] is one of the most promising technologies that will enable a higher capacity in wireless networks by effectively reducing multipath and channel interference. This is achieved by focusing the radiation only in the desired direction and adjusting itself to changing traffic conditions or signal environments.

The Butler matrix is one of passive beamforming network consisting of N input/output port and N input/output antenna elements produced N principal

H. Errifi (✉) · A. Baghdad · A. Badri · A. Sahel
EEA&TI Laboratory, Faculty of Sciences and Technology, Electrical Engineering Department, Hassan II Casablanca University, 28800 Mohammedia, Morocco
e-mail: errifi.hayat@live.fr

R. El-Azouzi et al. (eds.), *Advances in Ubiquitous Networking 2*, Lecture Notes in Electrical Engineering 397, DOI 10.1007/978-981-10-1627-1_30

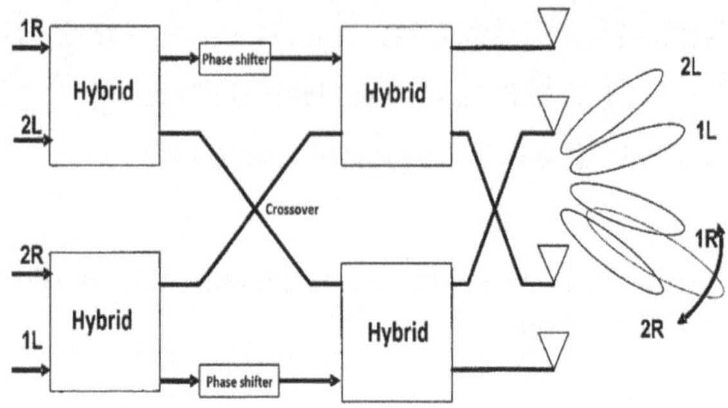

Fig. 1 The geometry and the desired beam-set of a four-element phased array

orthogonal beam at different location [1]. Butler matrix is easy to implement using the microstrip technique due to numerous advantages such as low profile and low cost [6, 7].

The conceptual operation of the matrix is described briefly as follows. First, the Radio Frequency (RF) signal excites each of the input ports, and, then, the signal goes through the output ports, thereby feeding the array elements. Then, the signal is distributed equally with a constant phase between them. As a result, beam radiations are generated at a certain angle. Figure 1 shows the topology of the 4 × 4 Butler matrix. The beam direction is illustrated with respect to each input port. Therefore, by feeding any of input port, user can select the direction of the radiation main beam as desired. In this paper, the design of 4 × 4-Butler matrix in X-band is presented. The design consists of four 90° Hybrids, two-crossovers and two-phase shifters. The butler matrix is then used to feed four micro-strip edge feed patch antennas [8]. The design and simulation are achieved using High Frequency Structures Simulator software [9].

2 Design and Simulation of 4 × 4 Butler Matrix

The design and simulation of the Butler matrix, operating at 10 GHz requires the study and characterization of its different components. They must be efficient with the minimum losses [10]. At first, the important mathematical calculations for finding the dimensions of all the individual components is done using MATLAB and then they are designed and simulated using HFSS. After getting good results, all the individual components are combined on a single substrate to implement the Butler matrix.

2.1 Choice of Substrate

RT/Duroid 5880 is used for the substrate of all matrix Butler components. Table 1 lists its characteristics. The choice of the substrate is based on the ease of availability, low cost and low radiation losses.

Table 1 Characteristics of the substrate

Parameter	
Substrate	RT Duroid 5880
Relative permittivity	2.2
Loss tangent	0.0027
Substrate height	0.79 mm

2.2 *Hybrid Coupler*

The most important element in the Butler matrix is the hybrid coupler [11]. The general structure of this type of coupler is defined in Fig. 2. This structure is an octopole allowing the division of an input signal into two output signals of equal amplitude and 90° phase shift at the frequency of operation. The main line of the coupler is coupled to a secondary line by two-quarter wavelength long sections spaced over one quarter wavelength [12]. Typically, the input is at port 1 and the output ports 2 and 3 while the isolated port 4 is terminated in a match load. In this study, 90° hybrid is designed by two ($Z_0 = 50\ \Omega$) and two ($Z_0 = 35\ \Omega$) transmission lines.

The simulated magnitudes (S_{11}, S_{12}, S_{13}, and S_{14}) for the hybrid coupler are −28 dB, −4 dB,−4 dB and −28 dB respectively as shown in Fig. 3. As expected, the phase difference between port 2 and port 3 is 87° (Fig. 4).

The simulation results of the coupler are encouraging. They have an equiamplitude and quadrature phase in the output. Therefore, it has good impedance matching.

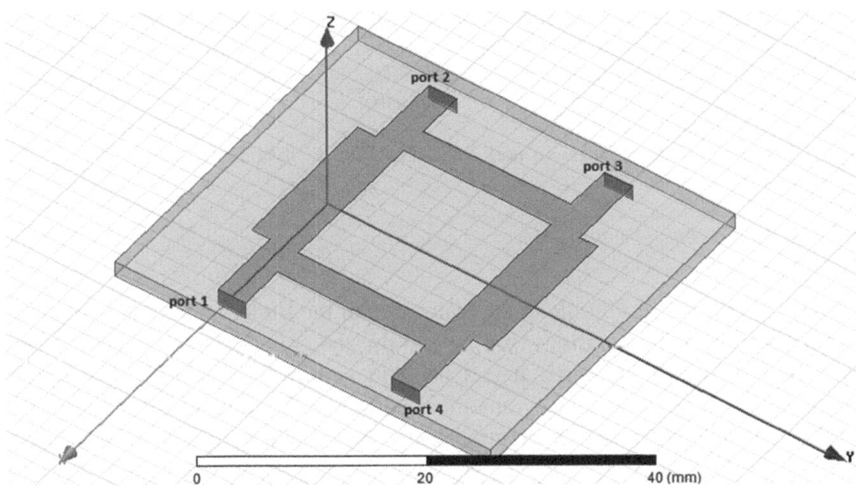

Fig. 2 Structure of hybrid coupler

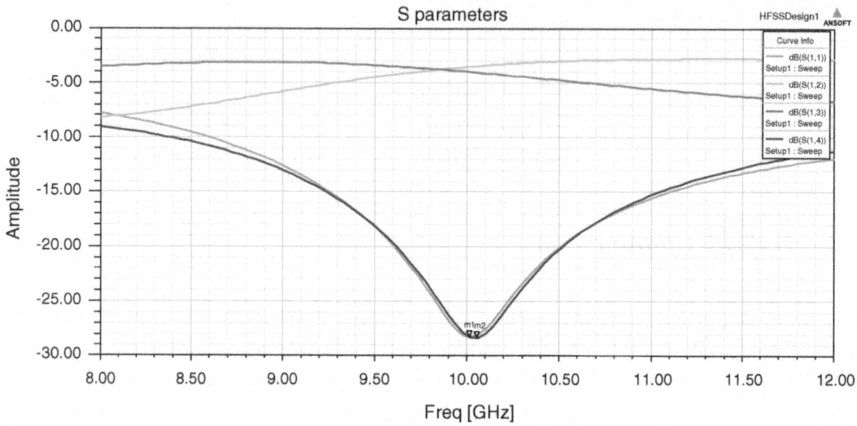

Fig. 3 S parameters versus frequency for the hybrid coupler

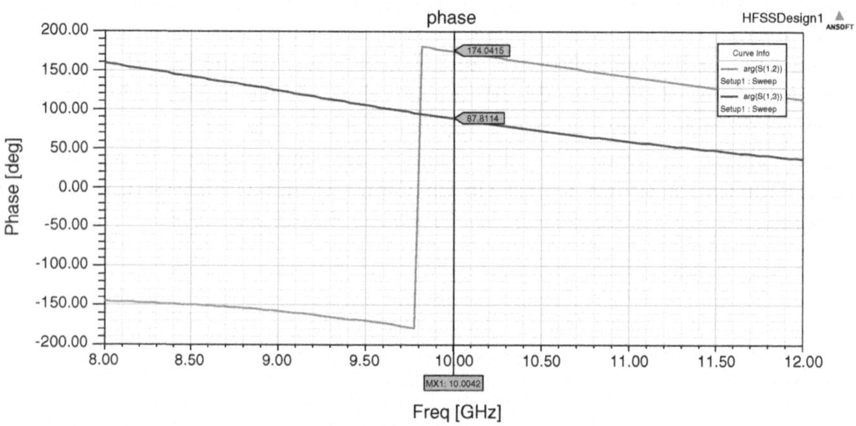

Fig. 4 The phase difference between port 2 and port 3 for the hybrid coupler

2.3 *Crossover*

Crossover is the biggest hurdle in the realization of the Butler matrix [11]. To avoid overlapping signals at crossings, we have to use the crossover with good isolation level between the input ports. It can be built simply by cascading two hybrid couplers [12]. It has also four symmetrical ports with 2 inputs and 2 outputs. The perfect design of crossover is accomplished if every adjacent ports are isolated.

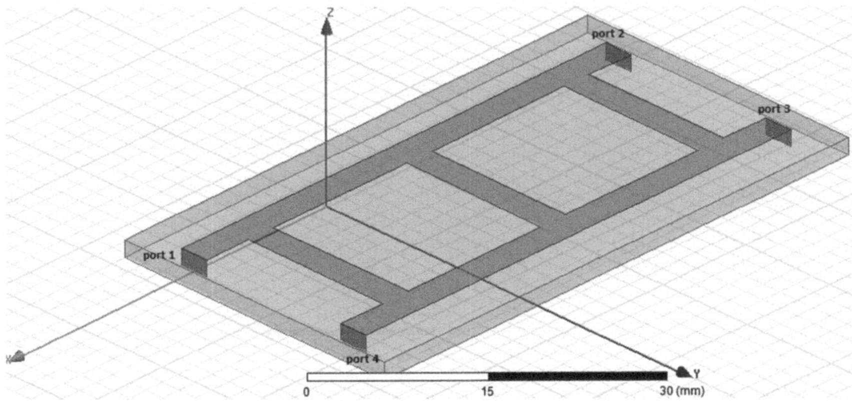

Fig. 5 Structure of crossover

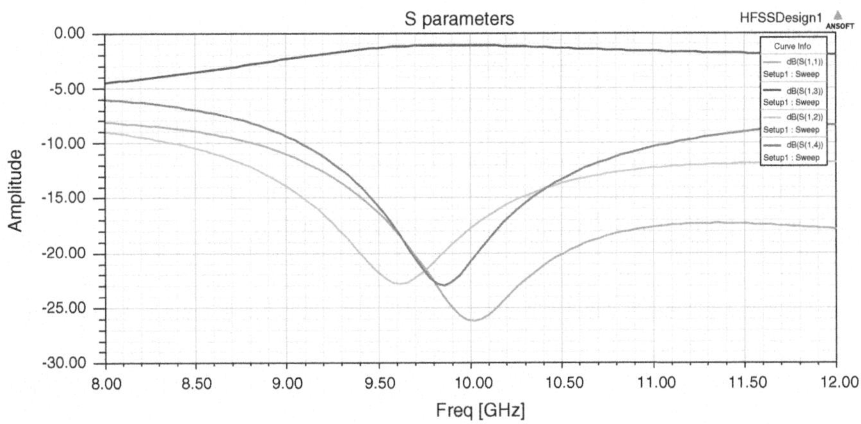

Fig. 6 S parameters versus frequency for the crossover

In the same way, the crossover has been designed using 50 Ω microstrip transmission lines as shown in Fig. 5. The insertion loss for the coupled port S_{13} is −1 dB while return loss S_{11} is −26 dB and the isolated ports S_{12} and S_{14} are −23 dB and −23 dB respectively for the frequency of interest (Fig. 6). These results are satisfactory in terms of reflection and isolation parameters.

2.4 *Phase Shifter*

Every line that is longer than a reference line L by a certain amount ΔL introduces a phase shift θ given by [12]:

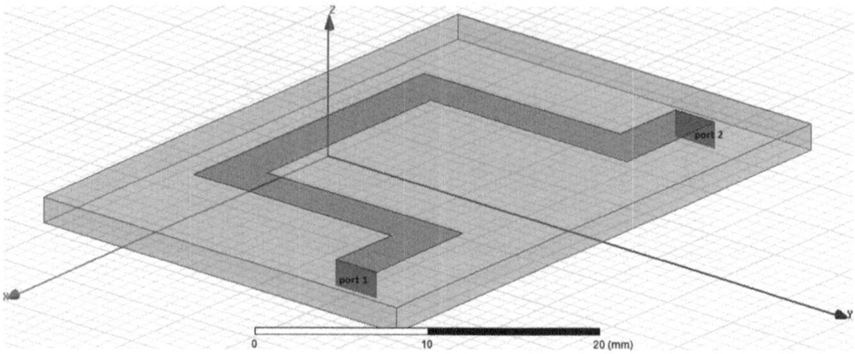

Fig. 7 Structure of phase shifter

$$\Delta L = \theta * \lambda_g / 360° \quad (1)$$

where the λ_g denotes the guided wavelength.

The optimum design for 50 Ω microstrip line with 45° phase shift was achieved as shown in Fig. 7.

As shown in Fig. 8, Insertions loss represented by S_{12} parameter is −0.5 dB, while the return loss S_{11} is −24 dB at 10 GHz.

For the phase shift, the simulation result is shown in Fig. 9. The phase difference between the ports 1 and 2 is 45° at the desired frequency.

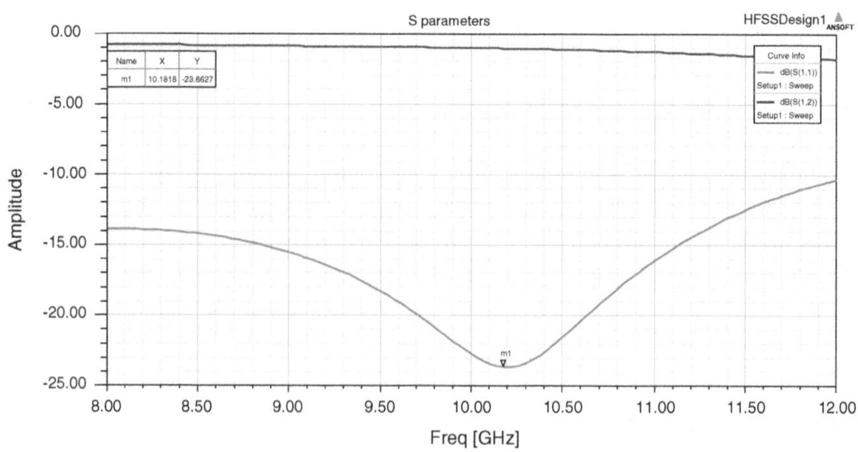

Fig. 8 S parameters versus frequency for the phase shifter

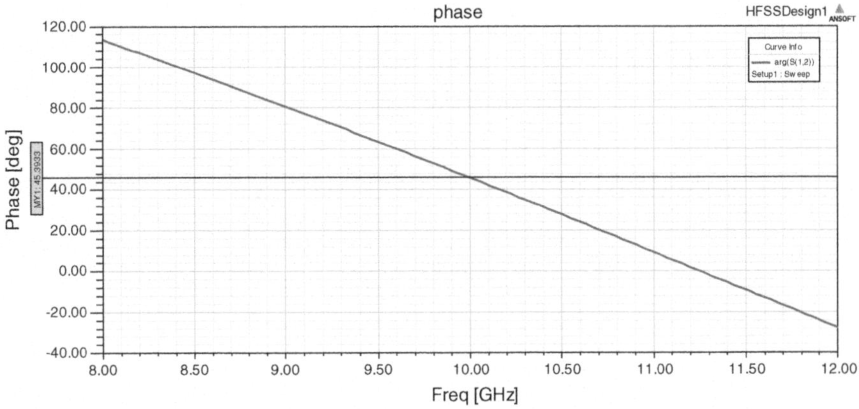

Fig. 9 Simulated phase difference between the ports 1 and 2 in 45° phase shifter

2.5 *Butler Matrix Design and Setup*

The layout of the proposed 4 × 4 Butler matrix is presented in Fig. 10. Combining the components presented in Sections B, C and D, the proposed Butler matrix was designed as a passive microstrip network on the same substrate RT-Duroid. When one of the input ports (port 1, port 2, port 3 or port 4) was excited by an RF signal, all the output ports (port 5, port 6, port 7 and port 8) were excited, though with equal amplitude and specified relative phase differences.

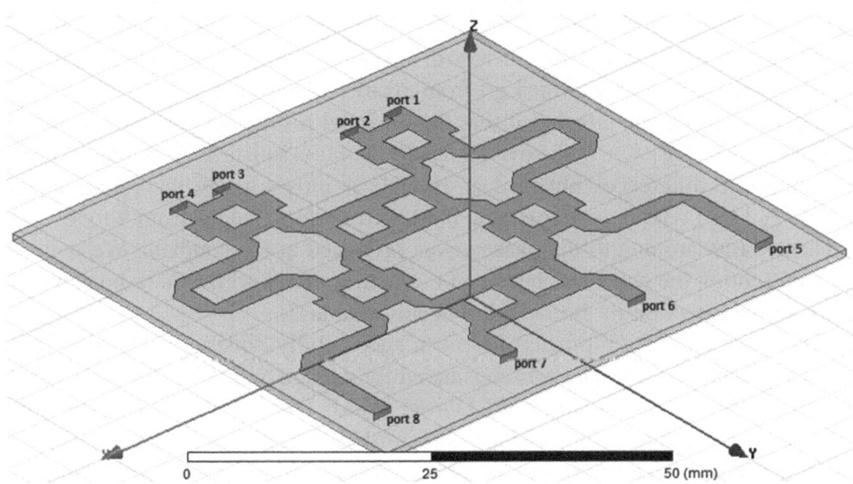

Fig. 10 Structure of the simulated Butler matrix

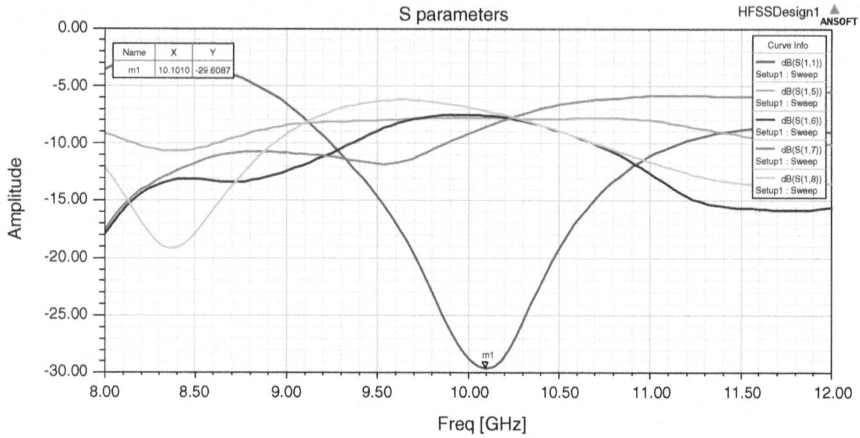

Fig. 11 Simulated results of magnitude S parameters of the 4 × 4 Butler matrix when port 1 is fed

As different input ports were excited, the proposed Butler matrix was treated as a beam forming network, which provided four output signals with equal power levels and with progressive phases of +54°, +144°, −123° and −52° at the center frequency of 10 GHz. Hence, the user can switch the direction of the main radiation beam by exciting the designated input port. The system is capable of producing multiple narrow beams in different directions and thereby selecting the strongest signal among all of the available signals.

Figure 11 show simulation results of the insertion loss and return loss for port 1 when the other ports are matched. These results illustrate that the return loss is better than −20 dB and the coupling to the output ports is well-equalized (around −7 dB). These losses due to the microstrip discontinuities have a minimal influence on the working frequency, which go from 10 to 10.1 GHz.

In general, it can be concluded that the obtained results are very promising.

According to Fig. 12a, b, the overall results are quite satisfactory, it may be noted that the phase differences obtained at 10 GHz is closer to the theoretical model (± 45°, ± 135°) with a quite tolerable phase error which ranges from 9 to 12°.

These errors are due to the phase errors produced at the couplers. We will see when presenting the generated beams of our matrix that this error is responsible for a shift in the direction of the radiation beams (Table 2).

The design simulation process has been successfully done using Ansoft HFSS 13.0. For simulation, the data was considered for a range of 8 to 12 GHz frequency. The final design is implanted on RT-Duroid board. The CorelDraw Graphic Suite 12 has been used to transform the circuit from HFSS 13.0 into layout printable on transparency as shown in Fig. 13.

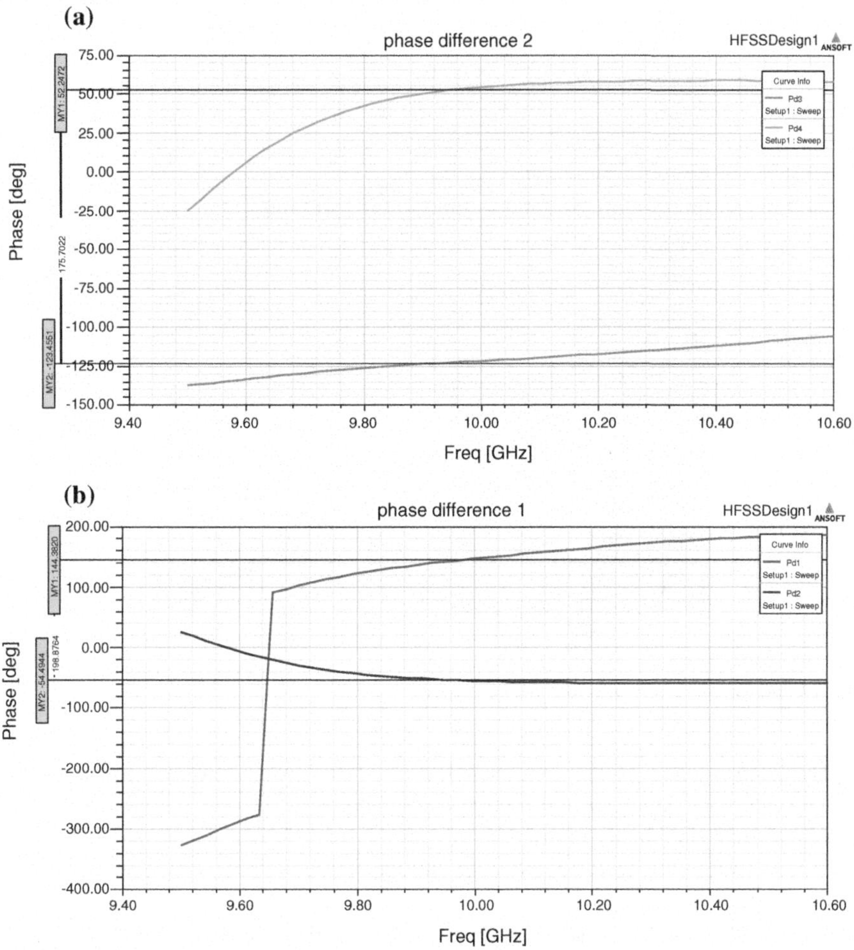

Fig. 12 **a** Simulated phase differences between two adjacent ports for port 1 and 2. **b** Simulated phase differences between two adjacent ports for port 2 and 3

Table 2 Summarizes various parameters of the proposed butler matrix when port 1 is excited

Parameters	
Input reflection coefficient	–30 dB
VSWR	1.06
Band width	100 MHz
Beam direction	10° right
Length of layout	60 mm
Width of layout	50 mm
Operating frequency	10 GHz

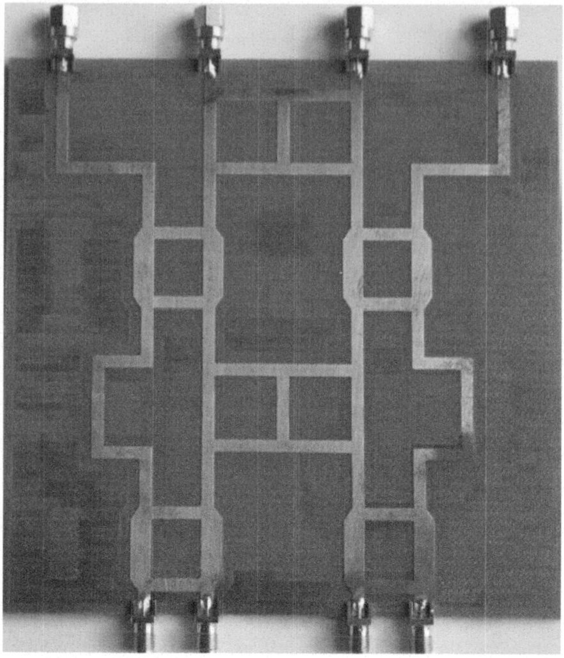

Fig. 13 Prototype of the proposed Butler matrix

The performance of the proposed Butler Matrix was measured by using the Agilent Network Analyzer and plotted using Microsoft's Excel. Simulated (in red) and measured (in black) return loss at port 1 for 10 GHz is shown in Fig. 14. Measured results are well agreed with simulated predictions with a small shift to higher frequencies.

To demonstrate the performance of 4×4 Butler matrix in terms of beam-forming, the proposed matrix is connected to four-patch antenna array [13, 14].

3 Simulation Results of Microstrip Smart Antenna System

Outputs of the butler matrix were fed to the array of four micro-strip edge feed patch antennas implemented on the same substrate as shown in Fig. 15. The inter element distance is 0.6 λ in order to obtain a minimum mutual coupling between the elements and thus preserve the orthogonality condition between the various radiation beams.

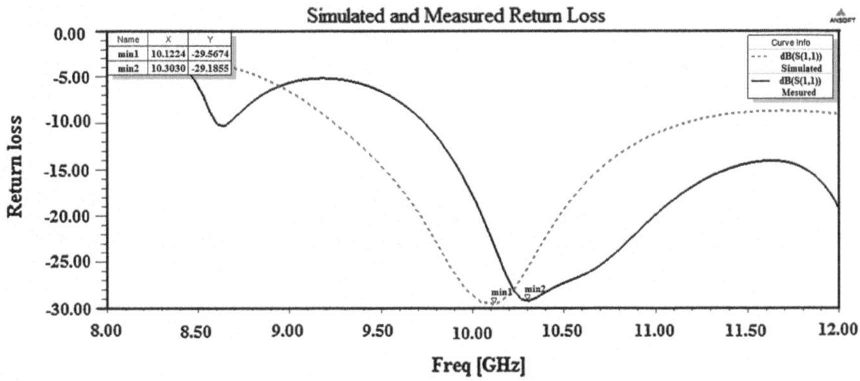

Fig. 14 Simulated and measured results of magnitude S_{11} parameters of the 4 × 4 Butler matrix when port 1 is fed

The far field radiation patterns were simulated using Ansoft HFSS at the center frequency of 10 GHz. The generated beams were found to be at ±10° and ±50° with 5° error in both cases. HPBW is 20° as well as the gain is 13 dB. This topology of multibeam antenna is suitable for beamforming applications; it can cover an area of 120°. Figure 16 shows the resulting beam directions.

According to these results, it is clear that the phase errors have a minimal effect on the side lobe levels of the generated beams, which proves that the proposed system is suitable for narrowband applications.

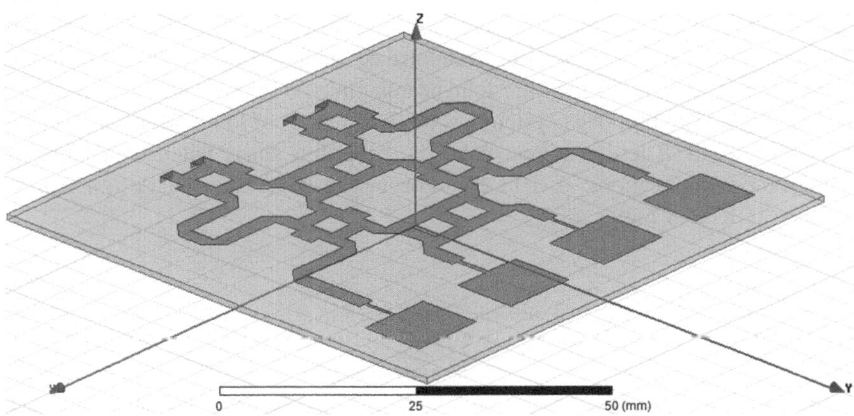

Fig. 15 Structure of microstrip smart antenna system

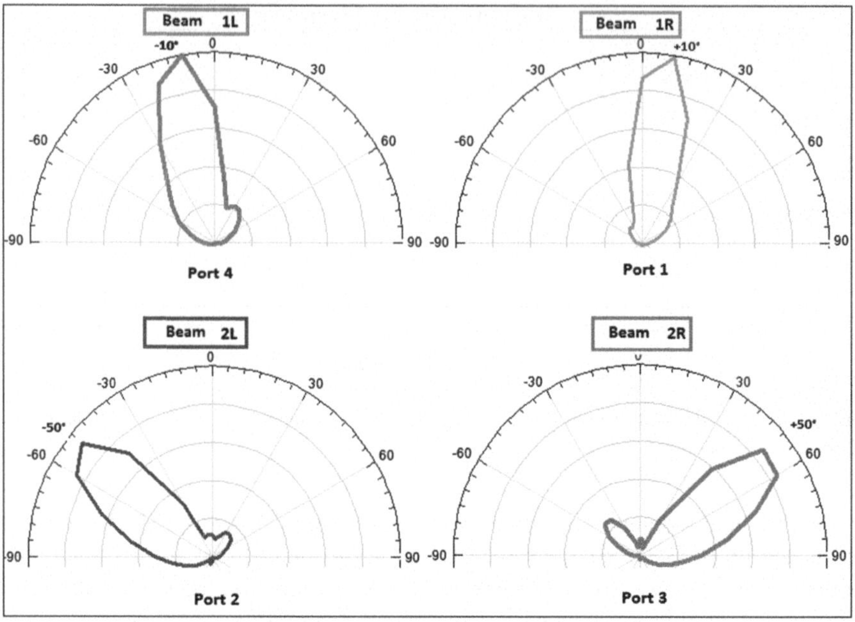

Fig. 16 Beam directions for microstrip smart antenna system

4 Conclusion

The 4 × 4 Butler matrix and its components were studied, designed and simulated successfully. The simulated results show that the proposed 4 × 4-butler matrix combined to 4 patch antenna array produces four orthogonal and narrow beams in the directions (±10°, ±50°) with an improved gain (13 dB). When taken into account the HPBW (20°) then the four beams effectively cover the required 120° azimuth angle. The proposed system has the advantages of low cost, small volume, and light weight. These features make the proposed smart antenna suitable for beamforming applications at 10 GHz. For larger coverage, the design can be extended to 8 × 8 butler matrix.

References

1. Desmond, N.C.T.: Smart antennas for wireless applications and switched beamforming. Department of Information Technology and Electrical Engineering, The University of Queensland (2001)
2. Corona, A., Lancaster, M.J.: A high-temperature superconducting Butler matrix. IEEE Trans. Appl. Supercond. **13**(4), 3867–3872 (2003)

3. Daneshmand, M., Mansour, R., Musavi, P., Choi, S., Yassini, B., Zybura, A., Yu, M.: Integrated interconnect networks for RF switch matrix applications. IEEE Trans. Microw. Theory Tech. **53**(1), 499–507 (2006)
4. Bona, M., Manholm, L., Straski, J.P., Svensson, B.: Low-loss compact Butler matrix. IEEE Trans. Microw. Theory Tech. **50**(9), 2069–2075 (2002)
5. Winters, J.: Smart antennas for wireless systems. IEEE Pers. Commun. 23–27 (1998)
6. Garg, R.: Microstrip Antenna Design Handbook. Artech House Books, Nov 2000
7. Li, W.R.: Switched-beam antenna based on modified butler matrix with Low sidelobe level. Electron. Lett. **40**(5) (2004)
8. Errifi, H., Baghdad, A., Badri, A.: Design and analysis of microstrip patch array antenna with serie, corporate and serie-corporate feed network. Int. J. Electron. Electr. Eng. (2015)
9. HFSS software user guide (2005). Ansoft Corporation
10. Denidni, T.A., Liber, T.E.: Wide band four-port butler matrix for switched multibeam antenna arrays. IEEE **14**, 2461–2464 (2003)
11. Srivastava, G., Gupta, V.: Microwave devices and circuit design. Prentice-Hall of India, New Delhi (2006)
12. Pozar, D.M.: Microwave Engineering, 3rd edn. John wily and sons, New York (2005)
13. Errifi, H., Baghdad, A., Badri, A., Sahel, A.: Radiation characteristics enhancement of microstrip triangular patch antenna using several array structures. Wirel. Microw. Technol. doi:10.5815/ijwmt.2015.03.01
14. Errifi, H., Baghdad, A., Badri, A., Sahel, A.: Conception et simulation d'une antenne réseau directive à deux patch triangulaires pour les applications dans la bande X. Colloque international Télécom'2015 & 9èmes JFMMA, 13–15 Mai 2015, Meknès, Maroc

Comparative Study of Radiation Performance Between Two Ultra Wide Band Planar Patch Array Antennas for Weather Radar Applications in C-Band

Abdellatif Slimani, Saad Dosse Bennani, Ali El Alami and Kaoutar Allabouche

Abstract This paper presents a comparative study of radiation performance between two Ultra Wide Band (UWB) microstrip planar array antennas for weather RADAR applications which operate in C band. These arrays are etched onto a FR4 printed circuit board with an overall size of $(162 \times 100 \times 1.58)$ mm^3 and dielectric constant $\varepsilon_r = 4.4$. The proposed arrays antennas are composed of a twenty radiating patch element with a T-junction power divider which has a role to divide the power equally to all antenna elements above a partial ground plane. Simulation results show that each array antennas has its own characteristics. The results show that the use of an array antennas with rectangular radiating elements give a bandwidth which is about 118 %, gain and high directivity can exceed 12 dB and a half power beam width of 10°, which are relatively closer to those obtained in array antennas with circular radiating elements.

Keywords Array antennas · UWB · Weather RADAR · VSWR · Gain

1 Introduction

Generally, to discover the climatic condition of the weather, and forecast the position of precipitation, the weather radars is used. At present, the radars which are able to detect the precipitations and determine its intensity are Doppler radars [1, 2],

A. Slimani (✉) · A. El Alami · K. Allabouche
Faculty of Sciences and Technics, University Sidi Mohamed Ben Abdellah, Fes, Morocco
e-mail: abdellatif.slimani@usmba.ac.ma

S.D. Bennani
Laboratory of Renewable Energy and Intelligent Systems, National School of Applied Sciences, Fes, Morocco

R. El-Azouzi et al. (eds.), *Advances in Ubiquitous Networking 2*, Lecture Notes in Electrical Engineering 397, DOI 10.1007/978-981-10-1627-1_31

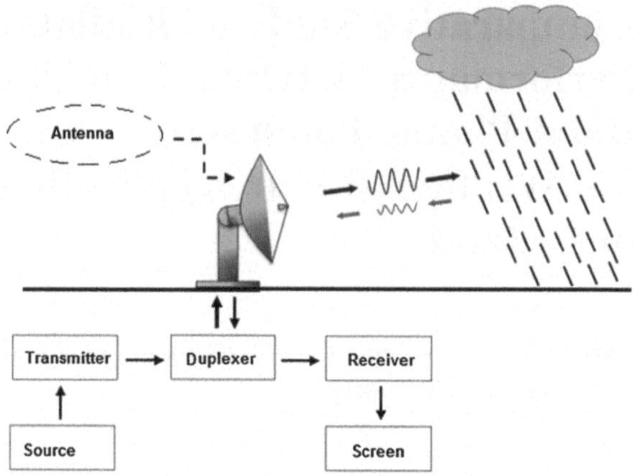

Fig. 1 Basic components of the weather radars

it is assigned several frequency bands for them, including the S band (2–4 GHz), C band (4–8 GHz) and X band (8–12 GHz). In this paper, the focused band is from 4 to 8 GHz which appertain to range frequency of IEEE 802.15a (3.1–10.6 GHz) covers the standard of the UWB [3, 4].

Radars are very complex systems, its design is based generally on their mission, the most of them are composed of the following components: receiver, transmitter, duplexer, antenna, power source and screen. In this paper, we focused our study only on the antenna element, its miniaturization and the optimization of its radiation performance, the following figure shows the basic components of the weather radars (Fig. 1) [5–7].

In this paper, we made a comparison results between an array antennas that consists of twenty circular radiating elements and an array antennas that consists of twenty rectangular radiating elements [8], to express the difference between the performance of these two types of arrays in terms of adaptation and radiation.

Design methodology of these arrays is described bluntly in the paper [9]. Performance simulations of patch arrays antennas were performed with Ansoft HFSS and CST Microwave Studio, which utilize different numerical methods for electromagnetic computations.

2 Methods and Materials

In this paper, we used the Quarter-wave transformer impedance technique to divide the power equally to all radiating elements (Fig. 2). Commonly, in microwave circuit design the impedance matching is very important, which is relatively simple at a single resonance frequency, but becomes very difficult if a wideband impedance matching is desired.

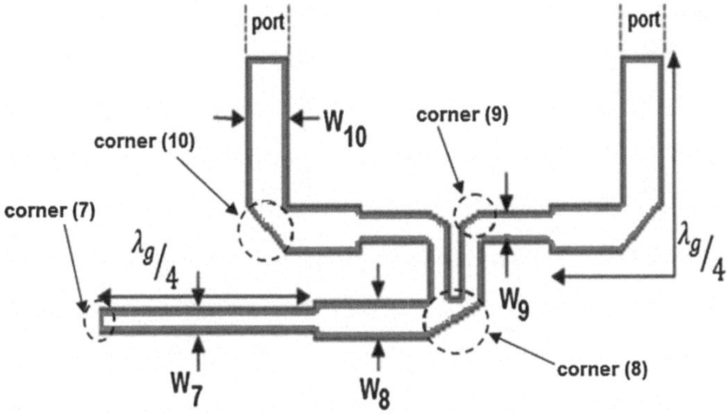

Fig. 2 Quarter wave-transformer

In the designing of the feed array we have to consider reflection levels at and electrical lengths of the bends. Removing a part of the area of metallization in the bend's corner can reduce the reflection level of the bend. The percentage mitre is the cut-away fraction of the diagonal between the inner and outer corners of the un-mitred bend (Fig. 3).

Fig. 3 Microstrip mitred bend

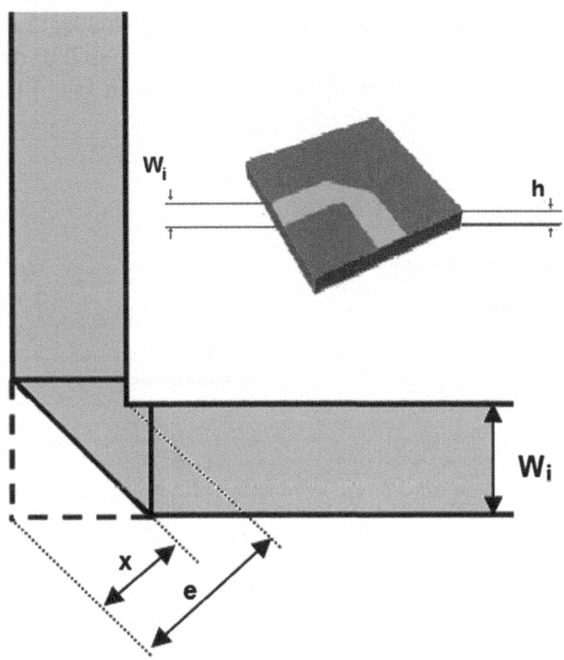

Table 1 Parameters corners of the proposed antenna font sizes of headings

Corner	e_i (mm)	x_i (mm)
7	0.67	0.4
8	3	1.7
9	1.1	0.6
10	2.9	1.65

Table 2 Microstrip line impedances

Impedance	Value (Ω)	Impedance	Value (Ω)
Z_1	44	Z_7	87.1
Z_3	33.9	Z_8	72
Z_4	54	Z_9	80.6
$Z_5 = Z_6$	40.4	Z_{10}	60

The optimum percentage mitre is given by [10]:

$$M = (100\,x/e)\% = \left(52 + 65exp^{(-1.35W_i/h)}\right)\% \tag{1}$$

where, h is the thickness substrate, $W_i/h \geq 0.25$ and dielectric constant $\varepsilon_r \leq 25$.

For our antenna design, we give e for each corner and calculate the value of x by applying Eq. (1) and the following Table 1 shows the results obtained.

The characteristic impedances of the microstrip lines being used for feeding elements of the arrays are given in the following Table 2.

A standard T-junction power divider is used to divide power equally to the all elements of patch arrays antennas shown in Fig. 4 [11]:

$$Z_3 = \sqrt{Z_1.Z_4/2} \tag{2}$$

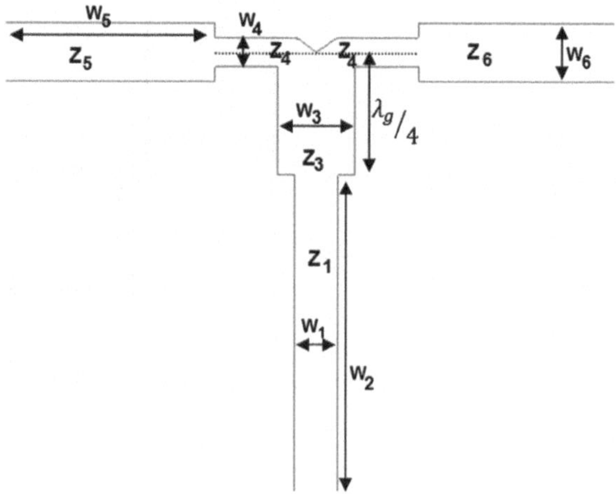

Fig. 4 T-junction power divider

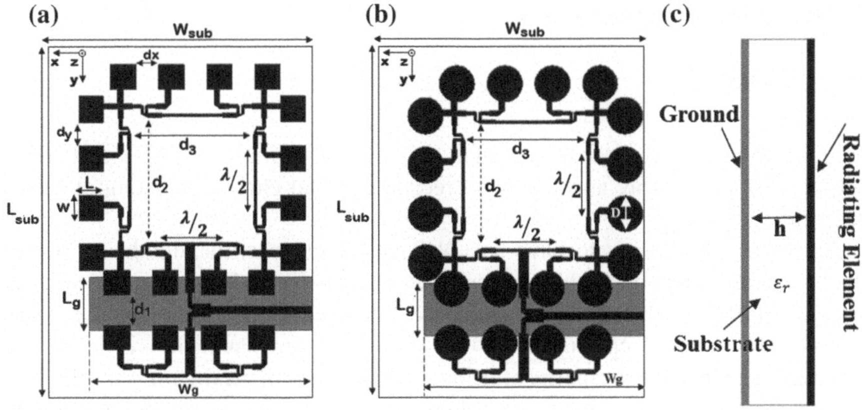

Fig. 5 Geometry of the proposed UWB arrays antennas. **a** and **b** the top view of the arrays antennas, **c** the side view of the arrays antennas

Table 3 Parameters of the proposed arrays antennas

Parameter	Value (mm)	Parameter	Value (mm)
W_1	3.5	W_{10}	2.2
W_2	28.7	W	10.5
W_3	5.2	L	10
W_4	2.5	W_{sub}	118
W_5	19	L_{sub}	148
W_6	4	W_g	82.4
W_7	1	L_g	21.2
W_8	1.5	$d_x = d_y$	9.5
W_9	1.2	$d_2 = d_3$	51
D	14.8	d_1	12

The geometry of the proposed UWB arrays antennas are depicted in Fig. 5 with their characterizing parameters. The antennas are located on the x-y plane and the normal direction is parallel to z-axis. There are printed on FR-4 substrate with a dielectric $\varepsilon_r = 4.4$, thickness $h = 1.58$ mm, a loss tangent $\delta = 0.02$ and there are excited by a 50 Ω coplanar waveguide (CPW) transmission line printed on a partial grounded substrate.

Table 3 shows the geometric parameters of our antenna that have been calculated by use of the relation [12, 13]:

3 Results and Discussion

3.1 Return Loss

Figure 6 shows the comparison simulation results of return loss in CST and HFSS between the patch array antennas with rectangular and circular radiating elements geometry.

Since our objective is to have an UWB array antennas, the simulation results shows that all reflection coefficients are lower than −10 dB in the band of interest. The only difference that, the array antennas with rectangular radiating elements has a peaks of resonance can reach −40 dB, while the array antennas with circular radiating elements, it has a peaks of resonance can reach in maximum −28 dB. This difference returns to the analytical method used to solve the equations of EM field in the array elements, such as the rectangular shape using the method of transmission lines while the circular shape using the method of cavity. Generally, the arrays antennas are well adapted in the C desired band.

3.2 Voltage Standing Wave Ratio (VSWR)

Figure 7 shows the VSWR simulated for both patch array antennas. The results simulations indicate that the VSWR remains less than 2 over the bandwidth range of 3.6–9 GHz which includes the C band.

So if we talk about adaptation, the both array antennas are well matching in the C band frequency range, while the difference appears only in the peaks of resonance frequency.

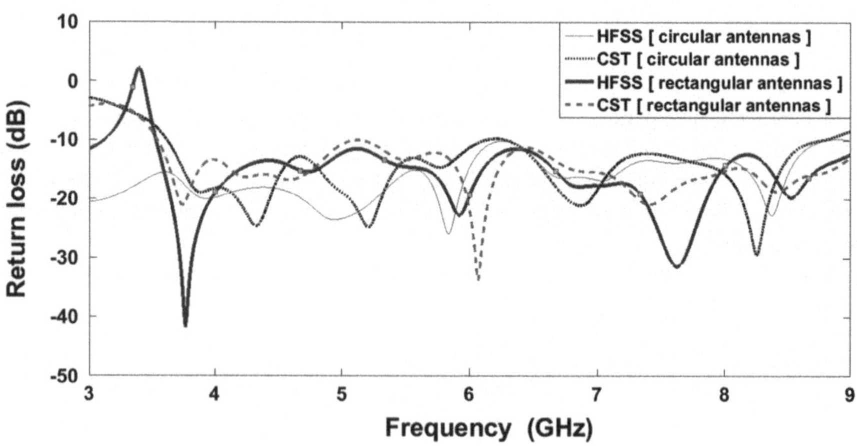

Fig. 6 Return loss comparison between HFSS and CST of arrays antennas

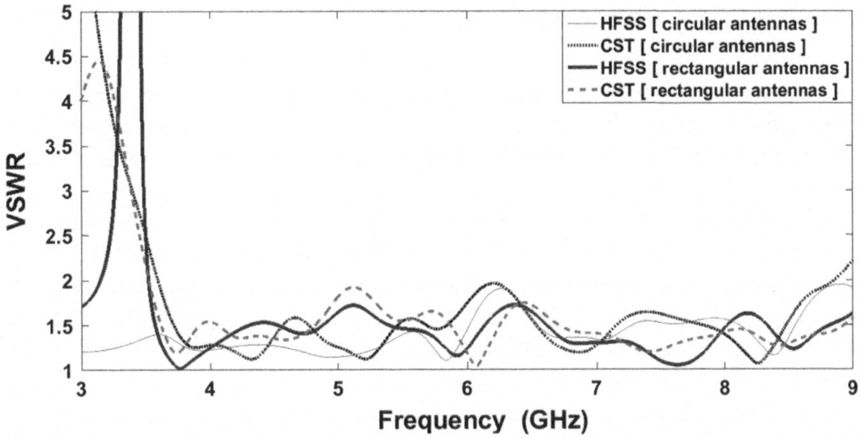

Fig. 7 VSWR comparison between HFSS and CST of arrays antennas

3.3 *Gain Versus Frequency*

Figure 8 shows the gain of different array antennas in the UWB frequency range. In most of the frequencies between 3 and 9 GHz, the gain of the arrays is increase between 10 and 20 dB. We can see that in the band of 3–4 GHz, both arrays antennas have the same gain variation, while in the band of 4–7.8 GHz, we can see that the level of the gain in array with circular radiating elements, decreases relative to gain level of the other array antennas, and in the band of 7.8–8 GHz, the gain of the circular elements increases a little, compared to the level gain of the array with rectangular elements.

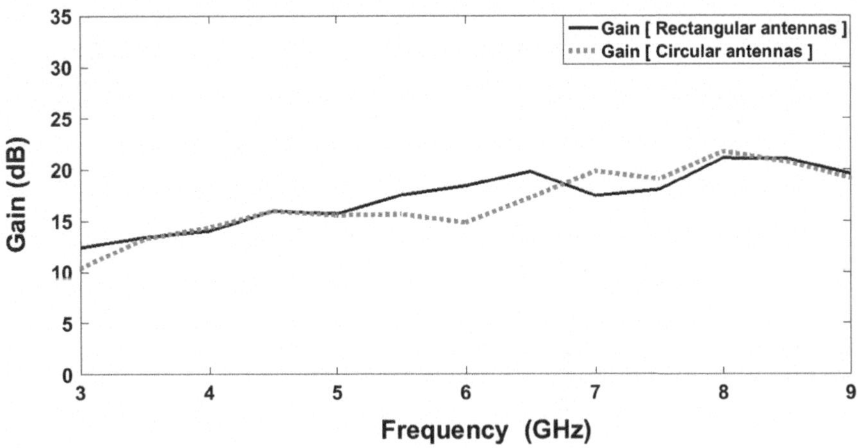

Fig. 8 Comparison gain between different arrays antennas

3.4 Far Field Radiation Pattern

Figure 9 shows the polar radiations of microstrip arrays antennas in 2D between HFSS and CST at 6–8 GHz. Figure 9(a), (b) show the polar gain radiation pattern in E-plane (x − z) and Fig. 9(c), (d) show the polar gain radiation pattern in H-plane (y − z).

According to the Fig. 9, we find that the arrays antennas have a bidirectional radiation pattern directed towards the desired directions (End-Fire angles).

In E-plane, it can be seen that the number of side lobes in the resonance frequency of 6 GHz is minimal, there are three secondary lobes, by against, in the case of the resonance frequency 8 GHz, there are four secondary lobes with a main lobe reaches a gain of 20 dB.

In H-plane, we can see that there are four side lobes at the resonance frequency 6 GHz, it is almost the same number in the case of the resonance frequency of 8 GHz. The passage between resonance frequencies of 6–8 GHz produces an elevated of secondary lobes, but at the same time an increase of gain level from 15–20 dB.

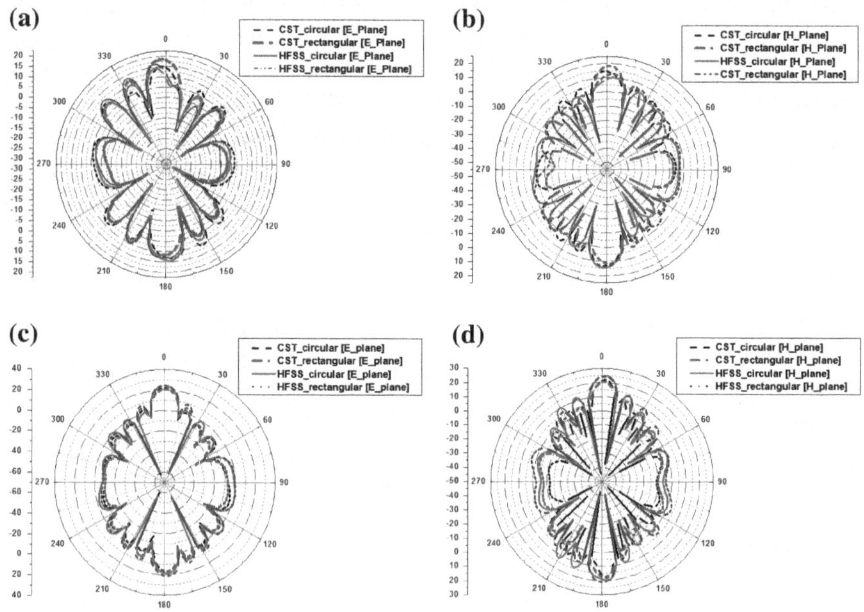

Fig. 9 Comparison of 2D Gain radiation pattern of the proposed arrays antennas, **a** for $f = 6$ GHz, **b** for $f = 6$ GHz, **c** for $f = 8$ GHz, **d** for $f = 8$ GHz

3.5 Surface Current Distributions

The surface current distributions on the top patch arrays antennas for the resonance frequency 6–8 GHz are depicted in Fig. 10, the blue color indicates the minimum current density while the red color indicates the maximum current density.

It can be seen that the current distribution density is higher in the case of the array antennas with rectangular radiating elements more than the case of circular elements in the both resonant frequencies.

As expected, strong surface current densities were present along the T-junction power divider region. When moving away from this cross-section, the current density decrease and the interaction vanishes rapidly because the distance increases from the point of excitation.

Finally, the compared results of the UWB arrays antennas present the best performance in terms of adaptation, bandwidth, gain and HPBW. These performances are summarized in Table 4.

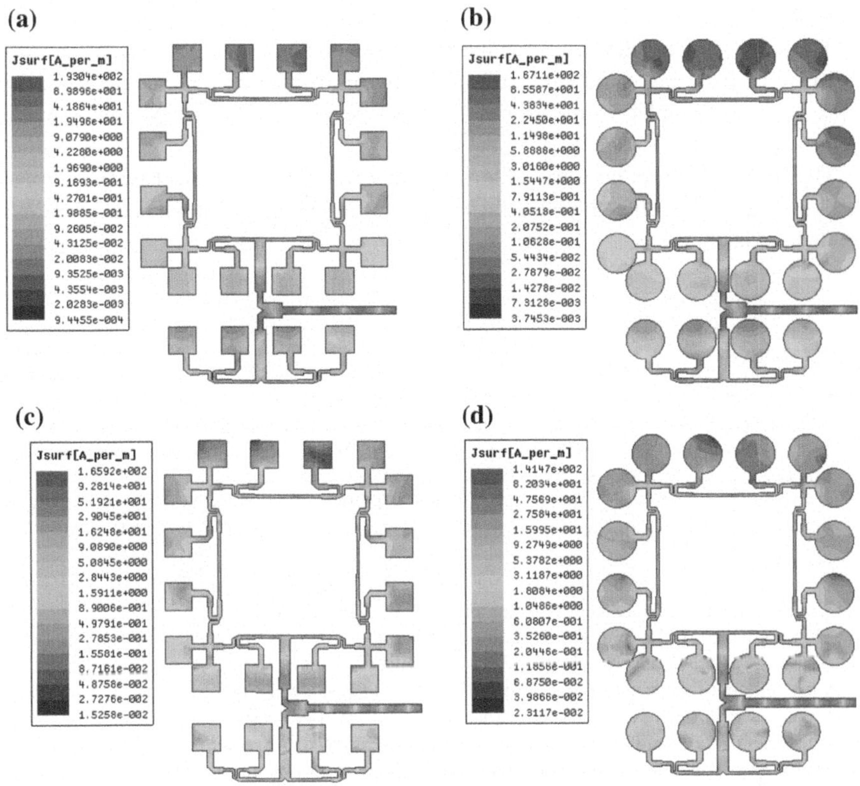

Fig. 10 The Surface current distribution of the different array antennas. **a** and **b** show the distribution in resonance frequency of 6 GHz while **c** and **d** show the distribution in resonance frequency of 8 GHz

Table 4 Comparison result between HFSS and CST

	HFSS				CST			
	6 GHz		8 GHz		6 GHz		8 GHz	
	Rect	Circ	Rect	Circ	Rect	Circ	Rect	Circ
Gain (dB)	16.46	14.66	20.69	22.25	18.6	15.1	21.2	22.6
HPBW (degree)	10.7	12.7	7.8	7.65	10.8	12.5	8	7.6
Bandwidth (%)	115	125	115	125	115.1	115.6	115.1	115.6

Note
Rec array antennas with rectangular elements
Circ array antennas with circular elements

4 Conclusion

In this paper, a comparison in terms of radiation performance between two ultra wide band planar array antennas for C-band weather radar applications has been presented.

The simulation results show that the both array antennas have approached results in terms of gain and bandwidth, but in terms of surface current distribution, the array with rectangular elements has a higher level than the array with circular elements.

The high gain over an ultra wide frequency range from about 4 GHz to upper than 8 GHz is benefits theses arrays antennas to be good candidates in UWB weather radar systems.

References

1. Office of the Federal Coordinator for Meteorology: Federal Research and Development Needs and Priorities for Phased Array Radar, FMC-R25-2006, Interdepartmental Committee for Meteorological Svcs. and Supporting Research, Committee (2006)
2. Heinselman, P.L., Priegnitz, D.L., Manross, K.L., Smith, T.M., Adams, R.W.: Rapid sampling of severe storms by the national weather radar testbed. Weath. Forecast. **23**(5), 808–824 (2008)
3. First Report and Order, Revision of Part 15 of the Commission's Rules Regarding Ultra-wideband Transmission Systems FCC, FCC 02-48 (2002)
4. Siriwongpairat, W.P., Liu, K.J.R.: Ultra-Wideband Communication Systems. Wiley (2008)
5. Karimkashi, S., Zhang, G., Kishk, A.A.: A-dual polarization frequency scanning microstrip array antenna for weather radar applications. In: 7th European Conference on Antennas and Propagation, Gothenburg, Sweden, 8–12 Apr 2013, pp. 1795–1798
6. Karimkashi, S., Zhang, G.: A dual-polarized series-fed microstrip antenna array with very high polarization purity for weather measurements. IEEE Trans. Antennas Propag. **61**(10), October 2013
7. Flashy, M.A., Shanthi, A.V.: Microstrip circular antenna array design for radar applications. In: International Conference on Information Communication and Embedded Systems (ICICES) (2014)

8. Slimani, A., Bennani, S.D., El Alami, A., Harkat, H.: Conception et optimisation d'un Nouveau Réseau d'Antennes ULB en Technologie Micro-ruban pour l'Évaluation des Changements Climatiques, Pôle de recherche Technologie de l'Information et de Communication, Systèmes et Modélisation (TICSM), FST Fès, Maroc (2015)
9. Slimani, A., Bennani, S.D., El Alami, A., Harkat, H.: Conception and optimization of a bidirectional ultra wide band planar array antennas for C-band weather radar applications. In: The 2nd International Conference on Information Technology for Organizations Development IT4OD in Fez, Morocco, March 30–April 1st 2016
10. Douville, R.J.P., James, D.S.: Experimental study of symmetric microstrip bends and their compensation. IEEE Trans. Microw. Theor. Technol. **26**(3), 175–182 (1978)
11. Chorfi, H.: Conception d'un Nouveau Système d'antenne Réseau Conforme en Onde Millimétrique, Université du Québec à Chicoutimi, Abitibi-Témiscamingue (2012)
12. Ibigbami, N.O., Adediran, Y.A.: Performance analysis of a patch antenna array feed for a satellite C-band dish antenna. Multi. J. Sci. Technol., J. Sel. Areas Telecommun. (JSAT), 24–30 (2011)
13. Slimani, A., Bennani, S.D., El Alami, A., Harkat, H.: Comparative study of the radiation performance between uniform and non-uniform excitation of linear patch antenna array for UWB radar applications. In: Wseas Books: Mathematical and Computational Methods in Electrical Engineering, pp. 89–95 (2015). ISSN: 1790-5117. ISBN: 978-1-61804-329-0

Performance Evaluation of MB-OFDM UWB Systems Based on Optimization Algorithm for CP Decomposition

Zakaria Mohammadi, Awatif Rouijel and Rachid Saadane

Abstract In this paper, a Canonical Polyadic (CP) tensor decomposition for Ultra WideBand (UWB) based MultiBand Orthogonal Frequency Division Multiplexing (MB-OFDM) systems is presented. Therefore, the conventional MB-OFDM, which is a multibanding technique of Ultra WideBand technology that differs in approach from the impulse Direct-Sequence (DS-UWB) is introduced. An application of the proposed Canonical Polyadic decomposition, with isolation of scaling matrix to MB-OFDM system is proposed. A simple blind receiver based on the enhanced gradient algorithm is then presented. For illustrating this application, computer simulations are provided to demonstrate the good behavior of these algorithm compared to others in the literature.

Keywords CP decomposition · Tensor modeling · UWB · MB-OFDM · Blind separation

1 Introduction

Nowadays, Ultra Wide-Band (UWB) has received mush interests from research and industrial community [1]. This technology was adopted as a reliable solution for the IEEE 802.15.4.*a* physical layer for robustness in low data-rate Wireless Personal Area Network (WPANs), offering large bandwidth under strict low-max effective isotropically radiated power (EIRP) spectral density of −41.3 (dBm/MHz) [2]. This limit gives to UWB signal a noise-like behavior that allows overcoming the performance impairment due to Narrowband devices. Several proposals have since then been proposed to realize a short-range high data-rate UWB-based

Z. Mohammadi · A. Rouijel
GSCM-LRIT Laboratory, Associate Unit to CNRST (URAC 29),
Mohammed V University, Rabat, Morocco

R. Saadane (✉)
LETI Laboratory, Hassania des travaux publiques,
KM.7 Route d'EL Jadida, B.P, 8108 Casablanca, Morocco
e-mail: Rachid.saadane@gmail.com

R. El-Azouzi et al. (eds.), *Advances in Ubiquitous Networking 2*, Lecture Notes in Electrical Engineering 397, DOI 10.1007/978-981-10-1627-1_32

communication link. Today, both impulse direct-sequence (DS-UWB) and MB-OFDM UWB systems are considered for standardization.

The standard proposed by the MultiBand OFDM alliance (MBOA) is based on the subdivision of the available UWB spectrum (3.1–10.6 GHz) into 14 subbands, each with 528 MHz bandwidth [3]. This Multi-Band configuration was justified previously in many works. The results carried over from [4], where the Degrees Of Freedom (DOF) of the UWB channel were estimated, show that the DOF number N_{DOF} exhibits a saturation behavior beyond a certain bandwidth value BW_C (By chance this value is typically around 500 MHz). The use of OFDM within the MB-OFDM approach is due to very high-data rate, spectral efficiency, immunity towards frequency selectivity but also the maturity that has acquired thenceforth.

Recently, the use of multi-linear algebra methods has attracted attention in several areas such as data mining, signal processing and particularly in wireless communication systems. Indeed, Wireless communication data can sometimes be viewed as components of a high order tensor (order larger than 3). Solving the problem of source separation would result in finding a decomposition of this tensor and determining its parameters. One of the most popular tensor decompositions is the Canonical Polyadic (CP) decomposition, also known as Parallel Factor Analysis (PARAFAC), which decomposes the tensor into a sum of rank-one components [5]. In the literature, typical algorithms for finding the CP components include alternating least squares (ALS) and descent algorithms [6, 7], which do not isolate the scaling factor matrix. Herein, an application of the new decomposition proposed in [8, 9] to MB-OFDM is presented. A new MB-OFDM transmission model is formulated by exploring the tensor modeling, while isolating the scaling factor. Based on the resulting tensor model of the received MB-OFDM signal, we study the blind separation using the proposed receiver "Algorithm 2", which is the enhanced version of the gradient algorithm [9]. This paper is organized as follows. Section 2 presents notations and properties of the 3rd order tensors, before formulating the exact CP decomposition problem. The conventional MB-OFDM, which is a multibanding technique of Ultra WideBand technology that differs in approach from the impulse Direct-Sequence (DS-UWB) is introduced in Sect. 3. In Sect. 4, we present a MB-OFDM transmission system exploiting the constrained structure of the proposed decomposition. Some simulation results are provided in Sect. 5 for bit-error-rate performance evaluation. We will conclude this work in the last section.

2 Notation and Preliminaries

Let us first introduce some essential notation. Vectors are denoted by boldface lowercase letters, e.g., $\mathbf{a}$; matrices are denoted by boldface capital letters, e.g., $\mathbf{A}$. Higher-order tensors are denoted by boldface Euler script letters, e.g., $\mathcal{T}$. The pth column of a matrix $\mathbf{A}$ is denoted $\mathbf{a}_p$, the (i,j) entry of a matrix $\mathbf{A}$ is denoted by A_{ij}, and element (i,j,k) of a third-order tensor $\mathcal{T}$ is denoted by T_{ijk}. The outer (tensor) product is represented by the symbol ∘, the Kronecker product by ⊗ and The Frobenius norm by $\|\mathcal{T}\|_F$.

Any tensor $\mathcal{T}$ admits a decomposition into a sum of rank-1 tensors, called CP decomposition. In the case of a 3rd order tensor, this decomposition takes the form below:

$$\mathcal{T} = \sum_{r=1}^{R} \lambda_r \mathbf{a}_r \circ \mathbf{b}_r \circ \mathbf{c}_r \tag{1}$$

Denote by $()^T$ matrix transposition, λ_r real positive scaling factors, and R the tensor rank. Vectors $\mathbf{a}_r$ (resp. $\mathbf{b}_r$ and $\mathbf{c}_r$) live in a linear space of dimension I (resp. dimension J and K).

As given in (1), the explicit writing of decomposable tensors is subject to scale indeterminacies. In the tensor literature, optimization of the CP decomposition (1) has been done without taking into account scaling factor $\mathbf{\Lambda}$. In [8], we propose to pull real positive factors λ_r outside the product, which permits to monitor the conditioning of the problem. Scaling indeterminacies are then clearly reduced to unit modulus but are not completely fixed.

3 MB-OFDM-Based UWB

Regarding the question of using multiband-UWB, interest was shown in the division of the available spectrum into many subband as shown in many works. It allows maximum exploitation of the channel capacity and the diversity order, based on the fact that degrees of freedom of the UWB channel tends to present a saturation value beyond certain bandwidth [4]. It is all the more suitable to use the MultiBand OFDM based UWB to fight against Multipath propagation and frequency selectivity, not to mention the optimum spectral efficiency introducing by the OFDM.

The MB-OFDM was proposed by the group IEEE 802.15 [3], and consists of dividing the available spectrum into 5 groups with 3 or 2 subbands (528 MHz). within each subband, the multicarrier modulation OFDM is used, hence the acronym MB-OFDM. A typical MB-OFDM system architecture is given in [3]. The information bits are multicarrier modulated as baseband signal, using the Inverse Fast Fourier Transform (IFFT). A total of 128 subcarriers are used per band (100 for data, 10 guard, 12 pilot and 6 Null subcarriers). To the symbol time $T_S = 242.4$ ns, a cyclic prefix with duration $T_{CP} = 60.6$ ns and a guard period $T_{GI} = 9.5$ ns are added ($T_0 = 312.5$ ns) as shown in the Fig. 1.

Although a wide subband of frequencies can be used from a theoretical viewpoint, certain practical considerations limit the frequencies that are normally used for MB-OFDM UWB system. The mandatory-mode is based on the utilization of the first 3 subbands, and by introducing a Frequency-Hopping Code (FH-Code) that define the one concerned by transmission within a certain time. This configuration is considered optimal for initial deployments and researches. Limiting the upper bound

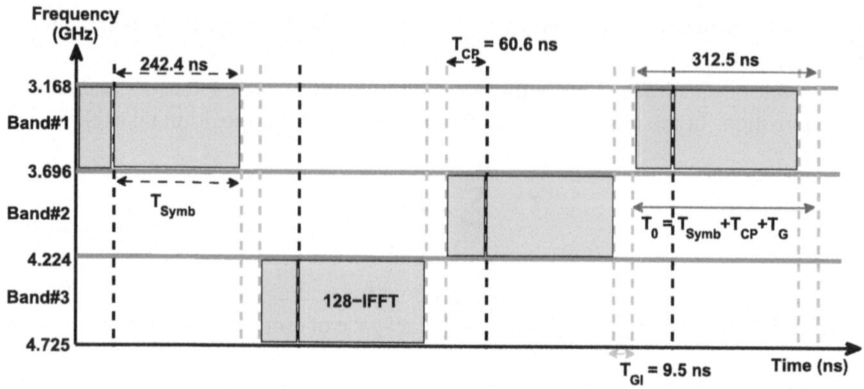

Fig. 1 Example of MB-OFDM signal using a Frequency-Hopping Code = 1, 3, 2, 1, …

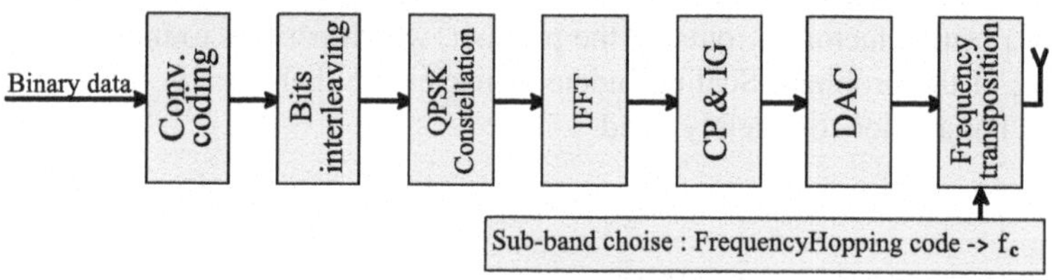

Fig. 2 The MB-OFDM-based UWB transmitter reported in [3]

to 4.8 GHz simplify the design of the radio and analogue Front-End circuits as well as reducing interference with other narrowband services.

As shown in the Fig. 2, the MB-OFDM based transmitter consists first of coding the binary-input using convolutional code, with a coding rate of $R_C = 1/3$, a constraint length $k = 7$ and a generator polynomial expressed in octal basis ($g_0 = 133_8, g_1 = 145_8, g_2 = 175_8$). Other coding rates ($R_c = 1/2, 3/4, 5/8$ and $11/32$) can be obtained from the initial code through the puncturing operation. Once completed, the coded bits are interleaved (Inter/Intra symbols) to provide a diversity for any possible transmission via multipath channels. The entrelaced bits are then coded using a QPSK contellation before using the IFFT to multicarrier modulating the input coded and interleaved bits. Thereafter, the resulted baseband signal is prolonged by a prefix cyclic and guard interval duration and Digital/Analogic converted before transposing it on a frequency carrier depending on the subband in use. Many throughputs can be then achieved depending on the coding rate R_C of the Forward Error Correction FEC, and diversity schemes (in Time or frequency) used [3].

On the receiver side, by knowing the used carrier frequency (via the FH-code), the received signal is expressed in baseband before using the Fast Fourier Transform (FFT) to retrieve the emitted bits. To mitigate the effects of propagation channel, the plot carriers are recovered for estimating the channel response, and for equalizing the received signal. Herein, the Least-Square channel estimation (LSCE) and Maximum Mean Square Equalization (MMSE) techniques are used.

Finally, the reciprocal operations (deinterleaving, convolutional decoding) are performed to find the original binary data.

4 MB-OFDM Tensor Model

4.1 Existence and Uniqueness

In the literature, the existing results in [10] showed that under certain conditions, a third-order tensor of rank R can be uniquely represented as sum of R rank-1 tensors. In practice, it is preferred to adapt to a lower multi-linear model of a fixed rank $R < rank\{\mathcal{T}\}$, so that we have to deal with a problem of approximation. The best rank-R approximate is defined by the minimum of the objective function:

$$Y(\mathbf{A}, \mathbf{B}, \mathbf{C}, \mathbf{\Lambda}) = \|\mathcal{T} - \sum_{r=1}^{R} \lambda_r A_{ir} B_{jr} C_{kr}\|_F^2 \tag{2}$$

By expanding the Frobenius norm in (2), and canceling the gradient with respect to $\mathbf{\Lambda}$, we will calculate the optimal value of $\mathbf{\Lambda}$, which satisfy the following linear system:

$$\mathbf{G}\lambda = \mathbf{f} \tag{3}$$

where $\mathbf{f}$ is R-dimensional vector and $\mathbf{G}$ represents the $R \times R$ Gram matrix defined by: $G_{pq} = (\mathbf{a}_p \otimes \mathbf{b}_p \otimes \mathbf{c}_p)^H (\mathbf{a}_q \otimes \mathbf{b}_q \otimes \mathbf{c}_q)$. In view of matrix $\mathbf{G}$, we can see that the coherence plays a role in the conditioning of the problem, and has deeper implications, particularly in existence and uniqueness of the solution to Problem (2). See [9] for further details.

4.2 Optimization Algorithm for CP Decomposition

Various optimization algorithms exist to compute CP decomposition without constraint, as ALS or descent algorithms [6, 7]. We subsequently present optimization algorithms to compute the CP decomposition (2), under the constraints of unit norm columns of loading matrices. Our optimization problem consists in minimizing the squared error Y under a collection of 3R constraints, namely:

$$\min_{\mathbf{A},\mathbf{B},\mathbf{C},\mathbf{\Lambda}} \|\mathcal{T} - \sum_{r=1}^{R} \lambda_r \mathbf{a}_r \circ \mathbf{b}_r \circ \mathbf{c}_r\|_F^2, \quad \|\mathbf{a}_r\| = \|\mathbf{b}_r\| = \|\mathbf{c}_r\| = 1, 1 \le r \le R \tag{4}$$

Therefore, we need to find three matrices **A**, **B** and **C** of unit norm columns which minimize (4). Stack these three matrices in a $I+J+K$ by R matrix denoted by X. The objective can now be also written $\Upsilon(\mathbf{X};\boldsymbol{\Lambda})$, for the sake of convenience. The solution we propose is to use a projected gradi- ent algorithm while $\boldsymbol{\Lambda}$ is considered as an additional variable. By cancelling the gradient of $\Upsilon(\mathbf{X};\boldsymbol{\Lambda})$ with respect to $\boldsymbol{\Lambda}$, one obtains Eq. (3), which can be solved for $\boldsymbol{\Lambda}$ when $\mathbf{X}$ is fixed. This gives the algorithms below.

ALGORITHM 2.

1. Initialize $(\mathbf{A}(0), \mathbf{B}(0), \mathbf{C}(0))$ to full-rank matrices with unit-norm columns.
2. Compute $\mathbf{G}(0)$ and $\mathbf{f}(0)$, and solve $\mathbf{G}(0)\,\lambda = \mathbf{f}(0)$ for λ, as defined in Sect. 4.1 by (3). Set $\boldsymbol{\Lambda}(0) = Diag\{\lambda\}$.
3. For $k \geq 1$ and subject to a stopping criterion, do
   (a) Compute the descent direction as the gradient w.r.t. $\mathbf{X}$: $\mathbf{D}(k) = -\nabla\Upsilon(\mathbf{X}(k-1);\boldsymbol{\Lambda}(k-1))$
   (b) Compute a stepsize $\ell(k)$ using the backtracking method [11] such as: $\Upsilon(\mathbf{X}(k-1)+\ell(k)\mathbf{D}(k);\boldsymbol{\Lambda}(k-1)) < \Upsilon(\mathbf{X}(k-1);\boldsymbol{\Lambda}(k-1))$.
   (c) Update $\mathbf{X}(k) = \mathbf{X}(k-1) + \ell(k)\,\mathbf{D}(k)$
   (d) Extract the 3 blocks of $\mathbf{X}(k)$: $\mathbf{A}(k)$, $\mathbf{B}(k)$ and $\mathbf{C}(k)$
   (e) Normalize the columns of $\mathbf{A}(k)$, $\mathbf{B}(k)$ and $\mathbf{C}(k)$
   (f) Compute $\mathbf{G}(k)$ and $\mathbf{f}(k)$, and solve $\mathbf{G}(k)\,\lambda = \mathbf{f}(k)$ for λ, according to (3). Set $\boldsymbol{\Lambda}(k) = Diag\lambda$.

The convergence at the kth iteration is declared when the error between tensor $\mathcal{T}$ and its reconstructed from the estimated loading matrices, does not significantly change between iterations k and $k+1$ (a change smaller than a predefined threshold).

4.3 Tensor Modeling

We assume that R subbands are available, and N_p subcarriers can be used per band. Let us define S_{ir} as the ith symbol transmitted over the rth subband. The received MB-OFDM signal at the time j over the ith symbol period can be written as follows:

$$y(i,j) = \sum_{r=1}^{R}\Big(\sum_{n=1}^{N_p} S_{ir}\exp(\jmath 2\pi(n-1)\Delta f)\exp(\jmath 2\pi f_r t_j)h_{nr}\Big) \qquad (5)$$

where Δf is the frequency step, f_r denote the central frequency of the rth sub-band, t_j is the jth time and h_{nr} is the complex channel gain. Now, rewriting $T_{inj} = y(i,j)$, $A_{ir} = S_{ir}$, $B_{nr} = \exp(\jmath 2\pi(n-1)\Delta f)h_{nr}$ and $C_{jr} = \exp(\jmath 2\pi f_r t_j)$, gives us the relation resembling the CP decomposition:

$$T_{inj} = \sum_{r=1}^{R} \lambda_r A_{ir} B_{nr} C_{jr} i \in [1, I] j \in [1, J] n \in [1, N_p]. \tag{6}$$

where A_{ir} (resp. B_{nr} and C_{jr}) denote the entries of vector $\mathbf{a}_r$ (resp. $\mathbf{b}_r$ and $\mathbf{c}_r$), $\mathbf{a}_r$, $\mathbf{b}_r$ and $\mathbf{c}_r$ are the normalized vectors; the scaling ambiguities on the estimated symbols are eliminated by normalizing each symbol sequence by its norm, and calculating the exact scaling factor Λ, with $\Lambda = diag[\lambda_1, \lambda_2, \dots, \lambda_r]$. Separating the received signals is then equivalent to decompose the tensor $\mathcal{T}$ into a sum of R contributions, where R represents the number of active sub-bands in the system.

To calculate the CP decomposition of $\mathcal{T}$, we resort to Algorithm 2 presented in the previous section. The detection and separation of the matrix $\mathbf{A}$ of transmitted symbols will be made, using the following objective function:

$$\Upsilon(\mathbf{A}, \mathbf{B}, \mathbf{C}, \mathbf{\Lambda}) = \|\mathcal{T} - \sum_{r=1}^{R} \lambda_r \mathbf{a}_r \circ \mathbf{b}_r \circ \mathbf{c}_r\|_F^2. \tag{7}$$

where $\mathbf{a}_r$, $\mathbf{b}_r$ and $\mathbf{c}_r$ are the normalized vectors; the scaling ambiguities on the estimated symbols are eliminated by normalizing each symbol sequence by its norm, and calculating the exact scaling factor $\mathbf{\Lambda}$.

5 Simulation Results

In this section, we present some simulation results for illustrating the performances of the proposed blind receiver for MB-OFDM system. In this experiment, we consider a MB-OFDM system with $R = 4$ sub-bands, each with 528 MHz, where $N_p = 128$ subcarriers are available and the length of the transmitter sequence is $I = 256$ consecutive symbols. The propagation channel is supposed to be time-invariant during transmission of one packet but changes from packet to packet. The symbols are generated from an i.i.d distribution and are modulated using a pseudo-random Quadrature Phase-Shift Keying (QPSK) sequence.

The Fig. 3 illustrates the Binary Error Rate (BER) performance of our blind receiver with Algorithm 2, and those of the Algorithm 0 receiver, which is the gradient descent algorithm. These results indicate that the proposed receiver Algorithm 2 gives a better performance than the Algorithm 0 one. To see a significant difference between Algorithms 2 and 0, it is necessary to look at the convergence speed, since the final error is about the same. In all our experiments, Algorithm 2 converged faster in terms of number of iterations; this is illustrated in Fig. 4.

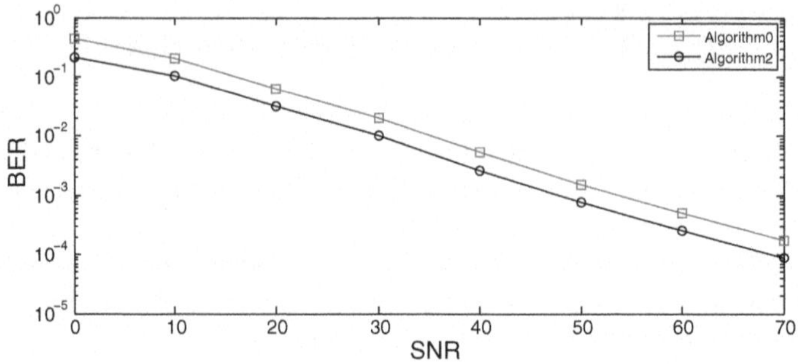

Fig. 3 Bit Error Rate (BER) versus Signal Noise Ratio (SNR) results for scenario: $R = 4, I = 256$, $J = size(t)$ and $N_p = 128$

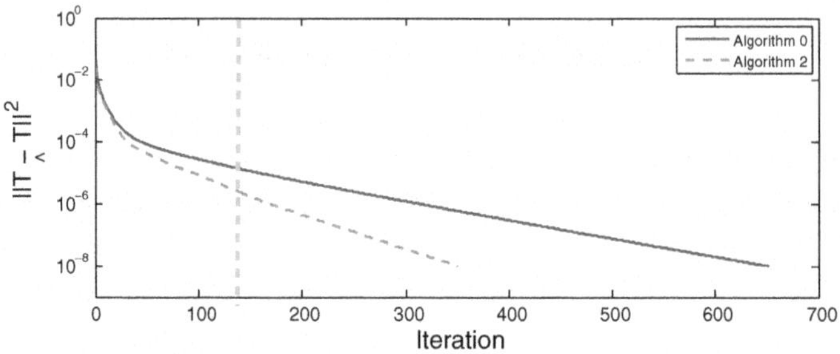

Fig. 4 Reconstruction error as a function of the number of iterations, for a tensor $\mathcal{T}$ of size $I \times N_p \times J$, rank R = 4 and SNR = 60 dB

6 Conclusion

In this paper, we have derived a new algebraic algorithm for the blind separation of UWB-based MB-OFDM signals. Various diversity scheme (Time/Frequency) introduced in MB-OFDM systems enabled formulating this problem as a Canonical Polyadic (CP) decomposition of a 3rd order tensor. This tensor-based approach with isolation of scaling matrix was the starting point for the proposition of a simple blind MB-OFDM receiver, based on the enhanced gradient algorithm. Some simulation results have illustrated the performance of the proposed algorithm. We have shown that the performance of the proposed algorithm converges better than the Gradient one, despite guaranteeing faster convergence.

References

1. Chong, C.C., Watanabe, F., Inamura, H.: Potential of UWB technology for the next generation wireless communications. In: Proceedings of the 9th IEEE International Symposium on Spread Spectrum Techniques and Applications (ISSSTA '06), pp. 422–429, Manaus, Amazon, Brazil, Aug 2006
2. Molisch, A.F., Balakrishnan, K., Chong, C.C., Emami, S., Fort, A., Karedal, J., Kunisch, J., Schantz, H., Schuster, U., Siwiak, K.: IEEE 802.15.4a channel model—final report. IEEE 802.15.4A Task Group (2004)
3. Batra, A., et al.: Multi-Band OFDM physical layer proposal for IEEE 802.15 task group 3a. In: Multi-Band OFDM Alliance SIG (2004)
4. Mohammadi, Z., Saadane, R., Aboutajdine, D.: Improving the estimation of the degrees of freedom for UWB channel using wavelet-based denoising. Eur. J. Sci. Res. **79**(4) (2012)
5. Harshman, R.A.: Foundations of the Parafac procedure: models and conditions for an "explanatory" multi-modal factor analysis. In: Phonetics, UCLA working papers, pp. 1–84 (1970)
6. Comon, P., Luciani, X., De Almeida, A.L.F.: Tensor decompositions, alternating least squares and other tales. J. Chemom. **23**, 393–405 (2009)
7. Sorber, L., Van Barel, M., De Lathauwer, L.: Optimization-based algorithms for tensor decompositions: canonical polyadic decomposition, decomposition in rank-s terms and a new generalization. SIAM J. Optim. **23**(2), 695–720 (2013)
8. Comon, P., Minaoui, K., Rouijel, A., Aboutajdine, D.: Performance index for tensor polyadic decompositions. In: 21th EUSIPCO, Marrakech, Morocco (2013)
9. Rouijel, A., Minaoui, K., Comon, P., Aboutajdine, D.: CP decomposition approach to blind separation for DS-CDMA system using a new performance index. In: EURASIP J. Adv. Sig. Process. **1** 128 (2014)
10. Kruskal, J.B.: Three-way arrays: rank and uniqueness of trilinear decompositions. Linear Algebra Appl. **18**, 95–138 (1977)
11. Boyd, S., Vandenberghe, L.: Convex Optimization. Cambridge University Press (2004). ISBN: 978-521-83378-3 hardback

CPW-Fed Dragon Fractal Antenna for UWB Applications

Abdelati Reha, Abdelkebir El Amri and Othmane Benhmammouch

Abstract This paper presents six iterations of CPW-Fed Dragon fractal antenna. The results show that more the number of iterations increases more the antenna has a broadband behavior. Also, Simulation results show that the 6th iteration of CPW-Fed Dragon fractal antenna have a bandwidth of 3.03 GHz (from 3.67 to 6.6 GHz) with important radiation gains (peak gains from 2 to 5.5 dB) which makes it a suitable solution for Ultra Wide Band (UWB) Applications. All the simulations were performed in CADFEKO, a Method of Moment (MoM) based solver.

Keywords Antenna · Design · Dragon · Fractal · Mom · UWB

1 Introduction

With the proliferation and miniaturization of telecommunications systems and their integration in restricted environments, such as Smart-phones, tablets, cars, airplanes, and other embedded systems. The design of compact multi-bands and Ultra Wide Band (UWB) antennas becomes a necessity.

A. Reha (✉) · A. El Amri
RITM Laboratory, CED Engineering Sciences, Ecole Superieure de Technologie, Hassan II University of Casablanca, Casablanca, Morocco
e-mail: reha.abdelati@gmail.com

A. El Amri
e-mail: elamri_abdelkebir@yahoo.fr

O. Benhmammouch
Université Internationale de CASABLANCA, Casablanca, Morocco
e-mail: othmane.benhmammouch@gmail.com

R. El-Azouzi et al. (eds.), *Advances in Ubiquitous Networking 2*, Lecture Notes in Electrical Engineering 397, DOI 10.1007/978-981-10-1627-1_33

Several techniques are used to design this kind of antennas:

1. Designing multi-bands antennas operating in several frequencies bands by using fractal geometries or adding slots to the radiating elements [1–4].
2. Designing UWB antennas operating in the frequencies bands exceeding 500 MHz or having a fractional bandwidth of at least 0.20. UWB wireless communication occupies a bandwidth from 3.1 to 10.6 GHz (based on the FCC "Federal Communication Commission") [5–8].

One of the interesting techniques used to have a broadband and multiband behavior is the fractal geometry, because it's a simple technique based on the auto-similarity, the most known techniques used are: MINKOWSKI, KOCH, TREE, HILBERT, SIERPINSKI, APOLLONIUS CIRCLES, CANTOR SET.

In this paper, a CPW-fed DRAGON Fractal Antenna is studied. The results show that more the number of iterations increases more the antenna has a broadband behavior. The proposed antenna is a good solution for UWB applications.

The simulations are done by CADFEKO based on the Method of the Moment (MoM).

2 Literature Review of Dragon Fractal Antennas

2.1 *The DRAGON Fractal Geometry*

The DRAGON's name comes from the fact that, for high iterations, the shape of the structure is close to that of the DRAGON (Fig. 2—iteration 20). The construction of this structure is made from a simple line and applying the following steps:

For the first iteration (Fig. 1)

- We move from the segment [AB] to the segment [AA1] by a rotation with the centre A and the angle $\frac{\pi}{4}$ followed by a scaling whose center is A and ratio $\frac{\sqrt{2}}{2}$;
- We move from the segment [AB] to the segment [A1B] by a rotation with the center B and the angle $-\frac{\pi}{4}$ followed by a scaling whose center is B and ratio $\frac{\sqrt{2}}{2}$;

For the other iterations, we apply the same procedure on each segment. Figure 2 shows the first four iterations and the 20th iteration of the DRAGON structure.

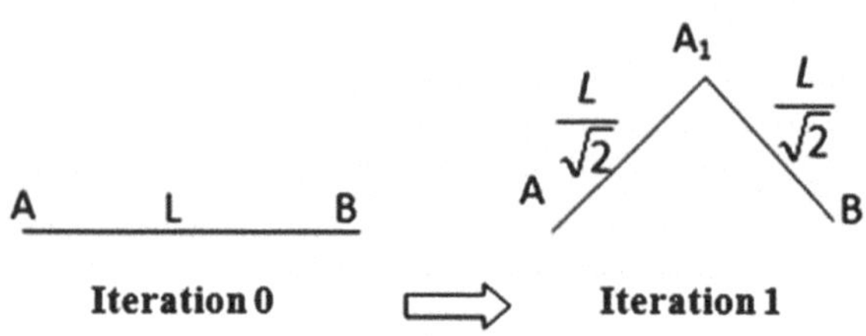

Fig. 1 Generation of the first iteration of the DRAGON structure

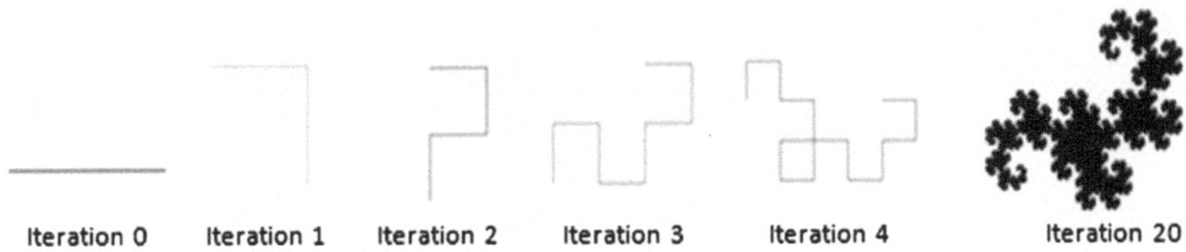

Fig. 2 The four first iterations and the 20th iteration of the DRAGON structure

In the fractal structures, we use another dimension concept known as "the HAUSDORFF dimension" which defined by the Eq. (1) [3, 9].

$$\dim = \frac{\ln(n)}{\ln(R)} \tag{1}$$

Where the fractal is formed of "n" copies whose size has been reduced by a factor of "R".

The HAUSDORFF dimension of the DRAGON structure is given by the Eq. (2).

$$\dim = \frac{\ln(n)}{\ln(R)} = \frac{\ln(2)}{\ln(\sqrt{2})} = 2 \tag{2}$$

2.2 *The Use of DRAGON Fractal Geometry in the Antenna Design*

According to our literature review published in a previous paper [3], it is clear that some fractal structures have been extensively used to design wire or planar antennas such as KOCH, SIERPINSKI, HILBERT and TREE structures, other ones were rarely used such as CIRCULAR structures, CANTOR Set, but some of them weren't studied such as DRAGON structures.

3 Antenna Design and Optimization

The configuration of the proposed antenna labeled is shown in Fig. 3. The antenna is constructed on a FR4 substrate with dielectric constant of 4.4, loss tangent 0.025, thickness of 1.6 mm, and size of 50 mm × 45 mm. The antenna has a coplanar configuration in which both the conductor and ground plane are on one side of the PCB. The parameters are optimized by FEKO, these parameters are: W_s = 50 mm, L_s = 45 mm, h = 1.6 mm, g = 0.5 mm, s = 1 mm, L_f = 16 mm and L_g = 6.55 mm. For the first iteration (Fig. 3), L_1 = 13.57 mm, for the iteration k, the length L_k is given

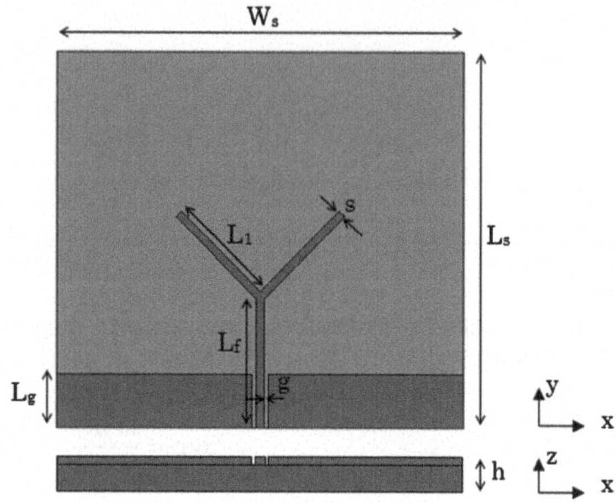

Fig. 3 The geometry of the proposed antenna (1st iteration)

by the Eq. (3). Figure 4 shows the six iterations of the CPW-fed DRAGON fractal antennas with the L_k parameters.

$$L_k = \frac{L_1}{(\sqrt{2})^{k-1}} \quad (3)$$

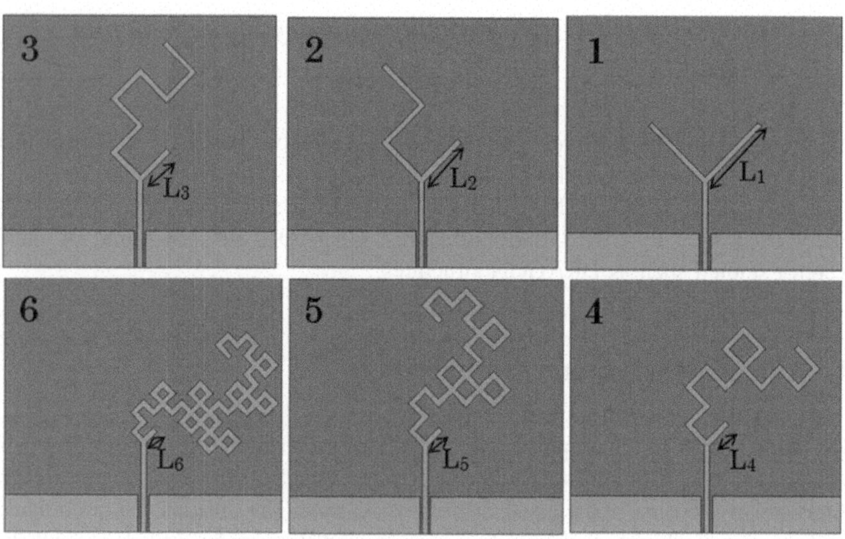

Fig. 4 The six iterations of the CPW-fed DRAGON fractal antennas

4 Results and Discussion

Figure 5 shows the variation of simulated S_{11} parameter versus the frequency for the studied antennas. Table 1 summarizes for each iteration, the resonant frequencies, the S_{11} values on the resonant frequencies, the −10 dB bandwidths, the maximum and minimum gains in the −10 dB bandwidths.

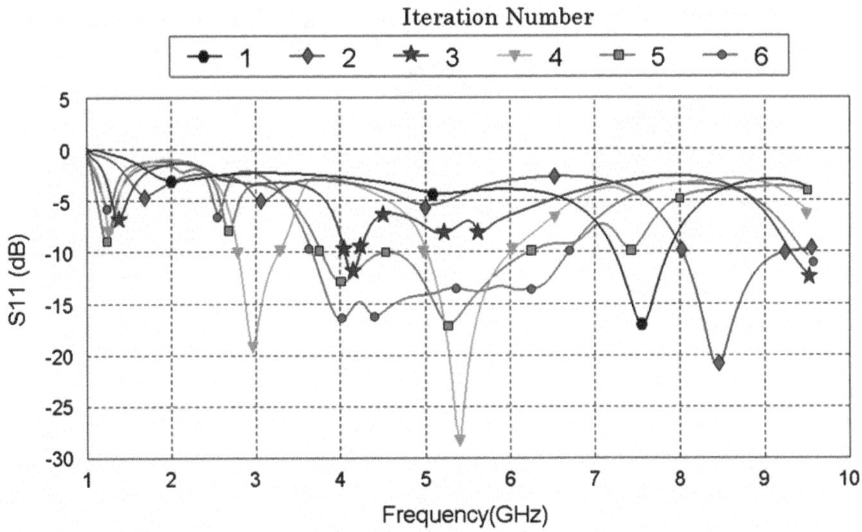

Fig. 5 Simulated S_{11} parameter versus frequencies and versus iteration numbers

Table 1 Simulated resonant frequencies, −10 dB bandwidths and the gains for the proposed antennas

Iteration number	First resonant frequency (GHz)	Resonant frequencies (GHz)/S_{11} (dB) (good matching)	(−10 dB) Bandwidth: from–to	Gain (dB)* min–max
1	2	$F_r =$ 7.56/−17.2	660 MHz: 7.2–7.86	3.7–4.2
2	1.7	$F_r =$ 8.43/−21.4	1.3 GHz: 8–9.3	3.7–4.2
3	1.38	$F_r =$ 4.13/−12	210 MHz: 4.04–4.25	2.1–2.9
4	1.3	$F_{r1} =$ 2.95/−19.4	600 MHz: 2.7–3.3	2.2–2.7
		$F_{r2} =$ 5.4/−28.6	1 GHz: 5–6	2.9–3.8
5	1.25	$F_{r1} =$ 4/−13.4	2.47 GHz: 3.75–6.22	2.5–4.4
		$F_{r2} =$ 5.26/−17.4		
6	1.25	$F_{r1} =$ 4/−17	3.03 GHz: 3.67–6.7	2–5.5
		$F_{r2} =$ 4.4/−16.5		

*The gain is simulated on the (−10 dB) bandwidth

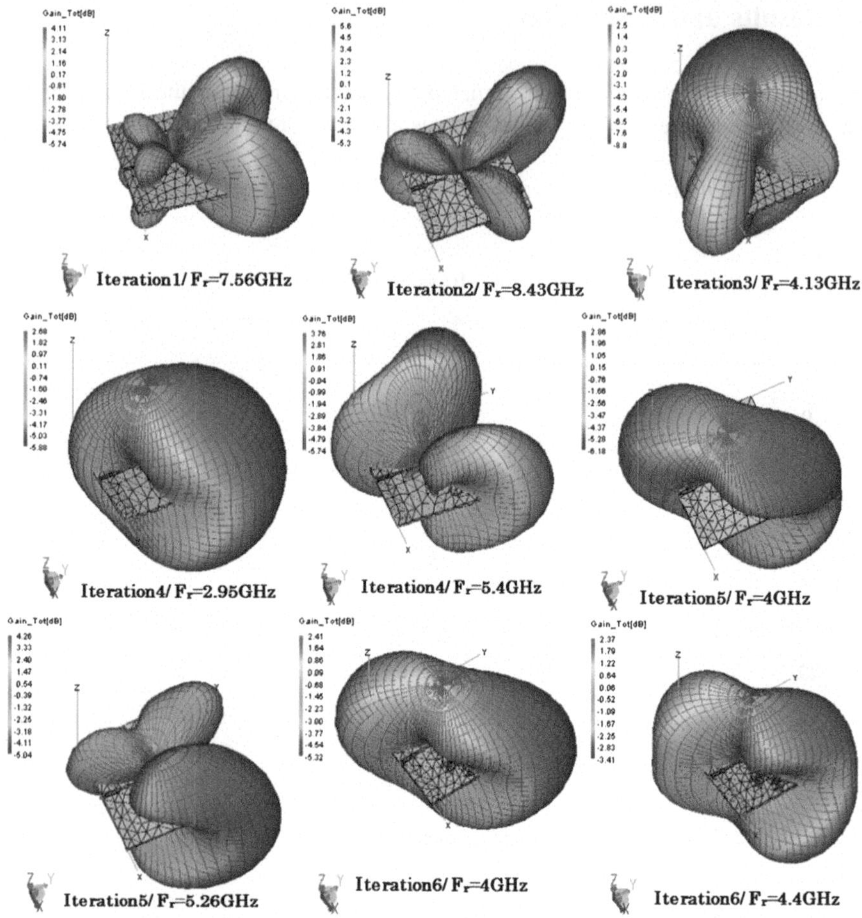

Fig. 6 The 3D-gain patterns for some frequencies

While the iteration number progressed, space-filling is achieved, the resonance is downward shifted and distinct multiband responses are observed. In the 4th iteration, two available broadbands with −10 dB bandwidth 20.3 % (2.7–3.3 GHz) and 18.5 % (5–6 GHz) are obtained. In the 5th iterations, one available broadband with −10 dB bandwidth 61 % (3.75–6.22 GHz) is obtained. In the 6th iteration, one available broadband with −10 dB bandwidth 75.7 % (3.67–6.7 GHz) is obtained. Figure 6 shows the 3D gain patterns in some resonant frequencies. We observe that this radiation patterns are stable, directional or bi-directional with low lobe sides.

5 Conclusion

The fractal concept is a one of the better solutions to design simple, low profile and miniaturized antennas. Six iterations of CPW-Fed DRAGON Fractal antenna are studied and show that more the number of iterations increase more the antenna has a broadband behavior. The 6th iteration of CPW-Fed Dragon fractal antenna have a bandwidth of 3.03 GHz (from 3.67–6.6 GHz) with important radiation gains (peak gains from 2–5.5 dB) which makes it a suitable solution for Ultra Wide Band (UWB) Applications.

Also, more refinements and more iterations can be studied to obtain antennas with better performances and good impedance matching. Manufacturing and measurement should be done to confirm these results.

References

1. Chen, S.-Y., Wang, P.-H., Hsu, P.: Uniplanar log-periodic slot antenna fed by a CPW for UWB applications. Antennas Wirel. Propag. Lett. **5**(1), 256–259 (2006)
2. Reha, A., Said, A.O.: Tri-band fractal antennas for RFID applications. Wirel. Eng. Technol. **04** (04), 171–176 (2013)
3. Reha, A., El Amri, A., Benhmammouch, O., Oulad Said, A.: Fractal antennas : a novel miniaturization technique for wireless networks. Trans. Netw. Commun. **2**(5) (2014)
4. Reha, A., El Amri, A., Benhmammouch, O., Oulad Said, A., El Ouadih, A., Bouchouirbat, M.: CPW-fed H-tree fractal antenna for WLAN, WIMAX, RFID, C-band, HiperLAN, and UWB applications. Int. J. Microw. Wirel. Technol. 1–8 (2015)
5. Abdelraheem, A.M., Abdalla, M.A.: Compact curved half circular disc-monopole UWB antenna. Int. J. Microw. Wirel. Technol. 1–8 (2015)
6. Angelopoulos, E., Anastopoulos, A., Kaklamani, D., Alexandridis, A., Lazarakis, F., Dangakis, K.: Circular and elliptical CPW-fed slot and microstrip-fed antennas for ultrawideband applications. Antennas Wirel. Propag. Lett. **5**(1), 294–297 (2006)
7. Islam, M.T., Samsuzzaman, M., Faruque, M.R.I., Islam, M.M.: Compact metamaterial antenna for UWB applications. Electron. Lett. **51**(16), 1222–1224 (2015)
8. Abd El-Hameed, A.S., Salem, D.A., Abdallah, E.A., Hashish, E.A.: Fractal quasi-self complimentary miniaturized UWB antenna, 15–16 (2013)
9. Falconer, K.J.: Fractal Geometry: Mathematical Foundations and Applications, 2nd edn. Wiley, Chichester, England (2003)

An Efficient Method of Improving Image Retrieval Using Combined Global and Local Features

Abderrahim Khatabi, Amal Tmiri and Ahmed Serhir

Abstract Nowadays, with the increased use of digital images it has become essential to find an efficient system for searching and indexing of images from large image collections. CBIR systems can be used for searching and retrieving different kinds of images from large databases on the bases of the visual content of the images. Currently, CBIR techniques work on combination of low level features i.e. color, shape and texture. In this paper we have designed a content based image retrieval system based on the combination of local and global features. The local features are obtained through local binary pattern (LBP) technique which is used to extract texture-based features from an image, while the global features are extracted using Angular Radial Transform (ART). To demonstrate the efficacy of this combination, experiments are conducted on Columbia Object Image Li-brary (COIL-100) and MPEG-7 shape-1 part B database. The result showed significant improvement in the retrieval accuracy when compared to the existing system.

Keywords ART descriptors · Local binary pattern (LBP) · Content-based image retrieval · Global features · Local features

A. Khatabi (✉) · A. Tmiri
Department of Computer Science, Chouaib Doukkali University,
El Jadida, Morocco
e-mail: khatabiabdo5@gmail.com

A. Tmiri
e-mail: B_tmiri@yahoo.fr

A. Serhir
Department of Mathematics, Chouaib Doukkali University, El Jadida, Morocco
e-mail: A.serhir@gmail.com

R. El-Azouzi et al. (eds.), *Advances in Ubiquitous Networking 2*, Lecture Notes in Electrical Engineering 397, DOI 10.1007/978-981-10-1627-1_34

1 Introduction

Digital images are one of the effective media to present a variety of information in graphic form, due to the increase of huge amount of images generated every day such as in medical diagnosis, military and registration, and geographical images. Therefore, the search for images by visual content has become an active area of research. Several techniques have been developed to search for images accurately and efficiently, based on specific image features such as color [1], texture [2] and shape [3].

Content Based Image Retrieval (CBIR) system is becoming an important tool with the advance of multimedia and imaging technology. The technical CBIR describes the automatic retrieval of images from a database using visual content such as color, texture and shape to represent and index the images [4].

Representation and description of shape based region and contour can exploit the limits of shape information [5]. Global approaches do not divide shape into sub-parts. Typically a feature vector, derived from the integral boundary, is used to describe the shape. Various shape descriptors have been developed in literature, which are divided into contour-based and region-based descriptors [6, 7]. Contour-based descriptors concentrate on boundary lines. Various contour-based descriptors are such as Fourier descriptor [8], curvature scale space [9], wavelet Fourier descriptor [10], and chain codes [11]. Region-based descriptors extract features from whole area of object, hence they are most suitable as descriptors for complex shapes like Zernike moments [12] and Angular radial transform (ART) [13].

Mostly region-based methods use moment descriptors to describe a shape. These descriptors represent global properties of an object. ART have certain desirable properties such as rotation invariance, robustness to noise, orthogonally and fast computation to extract the global features approach, which provides almost feature extraction accuracy as that of ZMs with extra advantage of being quite efficient in its computation time. It applies to a large number of objects, such as complex objects consisting of multiple discrete regions or simple objects with or without hole.

Even though an image represented through global features obtain more information, it is not sufficient enough. Local features need to be extracted, and this can be done through sparse or dense descriptors.

In the case of dense descriptors, local features are captured over complete image. Local binary pattern (LBP) [14] is one of the most widely used approaches due to its invariant to monotonic gray-level changes and ease in extraction of the local features which was primarily introduced for texture analysis and classification [15, 16].

Different researchers have used methods and techniques which considered both local and global Features to derive shape an image. Jain and Vailya [17] proposed a weight-based solution by using feature vector based on edge direction and invariant moments. Wei et al. [18] have used ZMs as global features and centroid distance along with contour curvatures as local features. Shu and Wu [19] have integrate contour points distribution histogram with earth mover distance scheme for shape matching. Singh and Pooja [20] have developed local and global feature-based image retrieval system.

In this paper, we propose the combination of global and local features to improve the performance of CBIR, where the global information of images is extracted by the ART descriptor, while the LBP descriptor captures the significant local information.

The rest of the paper is organized as follows: In Sect. 2, an overview of ART is given. Description of the techniques used for local feature extraction using Local Binary pattern (LBP) is presented in Sect. 3. The proposed method and the similarity measurement used to evaluate the matching score of these methods are in Sect. 4. Section 5 gives details of experimental analysis. Lastly, conclusion and concluding remarks are presented in Sect. 6.

2 Angular Radial Transform (ART)

ART is a complex orthogonal unitary transform defined on a unit disk based on complex orthogonal sinusoidal basis functions in polar co-ordinates [21–23].

The ART coefficients, F_{nm} of order n and m, are defined by:

$$F_{nm} = \int_{0}^{2\pi} \int_{0}^{1} V_{nm}(\rho, \theta) f(\rho, \theta) \rho d\rho d\theta \tag{1}$$

Where $f(r, \theta)$ is an image intensity function in polar co-ordinates and $V^{*}_{nm}(r, \theta)$ is a basis function, which is complex conjugate of $V_{nm}(r, \theta)$ defined in polar coordinates over a unit disk. These are expressed in a separable form of both radial and angular parts as follows:

$$V_{nm}(\rho, \theta) = A_m(\theta) R_n(\rho) \tag{2}$$

The indices n and m are non-negative integers. The real valued $R_n(r)$ radial polynomial and the angular basis function $A_m(\theta)$.

The radial component are defined as follows:

$$R_n(\rho) = \begin{cases} 1 & n = 0 \\ 2\cos(\pi n \rho) & n \neq 0 \end{cases}$$

In order to make the transformation invariant in rotation, an exponential is used in the angular basis function:

$$A_m(\theta) = \frac{1}{2\pi} e^{jm\theta}$$

Where $j = \sqrt{-1}$ the important characteristics of ART is the rotational invariance. The magnitude values of ART are unaffected and remain identical for image functions before and after rotation. The original image is represented by the intensity image function $f(r, \theta)$ having ART is rotated counterclockwise by angle α; the transformed image function is

$$g(r, \theta) = f(r, \theta - \alpha).$$

The ART coefficients of original and rotated images are F_{nm} and $F_{nm}^{rot} = e^{-jm\theta} F_{nm}$, the magnitude values are identical, where

$$\left\| e^{-jm\theta} \right\| = 1$$

The standard MPEG-7 recommends using 12 features angular and 3 radial functions (n = 3 and m = 12). The distance measure between two shapes described by ART is obtained by using the L2 norm:

$$D_{(ART)}\left(F_{ART}^{R}, F_{ART}^{M}\right) = \frac{1}{N_G} \sqrt{\sum_{i=1}^{N_G} \left(F_{ART}^{R} - F_{ART}^{M}\right)^2} \tag{3}$$

This measure is defined as the square root of the sum of the squared differences between two turning angle vectors, one belonging to the query's shape, R, and the other belonging to a shape in the contents (database), M, for which a description is available.

In [24] shows that if ART coefficients are calculated in the polar coordinate system eliminates the geometric and integral error.

3 Local Binary Pattern (Lbp)

The original LBP operator, introduced by Ojala et al. [25] is an operator which transforms an image into an array of integer labels describing small-scale appearance of the image. In this work we use the Local Binary Pattern (LBP) to represent textural properties of images and encode the appearance of the local region and the spatial structures.

Let's Consider image I(x, y) and g_c denote the gray level of an arbitrary pixel (x, y), i.e.

$$g_C = \mathrm{I}(\mathrm{x},\ \mathrm{y}) \text{ and } g_p = \mathrm{I}(x_p,\ y_p) \tag{4}$$

However, x_p denote the gray value of a sampling point in an evenly spaced circular neighborhood of P sampling points and radius R around point (x, y).

The coordinates of the center pixel are (xc, yc) then the coordinates of his P neighbors (xp, yp) on the edge of the circle with radius R can be calculated with the sinus and cosines:

$$\begin{aligned} x_p &= \mathrm{x} + \mathrm{R}\cos(2\pi \mathrm{p}/\mathrm{P}), \\ y_p &= \mathrm{y} - \mathrm{R}\sin(2\pi \mathrm{p}/\mathrm{P}). \end{aligned} \tag{5}$$

The local texture neighborhood T is defined as a distribution of gray levels I(x, y) of P + 1 pixels. It is expressed below by:

$$\mathrm{T} = \mathrm{t}\left(g_C,\ g_0,\ g_1,\ \ldots,\ g_{p-1}\right) \tag{6}$$

The center pixel value can be subtracted from the neighborhood, the set T can be expressed by:

$$\mathrm{T} = \mathrm{t}\left(g_C,\ g_0 - g_C,\ g_1 - g_C,\ \ldots\ g_{p-1} - g_C\right) \tag{7}$$

By assuming the center pixel to be statistically independent of the differences, which allows for factorization of the distribution:

$$\mathrm{T} \approx \mathrm{t}(g_C)\mathrm{t}\left(g_0 - g_C,\ g_1 - g_C\ \ldots\ g_{p-1} - g_C\right). \tag{8}$$

Where t (gc) is the intensity distribution over I(x, y). From the point of view of analyzing local textural patterns, it is independent of the local image texture. So, the T distribution of differences can be defined as follows:

$$\mathrm{t}\left(g_0 - g_C, g_1 - g_C,\ \ldots, g_{p-1} - g_C\right) \tag{9}$$

In order to alleviate these challenges, only the signs of the differences are considered:

$$\mathrm{t}\left(\mathrm{s}(g_0 - g_C), \mathrm{S}(g_1 - g_C),\ \ldots, \mathrm{s}\left(g_{p-1} - g_C\right)\right) \tag{10}$$

For each $s(g_p - g_c)$ sign, there is associated a binomial weight 2^p that transforms different neighborhoods in the LBP. As in the case of basic LBP, it is obtained by summing the thresholded differences. (Fig. 1).

$$LBP_{P,R}(x_c, y_c) = \sum_{p=0}^{p-1} s(g_p - g_c)2^p \tag{11}$$

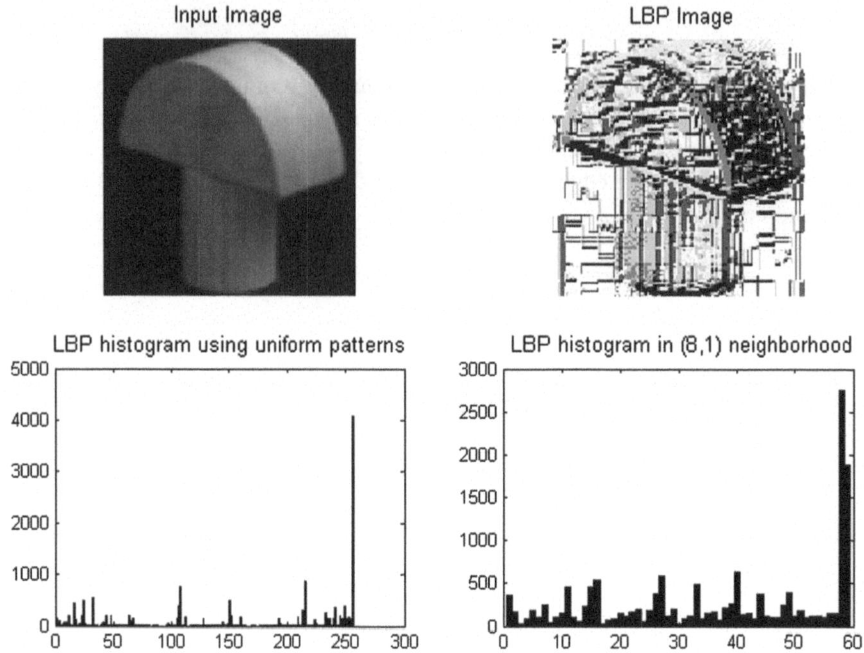

Fig. 1 The LBP image and LBP histogram in (8, 1) neighborhood

The local distribution of gray levels is approximately defined as follows:

$$T \approx t(LBP_{P,R}(x_c, y_c)) \tag{12}$$

However, the values of the gray levels gp depend on the image rotation, we tried to set this function to avoid this effect can be defined as follows:

$$LBP^{ri}_{P,R} = min\{ROR(LBP_{P,R},\ i) \qquad |i = 0,\ 1,\ \ldots,\ p-1\}$$

Where ROR(x, i) is responsible of the circular shift that takes place i times on the bit P having an x value.

And $LBP^{ri}_{P,R}$ is corresponds to the quantification of attributes occurrences that have an invariant rotation.

We use this $LBP^{riu2}_{P,R}$ instead of $LBP^{ri}_{P,R}$ to describe the texture invariant to rotation:

$$LBP^{riu2}_{P,R} = \begin{cases} \sum_{p=0}^{p-1} s(g_p - g_c) & if \quad u(LBP_{P,R}) \le 2 \\ P+1 & otherwise \end{cases} \tag{13}$$

And we use this operator $LBP_{P,R}^{riu2}$ can obtain an excellent performance. This kind of LBP texture representation only describes the change between the central pixel and neighbors.

4 The Proposed Method

In this paper, we propose a new method to build a new simple system by the content (CBIR) of images by using LBP and ART features.

ART are region-based descriptors; hence, they are capable of identifying the global similarity of all images with query image and LBP is applied on original image and all images in Database to capture the Local features.

With the proposed method, the original image is given as an input, ART is then applied for extracting the features and images are stored offline in the features database DB1. Similarly, the Local Dense descriptors described is applied on images, and the features are stored in database DB2.

The features of query image and all other images in the database are computed and the similarity of the query image and images from database is computed using any similarity metrics. The final retrieved images show the retrieval performance of the proposed method.

The steps taken to implement our system that combine Local and Global feature descriptors are as follow: step (1) is applied to obtain global features and step (2) is used to extract local features.

1. After segmentation of input image we can extract the global features based on ART (Magnitude) features
   (a) Calculate the shape features with ART (Magnitude) features.
   (b) Apply the process in step 1(a) on all images in the database for each class and store the features in database DB1.
2. Extract local features of each image using LBP descriptor.
   (a) Calculate features with LBP.
   (b) Apply the process in step 2(a) on all images in the database as per different and store the features in database DB2.
3. When the query image is submitted, the features of query image are calculated using step 1(a), and the similarity distance between ART (Magnitude)-based features of query image and features stored in database DB1 is computed. Similarly, use the same algorithm in step 2 to extract LBP features of query image and then find the similarity distance between local features of query image and features stored in database DB2. Afterward, the combined distance i.e., distance between features of ART and LBP are calculated and sorted in ascending order of similarity distance.

4. Presision and Recall are calculated from chosen retrieved images in step 3.

4.1 Precision–Recall (P–R)

The idea behind performance evaluation is to make a prediction on the retrieval performance of CBIR system. The common methods used to evaluate the performance of CBIR systems are User comparison, rank of best match, Average rank of retrieval images, Precision and Recall, Error rate etc.

Most of the researchers used precision (P) and recall (R) rate as a performance metric in CBIR systems.

Let I_q be the query image, precision and recall rate are defined as:

$$\begin{aligned} P(I_q) &= \frac{\text{(Total no. of Retrieval Relevant image)}}{\text{(Total no. of Retrieval image)}} \\ R(I_q) &= \frac{\text{(Total no. of Retrieval Relevant image)}}{\text{(Total no. of Relevant image)}} \end{aligned} \tag{14}$$

4.2 Euclidean Distance

Euclidean distance (L2 norm) is a straight line distance between any two points and used as a similarity metric.

$$D_E = \sqrt{\sum_{i=0}^{N-1} (A[i] - B[i])^2} \tag{15}$$

4.3 Combined Distance

The distance between global feature descriptors described by ART is defined:

$$D_{(ART)}(F_{ART}^{Q}, F_{ART}^{B}) = \frac{1}{N_G}\sqrt{\sum_{i=1}^{N_G} (F_{ART}^{Q} - F_{ART}^{B})^2}$$

The distance between local feature descriptors described by LBP is defined

$$D_{LBP}(F_{LBP}^{Q}, F_{LBP}^{B}) = \frac{1}{N_L}\sqrt{\sum_{i=1}^{N_L} (F_{LBP}^{Q} - F_{LBP}^{B})^2}$$

Where Q is the query image, B is an image from the database.

Where N_G and N_L are the maximum number of global and local features, respectively,

Finally, the combination takes place as follows:

$$D_{ART+LBP} = w_1 D_{ART} + w_2 D_{LBP} \quad (16)$$

In our work, we obtain w1 is 0.51 and w2 is 0.49.

5 Experimental Analysis

We conduct various experiments to demonstrate the effectiveness of the proposed method using different database images with different intensity, shape and texture.

The database used here are the Columbia Object Image Library (COIL-100) and MPEG-7 shape-1 part B.

5.1 Database

The proposed techniques have been implemented on a real dataset. We have used the Columbia Object Image Library (COIL-100) and MPEG-7 shape-1 part B database.

Figures 2 and 3 are used as a query image if the database images and the query image are described by using ART and LBP.

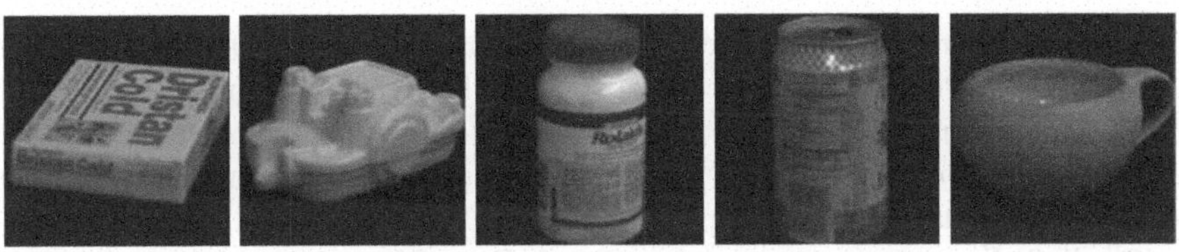

Fig. 2 Example images from the COIL-100 database

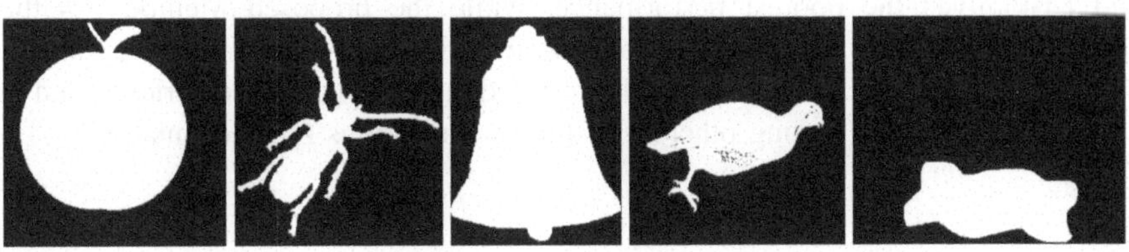

Fig. 3 Example images from the MPEG-7 shape-1 part B database

Table 1 Comparison of average of the proposed and recent methods

	FD	ZM	ART	LBP	Proposed
COIL-100	57.94	90.94	94.32	78.65	99.85
MPEG-7	40.54	70.35	98.52	82.50	99.78

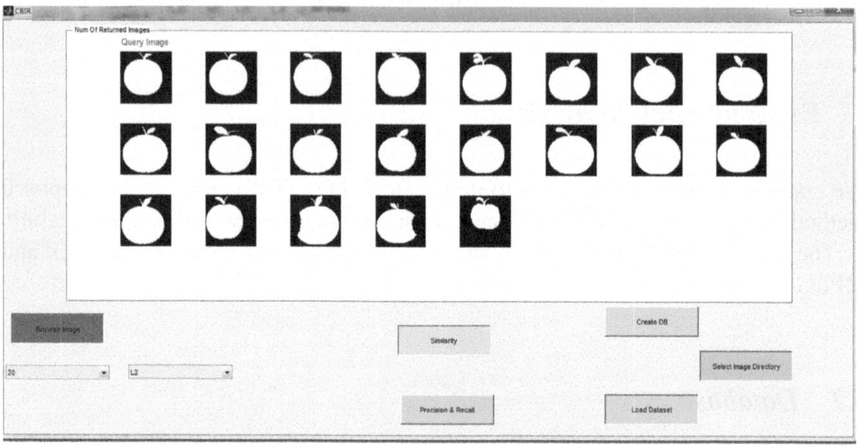

Fig. 4 Top 12 retrievals images taken from MPEG-7 using ART + LBP

- **COIL-100**: This database consists of 99 images categorized into 9 different classes where each class consists of 11 instances. The images are resized to 101 × 101 pixels.
- **MPEG-7 shape-1 part B**: set includes 1400 shape samples, 20 for each class. The images used for the experiments are resized to 101 × 101 pixels.

The Table 1 shows the average performance based on all the 10 categories from coill-100 and MPEG-7, showing comparison with few recent methods FD, LBP and ZM, where the proposed Method have performed better than the other descriptors.

The number of search results can change according to the number of similar images in the database Fig. 4.

The P–R diagrams are presented in Fig. 5. For subject change images, i.e., for MPEG-7 database, it is observed that Global features extracted by ART and ZMD provide similar performance. The performance of Global features extracted by FD and LBP gives the poorest performance, while the proposed Method has the superior performance. For COIL-100 database the P–R are given in Fig. 6, where it can be seen that the performance of proposed method has high retrieval and it attains its superiority among other descriptions for various types of images.

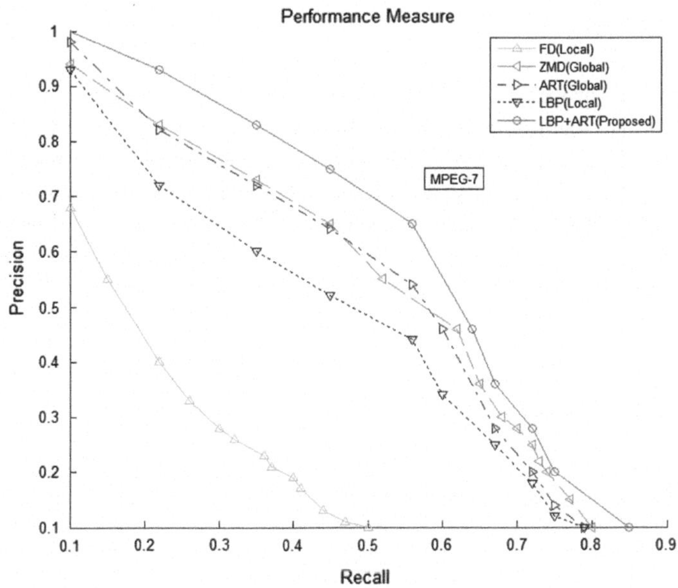

Fig. 5 P–R comparisons for MPEG-7 database of different methods and proposed

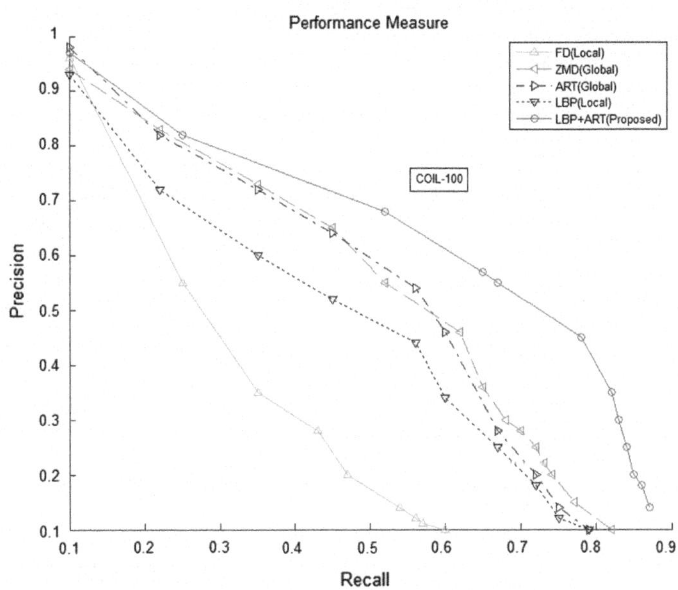

Fig. 6 P–R comparisons for Coill-100 database of different methods and proposed

6 Conclusion

In this paper, we propose a novel method for image retrieval system by combining local and global features using Local binary pattern (LBP) and Angular radial Trans-form (ART) respectively. Both global and local features are matched using Euclidean distance similarity measures. The extensive experimental results demonstrate the robustness and effectiveness of the proposed method as compared to the traditional ones for FD, LBP and ZM, which out performs the existing approaches of databases containing different kinds of image retrieval.

References

1. Swain, M., Ballard, D.: Color indexing. Int. J. Comput. Vis. **7**(1), 11–32 (1991)
2. Manjunath, B., Ma, W.: Texture features for browsing and retrieval of image data. IEEE Trans. Pattern Anal. Mach. Intell. **18**, 837–842 (1996)
3. Pentland, A., Picard, R.W., Sclaroff, S.: Photobook: content-based manipulation of image databases. Int. J. Comput. Vis. **18**(3), 233–254 (1996)
4. Shambharkar, S.A., Tirpude, S.C.: A comparative study on retrieved images by content based image retrieval system based on binary tree, color, texture and canny edge detection approach. In: IJACSA Special Issue on Selected Papers from International Conference & Workshop On Emerging Trends In Technology, pp. 47–51 (2012)
5. Khatabi, A., Tmiri, A., Serhir, A., Silkan, H.: Content-based shape retrieval (CBIR) using different shape descriptors. In: 2014 5th Workshop on Codes, Cryptography and Communication Systems (WCCCS), pp. 98–102. IEEE (2014)
6. Mehtre, B.M., Kankanhalli, M.S., Lee, W.F.: Shape measures for content based image retrieval: a comparison. Inf. Process. Manag. **33**(3), 319–337 (1997)
7. Zhang, D., Lu, G.: Review of shape representation and description techniques. Pattern Recognit. **37**, 1–19 (2004)
8. Zhang, D., Lu, G.: A comparative study of curvature scale space and Fourier descriptors for shape-based image retrieval. Vis. Commun. Image Represent. **14**(1), 41–60 (2003)
9. Flickner, M., Sawhney, H., Niblack, W., Ashley, J., Huang, Q., Dom, B., Gorkani, M., Hafine, J., Lee, D., Petkovic, D., Steele, D., Yanker, P.: Query by image and video content: the QBIC system. In: IEEE Computer (1995)
10. Dubois, S.R., Glanz, F.H.: An autoregressive model approach to two dimensional shape classification. IEEE Trans. Pattern Anal. Mach. Intell. **8**, 55–65 (1986)
11. Gevers, T., Smeulders, A.W.M.: Pictoseek: combining color and shape invariant features for image retrieval. IEEE Trans. Image Process. **9**(1), 102–119 (2000)
12. Kale, K.V., Deshmukh, P.D., Chavan, S.V., Kazi, M.M., Rode, Y.S.: Zernike moment feature extraction for handwritten Devanagari compound character recognition. In: Science and Information Conference (SAI), pp. 459–466. IEEE (2013)
13. Hwang, S., Kim, W.: Fast and efficient method for computing ART. IEEE Trans. Image Process. **15**, 112–117 (2006)
14. Suri, P.K., Verma, E.A.: Robust face detection using circular multi block local binary pattern and integral haar features. Int. J. Adv. Comput. Sci. Appl. Spec. Issue Artif. Intell. (IJACSA) (2010)
15. Liao, S., Law, M.W., Chung, A.: Dominant local binary patterns for texture classification. IEEE Trans. Image Process. **18**(5), 1107–1118 (2009)

16. Gho, Z., Zhang, L., Zhang, G.: A completed modeling of local binary pattern operator for texture classification. IEEE Trans. Image Process. **19**(6), 1657–1663 (2010)
17. Jain, A.K., Vailaya, A.: Shape-based retrieval: a case study with trademark image databases. Pattern Recognit. **31**(5), 1369–1390 (1998)
18. Wei, C.H., Li, Y., Chau, W.Y., Li, C.T.: Trademark image retrieval using synthetic features for describing global shape and interior structure. Pattern Recognit. **42**(3), 386–394 (2008)
19. Shu, X., Wu, X.J.: A novel contour descriptor for 2D shape matching and its application to image retrieval. Image Vis. Comput. **29**(4), 286–294 (2011)
20. Pooja, S.C.: Local and global features based image retrieval system using orthogonal radial moments. Opt. Lasers Eng. **50**(5), 655–667 (2012)
21. The Moving Picture Experts Group (MPEG) (2009). http://www.chiariglione.org/mpeg
22. Amanatiadis, A., Kaburlasos, V.G., Gasteratos, A., Papadakis, S.E.: Evaluation of shape descriptors for shape-based image retrieval. Image Process. **5**, 493–499 (2011)
23. Pooja, C.S.: An effective image retrieval system using region and contour based features. In: IJCA Proceedings on International Conference on Recent Advances and Future Trends in Information Technology, pp. 7–12 (2012)
24. Khatabi, A., Tmiri, A., Serhir, A.: A novel approach for computing the coefficient of ART descriptor using polar coordinates for gray-level and binary images. In: Advances in Ubiquitous Networking, pp. 391–401. Springer, Singapore (2016)
25. Ojala, T., Pietikäinen, M., Harwood, D.: A comparative study of texture measures with classification based on feature distributions. Pattern Recognit. **29**(1), 51–59 (1996)

A Comparative Experimental Study of Spectral Hashing

Loubna Karbil, Imane Daoudi and Hicham Medromi

Abstract Binary encoding methods that keep similarity in large scale data become very used for fast retrieval and effective storage. There have been many recent hashing technics that produce semantic binary codes. We are particularly interested in Spectral Hashing based methods which provide an efficient binary hash codes in a very simple way. This paper presents a comparative experimental study of Spectral Hashing to show the performance gain and the behaviour of this method on large scale Databases. In the best of our knowledge there is no experiments done on the evolution of the hamming matrix size on big data. Two large databases are used to show the limitation of Spectral Hashing and possible research tricks will be proposed.

1 Introduction

Similarity search, also known as approximate nearest neighbor search, has many applications such as content-based image retrieval (CBIR). Many real world applications of similarity search need to process a huge amount of data within a high dimensional space to answer a query. A simple linear scan in the original high dimensional data may be prohibitively expensive in term of storage and processing when data size grows.

Recently, hashing methods have been successfully used for approximating nearest neighbor search due to its fast query speed and low storage cost. Locality Sensitive Hashing (LSH) [1] is one of the most commonly used data-independent hashing methods. It uses random linear projections which are independent from

L. Karbil (✉) · I. Daoudi · H. Medromi
Systems Architecture, ENSEM, Hassan II University, Casablanca, Morocco
e-mail: l.karbil@gmail.com

I. Daoudi
e-mail: i.daoudi@ensem.ac.ma

H. Medromi
e-mail: h.medromi@yahoo.fr

R. El-Azouzi et al. (eds.), *Advances in Ubiquitous Networking 2*, Lecture Notes in Electrical Engineering 397, DOI 10.1007/978-981-10-1627-1_35

training data. Another class of hashing methods are called data dependent methods, which aim to learn hash functions from specific dataset. These data dependent methods include Spectral Hashing (SH) [2] which we focus our interest in, and where a subset of eigenvectors of a Laplacian graph is rounded to determine binary codes. Many other methods that give a compact binary codewords are based on Spectral Hashing, like Kernelized Spectral Hashing [3], Sparse Spectral Hashing [4], Weighted Hashing [5], Self-Taught Hashing [6], Multidimensional Spectral Hashing [7], HyperGraph Spectral Hashing [8], Spectral Hashing with Semantically Consistent Graph [9], Linear Spectral Hashing [10] and Robust Discrete Spectral Hashing [11].

The rest of this paper is organized as follows: Sect. 2 presents in detail the Spectral Hashing indicating its Shortcomings and provides a comparison to Spectral Based Hashing methods. Section 3 discusses the experiments and shows the behavior of Spectral Hashing in high dimensional space. Finally we conclude discussing results and opening future research directions.

2 Spectral Based Techniques

We will start the discussion with the basic method called Spectral Hashing (SH). Later, we will move to discuss different class of solutions that use spectral relaxation to obtain binary codes. These include Kernelized Spectral Hashing [3], Sparse Spectral Hashing [4] Weighted Hashing [5], Self-Taught Hashing [6], Multidimensional Spectral Hashing [7], HyperGraph Spectral Hashing [8], Spectral Hashing with Semantically Consistent Graph [9], Scalable Similarity Search with Optimized Kernel Hashing [12], Linear Spectral Hashing [10] and Robust Discrete Spectral Hashing [11].

2.1 Notations and Definitions

The following common notion of variables is used in this paper: there are n Images given for training. Each training item is represented with a feature vector x of dimension d and the data matrix is represented by $X \in R^{n \times d}$ where each row is a data point. Each binary code y is indicated by $y_i = [y_1, y_2, y_3, \ldots y_k]$ with k bits. $Y \in \{1, -1\}^{n \times k}$ represents the binary code matrix where each row is a binary code y_i of a vector x_i.

A good binary code is the one that satisfies the following properties: Balanced and Independence Properties. We require that each bit had 50 % chance to be one or zero, which is the balanced property. The code of k bits should be independent, i.e. each bit is obtained independently from the previous one. The two properties are given by:

- Balanced property: $\sum_i y_i = 0$
- Independence property: $\frac{1}{n}\sum_i y_i y_i^T = I$

2.2 Spectral Hashing

In Spectral Hashing Weiss et al. [2] formalized the problem of finding the best code for a given dataset and showed that these are equivalent to a particular of graph partitioning and it is NP-hard even for a single bit due to the balanced constraint. A spectral relaxation is applied whose solutions are simply a subset of thresholded eigenvectors of the graph Laplacien. By utilizing recent results on convergence of graph Laplacian eigenvectors to Laplace-Beltrami eigenfunctions of manifolds. They show how to efficiently calculate the code of a novel input. The optimization criteria is to minimize the expected Hamming distance between similar data points satisfying the independence and balanced properties:

$$minimize \sum_{ij} W_{ij}\left\|y_i - y_j\right\|^2 \tag{1}$$

$$\begin{aligned} subject\ to: \quad & y_i \in \{1, -1\}^k \\ & \sum_i y_i = 0 \\ & \frac{1}{n}\sum_i y_i y_i^T = I \end{aligned}$$

where $W_{ij} = e^{\left(-\left\|x_i - x_j\right\|^2 \setminus \in^2\right)}$ is the affinity matrix. Since we are assuming the inputs are embedded in R^d so that Euclidean distance correlates with similarity.

Representing the above optimization with matrices and adding a diagonal $n \times n$ matrix D where $D_{ii} = \sum_J W_{ij}$ gives a formulation like:

$$minimize\ trace\left(Y(D - W)Y^T\right) \tag{2}$$

$$\begin{aligned} subject\ to: \quad & Y(i,j) \in \{1, -1\} \\ & Y^T 1 = 0 \\ & YY^T = I \end{aligned}$$

This still of course a had problem; but by removing the constraint that Y(i, j) $\in$ {1, −1} we obtain an easy problem whose solutions are simply the k-eigenvectors of $D - W$ with minimal eigenvalue (after excluding the trivial eigenvector 1 which had eigenvalue 0).

But this would only tell us how to compute the code representation of items in the training set, for a new input the solution should be extended. Assuming the data points are samples from a probability distribution *p(x)*, and by utilizing recent results on convergence of graph Laplacian eigenvectors to the Laplace-Beltrami eigenfunctions of manifolds, we obtain a spectral problem whose solutions are eigenfunctions with small eigenvalue. For a case of uniform distribution on $[a, b]$ the eigenfunctions $\phi_k(x)$ of one dimensional L_p and eigenvalues λ_k are:

$$\phi_k(x) = \sin\left(\frac{\pi}{2} + \frac{k\pi}{b-a}x\right) \quad (3)$$

$$\lambda_k = 1 - e^{-\left(\frac{\epsilon}{2}\left|\frac{k\pi}{b-a}\right|^2\right)} \quad (4)$$

To summarize, given a training set of points $\{x_i\}$ and a desired number of bits k the spectral hashing algorithm works by:

- Finding the principal components of data using Principal Component Analysis (PCA).
- Calculation de k smallest single-dimension analytical eigenfunctions of L_p using a rectangular approximation along every PCA direction. This is done by evaluating the k smallest eigenvalues for each direction using Eq. (4), thus creating a list of dk eigenvalues, and then sorting this list to find the k smallest eigenvalues.
- Thresholding the analytical eigenfunctions at 0, to obtain binary codes.

2.3 Multidimensional Spectral Hashing (MDSH)

This method [7] is based on Spectral Hashing but the difference is that whereas the original SH uses only single-dimension eigenfunctions during retrieval, while this method expands the code to include the outer-product eigenfunctions as well.

The algorithm of the multidimensional Spectral Hashing is:

- Calculate the single-dimension eigenfunctions.
- Sort the λ_{ij} and find a set of k indices so that λ_{ij} are maximal.
- Encode each data point x with $y(x) = sign(\phi_{ij})$.
- Expand the code of x to include all outer-product eigenfunctions.
- Calculate the Hamming affinity between each x_i and x_j is given by:
- $$H(i,j) = \sum_l y_l(x_i)\lambda_l y_l(x_j) \quad (5)$$

Where the index l goes over the single-dimension bits as well as the outer-product bits.

2.4 Hyper Graph Spectral Hashing

This method [8] uses unified hypergraph [12] to model relationships between datasets and introduces hyperedges to represent the various similarities between data points. This technic extends the Spectral Hashing from an ordinary graph to a hypergraph and formulates the problem with the hypergraph Laplacian to get more precision than the original method.

2.5 Sparse Spectral Hashing

This method [4] uses boosting similarity and sparse principal component analysis under the original Spectral Hashing instead of the simple principal component analysis because in practice, generally semantics implied in an image is represented only with several distinctive features, rather than introduction other unrelated features in image expression. The method puts forward corresponding global optimization solution to establish explainable binary coding for large-scale image data and fulfill image index.

2.6 Weighted Hashing

This technique [5] assigns different weights on different hashing bits to capture their importance. The hashing codes and their corresponding weights are jointly learned in a unified framework by simultaneously preserving the similarities between data examples and balancing the variance of each hashing bits.

2.7 Spectral Hashing with Semantically Consistent Graph (SHSCG)

The method [9] gives a simple way to directly optimize the graph Laplacian. The method constructs a semantically consistent sparcified graph, which can better represent similarity between samples than the Euclidean distance. This learned graph is then applied to Spectral Hashing for effective binary code learning.

2.8 *Scalable Similarity Search with Optimized Kernel Hashing (SSOKH)*

This algorithm [12] follows the idea of generating data-dependent optimal hash codes similar to that used in Spectral Hashing. In this method, kernel hash function is explicitly represented and learned via optimization, which can be directly applied to novel inputs even in non-vector data format. In addition, several speed up techniques, such as those based on landmark points or Nystrom approximation, are incorporated to reduce the time complexity involved in the indexing and search stages. Finally, our method does not make any assumption about the similarity terms. Therefore, diverse types of similarities such as feature similarity, semantic category consistence, or other association relations can be easily handled.

2.9 *Linear Spectral Hashing*

This method [10] gives linear scalar products to overcome the problem of calculating hash codes for unseen data in Spectral hashing. This by connecting Spectral Hashing with Spectral Clustering and using normalized Laplacian instead of the simple Laplacian.

2.10 *Robust Discrete Spectral Hashing (RDSP)*

This method [11] propose a novel approach which targets at jointly learning discrete binary codes as well as robust hash functions. Unlike the Spectral Hashing which uses relaxation on discrete constraint to overcome on NP-hard difficulties for the discrete optimization, this method propose a unified hashing framework to directly output discrete binary codes.

3 Comparative Study

The table below shows different ameliorations and weakness attacked by these spectral based techniques (Table 1).

Despite the high performance and progress in similarity search made by all these methods, they have one important limitation: they use one hash table for retrieving neighbors, which means that a linear scan is done to find hash codes from the Hamming matrix that are similar to the query. In the best of our knowledge, there are no experiments done about the size evolution of the hamming matrix created by the Spectral Hashing in big data.

Table 1 A comparison of different methods based on the Spectral Hashing with solutions to different weakness of the basic method

Methods	Problem attacked	Solution proposed	Comparison with SH
Linear SH	Problem of Hashing unseen points	Change Laplacian with normalized Laplacian Connect Spectral Hashing with Spectral clustering	Better performance than SH for code length from 16 to 128 More training time
MDSH	Problem of low performance as the number of bits increases	Use of Kernel functions Change Hamming matrix with affinity Hamming matrix	Better performance than SH Less training time
Sparse SH	Problem of using all directions in PCA	Change PCA with Sparse PCA	The same performance of SH Less training time
SHSCG	Problem of using the euclidean distance for similarity	Generate an optimized graph Laplacian	Better performance than SH Less processing time
Weighted Hashing	Problem of considering hash bits carrying the same amount of information	Assign different weights on different Hashing bits	Better performance than SH The same processing time
RDSP	Problem of using relaxation to overcome the NP–Hard problems	Use discrete kernel hash functions and robust learning components	Better performance than SH in large scale databases
Hyper Graph SH	Problem of optimization by using an ordinary graph Laplacian	Change the graph Laplacian with a Hyper graph	Better performance than SH. The same processing time
SSOKH	Problem of using linear data problem of Hashing unseen points	Use optimized kernel functions provide new technique of Hashing new points	The same processing time. Better performance than SH for small code length

The next section will study the behavior of the Spectral Hashing in two large scale databases and see the evolution of the Hamming matrix size when data reach 7 million vectors and we will interpolate results until 10 million vectors.

4 Experimental Study

In this section, experimental results on two real multidimensional datasets are used to see the behavior of the basic Spectral Hashing in large scale databases. All our experiments are conducted on a computer with Intel Core i7 CPU 2.1 GHZ and 22 GB RAM.

4.1 Datasets

Two widely used datasets are adopted for evaluation: One is the Gist1 M [13] which is a public dataset containing 1 Million vectors with 960 dimensions. The other one is Cifar-10 [14] Dataset that contains 50.000 vectors, each one has 3072 dimensions. The behavior of the hamming matrix size and CPU time are evaluated on synthetic data of 7 million vectors and results are interpolated for 10 million vectors, each vector has 960 dimensions.

4.2 Evaluation

For a given database, we use for evaluation a test vectors that are not included in the original database and then we compute standard retrieval performance measures: precision and recall [15].

$$Precision = \frac{The\ number\ of\ retreived\ relevant\ points}{The\ number\ of\ all\ retreived\ points} \tag{6}$$

$$recall = \frac{The\ number\ of\ retreived\ relevant\ points}{The\ number\ of\ all\ relevant\ points} \tag{7}$$

The reported performance scores in the following Section are averaged over all test queries in the dataset.

To determine whether a retrieved vector in "relevant" to the given query, we proceed to a linear scan using the Euclidean distance. We vary the code length from 8 to 128-bit and the Hamming ball radius (i.e. the maximum Hamming distance between any retrieved vector and the query) from 0 to 20 in order to show their influences on the retrieval performance.

4.3 Results

The proposed evaluation of the Spectral Hashing aims to show the limitation of this method in large scale data bases. Figures 1 and 2 show the performance of Spectral Hashing in terms of its precision-recall curves (created by varying the code length while fixing the Hamming ball radius at 2. We see that Spectral Hashing does a bad job when database has a big size and when dimension is very high. Also Spectral Hashing gives bad performance when the length of the bit code grows and when it is very small. In addition the Spectral Hashing assumes that data is separable which explains the very bad performance that gives the method on Gist1 M which is skewed. The method does a poor job in approximating the far away neighbors.

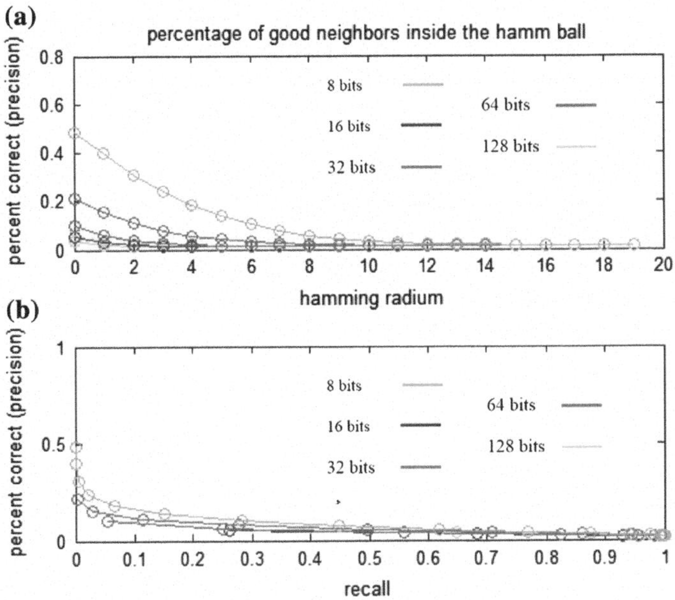

Fig. 1 **a** Precision results for different Hamming balls for Gist 1M and **b** results under the precision-recall curve for Gist 1 M

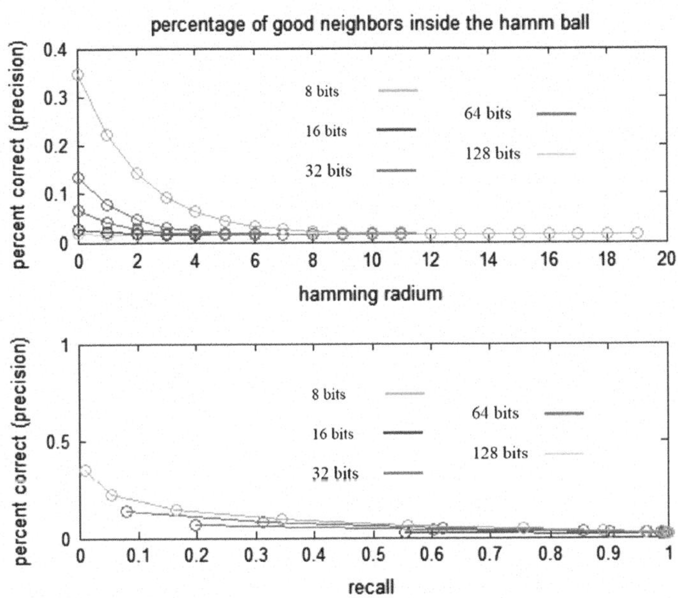

Fig. 2 Precision results for different Hamming balls and results under the precision-recall curve for Cifar-10

The Fig. 3 showed that the evolution of the Hamming matrix size is sublinear until 3.5 million vectors. Then we detect that the shape of the curve has changed. The evolution becomes linear after that number which is problematic. To zoom the evolution, we interpolate results until 10 million vectors. Figure 4 shows the shape of the curve.

This behavior of the size of the hamming matrix impacts the processing time as well. Figure 5 shows that the curve shape of the CPU time changes after 3.5 million vectors for 16 and 32 bits, which makes the linear scan over the hamming matrix becomes a problem in big data. To overcome this limitation, many tracks are possible: A hierarchical structure for example can give promising results–remains the objective of a future research.

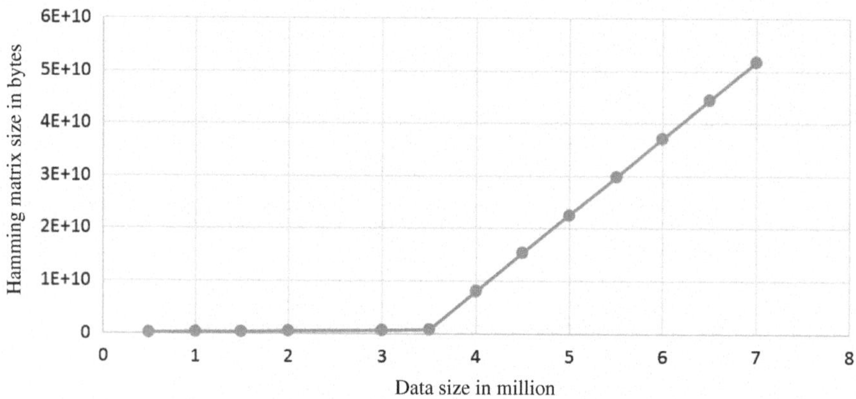

Fig. 3 The behavior of the Hamming Matrix size for different size of data until 7 million vectors

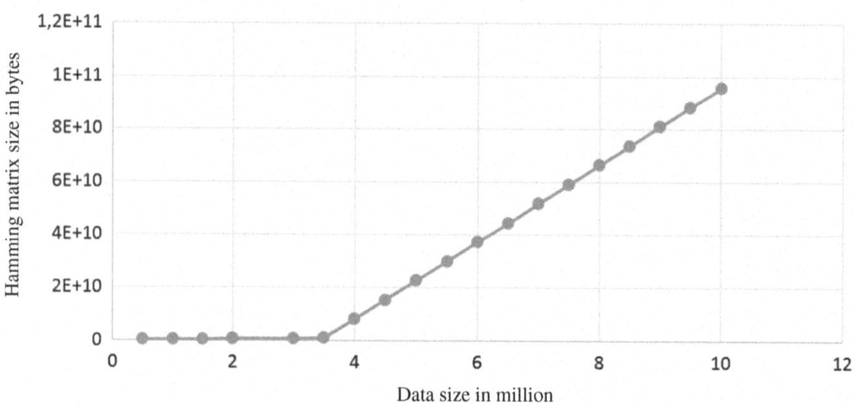

Fig. 4 The behavior of the Hamming Matrix size for different size of data until 7 million vectors with interpolation until 10 million vectors

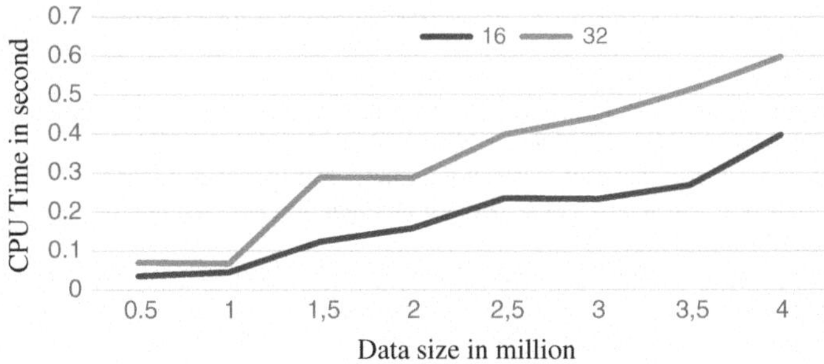

Fig. 5 The behavior of the CPU time for different data size for 16 and 32 bits code length

5 Conclusion

In this paper, we analyzed the performance in scaling of the Spectral Hashing method.

From the experiments, we conclude that the choice of the code length is crucial, and determines the search quality and performance. The major drawback of this approach is the required memory space to charge the Hamming matrix, for large databases such as video databases which is the main contribution of this paper. Several improvements should be made to the Spectral Hashing index to further reduce the computation complexity especially in large spaces.

References

1. Datar, M., Immorlica, N., Indyk, P., et al.: Locality-sensitive hashing scheme based on p-stable distributions. In: Proceedings of the Twentieth Annual Symposium on Computational Geometry, pp. 253–262. ACM, Study (2004)
2. Weiss, Y., Torralba, A., et al. Fergus, R.: Spectral hashing. In: Advances in Neural Information Processing Systems. pp. 1753–1760 (2009)
3. He, J., Liu, W., Chang, S.-F.: Scalable similarity search with optimized Kernel Hashing. In: Proceedings of the 16th ACM SIGKDD International Conference on Knowledge Discovery and Data Mining [Internet]. ACM, New York, NY, USA (2010)
4. Shao, J., WU, F., Ouyang, Chuanfei et al. Sparse spectral hashing. Pattern Recogn. Lett. **33**(3), 271–277 (2012)
5. Wang, Q., Zhang, D., Si, L.: Weighted Hashing for fast large scale similarity search. In: Proceedings of the 22Nd ACM International Conference on Information & Knowledge Management [Internet]. ACM, New York, NY, USA (2013)
6. Zhang, D., Wang, J., Cai, D., Lu, J.: Self-taught Hashing for fast similarity search. In: Proceedings of the 33rd International ACM SIGIR Conference on Research and Development in Information Retrieval [Internet]. ACM, New York, NY, USA (2010)
7. Weiss, Y., Fergus, R., Torralba, A.: Multidimensional Spectral Hashing. In: Fitzgibbon, A., Lazebnik, S., Perona, P., Sato, Y., Schmid, C. (eds.) Computer Vision ECCV 2012. Springer, Berlin (2012)

8. Zhuang Y, Liu Y, Wu F, Zhang Y, Shao J. Hypergaph Spectral Hashing for similarity search of social image. In: Proceedings of the 19th ACM International Conference on Multimedia [Internet]. ACM, New York, NY, USA (2011)
9. Li, P., Wang, M., Cheng, J., Xu, C., Lu, H.: Spectral Hashing with semantically consistent graph for image indexing. IEEE Trans Multimed. **15**(1), 141–152 (2013)
10. Bodó, Z., et al., Csató, L:. Linear spectral hashing. Neurocomputing **141**, 117–123 (2014)
11. Yang, Y., Shen, F., Shen, H.T., et al.: Robust Discrete Spectral Hashing for Large-Scale Image Semantic Indexing (2016)
12. Bu, J., Tan, S., He, X.: Music recommendation by unified hypergraph: combining social media information and music content. In: ACM Multimedia (2010)
13. http://horatio.cs.nyu.edu/mit/tiny/data/
14. https://www.cs.toronto.edu/~kriz/cifar.html
15. Manning, P.R., Sch¨utze, H.: Introduction to Information Retrieval. Cambridge University Press (2008)

Comparison of Feeding Modes for a Rectangular Microstrip Patch Antenna for 2.45 GHz Applications

Ouadiaa Barrou, Abdelkebir El Amri and Abdelati Reha

Abstract A microstrip patch antenna consists of a metal patch on a substrate on a ground plane. Different feeding modes are used such as: coaxial probe feed, microstrip line feed, proximity-coupled feed, and coplanar wave guide feed (CPW). The patch can take different shapes to meet various design requirements. The most known forms are rectangular, square, circular, hexagonal… The microstrip patch antenna is low-profile, conformable to planar and nonplanar surfaces, simple and cheap to manufacture using modern printed-circuit technology. There are many methods of analysis for the microstrip patch antennas. The most popular models are the transmission-line, cavity and full wave methods. In this paper, a microstrip patch antenna for 2.45 GHz applications is designed based on the transmission line method. The design is optimized with the Method of the Moments (frequency domain method) because it's one of the accurate methods for wire and planar antennas. Also, the four feeding modes are simulated and compared.

Keywords Antenna design · Feeding modes · Microstrip patch antenna

1 Introduction

With the development of wireless applications and their integration in restrict environment like smartphones, laptops and other embedded systems, the microstrip patch antennas are widely used because of their planer structure, low profile, light weight good efficiency, ease of manufacturing and integration with active devices.

There are many configurations that can be used to feed microstrip antennas. The most popular are the coaxial probe, microstrip line, proximity coupling, coplanar wave guide and others. Each feeding mode have some advantages and disadvantages and it was be used in depending on the requirements.

O. Barrou (✉) · A. El Amri · A. Reha
RITM Laboratory, CED Engineering Sciences, ESTC, Hassan II University of Casablanca, Casablanca, Morocco
e-mail: ouadiaa.barrou@gmail.com

R. El-Azouzi et al. (eds.), *Advances in Ubiquitous Networking 2*, Lecture Notes in Electrical Engineering 397, DOI 10.1007/978-981-10-1627-1_36

In this paper, first, a design methodology is presented. Next, the four feeding modes are simulated with CADFEKO, a Method of Moment (MoM) based solver. Finally, a comparison of the results is presented [1–7].

2 The Patch Antenna Design Methodology

2.1 The Design of Miscrostrip Patch Antenna

The microstrip patch antennas can be analyzed in various methods, the most popular are:

- Transmission-line method (TLM)
- Cavity method (CM)
- Full-wave methods: Are based on solving Maxwell's equations in differential or integral forms. The most popular are the Method of the Moments (MoM), the Finite element method (FEM), Finite-Difference Time Domain (FDTD).

Although the transmission line model has the least accuracy, it is the easiest method to implement and gives good physical insight. According to Balanis, the transmission-line model represents the microstrip antenna by two slots with a width of W and separated by a transmission line of length L (Fig. 1).

For the microstrip line shown in Fig. 2a, the field lines are inside the substrate and some of them are extended to outer space (Fig. 2b). For this, an effective dielectric constant (ε_{reff}) is introduced to account for fringing and the wave propagation in the line (Fig. 2c).

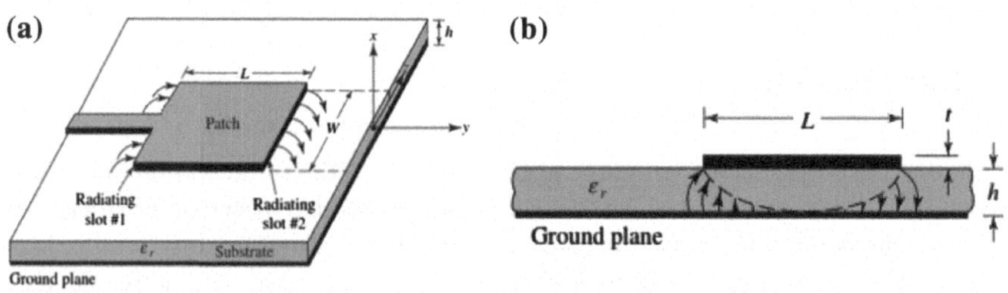

Fig. 1 Microstrip antenna [1]. **a** Microstrip antenna. **b** Side view

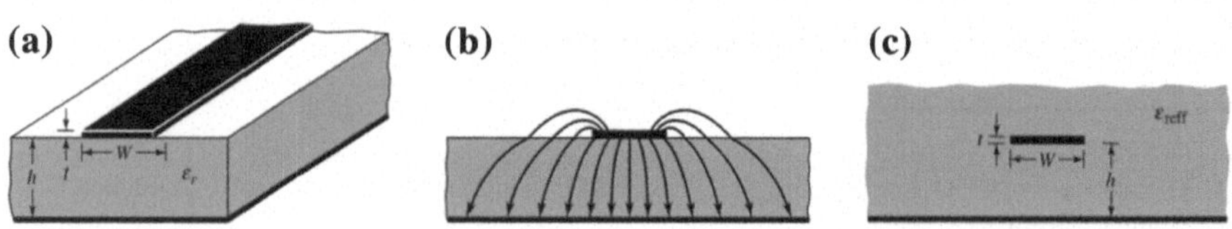

Fig. 2 Microstrip line and its electric field lines, and effective dielectric constant geometry [1]. **a** Microstrip line. **b** Electric field lines. **c** Effective dielectric constant

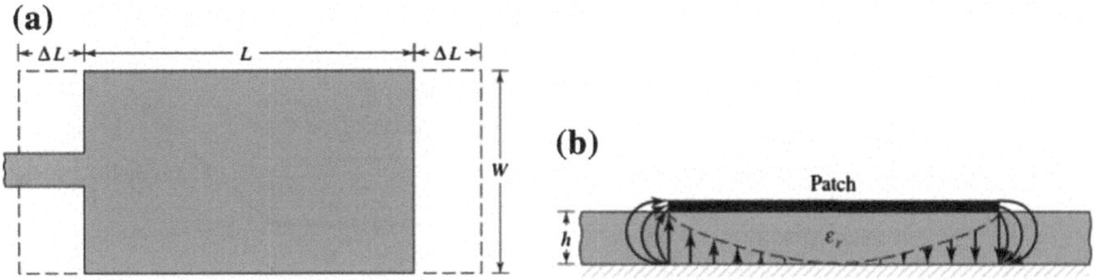

Fig. 3 Physical and effective lengths of rectangular microstrip patch [1]. **a** Top view. **b** Side view

ε_{reff} can be calculated from [1] by the formula (1)

$$\varepsilon_{reff} = \frac{\varepsilon_r + 1}{2} + \frac{\varepsilon_r \quad 1}{2} \times \left[1 + 12\frac{h}{W}\right]^{-\frac{1}{2}} \tag{1}$$

where, W/h > 1
ε_{reff} :Effective dielectric constant
ε_r: Dielectric constant of the substrate
W: Width of the radiating patch
h: Height of the substrate

As shown in Fig. 3, fringing effects looks greater than the microstrip patch dimensions. For the principal E-plane (xy-plane), the dimensions of the patch along its length have been extended on each end by a distance ΔL, which is a function of ε_{reff} and W/h given from [1] by the formula (2).

$$\frac{\Delta L}{h} = 0.412 \times \frac{(\varepsilon_{reff} + 0.3) \times (\frac{W}{h} + 0.264)}{(\varepsilon_{reff} - 0.258) \times (\frac{W}{h} + 0.8)} \tag{2}$$

The effective length of the patch is given by the Eq. (3).

$$L_{eff} = L + 2.\Delta L \tag{3}$$

It is also given by the Eq. (4).

$$L_{eff} = \frac{\lambda}{2\sqrt{\varepsilon_{reff}}} \tag{4}$$

The width of the patch is given from [1] by the Eq. (5)

$$W = \frac{\lambda}{2}\sqrt{\frac{2}{\varepsilon_r + 1}} \tag{5}$$

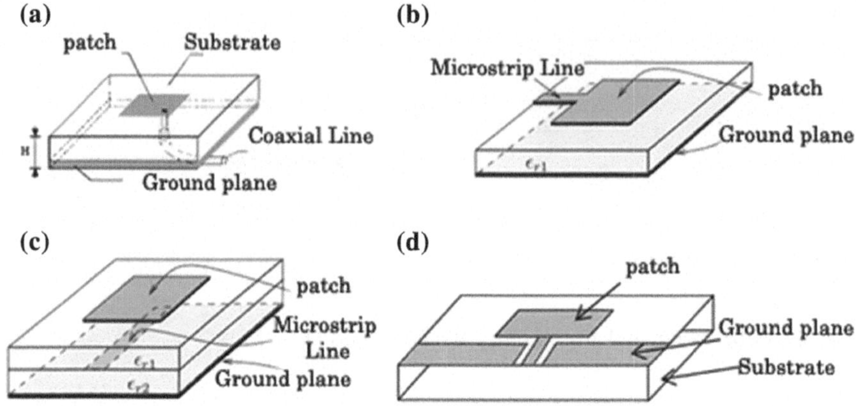

Fig. 4 Feeding Techniques. **a** Coaxial probe feed. **b** Microstrip line feed. **c** Proximity-coupled feed. **d** CoPlanar Wive guide feed (CPW)

Where, λ is the wavelength given by the Eq. (6)

$$\lambda = \frac{c}{f} \tag{6}$$

To design a microstrip patch antenna operating in the frequency of 2.45 GHz with the parameters: h = 1.6 mm and $\varepsilon_r = 4.4$ we follow the previous steps.

The results are: W = 37.26, L = 28.83 mm.

2.2 *Feeding Modes*

There are many configurations that can be used to feed microstrip antennas. The most popular are the microstrip line, coaxial probe, proximity coupling, and CPW (Fig. 4).

In the next section, a comparison of some patch antenna parameters will be done when we feed it with the four feeding techniques.

3 CADFEKO Simulation Results for Different Feeding Modes

To validate the previous design, the microstrip patch antenna was simulated with CADFEKO witch based on the Method of the Moments (MoM), one of the more accurate methods for wire and planar antennas [8–12]. Four feeding modes are simulated: coaxial probe, microstrip line, proximity coupling, and CPW.

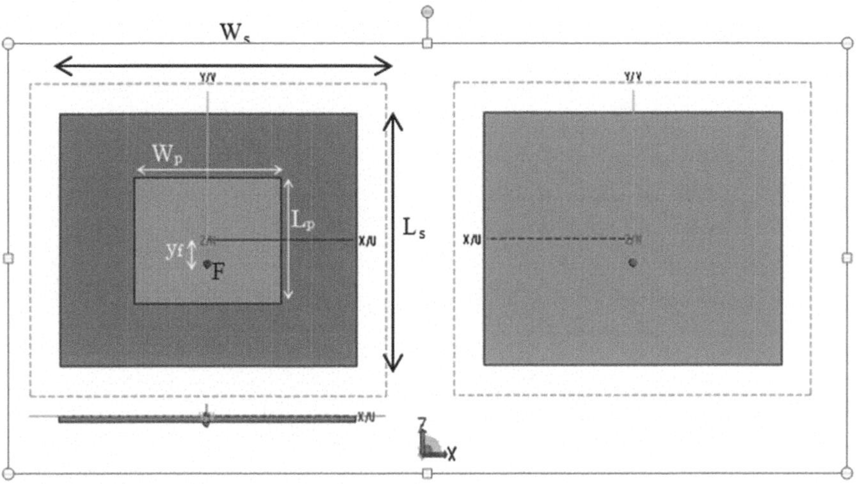

Fig. 5 Geometry of the patch antenna with coaxial probe fed

3.1 Coaxial Probe Feeding

Figure 5 illustrates the geometry of the patch antenna fed by a coaxial prob. The antenna is printed on a substrate EPOXY FR4 with relative permittivity $\varepsilon_r = 4.4$ and a thickness of 1.6 mm. The other parameters are: $W_p = 37.26$, $L_p = 28.83$ mm, $W_s = 2W_p$, $L_s = 2L_p$. The feeding probe is placed at the point F, placed at the y_f position from the center of the patch ($y_f = 4$ mm).

Figure 6 shows the S_{11} parameters. The simulated resonance frequency is 2.33 GHz, 130 MHz lower than the resonance frequency given by TLM. The design is optimized to have 2.45 GHz as the resonance frequency with MoM. The new dimensions of the patch are: $W_p = 35.66$ and $L_p = 27.56$ mm. Figure 7 shows the S_{11} for the optimized antennas.

To have a good impedance matching, the feeding point must be placed at a specific position. For that a parametric study is done. Fig 8 shows the behavior of S_{11} versus y_f. We observe that we have a good impedance matching for $y_f = 6$ mm. The 3D gain pattern is shown in Fig. 9, the maximum gain is 4.6 dB.

3.2 Microstrip Line Feed

The same patch antenna is fed by a microstrip line smaller in width as compared to the patch and having a characteristic impedance of 50 Ω. The width of this line is 2.95 mm based on the Eq. (7) [2].

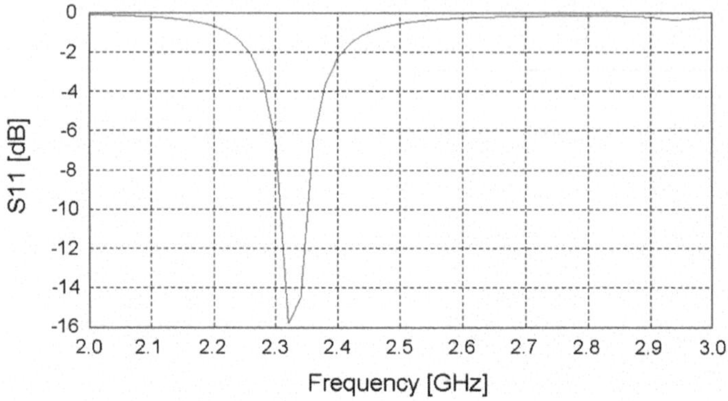

Fig. 6 S_{11} parameter for the patch antenna with coaxial probe fed

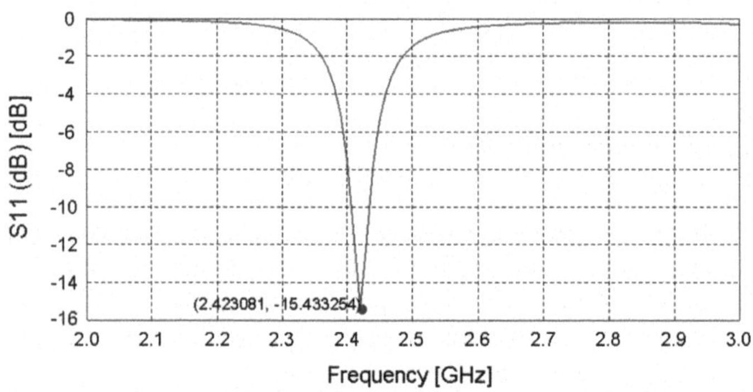

Fig. 7 S_{11} parameter for the optimized patch antenna with coaxial probe fed

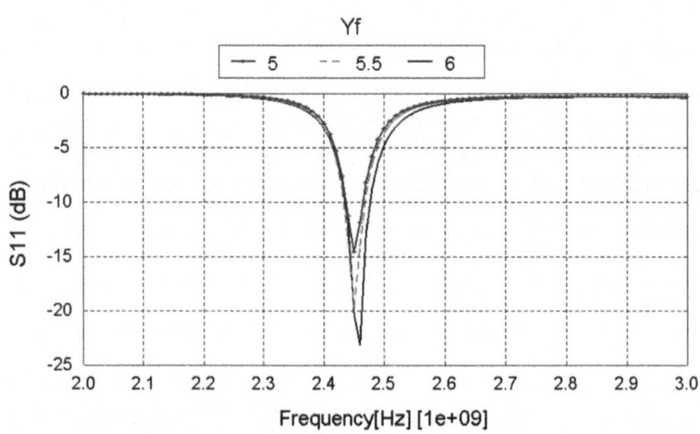

Fig. 8 Parametric study of S_{11} versus y_f

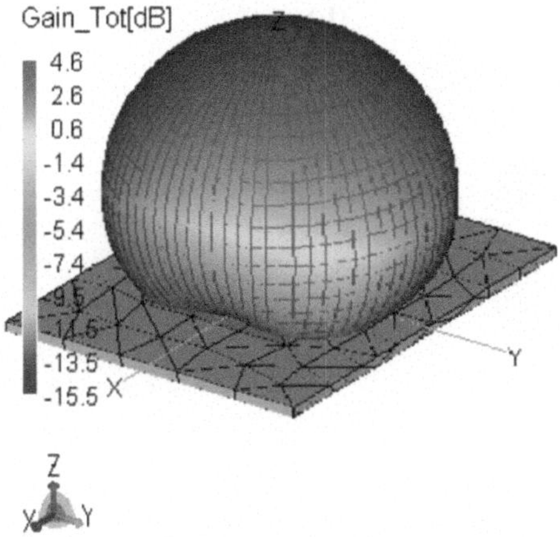

Fig. 9 3D gain pattern for the resonance frequency (fr = 2.45 GHz)

$$Z_0 = \frac{120\pi}{\sqrt{\varepsilon_r}\left(\frac{W_f}{h} + 1.393 + 0.667\ln\left(\frac{W_f}{h} + 1.44\right)\right)} \tag{7}$$

With Z_0: the characteristic impedance of the microstrip line.
W_f: the width of the microstrip line.
h: the high of the substrate.

Two configurations are studied, the first one without the inset feed point (Fig. 10a), the second one with the inset feed point (Fig. 10b). The S_{11} parameter and the 3D gain pattern for the two configurations are given by Fig. 11. We observe that when we set up the inset feed point; we obtain a good impedance matching and a better efficiency. Also a parametric study is done to know the effect of the length of the inset point (y_0). Figure 12 shows the variation of S_{11} versus y_0. A better impedance matching is obtained when $y_0 = 7.5$ mm.

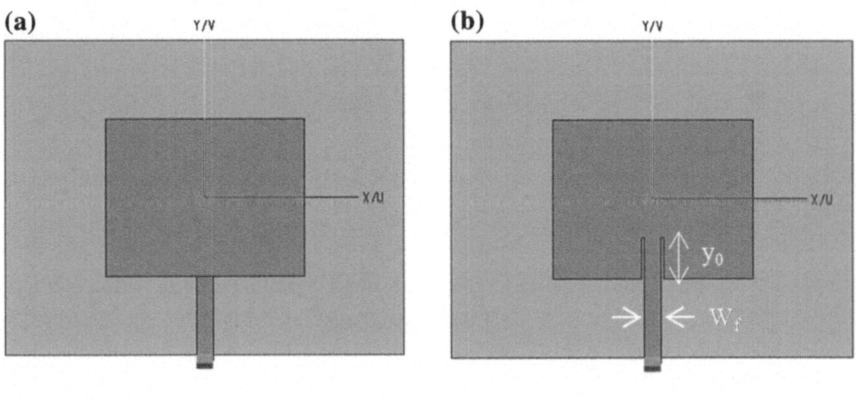

Fig. 10 Geometry of the patch antenna with Microstrip Line Feed

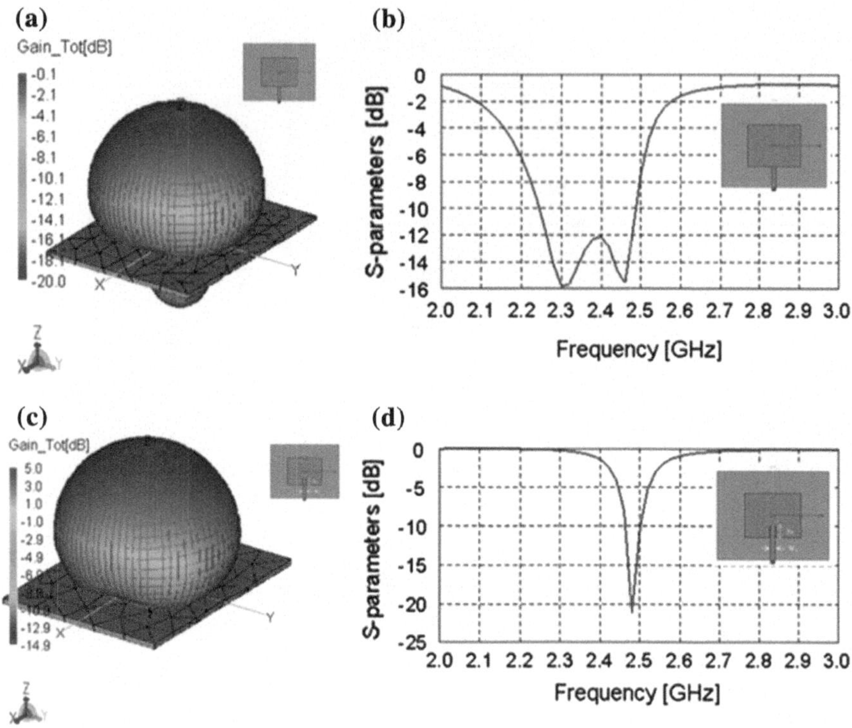

Fig. 11 3D gain pattern (**a**) and S_{11} parameter (**b**) without the inset feed point, and 3D gain pattern (**c**) and S_{11} parameter (**d**) for the second configuration (i.e., with the inset feed point)

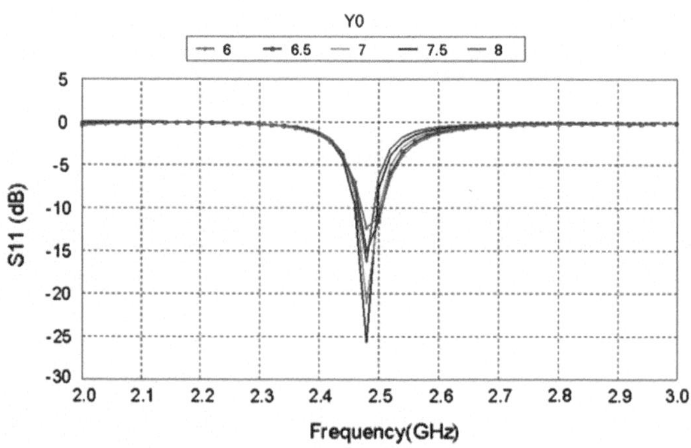

Fig. 12 Parametric study of S_{11} versus y_0

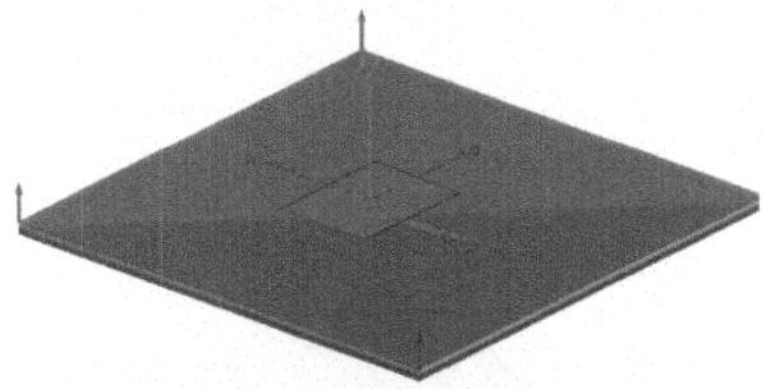

Fig. 13 Geometry of the patch antenna with Proximity coupled Feeding

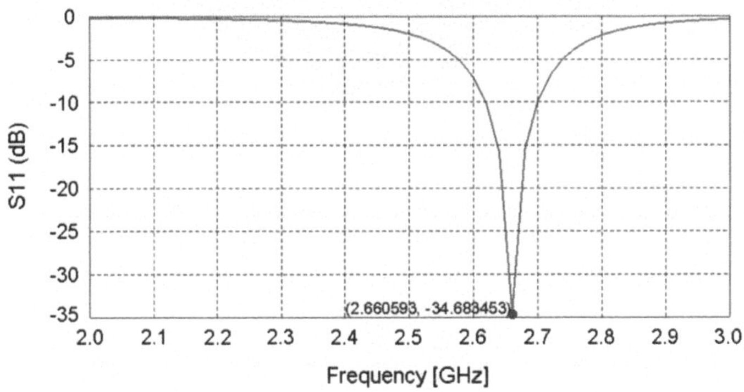

Fig. 14 S_{11} parameter for the patch antenna with Proximity coupled Feeding

3.3 Proximity Coupled Feed

For this feeding technique two dielectric substrates are used, not having necessarily the same electric characteristics and the microstrip line is placed between them (Fig. 13). The ground plane is placed on the bottom of the two substrates.

The first configuration studied using $\varepsilon_{r1} = 4.4$ and $\varepsilon_{r2} = 3.3$.
with ε_{r1}: the relative permittivity of the top layer
ε_{r2}: the relative permittivity of the bottom layer

The S_{11} parameter of the antenna is given by Fig. 14. We observe that the resonance frequency is 2.66 GHz, 210 MHz higher than 2.45 GHz, the resonance frequency obtained with the two first feeding modes. Figure 15 shows the 3D gain pattern of the antenna. We observe that the maximum gain is 5.7 dB, bigger compared to those obtained by the two first feeding modes. The S_{11} at the resonance frequency is the −10 dB bandwidth is 80 MHz (2.62–2.7 GHz). A good impedance matching is observed (48 Ω).

The variation ε_{r2} allow changing the resonance frequency. A parametric study is done and we observe that when ε_{r2} increases, the resonance frequency decreases (Fig. 16). This technique is one of the important ways to reduce the resonance frequency without increasing the dimensions of the antenna, so we can consider it

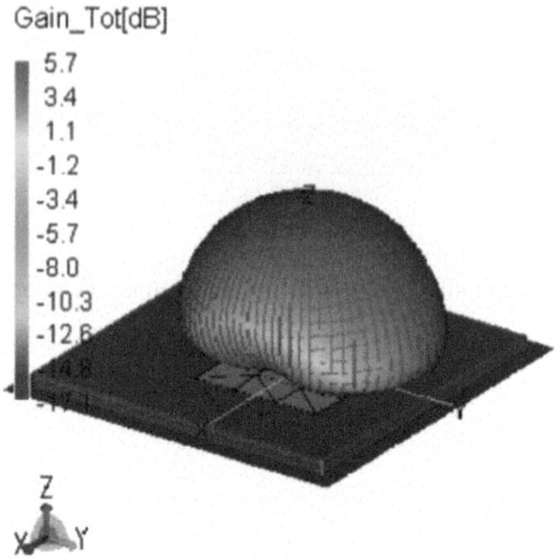

Fig. 15 3D gain pattern for the patch antenna with Proximity coupled Feed

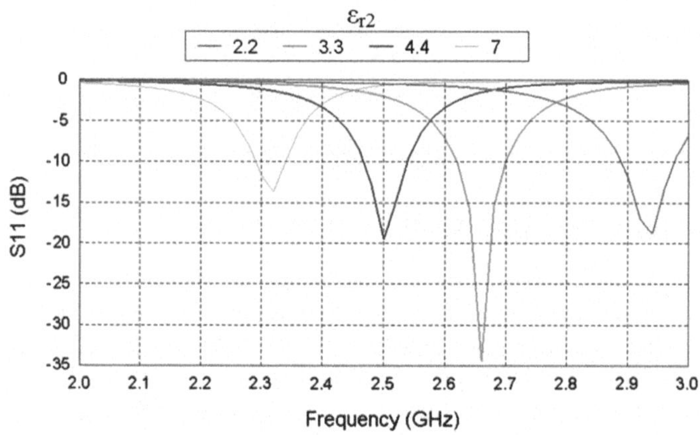

Fig. 16 Parametric study of S_{11} versus ε_{r2}

as a miniaturizing technique. Also, we observe that the −10 dB bandwidth decreases when ε_{r2} increases (Table 1).

3.4 CPW-Feeding Mode

This kind of feeding technique is also called CoPlanar Wave guide feeding (CPW-feeding). The ground plane is placed on the same plane as the patch as shown in Fig. 17. This antenna is easy to manufacture compared to the three first antennas using the Printed Circuit Board technique (PCB), in general it's used to

Table 1 Some results of the four feeding modes

Feeding mode		Resonance frequency (GHz)	S_{11} (dB)	Max. gain (dB)	Impedance	Bandwidth (From-To) (MHz)
Probe		2.44	−17.6	4.6	57−12i	45 (2.415–2.46)
Microstrip line		2.48	−26.2	5.1	54−2.6i	40 (2.46–2.5)
Proximity coupled	$\varepsilon_{r2} = 2.2$	2.93	−18	4.3	3+98i	100 (2.88–2.98)
	$\varepsilon_{r2} = 3.3$	2.65	−34.6	5.7	48	80 (2.62–2.7)
	$\varepsilon_{r2} = 4.4$	2.5	−20	5.4	40−1.5i	70 (2.46–2.53)
	$\varepsilon_{r2} = 7$	2.3	−13.5	5.1	29.5−1.4i	70 (2.29–2.35)
CPW		2.6	−11.4	1.3	85+7.2i	420 (2.4–2.82)

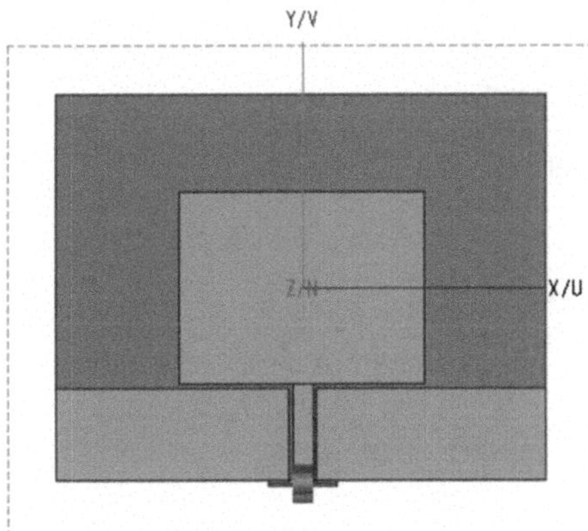

Fig. 17 Geometry of the patch antenna with CPW-feeding

obtain a large bandwidth, several studies used this technique to design antennas for Ultra Wide Band (UWB) and Broadband antennas [13–15].

The S_{11} parameter of the antenna is given by Fig. 18. We observe that the resonance frequency is 2.6 GHz with a large −10 dB bandwidth (420 MHz: 2.4 −2.82 GHz). The 3D gain pattern is given by Fig. 18, we observe that the maximum gain is 1.3 dB, also the antenna is omnidirectional Fig. 19.

4 Comparison of Different Feeding Modes

Each studied configuration has some advantages and disadvantages. The patch antenna with CPW feeding technique is simple to manufacture, omnidirectional, broadband but having a poor gain. The antenna with coaxial probe feed and microstrip line feed have the same behavior. There gain is important, directional but having a low bandwidth. The antenna with proximity-coupled feed is very complicated to manufacture, with medium bandwidth but the gain is very important and directional. Table 1 summarizes these results.

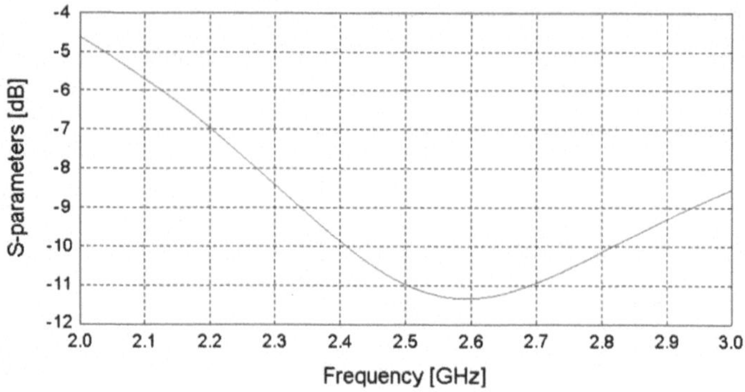

Fig. 18 S_{11} parameter for the patch antenna with CPW-feeding mode

Fig. 19 3D gain pattern for the patch antenna with CPW Feed

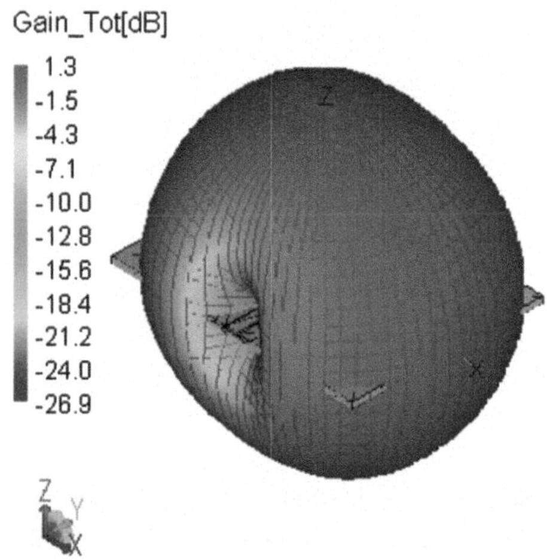

5 Conclusion

The microstrip patch is an adequate solution to design low profile antennas with important performances. It's also a good solution for designing embedded systems where the weight, cost and the ease of installation are the important requirements. The different feeding modes allow having some advantages: The proximity coupled allows having antennas with important gains also; the setup of substrate with high permittivity decreases the resonant frequencies. The CPW feeding mode increases the bandwidth of the antenna and having omnidirectional gain pattern. The coaxial probe and microstrip line feeding modes allow having antennas with short bandwidth and important gains.

To design microstrip patch antenna, the adopted feeding mode will be depend on the requirements performances.

As perspective of this work, manufacturing and measurements should be done to confirm these results with simulated ones.

References

1. Balanis, C.A.: Antenna Theory: Analysis and Design, 3rd edn. Wiley, Hoboken, NJ (2005)
2. Huang, Y., Boyle, K.: Antennas: From Theory to Practice. Wiley, Chichester, UK (2008)
3. Fang, D.G.: Antenna Theory and Microstrip Antennas. CRC Press/Taylor & Francis, Boca Raton, FL (2010)
4. Ta, S.X., Park, I.: A multiarm curl antenna for GPS applications. J. Electromagn. Waves Appl. 1–12, Nov 2014
5. Abraham, J., Mathew, T., Aanandan, C.K.: A novel proximity fed gap coupled microstrip patch array for wireless applications. Prog. Electromagn. Res. C **61**, 171–178 (2016)
6. Bakariya, P.S., Dwari, S., Sarkar, M., Mandal, M.K.: Proximity-coupled microstrip antenna for bluetooth, WiMAX, and WLAN applications. IEEE Antennas Wirel. Propag. Lett. **14**, 755–758 (2015)
7. Waterhouse, R.B.: Microstrip Patch Antennas: A Designer's Guide, Nachdruck der Ausgabe 2003. Kluwer Academic, Boston (2010)
8. Zhao, X.W., Liang, C.H.: Performance comparison between two commercial EM softwares using higher order and piecewise RWG basis functions. Microw. Opt. Technol. Lett. **51**(5), 1219–1225 (2009)
9. Clarke, S., Jakobus, U.: Dielectric material modeling in the MoM-based code FEKO. IEEE Antennas Propag. Mag. **47**(5), 140–147 (2005)
10. Reha, A., Said, A.O.: Tri-Band fractal antennas for RFID applications. Wirel. Eng. Technol. **04**(04), 171–176 (2013)
11. Sun, R.: The computer simulation of radiation pattern for cylindrical conformal microstrip antenna. Mod. Appl. Sci. **3**(10) (2009)
12. Davidson, D.B., Theron, I.P., Jakobus, U., Landstorfer, F.M., Meyer, F.J.C., Mostert, J., van Tonder, J.J.: Recent progress on the antenna simulation program FEKO 427–430 (1998)
13. Reha, A., El Amri, A., Benhmammouch, O., Oulad Said, A.: Compact dual-band Monopole Antenna for GPS/GALILEO/GLONASS and other wireless applications. In: Presented at the International Conference on Multimedia Computing and Systems, Marrakech, Apr 2014
14. A. Reha, El Amri, A., Benhmammouch, O., Oulad Said, A.: Dual-band antenna for 2.45/5.8 GHz RFID applications. In: Presented at the International Conference on Multimedia Computing and Systems, Marrakech, Apr 2014
15. Reha, A., El Amri, A., Benhmammouch, O., Oulad Said, A.: UWB compact monople antenna for LTE, UMTS and WIMAX applications. Rev. Méditerranéenne Télécommunications **4**(2), 95–98, Oct 2014

Online Signature Verification: A Survey on Authentication in Smartphones

Waseem Akram and Munam Ali Shah

Abstract Smart phones are advance generation of mobile phones. They enable us to access a large variety of services like data storages, voice communication, wireless connectivity etc. As the number of services increased the number of vulnerabilities and attacks has been increasing as well. There has been a corresponding rise of security solution proposed by researchers for authentication in smart phones like password, face and voice recognition, secret path and signature verification. The most popular authentication mechanism is online biometric signature verification. People adopt this mechanism due to its nature which is most fashionable, secure, trustable and difficult for unauthorized persons to breach privacy. With this work, we aim to provide a structured and broad over view of the research on signature verification process for authentication in smart phones. This paper surveys on signature verification process in smart phones, by focusing on different techniques used in signature verification process. We grouped existing approaches aimed to analyze performance and accuracy rate of different approaches adopted by signature verification process in smart phones. With this categorization, we aim to provide an easy and concise view of different approaches adopted by signature verification process in smart phones.

Keywords Smart phone · Authentication · Online signature verification

W. Akram (✉) · M.A. Shah
Department of Computer Science, COMSATS Institute of Information Technology, Islamabad, Pakistan
e-mail: imwaseem.khan@yahoo.com

M.A. Shah
e-mail: mshah@comsats.edu.pk

R. El-Azouzi et al. (eds.), *Advances in Ubiquitous Networking 2*, Lecture Notes in Electrical Engineering 397, DOI 10.1007/978-981-10-1627-1_37

1 Introduction

A Smart Phone has something more different than a mobile phone; it has more advanced ability of computing than mobile phones. Smart Phone is a telephone which is integrated with handheld computer. In the last 7 years, there is significant rise in the use of Smartphone devices [1]. Although, Smartphones provide several services like phone calls, Internet services, sharing data and keeping data or personal data to their users. As Smart Phone provides the enormous services; it is saddled with some challenges like security, since most of its operations are done on Internet. Therefore, it is necessary to guarantee the security of data in smart phones from authorized users [2]. Although Smart Phone has authentication like pattern password but this password is not secure on high percentage because if a person tries to breach the authentication, it is possible to guess and use it [2]. Critically, a lot of more malware malicious have been developed based on flexible smartphones APIs and most of them look like a safe software; some authentic applications collect user's information such as geographical Place without knowledge of the users [3].

To overcome authentication problems in Smart Phones, many solutions have been developed for the Smart Phones system authentication like password, voice recognition, images and signature [4]. Authentication is the process in which user provides his/her data and the system compares it with stored data in database, if verified then user becomes legitimate user to access his/her desired services [1] (Fig. 1).

The most popular authentication mechanism is online biometric signature verification [1]. People adopt this mechanism due to its nature which is most fashionable, secure, trustable and difficult for unauthorized persons to access privacy [2]. Online signature verification can also be used in Laptops and Mobile devices as well.

The signature mechanism is based on biometric rather than physical properties. Two approaches can be used for this mechanism: Static and Dynamic [2]. In Static process, signature is taken as input (bank check etc.). In Dynamic (online) process, signature is taken from sensitive screen. So, its dynamic features are used in the process of authentication like no. of strokes and orders, speed and pressure on each point etc. User gives sample of signature to the system and then signature features are extracted using some techniques and then comparing these information with stored signature by matching algorithms, if difference between test signature and template signature occurs upon threshold value, user will be rejected and if difference is less than threshold value then authorized, dissimilarity is based on min, max and average value of signals [2, 5].

In online signature verification process, two types of mechanisms are used that are function based and feature based. Function based approach uses time function from signal that are pen coordinates, pressure, speed etc. Features based approach is a holistic vector composed of global features [6–8]. Online signature verification process contains function based parameters like position, trajectories and pressure,

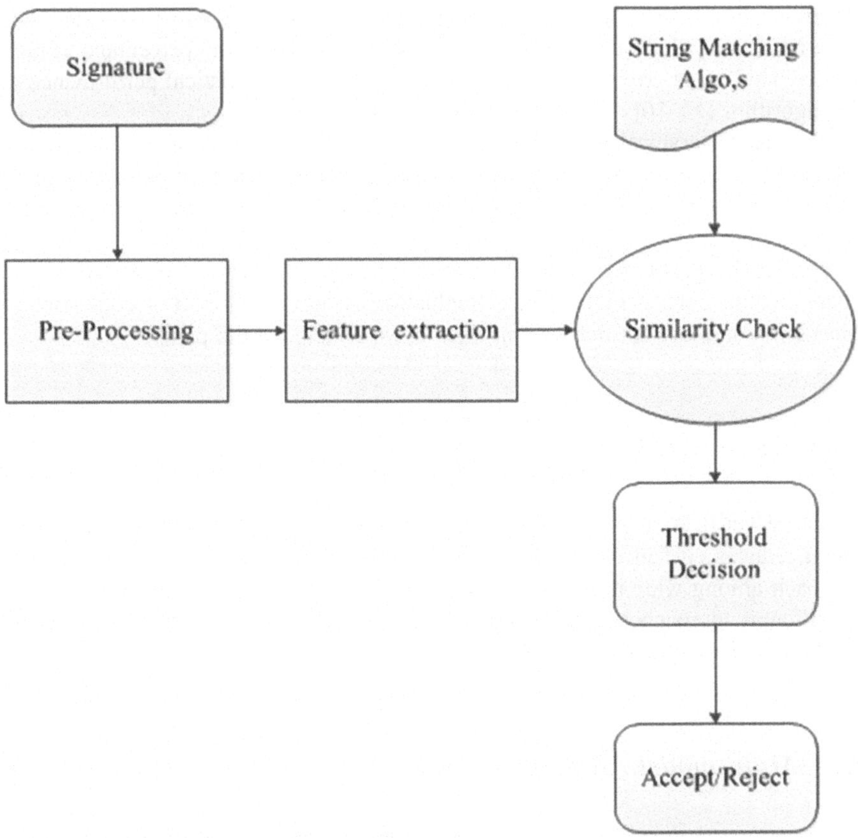

Fig. 1 Signature verification process

time etc. These properties are used for recognition [3]. Online signature is biometric behavioral process in which data is derived from action taken by a person. Data derivation and its analysis is a difficult task [9]. Quality of behavioral process is difficult to estimate [10]. Kinematic theory was used for quality measurement [11], for the analysis of these derived data, a Sigma-Lognormal analysis was used in [12].

The performance of algorithm is measure by two methods; False Rejection Rate (FRR), which measures the number of true signatures classified as forgeries, as a function of the classification threshold. False Acceptance Rate (FAR), it evaluates the number of false signatures classified as real ones as a function of the classification threshold. The performance of any technique is measured by this method as a function of the classification threshold. For example when we accept each signature, we will have a 0 % of FRR and a 100 % of FAR, and if we reject every signature, we will have a 100 % of FRR and a 0 % of FAR. The curve of FAR as a function of FRR, using the classification threshold as a parameter, is known by error trade-off curve. It shows the behavior and description of performance of the algorithm. In practice, this curve is often characterized by the equal error rate, i.e., the error rate at which the percentage of false accepts equal the percentage of false rejects. This equal error rate provides an estimate of the statistical performance of the algorithm [13–16].

This paper surveys online signature verification in Smart Phones. We grouped different mechanisms used in online signature verification process, and analyze different signature verification algorithms, its behavior, performance and its accuracy rate.

Rest of the paper is organized as follows: Sect. 2 presents the related works. Thereafter, in Sect. 3 Performance Evaluation is described, Sect. 4 comprises of Discussion and last but not least in Sect. 5 we concluded the paper.

2 Related Work

Signature verification process improves usability in many consumer applications and it reduces cost in cooperative environment. From many years, it is accepted approach among wide range of signature verification on handled devices.

Different approaches have been developed to the signature verification system in handled devices (Smart Phones), some of these approaches are following:

2.1 Multi-model Authentication

Verification system developed using Multi-Model approach which combines three factors i.e. signature, user name and password. Authentication process is done by which user gives their signature and other relevant information and the server check these information's by comparing with stored information. In this process, function based approach is used that abstracts time function from signature data, makes a string from these data. They used string matching algorithm's Lavenshtein distance, and Damerau-Lavenshatein distance and Sift3 [1]. String generation method is applied on three bases; Frequency string, Angle string, and Side Angle string. After string generation, the edit distance is calculated by comparing every signature of an object and finds the average edit distance. To be authenticating, edit distance score must be min or max. Test and experiment have been evaluated using 8 total reference signature set having 20 strings each. Result showed that among three algorithms, Shift Algorithm runs fast. This model is best among ever used methods due to using multi factors, like password, user name and signature, but some improvement required for accuracy [1].

2.2 Variant of Dynamic Time Warping

In [2, 17, 18], some contributions have been conducted by researchers that are pattern recognition; signature verification process is done by aligning client signature and its references signature using Dynamic Time Warping (DTW) technique and edit distance calculated on the bases of these parameters; nearest, farthest and template reference signature. The string is normalized by mean values using a three dimension feature vector [2, 17]. In this procedure, when users give their signature, dynamic feature is extracted; dynamic feature consists of no. of strokes, speed and pressure of pen on each point. Then this information is compared with reference signature to check similarity on threshold value. Test and experiment is evaluated on using Tablet 100 samples, first x and y coordinate and time recorded. Then align signature calculates edit distance by using DTW method. Experiment carried out using 94 people, total 619 signatures, result showed 1.4 % error rate.

2.3 Extreme Point Warping (EPW)

Another major improvement in online signature verification has been made by introducing extreme points warping technique (EPW) [3]. This technique is the enhancement of Dynamic warping technique. In DTW; Warp the whole features of signature, while EPW; warps some specific set of selected points. This is a functional based approach. These approaches improved error rate by 1.3 % and computation time reduced by 11 factors. In Extreme Warping technique [3], first of all extreme points are calculated from input signature signals (peaks and valley), then compares the extreme points with reference signature (peak to peak and valley to valley comparison) and decision is made on some threshold value. Result showed 1.3 % improvement in EER rate and computation time also reduced.

2.4 Class-Classifier Technique

Another approach for signature verification presented in [13] class classifier. In this process, global information like overall speed, velocity and pressure are used. This information's recognized by using class classifier. In previous work, DTW was used for extraction of global and local information. In class classifier, they extended the previous work. In this technique, they calculated the sketch of target set of objects [13]. The object is the signature and individual is called class. And build classifier for each individual using machine expert. Machine expert consist of complete set of global features (dimension features of vector). Then modify training data set which generates new sets, on the bases of new sets, class classifier is made. New set is the subset of all features [13]. Experiment carried out on 100 signatures 500 signatures, showed high performance by using this methodology.

2.5 Gaussian Mixture Model(GMM)

In [19] Gaussian mixture model approach was used for online signature verification. Gaussian components are used for representation of local features of signature. For features selection, MDL principle was used. GMM statistical method has been used in many type of verification system, it is single state of HMM model means that HMM can be used with few modifications. In this process, data were include pen movement, x and y coordinates, azimuth and elevation angle. After data acquisition, MDL principle was used for cost function and complexity. Result showed that this approach is user dependent model and produces significant result.

2.6 Hybrid Model

Another hybrid model is proposed in [20, 21], a model containing of both feature and function based approaches. In feature base, there were 100 feature categories on the base of time, speed, velocity, acceleration and direction. SFFS used for feature selection which reduce dimensionality of feature vector and increase optimization in computational complexity. Then normalization is done and similarity is check by Mahalanobis Distance technique [21]. In function based approach, two mechanisms were used; local and regional. In local, time function is matched by matching algorithm and in regional time function is segmented in region and vectors and matching is done by HMM. Similarity is checked by DTW. This is a hybrid method by combining of function and feature approach and result showed EER of 0.5 % [21].

2.7 Segment-Segment Comparison

In [22, 23], work is done on Geometric Extrema in signature verification process. Verification is done by comparing of segment to segment. Geometric Extrema is used for finding segment to segment correspondence [24]. Some rules defined from Geometric Extrema and these rules were used in evaluation of similarity. Geometric Extrema points are loci changes vertical and horizontal. Dynamic programming is used for mapping of optimal correspondence. On global parameters, result was 2.6 % EER and with the combination of both global and geometric extrema result was 0.94 % EER [23, 25].

2.8 Hyden Markaw Model

In [26–28] Hyden Markaw Model approach was used for signature verification process. It is a function based approach and experiment is carried out on MCYT bio model biometric database and Dynamic warping technique was first used in speech recognition problem but now days it can be used in signature verification process. In this process, first data is collected from signature that is force and inclination angles, then features are extracted by using Dynamic Warping Technique (DTW). Then find optimal alignment between two patterns that are reference and test signature. On the experiment result was 6.7 % and it was found that Mahabolanis algorithm in finding edit distance is not an optimal algorithm [29, 30], there were 145 subjects and total 3625 signatures were used, result showed that there is 0.05 % EER [28, 31, 32].

3 Performance Evaluation

In this section, we evaluate the performance of different techniques of signature verification process in smartphones. We compared performance on the base of string generating; string matching technique and equal error rate parameters, because they are the most common parameters in all techniques listed in related work section. The equal error rate shows the performance and accuracy of each technique. The result is shown in the following table (Table 1):

4 Discussion

Multi-Model authentication approach is more reliable and secure than other techniques, because it's using multi factor approach, it uses password, user name along with signature, and it is found that Sift algorithm runs fast so the computation time is less in this process. Variant of DTW shows little performance in case of when pressure information of signature is used. The EPW approach uses some specific point (extreme point), warps the signal linearly and achieves the goal of warping the whole signal, so computation time is reduced. When it is compared with DTW, DTW showed 415.7 ms runtime while EPW only 0.037 ms. This improvement in running time is too significant in case, if it runs on slowest system. The Class-Classifier technique makes different classes from signature information (IPD, NND, PWC and PCAD), by performance comparing, it showed that PWC is less reliable than other classes, while by fusion of class classifier (PCAD + PWC, NND + PWC, LPD + PCAD), showed good performance and EER is also reduced. When the performance of HMM and GMM is compared, it showed equal EER, but HMM is not good in decimation performance. The Hybrid approach adopts Function and Feature based approach making it Hybrid system. It is complex system and its running time is large but EER is reduced. HMM approach showed EER is 0.05 % which is very small as compared to other system ever used. The DTW (Dynamic Time Warping) technique showed EER is 6.7 %, which is very large as compared to the others, it has less performance and Mahabolanis

Table 1 Performance Evaluation

Model type	String generating technique	String matching technique	Equal error rate (EER %)	Limitation
Multi-model authentication	Frequency and angle	Lavenshtein and sift	2.5	• Reliable due to multi-factor • Fast
Variant DTW	Nearest, farthest and template	3-Dimension feature vector	1.4	• Little performance in case of Pressure info
EPW	Extreme point making	DP matching algo	1.3	• Small Running time (0.037 ms • Also Run on Slow device
Class-Classifier	Speed, velocity and pressure	Linear prog-discription		• PWC class less Reliable • Fusion of Classes Show good performance
GMM	Movement, xy coordinates and angle	MDL principal	3.1	
Hybrid model	Feature and function	SFFS and mahalanobis	0.5	• Runtime is large • Complex approach
HMM	Movement, xy coordinates and angle	HMM technique	0.05	• Not good in decimation performance
Segment comparing	Segment making	Geometric extrema	1.94	• Combination with Global parameters, EER 0.9 %

Algorithm is not optimal in signature verification process. By using Segment to Segment comparing approach, EER is 1.94 % and by combining with global parameter approach the EER becomes 0.9 %.

5 Conclusion

In this work we surveyed on Signature Verification for Authentication in Smart Phones, and classified different techniques used in signature verification process in Smart Phones, provided brief over view of these techniques. Also in this paper, we highlighted advantages and limitations of previous work then; we made a comparison between different techniques of signature verification process in Smart Phones.

References

1. Forhad, N., Poon, B., Amin, M.A., Yan, H.: Online Signature Verification for Multi-modal Authentication using Smart Phone, vol. I, pp. 18–21 (2015)
2. Kholmatov, A., Yanikoglu, B.: Identity authentication using improved online signature verification method. Pattern Recogn. Lett. **26**(15), 2400–2408 (2005)
3. Feng, H., Wah, C.C.: Online signature verification using a new extreme points warping technique. Pattern Recogn. Lett. **24**(16), 2943–2951 (2003)
4. Beton, M., Marie, V., Rosenberger, C.: Biometric secret path for mobile user authentication: A preliminary study. In: 2013 World Congress Computer Information Technology, pp. 1–6 (2013)
5. Karlesky, M., Sae, N.: Who You Are by way of What You Are: Behavioral Biometric Approaches to Authentication (2014)
6. Martinez-Diaz, M., Fierrez, J., Galbally, J., Ortega-Garcia, J.: Towards mobile authentication using dynamic signature verification: useful features and performance evaluation. In: ICPR'08 19th International Conference on Pattern Recognition, pp. 1–5 (2008)
7. Lei, H., Govindaraju, V.: A comparative study on the consistency of features in on-line signature verification. Pattern Recogn. Lett. **26**(15), 2483–2489 (2005)
8. Marcos Martinez-Diaz, J.G., Fierrez, J., Krish, R.P.: Mobile signature verification: feature robustness and performance comparison. In: IET Biometrics, no. October 2013, pp. 1–11 (2014)
9. Impedovo, D., Pirlo, G., Plamondon, R.: Handwritten signature verification: new advancements and open issues. In: proceedings of International Workshop on Frontiers in Handwriting Recognition, IWFHR, pp. 367–372 (2012)
10. Pen, T.F.: Automatic Signature Verification Using a Three Axis Force: e, no. 3, pp. 329–337 (2010)
11. Tolosana, R., Vera-Rodriguez, R., Ortega-Garcia, J., Fierrez, J.: Preprocessing and feature selection for improved sensor interoperability in online biometric signature verification. IEEE Access **3**, 478 (2015)
12. Galbally, J., Fierrez, J., Martinez-Diaz, M., Plamondon, R.: Quality analysis of dynamic signature based on the sigma-lognormal model. In: 2011 International Conference on Document Analysis and Recognition, pp. 633–637 (2011)
13. Nanni, L.: Experimental comparison of one-class classifiers for online signature verification. Neurocomputing **69**, 7–9 SPEC. ISS., 869–873 (2006)
14. Mendaza-Ormaza, A., Miguel-Hurtado, O., Blanco-Gonzalo, R., Diez-Jimeno, F.: Analysis of handwritten signature performances using mobile devices. In: 2011 International Carnahan Conference on Security Technology (ICCST), pp. 1–6 (2011)
15. Zheng, N., Bai, K., Huang, H., Wang, H.: You are how you touch: user verification on smartphones via tapping behaviors. In: 2014 IEEE 22nd International Conference on Network Protocols, pp. 221–232 (2014)
16. Tolosana, R., Vera-Rodriguez, R., Fierrez, J., Ortega-Garcia, J.: Feature-Based Dynamic Signature Verification Under Forensic Scenarios. Biometric Recognition Group—ATVS, Escuela Politecnica Superior Universidad Autonoma de Madrid Avda. F (2015)
17. Fierrez-Aguilar, J.F.-A.J., Ortega-Garcia, J.O.-G.J., Gonzalez-Rodriguez, J.G.-R.J.: Target dependent score normalization techniques and their application to signature verification. IEEE Trans. Syst. Man, Cybern. Part C (Appl. Rev.) **35**(3), 1556–1558 (2005)
18. Martens, R., Claesen, L: On-line signature verification by dynamic time-warping. In: Proceedings of International Conference on Pattern Recognition, vol. 3, pp. 38–42 (1996)
19. Richiardi, J., Drygajlo, A.: Gaussian mixture models for on-line signature verification. Proceedings of 2003 ACM SIGMM workshop on Biometrics Methods and Applications, pp. 115–122 (2003)

20. Richiardi, J., Ketabdar, H.K.H., Drygajlo, A.: Local and global feature selection for on-line signature verification. Eighth International Conference on Document Analysis and Recognition, pp. 0–4 (2005)
21. Krish, R., Fierrez, J.: Dynamic signature verification on smart phones. Highlights Pract. September 2015, 213–222 (2013)
22. Alonso, B., Ferrer, M.A., Travieso, C.M.: Offline geometric parameters for automatic signature verification using fixed-point arithmetic. IEEE Trans. Pattern Anal. Mach. Intell. **27** (6), 993–997 (2005)
23. Lee, J., Yoon, H.-S., Soh, J., Chun, B.T., Chung, Y.K.: Using geometric extrema for segment-to-segment characteristics comparison in online signature verification. Pattern Recogn. **37**(1), 93–103 (2004)
24. Gupta, G., Joyce, R.: Using position extrema points to capture shape in on-line handwritten signature verification. Pattern Recogn. **40**(10), 2811–2817 (2007)
25. Rhee, T., Cho, S., Kim, J.: On-line signature verification using model-guided segmentation anddiscriminative feature selection for skilled forgeries. Doc. Anal. Recogn. 645–649 (2001)
26. Martinez-Diaz, M., Fierrez, J., Galbally, J.: The DooDB graphical password database: data analysis and benchmark results. IEEE Access **1**, 596–605 (2013)
27. Houmani, N., Mayoue, A., Garcia-Salicetti, S., Dorizzi, B., Khalil, M.I., Moustafa, M.N., Abbas, H., Muramatsu, D., Yanikoglu, B., Kholmatov, A., Martinez-Diaz, M., Fierrez, J., Ortega-Garcia, J., Roure Alcobé, J., Fabregas, J., Faundez-Zanuy, M., Pascual-Gaspar, J.M., Cardeñoso-Payo, V., Vivaracho-Pascual, C.: BioSecure signature evaluation campaign (BSEC2009): evaluating online signature algorithms depending on the quality of signatures. Pattern Recogn. **45**(3), 993–1003 (2012)
28. Fierrez, J., Ortega-Garcia, J., Ramos, D., Gonzalez-Rodriguez, J.: HMM-based on-line signature verification: feature extraction and signature modeling. Pattern Recogn. Lett. **28**(16), 2325–2334 (2007)
29. Fuentes, M., Garcia-Salicetti, S., Dorizzi, B.: On line signature verification: fusion of a Hidden Markov Model and a neural network via a support vector machine. In: Proceedings of Eighth International Workshop on Frontiers in Handwriting Recognition (2002)
30. Parizeau, M., Plamondon, R.: Comparative analysis of regional correlation, dynamic time warping, and skeletal tree matching for signature verification. IEEE Trans. Pattern Anal. Mach. Intell. **12**(7), 710–716 (2009)
31. Dolfing, J.G.A., Aarts, E.H.L., van Oosterhout, J.J.G.M.: On-line signature verification with hidden Markov models. Proc. Fourteenth Int. Conf. Pattern **2**, 1309–1312 (2013)
32. Kashi, R., Hu, J., Nelson, W.L., Turin, W.: A Hidden Markov Model approach to online handwritten signature verification. Int. J. Doc. Anal. Recogn. **1**(2), 102–109 (2011)
33. Mittal, P., Dhruv, B., Kumar, P., Rawat, S.: Analysis of security trends and control methods in Android platform. In: 2014 Innovative Applications of Computational Intelligence on Power, Energy and Controls with their impact on Humanity, no. November, pp. 75–79 (2014)

Hybrid Approach for Moving Object Detection

Bouchra Honnit, Mohamed Nabil Saidi and Ahmed Tamtaoui

Abstract Moving object detection is a major step for video analysis. In this paper, we present a new approach for moving object detection. It is based on motion and edge detection technique. It makes use of the most three recent consecutive frames to detect moving area. Firstly, we compute the infimum gradient and we generate the motion saliency map. Then, we normalize both results to eliminate the parasitic elements. Finally, after applying point-by-point addition between the infimum gradient with the motion saliency map, a morphological closing operation is applied to complete the edge. The experimental results show the effectiveness of our approach for moving object detection with an accuracy rate of 92.49 %.

1 Introduction

Video analysis have become a hot issue due to its several usages in computer vision. It includes three major operations: moving object detection, tracking the object from frame to frame and recognition. There main areas of application are: traffic monitoring system, human identification and surveillance. Moving object detection consists on extracting, locating and identifying moving object of interest such as human and vehicle. The aim of the detection operation is to find a foreground moving object either at the first appearance of the target or at every video frame. However, real time moving object detection is a challenging task for researchers, on account of the following reasons: complex background, camera motion, object size variation, poorly textured objects, illumination condition and shadow.

B. Honnit (✉) · A. Tamtaoui
National Institute Of Posts and Telecommunication (INPT), Rabat, Morocco
e-mail: honnit@inpt.ac.ma

A. Tamtaoui
e-mail: tamtaoui@inpt.ac.ma

M.N. Saidi
National Institute of Statistic and Applied Economy (INSEA), Rabat, Morocco
e-mail: msaidi@insea.ac.ma

R. El-Azouzi et al. (eds.), *Advances in Ubiquitous Networking 2*, Lecture Notes in Electrical Engineering 397, DOI 10.1007/978-981-10-1627-1_38

Detection operation is based on at least one of these information color, texture or edge. In the literature, there are several methods proposed for moving object detection. They can be divided into three categories:

1. Background subtraction which extracts the background model from the input frames. It is considered as the most used and developed methods over the recent past. Some of the background subtraction approach [1–3] are: Gaussian Mixture, Running Gaussian average based background model, frame difference Region-based or spatial information, Wall flower based background model, Eigen Space decomposition, Sequential KD approximation and Temporal median filter. It is sensitive to dynamic background, illumination change and shadow.
2. inter-image difference has two main approaches. One is called temporal difference and it is based on pixel-wise difference among two successive frames [4]. The second one is called frame difference and it makes use of the subtraction operator to identify the presence of moving object. Frame difference can be divided into two sub categories:

   (a) Methods based on the subtraction operation between current frame and its neighbors [5–7].
   (b) Methods that compute the difference between current frame and another one called the reference frame; this last is initialized manually or extracted by a background subtraction [8, 9], but it can not obtain the complete moving object edge, as a result a morphological operation is used to improve the obtained result.

   Even if this category is robust to dynamic background but it can not detect slow motion due to the minor difference between two consecutive frames.
3. Optic flow methods calculate the optical flow field for an image or video frame. Then they perform clustering according to the characteristics of the optical flow distribution. However, it is sensitive to noise and it requires an extensive computational time.

Inspired by biological approach, researcher consider motion as a powerful feature to detect moving object. It can be categorized into three categories: temporal-based [10], spatial-based [11] and spatiotemporal-based [12]. First one is generally based on frame difference or background subtraction. The second one is applied in moving object detection in static camera and the third one combines temporal and spatial based methods.

We propose in this paper an hybrid approach for moving object detection based on temporal-based information. It combines motion saliency map generated in [10] and the infimum gradient described in [13] for moving object edge detection. It relies on three main steps. Firstly, motion saliency map is generated by the continues symmetry difference of the adjacent frame and the edge of moving object is extracted using the infimum. Secondly, both results are normalized to extract moving area. Finally, we apply point-by-point addition between motion saliency map and the infimum result. The contour is completed using morphological closing.

The rest of this paper is organized as follows. In Sect. 2, we present the proposed moving object detection method. In Sect. 3 we describe and interpret the obtained results. This paper is concluded in the last section.

2 The Proposed Approach

Wang and all proposed in [10] a new approach for moving object detection based on motion saliency map. They utilize the temporal information to generate motion saliency map. Then, the maximum entropy and fuzzy growing method are applied to extract the ground truth of the moving object. However, fuzzy growing is known as a parametric and complex method and it is extensive in computational time. In this method they consider the 5 recent images. Also, motion saliency map is obtained from the adjacent frame, so it does not manipulate the current frame information.

Dewan and all in [13] proposed a new method based on edge segment instead of edge pixel. It deals not only with an individual edge pixel independently, but the entire edge pixels are manipulated together. Generally, this method is based on the following steps: firstly, the difference between neighboring couples *PC* (Previous, Current) and *CN* (Current, Next) is computed. Secondly, after filtering *PC* and *CN* by a Gaussian mask, the edge map is generated by applying Canny edge detection algorithm. Thirdly, ROI is extracted using an edge-matching algorithm between *PC* and *CN*. Then, it is segmented by watershed algorithm. The infimum gradient image is computed from *PC* and *CN*. Finally, every pixel is classified as foreground or background according to a threshold obtained from ROI and infimum gradient image. The main drawback of this method is complicated and it is sensitive to slow motion.

In this paper we combine motion saliency map with the infimum gradient, in order to, reduce the computational time, make an efficient and suitable method for rel-time detection and make a robust method to slow motion by using motion information. The structure of the proposed model is shown in Fig. 1. Each step is described in the following subsection.

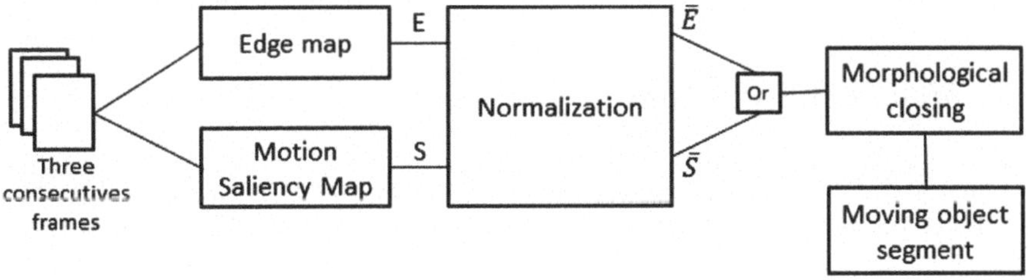

Fig. 1 Steps of our method

2.1 Edge Map Generation

Firstly, we compute the difference between the neighboring images (current, previous) and (current, next) with (1) and (2).

$$D_{n-1} = | I_n - I_{n-1} | \tag{1}$$

$$D_n = | I_n - I_{n+1} | \tag{2}$$

The difference is convoluted with a Gaussian mask to filter out the noise, which makes the proposed method robust to noise and illumination change. Secondly, the gradient magnitude is computed using a gradient operator. Then, an edge map is generated using canny edge detector, (3), (4).

$$DE_{n-1} = \varphi \nabla G * D_{n-1} \tag{3}$$

$$DE_n = \varphi \nabla G * D_n \tag{4}$$

Where φ, ∇ and G are respectively, canny edge detector, gradient operator and Gaussian mask for filtering on the difference images. Finally, since the current moving object exists in both difference images, its position will exist in both $\nabla^{n-1}$and∇^n with a higher gradient value in the boundary. Considering 3 consecutive frames, infimum gradient is computed using Eq. (7).

$$\nabla^{n-1} = \varphi G * D_{n-1} \tag{5}$$

$$\nabla^n = \varphi G * D_n \tag{6}$$

$$\nabla_{i,j}^{n-1} = \max_{i-1 \le k \le i+1, j-1 \le l \le j+1} \min(\nabla_{k,l}^{n-1}, \nabla_{k,l}^{n}) \to E \tag{7}$$

where $\nabla_{k,l}^{n-1}$ and $\nabla_{k,l}^{n}$ are respectively ∇^{n-1} and ∇^n of the pixel (k, l).

2.2 Motion Saliency Map

A frame set is constructed at a time t using three consecutive frames I_{t-1}, I_t, I_{t+1}. The difference between two consecutive frames is calculated by the Euclidean distance of the corresponding pixels in RGB color space. The pseudo-code proposed in [10] for motion saliency map generation is described as fellow:

1. Initialize the motion saliency matrix $S = 0$
2. For each i from 1 to n
   (a) Calculate the difference between two adjacent frames, i.e. $D_{t,t+i}$ and $D_{t,t-i}$
   (b) Take array multiplication of $D_{t,t+i}$ and $D_{t,t-i}$, i.e. $D_i = D_{t,t-i} \times D_{t,t+i}$
   (c) Perform morphological opening on the difference matrix $D_i' = imopen(D_i)$
3. The motion saliency matrix $S = \sum_{i=1}^{n} D_i'$

Morphological opening operation is used to eliminate the parasitic elements. Manipulation parameter is flat disk-shaped structure with radius $r = 1$ and n that is the number of frame is set to 3. $\bar{E}$ and $\bar{S}$ are calculated by normalizing E and S. At last, we compute the sum of motion saliency map and infimum gradient and we perform the morphological closing to complete the contour. Since our approach uses Gaussian filter, it is robust to noise, weather condition and illumination change. It does not require an extensive computational time and it is rapid and less complex than other methods. Also, it is robust to slow motion and it is suitable for human and traffic detection.

3 Experimental Results

Diverse comparative tests were carried out on various video sequences condition to validate the effectiveness and robustness of our method. They were performed using the public datasets of MIDAS desktop [14] and DataSet2014 [15]. The performance analysis of our method is simulated with Matlab in a processor Intel(R) Core(TM) i5, CPU 3.20 GHZ.

Figure 2 shows some example of moving object detection results. First column represents the input image (current frame), second column illustrates the result after summing the infimum gradient image and motion saliency map and the last column represents moving object localizing results. We tried in our video sequences selection to cover a large range of moving object detection conditions, such as backdoor, sunny day, dynamic background: vehicle passing in front of a tree shaken by the wind, street at night and shadow. As shown in Fig. 2 our approach is robust to weather condition and low contrast. It detects moving objects with high accuracy, where as, the shadow of the detected object is detected to. However there are many articles that deal with this problem [16]. In this paper we do not handle this issue.

In order to validate the efficiency of our method quantitatively using numeric experiments, we selected five datasets: backdoor, pedestrians, fall, street corner at night and high way from [15]. $F - measure$ with the parameters *precision* and *recall*

Fig. 2 Moving object detection results: **a** Input frames **b** Edge segment **c** Detection results

is one of the widely used metrics for measuring methods efficiency [17]. They are defined as follows:

$$F - measure = \frac{2 \times (precision \times recall)}{precision + recall} \tag{8}$$

$$precision = \frac{TP}{TP + FP} \tag{9}$$

$$recall = \frac{TP}{TP + FN} \tag{10}$$

TP, *FP* and *FN* are respectively, true positive, false positive and false negative. In our case, they are represented respectively as moving pixels, static pixels correctly detected and static pixels incorrectly detected. The higher the $F - measure$ is the good the accuracy of moving object is. Other metrics are used to evaluate our method performance such as *Re*(Recall), *Sp*(Specificity), *FNR* (False Negative Rate) and *PWC* (Percentage of Wrong Classification), which are described in [15].

Comparing our results shown in Table 1 with motion saliency map results in Table 2 and the infimum gradient in Table 3, our approach allows getting a higher $F - measure$ rate of 92.49 % than motion saliency map 84.23 % and infimum gradient 76.10 %.

Table 1 Detection results of our method for the selected datasets

Data-Set	Re	Sp	FNR	Precision	F-measure
Backdoor	0.88	0.9130	0.019	0.9130	0.8961
Pedestrians	0.9650	0.9031	0.183	0.9301	0.9472
Fall	0.8640	0.9280	0.0665	0.9280	0.8948
Street corner at night	0.6760	0.6903	0.0256	0.6903	0.9332
High way	0.9582	0.9490	0.0291	0.9490	0.9535
Average	0.8686	0.9964	0.3232	0.8820	0.9249

Table 2 Experiment results of motion saliency map [10]

Dataset	Re	Sp	FNR	Precision	F-measure
Backdoor	0.8886	0.9983	0.1132	0.9118	0.8992
Pedestrians	0.9603	0.9992	0.0397	0.9252	0.9424
Fall	0.8760	0.9860	0.1240	0.5603	0.6853
Average	0.9083	0.9921	0.0923	0.7991	0.8423

Table 3 Experiment results of the infimum

Dataset	Re	Sp	FNR	Precision	F-measure
Backdoor	0.8044	0.9070	0.1459	0.6886	0.7420
Pedestrians	0.9158	0.9180	1.5060	0.8141	0.8619
Fall	0.8730	0.9120	0.3370	0.5560	0.6793
Average	0.8644	0.9123	0.6629	0.6862	0.7610

Table 4 Average of computational time *(msec)*

Our method	Infimum	Temporal information
24.4	36.8	46.9

Table 5 Performance of our moving object detection method

Dataset	# of included object	Detection accuracy (%)	False alarm rate (%)
Backdoor	555	97.07	19.2
Pedestrians	405	97.27	19.65
Fall	836	98.11	17.89
Night corner street	948	94.78	20
High way	1280	98.43	15.04

Since our method is based on a simple operation such as subtraction, addition and morphological closing; results shown in Table 4 prove that our method is not extensive in term of computational task, which makes it suitable for real-time moving object detection.

Considering the mentioned results in [15] and our results in Table 5, our approach is able to emulate the other proposed methods. It shows a high accuracy rate of more than 94 % and a low false alarm rate of less than 20 % in all the datasets.

Also, our approach shows a relevant results compared to the infimum gradient and saliency motion map in respectively Tables 2 and 3. Except pedestrians dataset, the FNR is much lower than other methods. We consider two information: intensity and edge, as a result, neglected elements by the infimum gradient are extracted by motion saliency map and inversely. Even F-measure, precision and Sp proves that our approach detects effectively the moving object.

4 Conclusion

In this paper, we propose an hybrid approach for moving object detection. It is based on the infimum gradient and motion saliency map. It is characterized by its rapidity and simplicity. Thus, it is suitable for embedded system. Our moving object edge segment is more precise, so our approach can be extended to other moving object processing such as tracking, recognition and classification. Our experiment and numeric results proved that our method is efficient for real-time moving object detection.

As a future work, on the one hand, we want to validate our method against another video sequences. On the other hand, we want to validate it by using the classification and recognition of the detected shapes.

References

1. Patel, S.K., Mishra, A.: Moving object tracking techniques—a critical review. Indian J. Comput. Sci. **4**(2), 95–102 (2013)
2. Nayagam, M.G., Ramar, K.: A survey on real time object detection and tracking algorithms. Int. J. Appl. Eng. Res. **10**(9), 8290–8297 (2015)
3. Piccardi, M.: Background subtraction techniques: a review. In: IEEE International Conference on Systems, Man and Cybernetics, pp. 3099–3104 (2004)
4. Kulchandani, J.S., Dangarwala, K.J.: Moving object detection: review of recent research trends. In: International Conference on Pervasive Computing (ICPC) (2015)
5. Mostefai, M., Mechta, D., Chahir, Y.: Efficient real time face tracking operator study and implementation within virtex FPGA technology. Int. Arab J. Inf. Technol. **4**(1), 11–16 (2007)
6. Kanvar, K., Patil, S.A., More, S.A.: Real time application to generate the differential time lapse video with edge detection. In: Nirma University International Conference on Engineering (NUiCONE) (2012)
7. Zhu, S., Zhang, Q., Belloulata, K.: A novel spatio-temporal video object segmentation algorithm. In: IEEE International Conference on Industrial Technology (ICIT) (2008)
8. Mo, L., Liao, P., Liu, X.: A Motion Detection Algorithm Based on Background Subtraction and Three Frame Differencing. Microcomputer Information, Computer Simulation (2011)
9. Hoshen, Y., Arora, C., Poleg, Y., Peleg, S.: Efficient representation of distributions for background subtraction. In: IEEE International Conference on Advanced Video and Signal Based Surveillance, pp. 276–281 (2013)
10. Wang, Z., Liao, K., Xiong, J., Zhang, Q.: Moving object detection based on temporal information. IEEE Signal Process. Lett. **21**(11), 1403–1407 (2014)
11. Imamoglu, N., Lin, W., Fang, Y.: A saliency detection model using low-level features based on wavelet. IEEE Trans. Multimed. **15**(1), 96–105 (2013)
12. Hao, S., Shuxiao, L., Chengfei, Z., Hongxing, C., Jinglan, Z.: Moving object detection in aerial video based on spatiotemporal saliency map. **26**(5), 1211–1217 (2013)
13. Dewan, A., Hossain, J., Chae, O.: Background independent moving object segmentation using edge similarity measure. Image Analysis and Recognition. Lecture Notes in Computer Science, pp. 318–329. Springer, Berlin (2007)
14. The Multimedia Digital Archiving System. http://www.midasplatform.org/MIDAS/resources/software.html
15. IEEE Workshop on Change Detection in Conjunction with CVPR 2014. http://www.changedetection.net
16. Sanin, A., Sanderson, C., Lovell, B.: Shadow detection: a survey and comparative evaluation of recent methods. Pattern Recognit. **45**(4), 1684–1695 (2012)
17. Cai, J.F., Cands, E.J., Shen, Z.: A singular value thresholding algorithm for matrix completion. SIAM J. Optim. **20**(4), 1956–1982 (2010)

The Analysis of KDD-Parameters to Develop an Intrusion Detection System Based on Neural Network

Ilhame El Farissi, Sara Chadli, Mohamed Emharraf and Mohammed Saber

Abstract The intrusion Detection System (IDS) is designed to protect a computer or a network by detecting malicious attempts to storm the system. Therefore, it is important to develop a flexible IDS which is able to detect attacks with best performance. In recent researches, most of IDS are based on neural network and alimented by KDD data. Which means that the neural networks inputs correspond to the KDD-parameters. However, some of KDD-parameters are meaningless and can increase the detection rate. In order to optimize the IDS performance, it is primordial to exploit uniquely the most important and crucial parameters. In fact, there are three categories of KDD attributes; the basic attributes, the parameters relating to content and the time-based ones. The study carried out in this work consists on selecting the most efficient parameters in intrusion detection by demonstrating their utility and performance and neglecting the meaningless attributes. It has to be emphasized that MATLAB tool has been used to develop and put into practice the IDS in question.

Keywords Intrusion detection system · Artificial neural network for pattern recognition · KDD data · NSL-KDD data · KDD parameters · Attack categories · MATLAB

I. El Farissi · M. Emharraf · M. Saber
Laboratory LSE2I, National School of Applied Sciences, Oujda, Morocco
e-mail: ilhame.elfarissi@gmail.com
URL: http://wwwensa.ump.ma

M. Emharraf
e-mail: m.emharraf@gmail.com

M. Saber
e-mail: mosaber@gmail.com

S. Chadli (✉)
Laboratory Electronics and Systems, Sciences Faculty,
Mohammed First University, Oujda, Morocco
e-mail: chad.saraa@gmail.com

R. El-Azouzi et al. (eds.), *Advances in Ubiquitous Networking 2*, Lecture Notes in Electrical Engineering 397, DOI 10.1007/978-981-10-1627-1_39

1 Introduction

Intrusion Detection Systems (IDS) are widely employed in hosts and networks protection against malicious attempts. In recent years, most of researches in this area aim to develop a performant IDS by adopting an intelligent method such as Artificial Neural Network (ANN). It is due to the ability of the ANN to detect the known attacks and also the recently developed ones [1, 2].

The common objective of IDSs developers consists on increase the detection rate and decrease the attack success. Therefore, many of the recent neural network-based IDSs are based on KDD dataset.

KDD is a publically dataset that contains recorded attacks relying on 41 features. In the realized researches, all these features have been exploited. However, some of them can be unnecessary and may even decrease the systems performance. Thus, it seems interesting to filter and select the crucial features to design an optimum neural network-based IDS.

First of all, it is necessary to analyze the usefulness of each features category. In fact, there are four categories in the KDD dataset; basic attributes, attributes which are related to content, attributes which are based on the time using windows of two-second time and time-based attributes using windows of 100 connections time. In order to design powerful IDS, we have tested the performance of each category. For that, we have developed four scenarios; the first one consists on alimenting the neural network inputs by the basic attributes. In the second scenario, we have taken into consideration also the content-based attributes. Then, we have powered the neural network, in the third scenario, by the basic attributes and attributes that are based on the time using windows of two-second time. Concerning the last scenario, we have taken into account the basic attributes and features belonging on time-based attributes using windows of 100 connections time category.

In the first section of this paper, we present the realized works in this area. Then, we describe the KDD dataset architecture in the second section. As far as the third part of this article, it is dedicated to presentation of the Artificial Neural Network and its functioning and then we introduce the problem statement. According to the last section, it aims to present the proposed approach; its conception, realization and diagnostic of the obtained results in each scenario.

2 Related Works

The intrusion Detection System (IDS) is a device or an application that protects a system for malicious activities. Several studies are carried out on this area. The recent ones are based on artificial intelligence by exploiting the KDD 99 [3] or NSL-KDD datasets [4].

Firstly, [5] have proposed to select the optimum subset by using Linear Discriminant Analysis (LDA) algorithm and Genetic Algorithm (GA) and to exploit the

Radial Basic Functions to classify attacks. Ibrahim [6] has realized a comparison study for intrusion database (KDD99, NSL-KDD) based n Self Organization Map (SOM), the obtained detection rate for KDD99 database is 92.37 % while detection rate for NSL-KDD is 75.49 %. Concerning [1], they have exploited Multi Layer Perceptron(MLP), Generalized Feed Forward (GFF), Radial Basis Function (RBF), Self-Organizing Feature Map (SOFM) and Principal Component Analysis(PCA) to develop and IDS but they have taken into account all of the KDD features.

In order to evaluate the IDS performance, various approaches have been developed [7–10]. In fact, the proposed IDSs return an important percentage of false positive and false negative. In our opinion, this is due to the use of some no value-added parameters and which have no impact on attack and its detection.

The aim of our study consists on designing and developing an optimum and performant Intrusion Detection System by using the Neural Network and selecting uniquely the crucial parameters.

3 Description of KDD Dataset

KDD [3] is an available publically attacks dataset. It is composed of 42 features, the first 41 ones indicate the attacks attributes (characteristics) and the last one represents the attacks type.

As we are seeking to optimize the IDS by selecting the most significant parameters, it is necessary to describe the KDD architecture. Firstly, the KDD-features are divided into four categories:

- Basic attributes;
- Attributes which are related to content;
- Attributes based on the time using windows of two-second time;
- Time-based attributes using windows of 100 connections time;

Secondly, the recordings representing attacks are categorized in four types:

- **Denial of Service (DOS)**: The attackers conducting this type of attacks aim to prevent users from exploiting the service.
- **Probe**: It consists on collecting information concerning system vulnerabilities in order to employ them to launch attacks later.
- **Remote to Local (R2L)**: In order to send packets over the network, the attacker gains unauthorized local access in a remote machine.
- **User to Root (U2R)** : Through this type of attacks, the attacker gains the privileges of superuser root.

4 Presentation of Artificial Neural Network

The Biological Neural Network (BNN) such as the human one contains several interconnected neurons. The dendrites transport the electrical signals to set the connection between the neurons. The Artificial Neural Network (ANN) activity is inspired from the BNN functioning. ANN is composed of layers containing interconnected neurons, the connection between them acts as the dendrites and contribute in results calculating by using specific parameters, which are called the synaptic weights. There are three types of layers:

- **Input layer**: It contains input neurons which are alimented by external parameters. In fact, the received signals are summed through an activation function and the obtained results activate the neurons belonging to the next layer. In general, this last one is a hidden layer.
- **Hidden layers**: These layers are added to increase the neural network performance. They are also employed to activate the last layer that is called the output layer.
- **Output layer**: This layer contains neurons that allow returning the final result.

Given their learning capacity, the ANNs are employed in prevention [11, 12], classification, detection [13]. This is what allows us to benefit from the ANNs ability to develop a performant IDS. In order to develop a neural network, there is two main phases:

- **Training**: The ANN learning is categorized into two forms; the supervised learning and the unsupervised one. In our study, this last one interests us. Its particularity consists on presenting the estimated output during the training phase, to the network in order to compare it with the obtained one, calculate the difference between them and generate the error. Moreover, the network adapt the synaptic weights to obtain the minimum error.
- **Test**: In this phase, the synaptic weights values ate fixed and network capacity is tested.

In addition, there is a third step called validation. The validation step does not have any impact on synaptic weights but it is important to verify the learning rate before continuing the training. Furthermore, the three patterns sub-datasets, which are employed in the three phases, are different.

5 Problem Statement

An intrusion Detection System(IDS) has been designed with the objective of recognition of all types of network attacks. Detection precision and detection stability are two crucial factors allowing the IDS evaluation [14]. In order to enhance the detection precision and stability, many works have been performed [15]. In recent ones, the

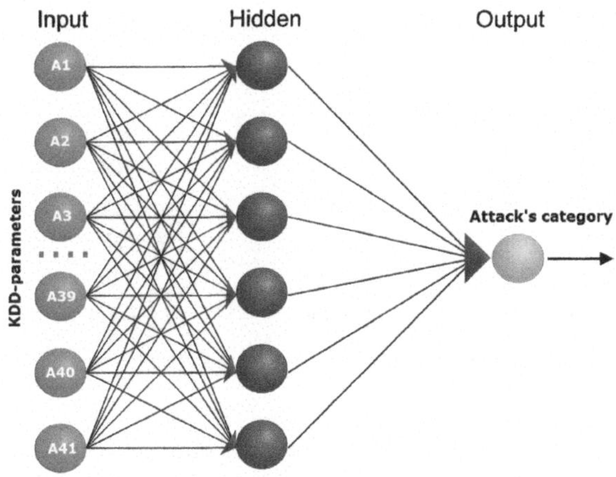

Fig. 1 Preview of the predicted Neural Network

intelligent methods belonging on ANN are the widely employed, they have proved their ability to detect the previous attacks and the recent ones.

Furthermore, the recent Neural Network-based IDSs are alimented by all the KDD features. However, the main weakness of these IDSs consists on the application of all the features. Indeed, there are some of them with no role in attack detection and may even decrease the detection rate. For this purpose, we suggest to develop an optimum neural network-based IDS alimented by the most crucial KDD-parameters. The Fig. 1 presents a preview of the predicted Neural Network.

6 Proposed Approach: Conception, Realization and Diagnostic of an Optimum Neural Network-Based IDS

In recent researches, most of the developed IDSs are based on ANN and alimented by KDD patterns. But KDD dataset contains 41 features, some of them are not effective in attack detection. However, they are integrated in IDSs design and development, which minimizes the detection rate. In order to optimize the IDS functioning, it seems interesting, for us, to filter and select the most effectives features that we intend to use as inputs of the ANN.

6.1 Dataset Selection

First of all, we have to choose the dataset to use. In fact KDD99 includes six datasets; three of them are not labeled, that is to say they do not contain the 42nd features corresponding to the attack type. Since we have opted for the supervised learning,

Table 1 Number of the employed patterns per attacks class

Dataset1		Dataset2	
Class	Patterns	Class	Patterns
Normal	60593	Normal	3883370
DOS	223298	DOS	972781
Probe	2377	Probe	41120
R2L	5993	R2L	1126
U2R	39	U2R	52
Total	**292300**	**Total**	**4898449**

we cannot employ them. The three others are labeled and includes the five classes Normal, DOS, Probe, R2L and U2R. The number of patterns in dataset is as follows:

- **Dataset 1**: It includes 4898431 patterns
- **Dataset 2**: It contains 292300 patterns
- **Dataset 3**: It represents 10 % of datatset1.

The patterns in dataset1 and dataset2 are classed per attacks category as presented in Table 1.

Thus, in order to avoid the redundant patterns we have imported the datasets in a DataBase Management System(DBMS) and we selected distinct recordings.

6.2 Conception of an Optimum Neural Network Based IDS

As mentioned above, the KDD features are divided into four categories basic attributes, attributes that are related to content, attributes based on the time using windows of two-second time and time-based attributes using windows of 100 connections time. The existing Neural network base-IDSs depend on the four categories. In this study, we aim to filter the KDD features and select the most effective ones in attacks detection.

Moreover, the ANNs model that we have employed is Neural Pattern Recognition proposed by Matlab tool . In addition, the data imported in the tool as input patterns are divided into three sub-datasets. The three generated sub-datasets are used in training, validation and test phases.

Therefore, to demonstrate the utility of features, we have designed four categories:

- **Scenario 1**: The first scenario consists on taking into consideration only the basic attributes.
- **Scenario 2**: In the second scenario, we have employed the basic attributes and the attributes that are based on content.
- **Scenario 3**: We have used the basic attributes and the third category of features.

- **Scenario 4**: This one consists on employing in addition of the basic attributes, the parameters belonging to time-based attributes using windows of 100 connections time category.

Furthermore, according to [2, 16] the features A7, A8, A9, A11, A14, A15, A17, A19, A20, A21 , A32, A40 have either no role in attack detection, or a minimum role. Consequently, we have two use cases in each of the first, second and fourth scenarios.

As a result, we will introduce the confusion matrix that is a performance representation. Each column of the matrix represents the instances in an estimated class while each row represents the instances in an output class. That allows us to preview the percentage of true positive, true negative, false positive and false negative results. In addition, for each scenario, we will present two confusion matrix corresponding respectively to the obtained results in training and test phases. The target and output classes are hosted by a number. In fact, the normal execution, U2R, R2L, DOS and Probe correspond respectively to numbers < **1** >, < **2** >, < **3** >, < **4** > and < **5** > as indicated in the confusion matrix below.

6.3 Scenario 1: Basic Attributes as the Neural Networks Input

There are two cases study in this scenario. In the first one, we have developed a neural network with nine neurons in the input layer matched to the nine basic attributes. By selecting distinct patterns from the dataset1 database, we have 317594 recordings, 55 % of them have been employed in training, 15 % in validation and 30 % of these patterns have been exploited during test phase. In the second part of this scenario, we have ignored the A7, A8 and A9 attributes. Thus, we have developed a network with seven input neurons that we call in the rest of the article the optimum basic attributes.

According to training and test confusion matrix that are illustrated in Fig. 2, we conclude that the basic attributes are crucial features for distinction between the normal execution and attack but not sufficient to detect attacks category. Furthermore, as presented in Fig. 3 we confirm that effectively the A7, A8 and A9 have no role in attacks detection.

6.4 Scenario 2: The Optimum Basic Attributes + Attributes Related on the Content as Networks Input

Firstly, we initiate this scenario by presenting the obtained results when the developed ANN is alimented by the optimum basic attributes + the attributes that are related on the content.

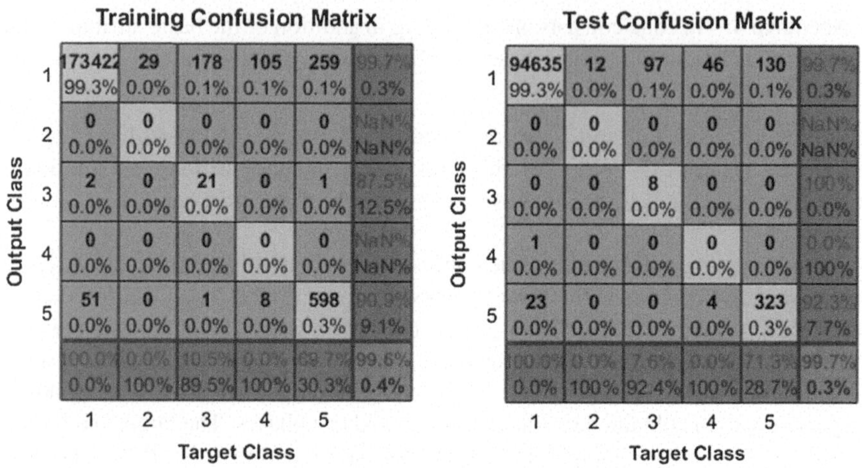

Fig. 2 Training and test confusion matrix of Neural Network alimented by the KDD basic attributes

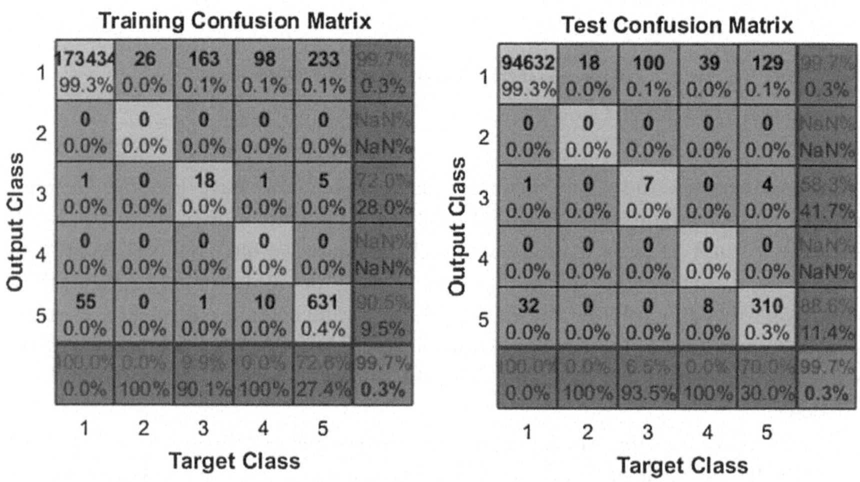

Fig. 3 Training and test confusion matrix of Neural Network alimented by the optimum basic attributes

Secondly, we have designed a neural network that is alimented by the optimum basic attributes + the attributes that are related on the content without the attributes A11, A14, A15, A17, A19, A20 and A21.

The Figs. 4 and 5 demonstrate that the attributes that are related on the content contribute to detect U2R attack and to increase detection rate concerning Probe attack. In addition, as illustrated in Fig. 5 we note that A11, A14, A15, A17, A19, A20 and A21 are not necessary features in detection of R2L, DOS and Probe attacks but they have an impact in detection of U2R category.

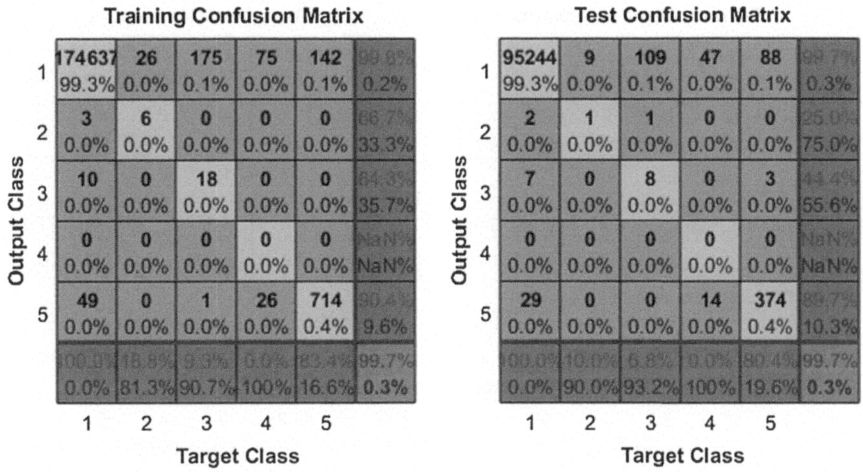

Fig. 4 Confusion matrix of Neural Network alimented by the optimum basic attributes + Attributes related on the content as networks input

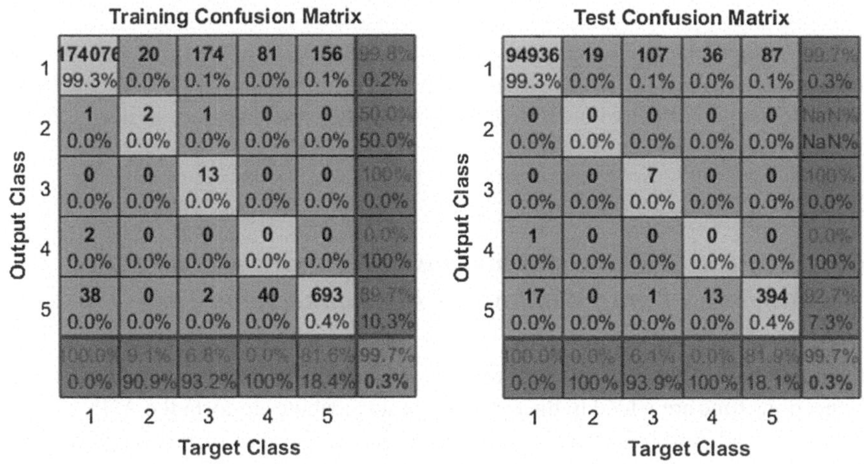

Fig. 5 Confusion matrix of Neural Network alimented by the optimum basic attributes + A10, A12, A13, A16, A18, A22 as networks input

6.5 Scenario 3: The Optimum Basic Attributes + Attributes Based on the Time Using Windows of Two-Second Time as Networks Input

The third scenario in this study consists on alimenting the ANN by the basic attributes + all the features belonging on the third category.

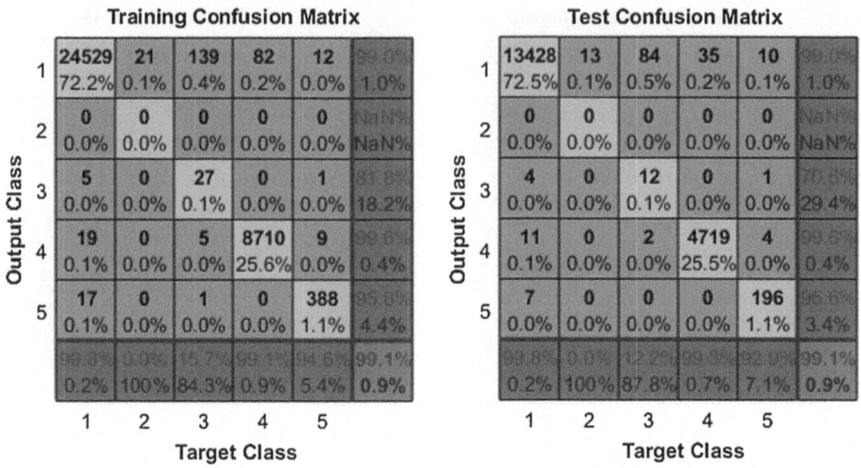

Fig. 6 Training and test confusion matrix of Neural Network alimented by the optimum basic attributes + Attributes based on the time using windows of two-second time

According to the Fig. 6 that presents the obtained results in the third scenario, we acknowledge that the attributes based on the time using windows of two-second time are sufficient and effective in DOS detection and they are necessary to detect Probe attack.

6.6 Scenario 4: The Optimum Basic Attributes + Time-Based Attributes Using Windows of 100 Connections Time

In the first part of this scenario, all the time-based attributes using windows of 100 connections time are added to the optimum basic attributes to form the ANN inputs. Thus, A32 and A40 have been eliminated in the second part of this scenario. The obtained results in this case are presented in Figs. 7 and 8.

As a result of this scenario, we have generated the Figs. 7 and 8 that allow us to confirm that the time-based attributes using windows of 100 connections time without A32 and A40 are less effective in some cases.

6.7 Diagnostic of the Proposed Approach

First of all, we initiate with the first scenario that consists on alimenting the neural network with the basic attributes. The first part of this scenario demonstrates that nearly 100 % of normal execution cases have been well classified. But, only 70.8 %

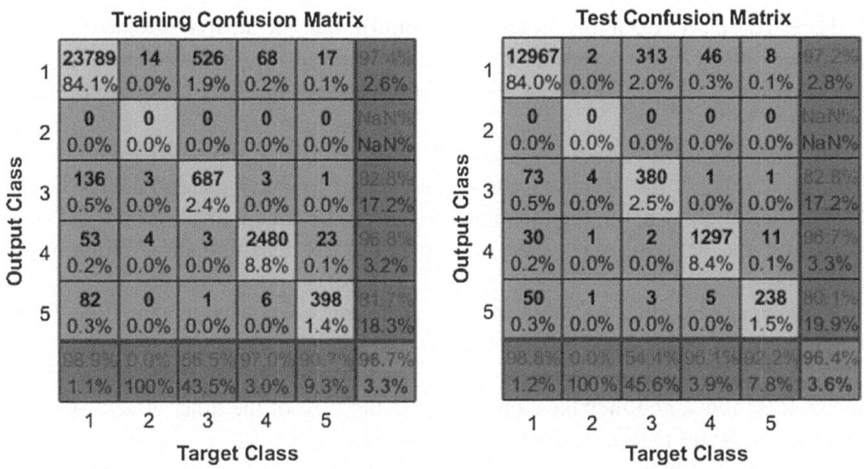

Fig. 7 Confusion matrix of Neural Network alimented by the optimum basic attributes + Time-based attributes using windows of 100 connections time

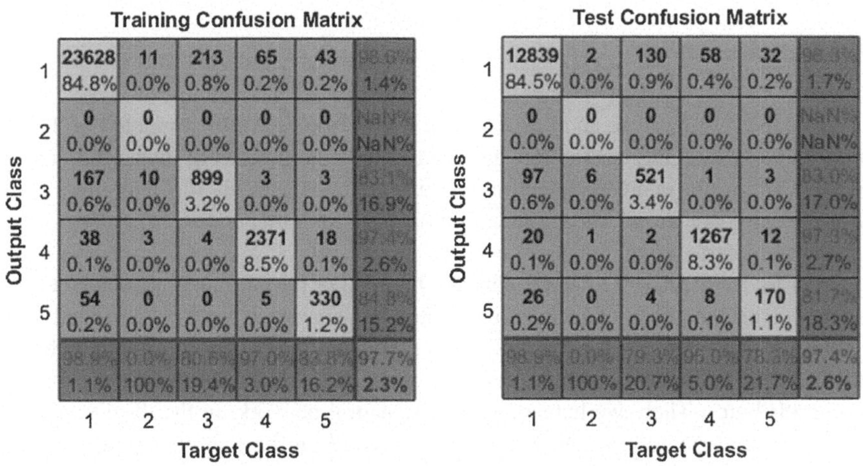

Fig. 8 Confusion matrix of Neural Network alimented by the optimum basic attributes + Time-based attributes using windows of 100 connections time without A32 and A40

of Probe attacks have been detected. According to the second part where we have neglected the A7, A8 and A9 attributes, we conclude that these last ones have no impact and no role in attack detection. Consequently, we have ignored them in the rest of the work.

Thus, we have employed the optimum basic attributes and the attributes that are related on the content as network inputs. Initially, we observe that Probe and U2R detection has increased respectively to 82.3 % and 16.3 %. However, by neglecting A11, A14, A15, A17, A19, A20 and A21, we note that U2R detection has decreased by 12 %. Therefore, from this scenario we confirm that the attributes related to content contribute in Probe detection and A11, A14, A15, A17, A19, A20 and A21 have no role in the case of R2L, DOS and Probe categories but they have an impact in U2R.

As far as the third scenario, we have obtained 99.1 % in True Positive and 0.9 % error. In one hand, the detection rate of DOS and Probe has increased to 99.2 and 93.9 %. In the other hand, the attributes based on the time using windows of two-second time as networks input have no impact in U2R detection and minimum role in R2L case.

Moreover, according to the fourth scenario we acknowledge that time-based attributes using windows of 100 connections time are efficient and necessary to classify DOS and Probe attacks. In addition, A32 and A40 features decrease the detection rate of R2L attack and they have no impact in the case of the other attacks. Consequently, it is better to eliminate them.

Finally, Since the U2R attacks are performed following to the launch of another attacks category, this explains the low detection rate in U2R case.

7 Conclusion

In this paper, we propose an optimum neural network-base IDS alimented by KDD data. Indeed, the KDD-parameters are used as the neural network inputs but not all of them are useful in attack detection, there some ones with no role and others with minimum impact. In addition, the strength of an ANN depends essentially on the input parameters.

In order to increase the neural network-based IDS performance, it seems necessary to verify the utility of each features category. Therefore, we have designed four scenarios; the first one consists on alimenting the neural network inputs by the basic attributes. In the second scenario, we have taken into consideration also the content-based attributes. Then, we have powered the neural network, in the third scenario, by the basic attributes and attributes that are based on the time using windows of two-second time. Concerning the last scenario, we have taken into account the basic attributes and features belonging to time-based attributes using windows of 100 connections time category.

According to the obtained results, we conclude that the optimum basic attributes are necessary to detect Probe attacks but not sufficient to recognize other ones. In addition, the attributes which rely on the content contribute to detect U2R attacks. Concerning the attributes which are based on the time using windows of two-second time, they are crucial to recognize the Probe and DOS attacks. Subsequently, by adding the time-based attributes using windows of 100 connections time to the optimum basic attributes we acknowledge that the detection rate of Probe, DOS and R2L has significantly increased.

References

1. Beghdad, R.: Critical study of neural networks in detecting intrusions. Comput. Secur. **27**(5), 168–175 (2008). doi:10.1016/j.cose.2008.06.001
2. Ingre, B., Yadav, A.: Performance analysis of NSL-KDD dataset using ANN. In: 2015 International Conference on Signal Processing And Communication Engineering Systems (SPACES), pp. 92–96, 2–3 Jan 2015. doi:10.1109/SPACES.2015.7058223
3. KDD data set. http://kdd.ics.uci.edu/databases/kddcup99/kddcup99.html
4. NSL-KDD data set. http://www.unb.ca/research/iscx/dataset/iscx-NSL-KDD-dataset.html
5. Imran, H.M., Abdullah, A.B., Hussain, M., Palaniappan, S., Ahmad, I.: Intrusions detection based on optimum features subset and efficient dataset selection. Int. J. Eng. Innovative Technol. **2**(6), 265–270 (2012)
6. Ibrahim, L.M., Basheer, D.T., Mahmod, M.S.: A comparison study for intrusion database (Kdd99, Nsl-Kdd) based on self organization map (SOM) artificial neural network. J. Eng. Sci. Technol. **8**(1), 107–119 (2013)
7. Saber, M., Chadli, S., Emharref, M., El farissi, A.: Modeling and implementation approach to evaluate the intrusion detection system. in: Networked Systems, pp. 1–5. Springer International Publishing (2015). doi:10.1007/978-3-319-26850-7_41
8. Akhlaq, M., Alserhani, F., Awan, I., Mellor, J., Cullen, A. J., Al-Dhelaan, A.: Implementation and evaluation of network intrusion detection systems. In: Network Performance Engineering, pp. 988-1016. Springer, Berlin, Heidelberg (2011). doi:10.1007/978-3-642-02742-0_42
9. Rastegari, S., Hingston, P., Lam, C.-P.: Evolving statistical rulesets for network intrusion detection. Appl. Soft Comput. **33**, 348–359 (2015). ISSN: 1568-4946. doi:10.1016/j.asoc.2015.04.041
10. Corchado, E., Herrero, A.: Neural visualization of network traffic data for intrusion detection. Appl. Soft Comput. **11**(2), 2042–2056 (2011). ISSN: 1568-4946, doi:10.1016/j.asoc.2010.07.002
11. Ji, S.-Y., Jeong, B.-K., Choi, S., Jeong, D.H.: A multi-level intrusion detection method for abnormal network behaviors. J. Netw. Comput. Appl. **62**, 9–17 (2016). ISSN: 1084-8045, doi:10.1016/j.jnca.2015.12.004
12. Slimani, I., El Farissi, I., Achchab, S..: Application of game theory and neural network to study the behavioral probabilities in supply chain. J. Theor. Appl. Inf. Technol. **82**(3) (2015)
13. El Farissi, A.I., Azizi, M., Moussaoui, M.: Detection of smart card attacks using neural networks. In: 2012 International Conference on Multimedia Computing and Systems (ICMCS), pp. 949–954, 10–12 May 2012. doi:10.1109/ICMCS.2012.6320286
14. de Sa Silva, L., dos Santos, A.C.F., Mancilha, T.D., da Silva, J.D.S., Montes, A.: Detecting attack signatures in the real network traffic with ANNIDA. Expert Syst. Appl. **34**(4), 2326–2333 (2008). doi:10.1016/j.eswa.2007.03.011
15. Patcha, A., Park, J.M.: An overview of anomaly detection techniques: existing solutions and latest technological trends. Comput. Netw. **51**(12), 3448–3470 (2007). doi:10.1016/j.comnet.2007.02.001
16. Bajaj, K., Arora, A.: Improving the intrusion detection using discriminative machine learning approach and improve the time complexity by data mining feature selection methods. Int. J. Comput. Appl. (0975-8887) **76**(1), 5–11 (2013). doi:10.5120/13209-0587

A CAD System for the Detection of Abnormalities in the Mammograms Using the Metaheuristic Algorithm Particle Swarm Optimization (PSO)

Khaoula Belhaj Soulami, Mohamed Nabil Saidi and Ahmed Tamtaoui

Abstract The discovery of a malignant mass in the breast is considered one of the most devastating and depressing health issue women can face. However an early detection can be so helpful and could bring hope to control the disease and even cure it. Nowadays In spite the fact that Digital mammograms have proven to be an efficient tool for the screening of breast cancer, an accurate detection of the abnormalities remains a challenging task for radiologists. In this paper, we propose an effective method for the detection and classification of the suspicious regions. In our proposed approach, we use Entropy thresholding for pectoral muscle removal, and we extract the region of interest (ROI) using the Metaheuristic algorithm Particle Swarm Optimization (PSO). Then we extract Shape and texture features from the abnormalities using Fourier transform and Gray Level Co-Occurrence Matrix (GLCM) respectively. The classification of the detected abnormalities is carried out through the Support Vector Machine, which classifies the segmented region into normal and abnormal based on the extracted features.

1 Introduction

Breast Cancer is the most common worldwide health issue that occurs among middle-aged women, and the leading cause of female cancer deaths. It starts in the tissue of the breast as a group of a dividing cells that forms an abnormal mass known as tumors. They can be cancerous (malignant) tumors or non-cancerous (benign) ones. Early detection plays a fundamental role in cancer prognosis since the death

K.B. Soulami (✉) · M.N. Saidi
National Institute of Posts and Telecommunications
(INPT, CEDOC 2TI, STRS), Rabat, Morocco
e-mail: kbelhajsoulami@gmail.com

M.N. Saidi
e-mail: msaidi@insea.ac.ma

A. Tamtaoui
National Institute of Statistic and Applied Economy (INSEA), Rabat, Morocco
e-mail: tamtaoui@inpt.ac.ma

R. El-Azouzi et al. (eds.), *Advances in Ubiquitous Networking 2*, Lecture Notes in Electrical Engineering 397, DOI 10.1007/978-981-10-1627-1_40

rate can be significantly reduced. Mammography is currently the most reliable technique for detecting breast abnormalities so the tumor can be treated at an early stage when the cancer would not has been spread yet. However, the identification of the suspicious masses is a tough task, because it is significantly subjective and relays on the radiologists expertise, and hence can lead to inaccurate predictions. That is why an automated detection using a computer Vision technique is highly recommended to assist radiologists in their diagnosis and give them a second opinion.

Before applying the identification and classification algorithms on the mammograms, a preprocessing task is required, which includes noise reduction, artifacts suppression, and pectoral muscle removal; this step mainly affects both detection and classification of the abnormalities and should be done first. The suppression of the pectoral muscle is highly recommended task in the preprocessing step, it helps in term of keeping only the breast profile of the mammogram, the removal of this muscle is necessary for the detection of the abnormalities, since it is a high intensity region that has similar features to the abnormal lesions. Many researches were conducted in order to remove the pectoral muscle Yanfeng et al. [1] used homogenous texture and high intensity deviation to identify the edge of the pectoral muscle, then a kalman Filter was applied to refine the roughness of the edge, the method attends 90 % of acceptance. A supervised technique was proposed by Arnau et al. [2], they used a model of three region in the breast (background, breast and pectoral muscle), and based on intensity, texture, and position information, they applied the training. The approach has shown an overlap between the automated and manual segmentation using 149 mammograms from the Mini-MIAS database. Jawad et al. [3] adopted an approach based on morphological operations and a Seeded Region Growing algorithm to automatically segment the breast profile and remove the pectoral muscle.

After the pectoral muscle removal, comes the step of the detection of abnormalities. There are several types of lesions in the beast, which can indicate cancer, such as microcalcifications, masses, architectural distortions and bilateral asymmetry. Particularly masses are often indistinguishable from the normal breast tissue due to their similar features, thus their detection and classification reveals to be so challenging. Several researchers focused their attention on different techniques to detect and classify abnormal region. An automated morphological operation based segmentation was proposed by [4] to find the suspicious masses in the breast then the features was extracted from the detected abnormalities using wavelet, and the classification was carried out using Support Vector Machine (SVM). Maitra et al. [5] proposed a Seeded Region Growing Algorithm to isolate normal and abnormal region in the breast after applying a Divide and Conquer algorithm for mammograms enhancement, followed by an edge detection algorithm, classification was performed using SVM. Anibou et al. [6] used SUSAN algorithm to detect the abnormalities in the high-density breasts, then they applied a Hierarchical watershed transform to detect the edge of the dense regions. They extract the shape features using Fourier Descriptor and an SVM classification was used based on the extracted descriptors and the rate of accuracy using this method achieved 78 %.

In this paper, we propose an automated method which detect and classify the suspicious regions using the metaheuristic algorithm Particle Swarm Optimization, then we analyze the extracted abnormalities using both shape and Texture features. The content of the paper is organized as follow: Sect. 2 gives an overview of the proposed approach; it describes the preprocessing step, and the techniques used for the segmentation of the breast. This section also illustrates the features extraction methods that we used and describes the procedure of classification. Section 3 presents the details of the image database and gives a highlight of the obtained experimental results using the proposed method. Finally, conclusion is given in the last section.

2 Proposed Method

Our approach is based on a CAD (computer Aided diagnosis) system that takes as an input the mammograms, removes the artifacts and the pectoral muscle in the first place so we can keep only the breast profile, and then we enhance the contrast of the image. We identify the region of interest (ROI) using the Particle swarm Optimization algorithm and we extract both shape and texture features, so we can classify the detected masses into abnormal (cancerous or non-cancerous) or normal ones.

The abnormalities detection in digital mammograms usually consists of the following steps: preprocessing (noise, Artifacts and pectoral muscle removal), segmentation (extraction of the region of interest), features extraction and the classification of the suspicious areas into normal and abnormal. Figure 1 shows the structure of the proposed approach. The following subsections describes each step.

2.1 Preprocessing

Preprocessing methods need to be performed on the mammogram images for the purpose of noise removal, background removal, radiopaque artifacts/label suppression and image contrast adjustment. As the breast profile should optimally be extracted from the background, the pectoral muscle needs also to be removed from the mammograms, since it could bias the process of the identification of abnormalities.

Artifacts and noise removal: This task is so crucial in the preprocessing step, since the radiopaque artifacts are usually sharply defined and bright regions of the mammograms background. It is one of the problems that bias the segmentation of the abnormalities. Generally mammograms contain different types of artifacts which is the case of the Mini-MIAS database images (High intensity labels, low intensity labels, scanning artifacts, Tape Artifacts). We managed to suppress the artifacts using a threshold of 0.16 and then we kept only the largest area which basically includes the breast and the pectoral muscle.

We used Two Dimensional-median filtering in a 3-by-3 connected neighborhood for the purpose of noise removal, since it suppress effectively scratches such as horizontal and vertical lines that tend to appear on most of the mammograms.

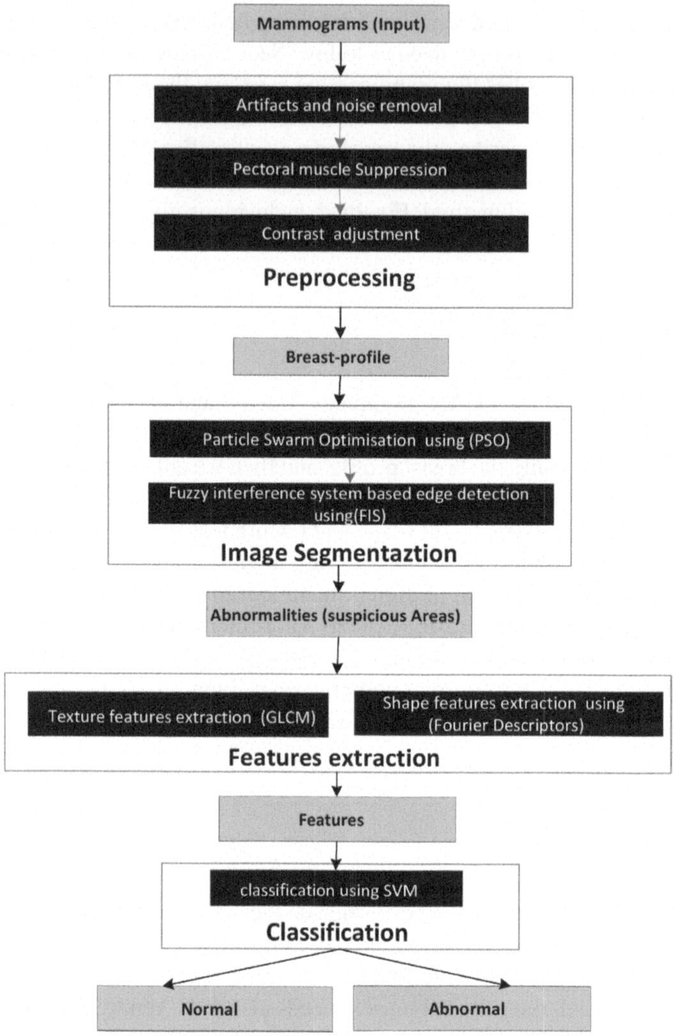

Fig. 1 The proposed CAD for the detection of abnormalities in mammograms

Pectoral muscle suppression: Pectoral muscle is localized in the upper right, left corner of the mammogram, it is a high intensity region that can influence the detection of the suspicious area due to their feature similarity to the abnormalities and hence need to be removed. For this purpose we used a multileveled Minimum Cross Entropy thresholding [8] which has been applied following three levels depending on the density of the mammogram, the more the breast is dense the more it requires a higher level of entropy thresholding because it contains a high intensity region that can be indistinguishable from the pectoral muscle.

Image contrast adjustment: Mammograms adjustment is achieved by performing contrast enhancement. Increasing the contrast of suspicious areas is very essential in mammograms, especially for dense breasts, where the contrast of abnormalities may not be discernable. As a result, differentiating between normal and abnormal regions could be so confusing.

The output of the preprocessing step, consists of the breast part, which will be used in the detection of the suspicious areas (malignant/benign masses).

Remark 1: We applied a morphological operation to refine the rough edges due to the pectoral muscle suppression, especially when it comes to dense breasts.

2.2 Breast Profile Segmentation

Detection of the abnormal masses: The segmentation of the breast profile is a fundamental step that leads to the detection of the lesions; in our method, we used the metaheuristic algorithm Particle Swarm Optimization (PSO).

Particle Swarm Optimization: is a robust stochastic optimization method and a Population-based search procedure that relays on the movement of swarms. It was proposed in 1995 by the social psychologist James Kennedy, from the U.S. Department of Labor Statistics and the electrical engineer and Russell Eberhart from the Purdue University. The particle swarm algorithm applies the concept of social interaction to solve problems, it mimes the principles of social psychology in a way that combines self-experiences with social experience. It was Inspired from the simulation of social behavior related to the dynamic movements and communications of insects, birds and fish [9].

PSO uses a number of agents or individuals called particles that constitute a flying around swarm, with a velocity $\vec{v}^t$, searching the best (optimal) solution in a multi-dimensional search space. Each particle is treated as a point in the space, which adjusts its velocity (1) according to its own flying experience as well as the flying experience of other particles (its neighbors). Which means A PSO system combines local search methods with global search methods, attempting to balance exploration and exploitation, that is why we used it in the detection of the abnormalities which requires both local and global information [10, 11].

$$\overrightarrow{v^{t+1}} = \overrightarrow{v^t} + c_1 * rand * (\overrightarrow{pBest} - \overrightarrow{p^t}) + c_2 * rand * (\overrightarrow{gBest} - \overrightarrow{p^t}) \tag{1}$$

The particle remembers the position where it had its best result. The best solution achieved so far by that particle, known as fitness, and it refers to its personal best (pbest). Particles need help in figuring out where to search, they exchange information about what they have discovered that is why there is another best value that is tracked by the PSO is the best value obtained so far by any particle in the neighborhood. This value is called (gbest) (cf. Algorithm 1). In basic, the co-operation in PSO uses the position of the neighbor with best fitness. This position is simply

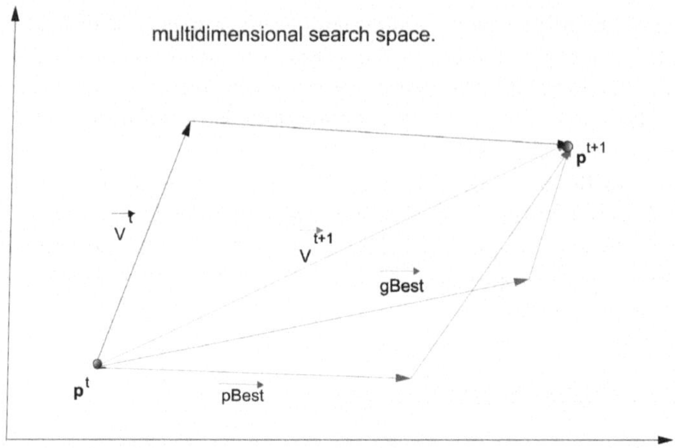

Fig. 2 PSO (particle swarm optimization)

used to adjust the particles velocity. In each iteration, a particle has to move to a new position (2), by adjusting its velocity (1). It relays on random weighted acceleration (c1, c2) to accelerate each particle toward its pbest and the gbest locations (Fig. 2).

$$\overrightarrow{p^{t+1}} = \overrightarrow{p^t} + \overrightarrow{v^{t+1}} \tag{2}$$

where p: particles position, v: particle' s velocity, c1: weight of local information (importance of personal best), it is the cognition parameter which represent how much the particle trusts its own past experience, c2: weight of global information (importance of neighborhood best), it is the social parameter which represents how much the particle trusts the swarm, pBest: best position of the particle, gBest: best position of the swarm, global best, rand: random variable (inertial weight)

for *each particle* **do**
  Initialize particle;
  Calculate fitness value;
  if *the fitness value is better than its peronal best* **then**
    set current value as the new pBest;
  end
  Choose the particle with the best fitness value of all as gBest;
  Calculate particle velocity(1);
  Update particle position(2)
end

Algorithm 1: PSO algorithm

Edge detection: consists of finding the boundaries of objects within images. It is used for image segmentation and data extraction. In order to identify the shape of abnormalities, we performed an edge detection algorithm on the extracted Region of Interest. This task plays an important role in keeping only the important structural properties of the lesions. For this purpose, we have chosen the Fuzzy Interface System based edge detection to detect the profile and shape of the extracted abnormalities. The FIS method was used from MATLAB Fuzzy Logic Image Processing Toolbox.

Fuzzy interference system based edge detection: A Fuzzy Inference System (FIS) is a way of mapping an input space to an output space using fuzzy logic. Instead of Boolean logic, the FIS uses rules and fuzzy membership functions, to reason about data. The membership functions define the degree to which a pixel belongs to an edge or not. The choice of membership function is problem dependent. But the most used function is "Triangular Membership function" (3), which is defined as:

$$f(a,b,c) = max\left(min\left(\frac{x-a}{b-a}, \frac{c-x}{c-b}\right), 0\right) \tag{3}$$

where a and c are the feet of the triangle and the parameter b defines the peak.

We have detected the edges of the abnormalities by comparing the gradient of every pixel in the x and y directions. If the gradient for a pixel is not zero, then the pixel belongs to an edge (white). We defined the gradient as zero using Gaussian membership functions for the FIS inputs.

2.3 *Features Extraction*

During feature extraction, the most important characteristics of the ROIs are studied and analyzed.

Shape feature extraction: The shape of the abnormalities is an important criterion which indicate whether the extract masses is abnormal (cancerous/non-cancerous) or not, so in order to extract the shape information from the abnormalities we used Fourier descriptor which is invariant to translation, rotation.

Fourier Descriptors: Fourier descriptors is a way of encoding the shape of a two-dimensional object by taking the Fourier transform of the boundary, where every point on the boundary is mapped to a complex number. To apply FD on the detected boundaries, two steps needs to be followed:

1. normalisation of the contour: In order to use the fast Fourier transform (FFT) properly we have to normalize the number of data set extracted from the edge, because the contours are different in shape and size.
2. calculation of the shape features using Fourier descriptor (4).

$$DF_n = \frac{1}{N}\sum_{k=0}^{N-1} r(k)exp(\frac{-i2\pi nk}{N}), n = 1, 2...N-1, \tag{4}$$

where N: is the number of normalized points, r(k) is the centroid distance function which represents the distance of the boundary points from the centroid (xc, yc)of the shape which is basically the average of the boundary coordinates.

Texture features extraction: The analysis of textures has proven a high efficiency in the detection of breast cancer, since texture is really outstanding when it comes to identifying specific characteristics of breast abnormalities. In our method, the texture-based features are extracted from the ROI region using Gray Level Co-Occurrence Matrices (GLCM).

The Grey-level Co-occurrence Matrix (GLCM): Level Co-occurrence Matrices (GLCMs) is one of the stunning texture analysis techniques. GLCM is a square matrix with dimension Ng (Number of Grey Levels) (5) that contains the occurrence of the combinations of grey level values. It gives an idea about the properties of the spatial distributions of the pixel intensity values in grayscale images. The parameters required for computing the GLCM are:

- Number of Grey Levels: usually it is 256 grey levels.
- Distance between Pixels: the matrix could be computed using non-neighbors pixels. Hence a distance between pixels is defined.
- Angle: the direction of the pair of pixels (0, 45, 90, 135).

$$G = \begin{bmatrix} p(1,1) & p(1,2) & \ldots & p(1,N_g) \\ p(2,1) & p(2,2) & \ldots & p(2,N_g) \\ . & . & . & . \\ . & . & . & . \\ . & . & . & . \\ p(N_g,1) & p(N_g,2) & \ldots & p(N_g,N_g) \end{bmatrix} \quad (5)$$

where p (i, j) is the sum of the occurrence of a pixel "i" in the specified spatial relationship to a pixel "j" in the input image.

In this paper apart from using 11 descriptors texture proposed by Haralick et al. [12], we used other recent texture descriptors [13, 14] and some features from the MATLAB Image Processing Toolbox.

2.4 *Classification*

Classification is a process related to categorization, the process in which objects are recognized, differentiated, and understood. In the classification step, the dataset is split into two disjoint sets: training and test. The training set is used to train the learning machine and the trained learning machine is then tested on the test set. In this paper the dataset sample was divided into two subsets from which one set was chosen as a training one and the other one was used for test.

In this work, the support vector machine (SVM) was performed using Sigmoid kernel [15]. SVM is basically a linear classification approach based on two classes. It separate individuals from two classes (+1 and −1) using the optimal hyperplane that separate the two sets, and guarantee a large margin between the two classes.

3 Experimental Results

3.1 Mini-Mias Database

Digital mammogram images were acquired from the mini-MIAS database [7] which consist of right and left breast images of dense, fatty-glandular and fatty breasts. The acquired mammogram images belongs to three categories: malignant, benign and normal. The abnormalities (benign and malignant) consists of five categories as follows: Ill-defined masses, architecturally distorted masses, Asymmetrical masses, Circumscribed masses and Spiculated masses. The size of the images is 1024 1024 pixels. The images are in grayscale with a pixel intensity of range [0, 255].

3.2 Preprocessing

The mammograms of Mini-MIAS database was preprocessed using the techniques described in Sect. 2.1 as the figure (Fig. 3) shows, the preprocess was applied on the three categories of the breast (fatty, fatty glandular, dense), this methods still have some drawbacks when it comes to the removal of pectoral muscle in dense mammograms. To avoid the over segmentation of the breast, we have chosen the level of Entropy thresholding manually, since in this case, the dense tissue of the breast is indistinguishable from the pectoral muscle.

3.3 Segmentation

The identification of the region of interest (abnormalities) was carried out using the segmentation methods described in Sect. 2.2. PSO algorithm was first applied on the preprocessed images followed by a fuzzy logic algorithm based edge detection, the figures (cf. Figs. 4, 5 and 6), show the experimental results of this step and it has been performed on the three different categories of the breast (dense Fig. 4, fatty glandular Fig. 5 and fatty Fig. 6). The majority of abnormalities was detected and there was cases where the output image was blank and thats describe a normal breast tissue, which supposed to not contain any abnormalities, this kind of results has fit our expectations.

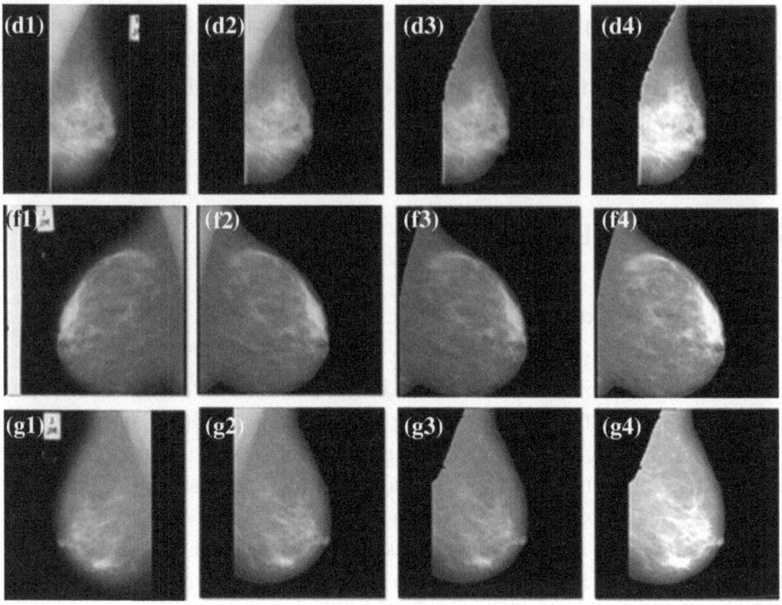

Fig. 3 The preprocessing step was performed on the three categories dense (d), fatty (f)and fatty glandular (g) respectively, **d1**, **f1**, **g1** original images **d2**, **f2**, **g2** refers to the suppression of artifacts and noise in the three categories. **d3**, **f3**, **g3** the removal of pectoral muscle, **d3**, **f3**, **g3** the contrast adjustment of the images, **d4**, **f4**, **g4**

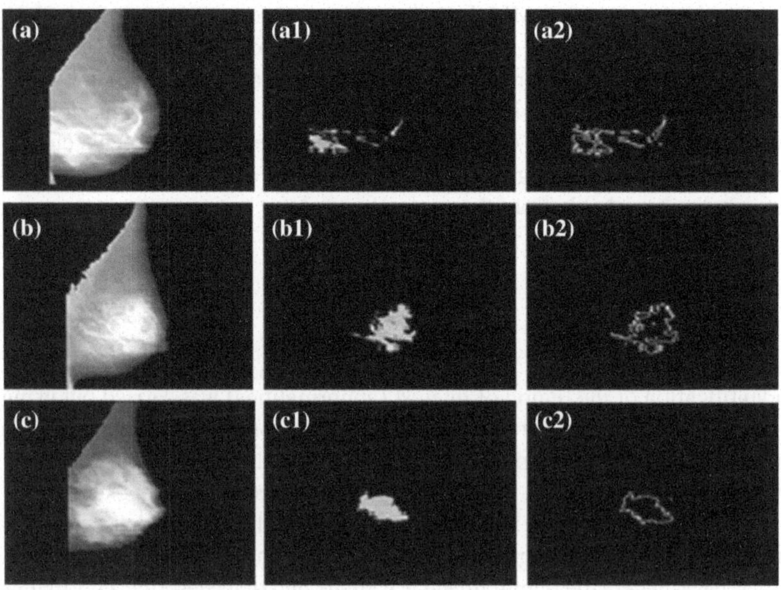

Fig. 4 **a**, **b**, **c** the preprocessed images of the dense breast category, which represents the normal malignant and benign cases respectively, **a1**, **b1**, **c1** the detected abnormalities using PSO, **a2**, **b2**, **c2** the edge of abnormalities using FIS

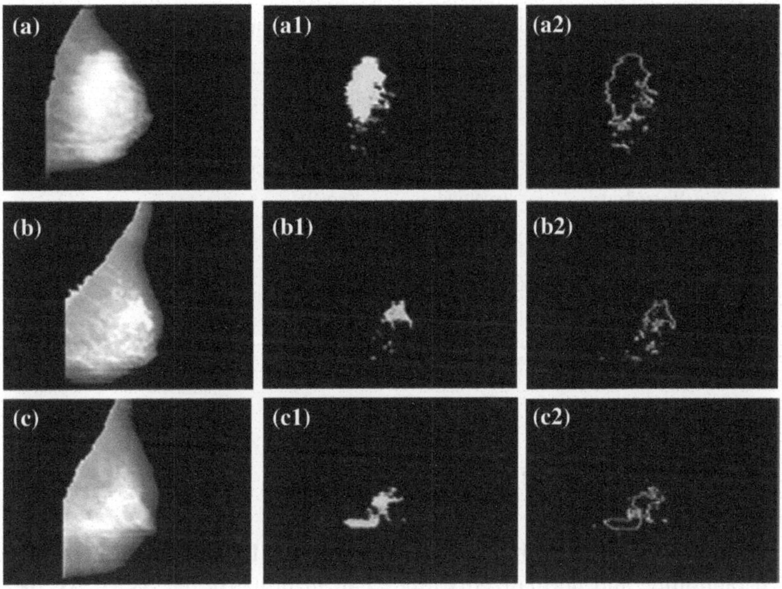

Fig. 5 **a**, **b**, **c** the preprocessed images of the fatty glandular breast category, which represents the normal malignant and benign cases respectively, **a1**, **b1**, **c1** the detected abnormalities using PSO, **a2**, **b2**, **c2** the edge of abnormalities using FIS

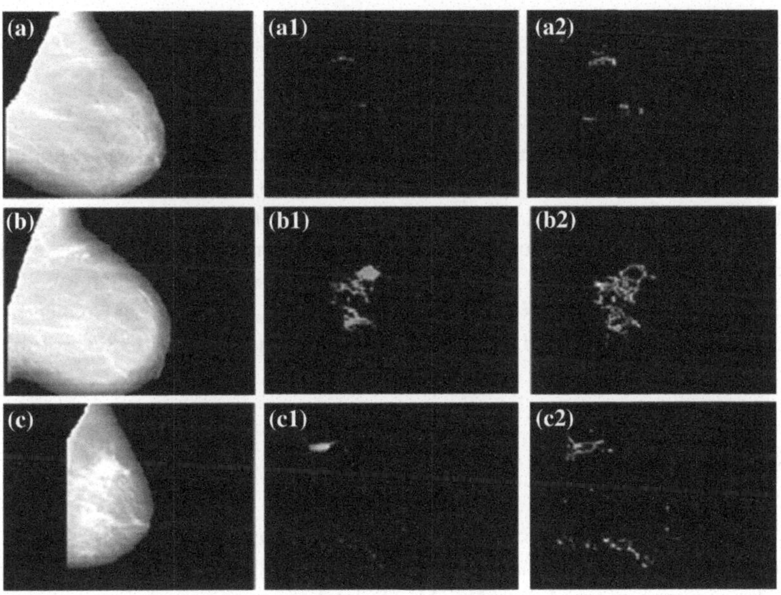

Fig. 6 **a**, **b**, **c** the preprocessed images of the fatty breast category, which represents the normal malignant and benign cases respectively, **a1**, **b1**, **c1** the detected abnormalities using PSO, **a2**, **b2**, **c2** the edge of abnormalities using FIS

Table 1 Comparaison with other techniques of detection of abnormalities in term of accuracy

Methods	Segmentation	Features	Classifier	Data-Set	Accuracy (%)
[16]	Small patches of 128 × 128 pixels	2DPCA	SVM	IRMA reference DDSM	80.07
[6]	Modified Susan and Watershed Transform	Fourier (centroid distance)	SVM(Sigmod Kernel)	Mini-MIAS	78.77
The proposed method	Particle Swarm Optimization	Shape (Fourier descriptors) + Texture (GLCM descriptors)	SVM(Sigmod Kernel)	Mini-MIAS	**83.87**

3.4 *Feature Extraction and Classification*

The obtained features from both methods FD and GLCM of the Sect. 2.3, were merged randomly, and normalized so they can fit properly the SVM. All the 107 features out of which 63 are shape features and the remaining describes the texture, were scaled (normalized) in the range between 0 and 1, the Feature normalization has been carried out using the following expression (6):

$$NF(x) = \frac{F(x) - min(F(x))}{max(F(x)) - min(F(x))} \tag{6}$$

where F(x) represents the feature of interest.

The normalized features are divided into two distinct sets, i.e. the training set and the testing set. The total number of ROI samples obtained from the acquired segmented breast data is 306, out of which 195 are normal samples and the remaining 111 are abnormal samples. where 80 % of the sample from both classes (normal/abnormal) was randomly allocated to the training set and the remaining 20 % of the sample from both classes was chosen as a testing set. The Performance of the proposed method is evaluated in terms of Accuracy which attend 83.87 % (Table 1).

4 Conclusion

In this paper, we have proposed a Computer Aided Diagnosis system that detect the abnormalities in digital mammogram and classifies them into normal and abnormal. The acquired images from Mini-MIAS database were preprocessed in order to remove noise, artifacts and pectoral muscle from the breast region so the segmentation algorithms could perform efficiently. Then we have extracted the suspicious regions using PSO algorithm, followed by an edge detection technique based on

FIS. We computed shape descriptors from the edge of abnormalities using Fourier Descriptors, then we extracted the texture-based features from the suspicious regions using GLCM. Both shape-based descriptors and texture-based ones were normalized and stored as feature vector. A support vector machine was carried out to classify suspicious regions into normal or abnormal. The proposed method was tested on Mini-Mias database. For further work, we want to evaluate our method on different private databases, and automate the entropy level thresholding for pectoral muscle removal, so we do not have to interfere manually, we will also try to detect the cancerous regions.

References

1. Lia, Y., Chena, H., Yangb, Y., Yanga, N.: Pectoral muscle segmentation in mammograms based on homogenous texture and intensity deviation. Pattern Recogn. **46**(3), 681–691 (2013)
2. Oliver, A., Llado, X., Torrent, A., Mart, J.: One-shot segmentation of breast, pectoral muscle, and background in digitised mammograms. In: 2014 IEEE International Conference on Image Processing (ICIP), pp. 912–916
3. Nagi, J., Kareem, S.A., Nagi, F., Ahmed, S.K.: Automated breast profile segmentation for ROI detection using digital mammograms. In: IEEE EMBS Conference on Biomedical Engineering & Sciences (IECBES 2010), pp. 87–92. Kuala Lumpur, Malaysia (2010)
4. Anitha, J., Peter, J.D.: A wavelet based morphological mass detection and classification in mammograms. In: International Conference on Machine Vision and Image Processing (MVIP), pp. 25–28 (2012)
5. Maitra, I.K., Nag, S., Bandyopadhyay, S.K.: Detection of abnormal masses using divide and conquer algorithmin digital mammogram. Int. J. Emerg. Sci. **1**(4), 767–786 (2011)
6. Anibou, C., Saidi, M.N., Aboutajdine, D.: Computer aid diagnostic in mammogram image using susan algorithm and hierarchical watershed transform. In: Lecture Notes in Computer Science, UNet 2015, pp. 355–366 (2016)
7. J. Suckling et al., The Mammographic Image Analysis Society digital mammogram database, Exerpta Medica **1069**, 375–378 (1994)
8. Brink, A.D., Pendock, N.E.: Minimum cross-entropy threshold selection. Pattern Recogn. **29**, 179–188 (1996)
9. Ait-Aoudia, S., Guerrout, E.-H., Mahiou, R.: Medical image segmentation using particle swarm optimization. In: 18th International Conference on Information Visualisation (IV), pp. 287–291 (2014)
10. Ghamisi, P., Couceiro, M.S., Martins, F.M.L., Benediktsson, A.: Multilevel image segmentation based on fractional-order darwinian particle swarm optimization. IEEE Trans. Geosci. Remote Sensing **99**, 1–13 (2013)
11. Raju, N.G., Rao, P.A.N.: Particle swarm optimization methods for image segmentation applied in mammography. Int. J. Eng. Res. Appl. **3**(6), 1572–1579 (2013)
12. Haralick, R.M., Shanmugam, K., Dinstein, I.: Textural features of image classification. IEEE Trans. Syst. Man Cybern. **SMC-3**(6) (1973)
13. Soh, L., Tsatsoulis, C.: Texture analysis of SAR sea ice imageryusing gray level co- occurrence matrices. IEEE Trans. Geosci. Remote Sens. **37**(2) (1999)
14. Clausi, D.A.: An analysis of co-occurrence texture statistics as afunction of grey level quantization. Can. J. Remote Sens. **28**(1), 45–62 (2002)
15. Sharma, S., Khanna, P.: Computer-aided diagnosis of malignant mammograms using zernike moments and svm. J. Digit. Imaging **28**(1), 77–90 (2015)
16. Deserno, T.M., Soiron, M., de Oliveira, J.E.E.: Computer-aided diagnostics of screening mammography using content-based image retrieval. Proc. SPIE **8315**, 271–279 (2012)

Texture Segmentation Based on Dual Tree Complex Wavelet Transform and Support Vector Machine

Amal Farress, Mohamed Nabil Saidi and Ahmed Tamtaoui

Abstract This paper presents a new approach for segmentation of the textured images that exploits properties of the dual-tree complex wavelet transform, shift invariance and six directional sub-bands at each scale, and uses a feature vector comprising of mean and standard deviation of the six directional sub-bands over a sliding window. The classification of each sliding window using Support Vector Machine (SVM) leads to a segmented image. Through experiments on a variety of synthetic images of texture data sets, we show that our algorithm yields significant performance improvements for texture segmentation, as compared with other state-of-the-art methods of feature extraction.

1 Introduction

Texture is one of the fundamental characteristics of images as it refers to surface characteristics and appearance of an object given by the size, shape, density, arrangement, proportion of its elementary parts. Although there is no precise definition of the notion texture, texture information is used in several computer-vision, remote-sensing [1], medical imaging [2], and content-based image retrieval applications. Major goals of texture research in computer vision are to understand, model and process texture, and ultimately to simulate human visual learning process using computer technologies. Texture segmentation is a fundamental problem in image processing, which is a prerequisite to high-level computer vision applications. It aims to partition an image into a set of disjoint regions based on texture properties, so that each region is homogeneous with respect to certain texture characteristics. Many studies have been performed for the segmentation of the textured images, such as Markov random field [3], probabilistic techniques [4], and wavelet transform [5]. A segmentation based classification is one of most technique used to segment a given

A. Farress (✉) · A. Tamtaoui
INPT, Rabat, Morocco
e-mail: amal.farres.1@gmail.com

M.N. Saidi
INSEA, Rabat, Morocco

R. El-Azouzi et al. (eds.), *Advances in Ubiquitous Networking 2*, Lecture Notes in Electrical Engineering 397, DOI 10.1007/978-981-10-1627-1_41

texture. However the texture classification process assigns a given texture to some texture classes. Classification can be divided into two families: supervised and unsupervised classification. Supervised classification is provided examples of each texture class as a training set. A supervised classifier is trained using the set to learn a characterization for each texture class. Unsupervised classification does not require prior knowledge, which is able to automatically discover different classes from input textures. In this paper we focus on a supervised classification method. We note that the majority of supervised classification methods involve a two-stage process: the learning phase and the recognition phase. In the learning phase, the target is to build a model for the texture content of each texture class present in the training data, which is generally comprised of images with known class labels, in which classifiers are trained to determine the classification for each input texture based on obtained measures of selected features. In this case, a classifier is a function that takes the selected features as inputs and texture classes as outputs. In the recognition phase, each pixel should belong to one class and only one.

A large list of techniques for modeling the texture have been proposed. Jain et al. [6] proposed a method by characterizing the channels by a bank of Gabor filters. Haralick et al. [7] proposed Gray level co-occurrence matrix and Wang et al. [8] used wavelet Transform as a feature descriptor. A comparative study of texture extraction has been presented by Al-Kadi et al. [9]. All the methods discussed above depend on feature evaluation set in spatial domain, therefore they have local advantages or disadvantages depending on the features used. Recently, Haghighat et al. [10] had used Gabor filters to extract features from the detected face region and mentioned that the most important advantage of Gabor filters is their invariance to rotation, scale, and translation. Anibou et al. [11] proposed discrete wavelet transform based method for texture classification. Method proposed by [11] uses real valued discrete wavelet transform, but its major problem is the lack of directionality and shift invariance. Following this work, in this paper, we propose a new method for textured images segmentation, which is based on Dual tree complex wavelet transform. The DT-CWT having advantages of shift invariance and better edge representation as compared to real valued wavelet transform. Thereby, to extract a feature vector, we use mean and standard deviation of the magnitude of DT-CWT complex coefficients. Each texture is after classified by using the SVM classifier. We evaluate in experimental study the proposed method on different number of samples. The results demonstrate that the proposed method give better texture characterizing rate. We also compare the proposed method with the state-of-the-art method proposed by [7, 10, 11]. Moreover, [11] had proposed several methods of fusion to improve the segmentation results. Two strategies based on information fusion were used in [11]: score and decision level fusion. In this paper we focus on the results obtained by decision level fusion using majority vote rule to demonstrate that our proposed method gives better results.

This paper is organized as follows: Sect. 2 describes basics of used feature extraction and feature learning. Section 3 gives experimental results. Finally, in Sect. 4, conclusion and suggest avenues for future work are given.

2 The Proposed Segmentation Approach

2.1 Feature Extraction

Feature selection is the key component in any classification algorithm. The correctness of any classification scheme lies on the selected feature. Castleman et al. [12] defined feature as, "A feature is a function of one or more measurement computed so that it quantifies some significant characteristics of objects". In our proposed method, we use the mean and standard deviation of the magnitude of the DT-CWT complex coefficients in six directional subbands as feature set. As mentioned above, the DWT suffers from the lack of shift sensitivity and the lack of strong edge detection. Kingsbury et al. [13, 14] proposed a solution to overcome these problems by using DT-CWT. As described in [13, 14], DT-CWT have the following properties:

- Good shift invariance = negligible aliasing. Hence, transfer function through each sub band is independent of shift and wavelet coefficients can be interpolated within each sub band, independent of all other sub bands;
- Good directional selectivity in 2-D, 3-D which derives from analyticity in 1-D (ability to separate positive from negative frequencies);
- Perfect reconstruction with short support filters;
- Limited redundancy: 2:1 in 1-D, 4:1 in 2-D etc.;
- Low computation: much less than the undecimated DWT.

As described by Kingsbury et al. in [15], DT-CWT uses complex valued filtering that decomposes the image into real and imaginary parts. The real and imaginary coefficients are used to compute magnitude and phase information. When the DT-CWT is applied to 2-D signals, it is performed separately, using 2 trees for the rows of the image and 2 trees for the columns. The 4 quad-tree components of each coefficient are combined by simple sum and difference operations to yield a pair of complex coefficients. These are part of two separate sub-bands in adjacent quadrants of the 2-D spectrum. This produces 6 directionally selective sub-bands at each level of the 2-D DT-CWT. The oriented and scale dependent sub-bands are visualized spatially in Fig. 1.

2.2 Feature Learning

Among the most popular classifiers, we choose the support vector machine (SVM), thanks to its performance in many image processing field. It was initiated by [16] and is primarily a linear classification approach to two classes. It tries to separate individuals from two classes (+1 and −1) seeking the optimal hyper plane that separates the two sets. This guarantees a large margin between the two classes. To extend SVM to the multi-class scenario, a number of models were proposed where typically a multi-class classifier is constructed by combining several binary classifiers. The major used

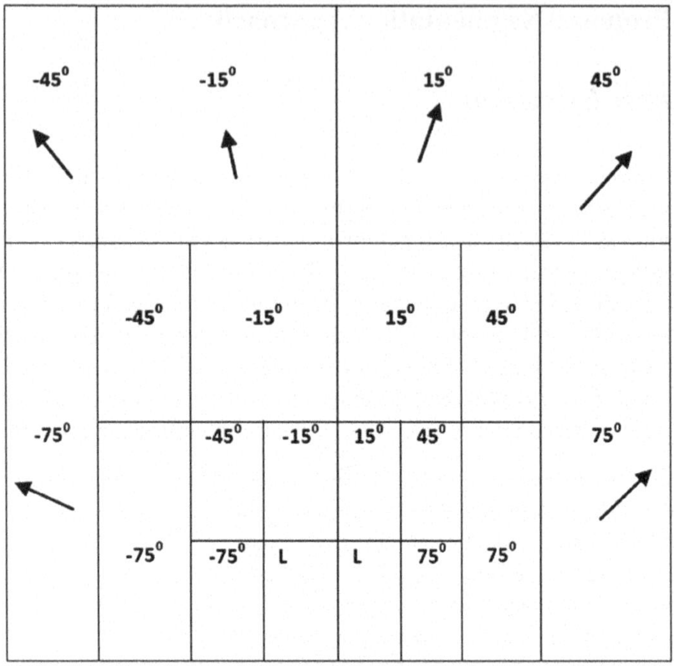

Fig. 1 Complex wavelet transform scale orientation labeled sub-bands

implementations for SVM multi-class classification are the one-against-all method and the one-against-one. Chih-Wei Hsu et al. have presented a comparative study of methods for multi-class Support Vector Machines in [17].

2.3 *Texture Classification Using a Sliding Window*

The precision obtained on the image can be improved using a sliding window with a recovery step, then the class found for this window is assigned to its central part. As described by Laanaya et al. in [18], to characterize the texture of an image, it is divided into tiles of size $L \times L$ pixels with a recovery step l (l < L) specified by the user in order to obtain an accurate classification (cf. Fig. 2). On each tile texture features are calculated using the mean and variance of DT-CWT coefficients. These features are used by the classifier. The classification of these tiles gives an image classified on homogeneous areas.

2.4 The Proposed Algorithm

The proposed method uses DT-CWT coefficient as a feature evaluation set and support vector machine as a classifier for classification of data. Using a sliding window with a recovery step, the class for this window is assigned to its central pixel. Steps of the proposed method are described below.

Step 1: The system applies DT-CWT of level 1 to the entire textured image (cf. Fig. 3) to extract features from each directionally selective filters denoted [+15°, +45°, +75°, −15°, −45°, −75°].

Step 2: The system compute the mean and standard deviation over the sliding window W from the corresponding channel as:

$$m_l = \frac{1}{N_W} \sum_{(i,j) \in W} (d_l(i,j)); \quad d_l \in \{+75°, -75°, +45°, -45°, +15°, -15°\} \tag{1}$$

$$std_l = \sqrt{\frac{1}{N_W} \sum_{(i,j) \in W} (d_l(i,j) - m_l)^2} \tag{2}$$

where, N_W denotes the number of pixel in the window W.

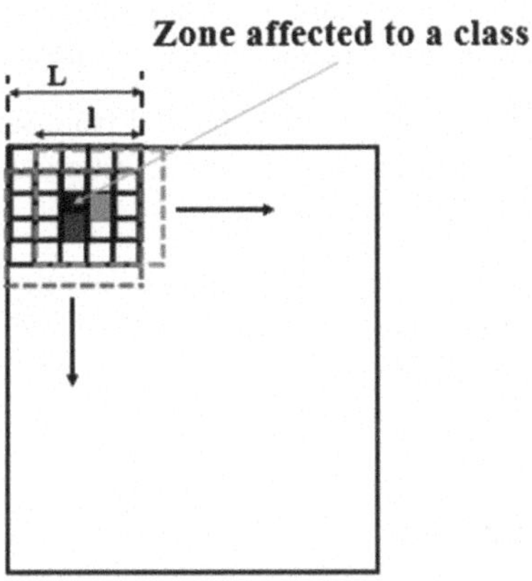

Fig. 2 Image classification using a sliding window

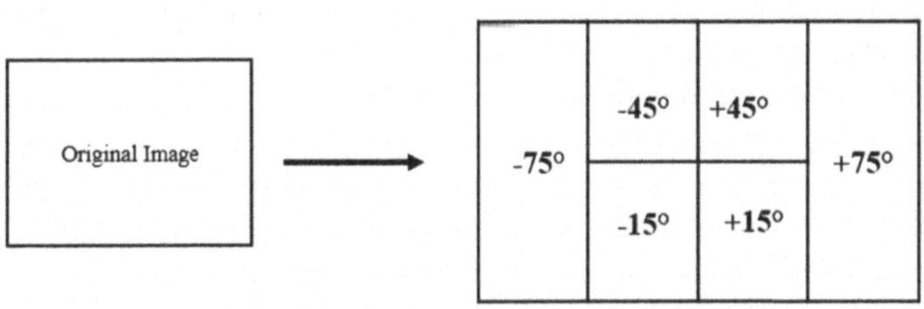

Fig. 3 Application of 1-level dual Tree Complex Wavelet Transform (DT-CWT)

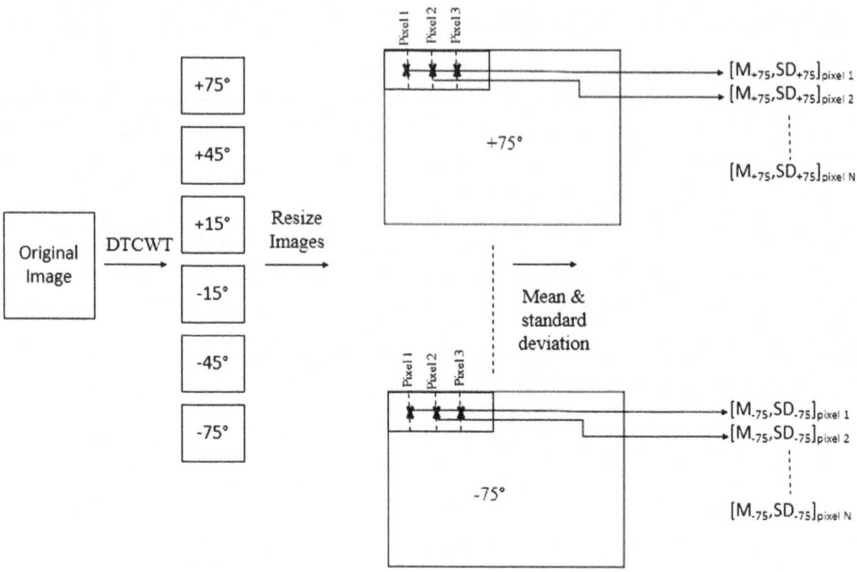

Fig. 4 Feature extraction process using Dual Tree Complex Wavelet Transform (DT-CWT)

Step 3: The feature vector corresponding to each pixel is composed of twelve parameters. Illustration of the feature extraction process is presented in Fig. 4.

Step 4: For the training data, features are extracted from samples of the remaining classes the same way as the testing data. Then the estimated feature vector of each pixel is sent to the SVM classifier for labeling.

Step 5: Different tests have been carried out on a range of window sizes from 7×7 to 23×23. The results of the experiments show that the block size of 15×15 is the proper one for texture discrimination.

3 Results and Discussion

To validate our aforementioned algorithm, we conducted experiments by segmenting a synthetic textured image Fig. 5a containing four natural textures images extracted from the Brodatz album of D16, D21, D32, and D77.

Figure 5a has been decomposed by DT-CWT at the first level. The features were extracted from the sub-bands by using a sliding window W 15×15 over each sub-band. For the training data, we extracted a number of samples with the same size of W for each class. Figure 6 shows the segmented results by SVM classifier by selecting 100, 200, and 400 samples, respectively, from each class.

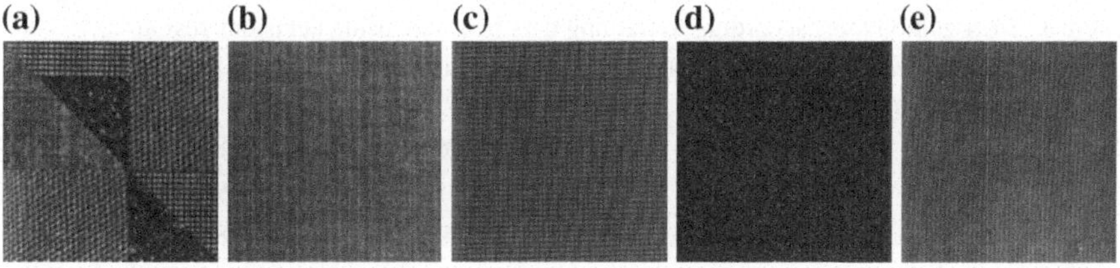

Fig. 5 **a** Textured image containing 4 textures from Brodatz, **b** D16, **c** D21, **d** D32, **e** D77

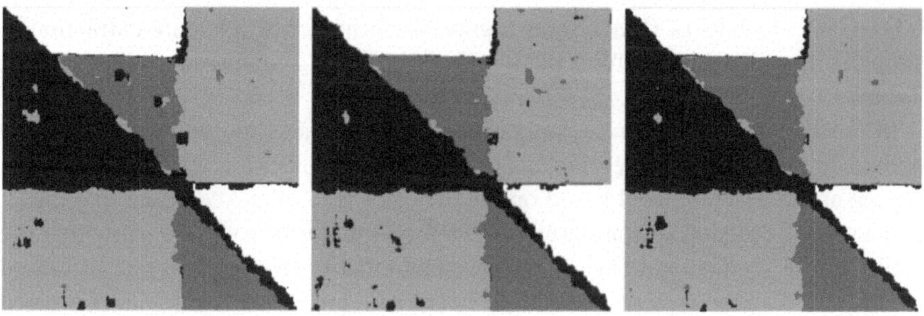

Fig. 6 Segmented image using the proposed approach for 100, 200, 400 samples, respectively

Table 1 Recognition rates according to training data base size

Training data base size (*samples* × *parameters*)	100 × 12	200 × 12	400 × 12
Recognition rates (%)	90.2722	91.4094	**92.3332**

For each experiment, we used the percentage of correct classification to evaluate the classification accuracy. Table 1 shows the recognition rates that were obtained by using different sizes of the training database with the DT-CWT and SVM classifier. The best result was obtained using 400 samples with recognition rates of 92 %.

The proposed method was compared with other state-of-the-art method proposed by [7, 10]. We have also evaluated classification by following the same approach in [11] by using DWT as a feature set. From Table 1, we can observe that the proposed method based on Dual tree complex wavelet transform gives better performance results in comparison to other state-of-the-art methods discussed above as feature, for texture segmentation. Due to improved directive and shift invariant properties of DT-CWT method outperforms the DWT method. Results of GLCM Method is more or less similar to discret wavelet transform method, but poor in comparison

Table 2 Recognition rates of different extraction methods using 400 samples per class

Methods name (%)	Accuracy
DT-CWT	**92.3332**
DWT	84.0175
GABOR	89.7104
GLCM	86.7137

Table 3 Recognition rates according to training data base size using decisions fusion

Training data base size (*samples × parameters*) (%)	100 × 12	200 × 12	400 × 12
DWT	73.2805	76.6529	84.0175
DT-CWT	90.2722	91.4094	**92.3332**
DWT using majority vote rule	78.0308	79.0811	89.3004
DT-CWT using majority vote rule	92.6020	93.1566	**93.3274**

to DT-CWT. Results of Gabor show a good performance as a feature extraction in comparison to GLCM and DWT, but it is time-consuming compared to other method (Table 2).

In order to give an accurate comparison between our proposed method and the method cited in [11], we integrate one of the strategies of fusion which is the fusion scheme at the decision level based on the majority vote rule. According to Table 3, the results of decision fusion based on DWT and majority vote rule show that the quality of the segmented image is improved and that the recognition rate increased by almost 5 % in the case of 400 samples per class, but still less efficient compared to our proposed method based on DT-CWT and majority vote rule.

4 Conclusion

In this paper, we proposed a new algorithm for texture segmentation using DT-CWT. The proposed texture classifier achieves an average correct classification rate of 92 % on a sets of texture by varying samples set. This performance shows that the proposed mean and standard deviation as features of the magnitude of the DT-CWT complex coefficients in six directional subbands for three scales are good candidates for texture classification. The proposed classifier exploits the benefits of shift invariance and six directional sub-bands at each scale of DT-CWT to give better description and discrimination of texture than using real valued coefficient. We also compared the performance of the proposed DT-CWT classifier with the DWT, GLCM and Gabor based classifier, and show that the former outperforms the latters by achieving high correct classification rate. In this study, we have only used mean and standard deviation of the magnitude of DT-CWT as an effective feature for texture segmentation. A statistic modelisation of the sub-bands can be used to further improve the performance of the classifier.

References

1. Yuan, J., Wang, D., Li, R.: Remote sensing image segmentation by combining spectral and texture features. IEEE Trans. Geosci. Remote Sens. **52**(1), 16–24 (2014)
2. Christodoulou, C.I., Pattichis, C.S., Pantziaris, M., et al.: Texture-based classification of atherosclerotic carotid plaques. IEEE Trans. Med. Imaging **22**(7), 902–912 (2003)

3. Deng, H., Clausi, D.A.: Unsupervised segmentation of synthetic aperture radar sea ice imagery using a novel Markov random field model. IEEE Trans. Geosci. Remote Sens. **43**(3), 528–538 (2005)
4. Alpert, S., Galun, M., Brandt, A., et al.: Image segmentation by probabilistic bottom-up aggregation and cue integration. IEEE Trans. Pattern Anal. Mach. Intell. **34**(2), 315–327 (2012)
5. Kim, S.C., Kang, T.J.: Texture classification and segmentation using wavelet packet frame and Gaussian mixture model. Pattern Recogn. **40**(4), 1207–1221 (2007)
6. Jain, A.K., Farrokhnia, F.: Unsupervised texture segmentation using Gabor filters. Pattern Recogn. **24**(12), 1167–1186 (1991)
7. Haralick, R.M., Shanmugam, K., Dinstein, I.: Textural features for image classification. IEEE Trans. Syst. Man Cybern. **3**(6), 610–621 (1973)
8. Wang, B., Zhang, L.: Supervised texture segmentation using wavelet transform. In: Proceedings of the 2003 International Conference on Neural Networks and Signal Processing, pp. 1078–1082. Nanjing, China (2003)
9. Al-Kadi, O.S.: Supervised texture segmentation: a comparative study. In: Proceedings of 2011 IEEE Jordan Conference on Applied Electrical Engineering and Computing Technologies (AEECT), pp. 1–5. Amman, Jordan (2011)
10. Haghighat, M., Zonouz, S., Abdel-Mottaleb, M.: CloudID: trustworthy cloud-based and cross-enterprise biometric identification. Expert Syst. Appl. **42**(21), 7905–7916 (2015)
11. Anibou, C., Saidi, M.N., Aboutajdine, D: Classification of textured images based on discrete wavelet transform and information fusion. J. Inf. Process. Syst. **11**(3), 421–437 (2015)
12. Castleman, K.R.: Digital Image Processing. Prentice Hall, Englewood Cliffs, NJ, USA (1996)
13. Kingsbury, N.G.: The dual-tree complex wavelet transform—a new technique for shift invariance and directional filters. In: proceeding 8th IEEE DSP Work-shop, Bryce Canyon (1998)
14. Selesnick, I.W., Baraniuk, R.G., Kingsbury, N.G.: The dual-tree complex wavelet transform. IEEE Sign. Process. Mag. **22**(6), 123–151 (2005)
15. Kingsbury, N.: Complex wavelets for shift invariant analysis and filtering of signals. Appl. Comput. Harmon. Anal. **10**, 234–253 (2001)
16. Cortes, C., Vapnik, V.: Support-vector network. Mach. Learn. **20**, 273–297 (1995)
17. Hsu, C.-W., Lin, C.-J.:. A comparison of methods for multiclass support vector machines. IEEE Trans. Neural Netw. **13**(2), 415–425 (2002)
18. Laanaya, H., Martin, A., Aboutajdine, D., Khenchaf, A.: Classifier fusion for post classification of textured images. Inf. Fusion **42**, 1–7 (2008)

Parallel and Reconfigurable Mesh Architecture for Low and Medium Level Image Processing Applications

Ihirri Soukaina, Errami Ahmed and Khaldoun Mohamed

Abstract Image processing and especially real time image processing is a very compute intensive task. Nowadays, with the high volume of data to be processed and the increasing size of images, the development of image processing architectures is very required, but most cases of architectures are mostly limited to one single task. This work introduces a parallel Reconfigurable Mesh architecture called RMC (Reconfigurable Mesh Computer) suitable for image processing applications. This architecture provides the flexibility of a programmable architecture and performance of a dedicated circuit, geared to the efficient parallel execution of low and medium level image processing operations. These processing operations derive abstractions from the image pixels so that it can help in further decision making about image. Before describing the proposed architecture, this paper reviews the criteria to be taken into consideration to compare image processing architecture, reinforced by an illustration of some hardware image processing architectures. We also identify some performed applications on RMC, to finally conclude with our future research directions for RMC architecture.

Keywords RMC · Low and Mid-level operations · Parallel processing · Image processing

I. Soukaina (✉) · E. Ahmed (✉) · K. Mohamed
RTSE, ENSEM, Hassan II University, Casablanca, Morocco
e-mail: Ihirri.soukaina@gmail.com

E. Ahmed
e-mail: aerrami@yahoo.fr

K. Mohamed
e-mail: m.khaldoun@ensem.ac.ma

R. El-Azouzi et al. (eds.), *Advances in Ubiquitous Networking 2*, Lecture Notes in Electrical Engineering 397, DOI 10.1007/978-981-10-1627-1_42

1 Introduction

Image processing is a method to perform some operations on an image in order to extract some useful information from it. This domain is gaining large importance in a variety of applications in science and engineering such as bio-medical science, space, agricultural and geological science, etc. Recent years have seen a flurry of activity in the area of parallel architecture especially for image processing because it usually demand solutions involving high performance and real time requirement. Current parallel structures comprise architectural solutions such as GPUs (Graphics Processing Unit), Systolic Array, and specific physical organization FPGA (Field-Programmable Gate Array). Through the brief history of digital image processing, special parallel processing architectures have been proposed and implemented. Among the architectures, the Reconfigurable Mesh Computer has emerged as a very attractive and powerful computational platform; Images can be naturally mapped onto an RMC so that neighboring pixels are mapped onto neighboring processing elements. Because of this, local operations on the image can be performed by local operations on the mesh and can execute low and mid-level image processing application at a high speed. Our objectives in this paper are as following: (1) presenting an analytical study of parallel image processing architecture (2) presenting the robustness of the proposed architecture in term of simplicity, ability to execute low and mid-level operations in parallel. The next section of the paper present a discussion on the criteria to distinguish the image processing architectures, then an illustration of some Hardware image processing architecture is presented, then we describe our architecture giving some implemented applications, to finally conclude by the perspective of our work.

2 Analytical Study of Parallel Image Processing Architectures

There has been a rapid growth in the number of proposed and constructed parallel architectures over the past 10 years, especially for image processing applications. But the relationship between this large number of parallel architectures are difficult to appreciate due to the lack of a sufficiently classifications. There are several criteria to be taken into consideration in order to compare image processing architectures such as: the nature of image processing problems, technological aspect, in a view of algorithmic point…etc. in this section we will describe some of the major criteria for classifying an image processing architectures.

2.1 Taxonomy Criterion

An initial step in designing a complete parallel image processing system is to decide which basic machine architecture to use. However, the various forms of parallel architectures can be distinguished under the Flynn's classification [1] which categorizes all computers according to how many instruction stream and data streams. The four possibilities defined are:

- SISD (Single Instruction Single Data): which means that a single data stream is being processed by one instruction stream, e.g.: The standard von Neumann Model.
- SIMD (Single Instruction Multiple Data): each instruction is executed on a different set of data by different processors [2].
- MISD (Multiple Instructions Single Data): where multiple processing units operate on one single data stream. This kind has never been used.
- MIMD (Multiple Instructions Multiple Data): where each instruction operates on different data and each processor has a separate program [3].

The advantage of this classification is that it is well established, but it doesn't discriminate clearly between multiprocessor architectures, that's why there are other criteria to be taken into consideration to classify image processing architectures.

2.2 Topology Shape Criterion

The topology of the interconnection network linking the processors to each other is an important criterion in parallel image processing architectures. There are varied kind of topologies, divided into two groups: direct and indirect interconnection networks; direct network have point to point and fixed communication links between neighboring processors such as ring, meshes, tori [4] and cubes, while the indirect one have no fixed neighbors and the communication changed dynamically, the well-known examples are bus, multistage and crossbar switches [5].

In view of image processing applications, we should mention systems based on mesh interconnection operate on the entire image and perform local operations on all the pixels of the image simultaneously [6]. Moreover, Pipeline systems carry out sequences of operations, scan and treat a small neighborhood at each stage [7]. Thirdly, Pyramid scheme permit the construction of global operations through divide and conquer technics [8]. Furthermore, architectures whose organization is not directly modeled on the structure of the image enable the implementation of several styles of parallelism [9].

2.3 Algorithmic Criterion

There is a close relationship between image processing algorithms and the type of architecture adopted for their execution. Image processing algorithms can broadly classified into two main groups: Pixel level operations [10] and Region level operations [11]. For the pixel level operations we find the point and local operations [12], geometries operations and measurement of properties. While region level operations determine the properties of regions in an image from the representation of the region in the image. To compare the architecture using the algorithm criterion we found that Pixel operations are well suited for execution in SIMD [13] stream machine. On the other hand, region level operations are well suited to ring type architectures [14] and SIMD with linear network, which is a special case of the two dimensional arrays, based on a geometric parallelism.

2.4 Technological Aspect

Parallel image processing architectures may also be classified in terms of the nature of the processing units comprising them. The traditional hardware implementation of image processing may use either DSP (Digital Signal Processor) or RISC/CISC (Reduced/complex instruction set computer) or even a GPU (graphics processing Unit). However, the growing need for fast and cost effective systems triggers a shift to FPGA, because of its ability to make a specialized circuit, flexibility and better program performance. We can also find a hybrid solution integrating both FPGA and one or several microprocessors in one system.

2.5 Image Processing Operations Criterion

A distinction has to be made between the various layers in image processing: low-level, intermediate-level and high-level. These levels can be described by the difference in the data types handled at each level.

Low level operators deals with pixel operations, usually convert an image data to image data [15]. This level can be divided into: Point operators, Neighborhood operators, Global operators and Recursive neighborhood operators. The intermediate level image processing operations transform image into some sort of symbolic description Such a segmentation tasks [16]. And finally the high level operates on the symbolic descriptor to do such tasks as: recognition decision and generate higher abstraction to interpret the image content [17].

3 Illustration of Some Image Processing Architecture

To make the discussion on architectures easier we will describe in this section some hardware architecture and their corresponding image processing algorithms to illustrate the criterions which have been discussed in the first part.

The first paper [18] proposes a parallel hardware architecture kind of pipeline which scan image and treat a small neighborhood at each stage. They use SIFT (scale invariant feature transform) and BoF (Bag Of features) algorithms for image representation and SVM (support vector machine) algorithm for classification. In order to access the classification, they use VIRTEX FPGA from Xilinx as system consisting MICROBLAZE soft core processor, DSP and memory. This system can be used also for the detection and recognition of abject.

The second paper [19] is a vision chip based on a dynamically reconfigurable hybrid architecture comprising a reconfigurable 64 * 64 Process Element (PE) array, 16 * 16 SOM (Self Organizing Map) neural network processor, a 64 * 1 RP (Row Processor) array processor and a dual core 32 bit RISC MPU. This vision chip carry out local and global operations and implement low, mid-level image processing such as filtering and morphology, and hand recognition, face detection as a high level operations.

The last paper [2] is a SIMD cellular Processor array Vision chip suitable for real time computer vision application such as surveillance cameras, industrial machine vision. Based on a massively parallel cellular array of PE, in which incorporate a photo sensor with an ADC (Analog to Digital Convertor), digital processing circuit, ALU (Arithmetic and Logic Unit), Flag, Register and communication unit. This architecture operate on two modes: synchronous mode for low level image processing based on local pixel data and asynchronous mode (continuous mode) for global operations.

4 Reconfigurable Mesh Computer Model (RMC)

The RMC (Reconfigurable Mesh Computer) has emerged as a very attractive and powerful computational platform; Images can be naturally mapped onto an RMC so that neighboring pixels are mapped onto neighboring processing elements. The use of RMC is well established, with several works in the literature, such as [20, 21, 22].

The concerned model in this work is a parallel Reconfigurable Mesh Computer, which exploits a massive amount of rather simple processing elements connected through a reconfigurable interconnection network. It uses a strict SIMD programming model, where all processors execute exactly the same instructions.

Normally, systems based on a mesh interconnection perform just pixel operations, which mean low-level tasks. However, our architecture is able to perform not only low-level tasks, but also mid-level ones with simple instructions and in a constant time, with the intension to use the results for high level applications. In this section, we will present our architecture in term of three aspects: the topology aspect, zoom on the basic processing unit, and then the basic sets of treatment defined in the architecture, to finish with a presentation of the applications which have been simulated using the RMC model.

4.1 Topological Aspects of the RMC

As mentioned above, the RMC is a set of PE (processing elements) arranged in squared matrix, having neighborhood connectivity of 4, 6 or 8. Figure 1 shows the different neighborhood connectivity. In our case each PE is connected to just 4 neighbors by communication channels through its four ports (East, West, North, South).

Each PE has a set of switches called the switching unit, allowing at each moment the local configuration of a PE, updated by the PE according to the configuration imposed by the algorithm currently processed in the machine. The structure of this matrix depends to the switching mode used. In our case, a direct switching mode is employed, where the matrix elements are defined according to two rules: if two ports are connected, we have 1 as a value and 0 elsewhere as shown in Fig. 2.

It can be defined three basic configuration operations kind of bridge, allowing to update the switching unit in order to ex-change data over the mesh and to enhance the algorithmic complexity.

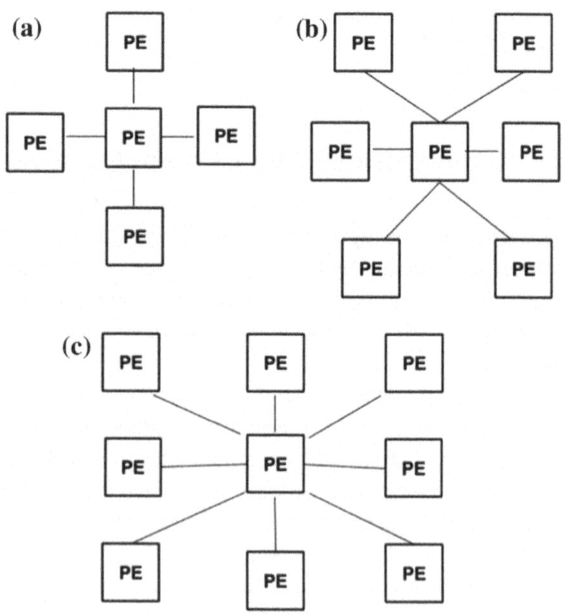

Fig. 1 **a** 4-connected PE, **b** 6-connected and **c** 8-connected PE

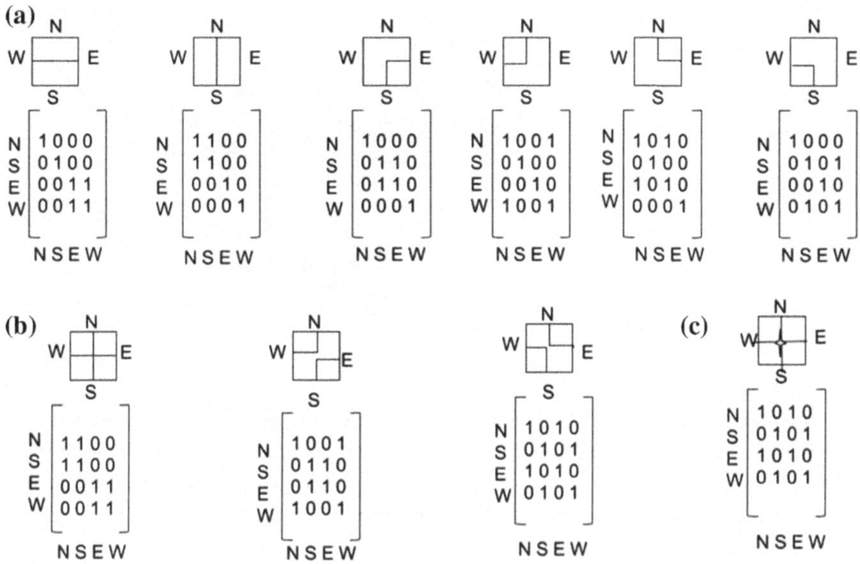

Fig. 2 Different configuration of a PE. And the corresponding Switching Matrix. **a** Simple bridge. **b** Double bridge. **c** Cross bridge

4.1.1 SB (Simple Bridge)

A PE is in SB state, when it connects only two of its four communication channels, in order to form a single bus. The remaining channels are in their initial state. Simple Bridge operation offers to the PE a set of reconfiguration states as shown in Fig. 2a.

4.1.2 DB (Double Bridge)

A PE is in DB state, when it connects two more of the remaining channels in the SB state. In this state the four channels of the PE are grouped into two pairs of channels, each pair constitutes a single bridge. The DB operation gives some reconfiguration state as in Fig. 2b.

4.1.3 CB (Crossed Bridge)

The crossed bridge operation is performed by a PE when it connects all its four channels to form a unique link between its four ports. Notice that according to the state of the ports of a PE (locked or unlocked), we can see other configurations as shown in Fig. 2c. Figure 2 summarizes the three kinds of bridging operations carried out by the PE's with the corresponding switching matrix.

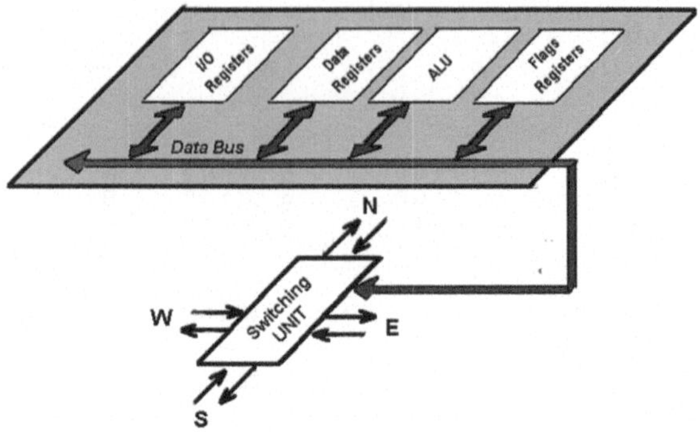

Fig. 3 Component of the PE

From the processing point of view, the PE's of the RMC are classified into 3 groups: PE's which are not concerned by the treatment, doesn't belong to the object of interest, and are in a disconnected mode. The second group is the set of PE which represent the object of interest are in a DB mode. And finally we found the PE which transport the information throughout the mesh is in SB mode.

4.2 Component of the PE

The PE of the Reconfigurable Mesh Computer can carry out arithmetic, logic and even reconfigurable operations in order to exchange data over the mesh. These operations are carried out by the basic components of a PE: ALU, Data Registers, Switching Unit, Flag Registers and Input/output Registers as shown in Fig. 3.

These components are classified according to two categories:

- Components in charge of treatment as the ALU unit, Data registers and the Flag registers.
- Components responsible for communication such the I/O registers and the Switching Unit.

4.3 Sets of Treatment of the RMC

4.3.1 Instruction Sets of the PE

For a processor to be able to process an instruction, it needs to determine what actions it may be asked to perform and have pre-determined available methods to carry out these actions, so the basic operations of a processor are the set of instructions that understand and may indicates what operation are expected to be performed. Like any standard processor, the PE's of the RMC have an instruction set relating firstly to the data processing operations including arithmetic, logical, shift and increment operations, secondly connection and comparison operation combine connection, blocking and comparison instructions, and finally we have transfer operations which are read and write instruction from a port, register and switching matrix. Table 1 summarizes the instruction set of each basic operation that can be done on a RMC Machine.

4.3.2 Elementary and Composed Procedures

Based on elementary operations that were mentioned above, we built a list of elementary procedures to be used in several image processing algorithms on the RMC machine. For example for the contour treatment, path planning, search for text in a document and for the determination of the convex hull. The main purpose of this procedure is to split a large program into small modules, make easy to write and test an algorithm and also avoid writing the same sequence of instructions again and again. Table 2. Summarizes the elementary operations of each basic procedure that can be done on a RMC Machine and that allows us to build a complex image processing algorithms.

And when a procedure calls another one, it is called composed procedure. These are also used to work with complex data structure and make easy the processing of complex algorithms. The Table 3 shows some composed procedures examples.

4.4 Classification of the Procedures According to the Type of Treatment

Because of the evident difference between the functionality of the data structures handled in the various image processing levels, almost image processing architectures are designed to be dedicated to one level tasks, low, medium or high level processing. The architecture proposed in our work is able to do low and mid-level tasks, with the intension to use the results for a high level applications.

Until this section, we have shown that a large number of procedures can be executed in RMC and many applications are performed in reducing complexity operations. This procedures can be categorized either treatment or configuration procedures.

For the PE configuration we have 3 famous procedures: identification of PE, Direct Broadcast and the determination of extreme PE of an object. In view of the low level procedures we have: Dilatation/ Erosion, Data flow, Smooth and Circular shift. While the rest of the procedures are directed towards a medium level processing tasks such as: contour orientation, Sorting, Ranking, Min/Max procedures…etc.

Table 1 Set of instruction of the PE's of the RMC

	Operations	Set of instructions
Data processing operations	Arithmetic operations	Addition of two operands
		Subtraction of two operands
		Multiplication of two operands
	Logic operations	'And' logic between two operands
		'Or' logic between two operands
		'Xor' logic between two operands
	Sift operations	Right shift of the bit
		left shift of the bit
	Increment/decrement/reset to 0 operations	Reset to 0 of a value
		Incrementing a value
		Decrementing a value
		No operation, incrementing the counter
Connection and comparison operations	Connection operations	Connection if equal to 0
		Connection if higher than 0
		Connection if less than 0
	Blocking operations	Blocking if equal to 0
		Blocking if higher than 0
		Blocking if less than 0
	Comparison (register operations)	Equal content
		Different content
		Upper content
		Lower content
	Marking (flog operations)	Marked if upper to a criterion
		Marked if less than a criterion
		Marked if equal to a criterion
		Marked if different from a criterion
Transfer operations	Read/write from a port	Read data from a port
		Write data in a port
	Read/write from a port	Read from a port
		Write in a port
	Read/write from a switching matrix	Read from a matrix
		Write in matrix
	Bit operations	Set a bit
		Clear a bit
		Read a bit

Table 2 Set of elementary procedure that can be used in a different image processing applications

Elementary procedure	Description	Elementary operations
Direct broadcast	A procedure of transmitting from a transmitter PE to a set of receivers simultaneously	Connections operations
		Transfer operations
Dilatation/erosion	A set of morphological operations used to clean an image of the background noise	Marking operations
		Transfer operations
		Arithmetic operations
		Logic operations
Determination of extreme PE of an object	A procedure that distinguishes the hollow point or peak	Transfer operations
		Arithmetic operations
		Comparison operations
Data flow	A procedure that determine the direction of flow of data for the characterization of the contour	Comparison operations
		Transfer operations
Identification of PE	An elementary procedure to identify each PE as its location in the matrix by its row and column coordinates	Arithmetic operations
Distance calculation	Procedure used to determine the nearest neighbor of a PE	Transfer operations
		Arithmetic operations
Contour orientation	A procedure to determine the direction of routing data	Connection operations
		Transfer operations
		Comparison operations
Smooth	A procedure that remove the extremes point	Transfer operations
		Comparison operations
		Connections operations
		Logic operations
		Increment/decrement operations
Circular shift	The procedure of rearranging the entries in the tuple	Shift operations
		Transfer operations
		Arithmetic operations
		Connection operations
Sorting	A procedure for sorting N elements in constant time	Comparison operations
		Arithmetic operations
		Logic operations
Hough transform	Procedure for detecting predefined features in digital image without loss of generality	Arithmetic operations
		Transfer operations
		Comparison operations
Calculation of width/height	Procedure used to determine the geometric characteristics of a component	Transfer operations
		Arithmetic operations
		Comparison operations

Table 3 Set of composed procedure used for image processing application on a RMC machine

Compound procedure	Description	Elementary operations
Min/Max	A procedure used to determine the PE having the largest or smallest value among all set of element	Transfer operations
		Arithmetic operations
		Connection operations
		Procedure of identifying a PE
		Direct broadcast procedure
Nearest neighbor	A procedure of determining the nearest neighbor of PE and Euclidean distance that separates it from is Nearest neighbors	Connection operations
		Transfer operations
		Distance calculation procedure
		Min/Max procedure
Segment extraction	Procedure allows isolating all PE located on the edge of a component to arrange the contour	Transfer operations
		Comparison operations
		Marking operations
		Logic operations
		Distance calculation procedure
Ranking	A procedure used for ranking the contour's pixels	Transfer operations
		Comparison operations
		Arithmetic operations
		Segment extraction procedure
Shrinking	A procedure involves reduction of pixels	Arithmetic operations
		Logic operations
		Min/Max procedure

This classification proves that the majority of procedures are kind of mid-level ones, which means that our architecture is oriented to medium processing tasks, something which is not usual for systems based on mesh which deals only with low-level tasks.

4.5 Applications

In this section, we illustrate the power of the reconfigurable mesh computer by giving some complex image processing application created from a small number of simple operations such as structural filtering, convex hull and path planning.

4.5.1 Structural Filtering

Structural filtering is used to eliminate all the high frequency space variations, or the irregularities of the shape hull, pattern recognition, mapping, analysis of scene, noisy contours.

The goal here is to achieve a structural filtering of multi-level pictures in order to remove any irregularities of high frequencies on the contours; the basic idea of our method relies on the fact that all the irregularities (noises) presented in the contour can be characterized by their structure and their surfaces.

The approach is as follows: from a noisy picture we extract the contours of the components of an image, and then applied a structural smoothing so that in the end we will have a spatial insulation of highly frequency variations. All this are achieved in a constant time with some simple procedures: Smooth procedure, Determination of extreme PE of an object and Transfer operations and erosion.

In terms of complexity, this application is executed in a constant time because there is no iterative procedure [23]. Figure 4a shows an example of a noisy synthetic image which is segmented into six connected components. The contour

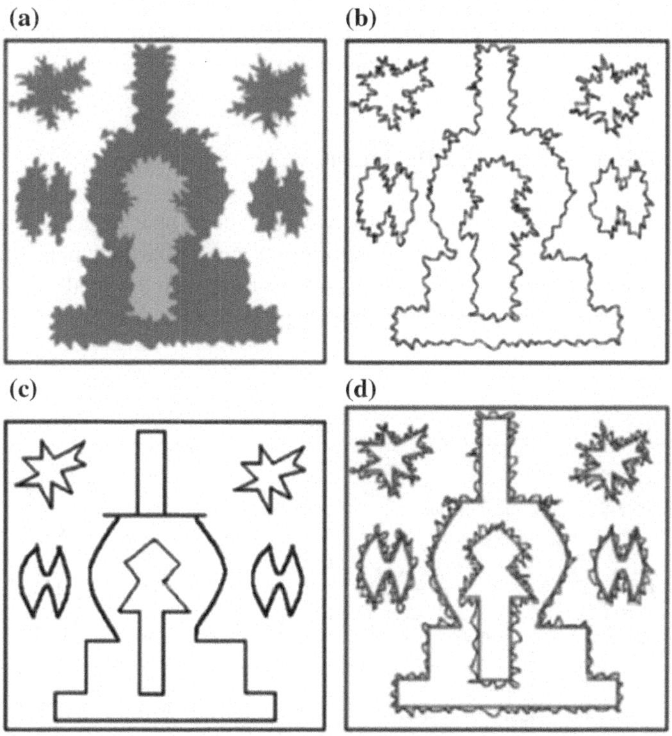

Fig. 4 Different stages of the filtering algorithm. **a** Example of multileveled connected component image. **b** Contour extraction of each component. **c** Filtered contours using external smoothing algorithm. **d** Superposition of noisy and smoothed contours

extraction of these components is given in Fig. 4b. Image structural smoothing and the spatial variations of the contour components are shown in Fig. 4c, d. The result of Fig. 4d correspond to the external smoothing procedure presented below.

4.5.2 Convex Hull

The convex hull is a very important problem especially for the standardization of a shape, for the decomposition of a complex shape. The aim here is to determine the convex hull of a multi-level image.

We use a geometric method that operate only on data extracted contours based on a calculation of distance and line segments to detect the extremes points, which form an hull contours. This method is based on the following procedure: Determination of extremes PEs of an object, Smooth procedure, Transfer operations and Comparison operations. Figure 5a shows the final extreme PEs of the convex hull of each component after removing the valley PEs. The convex polygons enclosing each component are presented in Fig. 5b.

From the standpoint complexity, to calculate the algorithm complexity, we took the worst case of a component that corresponds to an image whose shape is a circle. It was concluded that the complexity is O (log m) where m is the number of segments of the largest connected component of the image [23].

4.5.3 Search for a Text in a Document

For automatic processing of documents, we have been able to achieve a suitable algorithm to our RMC model for extracting text areas in a document following a bottom-up global structural analysis approach, starting with the smallest subject of a paragraph (Letter) to form the words that shape their turn paragraphs.

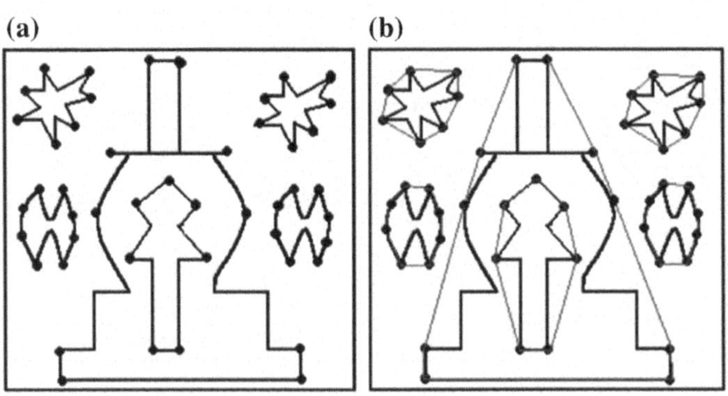

Fig. 5 Convex hull search. **a** Extreme vertices of each component. **b** The convex hull of each component

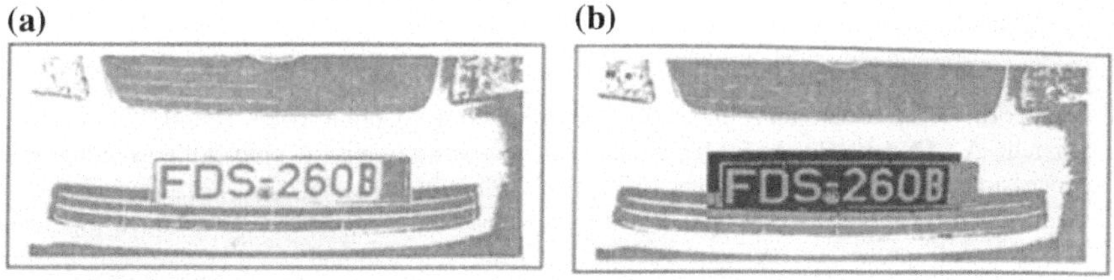

Fig. 6 Example of identifying a text area drowned in an image

To do so, this approach requires four analysis phases: (1) Extraction and characterization of the connected components of an image, (2) Extraction of text words by the criterion of a word is normally composed of components aligned representing a similarity in terms of width, height and spacing, (3) Concatenation of several words to build a text, (4) Extraction of subsections which are composed of more text sentences.

These steps were completed by simple procedures: Segment extraction procedure, Calculation of a height/width procedures, Direct broadcast procedure, Min/Max procedure, Nearest neighbors procedure, Dilatation operation and Marking operations.

By using the RMC model, we were able to achieve all phases of this algorithm in constant time (O(1) operations) regardless of the document structure. Figure 6a represents the original image; the 6b represents the text area that has been identified by our algorithm.

5 Conclusion and Perspectives

This paper introduces the reconfigurable mesh computer as a model of computation, able to do low and mid-level tasks. With the intension to use the results for a high level applications. We have presented the criteria to be taken into consideration in image processing architectures comparison. In fact, we have shown that the RMC is a powerful model due to the fact that a complex program can be split into a small and simple operations and executed in a constant time. Our contribution in this paper is to provide the list of set of instructions, elementary procedures and even composed ones, which open the door to build a large number of image processing algorithms on a reconfigurable mesh computer. The perspective task of this work is to design and implement these applications on RMC platform.

References

1. Skillicorn, D.B.: A taxonomy for computer architectures. Computer **21**(11), 46–57 (1988)
2. Lopich, A., Dudek, P.: A SIMD cellular processor array vision chip with asynchronous processing capabilities. Regul. Pap. IEEE Trans. Circuits Syst. **58**(10), 2420–2431 (2011)
3. Deguchi, K., Tago, K., Morishita, I.: Integrated parallel image processings on a pipelined MIMD multi-processor system PSM. In: 10th International Conference on Pattern Recognition, Proceedings, vol. 2, pp. 442–444 (1990)
4. Andújar-Muñoz, F.J., Villar-Ortiz, J.A., Sánchez, J.L., Alfaro, F.J., Duato, J.: N-Dimensional twin torus topology. IEEE Trans. Comput. **64**(10), 2847–2861 (2015)
5. Shen, H., Wang, J., Yuan, C., Wang, Z., Zheng, W.: A novel crossbar scheduling for multi-FPGA parallel sar imaging system. In: 2010 First International Conference on Pervasive Computing Signal Processing and Applications (PCSPA), pp. 394–397 (2010)
6. Miller, R., Prasanna, K., Stout, F., Dionisios, R.: Parallel computations on reconfigurable meshes (1993)
7. Kent, E.W., et al.: PIPE: pipeline image processing engine. J. Parallel Distrib. Comput. **2**, 50–78 (1985)
8. MéRIGOT, A., Ni, Y., Devos, F.: Architectures massivement paralleles pour la vision artificielle (2009)
9. Siegel, H.J., Siegel, L.J., Kemmerer, F.C., Smith, S.D.: PASM: a partitionable SIMD/MIMD system for image processing and pattern recognition (1982)
10. Hwang, J.-J., Liu, T.-L.: Pixel-wise deep learning for contour detection (2015)
11. Lee, C.-Y., Leou, J.-J., Hsiao, H.-H.: Saliency-directed color image segmentation using modified particle swarm optimization (2012)
12. Geisler, W.S., Perry, J.S., Super, B.J., Gallogly, D.P.: Edge co-occurence in natural images predicts contour grouping performance (2001)
13. Rosenfeld, A.: Parallel algorithms for image analysis. Mod. Sig. Process. (1985)
14. Kushner, T., Wu, A.Y., Rosenfeld, A.: Image processing on the ZMOB. IEEE Trans. Comput. (1982)
15. Persa, S., Nicolescu, C., Jonker, P.: Evaluation of two real time low level image processing architecture (2000)
16. Segmentation d'Images IRM Cérébrales sur Architecture Massivement Parallèle (GPU), Sept 2014
17. Liu, Y., Zhang, D., Lu, G., Ma, W.-Y.: A survey of content-based image retrieval with high-level semantics (2007)
18. Qasaimeh, M., Sagahyroon, A., Shanableh, T.: FPGA-based parallel hardware architecture for real-time image classification. Comput. Imaging IEEE Trans. **1**(1), 56–70 (2015)
19. Shi, C., Yang, J., Han, Y., Cao, Z., Qin, Q., Liu, L., Wu, N.-J., Wang, Z.: A 1000 fps vision chip based on a dynamically reconfigurable hybrid architecture comprising a PE array processor and self-organizing map neural network. Solid-State Circuits IEEE J. **49**(9), 2067–2082 (2014)
20. Moreira, A., York, B.W.: Matrix inversion in O (log n) on a scan-enhanced reconfigurable mesh computer, pp. 67–75 (1996)
21. Bouattane, O., Elmesbahi, J., Khaldoun, M., Rami, A.: A fast algorithm for k-nearest neighbor problem on a reconfigurable mesh computer. J. Intell. Robot. Syst. **32**(3), 347–360 (2001)
22. An efficient list-ranking algorithm on a reconfigurable mesh with shift switching (2007)
23. Errami, A., Khaldoun, M., Elmesbahi, J., Bouattane, O.: $\theta(1)$ Time algorithm for structural characterization of multi-leveled images and its applications on a reconfigurable mesh computer. J. Intell. Robot. Syst. **44**(4), 277–290 (2005)

A Survey on Segmentation Techniques of Mammogram Images

Ilhame Ait lbachir, Rachida Es-salhi, Imane Daoudi, Saida Tallal and Hicham Medromi

Abstract Mammogram images are important tools allowing visualization of various types of breast cancer. In fact, cancer detection refers to the extraction of region of interest ROI, which represents the tumor, in the mammogram image. In medical imaging field, Computer Aided Diagnosis systems (CAD) are used to analyze this type of images. To extract region of interest from mammograms, image segmentation methods have been wildly applied. These methods consist of partitioning the image on meaningful regions or segments easy to analyze. There are various techniques and methods of segmentation of mammogram images in the literature. In this paper, we present a survey of different approaches of segmentation that we compared theoretically in terms of advantages and drawbacks, particulary for mammogram images.

Keywords Mammogram images · Image segmentation · Region of interest ROI · Image analysis

---

I. Ait lbachir (✉) · R. Es-salhi · I. Daoudi · S. Tallal · H. Medromi
Systems Architecture, ENSEM, Hassan II University, Casablanca, Morocco
e-mail: aitlbachirilhame@gmail.com

R. Es-salhi
e-mail: rachida.es-salhi@ensem.ac.ma

I. Daoudi
e-mail: i.daoudi@ensem.ac.ma

S. Tallal
e-mail: s.tallal@ensem.ac.ma

H. Medromi
e-mail: h.medromi@yahoo.fr

R. El-Azouzi et al. (eds.), *Advances in Ubiquitous Networking 2*, Lecture Notes in Electrical Engineering 397, DOI 10.1007/978-981-10-1627-1_43

1 Introduction

Breast cancer is the second leading cause of mortality among women in the entire world, exceeded only by lung cancer. According to American Cancer Society, about 1 in 8 (12 %) women in the US will develop invasive breast cancer during their lifetime, and the foundation of Lalla Salma against the cancer affirms that 36.12 % of women in Morocco develop breast cancer. Only early detection can reduce the rate of mortality and increase recovery. Currently, mammography is the dominant tool to visualize and detect breast cancer, using low energy X-rays. The analysis of mammogram images remain a challenge among researchers, but in the few last decade, plenty of methods and Computer Aided Diagnosis systems (CAD) have been proposed. The goal of CAD systems is to assist and help radiologists in their interpretations and decision. We present in Fig. 1 the flowchart of mammogram image processing in CAD systems. In the first step, the image is pre-processed by removing noise and superfluous data, and then applying enhancement algorithms. The objective of this step is to prepare mammogram images to the following process. In the second step, titled segmentation, we aim to extract regions of Interest (ROI), which represent tumors, thereafter classified on benign or malignant masses. Once the mammogram segmentation is done, the goal of the third step is to describe the extracted ROI using different features like grey level histogram, intensity, size, texture and shape. The fourth step consist in selecting required features used in the final step to classify tumors as benign or malignant ones. In this paper, we focus on analyzing mammogram segmentation step. We will review different methods proposed in the literature, that we classified into two categories: Supervised and unsupervised methods. We discussed those methods in terms of advantages and disadvantages. This paper is organized as follows: Sect. 2 gives a review of image segmentation approaches. Section 3 presents Supervised segmentation techniques. Section 4 discusses about unsupervised techniques. Section 5 presents a brief discussion, and finally the conclusion.

2 Image Segmentation Approaches

Image segmentation is the process of finding groups of pixels that go together [14]. The aim of segmentation is to partition digital image into regions, sharing similar characteristics, easy to process. There are several image segmentation approaches in the literature categorized as follows:

Region-based approach: Region-based approach is one of the simplest as well as popular algorithms for segmentation. It is based on grouping pixels into homogenous regions [29]. The principle is expressed as follows:

$$\bigcup_{1}^{n} R_i = I,\ R_i \bigcap R_j = \emptyset \, for\, i \neq j \tag{1}$$

Fig. 1 Flowchart of mammogram image processing

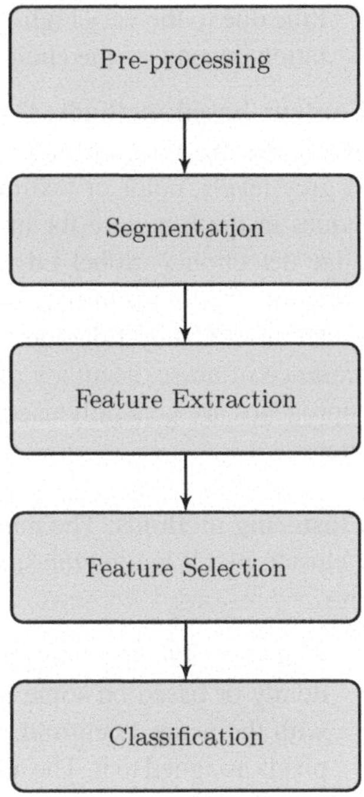

where: I is the image containing n regions Ri. Their intersection is the empty set and their union is the image I.

Region-based methods can be divided in three categories:

i. *Region growing*: Starting with a seed point, a region grows according to the homogeneity of neighboring pixels. This process is iterated until we get a homogenous and connected region.
ii. *Split and Merge method*: The image is represented in a quad tree, divided four by four squares satisfying a homogeneity criterion. Using the Region Adjacency Graph (RAG), the squares are merged according to the homogeneity of neighboring regions. This process is iterated until we get a homogenous and connected region.
iii. *Watershed transform* [7] is based on mathematical morphology. The image is considered as a topographic relief, where the height of each point is directly related to its gray level. Considering rain gradually falling on the terrain, the watersheds are the lines that separate the "lakes" called *catchment basins*. The watershed transform is usually computed on the gradient of the original image, so that the catchment basin boundaries are located at high gradient points. Watershed transform has the advantage of been simple, intuitive and fast. However, this method presents many drawbacks such as over segmentation due to the great number of non-meaningful small regions, in addition to sensitivity to noise.

The principal drawbacks of this approach is that it requires a large calculation time due to the very high-resolution of images. Therefore, the results of segmentation depend on the choice of the seed point.

Contour based methods: Contour-based approaches [21] consist on the detection of edges separating distinct regions. The detection is based on the discontinuities of grey levels, color or texture of the image. There are many edge detection techniques in the literature for image segmentation enumerated as follows: 1. Roberts Edge detection-2. Sobel Edge detection-3. Prewitt Edge detection-4. Kirsch Edge detection -5. Robinson edge detection-6. Marr-Hildreth edge detection-7. LoG edge detection-8. Canny Edge detection [21]. However, because of weak edges and the presence of noise, accuracy of results can be decreased. That is why contour based approaches are generally used in conjunction with region-based methods discussed above.

Clustering methods: The main purpose of clustering approaches is to resemble, in k cluster, pixels having the same properties [27]. Among clustering approaches, we cite:

i. *K-means methods*: This approach starts by placing K centroid set locations randomly or based on some heuristic [27]. Then, we assign each pixel to a cluster with the nearest centroid, we move thereafter each centroid to the mean of the pixels assigned to it. The algorithm continues until no pixels change cluster membership. K-means algorithm is simple. However, the performance depends on the initial positions of the centroids.
ii. *Fuzzy C-means (FCM) methods* [27]: The clustering method for both k means and FCM is same. The difference between them is that in the FCM method, each point has a degree of belonging to two or more clusters. Hence, FCM consumes more time than k-means algorithm, since it is doing more work. The performance of FCM algorithm depends also on the initial cluster centers. Moreover, FCM is sensitive to noise, which makes the identification of the initial positions difficult.

Thersholding methods: The methods based on thresholding are the simplest to implement, faster and inexpensive ones. They use the global information like histogram, to separate the object to be segmented (foreground) from the background. In order to do this, thresholding approaches transform an input image f into a binary image g as follows:

$$\begin{cases} g(i,j) = 1, f(i,j) \geq T \\ g(i,j) = 0, f(i,j) < T \end{cases} \quad (2)$$

where T is the threshold. Thresholding approach present many disadvantages such as: The difficulty of defining the thresholding T and the ambiguity imposed when two objects have the same color.

Energy function-based methods:

i. *Active contour/ deformable models* [13] are based on the evolution of a curve, in order to detect objects in an image. The basic idea is to start with a closed curve and iteratively modify it by applying a contour evolution, performed by the minimization of an energy function. There are two main approaches in active contours: snakes and level sets. Snakes require a predefined snake points that move explicitly depending on an energy minimization function, while level set method move contours implicitly as a particular level of a function. Hence, the use of level set theory provides more flexibility in the implementation of active contours. Compared to edge-based methods, discussed above, active contour methods have the advantage of estimating boundary with smooth curves or surfaces that bridge over boundary gaps.

ii. *Markov Random Field (MRF)* [18] MRF models are a graphical representation where the image is divided into sites, either at the pixel level or at the level of patches of predetermined spatial scale (size). Given an observation set corresponding to the feature vector of each site, image segmentation is defined as a labeling problem which consists on assigning a label to each site in the image, this label can be either region of interest or background and it depends on (i) the corresponding observation and (ii) the labels of its neighboring sites. The segmentation result is obtained as a global optimization problem, i.e., estimating the optimal label field, given the observations. In contrast to traditional deformable models that follow deterministic energy minimization approaches, MRF models are usually based on a probabilistic solution, i.e., they are driven by the maximization of a probability.

Mammogram segmentation [22] is partitioning the mammography into meaningful regions and segments having similar properties. It consists on finding the regions susceptible to contain anomalies. The aim of mammogram segmentation is to identify regions of interest (ROI) by measuring the volume and the size of the tumor.

General approaches of segmentation, already cited in Sect. 2, applied to mammography can either be supervised or unsupervised. Thus, the following sections will shell supervised mammogram segmentation techniques and unsupervised mammogram segmentation techniques. The following proposed methods are the most known and used ones.

3 Supervised Techniques

Supervised methods [23] are based on a training stage where knowledge about the object to be segmented is learnt. Thereafter, that knowledge is used to determine if a region is present in an image or not. These methods are interactive, since a user defines the information from which the algorithm of segmentation begins. Several semi-automatic methods where developed to segment mammographic images: Jennefer et al. [12] developed a semi-automated tumor segmentation method based on the watershed technique. The proposed method removes firstly noise and applies an image enhancement based on Speckle Noise Removal and EM algorithm. Secondly, masses are segmented using modified watershed approach. The idea of the modified method is to provide initial indicators using mouse clicks, which locate pixels belonging to lines considered as markers. Once the ROI segmented, features are extracted and classified using SVM classifier.

Martins et al. proposed in [5] a mass detection method in digital mammograms based on k-means method. This method uses k-means algorithm to segment the ROI and SVM classifier to classify the features extracted from the segmented ROI.

Hamdi et al. [9] developed a semi-automated mammography segmentation technique based on Watershed, Wavelet and Curvelet transform. The proposed algorithm uses Curvelet and Wavelet transform in the preprocessing step, in order to enhance the image contrast. Furthermore, background and foreground objects are marked. Finally, the watershed transform is applied to segment tumors.

4 Unsupervised Techniques

Unsupervised mammography segmentation methods [23] do not require human intervention or prior information about mammograms. Hence, masses are automatically segmented.

Zhanget al. [31] developed an automated breast masses segmentation able to extract suspicious regions automatically, without any user assistance. The proposed method starts with eliminating labels and artifacts scanned with the image, in order to focus on the breast region. Thereafter, local maxima are detected using pixels with maximum grey values. The extraction of ROI is done using a modified region growing algorithm, which controls the procedure of adding pixels.

Ball et al. [1] developed an automatic method for mammography segmentation based on Adaptive Level Set Segmentation, where an algorithm defines automatically a seed point. The segmentation here is done in the polar domain, using the seed point as the center of the polar image.

Eddaoudi et al. [6] proposed an automated masses detection technique. This method starts by pectoral muscle segmentation, using contour detection approach with an automatic initialization. Furthermore, Region of interest segmentation is performed using maxima thresholding-based approach. This method uses maximum intensity values as markers to detect suspicious objects. Once the ROI segmented, they are classified using SVM classifier. The tests of these methods were performed using the MIAS database [28], and showed better results compared to the existent methods.

Mahmood et al. [15] proposed an automated segmentation algorithm for breast profiling, based on a combination of thresholding technique and morphological preprocessing. The aim of this method is to separate background region from the breast region, and remove artifacts and labels. This method consists on converting the gray scale mammography to binary image, using a predefined threshold T. Artifacts, labels and noise are removed using morphological operations. The evaluation of the proposed algorithm is done using MIAS database [28], where results had proven accuracy.

Table 1 Center line of a mammogram analysis conditions

Background	Breast tissue	Noise block
0–89 gray level	90–230 gray level	>230

Raju et al. presented in [26] an automated mammogram segmentation based on region growing approach. The proposed method starts with an automated seed point identification for locating the center pixel of the abnormal regions. The identification is performed by dividing the center row of the mammogram image into blocks of 50 pixels. The median of gray level of each block is calculated and associated to three gray level conditions presented in Table 1 bellow:

Furthermore, adjacent blocks with the same gray levels are connected to form chains, where the longest one is attributed to a seed point. Therefore, every neighbor pixel, which satisfies a similarity condition, is added to the region, and acts as a next seed point for the next iteration. After segmenting the ROI, their boundaries are isolated using the gradient operators. This method showed better results where the time for extracting a ROI from a mammogram image is 3.7393 seconds, and 129 ROIs out of 149 where segmented, which contain abnormality content exactly as defined by the radiologist.

Xiang Lu et al. proposed an automated mammogram image segmentation methods based on improved VFC Snake model [14]. This method consists of three main processing steps: (1) Mammogram preprocessing, where Labels and artifacts are removed and the whole image is enhanced. (2) ROI extraction and location: edge detection operators applied to extract image edges. Then Linear Hough Transform LHT [11] is used to detect pectoral muscle, and Circular Hough transform CHT to detect mass edges. (3) Mass segmentation: the Snake model is used to perform accurate segmentation based on an the location of masses detected in the previous step.

5 Analysis and Discutions

The sections above cast some lights on the pros and cons of some commonly used mammogram segmentation techniques. A summary of the discussed approches is represented in Table 2 below:

Table 2 presents the advantages and drawbacks of image segmentation, particulary for mammograms. Region-based approach is wildely used for mammogram segmentation, but is time consuming because of the high resolution of mammogram images. Edges based approach can be used as a first process to detect principale edges of mammogram, but is insuffisant because of its sensitivity to noise. For clustering approach, the results depend on the initial identification of clusters number. Thresholding is the simplest and faster segmentation approach, but is sensitive to noise in mammograms. Compared to Edge-based approach, energy function-based

Table 2 Comparaison of mammogram segmentation techniques

Approach	Advantages	Drawbacks
Region based	1. Region growing method gives better results when the homogeneity criterion is well defined and is robust against noise	1. Depends on the seed point selection and is time consuming
	2. Starting segmentation in split and merge technique do not require any homogeneity criterion to be met	2.The resulting segments are too square because of the splitting process
	3. Watershed algorithm gives a closed boundaries	3. Over-segmentation problem
Edge based	1. Edge based techniques don't need prior information about the objects present in the image and perform best when dealing with high contrast images	1. A complicated process when the edges are ill-defined or there are too many edges
	2. They are fast	2. Detection of fake and weak edges provides erroneous segmentation results
Clustering	FCM is comparatively better than K-Means and gives good results for overlapped data	1. They are computationally expensive
		2. They are sensitive to outliers and initial number of clusters
Thresholding	Simple and computationally fast	1. Bad results when there is low contrast between objects and background and in the presence of noise
		2. Difficulty to fix the threshold when the number of regions increases
Energy function-based	1. Deformable models have less computational requirements	1. Sensitive to the initialization of the snake
	2. Flexible to represent complex shapes	

approach is flexible to represent shapes, and gives continuous contours as results. Yet, the results depend on the initialisation of the method.

It is worth mentioning that for best segmentation results, hybrid approach is used. It consists on combining two of the previous cited approaches, in order to benefit from their advantages and outshine their drawbacks.

Table 3 Breast segmentation methods according to the approaches used with accuracy of each method

Approach	Method	Preprocessing step	Dataset	Accuracy (%)
Energy function-based	Lu et al. [14]	• Label removal • Image enhancement • Pectoral muscle removal	• MIAS • DDSM	90.407
Region Growing	• Raba et al. [24] • Zhang et al. [32] • Chen et al. [3]	• Pectoral muscle segmentation • Noise and artifacts removal • Pectoral muscle segmentation	• mini-MIAS • mini-MIAS • mini-MIAS • EPIC	• 98 • N/A • 98.8 • 91.5
Thresholding	• Wei et al. [30] • Raba et al. [24] • Mustra et al. [19] • Maitra et al. [16] • Rahmati et al. [25]	• Pectoral muscle segmentation and artifacts removal • Pectoral muscle suppression • Pectoral muscle and artifacts removal • contrast enhancement and pectoral muscle separation • FCLAHE for image enhancement	• DDSM • MIAS • KBD-FER • mini-MIAS • DDSM	• 94.9 • 98 • 100 • 95.71 • 88.2

Otherwise, preprocessing aims to remove superflus data present in the image, including artifacts, labels, patient name, muscle part, etc. and enhance the region of interest which helps for efficient segmentation and detection of tumor [17]. Authors in [8] investigated the performance of different preprocessing techniques, and the results showed that the best detection performance is achieved when a preprocessing step is applied. Detection of pectoral muscle is one of the important steps in the preprocessing of medio-lateral oblique (MLO) views of mammograms [6, 20]. Previous works of mammogram segmentation show that the results are affected by the presence of the pectoral muscle in the mammogram image. That's why it is wise to segment the pectoral muscle as a preprocessing task.

The following Table 3 summarizes the different approaches and shows which one gives good results in terms of segmentation accuracy. Differences in segmentation accuracy are closely affected by the quality of the images used for testing. Mammogram segmentation algorithms were tested using different databases, publicly available for research purpose. Among them, we cite: MIAS [28], DDSM [10], KBD-FER [2] and EPIC [4].

Authors proposed in [25] a preprocessing filter called fuzzy contrast-limited adaptive histogram equalization (FCLAHE), improved on the contrast-limited adaptive histogram equalization (CLAHE) algorithm to remove noise and intensity inhomogeneities. Results showed an average increase of segmentation accuracy by 14.16 % when the new filter FCLAHE is applied. Thus, it is obvious to admit that a robust preprocessing algorithm gives better results of segmentation. On the other hand, removing only noise is not sufficient to achieve 100 % of accuracy. That's why it is important to develop suitable algorithms that remove noise, pectoral muscle and artifacts.

6 Conclusion

In this paper, we have presented a review of several mammogram image segmentation approaches, classified into supervised and unsupervised and compared in terms of advantages and drawbacks. Based on this theorical comparison, we concluded that combined approaches provide better segmentation accuracy. In addition, robust preprocessing methods improve obviously the segmentation results. It should be taking care of breast part when removing pectoral muscle, because breast details are sometimes also removed. A potential direction for future work is to concentrate on pectoral muscle and superflus data removal. Another direction is to use other types of breast images, such as thermography and ultrasound, having significantly different properties than mammography and proving to be better and less harmful.

References

1. Ball, J.E., Bruce, L.M.: Digital mammographic computer aided diagnosis (CAD) using adaptive level set segmentation. In: 29th Annual International Conference of the IEEE Engineering in Medicine and Biology Society, 2007. EMBS 2007, pp. 4973–4978. IEEE (2007)
2. Bozek, J., Grgic, M., Schnabel, J.A.: Validation of rigid registration of mammographic images. In: ELMAR, 2011 Proceedings, pp. 11–16. IEEE (2011)
3. Chen, Z., Zwiggelaar, R.: A combined method for automatic identification of the breast boundary in mammograms. In: 2012 5th International Conference on Biomedical Engineering and Informatics (BMEI), pp. 121–125. IEEE (2012)
4. Chen, Z., Zwiggelaar, R.: Segmentation of the breast region with pectoral muscle removal in mammograms. iN: Medical Image Understanding and Analysis (MIUA), pp. 71–76 (2010)
5. de Oliveira Martins, L., Junior, G.B., Silva, A.C., de Paiva, A.C., Gattass, M.: Detection of masses in digital mammograms using k-means and support vector machine. ELCVIA: Electron. Lett. Comput. Vis. Image Anal. **8**(2), 39–50 (2009)
6. Eddaoudi, F., Regragui, F., Mahmoudi, A., Lamouri, N.: Masses detection using svm classifier based on textures analysis. Appl. Math. Sci. **5**(8), 367–379 (2011)

7. Grau, V., Mewes, A., Alcaniz, M., Kikinis, R., Warfield, S.K.: Improved watershed transform for medical image segmentation using prior information. IEEE Trans. Med. Imaging **23**(4), 447–458 (2004)
8. Gulsrud, T.O., Mestad, E.: Perprocessing techniques for improved segmentation of clustered microcalcifications in digital mammograms. In: 2nd International GABOR Workshop in Vienna (2001)
9. Hamdi, M.A., Ettabaa, K.S., Harabi, M.L.: A new mammography segmentation technique based on watershed, wavelet and curvelet transform. In: Computers, Automatic Control Signal Processing and Systems Science
10. Heath, M., Bowyer, K., Kopans, D., Moore, R., Kegelmeyer, W.P.: The digital database for screening mammography. In: Proceedings of the 5th International Workshop on Digital Mammography, Citeseer, pp. 212–218 (2000)
11. Hough, P.V.: Method and means for recognizing complex patterns. Technical report (1962)
12. Jenefer, B.M., Cyrilraj, V.: An efficient image processing methods for mammogram breast cancer detection. J. Theor. Appl. Inf. Technol. **69**(1) (2014)
13. Kass, M., Witkin, A.: Terzopolous:snakes: active contour models. Int. J. Comput. Vis. (1987)
14. Ma, Y., Lu, X., Dong, M., Wang, K.: Automatic mass segmentation method in mammograms based on improved VFC snake model (2014)
15. Mahmood Mina, L., Isa, M., Ashidi, N.: A fully automated breast separation for mammographic images. In: 2015 International Conference on BioSignal Analysis, Processing and Systems (ICBAPS), pp. 37–41. IEEE (2015)
16. Maitra, I.K., Nag, S., Bandyopadhyay, S.K.: Technique for preprocessing of digital mammogram. Comput. Methods Progr. Biomed. **107**(2), 175–188 (2012)
17. Makandar, A., Halalli, B.: Breast cancer image enhancement using median filter and clahe. Int. J. Sci. Eng. Res. **6**(4), 462–465 (2015)
18. Manjunath, B., Chellappa, R.: Unsupervised texture segmentation using Markov random field models. IEEE Trans. Pattern Anal. Mach. Intell. **5**, 478–482 (1991)
19. Mustra, M., Bozek, J., Grgic, M.: Breast border extraction and pectoral muscle detection using wavelet decomposition. In: EUROCON 2009, EUROCON'09. IEEE, pp. 1426–1433. IEEE (2009)
20. Mustra, M., Grgic, M., Rangayyan, R.M.: Review of recent advances in segmentation of the breast boundary and the pectoral muscle in mammograms. Med. Biol. Eng. Comput. 1–22 (2015)
21. Muthukrishnan, R., Radha, M.: Edge detection techniques for image segmentation. Int. J. Comput. Sci. Inf. Technol. **3**(6), 259 (2011)
22. Nithya, R., Santhi, B.: Computer aided diagnosis system for mammogram analysis: a survey. J. Med. Imaging Health Inf. **5**(4), 653–674 (2015)
23. Oliver, A., Freixenet, J., Marti, J., Pérez, E., Pont, J., Denton, E.R., Zwiggelaar, R.: A review of automatic mass detection and segmentation in mammographic images. Med. Image Anal. **14**(2), 87–110 (2010)
24. Raba, D., Oliver, A., Martí, J., Peracaula, M., Espunya, J.: Breast segmentation with pectoral muscle suppression on digital mammograms. In: Pattern Recognition and Image Analysis, pp. 471–478. Springer (2005)
25. Rahmati, P., Hamarneh, G., Nussbaum, D., Adler, A.: A new preprocessing filter for digital mammograms. In: Image and Signal Processing, pp. 585–592. Springer (2010)
26. Rajkumar, K., Raju, G.: Automated mammogram segmentation using seed point identification and modified region growing algorithm. Br. J. Appl. Sci. Technol. **6**(4), 378 (2015)
27. Ramani, R., Valarmathy, S., Vanitha, N.S.: Breast cancer detection in mammograms based on clustering techniques-a survey. Int. J. Comput. Appl. **62**(11) (2013)
28. Suckling, J., Parker, J., Dance, D., Astley, S., Hutt, I., Boggis, C., Ricketts, I., Stamatakis, E., Cerneaz, N., Kok, S., et al.: The mammographic image analysis society digital mammogram database. In: Exerpta Medica. International Congress Series, vol. 1069, pp. 375–378 (1994)
29. Szeliski, R.: Computer Vision: Algorithms and Applications. Springer (2010)

30. Wei, K., Guangzhi, W., Hui, D.: Segmentation of the breast region in mammograms using watershed transformation. In: 27th Annual International Conference of the Engineering in Medicine and Biology Society, 2005. IEEE-EMBS 2005, pp. 6500–6503. IEEE (2006)
31. Zhang, H., Foo, S.W., Krishnan, S.M., Thng, C.H.: Automated breast masses segmentation in digitized mammograms. In: 2004 IEEE International Workshop on Biomedical Circuits and Systems, pp. S2–2. IEEE (2004)
32. Zhang, Z., Lu, J., Yip, Y.J.: Automatic segmentation for breast skin-line. In: 2010 10th IEEE International Conference on Computer and Information Technology (CIT 2010), pp. 1599–1604. IEEE (2010)

Part IV
Special Session 1: Smart Cities and Urban Informatics for Sustainable Development

A Hybrid Machine Learning Based Low Cost Approach for Real Time Vehicle Position Estimation in a Smart City

Ikram Belhajem, Yann Ben Maissa and Ahmed Tamtaoui

Abstract The Global Positioning System (GPS) enhanced with low cost Dead Reckoning (DR) sensors allows to estimate in real time a vehicle position with more accuracy while maintaining a low cost. The Extended Kalman Filter (EKF) is generally used to predict the position using the sensor's measures and the GPS position as a helper. However, the filter performance tails off during periods of GPS failure and may quickly diverge (e.g., in tunnels or due to multipath phenomenon). In this paper, we propose a novel hybrid approach based on neural networks (NN) and autoregressive integrated moving average (ARIMA) models to circumvent the EKF limitations and improve the accuracy of vehicle position estimation. While GPS signals are available, we train NN and ARIMA models to learn the non-linear and linear structures in the vehicle position; therefore they can provide good predictions during GPS signal outages. We obtain empirically an improvement of up to 95 % over the simple EKF predictions in case of GPS failures.

Keywords Global Positioning System · Intelligent transportation systems · Smart cities · Machine Learning · Extended Kalman Filter · Neural networks · Autoregressive integrated moving average · Low cost

1 Introduction

Context. A smart city takes advantage of the information and communication technologies to make the city's infrastructure and services more efficient for a better quality of life, with an optimal management of natural resources [1]. One of the

I. Belhajem (✉) · Y. Ben Maissa · A. Tamtaoui
Laboratory of Telecommunications, Networks and Service Systems,
National Institute of Posts and Telecommunications, Rabat, Morocco
e-mail: belhajem@inpt.ac.ma

Y. Ben Maissa
e-mail: benmaissa@inpt.ac.ma

A. Tamtaoui
e-mail: tamtaoui@inpt.ac.ma

R. El-Azouzi et al. (eds.), *Advances in Ubiquitous Networking 2*, Lecture Notes in Electrical Engineering 397, DOI 10.1007/978-981-10-1627-1_44

main fields of interest is the intelligent transportation systems (ITS) whose goal is to offer effective and energy sufficient transport services with a low cost. To fulfill these requirements, ITS use, inter alia, fleet management solutions to track the vehicles and to collect historical data of the taken paths in order to optimize the route and to conserve fuel. That is why many applications of the ITS rely on accurate real time vehicle positioning to enhance safety and comfort for drivers.

Usually, the current state of a vehicle is identified via the Global Positioning System (GPS) able to determine the location, altitude, velocity and direction based on satellite signals received. However, the GPS performance is limited due to atmospheric disturbances, signal masking and multipath errors in areas such as urban canyons (i.e., places where the street is flanked by buildings on both sides), dense foliage and tunnels [2]. One solution is the Differential GPS (DGPS) capable of reducing significantly some GPS errors. The other solution is to combine GPS and Inertial Navigation Systems (INS) to ensure a long term stable positioning accuracy and to overcome the limitations of using each sensor individually [3]. The INS integrate three accelerometers and three gyroscopes to provide the vehicle rotation rates and accelerations.

Data coming from different sensors is typically fused using the Kalman filter (KF). The filter is an optimal state estimator of linear dynamic systems that include noisy and erroneous measurements. For non-linear systems, the Extended Kalman Filter (EKF) is adopted through a linearization procedure using Taylor series expansions.

Problem. The multipath phenomenon remains even with the use of the DGPS high cost solution. Furthermore, the inertial sensors impose restrictions because of their computing complexity.

The KF position prediction tends to quickly diverge when the GPS signal is lost (i.e., vehicle goes through urban canyons, dense foliage areas or tunnels). This is due to imprecise modeling of the system and measurement dynamics or insufficient a priori information of system noises and measurement errors.

Contribution. This paper aims at introducing a new low cost hybrid approach combining neural networks (NN) and autoregressive integrated moving average (ARIMA) models in order to yield an optimal positioning solution by limiting the EKF shortcomings during GPS outages. This approach exploits the NN and ARIMA models strengths in non-linear and linear modeling. The sensors used are GPS and a low cost Dead Reckoning (DR) system (an odometer and a gyrometer) which is easy to use and keeps the calculations simple. When GPS information exists, the EKF estimates the vehicle position; meanwhile the NN are trained to learn the non-linear relationship of the vehicle position, then ARIMA models are used to capture the linear structure in residuals from the non-linear model. During periods of GPS signal blockage, the EKF works in the prediction phase, hence position errors can exceed the level of accuracy required. Here, the NN and ARIMA models compensate this lack of accuracy and provide more accurate vehicle position.

Contents. This paper is structured as follows. In Sect. 2, we discuss some selected works related to this topic. The essential background on EKF, NN and ARIMA models are presented in Sect. 3. Section 4 provides the formulation of the suggested approach. The experimental results are detailed in Sect. 5.

2 Related Works

Various research investigations related to the vehicle positioning suggest the integration of GPS and DR sensors like odometers and gyrometers, among which are [4, 5]. Usually the odometer (an anti-lock braking system (ABS) or an installed odometer) provides information about the distance travelled by the vehicle, thus its speed and the gyrometer measures the angular velocity. Despite the fact that they are autonomous and not subject to signal blockage, these proprioceptive sensors (i.e., sensors that measure values internally to the system) are prone to time growing errors such as bias drift and scale factor change. Consequently, such combined system GPS/DR takes advantage of DR short-term stability and GPS long-term reliability. It helps, when GPS is available, to keep down the odometer and gyrometer errors in real time and ensures a continuous positioning during the GPS signal outages.

Lucet et al. [5] employ the EKF for the fusion of data from GPS and a low cost DR system. However, the loss of GPS signal for long periods reduces the filter accuracy and even causes its divergence. In their work [4], Zamora-Izquierdo et al. compare the performance of the Kalman Filtering (KF and EKF) and the Particle Filter (PF); they conclude that superior results are obtained with PF even in case of GPS masks. Nevertheless, the increased power of PF, also known as Sequential Monte Carlo method, may require a large number of particles; this comes at the cost of higher computational complexity [7].

Chiang et al. [6] propose the use of multilayer feed-forward NN as the core algorithm for developing an intelligent scheme to reduce the EKF accumulated position errors. The sensors used are DGPS and low cost INS based on microelectromechanical systems (MEMS). The MEMS technology offers cost reduction coupled to small size and lower power consumption advantages. Unfortunately, these sensors suffer from a rapid accumulation of errors when operating in a stand-alone mode during GPS outages.

Goodall et al. [8] claim that the use of NN can bridge the gap in the EKF prediction mode based on data from GPS and INS. When GPS information exists, the EKF estimates the vehicle position while the NN learn the position errors. Then, the NN compensate the additional EKF drifts when no GPS signal is available. The general idea of NN is to build up the non-linear input-output mapping relationship by learning from given samples. However and despite their ability to approximate any continuous and differentiable function [9], NN perform inconsistently for linear relationships.

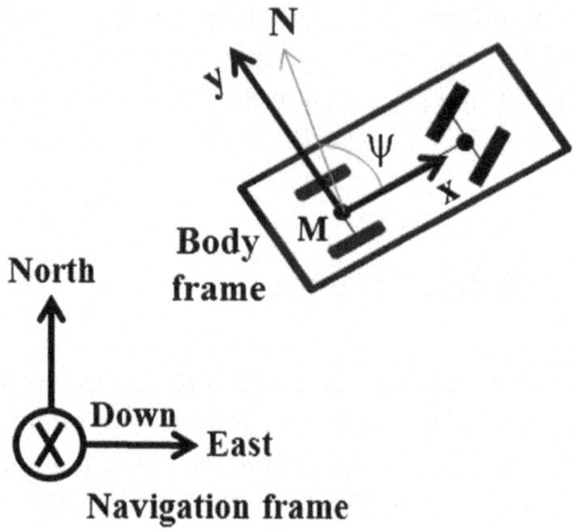

Fig. 1 Vehicle kinematic model

In [10], Chen et al. suggest a recursive methodology to solve the KF errors during GPS signal loss, the sensors include DR system and GPS. While GPS is available, the KF is used to estimate the velocity of the vehicle and its moving distance. During GPS outages, the autoregressive moving average (ARMA) model provides the status of the vehicle and gives better results than the KF predicted state estimate. In an ARMA model, the future value of a variable is generated as a linear function of current and past observations as well as a white noise; then it may be inadequate for complex non-linear problems.

3 Background

In this section, we describe the vehicle kinematic model and cover the background of the EKF, NN and ARIMA models.

3.1 Kinematic Model

Let us consider a car-like model of a front-wheel drive vehicle. The origin M of the body frame (rigidly attached to the vehicle) is located midway the rear axle while the x-axis is aligned with the vehicle longitudinal axis (see Fig. 1). For the vehicle dynamics analysis, the North-East-Down frame known also as a navigation frame is used; so any movement related to the body frame have to be converted to the navigation frame. Therefore, the vehicle position is denoted by (N, E, ψ) where (N, E) are the north and east components and ψ represents the heading.

The kinematic equations mentioned below describe the vehicle position at time epoch k + 1 [5]:

$$\begin{cases} N_{k+1} & = N_k + ds_{k+1}.sinc(\frac{d\psi_{k+1}}{2}).cos(\psi_{k+1} + \frac{d\psi_{k+1}}{2}) \\ & \quad - d\psi_{k+1}.(D_x.sin(\psi_k) + D_y.cos(\psi_k)) \ . \\ E_{k+1} & = E_k + ds_{k+1}.sinc(\frac{d\psi_{k+1}}{2}).sin(\psi_{k+1} + \frac{d\psi_{k+1}}{2}) \\ & \quad + d\psi_{k+1}.(D_x.cos(\psi_k) - D_y.sin(\psi_k)) \ . \\ \psi_{k+1} & = \psi_k + d\psi_{k+1} \ . \end{cases} \tag{1}$$

where

- ds_{k+1} is the distance traveled by the vehicle between k and k + 1;
- $d\psi_{k+1}$ represents the heading variation corresponding to the angular velocity between k and k + 1;
- D_x and D_y are the distances in the body frame between the GPS antenna and the middle of the rear axle.

3.2 *Extended Kalman Filter*

The EKF is a non-linear version of the KF that linearizes the process and measurement models about the current mean and covariance. The filter is a set of mathematical equations which uses the process model to estimate the current state of a system, then a correction of this estimate is performed using any available sensor measurements.

Figure 2 shows a scheme for the KF recursive procedure. A prior knowledge of the initial state $\hat{X}_0^+$ and the corresponding error covariance P_0^+ are required before starting the estimation process. In the prediction mode, the filter projects the state and the error covariance ahead to estimate $\hat{X}_k^-$ and P_k^-. When new measurements arrive at time epoch k, the filter starts the update mode. At this stage, the Kalman gain K_k, the updated state $\hat{X}_k^+$ and the error covariance P_k^+ are computed.

Fig. 2 Kalman filter model

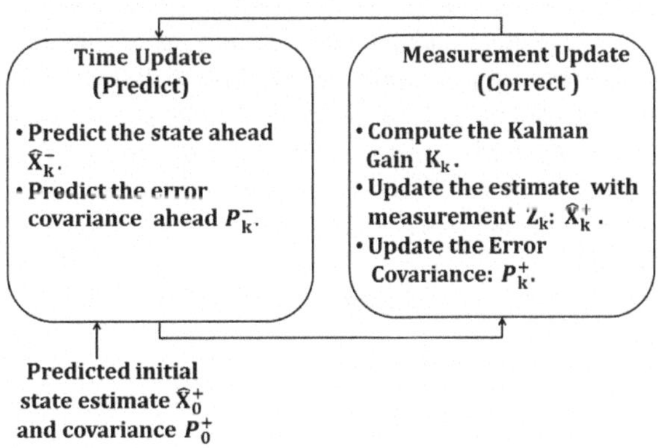

3.3 Neural Networks

NN are a subset of Machine Learning methods that use flexible computing paradigms attempting to mimic the structure and operation of the human brain to solve complex problems [11]. They are composed of many interconnected processing elements called neurons linked by synaptic weights. The arrangement of neurons into layers and the connection strengths within and between these layers refers to the network architecture.

In many practical applications, the most used model is the multilayer feed-forward NN in which all signals flow in one direction; from the input layer to the output layer passing by one or more hidden layers. In order to train the NN, weights of each unit are adjusted in accordance with learning rules. The training algorithm consists then of applying repeatedly small adjustments to the weights until the desired error between the expected output and the actual output is obtained [12].

3.4 Autoregressive Integrated Moving Average Models

The ARIMA models are characterized by their flexibility and ability to build a forecasting model for representing several different types of time series. The main advantage of ARIMA forecasting is that it requires data on the time series in question only. ARIMA models have been originated from the autoregressive (AR) models, the moving average (MA) models and the combination of the AR and MA, the ARMA models. Hence, the process ARIMA(p,d,q) generating the time series is formulated as a linear relationship related past values and random errors:

$$Y_k = \mu + \sum_{i=1}^{p} \phi_i Y_{k-i} + \epsilon_k + \sum_{i=1}^{q} \theta_i \epsilon_{k-i} \ . \tag{2}$$

where p is the number of autoregressive terms, d is the order of differencing, q is the number of moving-average terms, μ is a constant, ϕ_i $(i = 1, 2, \dots, p)$ and $\theta_i (i = 1, 2, \dots, q)$ are model parameters and ϵ_k is a white noise.

The first step in applying ARIMA methodology is to check the stationarity of the collected data. Stationarity implies that the statistical characteristics of the series remain at a fairly constant level over time. In case of nonstationarity, differencing is applied to data so to render the series stationary. The second step is to identify and estimate the appropriate ARMA model. For this, the Box-Jenkins procedure is performed, in which an iterative process of model identification, parameter estimation and diagnostic checking are included [13].

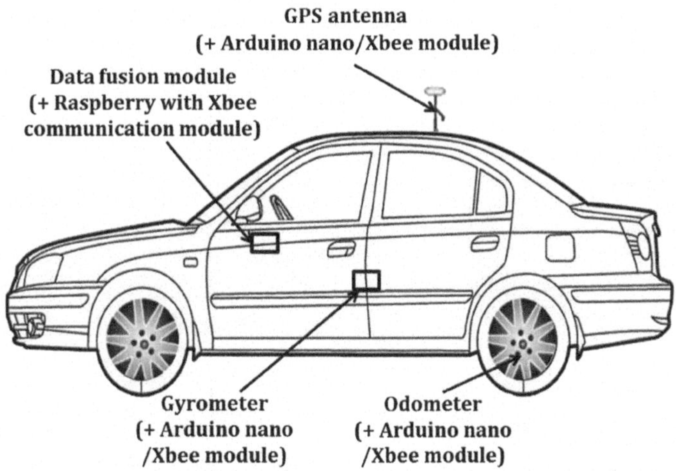

Fig. 3 Wireless sensor network

4 Formulation of the Hybrid EKF/NN/ARIMA Approach

In this section, we present a possible vehicle prototype and the formulation of our suggested approach in both the training and the prediction phases. The prototype is not yet implemented, however we conducted extensive simulations on relevant data sets with improvements over the EKF solution of up to 95 %.

4.1 Possible Vehicle Prototype

Figure 3 illustrates the positions of the GPS and the odometer/gyrometer sensors in our possible vehicle prototype; each one of them is coupled to an Arduino nano and a Xbee module. The Arduino nano is dedicated for the treatment while Xbee module ensures a Zigbee communication for the wireless sensor networks. For the data treatment, a Raspberry with Xbee communication module is mounted on the car's dashboard. Our vision then is to support a real implementation of this prototype.

4.2 Training Stage

Practically, the EKF receives the speed V_{odom} and the heading ψ_{gyro}; then computes the vehicle predicted position (N_{pred}, E_{pred}). When new GPS measurements arrive, the EKF updates the predicted position (N_{cor}, E_{cor}) as presented in Fig. 4. However, the EKF performance depends on how the sensor components are correctly modeled; though a perfect tuning of the filter is rarely achieved since vehicle dynamic variations and environment changes occur oftenly. Accordingly, the EKF performs badly during GPS signal blockage which may result in its divergence. To limit these

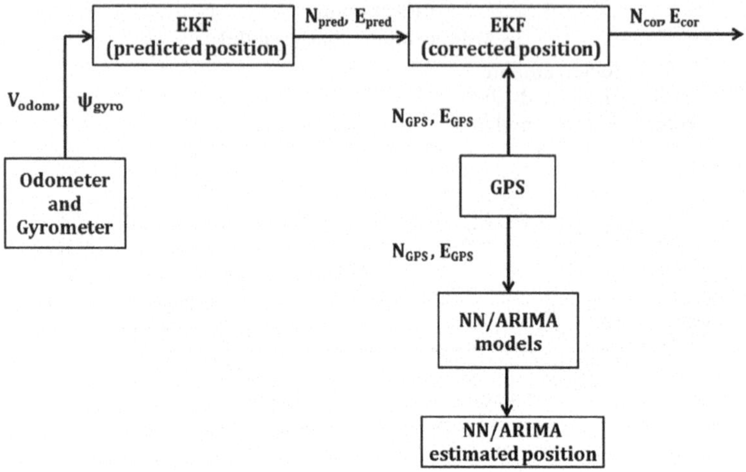

Fig. 4 NN/ARIMA training phase

deficiencies, the NN/ARIMA models are trained *in parallel* on different dynamics (see Fig. 4); so they can be used during the EKF prediction phase to estimate the vehicle position where no GPS signal exists.

The north position (resp. east) at time epoch k is considered as a function of a non-linear component $N_{north,k}$ (resp. $N_{east,k}$) and a linear component $L_{north,k}$ (resp. $L_{east,k}$):

$$\begin{cases} N_{GPS,k} = N_{north,k} + L_{north,k} \, . \\ E_{GPS,k} = N_{east,k} + L_{east,k} \, . \end{cases} \tag{3}$$

To model the non-linearity, two three-layer feedforward NN with a back propagation learning algorithm (given in Fig. 5) are proposed to model the north and east position components. For this purpose, the north NN inputs (resp. east) are the previous north GPS observations $(N_{GPS,k-1}, \ldots, N_{GPS,k-r})$ (resp. $(E_{GPS,k-1}, \ldots, E_{GPS,k-r})$). The networks outputs are the estimated north and east vehicle position at time epoch k.

By removing the non-linearity remains the linear information of the position components (north and east) modeled by two ARIMA models. The residuals are defined as:

$$\begin{cases} e_{north,k} = N_{GPS,k} - \hat{N}_{north,k} \, . \\ e_{east,k} = E_{GPS,k} - \hat{N}_{east,k} \, . \end{cases} \tag{4}$$

where $e_{north,k}$ and $e_{east,k}$ present the north and east residuals for time epoch k while $\hat{N}_{north,k}$ (resp. $\hat{N}_{east,k}$) denotes the estimated north NN output (resp. east). Then, these residuals are modeled using ARIMA models as follows:

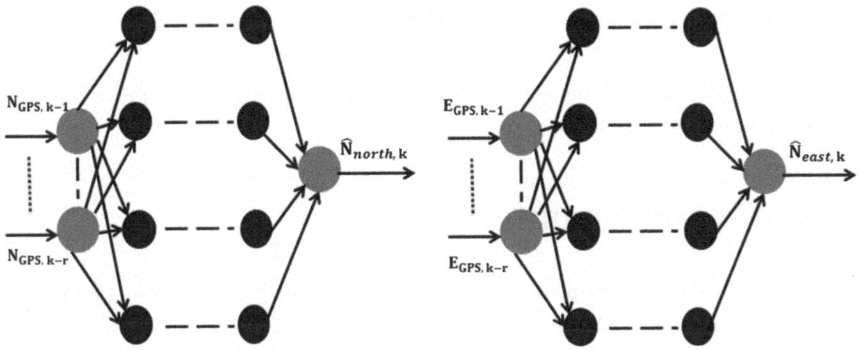

Fig. 5 North and East networks architecture

$$\begin{cases} e_{north,k} = \mu_{north} + \sum_{i=1}^{p} \phi_{north,i} e_{north,k-i} + \epsilon_{north,k} + \sum_{i=1}^{q} \theta_{north,i} \epsilon_{north,k-i} \\ e_{east,k} = \mu_{east} + \sum_{i=1}^{p} \phi_{east,i} e_{east,k-i} + \epsilon_{east,k} + \sum_{i=1}^{q} \theta_{east,i} \epsilon_{east,k-i} \end{cases} \quad (5)$$

Denoting $\hat{L}_{north,k}$ and $\hat{L}_{east,k}$ as the forecast linear north and east values from (5), the estimated position will be:

$$\begin{cases} \hat{N}_{GPS,k} = \hat{N}_{north,k} + \hat{L}_{north,k} \\ \hat{E}_{GPS,k} = \hat{N}_{east,k} + \hat{L}_{east,k} \end{cases} \quad (6)$$

4.3 Prediction Stage

After training on different examples, the NN/ARIMA models are used to provide the predicted vehicle position when GPS data is not available. During the prediction mode, the north NN inputs in the first iteration are $(N_{GPS,n}, \dots, N_{GPS,n-r})$ and the output is $\hat{N}_{north,n+1}$ given the fact that $N_{GPS,n}$ is the last north GPS observation before the signal blockage. In the second iteration, the inputs are $(\hat{N}_{north,n+1}, \dots, N_{GPS,n+1-r})$ while the output is $\hat{N}_{north,n+2}$. Finally, the inputs are $(\hat{N}_{north,n+m-1}, \dots, \hat{N}_{north,n+m-1-r})$ and the output is $\hat{N}_{north,n+m}$ in the last iteration supposing that m is the GPS outage duration before the signal recovery. It should be noted that the same procedure is followed for the east NN to predict the east vehicle position. For the ARIMA models, the predicted north and east residuals are computed based on the relation depicted in (5); the parameters are fixed according to the best fitted models chosen during the training stage.

5 Experimental Tests and Results

In this section, we present the test vehicle prototype and the simulation results of the suggested hybrid approach.

5.1 Test Vehicle Prototype

The performance of our proposed approach was examined with the Institute Pascal Data Sets [14]. The field test data were collected using VIPALAB, a platform equipped with multiple sensors. In our case, the test system comprises three sets of an uBlox-6T-0-001 GPS receiver, an odometer and a Melexis MLX90609-N2 gyrometer. GPS values were collected at the frequency rate of 1 Hz while the odometer/gyrometer data at 50 Hz. The road test trajectory used is CEZEAUX-Heko (given in Fig. 6) which spans over a distance of 4.2 km during 28 min.

5.2 Simulation Results

To test the proposed model performance, a total of five GPS outages were intentionally introduced at different locations over the whole test trajectory as shown in Fig. 6. These simulated outages last 100 s each; so to leave the position errors enough time to grow broadly without any EKF update phase.

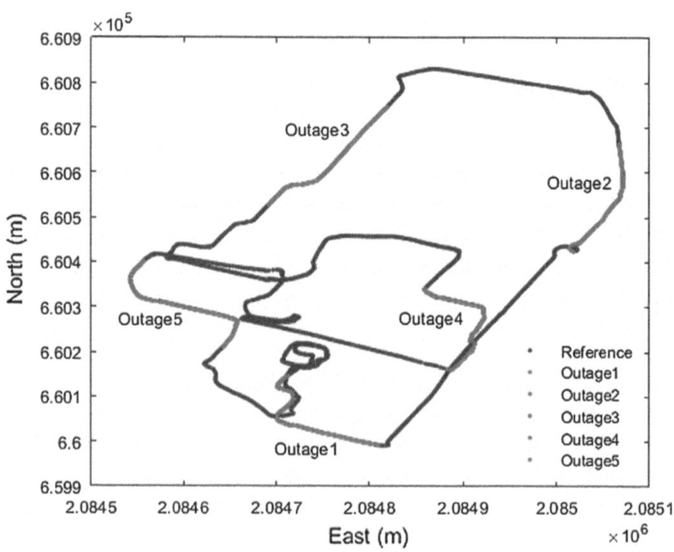

Fig. 6 Field test trajectory

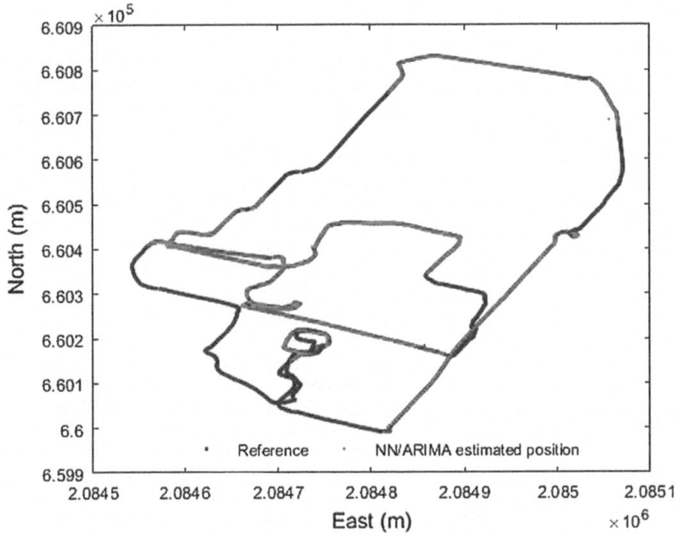

Fig. 7 NN/ARIMA training results

Before each outage, a set of new inputs/targets are collected with the intention of training the NN for at least 96 s at the GPS sampling rate. Then, the residuals, resulted as a difference between GPS target values and NN outputs, are modeled by the ARIMA models. Hence, the estimated vehicle position is an aggregation of predicted components from both NN and ARIMA models. In our case study, the north and east networks architecture chosen empirically consist of 4 inputs, 7 hidden neurons and one output while the training goal is to reach mean squared error (MSE) between outputs and desired values less than $5 \times 10^{-3}\,\text{m}^2$ given the real time constraints. Concerning the ARIMA models, ARIMA(1,1,2) and ARIMA(1,1,1) have been found to be the most adequate for representing the north and east position residuals. Figure 7 shows the results of the NN/ARIMA training stage before each outage. It appears clearly that the NN/ARIMA outputs are very close to the GPS target values.

The accuracy of the NN/ARIMA approach is examined during the prediction stage with simulated GPS failures since no natural ones are detected in the field test data. To compare the performance of the proposed model with the one of EKF, two different evaluation indicators are calculated for each outage: the root mean square error (RMSE) and the mean absolute deviation (MAD). They are expressed by the following formulas:

$$RMSE = \sqrt{\frac{1}{n}\sum_{k=1}^{n}(A_k - F_k)^2}\ ;\ MAD = \frac{1}{n}\sum_{k=1}^{n}\left|\frac{A_k - F_k}{A_k}\right| \ . \tag{7}$$

Table 1 GPS test outages north improvement

Outages	EKF		NN/ARIMA		Improvement (%)	
	RMSE (m)	MAD	RMSE (m)	MAD	RMSE (m)	MAD
Outage 1	836.07	728.27	123.59	110.29	85	84
Outage 2	577.07	492.95	339.61	303.12	41	38
Outage 3	828.65	721.82	150.06	128.82	81	82
Outage 4	815.72	704.43	180.23	164.76	77	76
Outage 5	864.47	750.64	76.62	61.79	91	91

Table 2 GPS test outages east improvement

Outages	EKF		NN/ARIMA		Improvement (%)	
	RMSE (m)	MAD	RMSE (m)	MAD	RMSE (m)	MAD
Outage 1	690.97	609.17	36.70	28.53	94	95
Outage 2	649.42	556.52	143.02	130.93	77	76
Outage 3	800.13	698.80	179.06	161.75	77	76
Outage 4	712.92	610.13	189.23	169.36	73	72
Outage 5	675.79	595.95	121.34	103.01	82	82

where A_k and F_k are the actual and predicted values at time epoch k while n is the total number of predictions. These results are listed in Tables 1 and 2 for north and east position components. By combining NN and ARIMA models together, the results show a significant decrease in RMSE and MAD over the EKF method. This hybrid approach enhances the vehicle position accuracy over the EKF predictions during GPS outages.

6 Conclusion

In this paper, we present a novel approach to continuously estimate the real time vehicle position based on data coming from a GPS and a low cost DR integrated sensors; so it can be used in different ITS applications. We propose the combination of NN and ARIMA models to overcome the EKF deficiencies since no GPS signal is detected. The use of these techniques jointly is motivated by their capability to capture the non-linear and linear relationships in data. The hybrid NN/ARIMA models are trained on different samples; so to provide more accurate positioning during GPS outages where EKF position errors grow largely. Experimental results with field test data demonstrate the ability of NN and ARIMA models to learn and make reasonable predictions during different periods of GPS blockage. Consequently, our hybrid approach ameliorates the positioning exactitude up to 95 % compared to the EKF results.

Future Work. Empirical results with simulated GPS outages showed very promising progress. Nonetheless, it is necessary to take into account the complexity of real GPS outages due to the GPS quality degradation unlike the case where the GPS signal is removed intentionally. Therefore, in addition to our promising real data set simulation results, we intend to implement our vehicle prototype in future related works.

Acknowledgments This work was financially supported by the National Center of Scientific and Technical Research (CNRST), Morocco.

References

1. Nam, T., Pardo, T.A.: Conceptualizing smart city with dimensions of technology, people, and institutions. In: Proceedings of the 12th Annual International Digital Government Research Conference: Digital Government Innovation in Challenging Times, pp. 282–291. ACM, New York (2011)
2. Easton, R.D., Frazier, E.F.: GPS Declassified: From Smart Bombs to Smartphones. Potomac Books Inc., USA (2013)
3. Grewal, M.S., Weill, L.R., Andrews, A.P.: Global Positioning Systems, Inertial Navigation, and Integration. Wiley-Interscience, New York (2007)
4. Zamora-Izquierdo, M.A., Betaille, D.F., Peyret, F., Joly, C.: Comparative study of Extended Kalman Filter, Linearised Kalman Filter and Particle Filter applied to low-cost GPS-based hybrid positioning system for land vehicles. In: International Journal of Intelligent Information and Database Systems, vol. 2, pp. 149–166. Inderscience Publishers, Geneva (2008)
5. Lucet, E., Betaille, D., Donnay Fleury, N., Ortiz, M., Salle, D., Canou, J.: Real-time 2D localization of a car-like mobile robot using dead reckoning and GPS, with satellite masking prediction. In: Accurate localization for land transportation Workshop, pp. 15–18. France (2009)
6. Chiang, K-W., Niu, X., El-Sheimy, N.: On the development of a conceptual intelligent navigator for a low cost DGPS/MEMS IMU integrated system. In: Proceedings of the 18th International Technical Meeting of the Satellite Division of The Institute of Navigation (ION GNSS 2005), pp. 494–502. Long Beach, CA (2005)
7. Aggarwal, P., Syed, Z., El-Sheimy, N.: Hybrid Extended Particle Filter (HEPF) for integrated civilian navigation system. In: Position, Location and Navigation Symposium, 2008 IEEE/ION, pp. 984–992. IEEE, USA (2008)
8. Goodall, C.L.: Improving usability of low-cost INS/GPS navigation systems using intelligent techniques. Ph.D. dissertation, Department of Geomatics Engineering. Universiry of Calgary, Canada (2009)
9. Zhang, G., Patuwo, B.E., Hu, M.Y.: Forecasting with artificial neural networks:: the state of the art. Int. J. Forecast. **14**, 35–62 (1998). Elsevier Science B.V., USA
10. Chen, S.H., Hsu, C.W., Huang, S.H.: Recursive estimation of vehicle position by using navigation sensor fusion. In: 12th International Conference on ITS Telecommunications, pp. 532–536. IEEE, Taipei (2012)
11. Yashchenko, V.: Neural-like growing networks the artificial intelligence basic structure. In: Intelligent Systems in Science and Information 2014, pp. 41–55. Springer International Publishing (2014)

12. Khoi, D.D., Murayama, Y.: Multi-layer perceptron neural networks in geospatial analysis. In: Progress in Geospatial Analysis, pp. 125–141. Springer, Japan (2012)
13. Box, P., Jenkins, G.M.: Time Series Analysis: Forecasting and Control. Holden-day Inc., San Francisco (1976)
14. Korrapati, H., Courbon, J., Alizon, S., Marmoiton, F.: "The Institut Pascal Data Sets": un jeu de données en extérieur, multicapteurs et datées avec réalité terrain, données d'étalonnage et outils logiciels. In: ORASIS 2013, Congrès des jeunes chercheurs en vision par ordinateur. France (2013)

Toward a Practical Method for Introducing and Evaluating Trust Learning Models in Open Multi-agent Systems

Youssef Mifrah, Abdeslam En-Nouaary and Mohamed Dahchour

Abstract In multi-agent systems, agents often interact with each others to achieves their own goals. In open dynamic systems, trust between agents become a critical challenge to make such interactions effective. Many trust models have been proposed to formalise this concept. These models are such good for dealing with trust by proposing components that present a computational form of this concept and a learning strategies to manage it. Components and learning strategies differs from one model to another. This diversity may influence the decision of a user about the best trust model to use in his system. A comparative study is needed to evaluate each trust model and to show the prediction quality of each one. Several testbeds for the evaluation of trust models have been proposed. However, those testbeds are not flexible enough to handle different scenarios in various contexts. In this paper we formulate a practical method based on a framework that introduce the trust concept into open distributed systems, and a testbed that can be used to evaluate trust models in different contexts.

Keywords Trust · Intelligent agent · Autonomous agent · Open distributed systems · Multi-agent systems · Testbed · JADE

1 Introduction

Multi-agent systems (MAS) are a kind of distributed systems that consist of multiple interacting intelligent agents, used together to solve problems which are difficult to be solved by an individual agent. In MAS context, intelligent agents are involved with some key features such as perception, reasoning, goal-driven and autonomy.

Y. Mifrah (✉) · A. En-Nouaary · M. Dahchour
Institut National des Postes et Télécommunications Rabat, Rabat, Morocco
e-mail: mifrah@inpt.ac.ma

A. En-Nouaary
e-mail: abdeslam@inpt.ac.ma

M. Dahchour
e-mail: dahchour@inpt.ac.ma

R. El-Azouzi et al. (eds.), *Advances in Ubiquitous Networking 2*, Lecture Notes in Electrical Engineering 397, DOI 10.1007/978-981-10-1627-1_45

In open MAS, an agent can join and leave the system without any constraint. Ad hoc Network such as VANET and MANET, Online auction websites and peer-to-peer Net are a kind of open MAS, where users present agents in those systems. In general, an agent in MAS has some limited competencies, and it often tries to rely on their peers to achieve their own goals. In an open MAS context, agents could not be sure that other agents expose accurate informations about themself. In this class of MAS, some agents may hide bad intentions and spreed false informations to mislead theirs peers. In such context, the trust concept become a critical challenge that is necessary to make interactions between agent more effective, and also to reduce the risk of being unsatisfied [1]. Trust management in agent systems is one of the most critical challenges that faces researchers in this area. In human societies, trust presents an ubiquitous concept involved implicitly in any interaction. Trust and reputation management has become a topic of research due to the fast growing of distributed architectures and multiagent systems. Intelligent agents are intended to achieves specific goals with a minimum or without human intervention. The Autonomy feature that characterize agents present an advantage for task automation and coordination. However, when an agent initiator delegate a task to an agent processor, the result obtained depend on the capacities, intention and willingness of the agent processor. To address this concern, researchers have looked for new techniques and concepts that would improve agents' interactions and help them achieve their goals in an efficient way, by proposing models that integrate trust and reputation management. With a trust model, agents can associate to their partners an evaluation value that estimate the level of trustworthiness of those partners. Each of the trust models, found in literature, presents its main components and the methodology to use for managing trust [2]. The multitude of trust models proposed and theirs various background theories make it difficult to choose the right model for a distributed system under development. The diversity of computational trust models make the user confused. Indeed, each model uses its specific terminology to describe elements that compose the model (trust, credibility, confidence..). Some models give a deep analysis about the trust management, and more elements that construct the trust than others. another limitation is that the proposed models are experimented using limited scenarios and a set of data decided by their authors. To address these limitations, comparative tools, called testbeds, were proposed by researchers to expose those computational trust models to different scenarios, and also to compare the robustness of each model with respect to different scenarios. Those testbeds have shown their utilities of verifying the perception level of trust models. But the lack in existing testbeds is that some of them do not handle trust per context, others do not handle diverse attack strategies.

To address the previous limitations, we propose in this paper an approach for introducing and evaluating trust learning models in MAS. this approach is based on our framework, called GeFMAT, for modeling trust within a MAS, with an implementation in the JADE platform. This framework defines in an abstract level how to design a multi-agent system in an open environment. GeFMAT does not only give a structure of the system, but also how the workflow information is done, and how agents interact between each other and how to apply their trust metrics while managing trust and preparing for the phase of decision making. the evaluating of trust learning model is done by our testbed. the key features of this testbed is that it offers the possibility to manages trust assessment per context while handling diverse services, and it introduces various attack strategies to evaluate the robustness of trust models against those attacks.

2 Trust Models and Testbeds: State of the Art

Computational trust model is to formalise the process of making an intelligent agent trust another agent and delegate a task to this trustee agent within the system. This decision is a consequence of trust and will help updating the value of this trust through cumulative experiences with that agent. The process of formalizing trust can be done from different perspectives. Researchers propose models that handle agents concepts and introduce trust management from a social approach. Human societies is the most important source of inspiration for researchers in this field. But the way the trust assessments is managed differs from one trust model to another. There are many computational trust models in the literature [3, 4]. Different classifications of trust models have been established based on different critereas such as the paradigm type [2], and the representation of trust [5]. When classifying by paradigm type, models are classified as cognitive [6] or probabilistic [7]. While for the trust representation criterea, model is classified following the composition of the trust assessment component. Some models use a single value that represent the trust assessment [8], while others propose more than one value, such as Beta Reputation System [4] that formulate a trust assessment as a vector of tree value : belief, disbelief, and uncertainty, in addition to some constraints and formulas used while calculating those values. Guha et al. also propose a model that handle an assessment of trust and distrust, and put forward an algorithm for the propagation of those values inside the system. Those classifications shows how many trust model exist in literature and how existing trust models differs from one another. To handle the diversity of existing trust models, researchers have proposed testbeds to evaluate and compare trust models. To the best of our knowledge, four testbeds could be found in the literature, namely the Agent Reputation and Trust (ART) [9], Trust and Reputation Experimentation and Evaluation Testbed (TREET) [10], Alpha testbed [11] and the one of Zhang [12].

In ART, agents play the role of art appraisers. Each agent appraises the value of paintings of a client by either using its own knowledge or by asking other agents for help. Agents can use trust models to evaluate their peers and their opinions. In ART, there is no variety of attack strategies that dishonest agents could use; there is also no collision attacks that a group of agents could perform. The ART testbed has inspired Kerr and Cohen [10] to create a more general and focused testbed, called TREET.

In TREET, the game scenario is a simulation of a marketplace where sellers and buyers exchange items. Buyers are motivated by the value of items, while sellers are motivated by the profit they make from sales. Sellers can cheat by not shipping items and so increasing their profits. With TREET, no specific format of trust value is imposed, and the user can implement new collision attacks easily. Both testbeds (i.e., ART and TREET) do not evaluate the quality of trust model classification, but just the quality of whether the selected partner is honest or not.

Another testbed, called Alpha, was proposed by Jelenc et al. [11]. It tries to distinct between the trust assessment phase where agents apply learning techniques to evaluate the trustworthiness of their peers, and the decision making phase. The authors confirm their research that the decision making mechanism influences the performance of the trust model.

The next section introduce an abstract framework that include trust management in multi-agent system. This framework helps users design and analyze multi-agent systems that integrate the concept of trust into the decision process.

3 Our Trust Management Framework

The process of setting up a framework for managing trust in MAS starts with a deep analysis of key features and common concepts used in MAS and components related to the trust concept. Several frameworks were proposed by researchers to design multi-agent systems by capturing common concepts used in such systems. Each of the proposed frameworks uses a specific domain model that design agents systems from different perspectives. Some of them are based on organizational and hierarchical perspectives, and include the notions of environment, hierarchy and role. Others focus on the interactional aspect of agent systems. There is also other works that combine between different perspectives. However no meta-model of the proposed framework handle explicitly the concept of trust. To introduce the concept of trust into multi-agent systems, we propose a meta-model for modeling multi-agent systems in open environment. This meta-model that captures the semantics of concepts of multi-agent systems involved in an open environment. It includes also a set of components related to the trust concept. Figure 1 present the components of the meta-model and theirs relationships.

As presented in Fig. 1, The meta-model is in the form of a related meta-classes. Each meta-class represent a common concept used in existing multi-agent systems. The meta-class Agent presents an autonomous entity that communicates with others agents and provides one or more services. There is no constraint on the internally specification of the agent model. While designing a system, we specify features that will characterize agents in the system. An agent will have a list of features values. For example, in an e-commerce platform where some users could be represented by agents, each user has a profile composed of features (user name, country, registration time, experiences…etc). Those features constitute the agent's profile, and they reflect its instance in the system. a detailed description of each of those meta-classes was published in a previous paper [13]. The presented meta-model reflect the structural part of the framework, the behavioral part is represented by a process model that define steps applied by agents to assess trust and improve the process of decision making. The framework adopts a workflow of information for the management of the

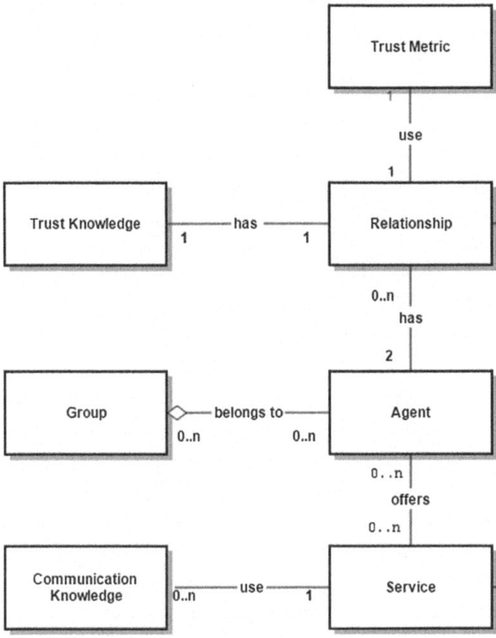

Fig. 1 The meta-model of GeFMAT

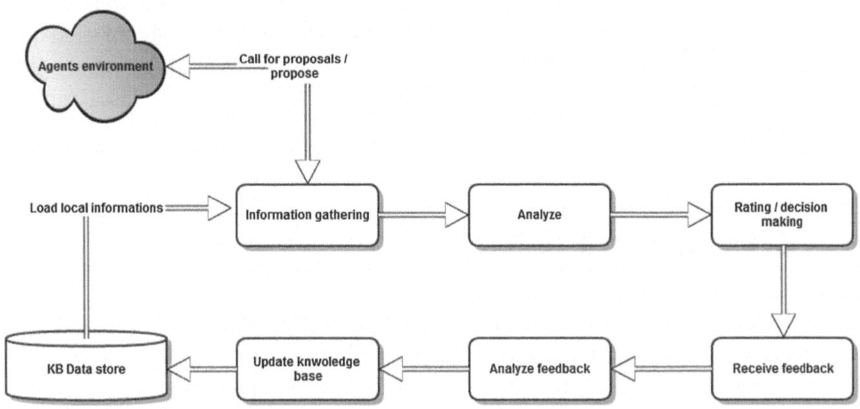

Fig. 2 GeFMAT framework workflow

trust. This workflow as presented in Fig. 2, captures and describes in an abstract way how agents should manage information about the environment during the decision process, and how to use feedbacks after the decision process.

To provide a proof of concept of our framework, an implementation and an experimentation of the designed framework is required. There are many development platforms dedicated to implement multi-agent systems [14–16]. some detailed study

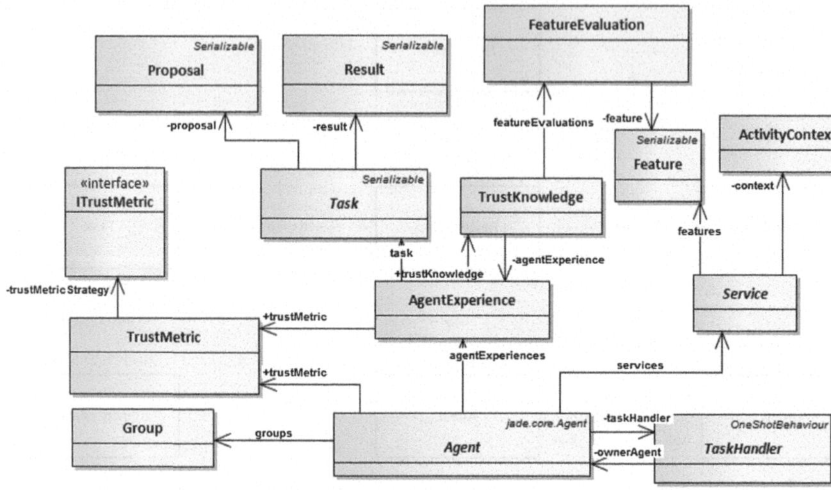

Fig. 3 GeFMAT meta-model implementation

have been done to classify existing platforms [17] and to show differences between those platforms. The Java Agent DEvelopment Framework (JADE) [14] is one of the featured platform of agents development. The advantage of JADE over other platforms is that it complies with the FIPA specification for interoperable intelligent MAS and represents an agent middleware providing a set of graphical tool used during the development process. JADE Framework provide a set of classes that could be directly extended and used in our framework. Figure 3 shows the classes implementation of GeFMAT.

4 A GeFMAT Based Testbed

We have presented in Sect. 2 some of the existing testbeds, their features, advantages, and limitations. Our contribution is a new testbed that provides some new features that do not exist in the existing testbeds. In our testbed, we simulate agents system as a group of students. Each student is qualified in one or more topics. Those topics are such as arithmetic operations, physics operations and literature knowledges. Each student has a set of homeworks to do, but he is not qualified in all of their topics. In this situation, a student (initiator) will seek help from his peers. Their peers will respond by a refuse or a proposal as a feedback for the request of the initiator student. Then, the initiator will use his history and received proposals to apply a trust metric strategy which present its social intelligence. then generate an evaluation of trustworthiness of each of his peers; such process will help the initiator choose the right student that will process the homework. Here, the right student does not mean the one who will certainly satisfy the initiator, but the student who is more likely qualified to satisfy the initiator from its point of view. The decision takes into consideration experiences acquired and the trust metric strategy used to evaluate each of his peers. Each student in this group is simulated as an agent in the testbed, and he can play the role of a requester and/or a proposer.

4.1 Testbed Architecture

As presented before, our testbed is based on the idea of the group students who try helping each other resolving their homeworks. There are two principal categories of students that we can distinguish: honest students and dishonest students. An honest student represents a student that has no intention to trick his peers, but he can sometime fail achieving some tasks with a specific probability. A dishonest student is a student with a bad intention while communicating with his peers. Every dishonest student follows an attack strategy. These attack strategies against trust and reputation systems differ in their natures and complexities from one to another. Some attacks are used against reputation systems where agents would ask for recommendation from other agents in the system such as Constant and Sybil attack [1]. Others are used to directly attack agents that acquired some positive/negative experience with the attacker agent [12]. We configure our testbed to handle the four known attacks: Constant or always dishonest attack, Camouflage attack, and Whitewashing attack. An agent that applies a constant dishonest attack always acts unfairly. This means that at each time the agent is selected by an initiator agent to perform a task, he will return a result different than what is expected by the initiator. This is the simplest attack because there is no timing strategy to trick peers over time or changing its identity. The second strategy that we introduce is the camouflage strategy. The idea of this attack is that the attacker gives fair result at the beginning of building its experience with his peers. The third attack strategy introduced in our testbed is Whitewashing attack. An attacker agent who applies this strategy will leave and rejoin the system each time that he builds a bad reputation with his peers. The fourth attack is the random attack, where the agent randomly decides to respond fairly or unfairly.

Recall that the purpose of evaluating a trust model is to check whether or not the model has the capability of assessing the trustworthiness for each agent. The process of evaluating computational trust model refers to statistical measure classifications. In statistics, we find methods that help evaluating the quality of a model prediction [18]. Several works have been done to compare such methods and to show their consistency and efficiency within different scenarios [19–21]. Jurman shows that the Matthews Correlation Coefficient (MCC) method is the best suited method for evaluating the quality of binary classifications [19], where we have to classify our dataset into two groups. Trust evaluation models are an example of a binary classification where agents are grouped by their trustworthiness. In case that the trust evaluation values belong to a range, we can first convert them to match a binary repartition, and then use measure classifiers to evaluate the model.

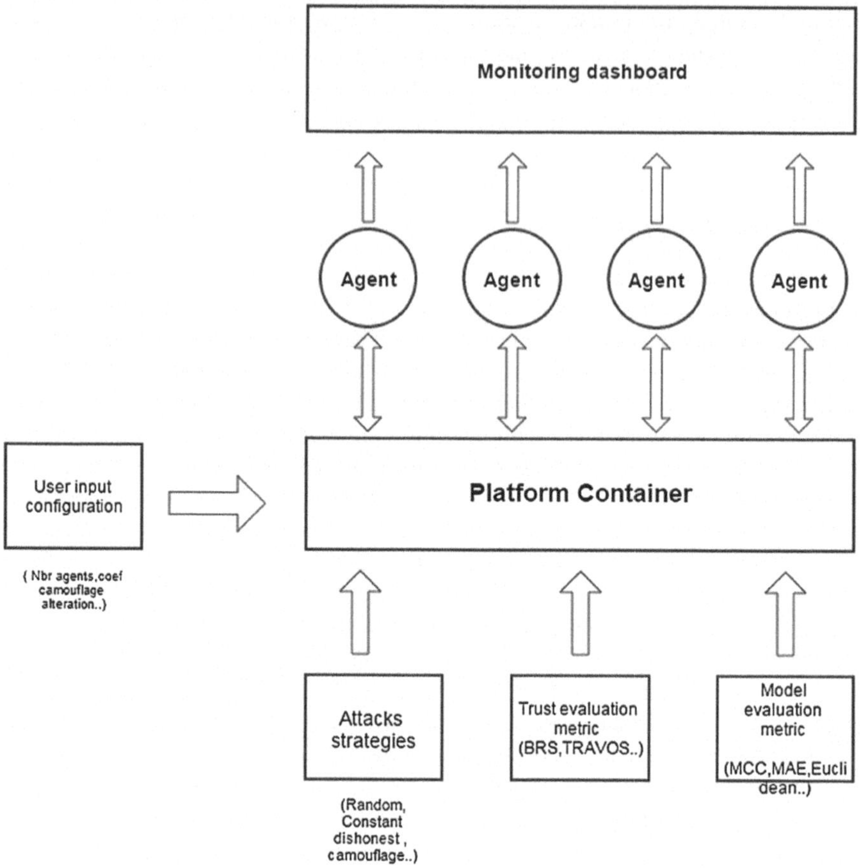

Fig. 4 The testbed architecture

Figure 4 illustrates the high level architecture of our testbed. The attack strategies and trust evaluation metrics are configured to be used by the students who are represented by agents in the figure. The user will configure a test case by setting the number of honest students, the number of dishonest students, and will also specify the nature of each one of them. Later, the user defines the trust evaluation metrics that each agent requestor will use to evaluate his peers. When a test case starts, the monitoring dashboard could be used to preview the trust assessment values that set each of the task requester agent about his peers. Model evaluation metrics are used to evaluate the quality of the prediction of the trust model being used.

Our testbed provide some features that are not available in existing testbeds. It can evaluate more than one trust models at the same time in the same test case scenario. It can also evaluate those models against different attack strategies. Table 1 compares the characteristics of our testbed and some of the existing ones.

Table 1 The comparison of our testbed and existing ones

Testbed	Parallel evaluation	Attacks strategies	Extensible for new attacks	Model evaluation metric	System architecture	Different services
Our testbed	Yes	Constant attack, Random attack White washing, Camouflage	Yes	Evaluates the quality of prediction using correlation functions MCC	Decentralized	Yes
ART	No	Random dishonest	No	Evaluates the accuracy and cooperation achievable by the system of appraiser agents	Decentralized	No
TREET	No	Random dishonest	Yes	The ratio of sales (profits) between honest and cheating sellers	Centralized decentralized	No
Zhang et al.	No	Constant attack, Sybil attack, Camouflage, Composite attack	Yes	Specific function proposed by authors	Decentralized	No

4.2 Implementation and Experimentation

The testbed was designed following the best practices applied in object oriented programming such as separation of concerns and design patterns. The importance of design patterns is that they represent proven solutions to commonly occurring problems in software design. They also help developers to describe and understand design solutions using well-known conventions. We will not discuss the implementation details of each component here; we will just give an overview of the design structure and the role of each part. The source code of our testbed is now available online.[1]

The testbed design is composed of five distinct components: student behaviors, operations, tasks, features and trust models. Each student could provide a set of operations to serve his peers. The default implementation of an operation results in correct answer. A student who provides an operation may alter the result depending on the nature of his behavior. Honest students use the default implementation without any changes. Dishonest students alter the logic applied in those services. Each dishonest student uses his own alteration strategy. Each of the available operations is intended to process specific tasks.

To evaluate the result and manage the trustworthiness of an agent's peers, the agent needs a trust metric, which is the implementation of a trust model. Those trust models are intended to be evaluated by the testbed. We have implemented four different trust models:

Jonker trust model [22], Beta Reputation System (BRS) [4], Forgive Factor model [23], and an empty trust model called NoModel. It is also possible to introduce new models by implementing the interface ITrustMetric. The communication between students could be monitored using a sniffer tool provided by JADE platform. Figure 5 shows the screenshot of the interaction result between the set of students.

To evaluate the prediction quality and the robustness of implemented trust models against dishonest agents, the user needs to setup a configuration for an evaluation scenario. The configuration used to experiment the testbed in this paper is presented in Table 2.

It is possible to execute one or many task requester agents that will send a call for proposals to delegate their tasks. The testbed gives the possibility to evaluate more than one trust models in the same test case. This is done by configuring a task requester agent for each trust model. This evaluation scenario uses four task requester agents to evaluate the four implemented trust models at the same time. We have launched two test cases using this configuration. In the first test case, each task requester agent prepares and delegates 25 tasks, which is the number of task handler agents. In this case, the results will show the initial acquired trust assessment obtained using each of configured trust models. In the second test case, each task requester agent prepares and delegates 500 tasks. This case will show the convergence of trust assessments of those trust models. Results are presented in

[1] https://github.com/mifmif/JADETrustTestbed.

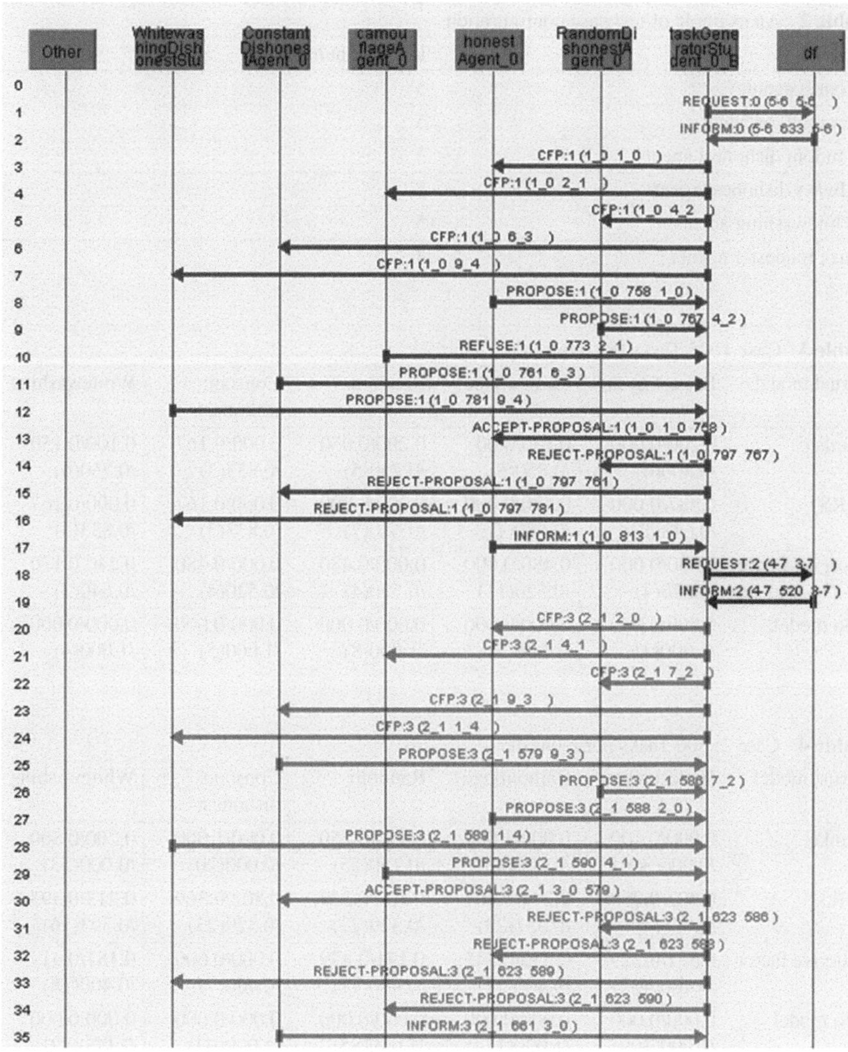

Fig. 5 Interactions between agents

Tables 3 and 4 that show the trust knowledge of a task requester about their peers grouped by category. The value of trustworthiness associated to an agent is in the form of *belief/disbelief/uncertainty (selection counter)*. The belief factor presents the degree of how an agent thinks that its peer is trustworthy. The disbelief factor presents the degree of untrustworthy that the agent thinks about his peer, and the uncertainty value presents the doubt that an agent still has about his peer. The *selection counter* parameter shows how many times the agent has selected a peer of the category.

Table 2 An example of test case configuration

Category	Instance number
Honest agents	5
Camouflage agents	5
Random dishonest agents	5
Always dishonest agents	5
Whitewashing agents	5
Task requester agents	4

Table 3 Case 1: 25 Tasks per requester

Trust model	Honest agent	Camouflage	Random	Constant dishonest	Whitewashing
Junker	0.200/0.000 /0.800(6)	0.150/0.00 /0.850(5)	0.200/0.050 /0.750(5)	0.000/0.167 /0.833(3)	0.100/0.150 /0.750(6)
BRS	0.387/0.000 /0.613(10)	0.320/0.000 /0.680(7)	0.320/0.000 /0.680(7)	0.000/0.167 /0.833(3)	0.000/0.167 /0.833(3)
Forgive factor	0.330/0.000 /0.670(4)	0.480/0.000 /0.520(11)	0.000/0.480 /0.520(4)	0.000/0.480 /0.520(4)	0.240/0.120 /0.640(3)
No model	0.000/0.000 /1.000(3)	0.000/0.000 /1.000(5)	0.000/0.000 /1.000(8)	0.000/0.000 /1.000(5)	0.000/0.000 /1.000(4)

Table 4 Case 2: 500 Tasks per requester

Trust model	Honest agent	Camouflage	Random	Constant dishonest	Whitewashing
Junker	0.900/0.100 /0.000(306)	0.000/1.000 /0.000(46)	0.150/0.750 /0.150(75)	0.000/1.000 /0.000(20)	0.200/0.800 /0.000(53)
BRS	0.922/0.050 /0.028(428)	0.244/0.404 /0.351(23)	0.302/0.369 /0.329(23)	0.302/0.369 /0.329(23)	0.213/0.393 /0.393(16)
Forgive factor	0.371/0.229 /0.400(437)	0.255/0.345 /0.400(21)	0.171/0.429 /0.400(17)	0.000/0.600 /0.400(7)	0.181/0.419 /0.400(18)
No model	0.000/0.000 /1.000(109)	0.000/0.000 /1.000(124)	0.000/0.000 /1.000(85)	0.000/0.000 /1.000(93)	0.000/0.000 /1.000(89)

Values within Tables 3 and 4 present means of evaluation values acquired by trust models about each category of students. For example, if the agent uses BRS as a trust metric, and delegates some tasks to five agents with the camouflage behavior (seven tasks in the case of Table 3), then the value shown in the table presents the mean evaluation of those five camouflage.

Values used in Tables 3 and 4 do not show the prediction quality of the used trust models. They just give a general idea about the defense of trust models against different attacks. To infer the prediction quality of each trust models, the results obtained by trust metrics (trust knowledge acquired during the task delegation process) need

Table 5 Trust model's correlation coefficient

Trust model	MCC	Satisfaction	Dissatisfaction
Junker	1	388	112
BRS	0.88	416	84
Forgive factor	0.79	461	39
No model	0	218	282

to be analyzed using measures discussed in Sect. 4 to extract the correlation between the prediction trustworthiness calculated by metrics, and the real trustworthiness. The MCC measure is used for this purpose. Table 5 shows the results obtained using those measures, and the number of satisfactions and dissatisfactions.

MCC values are between −1 and 1; when its value is close to 1, it means that the metric is positively correlated with the expected values, which are the trustworthiness of agents in the system, and therefore, it has a good prediction quality. If its value is close to −1, it means that the metric is negatively correlated with the expected values. For small values close to 0. they means that there is no correlation between the results obtained while using the metric and the expected results.

5 Conclusion

In this paper we presented a novel approach for introducing and evaluating trust learning models in open and dynamic distributed systems. This approach is conducted using our framework of trust management GeFMAT and a testbed for evaluating trust models. This framework is based on a meta-model that captures the semantics of concepts of MAS involved in an open environments. The testbed helps users compare the performance and efficiency of learning techniques proposed by existing trust and reputation models. The testbed was implemented and experimented using JADE platform. In our future work, we will work on improving the testbed for evaluating computational trust models. We also plan to focus on the scope of each agent, and how to manage this scope within the system.

References

1. Al-Mutaz, M., Malott, L., Chellappan, S.: Detecting Sybil attacks in vehicular networks. J. Trust Manage. 1–19 (2014)
2. Pinyol, I., Sabater, J.: Computational trust and reputation models for open multi-agent systems: a review. Artif. Intell. Rev. **40**, 1–25 (2013)
3. Teacy, W.T.L., Patel, J., Jennings, N.R., Luck, M.: Travos: Trust and reputation in the context of inaccurate information sources. Auton. Agents Multi-Agent Syst. (AAMAS), **12**(2), 183–198 (2006)

4. Audun, J., Roslan, I.: The beta reputation system. In: Proceedings of the 15th Bled Electronic Commerce Conference, pp. 324–337 (2002)
5. Victor, P., Cornelis, C., De Cock, M.: Trust Networks for Recommender Systems, vol. 4 (2011)
6. Sabater, J., Paolucci, M., Conte, R.: Repage: reputation and image among limited autonomous partners. J. Artif. Soc. Soc. Simul. **9**(2) (2006)
7. Patel, J., Teacy, W.L., Jennings, N.R., Luck, M.: A probabilistic trust model for handling inaccurate reputation sources. In: Trust Management, pp. 193–209. Springer, Berlin, Heidelberg (2005)
8. Massa, P., Avesani, P.: Trust-aware collaborative filtering for recommender systems. In: Proceedings of the Federated International Conference on the Move to Meaningful Internet, pp. 492–508 (2004)
9. Fullam, K., Klos, T., Muller, G., Sabater, J., Topol, Z., Barber, K.S., Rosenschein, J., Vercouter, L.: The agent reputation and trust (ART) testbed architecture. In: Proceedings of Workshop on Trust in Agent Societies, (AAMAS-05), pp. 50–62 (2005)
10. Kerr, R., Cohen, R.: TREET: The trust and reputation experimentation and evaluation testbed. Electron. Commerce Res. **10**(3–4), 271–290 (2010)
11. Jelenc, D., Hermoso, R., Sabater-Mir, J., Trček, D.: Decision making matters: a better way to evaluate trust models. Knowl.-Based Syst. 147–164 (2013)
12. Zhang, L., Jiang, S., Zhang, J., Ng, W.K.: Robustness of trust models and combinations for handling unfair ratings. In: Trust Management VI, pp. 36–51. Springer, Berlin, Heidelberg (2012)
13. Mifrah, Y., En-Nouaary, A., Dahchour, M.: An abstract framework for introducing computational trust models in JADE-based multi-agent systems. In: Advances in Ubiquitous Networking, pp. 513–523. Springer, Singapore (2015)
14. The JAVA Agent DEvelopment Framework. http://jade.tilab.com/
15. Multiagent Development Kit. http://www.madkit.net
16. Tryllian's Agent Development Kit. http://www.tryllian.com/adk.html
17. Kirn, S., Herzog, O., Lockemann, P., Spaniol, O.: Multiagent Engineering, Theory and Applications in Enterprises, pp. 503–530 (2006)
18. Parker, C.: An analysis of performance measures for binary classifiers. In: 2011 IEEE 11th International Conference on Data Mining (ICDM), pp. 517–526, Dec 2011
19. Jurman, G., Riccadonna, S., Furlanello, C.: A Comparison of MCC and CEN Error Measures in Multi-class Prediction (2012)
20. Huang, J., Ling, C.X.: Constructing new and better evaluation measures for machine learning. In: IJCAI, pp. 859–864, Jan 2007
21. Hernández-Orallo, J., Flach, P., Ferri, C.: A unified view of performance metrics: translating threshold choice into expected classification loss. J. Mach. Learn. Res. **13**(1), 2813–2869 (2012)
22. Jonker, C.M., Treur, J.: A dynamic perspective on an agent's mental states and interaction with its environment. In: Proceedings of the First International Joint Conference on Autonomous Agents and Multi-Agent Systems, AAMAS'02, 2002, pp. 865–872. ACM Press (2002)
23. Burete, R., Bădică, A., Bădică, C.: Reputation model with forgiveness factor for semi-competitive e-business agent societies. In: Networked Digital Technologies, pp. 402–416. Springer, Berlin, Heidelberg (2010)

Context-Aware Driving Assistance: An Approach for Monitoring-Based Modeling and Self-learning Cars

Afaf Bouhoute, Rachid Oucheikh and Ismail Berrada

Abstract Recent cars are equipped with a large number of sensors, electronic and communication devices that collect heterogeneous information about the vehicle, the environment and the driver. The use of the information coming from all these devices can highly contribute to the improvement of the vehicle safety as well as the driving experience. The last few years were marked by the development of a large number of in-vehicle intelligent systems that use driving behavior models to assist the driver ubiquitously. However, an important aspect to enhance driving experience is to make the provided assistance as close as possible to the behavior of the car owner, hence a need of personal models of drivers learned from their observed behavior. In this paper, the concept of intelligent and self-learning car is presented and examples of some car's embedded systems are given. Also, the role of modeling driver behavior in the design of driving assistance systems is emphasized. Further-more, the importance of monitoring-based driving behavior model construction to enable a personalized assistance is brought out together with some potential applications of formal driving behavior models.

Keywords Intelligent vehicles · Driving behavior model · Rectangular hybrid automata · Learning · In-vehicle monitoring · Driving safety · Formal verification

1 Introduction

Recent trends in automotive industry make cars look more like robots than physical systems. From navigational and informational systems to full automatic control, automakers are all oriented toward the application of computing and information

A. Bouhoute (✉) · R. Oucheikh · I. Berrada
USMBA University, LIMS, Fez, Morocco
e-mail: afaf.bouhoute@usmba.ac.ma

R. Oucheikh
e-mail: rachid.oucheikh@usmba.ac.ma

I. Berrada
e-mail: ismail.berrada@usmba.ac.ma

R. El-Azouzi et al. (eds.), *Advances in Ubiquitous Networking 2*, Lecture Notes in Electrical Engineering 397, DOI 10.1007/978-981-10-1627-1_46

technologies to transform cars into a Cyber Physical System (CPS). CPS is the name given to every system that has cyber technologies, both software and hardware, deeply embedded and interacting with physical components [22]. As automotive CPS, new cars are coming fitted with a set of powerful embedded sensors and communication technologies that make them capable of performing complex driving activities with advanced levels of autonomy. The developments in automotive CPS have been recently appearing at an accelerating pace, ranging from driver support systems (ABS, adaptive cruise control, lane departure warning…) to driverless systems (Audi's self-parking, Google's self-driving car…). Whether in semi-autonomous or full-autonomous systems, the role of the driver is very important and cannot be overlooked. Whereas monitoring the driver and his driving behavior is important in semi-autonomous cars, the driver preferences and his supervisory role are also to be considered in full autonomous cars. Hence, the need of integrating driver-in-the-loop considerations in the design of automotive CPS. Driver-in-the-loop systems integrate the driver with the vehicle's cyber systems through various degrees of autonomy, authority distribution, and systems interaction. From the monitoring of the driver interaction with the car CPS, we can extract a lot of information characterizing the driver, his driving style and how he reacts to real-time informational messages. This characterization of the driver enable assistance systems to provide a personalized assistance and can even make future cars more adapted to their drivers.

The contribution of this paper consists of an overview and examples of some car's embedded assistance systems are given, and also the role of modeling driver behavior in the design of advanced driving assistance systems is emphasized. Furthermore, the importance of monitoring-based driving behavior model construction to enable a personalized assistance is brought out together with some potential applications of formal driving behavior models.

The remainder of this paper is organized as follows: Sect. 2 gives an overview of what we mean by an intelligent car and of different type of driving assistance systems. In Sect. 3, we present briefly our approach for a formal and monitoring-based modeling. The potential applications of our approach are presented in Sect. 4. Section 5 concludes the paper and draws perspectives.

2 Related Work

In the last few years, the automotive industry is being considered as one of the promising area for the application of computing and networking technologies. Recent studies are considering cars as smart objects equipped with a powerful set of embedded technologies, and which are capable of perceiving and interacting with their external environment including other vehicles, roads, and pedestrians. This transformation has opened the floodgates for researches and developments in intelligent vehicular applications and semi-autonomous driving assistance systems. While some assistance systems focus merely on perception of the vehicle environment [8, 15], monitoring of the driver behavior [2, 5, 16] and displaying warnings, other systems excel this by removing the driver out the loop and performing corrective actions [6, 13]. With the aim of improving driving safety and comfort, developments in automotive systems were oriented to different directions that can be regrouped into three major categories: perception enhancement, driving behavior modeling and control systems.

In driving, perception is considered as an important issue as it forms the basis of all driving decisions. For perception enhancement of the driving environment, researches have focused on the use of information from the automotive sensory system to increase the awareness about the driving situation. Because of the complexity of the driving environment and the diversity/heterogeneity of its components (traffic signs, obstacles, environmental conditions...), different perception assistance systems have emerged that differ by their functionalities and the technologies on which they are based (vision/camera systems, Lidar/radar technologies, and vehicular communications...). For example in [8, 15], vision-based systems using image processing for traffic signs recognition were proposed. Both systems use robust and fast algorithms that achieve good performance, but the algorithm in [8] is unfortunately restricted to speed signs recognition, while an application as a speed regulator in an intelligent automated vehicle is proposed in [15] in addition the recognition system. Aside from traffic sign, obstacle (fixed or mobile) detection has been the focus of many papers like in [21] that proposes an approach for a visual processing system for pedestrian detection to be used in assistance applications. While perception enhancement is still an open research area, automotive industry is recently full of commercialized perception based assistance systems like: blind spot monitoring introduced by Volvo, lane departure warning systems firstly installed on Mercedes commercial trucks, intersection assistance first introduced by Toyota....

In addition to environment perception, understanding and modeling the driving behavior for application in assistance systems has been the topic of different studies. Some of researches in this field were oriented toward the prediction of driver intentions. Oliver and Pentland [18] used a trained models, that were created for different driving maneuvers based on collected data from the different sensors of a smart car, for the recognition and prediction of driver's maneuvers. The maneuvers that were modeled in [18] are passing another car, changing lane left and right, turning right and left, starting and stopping. The experiments have shown that the models are able to recognize and classify the driver maneuver one second before any significant change in the car signals. While the models in [18] use the information from all sensors (speed, throttle position, brake, gear, steering angle...), [20] presents prediction of car's future trajectories based primarily on the history of steering wheel angle. The goal of their approach is to create a probabilistic model of the driver behavior and use formal techniques to verify its properties that guaranty its safety and liveness. As a first step, a predictive model is created based on the future environment surrounding the car, the state of the driver and the history of steering maneuvers. The predictive model predicts for each environment and driver state a set of possible future trajectories, which are used as transitions probabilities within the driver model. A formal analysis of driver behavior is then carried out by the verification of quantitative properties of the constructed model.

The prediction and verification of driver behavior is considered as an important step for the realization of assistance systems. Based on the assistance provided to the driver, two categories are defined: passive assistance systems and active assistance systems. In passive assistance, different kinds of warnings (visual, auditory or tactile) are provided to the driver, alerting him about the current danger and/or possible corrective actions. Lane Departure Warning system (LDW) [9] is an example of passive assistance systems that, based on perception of the vehicle lateral movement, warns the driver if he is drifting out of the lane. However, [1] presents an assistance system as a driver behavior detection system that sends warnings to the driver as well as to the surrounding vehicles (via vehicular communication technology) when abnormal driving behaviors are detected. The main characteristic of passive assistance systems is that they always keep the human in the loop, which means that it always up to the driver to control it vehicle and the intervention of the system is nothing more that the provided warning. On the other hand, active assistance systems can in addition to warnings intervene on the physical level by automatically performing corrective actions if the driver does not respond to the given warnings. Partially similar to the lane departure warning, the lane keeping system [14] is an active assistance system that monitors the lateral movement of the car and automatically takes control of its steering to keep it on its lane, when the driver is not responding to the system's warnings. Another example of active safety is the system presented in [13] that provides the driver with brake assistance. The system analyzes the information about the vehicle, environment and driver, identifies the need for braking action, and based on the driver awareness decides if an automatic braking is needed.

Driving assistance systems either passives or actives always consider the driver as the main controller of the vehicle and intervene only in urgent cases, contrary to the recent vision of autonomous cars that keeps the driver in the loop for eventual intervention in case of system failures. Like in [10], where a formalism for the realization of a human-in-the-loop controller system that expect human intervention only for correct operation was presented. The paper focuses particularly on solving the problem of synthesizing a combination of human and autonomous controller from high level system specifications, and demonstrates its operation in the context of automated driving assistance.

After the investigation of these papers, we came to the following statements:

- The current autonomous and semi-autonomous systems focus on the safety aspect in a driving situation without considering the personal driving style of drivers to face this situation. Hence, the need of adjusting, while ensuring safety, the decisions made by these systems to meet drivers profiles.
- The modeling of the driver, vehicle and environment system is important to understand the behavior and preferences of the driver. Existing modeling approaches focus merely on modeling specific driving maneuvers, and predict the driver behavior based only on the current vehicle and driver state without referring to the history of the driver behavior in similar situations, or based on patterns of

driving behavior that were constructed and trained using different drivers and thus are not well-matched to the considered driver. In addition, we notice a lack of a formal framework for the description of driving behavior that captures the driver interactions with both the vehicle and environment.

To deal with these shortcomings, a formal framework that enables a holistic description of driving behavior and an algorithm to record the interactions of a driver with the vehicle and the driving contexts can be used. The behavior of a driver in a driving situation can then be predicted based on the collected records of its behavior in the same situation. And by combining these predictions with safety measures, autonomous and semi-autonomous cars can mimic the behavior of their owners while ensuring their safety.

3 Monitoring-Based Modeling for Self-learning Cars

Machine learning algorithms are called self-learning when performed by machine alone [23]. A car with self-learning skills is thus a car featuring advanced learning technologies that enable it to learn from the past driving experiences. This concept was lately brought to light by the car's company jaguar land rover, which developed a self-learning technology to offer a more personalized driving experience. What makes the jaguar land rover's car the first self-learning car is that the technology developed learns the driver preferences using a range of variables and can take care of multiple non driving tasks, unlike other technologies that focus only on navigation and traffic prediction. To provide a ubiquitous and personalized assistance, a learning of the driver behavior in all driving situations is required. In our previous works [3, 4], we have proposed a new formal framework to represent the driver behavior within the driver-vehicle-environment (DVE) system. In this section, The framework is shortly presented and the learning algorithm for the construction of the context-aware driving behavior model is given.

3.1 Description of the DVE System

The driving behavior is determined by several factors related to the driver, vehicle and driving environment. These three elements are often referred to as the Driver, Vehicle, Environment (DVE) system. This latter can be seen as a closed loop system: in which the driver act autonomously and make decisions based on of the perceived environment, on the other hand the vehicle as a passive component executes the driver decisions and move in the environment accordingly. The driver decisions in the next step are based on the feedback from the vehicle with the information from the environment [17]. Driving behavior can thus be defined as the activities performed by the driver to operate a vehicle, it represents the way the driver drives his car.

Formally, it can be seen as a change in the dynamic state of the vehicle made by the driver in response to a specific driving situation. A driving situation consists of a part of the traffic situation experienced by the driver in some unit of time and space [19], and which can be described as a combination of specific parameters that are categorized as:

- Dynamic parameters that refer to parameters about the behavior of other vehicles, pedestrians…;
- Static parameters, which are parameters regarding the constructional characteristics of the road (type of the road, curvature, intersection, type of intersection…) and the infrastructural characteristics (regulation type: implicit rules, road signs…).

A realistic representation of the driving behavior must thus takes into considerations the interactions of the driver within the DVE system. In our framework information about the ***dynamic state of the vehicle*** through the use of variables like velocity, acceleration…, ***the driver observable action*** like pressure on pedals, action on steering wheel…, and ***the driving environment*** (e.g. the authorized speed, GPS position…) are captured. And the driving behavior is represented as a transition system between the states of the vehicle enabled by the driver actions and the car environment situation. For this end, Probabilistic Rectangular Hybrid Input/Output Automata (PRHIOA) was proposed.

3.2 Modeling Framework: Probabilistic Rectangular Hybrid Input/Output Automata (PRHIOA)

PRHIOA is an adaptation of hybrid IO automata [11] and rectangular automata [7]. Hybrid I/O automata allows us to differentiate input and output actions, while rectangular hybrid automata allows us to define rectangular inequalities over variables. Formally, a Probabilistic Rectangular Hybrid I/O Automaton is defined as a tuple $(H, U, Y, X, Q, \Theta, inv, E, I, O, D, \mu)$, where:

- H is a set of internal variables, U is a set of input variables and Y is a set of output variables. H, U and Y are disjoint from each other and we write $X \triangleq H \cup U \cup Y$.
- $Q \subseteq val(H)$ is a set of states, where $val(H)$ is the set of valuations of H.
- Θ is a set of initial states.
- $inv : Q \rightarrow Rect(H)$ is an invariant function, where $Rect(H)$ is the set of all rectangular predicates over H. A rectangular predicate ϕ over H is a conjunction of rectangular inequalities; it defines the set of vectors $[\![\phi]\!] = \{z \in \mathbb{R}^n \mid \phi\ [H := z]\ is\ true\}$. A rectangular inequality over H is a formula $h_i \sim c$, where $h_i \in H$, c is an integer constant and $\sim$ is one of $<, \leq, >, \geq$. The function inv maps each state to its invariant condition.
- E, I and O are sets of internal, input and output actions, respectively. An internal action of E will be denoted later by "?".

- D is a set of discrete transitions. A discrete transition is labeled with an action, and is defined as a triple (q, o, g, q') where q is a source state, o is an action, $g \in Rect(H)$ is a guard on the transition, and q' is a target state. To simplify, if the guard is true we will only refer to a transition as a triple (q, o, q').
- μ is a transition probability over output actions O is defined such as: $\sum q' \in Q \sum_{a \in O} \mu(q, a, q') = 1$ for all $q \in Q$ and all $a \in O$.

The semantics of the PRHIOA is a transition system where states represents the state of the vehicle described by conjunctions of rectangular inequalities over a set of driving parameters (i.e. internal variables) such as velocity, lane, acceleration.... There are two types of transitions between states; one is triggered by a driver action (e.g. hit the gas/brake pedal, turn the wheel) also called output action and the other follows the change of the variables over time (e.g. contextual information) also called input action. The range of the driving variables values known already, the state space can be created and we can start the construction phase.

3.3 Algorithm of Construction of the Driving Behavior Model

In our approach, the driving behavior model describes how the driver behaved in all road encountered situations. Thus the construction, which consists of a learning of the transitions relating the states of the model, is performed through a continuous monitoring of the driver behavior reflected in the change of the variables values. The Algorithm 1 presents the steps of the construction algorithm. To learn and update the transitions probabilities a reinforcement scheme based on the algorithm of learning automata is used, the update scheme is illustrated in Algorithm 2.

Algorithm 1 Algorithm of construction

Require: Q: states set, $I=\{I_1, I_2, \ldots, I_n\}$: input actions sets, $O=\{O_1, O_2, .., O_m\}$: output actions set, $q_{initial}$: initial state
Ensure: D: transitions set, P : set of transitions probabilities
  $current\_state = previous\_state = q_{initial}$
  While the vehicle in movement Observe the DVE system
  if input I_j (i.e. driving environment requirement) **then**
    Add input transition
    Update the driving context
    Update transitions probabilities
  end if
  if input O_j (i.e. action of the driver) **then**
    Add output transition
    Update current state
    Update transitions probabilities
  end if

Algorithm 2 Transitions probabilities update

Require: $T = (q, a, q')$ transition to be added, D: transitions set, P : set of transitions probabilities, r: number of outgoing transition from q
Ensure: D: transitions set, P : set of transitions probabilities
  if T is not in D **then**
    $P(T) = 1$
    if $r \neq 0$ **then**
      $p(T) = \frac{1}{r}$
      $p(T') = (1 - \frac{1}{r})p(T')$ for $T' \neq T$
    end if
  else
    $p(T) = p(T) + \frac{1-p(T)}{r}$
    T' in the transitions set D
    $p(T') = (1 - \frac{1}{r})p(T')$ for $T' \neq T$
  end if

Running the algorithm of the model construction for many trips, we can have one model describing and reporting the driver behavior in the different encountered situations, from which we can retrieve information about his preferences, how he reacts behind the wheel and also the safety of his driving.

To enable a formal verification of the driving safety using logical properties, we have represented and captured the contextual information (e.g. input actions) as conditions on the internal variables of the model. A detailed study about how we define conditions for roads signs can be found in [4] and also about we update these conditions continuously within the states. Depending on the variables that were used we can define logical properties to verify in an automatic way, several driving behavior like the respect of the speed limit, the safe following distance…

4 Potential Applications of Formal Monitoring-Based Modeling

One of the important applications that we propose is a monitoring system for driver's education purposes. Monitoring systems have been recently an important trend for measuring driving styles. Monitoring systems have been mostly used in fleet management [12], and they have shown promising results thus con-firming the contributing role of monitoring in the improvement of a range of safety-related behaviors. Therefore for road safety purposes, monitoring systems must allow the assessment of the safety of the driver behavior in different traffic situations. Generally, the existing monitoring systems focus on the record of time-step driving data (acceleration, speed…) while the contextual details about the driving conditions are either not recorded or provided as videos of the external environment. Moreover, the large size of the generated files by these systems makes their analysis more complicated and increases their processing time. Comparing to the existing monitoring systems, driving data together with the contextual constraints of the environment are recorded as one mathematical model, and tools to analyze the convenience of recorded data to driving conditions are provided. In addition, smaller log files are generated, as only abstracted states of driver-vehicle with the emphasis on the input/output nature of the driving behavior are kept instead of storing the time step data of driving. As driving safety and performance has to be evaluated according to the contextual driving environment, focus has been put on modeling and recording of the driving behavior observed with the contextual information expressed as context constraints, which makes the analysis and verification process less complicated. Therefore, applications can be found in different other sectors with driving safety and driving analysis concerns can be found.

It can be used for example by insurance companies to reward/punish their customers according to the performance of their driving. Insurance companies are recently using driving monitoring to provide personal payment for car insurance based on driver habits, and to encourage their customer to improve their driving performance. The systems that are used until now track the driver's habits (braking and other data) to figure out if he will receive a good driver discount or not. The evaluation of driving habits is made primarily on data from the car engine computer. Yet the system that we propose here combines information from the vehicle with the information from environment to evaluate how the driver behaves in different driving situations. Information about environment consists of road types, traffic regulation, behavior of other vehicles, obstacles, etc. Insurers can then evaluate the driver habits in respect to the perceived environment instead of recognizing it from driving data. Examples of behavior insurers can evaluate are: the velocity of the car taking into account the road type or curvature and an overtaking maneuver considering the other vehicles on the road.

As a part of our approach focus on modeling traffic rules, traffic police can use it for the detection and generation of infringement notices. Recently, multiple intelligent solutions have been proposed to support traffic enforcement such as speed cameras that capture speeding vehicles, in-vehicle data recorders that record driving data to be used in a crash analysis,.... These solutions contribute to the improvement of road safety without requiring an increase in human police resources. Our approach of modeling offer a way to detect road rules infringement without the need of more intelligent road infrastructure as every vehicle will be able to auto-detect and record any disobedience to road rules and send data to police center for evaluation. The formal verification proposed will facilitate the evaluation of recorded data.

Another important application is to use the driver behavior model as a basis of a personalized driving assistance. The automotive technologies available in recent cars can be a source of a lot of information that can be used to learn driver preferences either in terms of driving, navigational or infotainment tasks. This information gathered from the in-vehicle technologies can be used to enrich the personal behavior model presented in this work. Because this probabilistic model can predict the driver behavior in the different situations that were experienced previously by the driver, we can anticipate and avoid risky behaviors and assistance functions can be implemented by the design of the controller unit to take control of the physical plant if a risky behavior is anticipated.

5 Conclusion

Technologies for smart cars are evolving rapidly and manufacturers are focusing more and more on the satisfaction of the customers hence the emergence of self-learning cars. This paper aims to elucidate the concept of smart self-learning cars and present examples of car's embedded technologies that were developed during the last years. We also emphasize, throughout this paper, the importance of the role driver in the design of advanced systems and a framework to formally represent the driving behavior together with a monitoring-based approach for the construction of personal driving behavior models are presented. In addition, the usefulness of the formal framework for an automatic verification of the safety of driver behavior is illustrated. Finally,an insight of some potential applications of formal and monitoring-based modeling is given.

To support the proposed approach, we are running two different simulation approaches. On one hand, some driving situations are simulated using the open driving simulator (OPENDS) and the verification tool PRISM is used for the automatic checking of some predefined safety properties. On the other hand, tests using real data from Green research Lab of The Nagoya University, Japan are considered. Details about the implementation, simulations and their results will be discussed in a future work.

References

1. Al-Sultan, S., Al-Bayatti, A.H., Zedan, H.: Context-aware driver behavior detection system in intelligent transportation systems. IEEE Trans. Veh. Technol. **62**(9), 4264–4275 (2013)
2. Aliane, N., Fernández, J., Bemposta, S., Mata, M.: Traffic violation alert and management. In: 2011 14th International IEEE Conference on Intelligent Transportation Systems (ITSC), pp. 1716–1720. IEEE (2011)
3. Bouhoute, A., Berrada, I., El Kamili, M.: A formal driving behavior model for intelligent transportation systems. In: Networked Systems, pp. 298–312. Springer (2014)
4. Bouhoute, A., Oucheikh, R., Zahraoui, Y., Berrada, I.: A holistic approach for modeling and verification of human driver behavior. In: 2015 International Conference on Wireless Networks and Mobile Communications (WINCOM), pp. 1–7. IEEE (2015)
5. Elmahalawy, A.M.: A car monitoring system for self recording traffic violations. World **4**, 5 (2014)
6. Enache, N.M., Mammar, S., Netto, M., Lusetti, B.: Driver steering assistance for lane-departure avoidance based on hybrid automata and composite lyapunov function. IEEE Trans. Intell. Transp. Syst. **11**(1), 28–39 (2010)
7. Henzinger, T.A., Kopke, P.W.: Discrete-time control for rectangular hybrid automata. Theoret. Comput. Sci. **221**(1), 369–392 (1999)

8. Johansson, B.: Road sign recognition from a moving vehicle. Technical report (2002)
9. Jung, C.R., Kelber, C.R.: A lane departure warning system based on a linear-parabolic lane model. In: Intelligent Vehicles Symposium, 2004 IEEE, pp. 891–895. IEEE (2004)
10. Li, W., Sadigh, D., Sastry, S.S., Seshia, S.A.: Synthesis for human-in-the-loop control systems. In: Tools and Algorithms for the Construction and Analysis of Systems, pp. 470–484. Springer (2014)
11. Lynch, N., Segala, R., Vaandrager, F.: Hybrid i/o automata. Inf. Comput. **185**(1), 105–157 (2003)
12. Marzooqi, A.: Road safety system for monitoring fleet drivers. In: Safer Driving, Reducing Risks, Crashes and Casualties. Proceedings of the 68th Road Safety Congress Held Blackpool, 3–5 Mar 2003 (2003)
13. McCall, J.C., Trivedi, M.M.: Driver behavior and situation aware brake assistance for intelligent vehicles. Proc. IEEE **95**(2), 374–387 (2007)
14. Enache, N.M., Netto, M., Mammar, S., Lusetti, B.: Driver steering assistance for lane departure avoidance. Control Eng. Pract. **17**(6), 642–651 (2009)
15. Monika, Y.D.D., Avinash, N., Jung, H.G., Na, H.: Real time traffic sign recognition system as speed regulator in IAV. In: IICAI, pp. 1936–1951 (2009)
16. Nejati, O.: Smart recording of traffic violations via M-RFID. In: 2011 7th International Conference on Wireless Communications, Networking and Mobile Computing (WiCOM), pp. 1–4. IEEE (2011)
17. Ni, D.: Traffic Flow Theory: Characteristics, Experimental Methods, and Numerical Techniques (2015)
18. Oliver, N., Pentland, A.P.: Driver behavior recognition and prediction in a smartcar. In: AeroSense 2000, pp. 280–290. International Society for Optics and Photonics (2000)
19. Plavsic, M.: Analysis and modeling of driver behavior for assistance systems at road intersections. Ph.D. thesis, Lehrstuhl für Ergonomie der Technischen Universität München (2010)
20. Sadigh, D., Driggs-Campbell, K., Puggelli, A., Li, W., Shia, V., Bajcsy, R., Sangiovanni-Vincentelli, A.L., Sastry, S.S., Seshia, S.A.: Data-driven probabilistic modeling and verification of human driver behavior. In: Formal Verification and Modeling in Human-Machine Systems (2014)
21. Shashua, A., Gdalyahu, Y., Hayun, G.: Pedestrian detection for driving assistance systems: single-frame classification and system level performance. In: Intelligent Vehicles Symposium, 2004 IEEE, pp. 1–6. IEEE (2004)
22. Sunder, S.: Foundations for innovation in cyber-physical systems. In: Proceedings of the NIST CPS Workshop, Chicago, IL, USA, vol. 13 (2012)
23. Vallverdu, J.: Handbook of Research on Synthesizing Human Emotion in Intelligent Systems and Robotics (2015)

ABE Based Raspberry Pi Secure Health Sensor (SHS)

Divyashikha Sethia, Suraj Singh and Vaibhav Singhal

Abstract Electronic health data collected from bio-medical sensors is having a profound and increasing impact on mobile health services. When transferred over the air this data should be accessible only to legitimate users such as doctors, nurses etc. In this paper, we propose a unique Secure Health Sensor (SHS) node which provides fine access control mechanism such that only legitimate users can access sensitive medical data. Medical data security is achieved through ensuring a secure communication by encrypting sensor data, preventing loss of sensor data by using wired connection between Raspberry Pi and sensors and using Ciphertext Policy Based Attribute Based Encryption (CP-ABE) to provide access control in multi-user enviroment. We propose storage of cyrptogrphic credentials on Raspbery Pi on hardware tamper resistant area such as Secure Element on form factors such as Go Trust microSD card. It comprises of java card applets used to store credentials and can be accessed by special securely compiled applications on the Raspbery Pi.

Keywords Health services · Bio-medical sensors · CP-ABE · Secure element

1 Introduction

Remote health services are a major requirement in both developed countries, where the cost of healthcare is high and security and privacy are critical issues and developing countries like India, having huge population to control in hospitals. A large population is estimated to reside in nursing homes and hospitals in near

D. Sethia (✉) · S. Singh · V. Singhal
Department of Computer Science & Engineering, Delhi Technological University, Main Bawana Road, Delhi, India
e-mail: divyashikha@dce.edu

S. Singh
e-mail: surajrider@gmail.com

V. Singhal
e-mail: singhalvaibhav28@gmail.com

R. El-Azouzi et al. (eds.), *Advances in Ubiquitous Networking 2*, Lecture Notes in Electrical Engineering 397, DOI 10.1007/978-981-10-1627-1_47

future. According to the study [1], within the next decade, people aged 65 and above will outnumber children under five in the world. Hence these health centres need to be equipped with continuous medical monitoring, medical data access and emergency communication services. An efficient, reliable, robust and secure health flow is important to manage patients, their health records securely and to make the right health service accessible to the patient at the right time. Privacy and security is a very important aspect of healthcare [2]. Secure Health Sensor (SHS) is useful in biometric and medical applications for real time monitoring of a patient's state or for acquiring sensitive data which can be used to provide the correct medical diagnosis.

The initial motivation of the work was to develop a secure health sensor (SHS) node for multi user environment such that medical data can be made available to number of users who have necessary attributes to decrypt sensitive data comprising of health sensor readings. Recently, few systems have been implemented that allow collection of patient data at any remote site and that can be shared with doctors, nurses etc. using cloud based services. Recently, a a few solutions like [3] provide remote health services using Remdi kit remote health services to patients, but lacks security measures. We have designed a system that not only encrypts this sensitive data but also makes it accessible to users having the appropriate attributes required to decrypt this medical data. With this design the sensors for temperature, blood pressure, pulse oximeter etc. can similarly be incorporated in the design to gather vital health parameters. Raspberry Pi has been chosen rightly as the single board computer for this application because it has the highest performance to cost ratio and is one of the smallest single board computers available in the market. This paper describes in detail, the components, design and functioning of SHS. The sensor information gathered can be communicated wirelessly to a remote device such as mobile phone using low energy wireless interfaces such as Bluetooth, Bluetooth lite or NFC. In this project, bluetooth technology is used. The rest of the paper consists of related work in Sect. 2, followed by Sect. 3 consisting of components used in SHS, Sect. 4 related to preliminaries of Attribute Based Encryption (ABE) and Sect. 5 consisting of the architecture. Results are discussed in Sect. 6, followed by conclusion and future work in Sect. 7.

2 Related Work

This paper extends our previous work [4] where we used encryption of sensor data using RC4 encryption algorithm. Here, we make use of AES for symmetric encryption of medical data and CP-ABE [5] for fine grained access control. The Plug-n-Trust module [6], where a mobile phone is responsible to collect the data from the various sensors (connected to the body) suffers from security issues [7]. The secure sensor node prototype designed with the Raspberry Pi [4] in our previous work makes use of similar architecture but does not make use of fine grained access control. Work in [8, 9] uses Raspberry Pi and sensors to gather the data but data storage and transmission do not incorporate any security. Saari [10] used

BeagleBone Black development board with an embedded Linux distribution in IoT based system but does not provide security of data transmission. Our system guarantees more security as compared to Plug-n-Trust module [6], Wireless Sensor network using Raspberry Pi and ZigBee [8], portable spectrometric sensor platform [9], data collector service [10] and secure sensor node using Raspberry Pi:

- Wired communication occurs between the Raspberry Pi and the sensor so there is no fear of information loss or security.
- The data transmission always happens in encrypted form.
- Fine grained access control is provided so that only users with valid attributes that match the access policy can decrypt the cipher text.

3 Components of SHS

A unique design and implementation of a secure sensor node has been carried out based on three major components: a single board computer, an accelerometer based sensor-ADXL345 [11] and heart-beat based sensor as shown in Fig. 1.

3.1 Major Components

1. **Single Board Computer**

We use Raspberry Pi [12] to design a secure health sensor interfacing with a number of sensors. Among other SBCs, it is the cheapest single board computer available with the best performance/cost and RAM/cost ratio used as wireless sensor node [13]. Its small size, low cost, low power consumption and high processing power makes it suitable for the design of SHS.

Fig. 1 Different components of SHS

2. **Accelerometer, ADXL345**
The ADXL345 [11] is a x, y and z axis accelerometer with a high resolution. It covers a range of ±16 g. The output data can be accessed through an SPI (4- or 3-wire) or a I2C digital interface. It is small, thin and has low-power, hence it is suitable for mobile device applications. It measures static acceleration due to gravity in tilt-sensing applications, and also dynamic acceleration resulting from motion or shock.
3. **Digital Heart Beat Sensor**
The Heart Beat Sensor provides digital output of heart beats when a finger is placed on it. Heart beat per minute (BPM) can be measured by connecting the output to Raspberry pi. The heartbeat count can be obtained serially (TTL) every minute. It works on the principle of light modulation by blood flow through the nerves of the finger at every pulse.

4 Preliminaries

4.1 Attribute Based Encryption

Attribute-based encryption (ABE) is a relatively new asymmetric key cryptography technique in which the private key of the user and cipher text are dependent on the attributes. Generally, in asymmetric key cryptography, data is enciphered for a specific receiver using the receiver's public-key. ABE on the other hand, defines the identity of users not as atomic but as a set of attributes, e.g., age, data of birth etc., and messages can be enciphered with a set of attributes (key-policy ABE—KP-ABE) or policies defined over a set of attributes (Ciphertext-policy ABE—CP-ABE).

4.2 Key-Policy ABE

Key-policy attribute-based encryption (KP-ABE) is a type of ABE, in which the access structure is incorporated into the secret key of the user and the encryption is done on messages under a set of attributes and private keys are associated with access structures that specify which ciphertexts the key holder is allowed to decrypt. Generally, in KP-ABE the cipher text size depends on the number of attributes incorporated into the cipher text. KP-ABE is dual to CP-ABE in a way because access policy is incorporated into the user's secret key, e.g., (A and C) or D, and a cipher text is decrypted with respect to a set of attributes, e.g., {A, B}.

4.3 Ciphertext-Policy ABE

In CP-ABE model [5], private keys are identified with a set S of descriptive attributes. A party that wishes to encrypt a message specify a policy (access tree) that private keys must satisfy in order to decrypt access tree T. Each non-leaf node of the tree represents a threshold gate, described by its children and a threshold value. If num_x is the number of children of a node x and k_x is its threshold value, then $0 < k_x \leq num_x$. Each leaf node x of the tree is described by an attribute and a threshold value $k_x = 1$. Figure 2 shows the access tree structure with 5 attributes with the policy as given below:

$$(patient\ or\ caretaker\ or\ doctor\ or\ (nurse\ and\ health\ professional))$$

The policy allows either patient, his/her caretaker, doctor and a nurse who is a health professional to decrypt the cipher text and obtain the sensitive medical data.

The cpabe toolkit [14] is used to implement the ciphertext policy based attribute based encryption. Below are the main functions provided by cpabe toolkit:

- *Setup*: It produces the public parameters public key (K_{CPABE_Pub}) and master key (master_key) as outputs using the cpabe-setup command.
- *Key-Generation (master_key, K_{CPABE_Pub}, S)*: This step takes three inputs, master secret key (master_key), public key (K_{CPABE_Pub}) and set of attributes of a user S. It uses cpabe-keygen command provided by cpabe-toolkit. It outputs the private key (S_K) for the user according to the set of attributes.
- *Encrypt(K_{CPABE_Pub}, M, w)*: This step uses the public key and policy access structure(w) defined by a set of attributes to encrypt the message (M). The cpabe-enc command is used to encrypt the message M. The output of this step is cipher text with .cpabe extension to input message file. For example, if the input file is directions.txt then the output file is directions.txt.cpabe.

Fig. 2 Example of access tree structure with 5 attributes

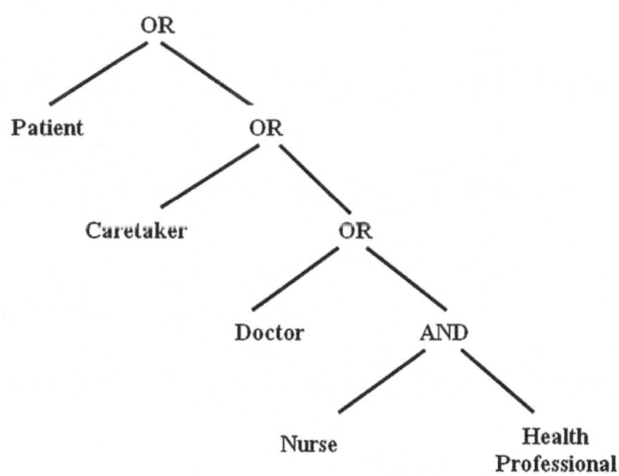

- *Decrypt* (K_{CPABE_Pub}, *C*, *SK*): The decryption step takes three inputs, public parameter (K_{CPABE_Pub}), a ciphertext (C) containing embedded access policy (w), and secret key(SK). If the attribute set on which secret key is defined, satisfies the access structure, the algorithm decipher the ciphertext and generate the plaintext, otherwise it returns error message that user attribute set does not satisfy the policy.

5 System Architecture

In SHS, the medical data is gathered at patient site through the bio-medical sensor and is sent over the air through bluetooth medium to remote site. The data is generally collected by medical professionals who are responsible for creating the access policy for the doctors, nurses etc. The transmitted data is made accessible to anyone having the required set of attributes; thus making it possible for multiple shareholders to access this sensitive medical data such as doctors, relatives of the patient etc. Thus, SHS is very flexible and provides fine grained access control over the medical data.

Figure 3 shows the architecture as well as the flow of medical data starting from reading collection through sensors and collected finally at remote site by different users. The bio-medical sensors are attached to patient to take the readings. Readings are securely transmitted through the wired interface to the raspberry pi device. Raspberry pi performs the encryption using CP-ABE to provide fine grained access control. This encrypted data is then transmitted through the Bluetooth device over the air to the remote user on demand. At remote site, the doctors, nurses, caretakers of the patient etc. can see this sensitive medical data. Any user without valid set of attributes cannot decrypt the encrypted sensor data file.

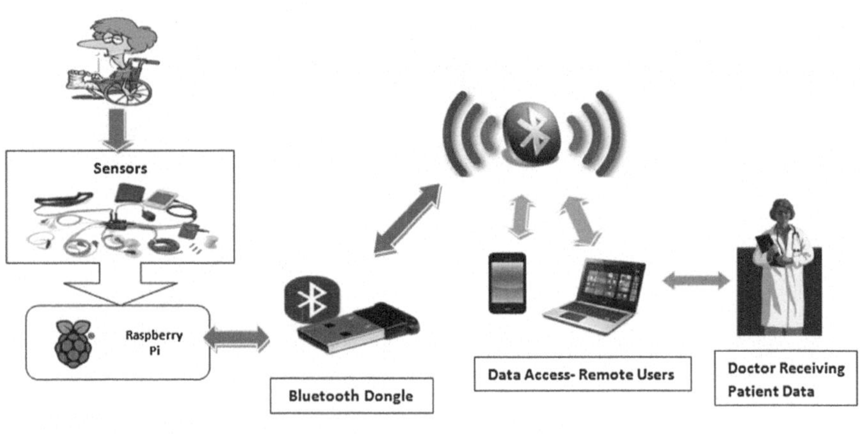

Fig. 3 SHS system architecture

6 Implementation

6.1 Connections of sensors in Secure Health Sensor Node

The accelerometer based sensor ADXL345 consists of eight pins, two of which are the power pin VCC and ground pin GND. Two Interrupt pins INT1, INT2 are available but are left unconnected. The CS and VCC pins are supplied with 3.3 V from the Raspberry Pi. The SDO and the GND pin of ADXL345 are connected to the GND pin of the Raspberry Pi. The SDA pin for data interchange is connected to the third pin and the SCL pin to the fifth pin of the Raspberry Pi i.e. to the corresponding SDA and SCL pins, Communication between the accelerometer based sensor and the Raspberry Pi can take place using either of the two serial protocols: SPI [11] or I2C [11]. The Digital Heart Beat Sensor has VCC, GND and OUT pins. VCC is connected to pin two of raspberry pi for providing +5v power. OUT pin can be connected to any GPIO pin and GND is connected to corresponding GND pin of pi.

6.2 Data collection from Sensors

The python library of ADXL345 is cloned and imported into the code for receiving the readings. The library is a basic implementation of the i2c protocol for the IC offering a simple way to get started with it on the Raspberry Pi. Along with this, libraries like python-smbus and i2c-tools were also installed on pi to facilitate the working of accelerometer. To measure the heartbeat, value of the GPIO pin to which the OUT pin of sensor was connected is noted periodically for around 500 times at an interval of 0.5 s. Heartbeat is the number of iterations out of total iterations for which the value of GPIO pin was found to be high.

6.3 Encryption

Before starting the encryption process in SHS, personalization step is performed in which the public key and master key is generated from the setup provided by cpabe toolkit as shown in the sequence diagram in Fig. 4. Table 1 shows the different notations used in the sequence diagram. Private keys of CP-ABE are also generated and distributed. CPABE-ENC is used to encrypt K_{AES_SYM} so that it can be securely distributed over the air. It requires the policy structure W as well as the K_{CPABE_Pub}. Finally, the private key (K_{PRIV}) of user and public key(K_{CPABE_Pub}) are transferred to the user through secure channel.

The readings are encrypted for secure transmission using AES (K_{AES_SYM} symmetric key). K_{AES_SYM} is again encrypted using CP-ABE encryption algorithm to generate EK1 which is transmitted over the air to user. We used symmetric key of length 16 byte (128-bit) in our program. To increase the strength of the encryption; AES key of length 24 byte (192-bit) or 32 byte (256-bit) can be used.

Figure 4 shows the sequence diagram of the SHS system. K_{AES_SYM} is used for encrypting the sensor data file. This results in generation of cipher text, say C1. The SHS sends C1 and EK1 over the transmission medium to the user.

6.4 Decryption

The decryption process is two step process which is reverse of encryption. User receives C1 and EK1. The user uses CP-ABE private key to decrypt EK1 to obtain the K_{AES_SYM} which is used to decrypt C1 (encrypted using AES at sender site) to finally obtain sensor data.

6.5 Data Transmission

The encrypted data can be transmitted to user over the bluetooth medium. To achieve this, bluetooth sockets and Client-Server architecture have been used. On SHS, a server application is initiated which listens for incoming connections and transmits the encrypted data on successful connection. Bluetooth client application at the user knows the MAC address of SHS's bluetooth device using QRcodes generated , and requests the data. Data can also be transmitted over the web using network sockets with the help of similar Client-Server architecture. For transmission over Bluetooth, Pybluez and bluetooth python packages [15] were used for making the server application at SHS.

7 Results

7.1 CP-ABE Private Key Generation Time

Figure 5, we found that CPABE private key generation time increases with number of attributes.

7.2 CPABE—AES Symmetric Key Encryption Time

Figure 6 above shows that time required for K_{AES_SYM} encryption depends upon the number of attributes used in the policy.

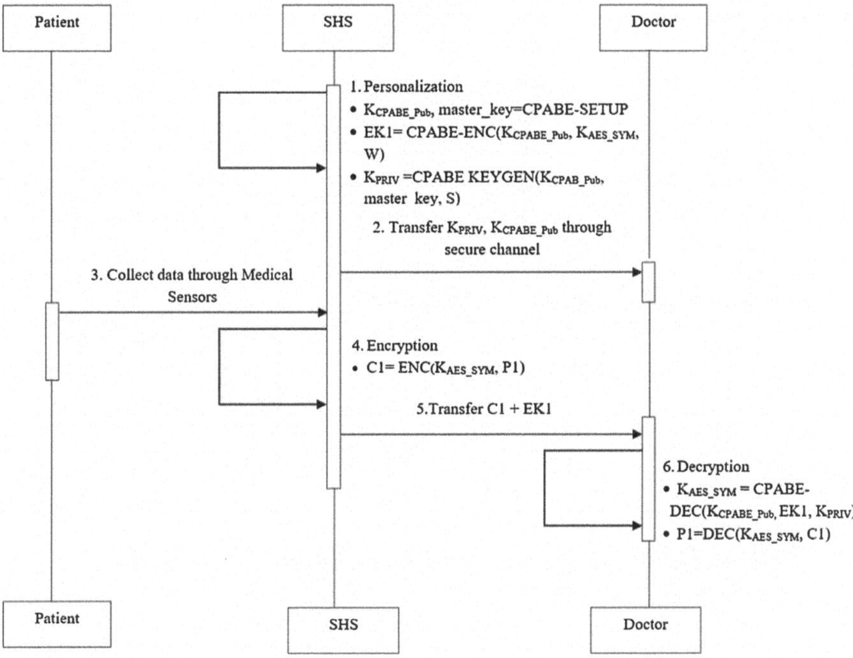

Fig. 4 Sequence diagram of SHS

Fig. 5 Key generation time against number of attributes

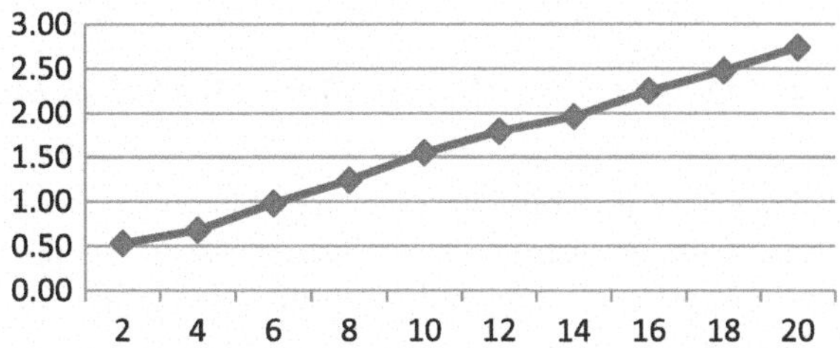

Table 1 Different notations used in sequence diagram

Key abbreviation	Description
K_{CPABE_Pub}	CPABE Public key
K_{AES_SYM}	AES symmetric key
K_{PRIV}	CPABE—User Private key
EK1	Encrypted AES key
C1	Encrypted Medical Data
P1	Patient Data
master_key	CPABE—Master Key

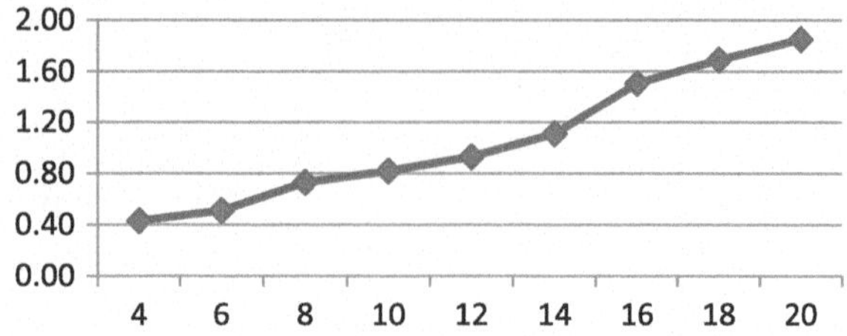

Fig. 6 CPABE—K_{AES_SYM} encryption time

Table 2 Sensor data Encryption and Decryption Time

Length of K_{AES_SYM} (bytes)	Average encryption time (millisecond)	Average decryption time (millisecond)
16	2.4229	20.6866
24	2.6389	21.4863
32	2.6252	22.0068

7.3 *Sensor Data Encryption and Decryption Time by AES Algorithm*

Table 2 shows the time taken for encryption and decryption on Raspberry Pi. We used different length AES key K_{AES_SYM} for encrypting the sensor readings and time was nearly same on all three cases. Decryption time increases along with key size K_{AES_SYM}.

7.4 *Data Transmission Time Over Bluetooth*

The average transmission time over bluetooth medium was found to be 4.756 s. The experiment was conducted several times and the average of all times is taken. The time taken is acceptable with the fine grained security access control provided by CPABE in multi stakeholder environment.

8 Conclusion and Future Work

In this work, we have designed a secure health sensor (SHS) node capable of securing sensitive medical data. SHS design is flexible and provides fine grained access control. The size of the Raspberry Pi being that of a mobile device provides compactness and portability to SHS. The use of bluetooth low energy dongle makes system more energy and cost efficient. The encryption, key generation and transmission time suggests that SHS can be incorporated in existing health care centers providing health services. The Raspberry Pi along with bluetooth low energy has interfaces which are easy to use, and can help designers in investigation and development of new sensors as well as encryption and compression algorithms for future sensors.

Java cards can be used as memory element for secure symmetric key storage on secure element (SE), thus enhancing the security of SHS. Instead of storing the symmetric key (K_{AES_SYM}) in encrypted form as EK2, it can be stored securely in java card and accessed by java card APIs, for encrypting sensor data. We could not incorporate it into the current design because java card interaction libraries with Raspberry pi (Arm platform) were not provided. We have implemented this on Linux based Intel 64-bit architecture systems. But, we could not replicate the same on raspberry pi since the libraries were not compatible with it and we couldn't get the compatible ones from the java card supplier. However, we can use Single Board Computer (SBC) like Intel Galileo 2 based on Intel architecture for implementing SHS using Java cards which can support java applet libraries. Raspberry Pi is based on ARM architecture.

Currently, our system includes sensors with wired connections. Wireless sensors can be incorporated into SHS by including appropriate wireless inventor kit for Raspberry Pi. We plan to improve upon the design of SHS by making it battery operated with switches to start the sensor. The limitation of data transmission through Bluetooth can be removed by making use of cloud servers to store the data. Users can request the data from the cloud server through appropriate user interface APIs. We are also planning to use QR codes for pairing between SHS and user mobile device used for accessing sensor information. This is an alternate to using NFC controller. We use Bluetooth since it has higher throughout and supports easy bidirectional security handshake.

References

1. World's elderly to overtake number of infants, an article in The Telegraph, UK, 18 June 2013
2. Avancha, S., Baxi, A., Kotz, D.: Privacy in mobile technology for personal healthcare. ACM Comput. Surv. (CSUR) **45**(1) (2012). Article 3
3. http://www.neurosynaptic.com/remedi-telemedicine-solution/
4. Banerjee, S., Sethia, D., Mittal, T., Arora, U., Chauhan, A.: Secure sensor node with Raspberry Pi. In: 2013 International Conference on Multimedia, Signal Processing and Communication Technologies (IMPACT), vol. 26, no. 30, pp. 23–25 (2013)
5. Bethencourt, J., Sahai, A., Waters, B.: Ciphertext policy attribute-based encryption. In: IEEE Symposium on Security and Privacy, pp. 321–334. IEEE Computer Society, Los Alamitos (2007)
6. Sorber, J., Shiny, M., Peterson, R., Kotz, D.: Plug-n-Trust: practical trusted sensing for mHealth. In: Institute for Security, Technology, and Society, Dartmouth College, Hanover, NH, USA Department of Computer Engineering, Myongji University, South Korea
7. Dimitriou, T., Ioannis, K.: Security issues in biomedical wireless sensor networks. In: Applied Sciences on Biomedical and Communication Technologies First International Symposium, Conference Publication (2008)

8. Nikhade, S.G.: Wireless sensor network system using Raspberry Pi and Zigbee for environmental monitoring applications. In: 2015 International Conference on Smart Technologies and Management for Computing, Communication, Controls, Energy and Materials (ICSTM), pp. 376–381, May 2015
9. Kar, A., Kar, A.: A novel design of a portable double beam-in-time spectrometric sensor platform with cloud connectivity for environmental monitoring applications
10. Saari, M., Sillberg, P., Rantanen, P., Soini, J., Fukai, H.: Data collector service—practical approach with embedded Linux. In: MIPRO 2015, 25–29 May 2015
11. Datasheet archives of various ICs(ADXL345). www.datasheetarchive.com/ADXL345-datasheet.html
12. Getting started with Raspberry Pi, Matt Richardson and Shawn Wallace, published by O'Reilly Media, First release December 2012
13. Vujović, V., Maksimović, M.: Raspberry Pi as a wireless sensor node: performances and constraints. In: 37th International Convention on Information and Communication Technology, Electronics and Microelectronics (MIPRO), pp. 1247–1252 (2014)
14. http://acsc.cs.utexas.edu/cpabe/
15. https://pypi.python.org/pypi/PyBluez

Towards Data-as-a-Service Provisioning with High-Quality Data

Elarbi Badidi, Hayat Routaib and Mohammed El Koutbi

Abstract Given the large amount of sensed data by IoT devices and various wireless sensor networks, traditional data services lack the necessary resources to store and process that data, as well as to disseminate high-quality data to a variety of potential consumers. In this paper, we propose a framework for Data as a Service (DaaS) provisioning, which relies on deploying DaaS services on the cloud and using a DaaS agency to mediate between data consumers and DaaS services using a publish/subscribe model. Furthermore, we describe a decision algorithm for the selection of appropriate DaaS services that can fulfill consumers' requests for high-quality data. One of the benefits of the proposed approach is the elasticity of the cloud resources used by DaaS services. Moreover, the selection algorithm allows ranking DaaS services by matching their quality-of-data (QoD) offers against the QoD needs of the data consumer.

Keywords Quality-of-data · DaaS selection · Smart cities · Cloud computing

1 Introduction

Over the last few years, the impressive progress in sensing and wireless technologies result in the proliferation of wireless sensor networks (WSNs) in many areas such as:

- Industrial machine surveillance for predictive maintenance.
- Intelligent buildings and smart homes applications.

E. Badidi (✉)
College of Information Technology, United Arab Emirates University, Al-Ain, United Arab Emirates
e-mail: ebadidi@uaeu.ac.ae

H. Routaib · M. El Koutbi
MIS Team, ENSIAS, Mohamed V University, Rabat, Morocco
e-mail: routaib.hayat@gmail.com

M. El Koutbi
e-mail: elkoutbi@ensias.ma

R. El-Azouzi et al. (eds.), *Advances in Ubiquitous Networking 2*, Lecture Notes in Electrical Engineering 397, DOI 10.1007/978-981-10-1627-1_48

- Military target tracking and surveillance.
- Government and environmental services like natural disaster relief.
- Seismic sensing.
- Long-term surveillance of elderly and chronically ill patients.
- Traffic management and garbage levels monitoring in smart cities.

A WSN typically contains many spatially distributed self-regulated sensors that cooperatively monitor the environmental conditions, like temperature, pressure, motion, sound, vibration, pollution, etc. Each node of a WSN usually contains an energy source most often cells/battery, a radio transceiver or some other wireless communication device, and a small microcontroller. These sensor nodes perform three main activities: sensing, processing raw sensed data and communicating that data. Some of the most common sensor devices, deployed in sensor network as sensor nodes, are camera sensor, accelerometer sensor, a thermal sensor, microphone sensor, and so forth. The above application areas are increasingly requiring real-time data to make decisions, deal with the changing environment and user conditions, and create value-added services. Data may be obtained at different quality levels. Section 2 describes the concept of data quality and the different attributes that characterize the quality of data. Data-as-a-Service provisioning raises challenges like aggregation of sensed data in a structured format, discovery, and selection of appropriate DaaS providers for data delivery to consumers.

The primary challenge in offering high-quality services like the ones we described above is the ability to store, process, and analyze sensed data obtained from different sources including environmental sensors, field sensors, IoT devices, and users' mobile devices. These sources typically generate massive amounts of data. To generate higher value information, aggregation of data from multiple sources is highly desirable. For instance, the initiative to transform many cities worldwide into smart cities includes transforming government services to smart services such as smart life, smart transportation, smart society, smart economy, smart governance, and smart environment. These smart services require aggregating data from different city stakeholders. One of the challenges of this initiative is raising awareness and changing people's perception towards smart services and launching more high-quality services to change people's lives for the better using high-quality data obtained from various data providers. Many of the existing data provisioning systems are suffering from the lack of scalability, extensibility, and interoperability. We believe that bringing data management and delivery to the cloud by deploying DaaS services on the cloud has the potential to deal with the above concerns as they will benefit from the power of the cloud regarding cloud storage abundance, computing power, up and down scalability, and elasticity.

To cope with the issues of DaaS delivery and DaaS providers' selection, we propose a framework for DaaS provisioning, which is relying on using a component called DaaS Agency, and deploying DaaS services on the cloud. The DaaS Agency mediates between data consumers and DaaS services using a publish/subscribe model. We believe that our approach will take advantage of the power of the cloud concerning elasticity, storage abundance, and scalability. Moreover, we describe an algorithm for the selection of DaaS services based on the QoD they can offer. The algorithm takes into account the QoD requirements of data consumers for each data to which they subscribe with the DaaS Agency.

The remainder of this paper is organized as follows. Section 2 provides background on cloud-sensor and data quality attributes. Section 3 provides a literature review identifying Data-as-a-Service and DaaS service provisioning. Section 4 describes our proposed framework, and Sect. 5 describes the proposed algorithm for DaaS service selection. Section 6 discusses the challenges of the approach and identifies implications for future research. Finally, Sect. 7 summarizes and concludes the paper while highlighting future work.

2 Background

2.1 Cloud Sensor Integration

Cloud computing is a fast-growing service-provisioning model for computing services delivered over the Internet and remote data centers [1]. It has established itself in the next generation of IT industry and business. Cloud computing promises reliable software, hardware, and services. This model permits the delivery of dynamically scalable and virtualized services called cloud services. Cloud services span a variety of IT functions including computation, storage, database, and application services. Users can access to cloud services through a Web interface or via an API. With cloud computing, organizations can focus on the core of their businesses without having to worry about the issues of infrastructure management and availability of resources [2].

The basic cloud services delivery models are IaaS, PaaS, and SaaS.

- Infrastructure-as-a-Service (IaaS), such as Amazon's EC2, allows organizations to rent storage space and computing resources that can be accessed over a private network or across the Internet.
- Platform-as-a-Service (PaaS), such as Google's Apps Engine and Microsoft Azure, allows organizations to develop their business applications in a cloud environment using software tools provided by the cloud provider. Management and maintenance of the cloud infrastructure including severs and their operating systems is the responsibility of the cloud provider.
- Software-as-a-Service (SaaS), such as Google Docs and Gmail, allows users to access the cloud service through a Web interface or via an API.

Cloud computing offers businesses many benefits. It allows them setting up a virtual office that gives them the flexibility of connecting to their business anywhere, anytime using a growing number of Internet-enabled devices (e.g. smartphones, tablets) and accessing to their data. There are many benefits to moving business operations to the cloud such as capital-expenditure free, reduced IT costs, scalability, disaster recovery and business continuity, collaboration efficiency, the flexibility of work practices, and access to automatic updates.

Given the significant amount of data sensed by sensors in a wireless sensor network, local servers can't cope with large volumes of data. Therefore, integrating sensor-based environment and cloud computing is increasingly becoming appealing to take advantage of the above benefits of cloud computing. Many efforts refer to this integration as a Sensor-Cloud platform [3, 4]. This platform aims to cope with many of the shortcomings of WSNs like storage capacity and the required processing power of data collected by the various sensor nodes of the network.

2.2 Data Quality Attributes

Concern for data quality has increased in recent years for the following reasons:

- Increased generation of data by sensor nodes, by mobile devices, and by social networks. Historically, production of large volumes of data was limited to governmental agencies and large corporations because of the high cost of its storage and processing.
- Increased use of data mining as a decision-support tool. Mining high-quality data has proven to be beneficial for organizations as it helps them identifying the needs of their customers to respond to them adequately.
- Increased proliferation of data providers offering Data-as-a-Service fuelled by the progress in cloud computing and the development of standards and protocols for data exchange.
- The diversity of the sensors used to sense different parameters (temperature, pressure, etc.) might be the subject of a reliability issue. Besides, the process by which raw data is aggregated might introduce additional biases.

To deal with the above matters, data has been characterized by some properties referred in the literature as quality-of-Data (QoD). QoD significantly impacts the behavior of DaaS services. Low-quality data increases the risk of making wrong decisions. QoD can be distinguished in space, time, and data type. For each of these dimensions, several indicators of quality including *precision*, *spatial resolution*, *temporal resolution*, *freshness*, *accuracy*, *consistency*, and *completeness* can be identified. Precision represents the granularity with which data describes a real world parameter. Spatial resolution characterizes the accuracy with which the physical area, to which an instance of data is applicable, is expressed. Temporal resolution is the period during which a single data instance is relevant. Freshness denotes the time that elapses between the sensing or the determination of data instance and its delivery to a requester.

3 Related Work

Over the last few years, smart cities' organizations, modern industry, and businesses create data at an unprecedented pace. Generated data is typically stored in multiple databases and accessed by users. Furthermore, this phenomenal growth in data generation is outpacing the ability of traditional software tools, techniques, and processes to use that data in an efficient manner. Data as a Service (DaaS) has the potential to provide cheaper, faster, and easier access to data. DaaS is an emerging cloud computing service model to making useful data available to users as a service through a network in a timely and cost efficient manner. DaaS can use new approaches for data access to corporate data centers, new architectural designs such as private clouds, or completely outsourced models within a public cloud. As data is the value of this model, it is essential to manage and process the massive amount of heterogeneous generated data to permit timely access to critical information.

In today businesses, getting copies of chunks of data stored in corporate databases by a business analyst, to assess the requirements of the organization on its operations and functions, might involve a process and several teams (storage, network, DBA). This process can last for days before getting the required data and might induce time and cost. By the time the business analyst gets her required data, the assessment results may no longer be relevant, requiring a data refresh and again repeating the process for several days. Of course, most managers are familiar with the time and cost of provisioning and updating non-production database copies, and so request approval either may not happen or will only occur after significant discussion and delay. In practice, the management approval/denial process adds yet another layer of annoyance, wasting time and delaying potential revenue benefit.

Several research efforts investigated several concerns of this emerging DaaS service model. Truong et al. [5] analyzed several concerns for DaaS, including data quality, auditing, business, IPR and legal, and service location, that should be well-specified and publishable so that DaaS can be searched, evaluated and selected. They considered that research efforts mainly focused on system perspective to make the data available via the service, but did not examine the concerns associated with the data offered by the service. They proposed and implemented a model for tackling the issues related to the selection and utilization of DaaSs. Mrissa et al. [6] described some DaaS-related problems that conventional service-oriented technologies do not handle in an appropriate way. They raised the issue of having a clear distinction between the roles of data providers and service providers and having better management of their privacy requirements. Also, they showed the limitations of the privacy models of traditional Web service concerning the privacy policies concerning data resources, and their limitations in dealing with user permissions and obligations and unstructured data resources. They proposed a model for representing privacy policies and annotating service descriptions formats (WSDL and REST) with privacy policies. Truong et al. [7] investigated the issue of DaaS agreements, which received little consideration from relevant stakeholders. They analyzed the necessary steps and interactions among data consumers, DaaS service providers, and data providers in the process of exchanging data agreements. They considered that given the complexity of data concerns and the rising trend of aggregating data from multiple sources, data agreements need to be associated with data discovery, its retrieval, and utilization. They described a service that allows composing, managing, and analyzing data agreements for DaaS in data marketplaces and cloud environments.

Zhang et al. [8] presented the design and a prototype of a Web 2.0 platform, which aims to provide sensor data as a service, permit users to discover reusable data and data analysis tools, and integrate them into value-added workflows. They considered that decoupling the storage and management of sensor data from platform-oriented metadata permits the handling of both discrete and streamed sensor data. Terzo et al. [9] proposed a DaaS approach to support intelligent sharing and processing of large data collections. The aim of the proposed approach is abstracting the data location, and fully decoupling the data and its processing. Besides, the authors aimed at building a cloud-based platform that implements DaaS to provide support to large communities of users who need data sharing, accessing, and processing.

Our work is in line with these efforts. It aims to have a framework to facilitate for data consumers the process of selecting DaaS services with high-quality data and permit to DaaS providers to advertise their services. The implications of this are that data quality should be considered by DaaS providers from the first phases of acquiring data either from sensor networks or open data sources.

4 A Framework for DaaS Provisioning

In every business, agencies emerge to mediate between consumers and providers. For data delivery, a DaaS agency may be used to decouple data consumers from DaaS services. Figure 1 depicts our framework for DaaS provisioning.

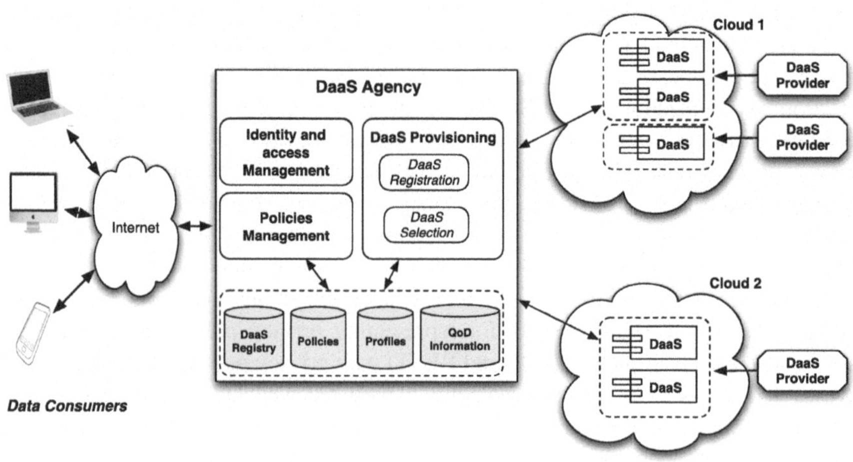

Fig. 1 Framework for cloud-based data provisioning

4.1 DaaS Agency

The DaaS agency is a mediator service that decouples data consumers from DaaS services. It is in charge of registering DaaS services and managing subscriptions of data consumers to receive some data types. DaaS services publish their newly acquired data to the DaaS agency, which notifies data consumers about the availability of newly acquired data. In parallel with this publish/subscribe model for data provisioning, the DaaS agency implements a regular on-demand request/response model wherein it requests up-to-date data from DaaS services once a data consumer requires information for a given data type. Therefore, the DaaS agency either pulls data from DaaS services or let them push updated data. DaaS services might deliver data to data consumers with various quality-of-data (QoD). Therefore, the DaaS Agency is in charge of selecting appropriate DaaS services to provide data to which a data consumer has subscribed. Data may be delivered to the same consumer by several DaaS services. Each one may provide a piece of data (a data type) that the consumer requires.

4.2 DaaS Services

As we have mentioned earlier in the related work section, high-level data is typically obtained from DaaS services that aggregate raw data sensed by sensors and mobile devices. Given the massive amount of data processed and stored by DaaS services and the broad acceptance of the cloud computing technology, data providers now can leverage their services by deploying them on the cloud.

Figure 2 depicts the process of data acquisition and the deployment of DaaS services on the cloud to provide high-quality data to data consumers. Raw data sensed by various devices and sensors is cleaned, processed, and aggregated by the Data Aggregator components in a structured format, and then uploaded to the cloud-based DaaS services. Another source of data, which is becoming an increasing trend in smart cities, is open data providers. Open Data is a new paradigm that defines the publication of government or private sector data without any copyright restrictions. Data can be freely accessed, used, and shared. The data is formatted so that citizens and enterprises can reuse it to create new innovative services or applications. Today, in many countries the Open Data movement is driving a significant change in the relationships between governments, citizens, and businesses. Their primary objective is keeping increased innovation in the public sector [10]. The way to catalyze this change is by releasing information on Open Data portals. Mobile applications are becoming the most promising tools for using such published data as well as real-time data generated by sensors.

The most pertinent Open Data portals, like those from the English, American, Austrian, and Dutch governments or the PublicData.eu portal, use the Comprehensive Knowledge Archive Network—CKAN—as the Open Source software

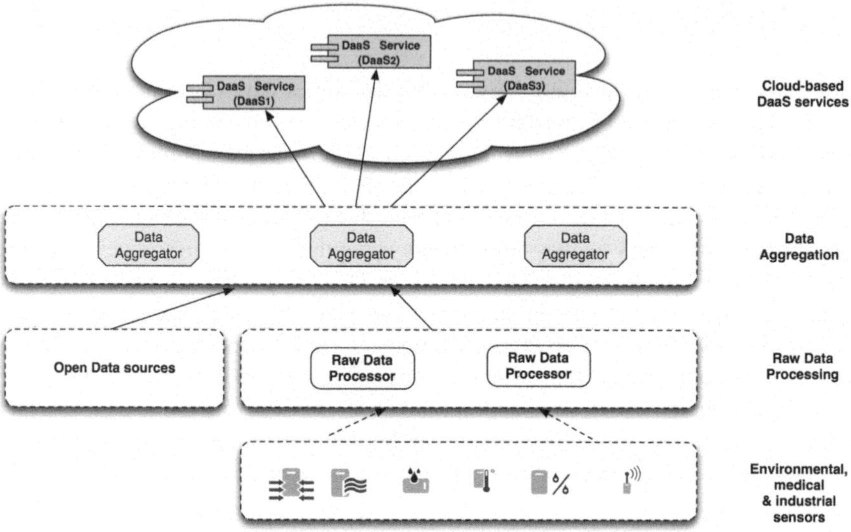

Fig. 2 Deployment of high-quality data on cloud-based DaaS services

platform for support [11]. This platform is a public registry of datasets and metadata, developed by the Open Knowledge Foundation (OKFn). A CKAN platform offers tools to streamline publishing and finding datasets, storing and managing data and metadata, and adapting and extending them through APIs.

One of the advantages of deploying DaaS services on the cloud is the economy of scale. By using the cloud infrastructure provided by a cloud vendor, a data provider can offer better, cheaper, and more elastic and reliable services than is possible within its premises. The net benefit for data consumers is the ability to receive high-quality data in a cost effective manner.

4.3 Interaction Model

Figure 3 shows the interactions among the components of the framework. The DaaS Agency acts as an intermediary between publishers (DaaS services) and subscribers (data consumers) on a collection of data types. A data consumer invokes the *subscribe()* method of the DaaS Agency to register its interest in receiving updates on some data types (such as location and temperature). If the processing of *subscribe()* is successful, the DaaS agency returns a subscription ID to the data consumer. Similarly, a DaaS service invokes *registerDaaSService()* of the DaaS Agency to register its interest to publish some types of data through the DaaS agency. If the processing of that method is successful, the DaaS Agency returns a registration ID to the DaaS service.

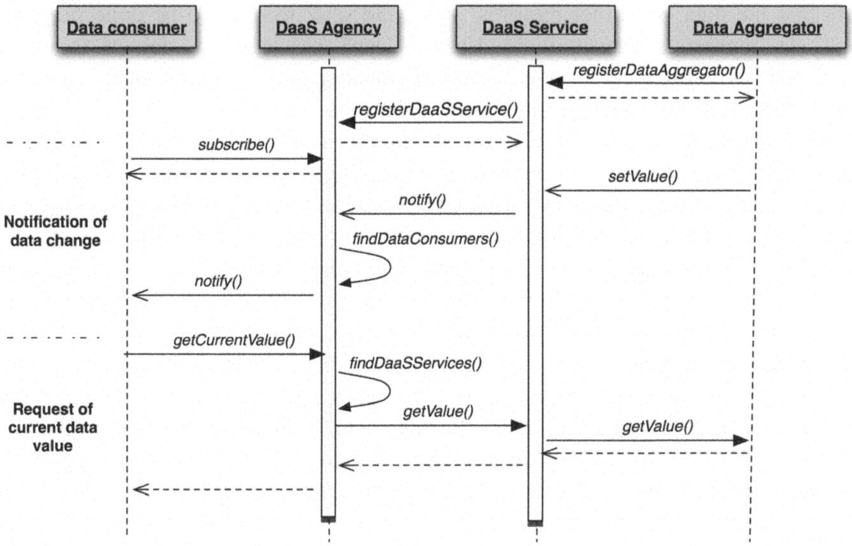

Fig. 3 Diagram of interactions among the framework components

The DaaS Agency receives notifications of data change through its *notify()* method that a DaaS service invokes. It, then, informs a data consumer about data change by invoking its *notify()* method. Furthermore, a data consumer may request the current value of a given data by invoking *getCurrentValue()* of the DaaS Agency. The DaaS Agency forwards the request to DaaS services that are providing that data requested by the data consumer. The DaaS Agency has also two additional methods *findDataConsumers()* and *findDaaSServices()* that are self-invoked. The first one is invoked to get the list of data consumers that have subscribed to a given data once a notification of data change has been received for that data. The second one is invoked to get the list of DaaS services that are publishing the data requested by a data consumer that has invoked *getCurrentValue()*. A Data Aggregator can register at a DaaS service by specifying the data it aggregates. Once registered, a Data Aggregator can submit the current value for a given data by invoking the *setDataValue()* method of the DaaS service. When the data value is changed in the DaaS service, the *notify()* method of the DaaS Agency is triggered to notify all subscribers of that data.

5 DaaS Services Selection

As we have stated earlier, the DaaS Agency is in charge of selecting suitable DaaS services to deliver data to which a data consumer subscribed. Data may be provided to the same consumer by several DaaS services. Each one may provide a piece of data that the data consumer requires. Thus, the selection has to be done per data type. As numerous potential cloud-based DaaS services can deliver the data requested by a consumer, it is essential to consider only potential DaaS services, which can fulfill the QoD requirements of the data consumer.

Let $D = \{d_1, d_2, \ldots, d_C\}$ be the list of data types to which a data consumer has subscribed by showing its interest in receiving such data. Let $DS = \{DS_1, DS_2, \ldots, DS_K\}$ be the list of cloud-based DaaS services, which subscribed with the *DaaS Agency*. These services typically provide data with different QoD. We assume that QoD indicators are in normalized form with values between 0 and 1. A value of 1 means highest quality and 0 means the lowest quality. For example for the "freshness" quality indicator, one means that data sources have sensed the information in the last minute, and zero means that they have sensed it in the last 10 min.

When subscribing to data type, a data consumer specifies the min values of the QoD indicators that can be tolerated. For example, a data consumer subscribed to the "location" data might require a minimum value of 80 % for the "freshness" quality indicator, 93 % for the "temporal resolution" quality indicator. Let $Q = \{Q_1, Q_2, \ldots, Q_N\}$ be the list of QoD indicators that are of interest to the data consumer. The minimum QoD requirements that the data consumer tolerates for a given data type d_j, with $1 \le j \le C$, are expressed by the following vector: $M_j = \{m_{1,j}, m_{2,j}, \ldots, m_{N,j}\}$ $0 \le m_{i,j} \le 1$, with $1 \le j \le C$ and N is the cardinality of Q. Therefore, the matrix T expresses the total QoD requirements of the data consumer, for all its subscribed data types and all QoD indicators considered in the system.

$$T = \begin{matrix} d_1 \\ \\ d_2 \\ \cdots \\ \cdots \\ d_C \end{matrix} \begin{bmatrix} m_{1,1} & m_{2,1}\cdots & \cdots & m_{N,1} \\ \vdots & & & \vdots \\ m_{1,2} & m_{2,2}\cdots & \cdots & m_{N,2} \\ \vdots & & & \\ m_{1,C} & m_{2,C}\cdots & \cdots & m_{N,C} \end{bmatrix} \qquad D_r = \begin{matrix} d_1 \\ \\ d_2 \\ \cdots \\ \cdots \\ d_C \end{matrix} \begin{bmatrix} q^r_{1,1} & q^r_{2,1}\cdots & \cdots & q^r_{N,1} \\ \vdots & & & \vdots \\ q^r_{1,2} & q^r_{2,2}\cdots & \cdots & q^r_{N,2} \\ \vdots & & & \\ q^r_{1,C} & q^r_{2,C}\cdots & \cdots & q^r_{N,C} \end{bmatrix}$$

A zero value in any value of the matrix T means that the data consumer has no constraint on the corresponding QoD indicator. The goal of our proposed selection algorithm is to find for each data type d_j, to which the data consumer subscribed, a suitable DaaS service from the set DS capable of fulfilling the minimum quality requirements of the data consumer. The matrix D_r expresses the QoD offer of a DaaS service DS_r. DS_r is suitable for a data type d_j if the following condition holds:

$$0 \le m_{i,j} \le q^r_{i,j} \le 1 \text{ for } 1 \le i \le N \text{ and } 1 \le j \le C$$

The data consumer might set relative weights for the QoD indicators. He may even set weights for each data type to which it subscribed. For example, for the "location" data, more weight may be given to the "spatial resolution" indicator than to the other QoD indicators. W is the weight matrix.

$$W = \begin{matrix} d_1 \\ d_2 \\ \dots \\ \dots \\ d_C \end{matrix} \begin{bmatrix} w_{1,1} & w_{2,1}\cdots & \cdots & w_{N,1} \\ \vdots & & & \vdots \\ w_{1,2} & w_{2,2}\cdots & \cdots & w_{N,2} \\ \vdots & & & \\ w_{1,C} & w_{2,C}\cdots & \cdots & w_{N,C} \end{bmatrix} \quad U_r = \begin{matrix} d_1 \\ d_2 \\ \dots \\ \dots \\ d_C \end{matrix} \begin{bmatrix} u^r_{1,1} & u^r_{2,1}\cdots & \cdots & u^r_{N,1} \\ \vdots & & & \vdots \\ u^r_{2,1} & u^r_{2,2}\cdots & \cdots & u^r_{N,2} \\ \vdots & & & \\ u^r_{1,C} & u^r_{2,C}\cdots & \cdots & u^r_{N,C} \end{bmatrix}$$

The utility of a given QoD indicator Q_i for a given data type d_j by the DS_r offer is:

$$u^r_{i,j} = w_{i,j} \times q^r_{i,j} \text{ for } 1 \le i \le N \text{ and } 1 \le j \le C \tag{1}$$

U_r is the utility matrix of the DS_r offer, for all QoD indicators and all data types.

Given the data consumer's weight matrix W and her minimum QoD requirements matrix T, the minimum utility matrix is U_{min}.

$$U_{min} = \begin{matrix} d_1 \\ d_2 \\ \dots \\ \dots \\ d_C \end{matrix} \begin{bmatrix} \propto_{1,1} & \propto_{2,1}\cdots & \cdots & \propto_{N,1} \\ \vdots & & & \vdots \\ \propto_{1,2} & \propto_{2,2}\cdots & \cdots & \propto_{N,2} \\ \vdots & & & \\ \propto_{1,C} & \propto_{2,C}\cdots & \cdots & \propto_{N,C} \end{bmatrix}$$

where $\propto_{i,j} = w_{i,j} \times m_{i,j}$ for $1 \le i \le N$ and $1 \le j \le C$

The difference matrix, $U_r - U_{min}$, shows whether DS_r may satisfy or not all QoD requirements for all data types to which the data consumer has subscribed to. A value that is less than zero in this matrix means that DS_r cannot satisfy the QoD requirements for the associated data type and QoD indicator. By reasoning per data type, we consider only DaaS services that can meet the QoD requirements for that data type. The utility per data type d_j for a potential DaaS service DS_r offer is:

$$v^r_j = \sum_{i=1}^{N} u^r_{i,j}. \tag{2}$$

Considering the utility values of all potential DaaS services, we get the following decision matrix:

	DS_1	DS_2	.	DS_K	Max utility	Selected DaaS
d_1	v^1_1	v^2_1	.	v^K_1	…	…
d_2	v^1_2	v^2_2	.	v^K_2	…	…
…	…	…	.	…	…	…
d_C	v^1_C	v^2_C	.	v^K_C	…	…

A utility value in the decision matrix is zero if the DaaS service cannot meet the QoD requirements for a given data type. The maximum utility value of each row j corresponds to the best QoD offer that can fulfill the QoD needs of the data consumer for the data type d_j. The most suitable DaaS service for data type d_j, that we call here $Best_j$, will be the one that maximizes the above utility value, that is:

$$Best_j \leftarrow \max_{1 \le r \le K} \left(v_j^r \right) \tag{3}$$

If no DaaS service satisfies the data consumer QoD requirements for a given data type, then the DaaS Agency might ask the data consumer to lower its QoD expectations.

6 Discussion

Deploying DaaS services on the cloud provide several benefits we discussed in previous sections. However, it also raises numerous issues for data providers including interoperability, security, and performance concerns. The interaction model described in Sect. 4 provides the basis for the development of a DaaS service API that will be used by both the DaaS Agency and data consumers to interact with DaaS services. Heterogeneity of the APIs offered by various DaaS services will be one of the challenges of the approach. The DaaS Agency should, then, be able to interoperate with all these various DaaS services. Security is a significant issue in cloud computing. Care must be taken when designing and implementing a security solution for a DaaS service to keep it as straightforward and efficient as possible. For instance, the DaaS service may have to be integrated with an identity management service. Billing, performance monitoring, managing customers' expectations are also significant concerns among others that a DaaS provider has to handle. The DaaS provider must ensure that its DaaS service is highly available and that its customers can access it. One outage or crash of the service can affect all its customers.

7 Conclusion

High-level data is typically obtained from DaaS services that aggregate raw data sensed by sensors and mobile devices. Given the enormous amount of data processed and stored by DaaS services and the broad acceptance of the cloud computing technology, data providers now can leverage their services by deploying them on the cloud. In this paper, we have presented our proposed framework for cloud-based DaaS provisioning. The framework relies on a DaaS Agency for data dissemination using a publish/subscribe model. DaaS services, deployed on the cloud, can scale up and down, regarding cloud resources they use, according to the demand for data. We have described a preliminary model of interactions, among the components of the framework, and that could be the basis for a DaaS service API. As a future work, we intend to implement a prototype of the framework by considering some real scenarios for data provisioning and implementing a DaaS Agency and few similar cloud-based DaaS services using open-source software tools.

References

1. Armbrust, M., Fox, A., Griffith, R., Joseph, A.D., Katz, R., Konwinski, A., Lee, G., Patterson, D., Rabkin, A., Stoica, I., Zaharia, M.: A view of cloud computing. Commun. ACM **53**, 50–58 (2010)
2. Aceto, G., Botta, A., de Donato, W., Pescapè, A.: Cloud monitoring: a survey. Comput. Netw. **57**(9), 2093–2115 (2013)
3. Alamri, A., Ansari, W.S., Hassan, M.M., Hossain, M.S., Alelaiwi, A., Hossain, M.A.: A survey on sensor-cloud: architecture, applications, and approaches. Int. J. Distrib. Sens. Netw. **2013**, Article ID 917923 (2013)
4. Eggert, M., Häußling, R., Henze, M., Hermerschmidt, L., Hummen, R., Kerpen, D., Pérez, A. N., Rumpe, B., Thißen, D., Wehrle, K.: SensorCloud: towards the interdisciplinary development of a trustworthy platform for globally interconnected sensors and actuators. In: Krcmar, H., Reussner, R., Rumpe, B. (eds.) Trusted Cloud Computing, pp. 203–218. Springer International Publishing, Cham (2014)
5. Truong, H.L., Dustdar, S.: On analyzing and specifying concerns for data as a service. In: Proceedings of the Asia-Pacific Services Computing Conference (APSCC 2009), pp. 87–94 (2009)
6. Mrissa, M., Tbahriti, S.E., Truong, H.L.: Privacy model and annotation for DaaS. In: Proceedings of the IEEE 8th European Conference on the Web Services (ECOWS 2010), pp. 3–10 (2010)
7. Truong, H.L., Dustdar, S., Gotze, J., Fleuren, T., Muller, P., Tbahriti, S.E., Mrissa, M., Ghedira, C.: Exchanging data agreements in the DaaS model. In: Proceedings of the Asia-Pacific Services Computing Conference (APSCC 2011), pp. 153–160 (2011)
8. Zhang, J., Iannucci, B., Hennessy, M., Gopal, K., Xiao, S., Kumar, S., Pfeffer, D., Aljedia, B., Ren, Y., Griss, M., Rosenberg, S., Cao, J., Rowe, A.: Sensor data as a service—a federated platform for mobile data-centric service development and sharing. In: Proceedings of the IEEE International Conference on Services Computing (SCC 2013), pp. 446–453 (2013)
9. Terzo, O., Ruiu, P., Bucci, E., Xhafa, F.: Data as a Service (DaaS) for sharing and processing of large data collections in the cloud. In: Proceedings of the Seventh International Conference on Complex, Intelligent, and Software Intensive Systems (CISIS 2013), pp. 475–480 (2013)
10. De Jong, J.P.J., Vanhaverbeke, W., Kalvet, T., Chesbrough, H.: Policies for Open Innovation: Theory, Framework and Cases. VISION Era-Net, res. Project (2008)
11. Open Knowledge Foundation: Comprehensive Knowledge Archive Network—CKAN. http://ckan.org (2016)

Part V
Special Session 2: Unmanned Aerial Vehicles From Theory to Applications

Coverage and Power Gain of Aerial Versus Terrestrial Base Stations

Mohammad Mahdi Azari, Fernando Rosas, Alessandro Chiumento, Kwang-Cheng Chen and Sofie Pollin

Abstract Aerial stations have been recently recognized as an attractive alternative to provide wireless services to terrestrial users thanks to their superior coverage capability. In this paper, the coverage and power gain that can be achieved by a drone with respect to a terrestrial base station are studied. We address the problem by characterizing the coverage area based on the network outage probability, taking into account the height depending fading and path loss exponent that characterize air-to-ground wireless links. Results show that there exist a unique optimal altitude that provides the largest coverage and power gain, which strikes a fine balance between the path loss, due to the higher altitude, and a reduced influence of the multipath scattering. While numerical evaluations show that even at low altitudes the network gains up to 4x coverage or 20 dB power, the gain achieved at optimal altitude can be higher

Keywords Drone · Base station · Rician fading · Outage probability

1 Introduction

Aerial communication platforms have been increasingly used as an innovative method to develop robust and reliable communication networks. The growing demand for higher data rate and inherent limitations on the legacy terrestrial infrastructure have turned the aerial stations in an eminent technology that urges for prompt develpment. Facebook Aquila Drone [1], Google Loon [2], and the ABSOLUTE [3] are important projects exploiting these solutions. The two former projects provide internet access for users in remote areas, whereas the latter pursues the enhancement of network capacity while tackling public safety issue.

M.M. Azari (✉) · A. Chiumento · S. Pollin
Departement of Electrical Engineering, KU Leuven, Leuven, Belgium
e-mail: mazari@esat.kuleuven.be

F. Rosas · K.-C. Chen
Graduate Institute of Communication Engineering,
National Taiwan University, Taipei, Taiwan

R. El-Azouzi et al. (eds.), *Advances in Ubiquitous Networking 2*, Lecture Notes in Electrical Engineering 397, DOI 10.1007/978-981-10-1627-1_49

Drones, as an aerial platform, can be used in many applications including urban traffic surveillance [4], earth observation and agricultural purposes [5], and disaster recovery scenarios [6]. Moreover, a drone can be efficiently integrated into cellular heterogeneous networks as an aerial base station or a relay [7, 8]. In [7] the airborne relays are utilized to assist the existing cellular networks by providing the emergency coverage. The results are based on the experimental 3G field test and show that the local traffic imbalances can be tackled by improving the throughput in poor coverage zones. The problem of temporary site outage or overload is also addressed in [8] where using a swarm of drones is proposed to offload the traffic to the neighboring cells.

Drones altitudes can significantly influence their performance within a wireless network. In effect, the optimal positioning of a drone has been studied in [9] based on numerical simulations, without providing an analytical approach which could lead to the generalization of the results. Furthermore, the impact of altitude on the coverage area was explored in [10, 11], where unfortunately the effect of multipath and random fading is not taken into account. To the best of our knowledge, there is no study on the relationship between the altitude of a drone and its performance as a wireless base station that takes the impact of the small-scale fading into account, which corresponds to one of the most fundamental features of wireless channels.

In order to address this problem, the Rician fading model is a suitable choice to study the wireless link between a drone and ground terminals [12, 13]. However, it has been reported that the Rician factor is affected by the drone altitude [14]. In fact, it has been shown in [14] that the Rician factor is mainly dependent on the elevation angle, being larger at higher altitudes due to the presence of fewer multipath scatters. As the channel suffers from more path loss when the drone goes higher, this suggests there should exist an optimal point where the effects of fading and pathloss are balanced.

In this paper, we consider an air-to-ground communication network where the channel suffers from both the height-dependent path loss and small-scale fading. We study the coverage area of a drone, which provides wireless access to terrestrial nodes, based on the system outage probability. We analyze the coverage and power gain of the drone at different altitudes and show that there exist unique optimum elevation angle and altitude of the drone which maximize the network's gain. Moreover, the dependency of the coverage gain to the system parameters such as transmit power and SNR requirement are discussed.

The rest of this paper is organized as follows. In Sect. 2 the system model is proposed. The problem is defined in Sect. 3 following with Sect. 4 in which the coverage gain is analyzed. In Sect. 5 the numerical results are discussed. Finally the conclusion is presented in Sect. 6.

2 System Model

We consider a drone as an aerial base station serving terrestrial nodes within its coverage area that is defined by the target outage probability, transmit power and channel properties. The drone is located at an altitude h with the elevation angle θ with regard to a terrestrial node T as illustrated in Fig. 1. The node T is placed at the distance r_{T} from the projection of the drone on the ground O and hence $\theta = \tan^{-1}(h/r_{\mathrm{T}})$. It is to be noted that the special case with $h = 0$, and hence $\theta = 0$, corresponds to a ground-to-ground communication network where the drone plays the role of a terrestrial base station (see Fig. 1).

Let us consider the downlink channel between the drone and the terminal T which is assumed to include both the large scale path loss and small-scale fading effect. Thus the received SNR at node T can be expressed as

$$\Gamma_{\mathrm{T}} = \frac{AP_{\mathrm{D}}}{N_0 {\ell_{\mathrm{T}}}^{\alpha}} \Omega_{\mathrm{T}}, \tag{1}$$

where P_{D} is the transmit power, A is a constant containing the impacts of system parameters such as antenna gain and operating frequency, N_0 is the noise power, ℓ_{T} is the distance between the drone and the terminal T, α is the path loss exponent, and $\Omega_{\mathrm{T}} \in [0, \infty)$ is the small-scale fading power which is a variable such that $\bar{\Omega}_{\mathrm{T}} = 1$. Since A, P_{D} and N_0 are independent of θ, we define a θ-independent link budget γ_{D} as

$$\gamma_{\mathrm{D}} \triangleq \frac{AP_{\mathrm{D}}}{N_0}. \tag{2}$$

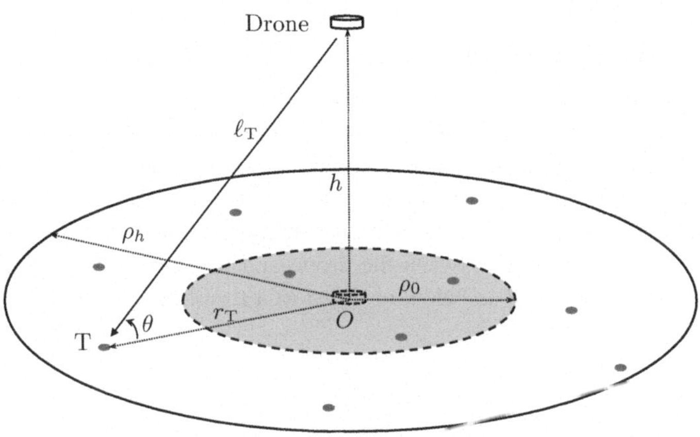

Fig. 1 A typical air-to-ground communication system with a number of ground nodes and a drone. The coverage radius of the drone on the ground is ρ_0 while its coverage at an altitude h is ρ_h

In presence of both LoS and multipath scatters at the receiver, the Rician distribution is an appropriate choice to model the small-scale fading. Using this model, Ω_T follows a non-central chi-squared probability distribution function (PDF) given by

$$f_{\Omega_T}(\omega) = \frac{(K+1)e^{-K}}{\overline{\Omega}_T} e^{\frac{-(K+1)\omega}{\overline{\Omega}_T}} I_0\left(2\sqrt{\frac{K(K+1)\omega}{\overline{\Omega}_T}}\right), \quad \omega \geq 0. \tag{3}$$

Above, $I_0(\cdot)$ is the zero order modified Bessel function of the first kind, and K is the Rician factor which is defined as the ratio of the LoS power to the power of multipath components at T. Indeed, the Rician factor represents the severity of the fading in which lower K corresponds to more severe one. With $K = 0$ the fading is reduced to a Rayleigh distribution and with $K \to \infty$ the channel converges to an AWGN.

Following the above-mentioned fact, since the predominance of the LoS might vary across different drone elevation angles, the Rician factor corresponding to a terrestrial node at different locations might not be the same and should be modeled as function of θ. To this end, we introduce a general function of $K = K(\theta)$ showing the dependency of the Rician factor on the elevation angle. The results presented in this paper are valid for any increasing function $K(\cdot)$.

Similar to the Rician factor, one can observe that the path loss exponent is also different for diverse elevation angles. In fact, the path loss exponent at $\theta = 0$ corresponding to a ground-to-ground communication link is the largest, while a free space communication link can be considered at $\theta = \pi/2$ associated to the smallest path loss exponent. Therefore, we introduce a decreasing function $\alpha = \alpha(\theta)$ showing the dependency between path loss exponent and the elevation angle.

3 Problem Formulation

In this section we characterize the problem of the coverage and power gain of a drone with respect to a terrestrial base station. In this regard, the optimum height of the drone which yields the maximum coverage or power gain is investigated by characterizing the coverage area based on the outage probability.

The communication link between the drone and the node T is in outage if the instantaneous channel SNR is below or equal to a threshold ξ. Therefore, the link outage probability is defined as

$$\mathcal{P}_{out} \triangleq Pr\{\Gamma_T \leq \xi\}, \tag{4}$$

where $Pr\{\cdot\}$ indicates the probability and Γ_T is as given in (1). Using (1) and (3), and noting that $\theta = \tan^{-1}(h/r_T)$ and $\ell_T = \sqrt{r_T^2 + h^2}$, the outage probability will be a function of h and r_T which can be rewritten as

$$\mathcal{P}_{out}(h, r_{\mathrm{T}}) = \mathbb{P}\left(\frac{AP_{\mathrm{D}}}{N_0 {\ell_{\mathrm{T}}}^{\alpha(\theta)}} \Omega_{\mathrm{T}} \leq \xi\right) \tag{5}$$

$$= 1 - Q\left(\sqrt{2K(\theta)}, \sqrt{2\xi\,[1 + K(\theta)]\,{\ell_{\mathrm{T}}}^{\alpha(\theta)}/\gamma_{\mathrm{D}}}\right), \tag{6}$$

where γ_{D} is defined in (2), and $Q(\cdot, \cdot)$ is the first order Marcum Q–function.

The coverage area of the drone at an altitude h is a geographical region characterized by the following relation

$$\mathcal{P}_{out}(h, r_{\mathrm{T}}) \leq \varepsilon, \tag{7}$$

where ε is the target outage constraint determined by the required quality of service. This region is a disk centered at O whose radius, denoted as ρ_h, can be obtained by solving the equation

$$\mathcal{P}_{out}(h, \rho_h) = \varepsilon. \tag{8}$$

By considering $h = 0$, ρ_0 is the radius of the coverage area that can be supported by the terrestrial base station. At any altitude of h, the coverage gain of the drone is defined as

$$\mathcal{G}_h^C = \frac{\rho_h}{\rho_0}. \tag{9}$$

Assuming that the maximum coverage radius is $\tilde{\rho}_h$, which is achieved at the optimal height $\tilde{h}$, the maximum gain $\tilde{\mathcal{G}}_h^C$ is defined as

$$\tilde{\mathcal{G}}_h^C = \frac{\tilde{\rho}_h}{\rho_0}. \tag{10}$$

Similarly, the power gain $\mathcal{G}_h^P$ is defined as the additional power (in dB) needed for the terrestrial base station to achieve the same coverage radius as that of the drone at altitude h, i.e. ρ_h. The power gain is numerically studied in Sect. 5.

4 Coverage Gain Analysis

In order to find analytical expression for the coverage gain, we investigate Eq. (8) to find the radius ρ_h for every altitude. Using (6) and (8) can be expressed as

$$Q\left(\sqrt{2K(\theta)}, \sqrt{2\xi\,[1 + K(\theta)]\,{\ell_{\mathrm{T}}}^{\alpha(\theta)}/\gamma_{\mathrm{D}}}\right) = \tau, \tag{11}$$

where $\tau = 1 - \epsilon$. By taking an auxiliary parameter defined as

$$x_\theta = \sqrt{2K(\theta)}, \tag{12}$$

Equation (11) is equivalent to

$$\sqrt{\frac{2\xi\,[1 + K(\theta)]\,\ell_{\mathrm{T}}{}^{\alpha(\theta)}}{\gamma_{\mathrm{D}}}} = Q^{-1}(x_\theta, \tau) \triangleq y_\theta, \tag{13}$$

where $Q^{-1}(x_\theta, \tau)$ is the inverse Marcum Q–function with respect to its second argument y_θ. Now, using (12) and (13) one can write

$$\ell_{\mathrm{T}} = \left(\frac{\gamma_{\mathrm{D}}\, y_\theta^2}{\xi\,[2 + x_\theta^2]}\right)^{\frac{1}{\alpha(\theta)}} \triangleq \Psi(\theta). \tag{14}$$

Thus, knowing that $h = \ell_{\mathrm{T}} \sin(\theta)$ and $\rho_h = \ell_{\mathrm{T}} \cos(\theta)$, one obtains

$$h = \Psi(\theta) \cdot \sin(\theta), \tag{15}$$

and

$$\rho_h = \Psi(\theta) \cdot \cos(\theta). \tag{16}$$

From (15) and (16), by varying θ from 0 to $\pi/2$, the coverage radius in every altitude is attained. For instance, at $\theta = 0$ one sees that $h = 0$ and $\rho_h = \Psi(0)$, and $\theta = \pi/2$ yields $h = \Psi(\pi/2)$ and $\rho_h = 0$. The optimum value of θ at which the coverage radius ρ_h reaches its maximum is denoted as $\tilde{\theta}$, so that $\tilde{h} = \Psi(\tilde{\theta}) \sin(\tilde{\theta})$ and $\tilde{\rho}_h = \Psi(\tilde{\theta}) \cos(\tilde{\theta})$. Therefore, the maximum coverage gain in (10) is obtained as

$$\tilde{\mathcal{G}}_h^C = \frac{\Psi(\tilde{\theta})}{\Psi(0)} \cos(\tilde{\theta}). \tag{17}$$

Using (14) and (17), provided that the dependency of $\tilde{\theta}$ on γ_{D} and ξ is negligible, we can write

$$\tilde{\mathcal{G}}_h^C \propto \left(\frac{\gamma_{\mathrm{D}}}{\xi}\right)^{\left[\alpha(\tilde{\theta})^{-1} - \alpha(0)^{-1}\right]}. \tag{18}$$

This result shows that the maximum coverage gain increases with the transmit power, however it decreases as the SNR threshold increases.

5 Numerical Results

In this section we provide the numerical evaluations by adopting the dependency of the Rician factor to the elevation angle as

$$K(\theta) = ae^{b\theta}, \tag{19}$$

where a and b are determined by the environment and system parameters such as carrier frequency. Denoting $\kappa_0 = K(0)$ and $\kappa_{\pi/2} = K(\pi/2)$, a and b can be written as

$$a = \kappa_0, \quad b = \frac{2}{\pi} \ln\left(\frac{\kappa_{\frac{\pi}{2}}}{\kappa_0}\right). \tag{20}$$

We also consider a linear dependency of the path loss exponent to the elevation angle as

$$\alpha(\theta) = c\theta + d, \tag{21}$$

where c and d are related to $\alpha_0 = \alpha(0)$ and $\alpha_{\pi/2} = \alpha(\pi/2)$ by

$$c = \frac{2}{\pi}\left(\alpha_{\frac{\pi}{2}} - \alpha_0\right), \quad d = \alpha_0. \tag{22}$$

We performed the numerical evaluations by assuming $\kappa_0 = 4$ dB, $\kappa_{\pi/2} = 15$ dB, $\alpha_0 = 3$, $\alpha_{\pi/2} = 2$, and $\epsilon = 0.01$ unless otherwise is mentioned.

Figure 2 shows that there exists an optimum elevation angle, i.e. $\tilde{\theta}$, which maximizes the coverage radius. From the figure, $\tilde{\theta} \simeq 72°$ for all curves meaning that the optimum elevation angle is almost independent from the transmit power and the SNR threshold. As can be seen, the coverage radius increases with γ_{D} and decreases with ξ. Note that by varying θ, the drone altitude is also changing as is expressed in (15).

The coverage gain achieved in every altitude is demonstrated for $\xi = 3$ dB in Fig. 3, which shows the existence of an optimum altitude $\tilde{h}$ with the maximum gain $\mathcal{G}_{\tilde{h}}^C$. Indeed, increasing the drone altitude leads to larger path loss since the link length grows. However, increasing h results in larger Rician factor, which is interpreted as less negative effect of multipath scatters at the receiver. The figure shows that at some point, i.e. $h = \tilde{h}$, these two effects are balanced which gives the maximum coverage gain. Although $\tilde{h}$ is relatively high at which a drone could be deployed, the coverage gains at lower altitudes are still significant such that a linear dependency between $\mathcal{G}_h^C$ and h can be found from the figure for all examined γ_{D}. For instance, at $h = 500\ m$, a gain of 3x can be obtained.

Figure 4 confirms the result of (18) where it is claimed that the maximum coverage gain is proportional to $\gamma_{\text{D}}{}^{[\alpha(\tilde{\theta})^{-1} - \alpha(0)^{-1}]}$ and $\xi^{-[\alpha(\tilde{\theta})^{-1} - \alpha(0)^{-1}]}$. In this plot, we have $\tilde{\theta} \simeq 72°$, $\alpha(\tilde{\theta}) \simeq 2.2$, and

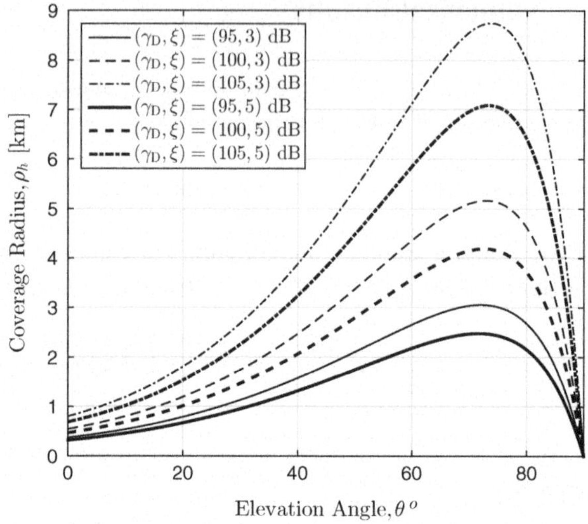

Fig. 2 There is an optimum elevation angle $\tilde{\theta}$ which maximizes the coverage radius ρ_h at $h = \tilde{h}$. The examined cases show that $\tilde{\theta}$ is independent from the link budget γ_D and SNR requirement ξ

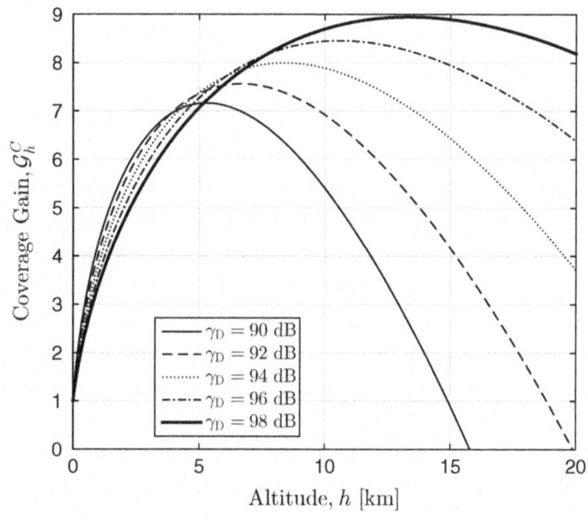

Fig. 3 The coverage gain $\mathcal{G}_h^C$ of an aerial base station at different altitudes h, for various link budgets γ_D defining the terrestrial coverage area ρ_0

$$\tilde{\mathcal{G}}_h^C \simeq 0.66 \left(\frac{\gamma_D}{\xi}\right)^{\left[\alpha(\tilde{\theta})^{-1} - \alpha(0)^{-1}\right]} \tag{23}$$

for all examined cases.

The power gain illustrated in Fig. 5 for $\xi = 5$ dB shows the profound role of a drone's altitude in terms of power saving. This figure illustrates the extra amount of transmit power that a terrestrial base station needs in order to reach the same coverage area as that of a drone at the height h. The optimum height at which the

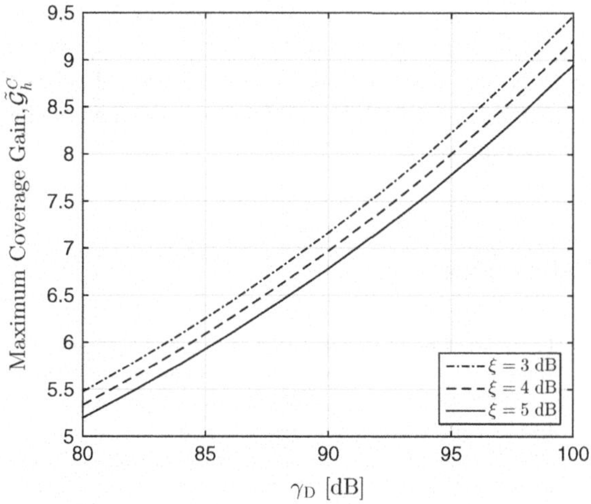

Fig. 4 The maximum coverage gain $\tilde{\mathcal{G}}_h^C$ of an aerial base station is proportional to $\gamma_D^{[\alpha(\tilde{\theta})^{-1}-\alpha(0)^{-1}]}$ and $\xi^{-[\alpha(\tilde{\theta})^{-1}-\alpha(0)^{-1}]}$ as is shown in Eq. (18)

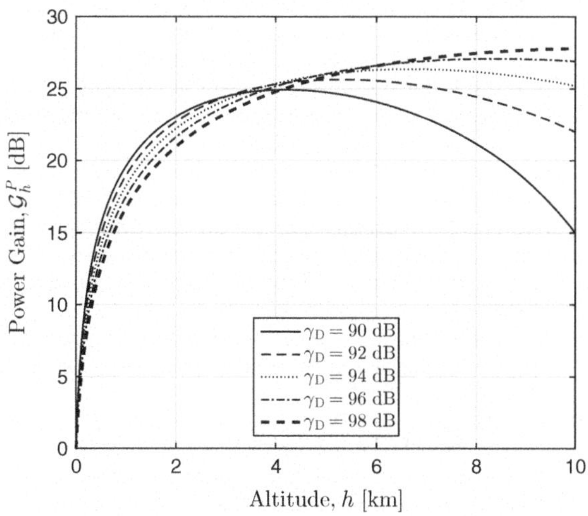

Fig. 5 Power gain of an aerial base station compared to a terrestrial base station with the same coverage ρ_h

maximum power gain is achieved is the same as the altitude of a drone yielding the maximum coverage gain. The aerial base station can gain up to 20 dB at $h = 1$ km in the examined cases.

6 Conclusion

We studied the achievable performance of a drone that acts as an aerial base station in a cellular system, comparing it with a terrestrial base station. Our results suggest that important gains in terms of coverage area or power savings can be achieved by positioning the drone at an adequate height. Numerical evaluations showed that these gains are a consequence of the reduced multipath fading and path loss exponent of air-to-ground wireless links. It is to be noted that drones can also be more vulnerable to interference, which is an effect that was not considered in this work and is part of our ongoing work.

References

1. Facebook: Connecting the World from the Sky. Facebook, Technical report (2014)
2. Katikala, S.: Google project loon. InSight: Rivier Acad. J. **10**(2) (2014)
3. ABSOLUTE (Aerial Base Stations with Opportunistic Links for Unexpected and Temporary Events). http://www.absolute-project.eu
4. Wu, C., Cao, X., Lin, R., Wang, F.: Registration-based moving vehicle detection for low-altitude urban traffic surveillance. In: 8th World Congress on Intelligent Control and Automation (WCICA), pp. 373–378. IEEE (2010)
5. Berni, J.A., Zarco-Tejada, P.J., Suarez, L., Fereres, E.: Thermal and narrowband multispectral remote sensing for vegetation monitoring from an unmanned aerial vehicle. J. IEEE Trans. Geosci. Remote Sens. **47**(3), 722–738 (2009)
6. Qiantori, A., Sutiono, A.B., Hariyanto, H., Suwa, H., Ohta, T.: An emergency medical communications system by low altitude platform at the early stages of a natural disaster in indonesia. J. Med. Syst. **36**(1), 41–52 (2012)
7. Guo, W., Devine, C., Wang, S.: Performance analysis of micro unmanned airborne communication relays for cellular networks. In: 9th International Symposium on Communication Systems, Networks & Digital Signal Processing (CSNDSP), pp. 658–663. IEEE Press (2014)
8. Rohde, S., Wietfeld, C.: Interference aware positioning of aerial relays for cell overload and outage compensation. In: Vehicular Technology Conference (VTC Fall), pp. 1–5. IEEE Press (2012)
9. Kosmerl, J., Vilhar, A.: Base stations placement optimization in wireless networks for emergency communications. In: IEEE International Conference on Communications Workshops (ICC), pp. 200–205 (2014)
10. Mozaffari, M., Saad, W., Bennis, M., Debbah, M.: Drone small cells in the clouds: design, deployment and performance analysis (2015). arXiv:1509.01655
11. Al-Hourani, A., Kandeepan, S., Lardner, S.: Optimal LAP altitude for maximum coverage. J. IEEE Wirel. Commun. Lett. **3**(6), 569–572 (2014)
12. Kandeepan, S., Gomez, K., Reynaud, L., Rasheed, T.: Aerialterrestrial communications: terrestrial cooperation and energy-efficient transmissions to aerial base stations. J. IEEE Trans. Aerosp. Electron. Syst. **50**(4), 2715–2735 (2014)
13. Matolak, D.W.: Air-ground channels & models: comprehensive review and considerations for unmanned aircraft systems. In: IEEE Aerospace Conference, pp. 1–17 (2012)
14. Shimamoto, S., et al.: Channel characterization and performance evaluation of mobile communication employing stratospheric platforms. J. IEICE Trans. Commun. **89**(3), 937–944 (2006)

Ultra-Reliable IEEE 802.11 for UAV Video Streaming: From Network to Application

Bertold Van den Bergh, Alessandro Chiumento and Sofie Pollin

Abstract Civilian application of Unmanned Aerial Vehicles (UAVs) are becoming more and more widespread. An important question is how ultra-reliable communication to and from the drone will be organised. At the moment complex and difficult to deploy point-to-point proprietary wireless links are often used. To enable ubiquitous usage of UAVs it is necessary to have a simple, reliable and widely available data link, such as IEEE 802.11. In this work we examine if infrastructure to control UAVs could be built from IEEE 802.11 access points already deployed for other applications. Our conclusions are based on a combination of measurements and simulations. The analysis presented assumes, but is not limited to, a representative UAV mission that involved streaming video to the ground. The proposed framework then significantly improves reliability by allowing the UAV to broadcast to multiple ground receivers and solves the limited acknowledgment available to the aerial node by applying FEC at the application layer.

Keywords IEEE 802.11 · Video streaming · UAV · Ultra-high reliability

1 Introduction

Over the last few years the popularity of Unmanned Aerial Vehicles (UAVs) has exploded. Novel uses are constantly being invented across many domains, such as: shipping, precision agriculture and remote sensing. A wide variety of applications, including surveillance, traffic and crowd monitoring require a constant video stream from the UAV. Extremely high reliability is then necessary to ensure both real-time control and efficient delivery of high throughput content from the UAV to the ground. While point-to-point microwave links can be used, they are not desirable for several reasons. Primarily, the required ground station equipment is often large and time consuming to deploy. Secondly, to guarantee protection from interference, it

B. Van den Bergh (✉) · A. Chiumento · S. Pollin
KU Leuven, Leuven, Belgium
e-mail: vandenbergh@bertold.org
URL: www.esat.kuleuven.be

R. El-Azouzi et al. (eds.), *Advances in Ubiquitous Networking 2*, Lecture Notes in Electrical Engineering 397, DOI 10.1007/978-981-10-1627-1_50

may be necessary to use licensed frequencies, which presents an administrative and cost overhead as these networks require the use of a subscription. Also, requests for licensed spectrum have to be placed well in advance, which, in turn, make applications that depend on quick turnaround difficult or impossible. Finally, it is possible that applications are developed where line-of-sight is not guaranteed, for example for law enforcement. Microwave links experience significant attenuation when the Fresnel zone is violated, making them impossible to use without almost perfect line-of-sight.

Several works have shown that the IEEE 802.11 protocol is suited for advanced high-bandwidth UAV communication [1–4]. Other alternatives such as the well-known XBee [5] cannot provide the required throughput.

The ubiquitous presence of 802.11 access points, especially in urban environments, makes this technology a perfect candidate for civil UAV payload and control communication. The usage of unlicensed spectrum brings the great benefit of mature and wide spread technology, at the risk of increased interference.

Previous work on the usage of 802.11 links for UAV communication has highlighted that the transition from the terrestrial to aerial context might be challenged by the elevated interference due to the improved propagation between ground based and airborne nodes [1].

Several elements will impact the reliability of an 802.11 link. Firstly, there is the actual radio channel between the transmitter and the receiver. Bad propagation conditions may be caused by obstacles in the transmission path. Furthermore, interference can reduce receiver sensitivity to the point where communication with a distant node is no longer possible. Finally, it is important that the firmware in the access point and radio chipset properly manages the link according to the nature of the payload.

As shown in [1], an airborne node is able to receive data from more WiFi access points as its altitude increases. This is shown to be caused by reduced shadowing. Since the wireless channel is reciprocal this also means that more ground equipment will be able to receive transmissions from the UAV. This is exploited in this paper.

In this work, a novel comprehensive solution for an efficient and ultra-reliable UAV video streaming application in the presence of multiple ground access points is presented. The proposed solution exploits the enhanced propagation conditions between an airborne transmitter and the ground infrastructure. In order to guarantee system performance and reliability multiple access points are used simultaneously. Furthermore, to ensure glitch free video streaming, an error correction solution is applied at the application layer, which allows trading off excess bandwidth for lower latency.

This paper is split into four main sections. In the next section a system model and a concrete example wide-area surveillance mission is described. Images from one or more cameras are streamed in real-time to a central observation post. It is assumed that the UAV is controlled remotely in line with regulatory requirements. In Sect. 3, the connectivity problem is analyzed, both as function of the link quality and of the application requirements. In Sect. 4, a model to trade-off application errors, or data code rate, for latency is developed and discussed. Finally, in Sect. 5 the conclusions are drawn.

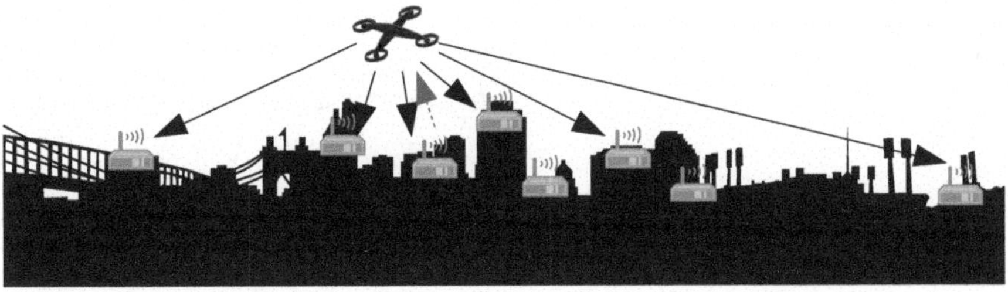

Fig. 1 System model: a UAV communicating with ground based 802.11 base stations

2 System Model and Reference Mission

In this paper we assume a UAV equipped with a standard compliant 802.11 WiFi card. It is operated in an area where standard WiFi access points are deployed on the ground. The UAV aims to transmit payload data using the existing ground infrastructure. Conversely, control data may be uplinked to the UAV. The scheme is shown in Fig. 1.

It is assumed that the ground infrastructure is willing to transfer the data for the UAV user. There are several ways in which this could be implemented in practice. Many ISPs already provide a WiFi service using consumer modems. These schemes could be extended to cater also to UAV users. Note that adding this functionality will not impact the normal wireless LAN operation. Alternatively, in corporate environments an elaborate WiFi deployment is likely already in place. This infrastructure can also be leveraged, greatly simplifying business park surveillance. Finally, the UAV operators may deploy 802.11 base stations for their dedicated private use.

This paper is structured around a video surveillance mission, in which an UAV should cover an urban or semi-urban area while providing constant video feed. To provide effective images a Full HD camera providing 30 frames per second is employed. To limit the bandwidth requirement an intra-frame video codec such as H.264 AVC or H.265 needs to be used. This codec will only send the differences between frames, which is significantly less data compared to sending every frame individually. The obtained data is then streamed over the 802.11 connection to the monitoring post. Commands for controlling the UAV are sent in the opposite direction.

3 Reliable Communication Strategy

It is assumed that the UAV will opportunistically use available ground infrastructure. At first, a single link between the UAV and single ground station will be examined. Such link, from the UAV to a single station, can be modeled using a 3-state markov model shown in Fig. 2 [6]:

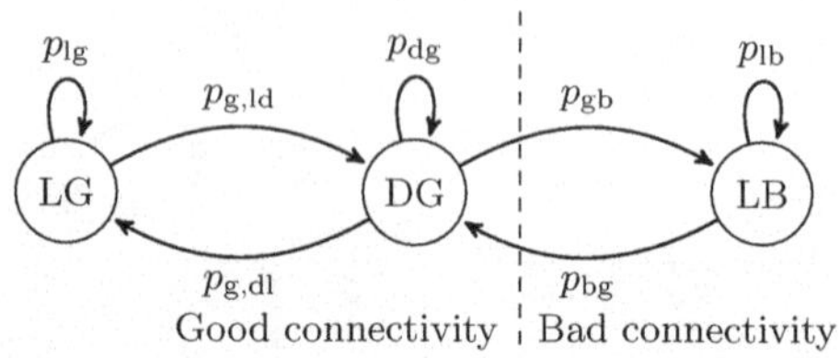

Fig. 2 Three state markov model

- DG: The frame is Delivered during Good conditions.
- LG: The frame is one of the few Lost altough conditions are Good.
- LB: The frame is Lost during an error burst.

The system is in the states LG and DG when conditions are good, this means that normally the transmitted data should be received by the node. Of course, packets can still be lost due to collisions and interference, but in general the losses are low. However, in the LB state, bad conditions are assumed. This could be because the UAV is outside of the communication range, or due to obstructions in the propagation path. In practice a few packets may still arrive, but for simplicity this has been left out of the model.

To analyze the typical packet loss performance of an individual, single frame, IEEE 802.11 transmission a measurement was performed. An AR9280 based transmitter, representing the UAV, was installed on a 17 m high tower. Packets (1500 bytes long) are transmitted every 10 ms to an AR9342 chip that is being carried around. This device measures what a ground node would see. The transmission rate was fixed at 6 mbps (MCS0) with STBC. Two vertically polarized dipoles were used, the transmission power was 100 mW and the frequency 2412 MHz. Figures 3 and 4 show the packet loss over time (computed over one second).

As can be seen on Fig. 4 there are significant time periods where the signal is not received by the ground node at all due to shadowing. Using just a single access point is not a reliable choice which greatly impacts the communication's Quality of Service. Furthermore, it is clear that the performance is strongly time varying. This

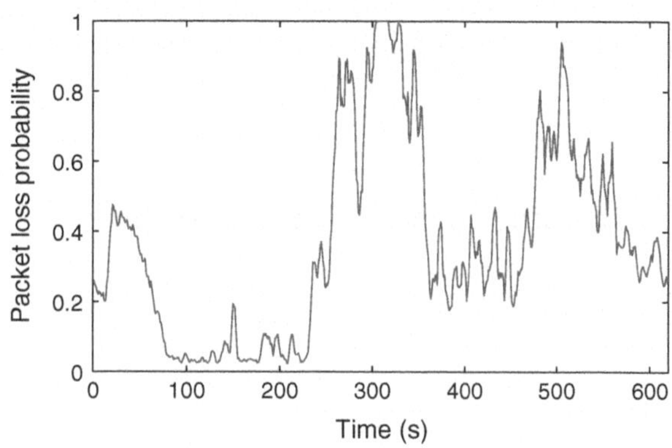

Fig. 3 Packet loss probability in an area with few buildings

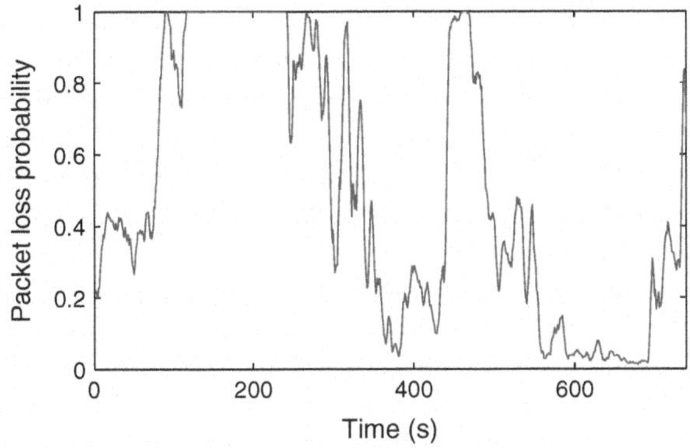

Fig. 4 Packet loss probability in an area with many buildings

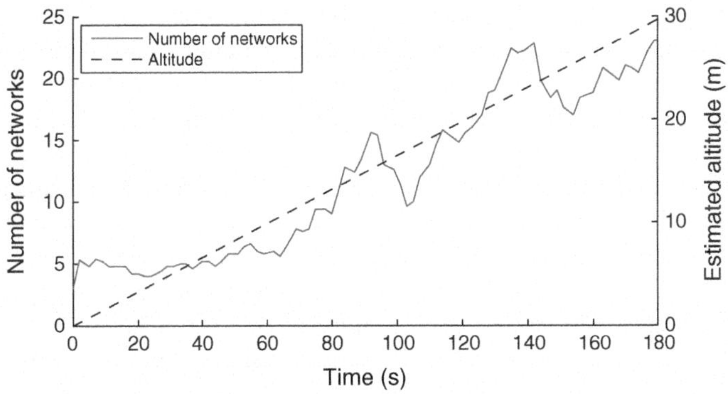

Fig. 5 Number of networks seen versus altitude. *Source* [1]

is an expected property of mobile wireless channels. It is therefore important to note that the parameters of the markov model are time varying.

Previous work [1] has shown that the UAV is able to communicate with an increasing number of ground base stations as its altitude raises. This is shown in Fig. 5.

The usual problem encountered when multiple links are available is the choice of which base station to communicate with in order to improve the overall reliability. This is particularly challenging in the case of high mobility nodes such as UAVs. We propose not to choose any station. Instead, the UAV can broadcast its data to all receiving ground nodes as this requires exactly the same amount of airtime compared to a targeted transmission. This solution can be implemented without making any changes to the 802.11 hardware. Apart from receiving data via its own BSSID, any WiFi device is capable of receiving packets sent using the broadcast BSSID (FF:FF:FF:FF:FF:FF). This is mandated by the standard to handle management frames, for example a probe request. A frame sent using both the broadcast destination address and BSSID would indeed be received by all listening nodes [7]. Support for the broadcast BSSID is also required by other standards, for example IEEE 802.11p for car-to-car communication. Current access point firmware would silently discard the packet, but a simple program could be developed that when installed in the router collects and forwards this kind of packets to the UAVs central infrastructure. This simple solution would allow the UAV to deliver its payload to multiple access points concurrently.

The main limiting factor to the application of the proposed method is that the access point will not transmit an acknowledgement to a frame formatted in this way. Even if it did, it would likely not be usable as the ACKs from many routers would collide at the UAV receiver. Furthermore, according to [1] the signal to interference ratio decreases significantly with increasing altitude. Even though the ground node likely received the packet, the ACK would have a high probability of being lost, resulting in needless retransmissions. A solution to both the problem of missing acknowledgments and to limit unnecessary retransmissions is presented in Sect. 4.

The final question is how to make distributed uplink transmissions from the ground control station to the UAV work. Simply transmitting them is unlikely to work due to the high level of interference experienced by the UAV. There are several solutions, for example on the PHY layer beamforming could be employed. Here, we present a scheme that can be applied on the MAC layer without requiring any hardware changes to the wireless access points. For a ground node to be able to interfere with the UAV means there is also the possibility of communication with this node. As such, the UAV can easily silence interfering ground transmissions by injecting a clear-to-send frame with its own MAC address as destination and a bogus duration. The 'locally administered' bit in the destination address could be set to differentiate the uplink request from normal 802.11 CTS transmissions that may also be used by the UAV. The control infrastructure can then deliver the next uplink packet to the ground node that received the last downlink packet with the highest signal level. Upon reception of the special CTS frame the station having the uplink packet will reset its NAV timer to zero and transmit during the CTS reserved window. Care should be taken to limit the rate and duration of these CTS requests as otherwise they could easily result in a wide-area denial-of-service attack against the ground networks.

4 Link Reliability

For many UAV applications, a stream of data has to be delivered to the ground. In general, no or very few errors are tolerated. For example, in the video streaming case, the amount of packet loss should be kept very low. Lost packets will cause the reference picture in the encoder and decoder to diverge. Thus, reference picture updates will have to be sent much more often. This consumes a significantly higher amount of bandwidth compared to only transmitting motion updates.

As shown in the previous section, the used scheme does not allow verifying whether a frame was received, since no ACK is transmitted by the receiver. This acknowledgement would then have to be implemented via the uplink mechanism and via the central control center. This would replace the normally instant 802.11 ACK with an acknowledgement delayed by a few internet route delays.

Because total latency is severely influenced by the average number of retransmissions necessary to ensure correct communication, this would have a catastrophic effect on latency. If lost packets are compensated by resending packets, the total latency can become several times the base link latency to allow for the ACK/NACK and retransmissions. The transmitter also needs a long buffer to be able to retransmit old packets.

Therefore, as an alternative to retransmissions, lost packets can be preempted at the source by employing Forward Error Correction (FEC). Here we assume that we have only application level access to the socket interface. Note that we are talking about datagram sockets. This could be UDP, or even a raw 802.11 socket. As such, a packet is either delivered intact or is lost completely. This channel is called a packet erasure channel. Packet erasure codes have been developed in information theory to ensure correct communication at the cost of increased bandwidth. A series of k packets is transformed into a sequence of $k + m$ packets where, for optimal codes, the original packets can be recovered as long as at least k packets are received. Initially these codes were applied to data storage and archival, but they can also be used in communication [8].

The main benefit of employing FEC is that the latency is only very slightly increased due to the block nature of the error correcting code. This is in stark contrast with a retransmission scheme where several times the base round trip time may be needed. The disadvantage is that the additional packets use extra bandwidth, even when they are not needed for correct reception.

The most basic packet loss model assumes each packet has a certain probability p of getting discarded. As seen above, to decode a block of packets correctly at least k packets out of $n = k + m$ transmitted packets must be received. Assuming a loss probability p, the probability that exactly l packets will be delivered is given by:

$$\begin{aligned} P(X = l) &= \binom{n}{l} \cdot (1-p)^l \cdot p^{n-l}, \\ &= \frac{n!}{(n-l)!\, l!} \cdot (1-p)^l \cdot p^{n-l}, \end{aligned}$$

where X is a random variable representing the number of delivered packets. Thus, the probability of at least k packets being delivered is:

$$P(X \geq k) = \sum_{l=k}^{n} [P(X = l)].$$

Finally, the probability of a block not being correctly decoded is:

$$p_{\text{fail}} = P(X < k) = 1 - \sum_{l=k}^{n}[P(X = l)] = \sum_{l=0}^{k-1}[P(X = l)]. \tag{1}$$

As seen in Eq. 1, the block error rate can be made arbitrarily small, as long as sufficient redundant packets are added.

This development outlines an interesting trade-off. In order to reduce latency the block size should be as small as possible. This, however, greatly increases the risk of a lost frame since the lost packets will be less spread out over different channel realisations. To compensate, a very high code rate will be required. For example, n packets are needed for a simple repetition code and a desired decoded error rate of p_{fail}:

$$p_{\text{fail}} = \sum_{l=0}^{0}[P(X = l)] = P(X = 0) = p^n, \tag{2}$$

$$n = \left\lceil \frac{\log p_{\text{fail}}}{\log p} \right\rceil. \tag{3}$$

Since only a single original frame is used, the code rate is exactly equal to n.

By using a larger frame size this effect is reduced, leading to the possibility of trading a small amount of latency for much better bandwidth efficiency. Figure 6 shows this trade-off between bandwidth efficiency and delay for different expected packet loss rates. The target error rate is set to one uncorrectable error every three minutes assuming one packet is transmitted per millisecond. In Fig. 6 it is assumed that the block transmission is subject to a delay constraint, this means that all redundant packets should be sent during the next block duration. This scheme is shown in Fig. 7.

As said before, another approach to reduce the probability of lost packets is simply retransmitting the packets when they are lost. This scheme is employed in the

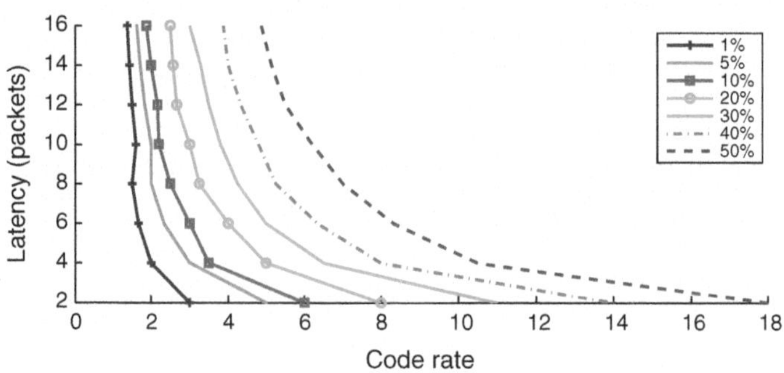

Fig. 6 Latency induced by FEC scheme if block transmission is strictly timed

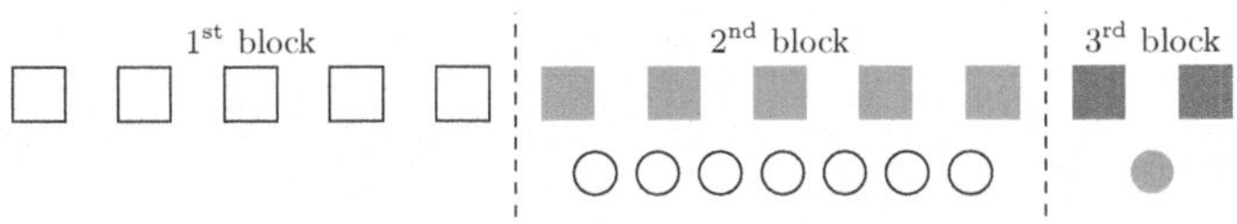

Fig. 7 In the block-timed mode redundant packets are all sent in the next block

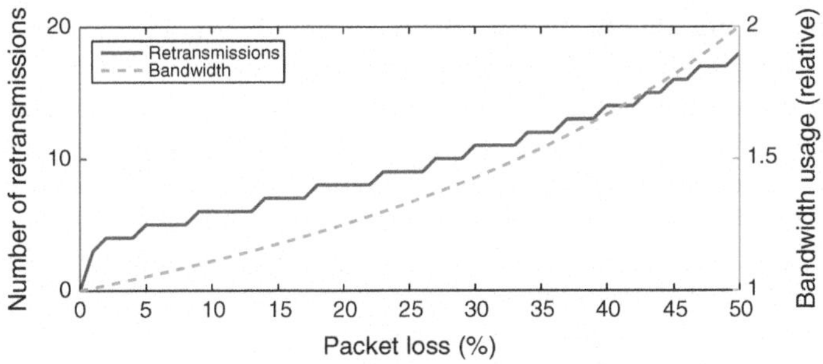

Fig. 8 Worst-case latency and bandwidth usage of a retransmission scheme

standard RTP/RTCP video streaming protocol. In order to compare both schemes a figure has been made. Figure 8 shows that the retransmission scheme is more bandwidth effective (dashed line). Indeed, extra bandwidth is consumed only when a frame loss event has happened. This is in contrast to FEC schemes which always send redundant packets because the source cannot know whether they will be needed. The average number of retransmissions with a packet loss probability p is:

$$n = \frac{p}{1-p}.$$

Even in the case of a 50 % frame drop probability the consumed bandwidth is, on average, only doubled. This is in contrast to the FEC scheme that uses bandwidth expansion ranging from 1.1 to 18 in the given example. However, to estimate the required latency, it is necessary to look at the worst-case expected number of retransmissions required to reduce the failure probability to a certain target. In the figure, the target is assumed to be one error per three minutes, with one source packet per millisecond, as above. This result is obtained using Eq. 3.

It should be noted that the unit of latency is round-trip times as the frame needs to be delivered from the sender to the receiver and the receiver needs to send an acknowledgement. Before the sender will repeat a packet, it needs to wait at least one round-trip time to verify if the packet arrived. This represents a major trade-off between retransmission and FEC schemes. If one wants to send a large amount of data without time constraints, retransmissions are better since the bandwidth overhead is lower. However, for real-time applications FEC schemes are significantly better for most links. The added latency of FEC is a number of frame transmission

Table 1 Comparison between FEC and retransmission scheme

	FEC	Retransmission
Bandwidth usage	4 Mbps	2.22 Mbps
Latency	21 ms	150 ms

The FEC system is the clear winner when it comes to latency!

times. However, the additional latency for retransmissions is a number of round-trip times. According to the model in Sect. 3, this delay is orders of magnitude larger.

Finally, to show the merits of this scheme we present a table with representative performance. A 2 Mbps stream is assumed. When sending 100 source packets per second this corresponds to approximately 300 bytes per frame. Assuming usage of MCS0 this results in a transmission duration of 600 µs, including overhead.

Since hardware level retransmissions are disabled it is assumed that 10 % packet loss occurs on a per frame per link level ($p = 0.1$). To achieve the stated performance of one uncorrectable error per three minutes the worst-case number of retransmissions is 5. The average bandwidth consumption would be 2.22 Mbps. As said before, we cannot use the standard IEEE 802.11 ACK mechanism. The ACK has to be delivered from the control plane over the internet. Therefore we assume that the minimum retransmission delay should be around 30 ms, resulting in a latency of 150 ms.

To compare with the FEC scheme it was decided to only allow a code rate of up to two. This doubles the required bandwidth to 4 Mbps. To achieve this a block size of 5 packets is required. Since in the worst case we will also need all 5 redundant packets the total latency will be 10 packets. As said before, the transmission time is 600 µs and the network transfer delay is 15 ms, corresponding to a total latency of 21 ms. These results are summarized in Table 1.

5 Conclusion

In this work, a case for real-time video streaming from an UAV in a urban setting has been analyzed. A framework allowing very reliable communication from one UAV to multiple ground nodes has been proposed and discussed. Furthermore, in order to allow the feasibility of such framework, to overcome the shortcomings of traditional retransmission schemes and to reduce communication latency a forward error correction scheme, acting at the application layer has been introduced.

The proposed framework shows that, indeed, broadcasting messages from an UAV to multiple access points, is not only advantageous in terms of reliability but also already possible without any modification to the 802.11 standard. The FEC scheme also removes the necessity of retransmissions by eliminating the necessity of recuperating the lost packets via the inclusion of redundancy in the signal itself. At a moderate bandwidth expansion, the proposed method is then able to reduce the latency by a large factor, making thus real-time constrained applications very possible.

References

1. Van den Bergh, B., Vermeulen, T., Pollin, S.: Analysis of harmful interference to and from aerial ieee 802.11 systems. In: Proceedings of the First Workshop on Micro Aerial Vehicle Networks, Systems, and Applications for Civilian Use, DroNet '15, pp. 15–19, New York, NY, USA. ACM (2015)
2. Jimenez-Pacheco, A., Bouhired, D., Gasser, Y., Zufferey, J.C., Floreano, D., Rimoldi, B.: Implementation of a wireless mesh network of ultra light mavs with dynamic routing. In: 2012 IEEE Globecom Workshops, pp. 1591–1596, Dec 2012
3. Andre, T., Hummel, K.A., Schoellig, A.P., Yanmaz, E., Asadpour, M., Bettstetter, C., Grippa, P., Hellwagner, H., Sand, S., Zhang, S.: Application-driven design of aerial communication networks. IEEE Commun. Mag. **52**(5), 129–137 (2014)
4. Yanmaz, E., Kuschnig, R., Bettstetter, C.: Achieving air-ground communications in 802.11 networks with three-dimensional aerial mobility. In: INFOCOM, 2013 Proceedings IEEE, pp. 120–124, Apr 2013
5. Allred, J., Hasan, A.-B., Panichsakul, S., Pisano, W., Gray, P., Huang, J., Han, R., Lawrence, D., Mohseni, K.: Sensorflock: an airborne wireless sensor network of micro-air vehicles. In: Proceedings of the 5th International Conference on Embedded Networked Sensor Systems, SenSys '07, pp. 117–129, New York, NY, USA. ACM (2007)
6. Norris, J.R.: Markov Chains (Cambridge Series in Statistical and Probabilistic Mathematics). Cambridge University Press (1997)
7. IEEE Standard for Information technology: Telecommunications and information exchange between systems Local and metropolitan area networks. IEEE Std 802.11-2012 (Revision of IEEE Std 802.11-2007)
8. Rizzo, L.: Effective Erasure codes for reliable computer communication protocols. In: SIGCOMM Comput. Commun. Rev. 24–36 (1997)

A New Adaptative Security Protocol for UAV Network

Oumhany Zouhri, Siham Benhadou and Hicham Medromi

Abstract An Unmanned Aerial Vehicle (UAV) is a pilotless aerial vehicle which can be controlled either autonomously by onboard computers or remotely controlled by a pilot at the Ground Control Station (GCS). UAV and Ground Control Station (GCS) define a new form network, this kind of network persuade some specific characteristics as sufficient energy, network connectivity, mobility and network security. These specifications persuade difficult challenges for building a trustworthy and secure communication architecture solution. In this paper we present our new secure communication protocol taking into account the specifications of UAV network. This new architecture depends mostly on the definition of a secure protocol which provides authentication, confidentiality and integrity, in preserving network resources for effective data exchanged between UAV and GCS.

Keywords Security architecture · Security protocol · Cryptography · UAV system

1 Introduction

Over recent years, UAVs are increasingly being used not only for military tasks, but also for civilian tasks, such as environment and traffic monitoring, delivery services, and aerial surveys [1]. UAV describe an aircraft that fly under no person aboard. An unmanned aerial systems (UAS) is formed when UAV is associated with ground control stations (GCS) that includes mission planning and monitoring software using a two-way data link for control and telemetry. When UAV communicate with GCS they form a temporary network called UAV Network (UAVNET). Communication between UAV and the GCS is handled by a communication protocol through a wireless link [2, 3]. UAVNET introduce several challenge regarding their high mobility and frequently changing topology, network connectivity.

O. Zouhri (✉) · S. Benhadou · H. Medromi
National High School of Electricity and Mechanics, ENSEM, 8118 Casablanca, Morocco
e-mail: oumhanyzouhri@gmail.com

R. El-Azouzi et al. (eds.), *Advances in Ubiquitous Networking 2*, Lecture Notes in Electrical Engineering 397, DOI 10.1007/978-981-10-1627-1_51

Unfortunately, the security protocols that currently exist are not designed for this environment [4]. They are not taking the constraint [5] of resources into account because not only resources are limited but the environment is dynamic. Which further complicates the problem, because we know that security solutions are inadequate in terms of resources. The limited energy, communication bandwidth and computational capacities of drone make protocols such as TLS [6] and Kerberos proposed for wired networks impractical for UAV networks.

In addition, human lives can be at stake when the drones can be used as a weapon [7, 8]. Therefore, it is essential to ensure that control traffic cannot be modified or deleted by unauthorized entity during the mission. It must be secretly exchanged and on time, due to a significant delay in the exchanged traffic can cause considerable damage. This is the challenge that we intend to solve within our research. A large part of the research related to security and threat modeling focuses on the causes and methods of computer security breaches. Through, these works are focused on answering the same questions: what are the vulnerabilities of the system in question, how can attacks be prevented and how can the threats be mitigated. A better approach is to perform a cause-effect analysis defining the cause of unintended or degraded subsystem functionality and evaluating the attack severity on mission/task performance [9]. A model of UAV communication scenario, which consists of various components and different types of communication links. Each of these links carries different types of information and data [10], as illustrated in Fig. 1.

This kind of network rely on wireless communication channels for communicating with each other. Also it has three types of links based on the type of information being transmitted, namely, radio communication, and Satellite link. Radio communication links carry telemetry data, audio, video, and control information. Although the various communication channels seem similar, there is a lot of difference between the channels with respect to security [11]. Each communication links have different security requirements. Components like Satellite link might have certain threats but may not be too vulnerable due to the existing security measures in place [12].

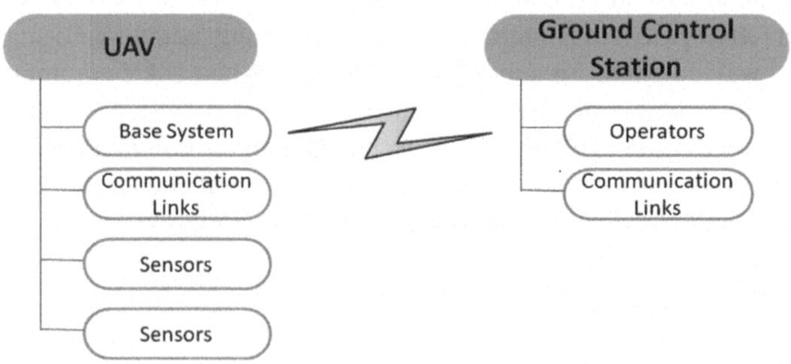

Fig. 1 Communication links in UAV system

Our main objective is to propose a new architecture that allows for secured communication between the UAV and the Ground Control Station (GCS). This architecture must support the different features of these networks (dynamic topology, limited resources, wireless communication…). Moreover, in the present work we find to offer detailed security architecture for controlling communication in UAV network that provide data security by cryptographic means.

The rest of this paper is organized as follows. The methodology and security requirements of UAVNET are described in Sect. 2. The Sect. 3 describes the state of art of UAVnet. Then, we present the proposed UAVnet protocols and system architecture in Sect. 4, and the conclusion is drawn in Sect. 5.

2 Methodology and Security Requirements of UAVNET

In this section, we first present some basic specifications of UAVNET after which the methodology and requirements of UAVs such systems are introduced.

2.1 UAV Network Specifications

Technically, UAVNET present a new mobile network, this type of network owns some specifications that must be taken when designing a new security solution. The first specification is mobility, sight connectivity between the UAV and GCS is ensured by a radio link, in that the communication between the two entities cannot be always available. Because of the UAV movements and variations of distance, link quality fluctuates and may cause loss of connectivity and performance degeneration. Nonetheless, the radio links are also easier to be interfering with an attacker. This makes UAVNET more susceptible to security breaches. According to the mobility, the connectivity between UAV and GCS could be lost while they are transmitting critical information.

The second specification is Efficiency is the most important issue for UAV, since the power and storage may limit network life. So the security protocol should impose a minimum calculation overhead and support simultaneous cryptographic operations.

2.2 Security Analysis for UAVNET

System security should never be considered as a state, but rather as a process. In order to support this process, it is important to be capable to describing and judging the current security status. Furthermore, it is desirable to be able to compare system configurations in terms of security levels. In order to fulfill these tasks, we are confronted with the question: What is security and how is it measured? Focusing on the technical aspect of the question, information security is defined in [13] as "protecting information and information systems from unauthorized access, use, disclosure, disruption, modification, or destruction". Hence, security is a value describing how good system is protected.

2.3 Security Requirements

We know that, security can be compromised by operating vulnerabilities via inserting undesirable changes in one of the three properties of the system, Confidentiality, Integrity and Availability. Through generation of policy rules, most of the existing security models try to address only one property of the system. All three property are very sensitive, and none of them can be compromised in this case [14, 15].

Also It is required to ensure that no malicious entity can disrupt the transmission of data messages. A secure UAVNET architecture should ensure the following security service: authentication, integrity, confidentiality and non-repudiation. Authentication, regarded as the first line of defense versus intruders. It allows to verify the identity of an entity in the network. Within the wireless networks, the authentication process is based on a trusted third party in which all network entities have confidence. The trusted third party is only Certificate Authority distributes certificates to the entities that have the right of access to a network service. This authentication scheme is centralized, and is known as Public Key Infrastructure (PKI) [16]. Apply the PKI model directly in UAVnet is not possible for reasons that the UAVnet is dynamic and frequently changing network topology.

Confidentiality is an essential service to ensure private communication between entities. It is a protection against threats that could cause unauthorized disclosure of information that should be protected. It is mainly based on cryptography, particularly encryption algorithms.

Integrity this service ensures that traffic from source to destination has not been altered or modified without authorization during transmission. It is the protection against threats that could cause unauthorized modification of system configuration or data. The Integrity Services are designed to ensure the proper functioning of resources and transmission.

Non-repudiation: is the ability to verify that the sender and recipient are parties that say they have sent or received the message. In other words, the non-repudiation of origin shows that the data was sent, and non-repudiation of arrival proves that they were received.

2.4 Cryptographic Primitives

The diversity of reliable protocol means it is challenging to consumption in the calculation. Hence in the most security protocols, cryptography can be divided in two categories that depend on the initialization phase of the protocol and data protection phase. The first phase is usually based on asymmetric encryption (public/private key) while the second is based on a symmetric encryption (secret key). In recent years, a new direction in cryptography is being developed, which is associated with the use of Signcryption. It is a new paradigm in public key cryptography that simultaneously fulfils both the functions of digital signature and public key encryption in a logically single step, and with a cost significantly lower than that required by the traditional signature followed by encryption. Signcryption costs 58 % less in average computation time and 70 % less in message expansion than does signature-then- encryption [17].

The high mobility of UAV and the low network resources require to develop a specific solution for UAANETs. Moreover, security should be taken into account at an early stage of the UAV network design. So far, communication architectures have been proposed for UAV [18].

3 State-of-the-Art

3.1 Related Works

Most of the existing works related to the area of UAV shows that security communication issues are not completely addressed in the literature. From a security and threat analysis perspective, it is necessary to understand that a typical UAV network is not similar to the traditional computer network. Some researchers have compared it to wireless sensor networks (WSNs) [19] and mobile ad hoc networks (MANETs). Although this network assume near similarity to WSNs, as both of them use wireless communication protocols [20], there are other aspects in which they differ. For instance, power requirements, amount of information being carried by channels, and the number of nodes in a WSN are much lower than in a UAVNET. Moreover, the coverage area for a UAVNET is almost 1000 times bigger than that of a WSN. Additionally, while in WSNs, all nodes usually transmit their sensor data to one central node which communicates with external systems, in the UAVNET, the UAVs communicate to the Ground Control Station (GCS) independently. Some researchers have combined the application of UAVs in sensor networks so as to utilize the bigger coverage area of UAVs [21].

3.2 Attacks on UAVNETs

A radio communication link has been used, for UAV management, telemetry, and other data transmission. It is inevitably confronted with threats [22]. Firstly, the principal components that can be vulnerable to attack are, GCS, UAV communication link. For instance, threats to the GCS are most often based [23], viruses, malware, keyloggers, trojans, etc. Hacking is a major threat for a UAV [24]. Eavesdropping, hacking, identity spoofing, cross-layer attacks [23] and multi-protocol attacks [24] are the major of compromising security of communication links.

On the other hand, system could be compromised by using, changing information and making new information [25]. We mention other types of attacks which have three categories: jamming, compromising signal and capturing the feed signal. Jamming aimed at disrupting communication through interference or collision before the reception [26]. Denial of Service attacks (DoS) [27, 28] is attacks which might affect availability, in particular based on network congestion or overflow in the system network card so that the system appears to be unavailable. This may be the major threat to the availability of the UAV system as wrong signals.

4 A New Security Communication Protocol Between UAV and GCS(SPUAV)

In this section we present the proposed architecture that would support the specifications of UAVNET and the main cryptographic primitive, it satisfies multiple security requirements, such as authentication, integrity, confidentiality. This architecture is divided into three module and a database. The interaction between the module is an automatic [29] and autonomous way, allowing self protection system and a multi-layer defense against attacks [30] (Fig. 2).

Authentication Protocol is the initialization phase of the Protocol, it generates a series of messages exchanged between entities to negotiate security settings and the and resources that nodes want to use. During this phase, communicators traded for

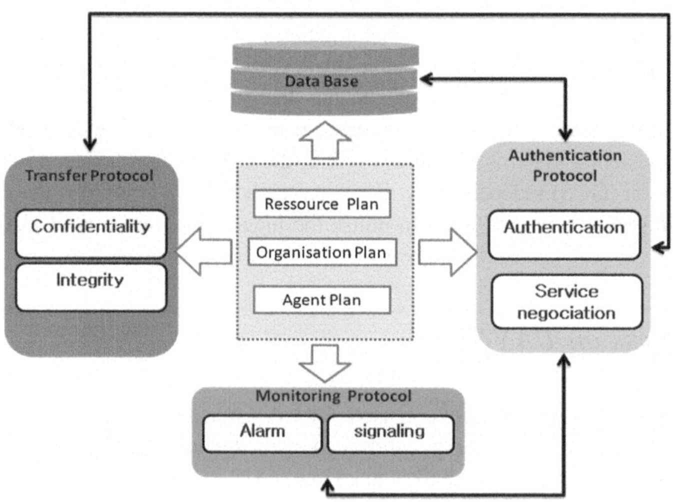

Fig. 2 Security Communication Protocol between UAV and GCS(SPUAV)

each service security settings and exchange information needed to encrypt their data traffic. In the other hand it negotiate the exchange of security parameters (Random Numbers, list of ciphers, hash) between the UAV and GCS before the application data is transmitted.

Transfer Protocol provides data transfer service between the UAV and GCS. It then allows of providing the services of confidentiality and data integrity. To ensure data confidentiality, it encrypts user data with key Kc, derived during the initialization phase of SPUAV. To provide message integrity, it uses a MAC function with a second key derived from the master key Kc.

Monitoring Protocol is divided into two module: Attacks alarm is responsible for detecting attacks and archives each type of attack for the prevention in future. Signaling Module is responsible to report errors encountered while checking messages and any contradiction that may occur during the initialization phase of the protocol. It also helps point out some session parameters such as the expiration of a session or the closing of a connection. also it offer the detection and prevention of attacks. In case of insufficient resources, which notifies the application takes a decision concerning the degradation of the quality required or cancellation of issuance.

Security Database resembles a relational database that interfaces with all other protocols, to record all the event log or session key. The security policy of a computer system to specify the actions allowed in this system. To control access to sensitive information of a system, a security policy must identify the resources of the system containing the sensitive information and transactions to access this information.

For a flexible coordination between different modules of our proposed architecture a set plane architecture is used. Also the commodity of detection and reaction components is enables. Which allows an effective and flexible self-protection of the communication between UAV and GCS.

5 Conclusion and Future Research

With the rapid development of UAV technology, this systems have emerged to manage and process data with high requirements of communication security. These emerging systems have given rise to various new challenges about how to develop a new communication security architecture. Furthermore, since the secure communication architecture has to provide confidentiality, authentication and integrity. In this paper, we have presented a new secure communication protocol between UAV and ground control station.

The objective of the design of this architecture does not stop at the proposal in the literature. One of the goal of this architecture is the realization and integration in the real drone. For this reason, the next work we try to simplify to the maximum the operation and the combination of these mechanisms to facilitate real implementation. In future work will involve working on some threats already cited and using mission information to model the threats more accurately.

References

1. John, D.B.: Unmanned Aerial Systems: A Historical Perspective, CreateSpace Independent Publishing Platform, pp. 1–12–45–49 (2010)
2. Nirmala, S., Ananda, C.M.: Communication methodology in formation of flying of micro air vehicles. In: The 3rd International Conference on Recent Advances in Design, Development and Operation of Micro Air Vehicle, pp. 63–67 (2014)
3. Jun, L,. Yifeng, Z., Louise, L.: Communication Architectures and Protocols for Networking Unmanned Aerial Vehicles, Globecom Workshop—Wireless Networking and Control for Unmanned Autonomous Vehicles (2013)
4. Kaps, J.-P.: Cryptography for ultra-low power devices, Ph.D. thesis, at Worcester Polytechnic Institute (2006)
5. Nicolas, L.: How can model driven development approaches improve the certification process for uas. In: International Conference on Unmanned Aircraft Systems (ICUAS), pp. 253–260. IEEE (2014)
6. Dierks, T., Aallen, C.: The TLS Protocol Version 1.0. IETF RFC No. 2246, Jan 1999
7. Maxa, J., Mahmoud, M., Larrieu, N.: Secure routing protocol design for UAV Ad Hoc networks, DASC'2015. In: IEEE/AIAA 34th Digital Avionics Systems Conference, Sep, Prague, Czech Republic (2015)
8. Hartmann, K., Steup, C.: The vulnerability of uavs to cyber attacks an approach to the risk assessment, in Cyber Conflict (CyCon). In: 2013 5th International Conference on IEEE, pp. 1–23 (2013)
9. Ethan, M.P,. Daniela, A.D.: A performance study of UAV-based sensor networks under cyber attack. In: Proceedings of the 6th International Conference on System of Systems Engineering, Albuquerque, New Mexico, USA, pp. 27–30 (2011)
10. Goraj, Z.: UAV Platform designed in WUT for border surveillance, AIAA paper 2965 (2007)
11. Rudniskas, D., Goraj, Z., Stankunas, J.: Security analysis of UAV radio communication system. Taylor & Francis Int. Res. J. Aviat. **13**(4) (2009)
12. Anjum, F., Mouchtaris, P.: Security for Wireless Ad Hoc Networks. Wiley (2007)
13. Bishop, M.: Introduction to Computer Security, 1st edn., Addison-Wesley, Ed. Boston, USA: Pearson Education (2004)
14. Bishop. M.: Introduction to Computer Security, Addison-Wesley, 5 Nov 2004. ISBN: 0-321-24744-2
15. Pfleeger, P.: Security in Computing, 4th edn., Prentice Hall of India (2008). ISBN-13: 978-8120334151
16. . Simplício, M.A., Barreto, P.S., Margi, C.B., Carvalho, T.C.: A survey on key management mechanisms for distributed. Wirel. Sens. Netw. Comput. Netw. 2591–2612 (2010)
17. Zheng, Y.: Signcryption and Its applications in efficient public key solutions. In: Proceedings of Information Security Workshop (1997)
18. Sahingoz, O.K.: Networking models in flying ad-hoc networks (fanets): concepts and challenges. J. Intell. Rob. Syst. **74**, 1–2–513–527 (2014)
19. Ian, F.A., Ismail, H.K.: Wireless sensor and actor networks: research challenges. J. Ad Hoc Netw. **2**, 351–367. ELSEVIER (2004)
20. Theodore, S., Rappaport, A., Annamalai, Buehrer, R.M., William, H.T.: Wireless communications: past events and a future perspective. In: IEEE Communications Magazine 50th Anniversary Commemorative Issue (2002)
21. Kam, M. L., Gerard, L.: Sensors and Communications Range Relations for UAV Operations, New Challenges in Aerospace and Technology Maintenance Conference, Singapore (2006)
22. Boukhdir, K., Marzouk, F., Medromi, H., Benhadou, S.: Secured UAV based on multi-agent systems and embedded intrusion detection and prevention systems. Am. J. Eng. Res. (AJER) **4** (8), 186–190 (2015)
23. Wang, W., Sun, Y., Li, H., Han, Z.: Cross-Layer Attack and Defense in Cognitive Radio Networks, IEEE Globe Communication Conference (Globecom), Miami, FL, Nov-Dec 2010

24. Jim, A.F.: Multi-Protocol Attacks and the Public Key Infrastructure, 21 st National Information Systems Security Conference, Arlington, Virgina, USA, Sept 1998
25. Sen, S., Clark, J.A., Tapiador, J.E.: Security threats in mobile ad hoc networks. In: Security of Self-Organizing Networks: MANET, WSN, WMN, VANET, Auerbach Publications, pp. 127–147 (2010)
26. Bhattacharya, S., Basar, T.: Game-theoretic analysis of an aerial jamming attack on a UAV communication network. In: American Control Conference Marriott Waterfront, Baltimore, MD (2010)
27. Carle, G., Dressler, F., Kemmerer, R.A.,. Koenig, H., Kruege, C., Laskov, P.: Network attack detection and defense. In: Manifesto of the Dagstuhl Perspective Workshop, 2nd–6th March 2008
28. Michel, B.: WiMax802.16 Threat Analysis, International Symposium on QoS and Security for Wireless and Mobile Networks, Q2SWinet'05, Montreal, Quebec, Canada 13 Oct 2005
29. Jeffrey, K., William, W.: An artificial intelligence perspective on autonomic computing policies. In: Fifth IEEE International Workshop on Policies for Distributed Systems and Networks (2004)
30. Chess, D., Palmer, C., White, S.: Security in an autonomic computing environment. IBM Syst. J. 107–118 (2003)

Optimal Beaconing Policy for Tactical Unmanned Aerial Vehicles

Sara Koulali, Mostafa Azizi, Essaïd Sabir and Rim Koulali

Abstract Unmanned Aerial Vehicles (UAV) were initially developed for military monitoring and surveillance tasks. However, they recently found several interesting applications in the civilian domain. One promising application is to use UAV for military operations behind enmy lines. Rapid deployment along with limited operating costs are key factors that boost the development of UAVs for both military and civilian utilizations. UAVs are battery-powered which makes energy consumption optimization a critical issue for acceptable performance, high availability and an economically viable UAVs deployment. In this paper, we focus on tuning the beaconing probability as an efficient mean of energy consumption optimization. The conducted study provides markov decision process perspective of the problem. Also, we conduct extensive numerical investigations to assist our claims about the energy efficiency of the optimal beaconing policy.

Keywords Unmanned aerial vehicles · Markov decision process · Delay/Disruption tolerant networks · Optimal beaconig policy · Activity schedule

S. Koulali (✉) · M. Azizi
MATSI Laboratory, Mohammed First University of Oujda, Oujda, Morocco
e-mail: s.koulali@ump.ac.ma

M. Azizi
e-mail: azizi.mos@ump.ma

E. Sabir
NEST Research Group, ENSEM, Hassan II University of Casablanca, Casablanca, Morocco
e-mail: e.sabir@ensem.ac.ma

R. Koulali
LARI Laboratory, Mohammed First University of Oujda, Oujda, Morocco
e-mail: r.koulali@ump.ac.ma

R. El-Azouzi et al. (eds.), *Advances in Ubiquitous Networking 2*, Lecture Notes in Electrical Engineering 397, DOI 10.1007/978-981-10-1627-1_52

1 Introduction

Unmanned aerial vehicles (UAVs) have been commonly associated with military technology suited for tactical offensive/defensive missions. Though, there has been a growing interest in broadening their usage range to cover civil applications such as monitoring traffic congestion, network coverage extension and disaster management. The Google Loon project [1] is based on the balloon deployment to provide ubiquitous networking. The balloon will be deployed in High Altitude in the stratosphere to provide internet access, especially in rural and poorly covered areas. Internet coverage will be provided for LTE-enabled devices by balloons relying on wind to relocate. The balloons form one large communications network. Facebook has its own vision for providing internet access named Drone project [2], the proposed architecture is a mixture of Low Earth Orbit, Geosynchronous Earth Orbit and stationary drones depending on the density of the target population. This could potentially lead content providers such as Google and Facebook to become independent Internet Service Providers (ISP) and circumvents existing ISPs to distribute their content.

Nevertheless, most UAVs successful applications are mainly in the defense and law enforcement fields. One can cite intelligence gathering, border surveillance and smuggling fight as concrete examples.

Rapidly deployed UAVs at low altitude could fulfill tactical missions behind enemy lines. Thus, they will form a temporary communication backbone between Control and Command Center (CCC) and deployed tactical teams. Such deployment, offers reliable communication infrastructure to coordinate military missions and provide timely guidance to on the ground operationals. Fast deployment and effective relocation on response to demand is one major asset of UAVs without being hampered by geographical constraints inherent to on the ground deployed communication networks. This ability to relocate allows great responsiveness and represents a sought-after characteristic. Figure 1 illustrates a UAV fulfilling a guidance mission for a tactical team on behalf of a CCC.

In this work, we examine the problem of optimal beaconing from single UAV perspective. To achieve the maximum system performance in terms of encounter probability and energy efficiency, we propose to carefully fix the beaconing probability. First, we introduce a Markov Decision Process (MDP) model for optimal beaconing period duration choice. Second, we compute the optimal beaconing policies. Finally, we show the efficiency of our proposed beaconing strategy through extensive numerical results.

The rest of this article is organized as follows: Sect. 2 surveys related work. In Sect. 3 we describe a markov decision process based framework to model UAV engaged in tactical missions. We examine a representative case study through extensive numerical investigations in Sect. 4. Finally, Sect. 5 draws some conclusions and future work.

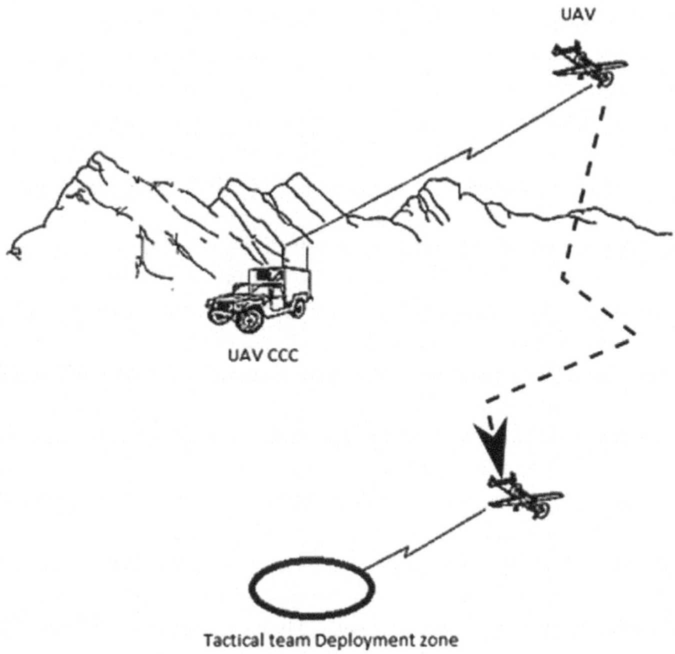

Fig. 1 UAV for tactical missions

2 Related Works

In [3], the location and movement of UAVs are optimized to improve the connectivity of a wireless network. Authors formulated deployment and movement problems for the UAV and developed adaptive algorithms to increase the network performance in terms of global message connectivity. They showed that network bisection and k-connectivity are improved by the addition of a UAV to the network. The work [4] proposed a novel usage model for a UAV network, where a number of UAVs are required to collect information from randomly located areas and transmit it wirelessly to a common receiver. The authors of [5] consider energy-efficiency maximization for UAV-based relay architectures. In this work a fixed-wing UAV relays data between stationary source and destination nodes. Thus, circular maneuvering is optimized through tuning the turning radius parameter. Energy efficiency is defined as the ratio of network capacity to the power consumption of both maneuvering and communication. The authors provide a closed form for a suboptimal solution for an approximate energy efficiency formula.

The authors of [6] propose a distributed framework for UAV-based disaster sensing. The presented framework comprises a client unit hosted by the UAV on board system and a server unit hosted by the remote computing cloud infrastructure that provides service-oriented resource support. To address the processing and storage limitations inherent to small civilian UAV they propose in-cloud selective data offloading and processing. The selection process on the UAV filters acquired video only offloads essentials frames power-hunger advanced processing. The work in [1] investigates UAV based relaying both for single and multiple relays UAV over test-beds. Performance bounds are derived based on stochastic geometry formulation. The proposed UAV-based relay is compared to load balancing and traffic management techniques.

The authors of [7] model a UAV system power requirement using energy requirements from all composing sub-systems. It shows that an efficient UAV system can only be improved by the use of energy efficient components. They propose to optimize the maximum operating range and frequency band for data-transfer to a ground station. Besides complex tasks are distributed among multiple UAV working as a fleet. Optimal beaconing control for Epidemic routing in Delay Tolerant Networks for energy efficiency is proposed in [8]. The authors propose a continuous Markov and derive a threshold beaconing policy that maximizes the delivery ratio within an energy constraint.

In [9], we studied the activity scheduling of unmanned aerial vehicles acting as Drone Small Cells for temporary events and disaster-relief activities. We formulated a model based on a non-cooperative game theory and characterized equilibrium beaconing period duration for competing drones. Then, we introduced a fully distributed learning algorithm that allows each drone to discover its optimal beaconing strategy without any knowledge of its opponent schedule. The equilibrium operating point allows drones to efficiently optimize their energy consumption while maximizing the likelihood of getting in contact with the mobile users on the ground.

3 Problem Formulation & Mathematical Model

We consider a single UAV that fulfills tactical missions on behalf of a CCC. The UAV mission is to collect data from a team deployed on the ground then to deliver it to the CCC. The UAV advertises its presence to both CCC and on the ground team by sending beacons according to a random strategy to avoid detection. Both CCC and tactical team are mobile to avoid detection by the enemy. Time is discrete, where in each time step (or slot) the UAV state (on a mission/not on a mission) changes according to the Markov chain depicted in Fig. 2.

The number of backlogged missions at slot t is denoted by N_t. The UAV, CCC and tactical team are moving according to different random mobility patterns to avoid

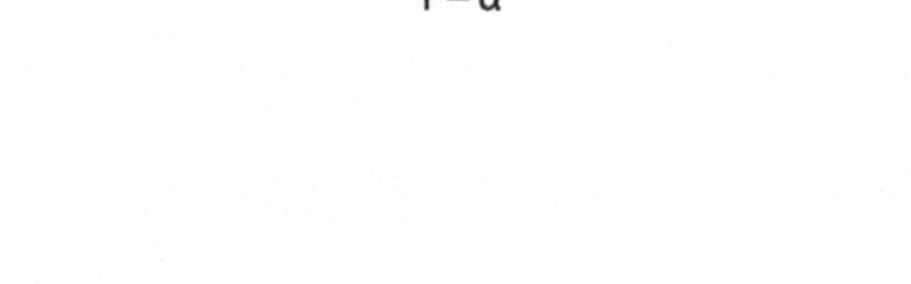

Fig. 2 Makovien model for the UAV's state evolution

detection. It has been proven that several random mobility models exhibit exponentially distributed inter-contact times [10]. In this work we denote by λ_s (resp. λ_d) the encounter rate between the UAV and the CCC (resp. the tactical team). The UAV chooses its optimal beaconing probability $P_{b,t}$ for each slot to increase the successful encounter probability with the CCC or tactical team depending on the number of backlogged mission (state). Indeed, if the UAV has already encountered the CCC then, it will make more sense to adjust its beaconing probability such that the likelihood of getting in contact with the tactical team increases. However, if the UAV has zero backlogged missions, then it will carefully choose its beaconing probability to increase the chances of being in contact with the CCC. Beaconing will reduce the power budget of the UAV (C_b unit per slot). Hence, it has to be incentivized to fulfill tactical missions. Whenever the UAV meets the CCC or the tactical team it is rewarded by an amount of energy transferred wirelessly C_e [11]. The optimal beaconing strategy has to ensure a delicate tradeoff between the amount of depleted power and the expected reward obtained when meeting either the CCC or the tactical team.

Then, the successful encounter probability between the UAV and the CCC/tactical team is expressed as follow:

$$\forall y \in \{s, d\}, P^{succ}_{y,\Delta t} = \int_t^{t+\Delta t} \lambda_y e^{-\lambda_y s} ds = e^{-\lambda_y t} \times (1 - e^{-\lambda_y \Delta t}) \tag{1}$$

Consequently, the UAV state evolution probabilities are given by:

$$\beta \triangleq P(N_{m,t+1} = 1 | N_t = 1, P_b) = \left(1 - P^{succ}_{d,\Delta t}\right) + P^{succ}_{d,\Delta t} \times \left(1 - P_{b,t}\right). \tag{2}$$

and

$$\alpha \triangleq P(N_{m,t+1} = 0 | N_t = 0, P_b) = \left(1 - P^{succ}_{s,\Delta t}\right) + P^{succ}_{s,\Delta t} \times \left(1 - P_{b,t}\right). \tag{3}$$

Each UAV faces a sequential decision problem for choosing optimal beaconing probability $P_{b,t}$. by [12] there exists an optimal policy which is deterministic and Markovian. Indeed, if an optimal control policy exists, there is no loss in restricting policies to be Markov, that is a policy which only uses the current state. We formulate the UAV decision making as a Markov Decision Process with finite horizon μ; $\Omega = (\mathcal{X}, \mathcal{A}, P, \upsilon, \mu)$, where:

- $\mathcal{X}$ is a finite set of UAV states corresponding to the number of backlogged missions.
- $\mathcal{A} = [0, 1]$ represents the set of available actions to each UAV. For a given UAV, the decision is the probability of being active (in the probing period) $P_{b,t}$.
- $P : \mathcal{X} \times A \to \mathcal{X}$: The transition probability matrix P represents the dynamics governing the UAV data buffer evolution.
- $\upsilon : \chi \times A \to \mathbb{R}$ represents at a given state the immediate payment of the UAV after choosing an action, the payment value at slot m is null. A UAV pays a cost C_b per

beaconing slot and is rewarded by C_e energy units upon mission completion. The utility of the UAV is expressed as follows:

$$v_t(N_{t+1}|N_t, P_{b,t}) = \begin{cases} -C_b \times P_{b,t} \times \Delta t, & N_{t+1} = N_t \\ C_e - C_b \times P_{b,t} \times \Delta t, & N_{t+1} \neq N_t \end{cases} \tag{4}$$

The expected total discounted reward from policy π starting in state N_1 is given by:

$$v^{\pi}(N_t) = E^{\pi}_{N_t}\left[\sum_{t=1}^{m-1} \delta^t v_t(N_t, P_{b,t}) + v_m(P_{b,m})\right] \tag{5}$$

4 Numerical Results

For our simulations, we set the beaconing cost and energy reward parameters such that: $C_e = 100 \times C_b = 20\,\text{J}$. First, we numerically compute optimal beaconing policies for the following scenarios: The mission deadline is $m = 100$ and the discount factor takes the value $\delta = 0.9$.

We compute numerically the optimal beaconing policy through Value Iteration algorithm [12]. The obtained policies were compared to the beacon all the time policy (BAT) corresponding to $P_{b,t} = 1$. The optimal beaconing probabilities (resp. value function) for the First scenario are illustrated in Fig. 3 (resp. Fig. 4). When the UAV has a backlogged mission ($N_m = 1$), it beacons with probability one for 3 slots then stops. However, when the UAV has no backlogged mission ($N_m = 0$) it beacons with probability one for 5 slots then stops beaconing. This is justified by the fact that its is more likely for the UAV to meet the tactical team than the CCC. Indeed, λ_d is higher than λ_s. Besides, as the mission deadline tends to expire, beaconing costs will be greater than the expected reward. Hence, it becomes useless to send beacons. From, Fig. 4, it is clear that the cumulative discounted reward for the optimal policy outperforms the one associated with the BAT policy.

In order to evaluate the effect of the beaconing cost on the optimal beaconing policy we choose $C_e = 40 \times C_b = 20\text{J}$. Also, we consider different encounter rate values to investigate their impact on the optimal policy. Indeed, we investigate the following scenarios:

- $\lambda_s = 0.4$ and $\lambda_d = 1$.
- $\lambda_s = 0.1$ and $\lambda_d = 0.25$.

As the beaconing cost increases, the UAV optimal policy is to beacon for a shorter period. Indeed, as illustrated in Fig. 5, UAV beacons only for the 2 (resp. 5) slots when it has a backlogged mission (resp. has no backlogged mission). Also, as the encounter rates decrease ($\lambda_s = 0.1$ and $\lambda_d = 0.25$), the UAV will beacon for longer periods to increase the encounter probability.

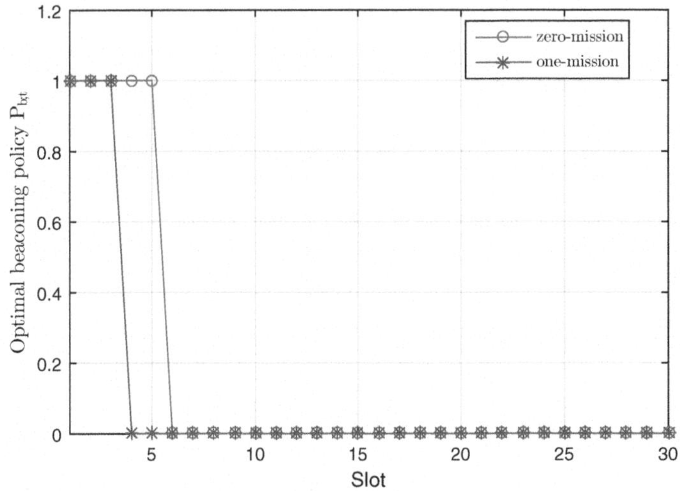

Fig. 3 Optimal beaconing policies ($\lambda_s = 4 \times 10^{-1}, \lambda_d = 1$)

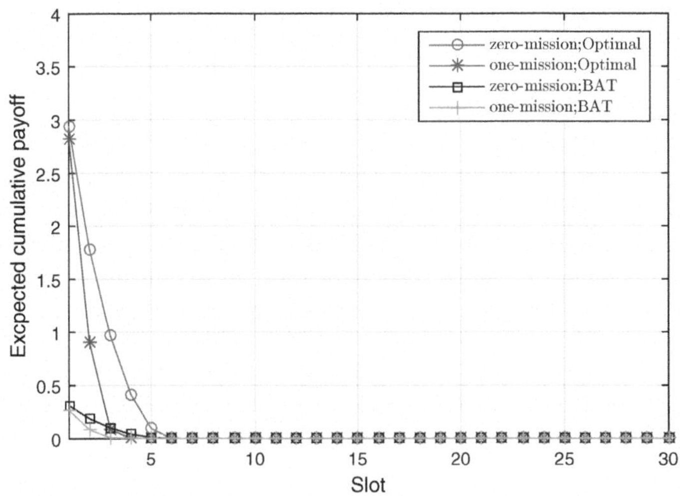

Fig. 4 Expected payoff for different policies

Once can notice that the optimal beaconing policies for both scenarios are of monotonic (i.e., decreasing in time). Particularly, the optimal policy is of threshold type. Such policies are easy to implement since the only required parameter is the threshold value.

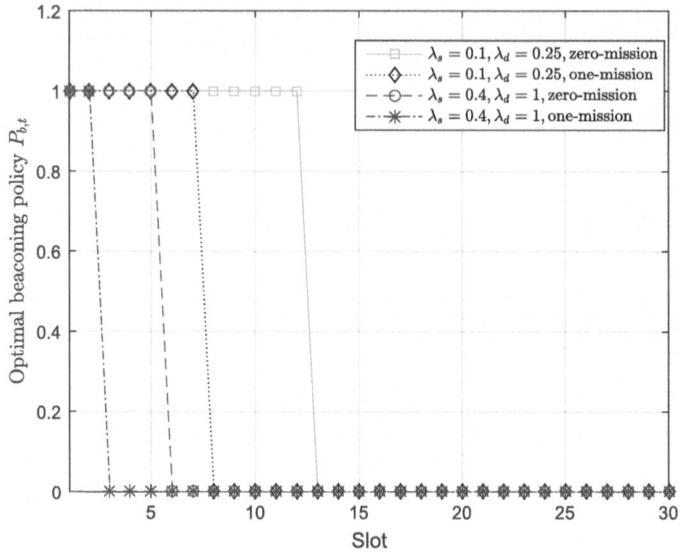

Fig. 5 Optimal beaconing policies for different encounter rates

5 Conclusion and Perspectives

In this paper, we studied the activity scheduling of unmanned aerial vehicles in military operations. We formulated a model based on a markov decision process and characterized optimal beaconing policy. The optimal operating point allows the UAV to efficiently optimize its energy consumption while maximizing the likelihood of getting in contact with the CCC and/or the deployed tactical teams.

As a future work we are working towards generalizing the problem for a large number of UAVs and missions. We are also interested in implementing such a distributed mechanism in a real UAV network. The case where energy harvesting is possible is also a very attractive open issue we would like to deal with.

References

1. Guo, W., Devine, C., Wang, S.: Performance analysis of micro unmanned airborne communication relays for cellular networks. In: Proceedings of 9th International Symposium on Communication Systems, Networks & Digital Signal Processing (CSNDSP) 2014, pp. 658–663 (2014)
2. Facebook, Connecting the World from the Sky. Technical report (2014)
3. Han, Z., Liu, K., et al.: Optimization of manet connectivity via smart deployment/movement of unmanned air vehicles. IEEE Trans. Vehicular Technol. **58**(7), 3533–3546 (2009)
4. Saad, W., Han, Z., Başar, T., Debbah, M., Hjørungnes, A.: A selfish approach to coalition formation among unmanned air vehicles in wireless networks. In: Proceedings of International

Conference on Game Theory for Networks, 2009 (GameNets'09), pp. 259–267 (2009)
5. Choi, D.H., Kim, S.H., Sung, D.K.: Energy-efficient maneuvering and communication of a single UAV-based relay. IEEE Trans. Aerosp. Electron. Syst. **50**(3), 2320–2327 (2014)
6. Luo, C., Nightingale, J., Asemota, E., Grecos, C.: A UAV-cloud system for disaster sensing applications. In: Proceedings of IEEE 81st Vehicular Technology Conference (VTC Spring) 2015, pp. 1–5 (2015)
7. Uragun, B.: Energy efficiency for unmanned aerial vehicles. In: Proceedings of 10th International Conference on Machine Learning and Applications and Workshops (ICMLA) 2011, vol. 2, pp. 316–320 (2011)
8. Li, Y., Wang, Z., Jin, D., Su, L., Zeng, L., Chen, S.: Optimal beaconing control for epidemic routing in delay-tolerant networks. IEEE Trans. Vehicular Technol. **61**(1), 311–320
9. Koulali, S., Sabir, E., Taleb, T., Azizi, M.: A green strategic activity scheduling for UAV networks: a sub-modular game perspective. IEEE Commun. Mag. (to appear)
10. Biondi, E., Boldrini, C., Passarella, A., Conti, M.: Duty cycling in opportunistic networks: the effect on intercontact times. In: Proceedings of the 17th ACM International Conference on Modeling, Analysis and Simulation of Wireless and Mobile Systems 2014, pp. 197–201 (2014)
11. Niyato, D., Wang, P., Tan, H.P., Saad, W., Kim, D.I.: Cooperation in delay-tolerant networks with wireless energy transfer: performance analysis and optimization. IEEE Trans. Vehicular Technol. **64**(8), 3740–3754
12. Puterman, M.L.: Markov Decision Processes: Discrete Stochastic Dynamic Programming. Wiley (2014)
13. Li, Y., Wang, Z., Jin, D., Su, L., Zeng, L., Chen, S.: Optimal beaconing control for epidemic routing in delay-tolerant networks. IEEE Trans. Vehicular Technol. **61**(1), 311–320

Part VI
Special Session 3: From Data to Knowledge: Big Data Applications and Solutions

Document-Oriented Data Warehouses: Complex Hierarchies and Summarizability

Max Chevalier, Mohammed El Malki, Arlind Kopliku, Olivier Teste and Ronan Tournier

Abstract There is an increasing interest in implementing data warehouses with NoSQL document-oriented systems. In the ideal case, data can be analysed on different dimensions. These dimensions follow strict hierarchies that we can use to roll-up and drill-down on analysis axes. In this paper, we deal with non-strict and non-covering hierarchies, common issues in data warehousing a.k.a. summarizability issues. We show how to model these hierarchies in document-oriented systems and we propose an algorithm that can deal with summarizability issues. The new approach is tested and compared to existing approaches.

Keywords Data warehouses · Document-oriented systems · NoSQL · Summarizability

---

M. Chevalier · M. El Malki (✉) · A. Kopliku · O. Teste · R. Tournier
Université de Toulouse, IRIT (UMR 5505), 118 route de Narbonne, Toulouse, France
e-mail: Mohammed.ElMalki@irit.fr

M. Chevalier
e-mail: Max.Chevalier@irit.fr

A. Kopliku
e-mail: Arlind.Kopliku@irit.fr

O. Teste
e-mail: Olvier.Teste@irit.fr

R. Tournier
e-mail: Ronan.Tournier@irit.fr

M. El Malki
Capgemini, 109 avenue du Général Eisenhower, BP 53655, 31036 Toulouse, France

R. El-Azouzi et al. (eds.), *Advances in Ubiquitous Networking 2*, Lecture Notes in Electrical Engineering 397, DOI 10.1007/978-981-10-1627-1_53

1 Introduction

There is an increasing interest in implementing data warehouses with NoSQL systems [20] including document-oriented systems such as MongoDB [5]. NoSQL systems are an interesting alternative to relational databases (RDBMS), because they offer interesting scaling, replication and flexibility features. Until now, the different studies have focused on modelling issues, instantiation and OLAP cuboids (On-Line Analytical Processing [10]). The management of complex hierarchies [6, 13, 17, 21] is an important issue in data warehousing. We introduce in this paper the management of complex hierarchies and summarizability issues with document-oriented data warehouses.

In OLAP settings, it is common to analyse data on different dimension combinations. During analysis, we can drill-down or roll-up at different levels of detail using the hierarchy of dimensions. It is common to have irregularities in these hierarchies such as non-strict hierarchies and non-covering hierarchies. The latter are also the cause of summarizability issues i.e. we cannot drill-down or roll-up in data. Several solutions have been proposed for summarizability issues, but these solutions are adapted to the relational model [1, 6, 8, 9, 12, 19] With these solutions, it is necessary to alter original schemas and to override attribute values to act as arrays. NoSQL systems have interesting features that can useful for dealing with complex hierarchies. This is the scope of this paper.

In particular, document-oriented systems are an interesting case study for managing complex hierarchies. They support atomic attributes as well as the complex attributes (nested records, arrays,...) for storing the data. Document-oriented systems are one of the most popular classes of NoSQL approaches [5]. Data is stored in documents and documents are grouped in collections [3, 5]. Documents have a flexible schema. They contain key-value pairs where keys act as metadata (they represent the data structure). Values can be of simple data type (strings, numbers, dates…), but they can also be arrays or sub-documents. Documents within the same collection can have different schemas. Document-oriented systems have been shown to work well for implementing data warehouses. They can scale horizontally and exploit parallel computation for faster querying. However, until now, the management of complex hierarchies and summarizability issues have not been treated with NoSQL systems in an OLAP setting.

In this context, we extend our previous work on data warehouses implementation with document-oriented systems. We introduce support for storing complex hierarchies and support for data summarization on the complex hierarchies. Our new contribution can be summarized as follows:

We show how we can easily store complex hierarchies in documents
We propose an algorithm for summarizability issues in document-oriented data warehouses. We compare our algorithm to other state-of-the-art algorithms

The rest of this paper is structured as follows. In the next section, we introduce the data warehouse basic notions, the multidimensional data model and the complex hierarchy issues. Then, we propose our approach for modelling, storing and dealing with complex hierarchies. In the following section, we propose experimental work to validate our work. We summarize related work and we end with conclusions. Data warehouses and complex hierarchies.

2 Data Warehouse Design

2.1 Multidimensional Modeling

To ease data analysis and decision making, it is common to centralize them in data warehouses [4]. These latter are suitable for on-line analysis called OLAP. In this setting, data is modelled with a multidimensional model composed of measurable facts and analysis dimensions. Several analysis topics (called facts) regroup a set of indicators (called measures). The values of these indicators are observed by different analytical axes, also called dimensions. These dimensions are composed by attributes, which represent different levels of detail, which are themselves organized into hierarchies.

The traditional example in data warehouses concerns sales as the fact and dimensions like customer, date, and supplier. For the sake of change, we will use another example from social media, more precisely the analysis of tweets (microblogs). In Fig. 1, we show the multidimensional schema. The *tweet* is analysed according to three dimensions: *Time*, *User* and *Subject*. One of the analysis measures is the popularity of a tweet (the number of times a tweet has been retweeted). At different analysis levels, we may wish to have the total amount of retweets grouped by topic or by category or by month or year. The measures can be observed, for example based on the "time" dimension with three detail levels (day, month, year) organized in a hierarchy with "day" the lower detail level, "month" at

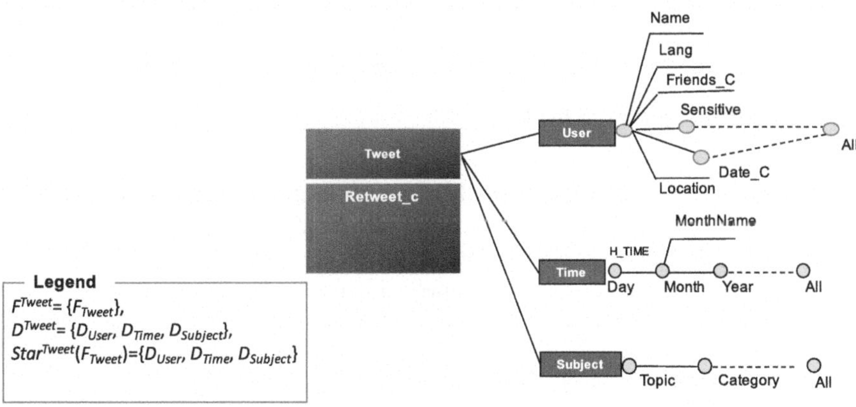

Fig. 1 A multidimensional conceptual schema allowing the analysis of Tweets

a higher level and so on. The hierarchies are useful structures that are employed to ease the pre-calculation of induced agglomeration (for example, calculate the annual sales from the weekly values). Generally, the situations in the real world are modelled according to the simple hierarchies. The associations between the different levels of one simple hierarchy are the type "one-to-many", e.g. one category is divided into many sub-categories.

Below, we provide some formalization on the multidimensional data model and OLAP cuboids [15]:

A **multidimensional schema**, namely E, is defined by $(F^E, D^E, Star^E)$ where: $F^E = \{F_1, \ldots, F_n\}$ is a finite set of facts, $D^E = \{D_1, \ldots, D_m\}$ is a finite set of dimensions, $Star^E: F^E \rightarrow 2^{D^E}$ is a function that associates facts of F^E to sets of dimensions along which it can be analyzed (2^{D^E} is the *power set* of D^E).

A **fact**, $F \in F^E$, is defined by (N^F, M^F) where: N^F is the name of the fact, $M^F = \{f_1(m_1), \ldots, f_v(m_v)\}$ is a set of measures, each associated with an aggregation function f_i.

A **dimension**, denoted $D_i \in D^E$ (abusively noted as D), is defined by (N^D, A^D, H^D) where: N^D is the name of the dimension, $A^D = \{a_1^D, \ldots, a_u^D\} \cup \{id^D, All^D\}$ is a set of dimension attributes, $H^D = \{H_1^D, \ldots, D_v^D\}$ is a set hierarchies.

A **hierarchy** of the dimension D, denoted $H_i \in H^D$, is defined by $(N^{Hi}, Param^{Hi}, Weak^{Hi})$ where: N^{Hi} is the name of the hierarchy; $\mathrm{Param}^{H_i} = \langle \mathrm{id}^D, \mathrm{p}_1^{H_i}, \ldots, \mathrm{p}_{v_i}^{H_i}, \mathrm{All}^D \rangle$ is an ordered set of $v_i + 2$ attributes which are called **parameters** of the relevant graduation scale of the hierarchy, $\forall k \in [1, \ldots, v_i]$, $p_k^{H_i} \in A^D$; $Weak^{Hi}: Param^{Hi} \rightarrow 2^{A^D - Param^{H_i}}$ is a function associating with each parameter possibly one or more weak attributes.

An **OLAP cuboid** O is derived from E, $O = (F^O, D^O)$ such that: F^O is a fact derived from F ($F \in F^E$) with a subset of measures, $M^O \subseteq M^F$; $D^O \subseteq 2^{Star^E(F)} \subseteq D^E$ is a subset of dimensions of D^E. More precisely, D^O is one of the combinations of the dimensions associated to the fact F ($Star^E(F)$).

If we generate OLAP cuboids using all dimension combinations of one fact, we have an OLAP cuboid lattice (also called a pre-computed aggregate lattice or cube).

2.2 *Complex Hierarchies*

In the real world, it is often the case when hierarchies are irregular. We say that the hierarchy is complex when it is a non-strict hierarchy and/or a non-covering hierarchy [6]. We will illustrate and define the above.

In Fig. 2, we show an example of complex hierarchy. The example is taken from an OLAP application on Twitter. The *subject* is one of the analysis dimensions and its attributes form a hierarchy *id-topic-category-all*. We can see that the tweet "*P1*" has two topics "*Foot*" and "*Tennis*"; the topic "*Tennis*" falls within two categories

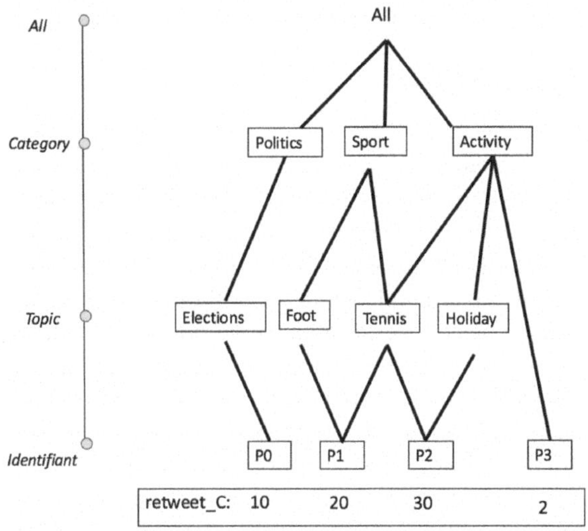

Aggregate by Category

category	retweet_c
Politics	10
Sport	70
Activity	82
Sum	**162**

Aggregate by Topic

topic	retweet_c
Elections	10
Foot	10
Tennis	50
Holidays	30
Sum	**110**

Fig. 2 Example of non-strict and non-covering hierarchy and summarizability issues

"*Sport*" and "*Activity*". This corresponds to a *many-to-many* relationship on *tweet-topic* and *topic-category*. This is called non-strict hierarchy.

The tweet P3 has no topic, but it falls within the category "*Activity*". This corresponds to a *one-to-any* relationship ([1…0–*]) on *tweet-topic*. This is called non-covering hierarchy. Now, we can define:

- A hierarchy is said to be non-strict when a child of a given level can have more than a parent of the superior level [12, 16].
- A hierarchy is said to be non-covering if a dimension value can have no direct upper parent [12, 16].

The complex hierarchies cause summarizability issues [11, 14] i.e. it is not easy to perform drill-down and roll-up analysis on data, because of potential missing or redundant information. One element can be considered several times or none when computing a pre-aggregate (for example the sum of measures by category when a product appears in multiple categories).

Let us illustrate the summarizability issues with our example from Fig. 2. If we count re-tweets by topic (Fig. 2), we obtain a total of 110 while the exact total is 62. The tweets *P1* and *P2* have been counted several times (twice each) which distorts the calculation of aggregates. If we wish to have the amount of re-tweets by category for the aggregate results at the level of topics (Fig. 2), we obtain a result of 162 in place of 62. The topic *Tennis* is attached to two categories. Furthermore, the erroneous aggregate results at the level of topics are reflected in the superior hierarchical levels.

3 Complex Hierarchies and Document-Oriented Systems

3.1 Document-Oriented Data Model

Document-oriented systems store documents in collections and are key-value stores. A unique key identifies every document (the value) that will be called identifier. The document is itself a set of key-value pairs. Keys define the structure of the document and act as meta-data. Each value can be an atomic value (number, string, date…), a sub-document or an array. Documents within documents are called sub-documents or nested documents. We distinguish the document instance from the document structure/schema. The document structure/schema corresponds to a generic document without atomic values i.e. only keys. A document instance belongs to a collection *C* and has an identifier, *id*. We refer to this document as *C* (*id*). We use the following symbols: ":" separates keys from values, "[ ]" denotes arrays, "{ }" denotes documents and a comma "," is used to separate key-value pairs from each other. Using this notation, we provide an example of a document instance:

```
User (30001): {

      name: "John Smith",
      addresses: [{city: "London", country: "UK"},
           {city: "Paris", country: "France"}],
      phone: {prefix: "0033", number: "61234567"}}
```

This example document belongs to the "User" collection, it has 30001 as identifier and it contains keys such as "name", "addresses", "phone". The addresses value is an array of sub documents and the phone value is a sub-document.

3.2 Mapping the Multidimensional Model and Complex Hierarchies

The formalism that we have defined earlier allows us to define a mapping from the conceptual multidimensional model to each of the logical models defined above.

The data model that we will propose is inspired by our previous work [3]. It takes into account that document-oriented implementations of data warehouses work better with flat models i.e. one fact and its dimensions are stored in one collection. This is different from RDBMS where we normalize data and we have one table for the fact and one table per dimension.

Our mapping can be explained in two steps:

(i) For a given fact, all dimension attributes are nested under the respective attribute name and all measures are nested in a sub-document with key "measures". This model is inspired from our work. This corresponds to the following mapping:

- Each conceptual star schema (one F^i and their dimensions $Star^E(F^i)$) is translated in a collection C.
- The fact F_i is translated in a compound attribute Att^{CF}. Each measure m_i is translated into a simple attribute Att^{SM}.
- Each dimension $D_i \in Star^E(F^i)$ is converted into a compound attribute Att^{CD} (i.e. a nested document). Each attribute $A_i \in A^D$ (parameters and weak attributes) of the dimension D_i is converted into a simple attribute Att^A contained in Att^{CD}.

(ii) For attributes within complex hierarchies, we use arrays. There are three cases:

In this case, the attribute can have no values (non-strict hierarchy)

- The attribute value has no value i.e. non-covering hierarchy
- The attribute value has one value i.e. normal behavior
- The attribute value has many values i.e. non-strict hierarchy

Below, there is an example from the Twitter case study. A combination of fact and dimensions will be stored in one document that looks as the following:

```
{
"User": {
        "user_id": "1704005545",
        "user_screen_name": "ann2thingelse",
        "user_friends_count": "150",
        "user_utc_offset": "28800",
        "user_time_zone": "Irkutsk",
        "user_created_at": "Tue Aug 27 07:16:41 +0000 2013",
        "user_lang": "ko",
        "user_location": ""
        },
"Time": {
        "id": "619883842770370560",
        "created": "Sat Jul 11 14:59:59 +0000 2015",
        "timestamp": "1436626799658",
        "day": "11",
        "month": "6",
        "year": "2015"
        },
"Subject": [{"topic": "football",
               "category":[ "football", "Activity]"},
             {"topic": "senat", "category": ["Politycs"]}],
Fact": {
      "Retweet_c": "15"
      }
}
```

3.3 Algorithm for Managing Complex Hierarchies

In this section, we propose an algorithm that can deal with non-strict and non-covering hierarchies.

Let C be a collection corresponding to an OLAP cuboids or detailed data. We will interest to one dimension d and a potentially complex hierarchy H. The data in C is described at some level of granularity; we suppose the lowest level of granularity corresponds to some attribute a. Our goal is to group data on another dimension attribute from H that stands higher in the hierarchy, say attribute b.

Furing aggregation, we suppose we want to apply *sum*(m) an aggregation function on one measure m.

We suppose data is modelled with the mapping we have defined earlier i.e. dimension attribute values within complex hierarchies will be stored with arrays.

To preserve summarizability, we propose on the data model we have proposed the following:

Resolution of non-strict hierarchies: The problem with non-strict hierarchies is that we aggregate measures multiple times when we have multiple values in the group_by dimension attribute. To deal with this issue we propose the use of two variables/fields:

- The ***real value***, which will be displayed for analysis. The real aggregation value is obtained, while aggregating all the measures m from the parent attributes of a^H in b^{H}. This value is calculated without taking into account the number of parents for each child attribute.
- The ***aggregate value***: which, it, will be used uniquely for calculating the superior hierarchical level. The ***aggregate value*** is calculated differently. For each attribute a, the algorithm calculates the number of parents it has in b^H that we call *parents*(a). If the child attribute has a single parent ($|parents(a)| = 1$) the measure will be aggregated one time. If the attribute has several parents ($|parents(a)| > 1$), the algorithm will count the number of parents P (the number of elements in the array) then add the measure aggregated value *sum*(m) will be divided by the number of parents $|parents(a)|$. In this way the measure will not be aggregated as many times as that of the parents.

Resolution of non-covering hierarchies: For treating non-covering hierarchies, we use classical approach, that regroup all the orphan values in an artificial value called *others*. For example, for an aggregation hierarchical level b^{H} a *others* value is created and contains all the orphan values of the hierarchical level a^H. This solution is used already in the relational model [8, 9].

Algorithm SCHS: Algorithm pseudocode for aggregating data (summing) on a measure groupig by a dimension attribute of potential complex hiearchy

```
Input: C // Collection of documents to a cuboid of dat
Param : Prameter used for aggregation
For doc in C do
        If doc.b= ∅ then
                doc.b ← NewParam(other)
                SumAgg[doc.b]+=doc.m
                SumReal[doc.b]+=doc.m
        Else
                For v in doc.b
                        SumAgg[v]+= doc.m / |doc.param|
                        SumReal[v]+= doc.m
        Endif
End
```

4 Experiments

4.1 Experimental Setup

The first experiments are about instantiating a data warehouse with the data model we proposed. We use for this purpose data from the Twitter case study. We load data and we study performance on a set of OLAP queries.

The second experiments are about validating our algorithm for data summarization with complex hierarchies. We also compare our approach to two approaches from state-of-the-art namely:

- The approach of Pederson et al. [12]: an approach that is considered as a reference approach for the summarizability issues
- The approach of Hachicha et al. [6]; that also uses a correction strategy when aggregating.

These two approaches are meant for the relational model; we have adapted them for document-oriented systems.

Hardware: The experiments are done on a cluster composed of 6 PCs, (4 core-i5, 8 GB RAM, 2 TB disks, 1 Gb/s network), each being a worker node and one node acts as dispatcher.

Dataset: The data is obtained with the Twitter API for data streaming [18]. Tweets are returned in JSON data format with each tweet having 67 data fields. We process tweets to follow the data model we have defined earlier. We also add a dimension called *subject* that has as attributes *topic* and *category*. These extra data is fictional and we introduce here arbitrarily non-strict hierarchy issues and non-covering hierarchy issues [18].

Queries: We test our approach to implemented the conceptual model to logical model, on 3 query sets. Three query sets are created with 3 queries per set. The query complexity increases from Q1 to Q3. Q1 involves 1 dimension, Q2 involves 2 dimensions and Q3 involves 3 dimensions.

4.2 Experimental Results: Data Warehouse Instantiation and Validation

In the first set of experiments, we have concentrated in transforming and loading data into MongoDB with the pre-defined model of data.

After loading data, we focus on interrogation. In the following table, we show query execution times on 9 queries on 5 different settings: 1 shard, 2 shards, 3 shards, 4, shards, 5 shards. We can observe that augmenting the number of shards reduces the query time. This is easy to explain. The query is executed in parallel across shards (Table 1).

4.3 Experimental Results 2: Data Summarization with Complex Hierarchies

In this section, we show results on data summarization (aggregation) using algorithms that fix summarizability issues on complex hierarchies. We compare our approach to the approaches of Hachicha and Pedersen. Results are shown in Tables 2 and 3. We use two different settings. In the first setting, we consider one configuration server and one data shard (Table 2). In the second setting, we consider one configuration server and 5 data shards (Table 3).

Table 1 Query execution times at different configuration with 400 millions documents, in seconds

# shards/query	1 shard	2 shards	3 shards	4 shards	5 shards
Q1.1	1070	1042	824	598	497
Q1.2	702	658	433	402	326
Q1.3	697	655	433	408	324
Q2.1	687	656	433	351	286
Q2.2	687	656	433	351	286
Q2.3	687	656	433	352	285
Q3.1	695	676	433	360	285
Q3.2	693	675	433	352	285
Q3.3	693	676	432	353	285

Table 2 Cuboids computation times (in seconds) compared on different approaches on single shard setting with 400 millions documents

Aggregate	Pedersen	Hachicha	SCHC
Topic-day-location	2107	1907	1889
Topic-month	1206	1201	1185
Category-month-location	167	63	64
Year-category	109	48	51
Avg	**3589**	**3219**	**3189**

Table 3 Cuboids computation times (in seconds) compared on different approaches on 5 shards setting with 400 millions documents

Aggregate	Pedersen	Hachicha	SCHC
Topic-day-location	903	808	604
Topic-month	597	534	486
Category-month-location	36	23	12
Year-category	34	15	10
Avg	**1570**	**1308**	**1112**

We show in the tables, the execution time to compute a pre-aggregate (OLAP cuboid) on given dimension combinations. We build cuboids on top of each other i.e. we will compute a cuboid from another existing cuboid that is closer to its granularity of data.

We observe the following results. In the average case, our approach works faster than the other approaches from state-of-the-art. We also observe that it is faster to compute top-level cuboids i.e. cuboids that group on few dimensions and top-level attributes. This is easy to explain, because there is less data. In this case, our approach performance is comparable with state-of-the-art approaches.

The above observations are true on both settings: single shard and multiple shards. We can confirm once again that sharding makes computation faster.

5 Related Work

In 1997, the summarizability has studied for the first time on multidimensional data by Lenz and Shoshani [11]. Since then, three approaches for treating the complex hierarchies have been proposed.

The first approach involves schema normalization. In this solution, two solutions are proposed. For the first, the authors propose to resolve the problem at the conceptual level while defining the rules of constraint and of implementation of the conceptual model towards the logic model [8]. In the second solution of normalization, the principle is to separate the correct hierarchies from the hierarchies susceptible to cause aggregate calculation errors. In this context, [12] propose to put the non-strict hierarchies in the new tables, called joint tables also called separated tables by Malinowski and Zimanyi [12]. In 2008, Mazon et al., proposed a conceptual model normalized UML, separating the different associations [14].

In the second approach, data is transformed for treatment of the complex hierarchies. This approach requires the modification of the fact-dimension instances. Pederson et al. were the first to propose a solution for this perspective [16]. Three algorithms were thus proposed, Makecover which is responsible for making the covered data. Makestrict, is responsible for transforming the multiple hierarchies to the simple hierarchies. For each element having multiple parents, a parent composed from the fusion of its parents is created and inserted between the two. The last algorithm Makeonto is used to manage onto hierarchies [12]. In similar work based on the solution from Pederson, Mansmann and Scholl [11] present a visual tool OLAP which allows for normalizing, browsing and visualizing the different levels of a hierarchy. In their graphic structure, each level of the hierarchy is modeled by a directory.

The third solution has to detect the non-strict hierarchies and non-covering hierarchies, and resolve them at the moment of aggregate computation. These solutions are often accompanied by implementation of operators. In 2005, Horner and Song [8] suggest a script to detect the measures already computed but without ever implementing them. Hachicha and Darmont [6], consider the managing of the summarazibility issues in the documents XML and propose a projection operator which returns a zero result in the case of non-strict hierarchies. In a similar way, Hachicha and Darmont, while drawing from the work of Pederson, propose an operator which operates in multidimensional data XML, by grouping the parents of an element for a single hybrid parent.

The representation of complex hierarchies in conventional relational DBMS turns out to be very complicated, even more so with the explosion of massive data, the bases of relational data shows the benefits of difficulty in management of such massive data. This is why, in this article, we take an interest in a new solution, the systems NoSQL [2], which seems to respond to the problem of massive data [20] in particular the system document-oriented.

6 Conclusions

In this paper, we have studied complex hierarchies and summarization issues in the context of document-oriented implementations of data warehouses.

First, we have proposed a set of rules to automatically translate the conceptual multidimensional schema at the level of logic NoSQL document oriented systems. Furthermore we have conducted a set of experiments to study the loading processes and interrogation. Then, we have tested our approach on datasets with Twitter tweets. Different volumes have been used. We have used MongoDB as the data base NoSQL. The first results show that our approach offers the best results and the best analysis.

As for future work, we hope to conduct investigations at the level of columns oriented models. This latter uses the versioned values (timestamp [7]), a very interesting point for updating data warehouses.

References

1. Chevalier, M., El Malki, M., Kopliku, A., Teste, O., Tournier, R.: Implementation of Multidimensional Databases in Column-Oriented NoSQL Systems (ADBIS 2015), pp. 79–91. Springer (2015)
2. Chevalier, M., El Malki, M., Kopliku, A., Teste, O., Tournier, R.: Implementing multidimensional data warehouses into NoSQL. In: 18th International Conference on Enterprise Information Systems (ICEIS 2016)
3. Chevalier, M., El Malki, M., Kopliku, A., Teste, O., Tournier, R.: Implementing Multidimensional Data Warehouses into NoSQL (ICEIS 2015)
4. Colliat, G.: OLAP, relational, and multidimensional database systems. SIGMOD Rec. **25**(3), 64–69 (1996)
5. Dede, E., Govindaraju, M., Gunter, D., Canon, R.S., Ramakrishnan, L.: Performance evaluation of a mongodb and hadoop platform for scientific data analysis. In: 4th Workshop on Scientific Cloud Computing, pp. 13–20. ACM (2013)
6. Hachicha, M., Kit, C., Darmont, J.: A novel query-based approach for addressing summarizability issues in XOLAP. In: COMAD'12, Pune, India, pp. 56–67. CSI (2012)
7. Hbase: https://hbase.apache.org/
8. Horner, J., Song, I.-Y., Chen, P.P.: An analysis of additivity in OLAP systems. In: DOLAP'04, Washington, DC, USA, pp. 83–91. ACM (2004)
9. Hurtado, C.A., Gutiérrez, C., Mendelzon, A.O.: Capturing summarizability with integrity constraints in OLAP. ACM Trans. Database Syst. **30**(3), 854–886 (2005)
10. Kimball, R., Ross, M.: The Data Warehouse Toolkit: The Definitive Guide to Dimensional Modeling, 3rd edn. Wiley (2013)
11. Lenz, H.-J., Shoshani, A.: Summarizability in OLAP and statistical data bases. In: SSDBM'97, Olympia, Washington, USA, pp. 132–143. IEEE Computer Society (1997)
12. Malinowski, E., Zimányi, E.: Hierarchies in a multidimensional model: from conceptual modeling to logical representation. Data Knowl. Eng. 348–377 (2006)
13. Mansmann, S., Scholl, M.H.: Empowering the OLAP technology to support complex dimension hierarchies. Int. J. Data Warehouse. Min. 31–50 (2007)
14. Mazón, J.-N., Lechtenbörger, J., Trujillo, J.: A survey on summarizability issues in multidimensional modeling. Data Knowl. Eng. **68**(12), 1452–1469 (2009)
15. Morfonios, K., Konakas, S., Ioannidis, Y., Kotsis, N.: R-OLAP implementations of the data cube. ACM Comput. Survey **39**(4), 12 (2007)
16. Pedersen, T., Jensen, Ch., Dyreson, C.: A foundation for capturing and quering complex multidimensional data. Inf. Syst. **26**(5), 383–423 (2001)
17. Rafanelli, M., Shoshani, A.: STORM: a statistical object representation model. In: SSDBM'90, LNCS, Charlotte, USA, vol. 420. Springer (1990)
18. Ben Kraiem, M., Feki, J., Khrouf, K., Ravat, F., Teste, O.: OLAP of the tweets: from modeling toward exploitation. In: IEEE 8th International Conference on Research Challenges in Information Science, RCIS'14, Marrakech, Morocco, pp. 1–10 (2014)
19. Stonebraker, M.: New opportunities for new SQL. Commun. ACM **55**(11), 10–11 (2012)
20. Stonebraker, M., Madden, S., Abadi, D.J., Harizopoulos, S., Hachem, N., Helland, P.: The end of an architectural era. In: 33rd (VLDB), pp. 1150–1160. ACM (2007)
21. Timko, I., Dyreson, C., Pedersen, T.B.: A probabilistic data model and algebra for location-based data warehouses and their implementation. Geoinformatica **18**(2), 357–403 (2014)

Application of APSIS on a Card Payment Solution

Hassan El Alloussi, Karim Benzidane, Othman El Warrak, Leila Fetjah, Said Jai-Andaloussi and Abderrahim Sekkaki

Abstract As the adoption of Cloud Computing is growing exponentially, many issues linked to security and lack of governance have been noted increasingly. In the domain of payment, other than coins and banknotes, the security of digital transaction is a big concern. In this paper, we extend the work done on APSIS (Advanced Persistent Security Insights System) by applying it on a real Cloud based Card Payment Solution. In next steps, we will focus on evaluating Risk Management of deployed Card Transaction Platform on a Public Cloud and all the strategies to reduce impacts of all potential risks.

Keywords Cloud computing · PCI-DSS · Card industry · PCI-SSC · Security Information and Event Management (SIEM) · Security intelligence · Big data · Hadoop · Spark · ITIL · SKMS

H. El Alloussi (✉) · K. Benzidane · O. El Warrak · L. Fetjah · S. Jai-Andaloussi · A. Sekkaki
Laboratory of Research and Innovation in Computer Science, Department of Computer Science, Faculty of Sciences Ain Chock, University Hassan II, Casablanca, Morocco
e-mail: halloussi@gmail.com

K. Benzidane
e-mail: k.benzidane@live.fr

O. El Warrak
e-mail: othmanelwarrak@gmail.com

L. Fetjah
e-mail: l.fetjah@fsac.ac.ma

S. Jai-Andaloussi
e-mail: andaloussi.said@gmail.com

A. Sekkaki
e-mail: a_sekkaki@yahoo.fr

R. El-Azouzi et al. (eds.), *Advances in Ubiquitous Networking 2*, Lecture Notes in Electrical Engineering 397, DOI 10.1007/978-981-10-1627-1_54

1 Introduction

As the competition puts pressure on companies to increase productivity and decrease capital investments, solutions like distributed computing, that offer scalable systems with low fees, are attractive options for management to take under consideration. However, when you are responsible of the overall security aspects of an IT infrastructure, the idea of migrating everything to an environment that is not controlled and even owned, probably makes the decision more difficult.

Therefore, many banks and card transactions companies, which are attracted to outsourcing card solution outside their premises, encounter several hurdles, mainly related to security and data governance. The client has the right to know where its data is and where it is going. This concept is the basis to data security, and plays a significant role in achieving and maintaining compliance with security norms.

Herein, we focus on applying a real Card Payment Solution use case of the platform APSIS developed in [1], for data aggregation, correlation, alerting, dashboard, compliance, retention and forensic analysis, then exposing it to big data with the application of security intelligence in order to have more accurate view on what is happing transaction wise. In the next section, we discuss some basic aspects and definitions around Cloud Computing, Big Data, the Payment Card Industry. In the Sect. 4, we present the Card Data processing solution, object of study. In the Sect. 3, a brief on APSIS [1]. And finally, the use case is illustrated in Sect. 5.

2 Background

2.1 Cloud Computing

Cloud Computing means outsourcing your data and its processing on remote servers, which eliminates the need to store them on premises. The interest is to access that data from any Internet-connected computer and synchronization across multiple devices.

The benefits are many; including a gain of space, resources, time and money. The user can freely access documents without worrying about the machine he uses. Cloud computing is, essentially, an subscription based offer to external services.

However, to adopt the Cloud, the customer should manage security issues at all levels. Indeed, the advent of cloud computing brings new solutions to significant improvement in security. The data are stored in the cloud and should be always accessible no matter what happens to all accessing devices (laptop, Tablet, Smartphone).

2.2 Big Data

Nowadays big data is playing a big role in mutating the world. It is an evolving term that describes any voluminous amount of structured, semi-structured and unstructured data that has the potential to be mined for information. Big Data can be characterized by definition with a several of Vs, but those of interested in regards to data are; the extreme Volume of data, their Variety and their Velocity at which they must be quickly processed. The term is often used when speaking about Petabytes and Exabyte of data, much of which cannot be integrated easily.

2.3 The Payment Card Industry

Electronic payment means all electronic flows of information and treatment needed to manage credit cards and associated transactions. Electronic money transfers have been conducted by banks since the 1960s and bank customers have been able to draw cash from ATM's since the 1970s (NCR, Diebold, Wincor...).

Historically, the first credit Cards, were existed before 1970, and were equipped with only "Embossing" (i.e. customer data printed in relief on the physical media). Information is the number of the card (backed by a bank account), the name and surname of the owner, date of expiry, etc.

In the mid-90s, electronic banking has evolved to include a new fully electronic channel and e-Commerce which is buying and selling of products or services via the web, Internet or other computer networks while M-commerce (or mobile commerce) is the buying of products or services via a device like Smartphone, PDA, etc.

(1) The stakeholders involved with payment card transactions:

Card holder: a person holding a payment card (the consumer in B2C).
Merchant: the business organization selling the goods and services (The merchant sets up a contract known as a merchant account with an acquirer).
Service provider: this could be the merchant itself (Merchant service provider (MSP)) or an independent sales organization providing some or all of the payment services for the merchant.
Acquirer or acquiring bank: this connects to a card brand network for payment processing and also has a contract for payment services with a merchant.
Issuing bank: this entity issues the payment cards to the payment card holders.
Card brand: this is a payment system (called association network) with its own processors and acquirers (such as Visa, MasterCard or CMI card in Morocco) (Fig. 1).

(2) Payment cards flowchart (Fig. 2): Basically payment cards work using two components. The first one, the 'transaction authorization', is where a message containing the transaction details is sent to the card issuer requesting authorization for the payment. The card issuer then authorizes the payment. This guarantees payment to the merchant. The second component known as 'clearing' is where the merchant submits the authorized transaction for payment (automatically or manually; daily or periodically) to Service Provider. The transaction then appears in the card holder's statement.

However, in e-commerce/m-commerce, the payment methods are slightly different.

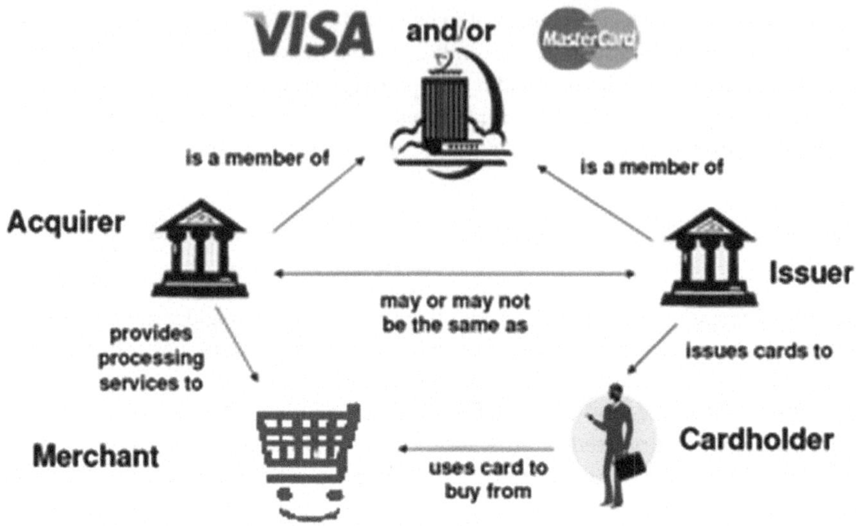

Fig. 1 Payment card stakeholders

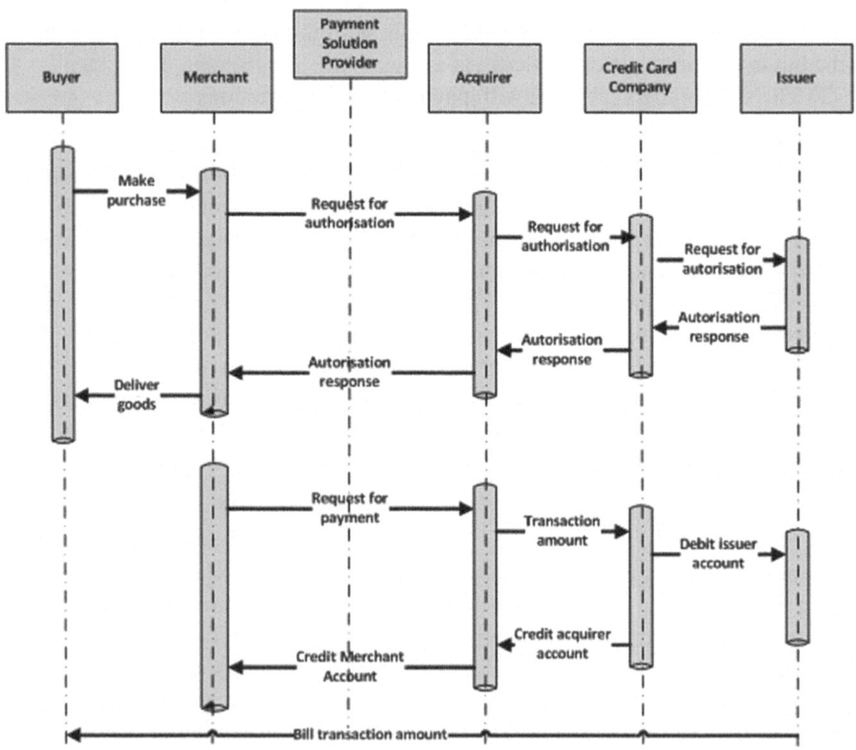

Fig. 2 Payment card flowchart

(3) The e-commerce/m-commerce system model: Generally most e-commerce/m-commerce systems can be designed as a three tier model. The three component parts are the client side, the service system and the back end system. These two last components are commonly known as Server Side. The client side connects users to the Server Side, which deals the users requests. From a business perspective the client side provides the customer interface, the service system provides the business logic and the back-end provides the required data to complete a transaction to its fate.

(4) Card Payment Vulnerabilities: Transaction security

The transaction process highlights the requirement for communication between the users, merchant, card issuer and may be the service provider. These communications must be protected to ensure confidentiality and integrity of the transaction details. This will prevent spying and data manipulation of the transaction details. By understanding the Payment Card system architecture it becomes apparent that the payment card data will be vulnerable if someone having obtained the payment card information details or can access the component parts of the server side system. Additionally, the communications between the component parts of the server side must be protected to ensure confidentiality and integrity of the transaction details.

2.4 Hadoop Ecosystem

Talking about big data would always lead us to think about Hadoop, which is an open-source software framework for distributed storage and distributed processing of very large data sets on computer clusters.

Hadoop relies on a dedicated file system called Hadoop File System (HDFS), which was developed using distributed file system design. It is run on commodity hardware. Unlike other distributed systems, HDFS is highly fault tolerant and de-signed using low-cost hardware. It holds very large amount of data and provides easier access. To store such huge data, the files are stored across multiple machines. These files are stored in redundant fashion to rescue the system from possible data losses in case of failure. HDFS also makes applications available to parallel processing. The goals of HDFS:

- Fault detection and recovery: Since HDFS includes a large number of commodity hardware, failure of components is frequent. Therefore HDFS should have mechanisms for quick and automatic fault detection and recovery.
- Huge datasets: HDFS should have hundreds of nodes per cluster to manage the applications having huge datasets.
- Hardware at data: A requested task can be done efficiently, when the computation takes place near the data. Especially where huge datasets are involved, it reduces the network traffic and increases the throughput.

3 APSIS: Advanced Persistent Security Insights System

In [1] we have described our solution' concept and the landscape of security detection as a new take on a SIEM. As the threats become more sophisticated with the advance of technology, organizations must get sophisticated as well with their IT security measures by acquiring better analytics with swift responses.

As aforementioned, each data entry is a source of information that could play a crucial role in detecting or predicting a poignant attack. That being said, implementing security intelligence at an organization's infrastructure, and coupled with Big Data capabilities would perform way better than any security appliance. Thus making stringent decisions and actions from the most comprehensive set of data gathered across the infrastructure.

Incorporating Big data with a security appliance from our approach' point of view needs to be within the DIKW (Data Information Knowledge Wisdom) model. The underutilization of data is a big issue in data centres today, leaving these facts and figures unprocessed without any form of interpretation or analysis. Processing data would lead us to have information, which would be much more meaningful in a security context. Hence, analyzing and structuring information would lead to an insightful knowledge that can be put or applied into action (preventing, blocking an attack, etc.). This model doesn't only stop at providing knowledge, but it goes far beyond that by integrating these elements and make them useful into bringing wisdom. Which is the ability to make sound judgements and decisions by increasing effectiveness and added value, and also characterized as knowing the right things to do.

Therefore, we have presented a solution called Advanced Persistent Security Insights System (APSIS), in order to achieve each pit stop of the DIKW model within a security context. APSIS is a new take on a comprehensive solution that provides greater visibility into the entire organization' environment in terms of security. The major asset of this approach is delivering comprehensive insights fuelled from various and distributed data entries, resulting into identifying and responding to threats by actionable decisions, thus, reducing security risks and also the rate of false positives. Our approach is a blended integration between in-house modules and open source technologies providing a massively scalable and elastic security intelligence analytics.

The analytics in APSIS relies upon processing events that are accruing across the various layers of the organization such as internal infrastructure, BYOD infrastructure, or even the Cloud. The concept of event processing is about tracking and analyzing streams of data from events to support better insight and decision making. As it is stated in the Vs definition of Big Data, volume and variety of data can be quite challenging to handle. Therefore, the use of Complex Event Processing (CEP) is the best fit for our purpose, since it is a type of event processing that combines data from multiple sources to identify patterns and complex relationships across various events [ref]. It helps also identify threats across many data sources and provides real-time alerts to act on them swiftly.

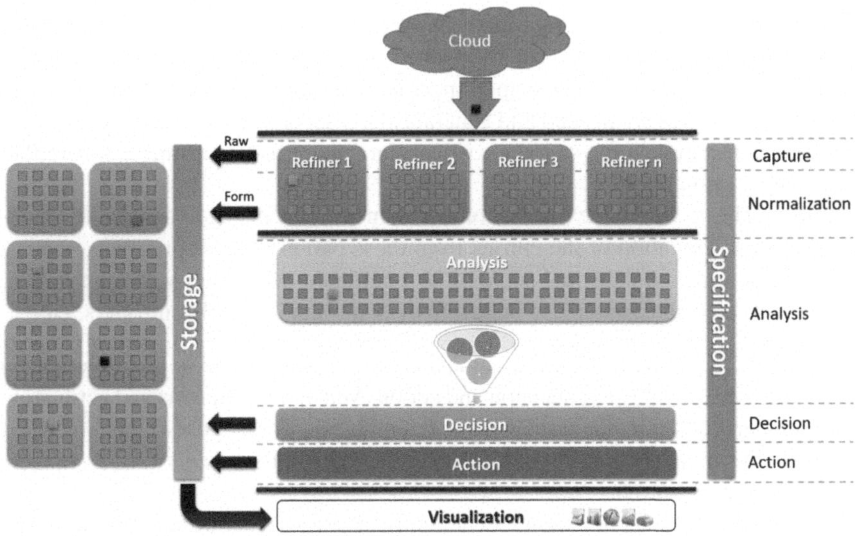

Fig. 3 The logical architecture of APSIS

There are currently seven modules in APSIS, where they all communicate with each other via REST-based API requests. Figure 3, represents a logical architecture which can be one of many ways to architect APSIS. With its pluggable and modular architecture, other tools can easily be swapped and plugged in.

- **Core (Aker)**: The core module is the kernel of APSIS. Its main task is to manage the modules and their communication. The Core provides a central hub for the authentication and authorization of (much like Keystone in OpenStack) each module, delivering them the proper credentials so that they can communicate in a secure channel. It has a SQL database where it is stored most of the overall information about the functioning of the modules; their IP address, timestamps, hardware resources, etc.

- **Refiner**: a component that is responsible for data collection and normalization with real-time pipelining capabilities. It is decoupled into two modules; capture and normalization. The capture module is responsible for the management of the overall assets. The captured entries can be in a pull, push or promiscuous mode. After being captured, the raw entries are archived in a distributed files system, then gone through a normalization process so they can be homogeneous within APSIS. The capture phase can be considered by the normalization module as an input that needs to be filtered.
- **Analyzer**: is a module responsible for real-time downstream analysis of data using policies defined in the PD (Policy Directory) module, then sending results to the decision module for alerting and action purposes. Its main objective is to find all the entries leading to a one single event (malicious activity for example).

- **Decision**: After being under the analysis process, an informative entry is generated then passed on to the decision module. The main purpose of this phase is to classify ingest entries and decide based upon specifications defined in the PD module when reaction measures are needed and generate alerts according to severity levels. This can go from simple information to critical alerts that needs real-time action (either human or automatic).
- **Action**: This is where the interaction between both the decision and the PD modules. The aggregation of the entries from these two modules represents a request that the action module would act upon by interfacing with other solutions via plugins such firewalls, routers, switches, SDN controllers, Cloud controllers, etc.

Therefore an incident management and response process had to be implemented for these modules; analyzer, decision and action. Managing decisions and actions are part of the process of incident management and response. The goal of this process is to contain the impact of unexpected and potentially disrupting events to an acceptable level for an organization.

To ensure that incident management and response will be inline with SLAs and the security policy, it is necessary to (Fig. 4):

- Ensure that incidents can reliably be managed and their impact contained. There must be a formal process in place to detect, identify, assess, and respond to incidents. This should be detailed in a standard or formal process on the PD module, and must be tested periodically.
- Ensure that incident management includes clear and reliable means for the administrator to mitigate events and problems detected by APSIS.
- The incident management process should include periodic reviews and re-porting.

In order to generate accurate decisions and actions, it is necessary to depend on ITIL framework, incident management processes and its SKMS Database Fig. 5. In fact, Service Knowledge Management System (SKMS) is a set of tools and databases that are used to manage knowledge and information. The SKMS includes the

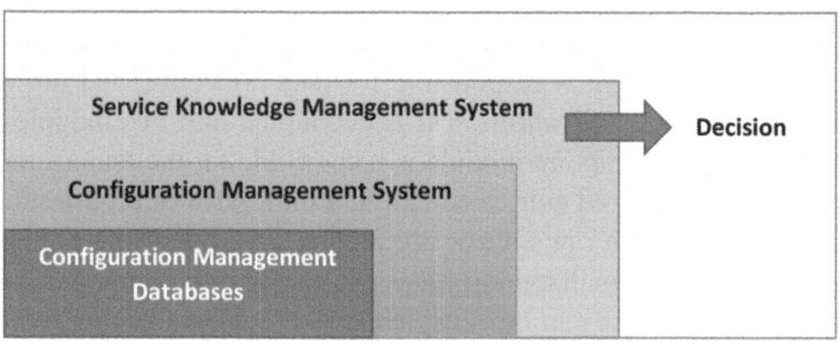

Fig. 4 DIKW model: toward a wiser decision

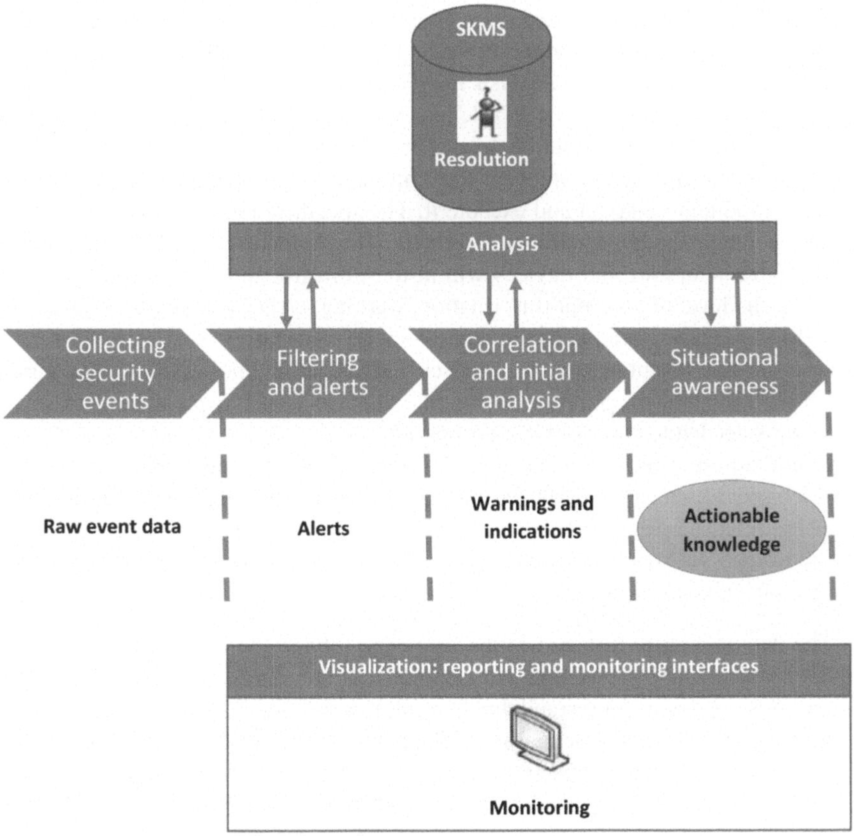

Fig. 5 Conceptual pipeline for events management

Configuration Management System, as well as other tools and databases. The SKMS stores, manages, updates, and presents all information that APSIS needs to manage the full Lifecycle of organization' infrastructure.

- **Policies Directory (specifications)**: Policies are operating rules that can be referred to as a way to maintain order, security, consistency, or otherwise further a goal or mission. From the assets to the action module, the policies directory (PD) specifies business rules-alike in the form of a YAML or JSON files, which each and every module should abide by in order to ensure a smooth interaction between them.

  For instance, each asset is defined and known to APSIS for its integration by an IP address, distinguished name, state (active or inactive), zone, group, deliverance method (push, pull or promiscuous), and Log format to know what filter should be applied, etc. Zones are a collection of assets that defines if a particular analysis

should be applied upon these particular assets. Groups can be found within zones, where filters are applied upon these group of assets that defines their log format so it can be normalized for analysis in APSIS.
For the analysis module, the PD defines what should be analyzed and detected and also what zone it should be upon. For the decision module, we define everything needed to generate an event that can then be a decision carried out to the action module to take on and execute it. The specifications for these latter modules encompasses information and directives about the event and its interactions with other solutions and devices within the infrastructure. These specifications can be the level of an event (information, warning, error), and also what needs to be done if certain parameters are met (block IP, close port), etc.
The PD module can be seen as a parser that takes these policy files as an input, and send them to the appropriate module via REST API as parameters to fill in the gaps needed to carry out its appropriate task.

- **Visualization**: This module is an implementation of a comprehensive dashboard, which provides a web based user interface to the modules for administration purposes, but also gives a visualization for the overall collected data in a much easier way to read via charts show casing all the prominent information such as alerts, warnings, taken actions, etc.

The data stream is gathered inline using the refiner and transformed as alerts, other events are purely informational, and many events can be examined in multiple dimensionsalone or in contextand thus serve as raw data for analysis. Once events are generated and collected, it is analyzed. The analysis consist of illuminating, assessing, and escalating indications and warnings that represent situationsrather than just reporting events. In doing so, we also need to minimize false positives and false negatives.

Real time analysis fits the need for time sensitive detection and response. It is based on simple alerts, which basically requires human review and a solid SKMS database. For example, an Attack signatures can be used to match events of interest against known scenarios, but this matching becomes very complex when detection or analysis time window take a bit longer time to detect persistent attacks. Further analysis involves assessment of complex events on a broader window of time and focuses on establishing context among seemingly unrelated individual events or changes over a long period of time.

As depicted in Fig. 5, we illustrate the event stream from generation, collection, analysis, and up to situational awareness. It also link up an event with the appropriate level, starting from a simple informative alert to a critical alert based upon complex events to improve the degrees of notoriety in terms of detection and action in APSIS (self-aware).

4 Use Case

4.1 Card Data Processing Solution on Cloud

HibaPay is an electronic Card Transaction Platform that allows banks to convert the opportunities offered by the development of smartphones and increase in revenue by setting up financial services that are simple and powerful.

The design of the platform HibaPay considers integrated manner the interests and constraints of the various stakeholders: the Client, the Merchant and the Bank. It also includes a prospective view of the state of the art either in the world of Mobile Phones or that of Electronic Card payment.

The platform HibaPay allows banks and operators to explore all business development opportunities offered by the financial Solutions in meeting with the specific needs of each market. HibaPay includes modules able to realize synergies between market players of mobile financial services.

In the Fig. 6, we illustrate the architecture of the solution HibaPay:

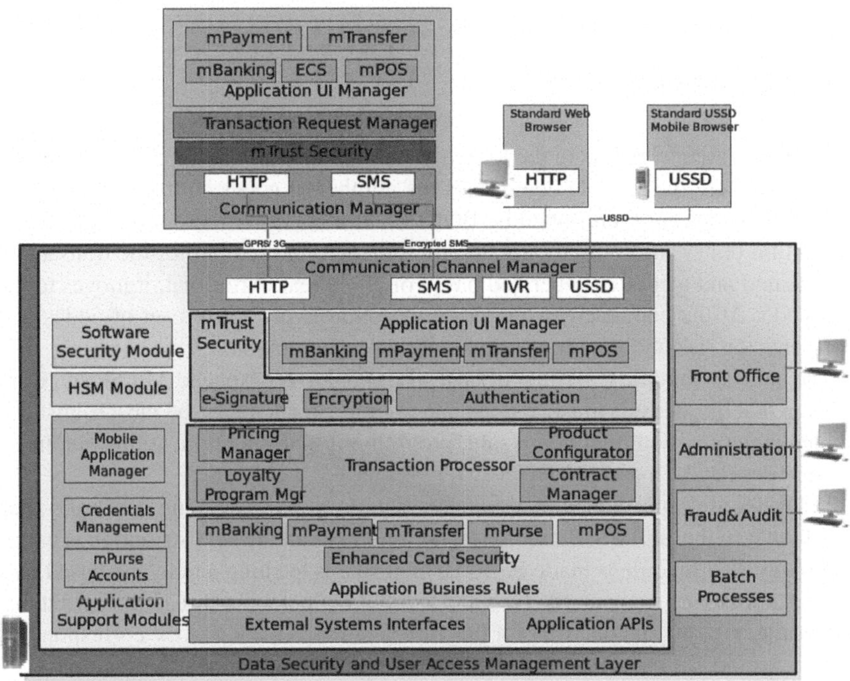

Fig. 6 The card processing solution architecture

The HibaPay main modules are:
The server Modules:

- mTrust Security
- Software Security Module
- Transaction Manager
- Credentials Manager
- Applications UI Manager
- Communication Channels Manager
- Mobile Application Manager
- mPurse Accounts Manager

The application Modules:

- mBanking
- mPayment
- mTransfer
- mPurse
- mPoS
- Enhanced Card Security

4.2 Application of APSIS

The module' architecture is in the form of a controller as described above, in the initialization phase the module determines the number of nodes constituting the controller (the nodes are represented by BOLTs), after being received, every payement transaction (entry log) is analyzed step by step, and on each hop of the transaction is evaluated and a code is generated based on the threat. After that, it moves to the next node. At the end, each suspicious transaction is paired to a set of codes that represents the risk and then transmitted to the decision module.

The analysis module is written in Java and paired to the Apache API Storm. Considering that the analysis must be done in real time, Storm offers the most robust and efficient solution, with that being said, programming Storm applications requires a set of nodes: BOLTS and SPOUTS.

SPOUTS are input nodes that recovers data by a pull (fetching data) and then distribute it to the BOLTS; these nodes, their number and their distribution is called a topology. Our module is made of Bolts, their role is fetching a field and based on a set of patterns they evaluate the type and level of the potential threat, the action that comes after is to use an identifier for that threat and send the log to the correspondent BOLT.

At the end of the program cycle we end up with safe transactions on one hand and a log of suspicious ones on the other, results are then sent to the decision module.

Regarding the IO Module, it has in addition more input specification in a stream of JSON code that recovers refiners who one analyzed returns as a result a JSON enriched by the results of the analyzer.

The architecture and organization of the module components (Bolts) is decided at the initiation of the topology. The topology retrieves the patterns and launches the necessary numbers of bolts and their organization so as to form a controller with multiples inputs (refiners) and two outputs (safe log and suspicious log). The main gain from this procedure is real time data analysis therefore the different transactions must be evaluated at the hover of a mouse.


```
INFO [2016-03-18 19:19:39,691] (MakeTransactionController.java,119) - Received Request :{"schemaVersion":"1.0",
"serviceparams":{"payerinfo":{"payment_account_info":{"accountNumber":"2","accountPIN":"test","subscriptionId":"2"
},"payment_method":"mwallet_account","payment_type":"SWAP_CARD"},"sessionId":"session"},"timestamp":"1458326591",
"transactioninfo":{"amount":"4500","currency":"USD","description":"comm","invoiceId":"3"},"deviceInformation":{
"deviceIdType":"ANDROID_ID","deviceId":"547ccb2dd363824c"},"locationInformation":{"lacid":"0","latitude":"",
"longitude":"","mac":"","mcc":"310","mnc":"260","cellId":"0","locationInfoType":"gps"},"merchantInfo":{"cashierId":
"28","merchantId":"1"},"requestId":"requestID"}
 INFO [2016-03-18 19:19:39,692] (MakeTransactionController.java,121) - Operation Response :{"schemaVersion":"1.0",
 "requestId":"requestId","timestamp":"2016-03-18 19:19:39.277","create_time":"2016-03-18 19:19:39.277",
 "update_time":"2016-03-18 19:19:39.277","state":"Approved","transactionId":"47","description":"comm"}
```


Fig. 7 Logs collected from payment card server


```
{
    "logInformation":
        {
            "time":"2016-03-18 19:18:56,642",
            "operation":"request",
            "requestId":"requestID",
            "type":"INFO",
            "otherInformation":"BalanceQueryController.java,59",
            "schemaVersion":"1.0"
        },
    "timestamp":"1458326555",
    "deviceInformation":
        {
            "deviceIdType":"ANDROID_ID",
            "deviceId":"547ccb2dd363824c"
        },
    "locationInformation":
        {
            "lacid":"0",
            "latitude":"",
            "longitude":"",
            "mac":"",
            "mcc":"310",
            "mnc":"260",
            "cellId":"0",
            "locationInfoType":"gps"
        },
    "merchantInformation":
        {
            "cashierId":"28",
            "merchantId":"1"
        }
}
```


Fig. 8 Logs transformed as standard format

```
{
    "logInformation":
        {
            "time":"2016-03-18 19:18:56,642",
            "operation":"request",
            "requestId":"requestID",
            "type":"INFO",
            "otherInformation":"BalanceQueryController.java,59",
            "schemaVersion":"1.0"
        },
    "timestamp":"1458326555",
    "deviceInformation":
        {
            "deviceIdType":"ANDROID_ID",
            "deviceId":"547ccb2dd363824c"
        },
    "locationInformation":
        {
            "lacid":"0",
            "latitude":"",
            "longitude":"",
            "mac":"",
            "mcc":"310",
            "mnc":"260",
            "cellId":"0",
            "locationInfoType":"gps"
        },
    "merchantInformation":
        {
            "cashierId":"28",
            "merchantId":"1"
        }
    "errorInfo"
        {
            "codeError":"D00"
            "nameError":"invalid deviceId"
        }
}
```


Fig. 9 Event classification

In the Fig. 7, we have a sample of codes retrieved from the logs generated by the application:

These logs are transformed by the refiner' normalization process to be more in standard as below (Fig. 8):

And after analyzing and adding information about the occurred event, in this case an error with an invalid device ID (Fig. 9):

It is passed to the display module through Kibana to give a dashboard experience and view all what is happening (Fig. 10):

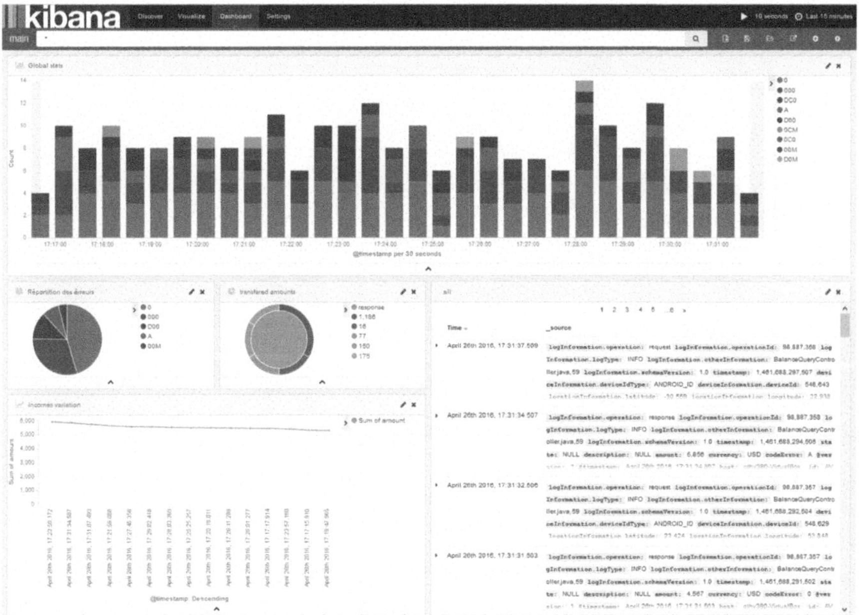

Fig. 10 Visualization of the event on a dashboard

5 Conclusion

The objective of our work is to implement the APSIS approach (a complete solution for Security Intelligence), that relies on taking advantage of the traditional capabilities of a SIEM system; which are data aggregation, correlation, alerting, dashboards, compliance, retention and forensic analysis, then exposing it to big data with the application of security intelligence in order to have more accurate view on what is happing on the infrastructure, on a real environment of Card Payment in the Cloud.

In this paper, we have explained our approach exhaustively and we argued it by a practical use case. In the next steps of our work, we will focus on applying the APSIS on other industries and improve it to be more convenient.

Reference

1. Benzidane, K., El Alloussi, H., El Warrak, O., Fetjah, L., Andaloussi, S.J., Sekkaki, A.: Toward a cloud-based security intelligence with big data processing. In: IEEE/IFIP Network Operations and Management Symposium, Apr 2016

Predicting Chronic Kidney Failure Disease Using Data Mining Techniques

Basma Boukenze, Abdelkrim Haqiq and Hajar Mousannif

Abstract Kidney failure disease is being observed as a serious challenge to the medical field with its impact on a massive population of the world. Devoid of symptoms, kidney diseases are often identified too late when dialysis is needed urgently. Advanced data mining technologies can help provide alternatives to handle this situation by discovering hidden patterns and relationships in medical data. The objective of this research work is to predict kidney disease by using multiple machine learning algorithms that are Support Vector Machine (SVM), Multilayer Perceptron (MLP), Decision Tree (C4.5), Bayesian Network (BN) and K-Nearest Neighbour (K-NN). The aim of this work is to compare those algorithms and define the most efficient one(s) on the basis of multiple criteria. The database used is "Chronic Kidney Disease" implemented on the WEKA platform. From the experimental results, it is observed that MLP and C4.5 have the best rates. However, when compared with Receiver Operating Characteristic (ROC) curve, C4.5 appears to be the most efficient.

Keywords Data mining · Kidney disease · SVM · ANN · C4.5 · BN · KNN · Performance

B. Boukenze (✉) · A. Haqiq
Computer, Networks, Mobility and Modeling Laboratory, NGN Research Group, Africa and Middle East, FST, Hassan 1st University, Settat, Morocco
e-mail: basma.boukenze@gmail.com

A. Haqiq
e-mail: Ahaqiq@gmail.com

H. Mousannif
LISI Laboratory, FSSM Cadi Ayyad University, 40000 Marrakesh, Morocco
e-mail: mousannif@uca.ac.ma

R. El-Azouzi et al. (eds.), *Advances in Ubiquitous Networking 2*, Lecture Notes in Electrical Engineering 397, DOI 10.1007/978-981-10-1627-1_55

1 Introduction

Kidneys are the seat of different ailments. Unlike painful renal colic, chronic renal failure often develops without symptoms. It is called "terminal" because it develops without warning signs. For severe cases, the only remedy option is either dialysis or organ transplant [1].

According to the 2010 Global Burden of Disease study, chronic kidney disease was ranked 27th in the list of causes of the total number of deaths worldwide in 1990. But it rose to 18th in 2010. 10 % of the population worldwide is affected by chronic kidney disease (CKD), and millions die each year because they do not have access to affordable treatment.

The British Medical Journal published Saturday, March 11, 2006 an editorial explaining that in France, the annual incidence (number of new cases of kidney disease) is higher than 100 per million inhabitants. It has doubled in this country in ten years, and will continue to grow by 5 to 8 % per year [2]. Chronic kidney disease can be treated. With early diagnosis, it is possible stop or at least slows its progression.

Information and Communication Technologies (ICT) play a very important role because they are the basis of the knowledge economy. They are used to store, process an increasing volume of data, and are a source of productivity gains and decision making for many domains, like the medical field and especially information about chronic kidney failure disease as a treatable disease if it is diagnosed early [3]. And when we talk about big data in the medical sector, data mining techniques such as classification, clustering and combination rules, play a great role in extracting unknown knowledge from databases. Classification is a data mining technology used to predict group membership for data instances [4].

Big data can exploit medical data to extract value and hidden knowledge that can help in decision making, and reduce the rate of false diagnosis. The goal is to reduce the cost of treatment, predict diseases in the first stage and anticipate the treatment to relieve suffering of sickness or even save people's lives.

Several classification methods are used to predict kidney failure disease, but in this study we focus on SVM [5], C4.5 [6], NB [7], K-NN [8], as they are classified currently among the top 10 classification methods Identified by IEEE and Related Python Resources [9]; they constitute a very active area of research and their classification proved a success on different domains.

The reminder the paper is organized as follows. Related works are discussed in Sect. 2. The proposed methodology is given in Sect. 3. Section 4 describes experimental results. Section 5 discusses the experimental results. Section 6 concludes the paper.

2 Related Works

Ashfaq Ahmed et al. [10] have presented a work using machine learning techniques, namely Support Vector Machine [SVM] and Random Forest [RF]. These were used to study, classify and compare cancer, liver and heart disease data sets with varying kernels and kernel parameters. Results of Random Forest and Support Vector Machines were compared for different data sets such as breast cancer disease dataset, liver disease dataset and heart disease dataset. It is concluded that varying results were observed with SVM classification technique with different kernel functions.

Vijayarani and Dhayanand [11] have presented a work to predict kidney disease by classifying four types of kidney diseases: Acute Nephritic Syndrome, Chronic Kidney disease, Acute Renal Failure and Chronic Glomerulonephritis, using Support Vector Machine (SVM) and Artificial Neural Network (ANN), then comparing the performance of those two algorithms on the basis of accuracy and execution time. The results show that the performance of the ANN is better than the SVM algorithm.

Palaniappan and Awang [12], developed a prototype called Intelligent Heart Disease Prediction System (IHDPS) using data mining techniques, namely, Decision Trees, Naïve Bayes and Neural Network. Results show that each technique has its unique strength in realizing the objectives of the defined mining goals. The effectiveness of models was tested using two methods: Lift Chart and Classification Matrix. The most effective model to predict patients with heart disease appears to be Naïve Bayes followed by Neural Network and Decision Trees.

Fan et al. [13] discovered the information of breast cancer recurrence of SEER data. After preprocessing the dataset, they investigated several algorithms like C 5.0, CHAID, CART, and QUEST. As a result, C5 algorithm showed to have the best performance of accuracy.

In Lakshmi et al. [14], three data mining techniques (Artificial Neural Networks, Decision tree and Logical Regression) are used to elicit knowledge about the interaction between variables and patient survival. A performance comparison of three data mining techniques is employed for extracting knowledge from data collected at different dialysis sites. ANN is suggested for Kidney dialysis to get better results with accuracy and performance.

In Vijayarani and Dhayanand [15], classification process is used to classify four types of kidney diseases. Comparisons of Support Vector Machine (SVM) and Naïve Bayes classification algorithms are done based on the performance factors, classification, accuracy and execution time. As results, the SVM achieves increased classification performance. Hence it is considered as the best classifier when compared with Naïve Bayes classifier algorithm. However, Naïve Bayes classifier classifies the data with minimum execution time.

In this study, we apply data mining techniques, recently ranked among the top 10 as best classifiers, to predict chronic kidney disease on the basis of the information attributes in the database used in order to categorize patients who are

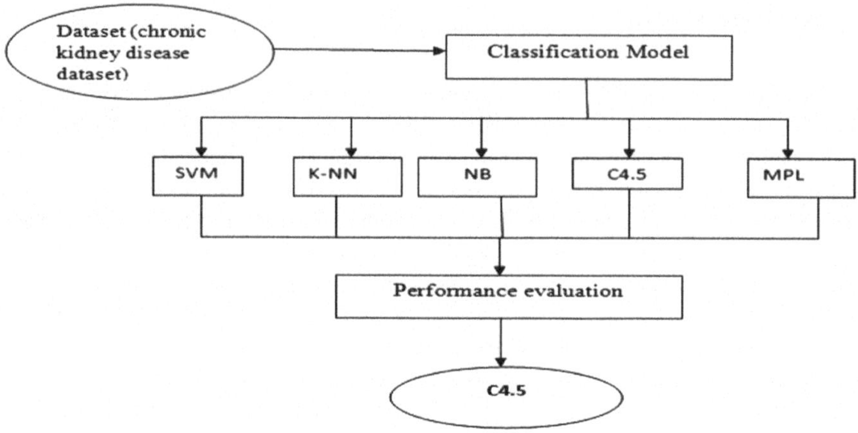

Fig. 1 System architecture

suffering from the chronic kidney disease (ckd) and patients who are not suffering from it (notckd).

3 Methodology

See Fig. 1

4 Environment

In this study, we use the Waikato Environment for Knowledge Analysis (Weka). It is a comprehensive suite of Java class libraries that implement many algorithms for data mining clustering, classification, regression, and analysis of results. This platform offers researchers a perfect environment to implement and evaluate their classification models compared to TANAGRA or ORANGE [16].

4.1 Data Set

We used the database Chronic Kidney Disease Dataset from UCI Machine Learning Repository [17]. This database contains 400 instances and 24 Integer attributes, with two classes (chronic kidney disease (ckd), non chronic kidney disease (notckd)). Table 1 describes the attributes of the database, while Table 2 describes the distribution of classes.

Table 1 Information attributes

Attribute	Representation	Information attribute	Description
Age	Age	Numerical	Years
Blood pressure	Bp	Numerical	Mm/Hg
Specific gravity	Sg	Nominal	1.005, 1.010, 1.015, 1.020, 1.025
Albumin	Al	Nominal	0.1.2.3.4.5
Sugar	Su	Nominal	0.1.2.3.4.5
Red blood cells	Rbc	Nominal	Normal, abnormal
Pus cell	Pc	Nominal	Normal, abnormal
Pus cell clumps	Pcc	Nominal	Present, not present
Bacteria	Ba	Nominal	Present, not present
Blood glucose random	Bgr	Numerical	Mgs/dl
Blood urea	Bu	Numerical	Mgs/dl
Serum creatinin	Sc	Numerical	Mgs/dl
Sodium	Sod	Numerical	mEq/L
Potassium	Pot	Numerical	mEq/L
Haemoglobin	Hemo	Numerical	Gms
Packed cell volume	Pcv	Numerical	
White blood cell count	Wc	Numerical	Cells/cumm
Red blood cell count	Rc	Numerical	Millions/cmm
Hypertension	Htn	Nominal	Yes, no
Diabetes mellitus	Dm	Nominal	Yes, no
Coronary artery disease	Cad	Nominal	Yes, no
Appetite	Appet	Nominal	Good, poor
Pedal edema	Pe	Nominal	Yes, no
Anemia	Ane	Nominal	Yes, no
Class	Classe	Nominal	Ckd notckd

Table 2 Class distribution

	Class	Distribution
1	Ckd	250 (62.5 %)
2	Notckd	150 (37.5 %)

4.2 Measures of Performance

Evaluation of classification algorithms is one of the key points in any data mining process. The most commonly tools used in classification algorithms results analysis are: confusion matrix and receiver operating Characteristic curve (ROC) [18].

In this study, we use a confusion matrix then we calculate different measures of performance, and we focus on the most important criteria identified in [19]. The following section defines the most useful performance indicators we used to compare our classifiers.

Precision: Percentage of correctly classified elements for a given class

$$\text{Precision} = \text{TP}/(\text{TP} + \text{FP}) \tag{1}$$

Sensitivity (recall): The probability of correctly detecting that the subject suffers from the disease. A higher sensitivity means that the predictive model can easily detect the disease.

$$\text{Sensitivity(recall)} = \text{TP}/\text{TP} + \text{FN} \tag{2}$$

F-measures: is the harmonic mean of precision and recall, used to evaluate each classifier.

$$\text{F} - \text{measures} = 2 * \text{precision.recall}/\text{prescision} + \text{recall} \tag{3}$$

Accuracy: The accuracy represents the total accuracy rate of classifying each subject into the correct group. This index not only represents the probability of accurately classifying the subject as healthy or not, but also correctly classifying each patient into the correct disease group.

$$\text{Accuracy} = (\text{TP} + \text{TN})/\text{TP} + \text{FP} + \text{TN} + \text{FN} \tag{4}$$

Error: indicates the proportion of cases classified incorrectly.

$$\text{Error} = 1 - \text{accuracy} \tag{5}$$

Metrics:

TP: True positive Rate is the fraction of positive cases predicted as positive.
FP: False positive Rate is the fraction of negative cases predicted as positive.
TN: True negative Rate is the fraction of negative cases was correctly classified as negative.
FN: False negative Rate is the fraction of positive cases that were incorrectly classified as negative (Table 3).

Table 3 Confusion matrix

		Predicted class	
		True class	False class
Actual class	True class	True positive (TP)	False positive (FP)
	False class	True negative (TN)	False negative (FN)

To visualize the classifiers' performance, we use another tool: ROC curve (Receiver Operating Characteristic). The ROC curve is a representation of the true positive rate according of the false positive rate.

5 Experimental Results

In order to evaluate our classifiers' performance, we implement them on Weka and apply them to our chronic kidney disease database. To do so, we go through three steps:

Pre-processing: This consists in selecting and loading data from the database. It shows us the distribution of every attribute.

Classifying: Once our data is loaded, we apply the chosen algorithm. To do so, we must choose a test option. In this study, we use cross validation with tenfold because it is the most appropriate option that gives us the possibility to evaluate the classifiers in term of performance.

Evaluating: The following table shows the results obtained after the application of each classifier on the database. The classifiers' outputs are shown in Table 4.

6 Discussion

In this work, we applied machine learning algorithms to predict patients who have chronic kidney disease, and those who are not sick, based on the data of each attribute for each patient. Our goal was to compare different classification models

Table 4 Classifiers performance

Evaluation criteria	C4.5	SVM (SMO)	K-NN (IBK)	NB	MLP
Time to build model (s)	0.08	0.41	0.01	0.03	23.95
Correctly classified instances	396	391	383	380	399
Incorrectly classified instance	4	9	17	20	1
Accuracy (%)	63	60.25	58.25	57.5	62.25
Error	0.37	0.39	0.41	0.42	0.37

and define the most efficient one. Our comparison was made on the basis of four algorithms ranked among the top 10; SVM, NB, K-NN and C4.5, and another algorithm that showed its performance in many predictions in healthcare: Multi-layer Perceptron as a network model of artificial neuron (see Fig. 1). The results after the implementation of classifiers on Weka show that:

K-NN is the fastest classifier because it spent the shortest time to build the classification model (0.01 s) followed by NB, C4.5 and SVM. MLP is the slowest because it took 23.95 s to build its model (see Fig. 2).

Regarding accuracy, which represents the percentage of instances classified correctly, we notice a variation between 50 and 60 %. This has no relationship with the classifiers, but it has it with the application domain and type of data. In our study, C4.5 scored a good accuracy (63 %) followed by MLP (62.25 %). Since Accuracy alone is not enough to define classifiers performance, we used many other criteria (see Fig. 2).

Regarding error rate, MLP and C4.5 marked the smallest error rate (0.37) and the largest one was scored by NB (0.42) (see Table 4).

The kappa statistic value (Table 5) shows that the value of all predictors is above 0.81. This means that our classifiers are excellent according to degree scale proposed by (Landis and Koch) [20], except that MLP scored the best prediction agreement followed by C4.5.

Regarding the measurement of predictors, the values of MAE, RMSE, RAE, RRSE showed that MLP predictors scored the lowest values (MAE = 0.00) (RMSE = 0.06, SER = 1.8, RRSE = 12.85) followed by C4.5, SVM, K-NN and NB (see Fig. 3).

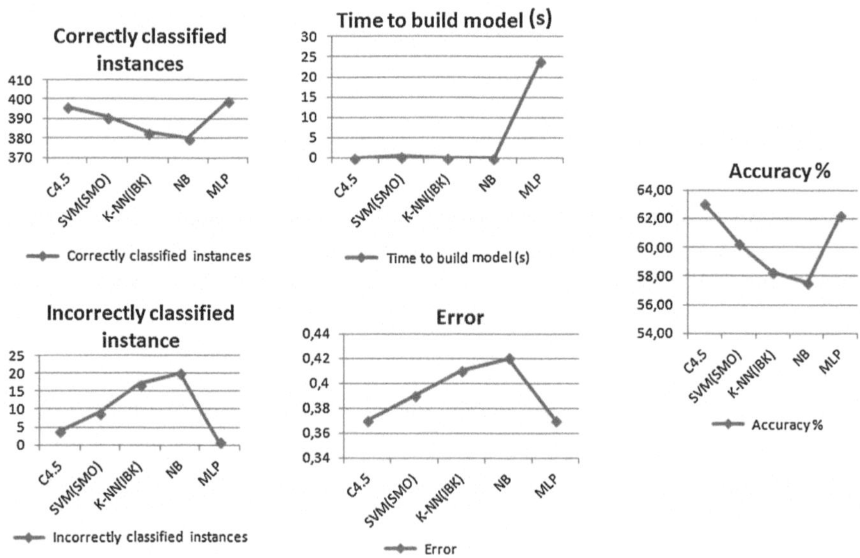

Fig. 2 Comparative graph of classifiers performance

Table 5 Simulation error

Evaluation criteria	C4.5	SVM(SMO)	K-NN	NB	MLP
Kappa statistic	0.97	0.95	0.91	0.89	0.99
Mean absolute error	0.02	0.02	0.04	0.04	0.00
Root mean squared error	0.08	0.15	0.20	0.20	0.06
Relative absolute error %	4.79	4.79	9.60	10.21	1.80
Root relative squared error %	16.66	30.98	42.46	42.25	12.85

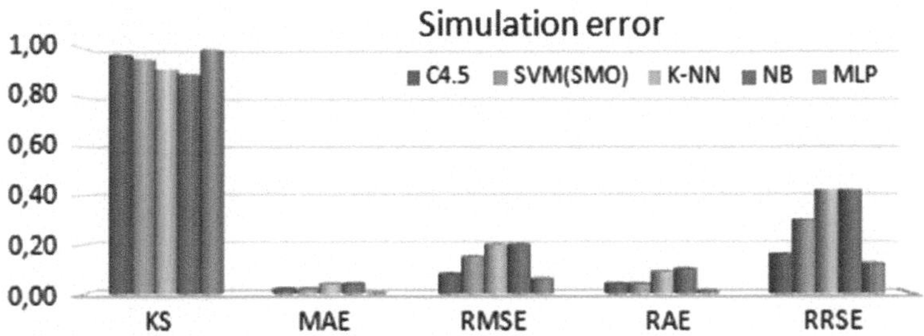

Fig. 3 Comparative diagram of learning algorithms

Another important measure is F-Measures which combines two performance measures: precision and recall. If we take the case of predicted patient with the disease (ckd), C4.5 and MLP marked the best and the same rate (0.99), and in the case of non disease (notckd) MLP was more reliable (Table 6). The confusion matrix (Table 7) shows us that MLP classified 399 instances correctly with 1 misclassified instance followed by C4.5 (396) and 4 instances as misclassified, then SVM, K-NN and NB (see Fig. 4).

Table 6 Accuracy measures by class 1

	TP	FP	Precision	Recall	F-measure	Class
C4.5	0.99	0.02	0.98	0.99	0.99	Ckd
	0.98	0.004	0.99	0.98	0.98	Notckd
SVM	0.96	0	1	0.96	0.98	Ckd
	1	0.03	0.94	1	0.97	Notckd
K-NN	0.93	0	1	0.93	0.96	Ckd
	1	0.06	0.89	1	0.94	Notckd
NB	0.92	0	1	0.92	0.95	Ckd
	1	0.08	0.88	1	0.93	Notckd
MLP	0.99	0	1	0.99	0.99	Ckd
	1	0.00	0.99	1	0.99	Notckd

Table 7 Diffusion Matrix

	Ckd	NotCkd	
C4.5 (J48)	249	1	Ckd
	3	147	Notckd
SVM (SMO)	241	9	Ckd
	0	150	Notckd
K-NN	233	17	Ckd
	0	150	Notckd
NB	230	20	Ckd
	0	150	Notckd
NN (MLP)	249	1	Ckd
	0	150	Notckd

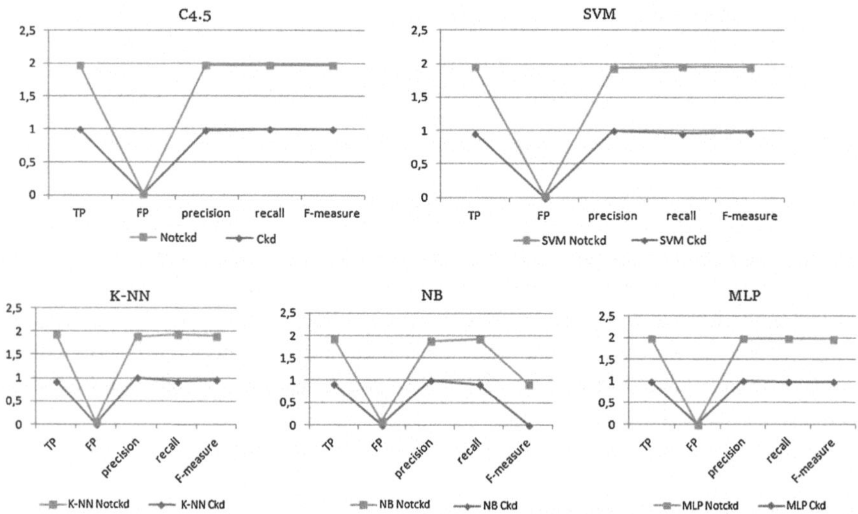

Fig. 4 Comparative graph of accuracy measures of used classifiers

MLP is deducted as the most efficient in terms of greatest number of instances correctly classified and the lowest error rate at the prediction. It is also the second one in accuracy and has the best f-Measures rate, but with the highest time of execution.

C4.5 is ranked as the second one after MLP, but it outperforms in building time of the classification and accuracy, and has the same error rate. The other algorithms are classified as follows: SVM, K-NN and NB.

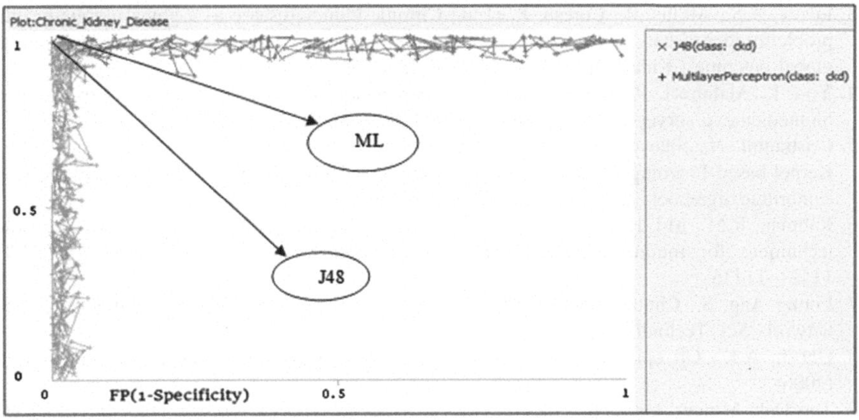

Fig. 5 ROC graph between MLP and C4.5

In conclusion, we can deduce that MLP and C4.5 are the most efficient algorithms for kidney failure prediction. The performance of MLP is higher than that of C4.5 when we refer to ROC graph (Graph 1), but it takes more time to execute than C4.5 (Fig. 5).

7 Conclusion

Applying data mining technologies in the medical field is very important because they definitely help in the decision making process. But to do so, such algorithms demand high performance with high accuracy, and a right choice of methods according to the context of work and the data handled. In this study, we employed five learning algorithms: C4.5, SVM, MLP, NB and K-NN, and C4.5 applied on the chronic kidney failure dataset, and we tried to compare them in terms of many criteria: accuracy, execution time, sensitivity and specificity. C4.5 has proved its performance on several levels ahead MLP, especially by the lowest error rate, and shortest execution time.

References

1. World Kidney Day: Chronic Kidney Disease. http://www.worldkidneyday.org/faqs/chronic-kidney-disease/(2015)
2. Jha, V., Garcia-Garcia, G., Iseki, K., et al.: Chronic kidney disease: global dimension and perspectives. Lancet **382**(9888), 260–272 (2013)

3. Levey, A.S., Atkins, R., Coresh, J., et al.: Chronic kidney disease as a global public health problem: approaches and initiatives—a position statement from Kidney disease improving global outcomes. Kidney Int. **72**(3), 247–259 (2007)
4. Yoo, I., Alafaireet, P., Marinov, M., Pena-Hernandez, K.: Data mining in healthcare and biomedicine: a survey of the literature. J. Med. Syst. **36**(4), 2431–2448 (2012)
5. Cristianini, N., Shawe-Taylor, J.: An Introduction to Support Vector Machines and Other Kernel-based Learning Methods, 200 pp. Cambridge University Press (2013). http://ebooks.cambridge.org/ebook.jsf?bid=CBO9780511801389]
6. Rahman, R.M., Md. Hasan, F.R.: Using and comparing different decision tree classification techniques for mining ICDDR, B Hospital Surveillance data. Expert Syst. Appl. **38**, 11421–11436
7. Loong Ang, S., Choon Ong, H., Chin Low, H.: Classification using the general Bayesian network. Sci. Technol. **24** (1) 205–211(2016)
8. Ghosh, A.K.: On optimum choice of k in nearest neighbor classification. **50**, 3113–3123 (2006)
9. Top Data Mining Algorithms Identified by IEEE & Related Python Resources. http://www.datasciencecentral.com/profiles/blogs/python-resources-for-top-data-mining-algorithms
10. Ashfaq Ahmed, K., Aljahdali, S., Hussain, S.N.: Comparative prediction performance with support vector machine and random forest classification techniques. Int. J. Comput. Appl. **69** (11),12–16 (2013)
11. Vijayarani, S., Dhayanand, S.: Kidney disease prediction using SVM and ANN algorithms. Int. J. Comput. Business Res. **6**(2), (2015)
12. Palaniappan, S., Awang, R.: Intelligent heart disease prediction system using data mining techniques. IEEE **1**(8), 108–115 (2012)
13. Fan, Q., Zhu, C.J., Yin, L.: Predicting breast cancer recurrence using data mining techniques. IEEE **1**(10), 310–311 (2010)
14. Lakshmi, K.R., Nagesh,Y., VeeraKrishna, M.: Performance comparison of three data mining techniques for predicting kidney dialysis survivability. Int. J. Adv. Eng. Technol. **7**(1), 242–254 (2014)
15. Vijayarani, S., Dhayanand, S.: Data mining classification algorithms for kidney disease prediction. Int. J. Cybern. Inf. (IJCI) **4**(4), 13–25 (2015)
16. Hall, M., Frank, E., Holmes, G., Pfahringer, B.: The WEKA data mining software: an update. **11**(1), 10–18 (2009)
17. Chronic kideney Data Set. https://archive.ics.uci.edu/ml/datasets/Chronic_Kidney_Disease#
18. Ma, H., Bandos, A.I., Rockette, H.I., Gur, D.: On use of partial area under the ROC curve for evaluation of diagnostic performance, static in medicine. Statist. Med. **32**, 3449–3458 (2013)
19. Sokolova, M., Lapalme, G.: A systematic analysis of performance measures for classification tasks. **45**, 427–437 (2009)
20. Santos, F.: The Kappa Cohen: a tool to measure the inter-rater agreement on qualitative characters. http://www.pacea.u-bordeaux1.fr/IMG/pdf/Kappa_Cohen.pdf (2015)

The manufacturer's authorised representative in the EU is Springer Nature Customer Service Centre GmbH, Europaplatz 3, 69115 Heidelberg, Germany. If you have any concerns regarding our products, please contact ProductSafety@springernature.com

Printed and bound by CPI Group (UK) Ltd, Croydon, CR0 4YY
25/09/2026
02222839-0018